MARINE BIOLOGY

AN ECOLOGICAL APPROACH

SIXTH EDITION

JAMES W. NYBAKKEN
Moss Landing Marine Laboratories

MARK D. BERTNESS
Brown University

PEARSON
Benjamin
Cummings

SAN FRANCISCO   BOSTON   NEW YORK
CAPETOWN   HONG KONG   LONDON   MADRID
MEXICO CITY   MONTREAL   MUNICH   PARIS
SINGAPORE   SYDNEY   TOKYO   TORONTO

ACQUISITIONS EDITOR: *Chalon Bridges*
EDITOR-IN-CHIEF: *Beth Wilbur*
ASSISTANT EDITOR: *Nora Lally-Graves*
MANAGING EDITOR: *Erin Gregg*
SENIOR PRODUCTION SUPERVISOR: *Corinne Benson*
MARKETING MANAGER: *Jeff Hester*
PRODUCTION EDITOR: *Heather Meledin,*
Progressive Publishing Alternatives
COMPOSITION: *Progressive Information Technologies*
MANUFACTURING MANAGER: *Pam Augspurger*
PHOTORESEARCHER: *Kristin Piljay*
TEXT AND COVER DESIGNER: *Seventeenth Street Studios*
COVER PRINTER: *Coral Graphics*
TEXT PRINTER: *R.R. Donnelley, Willard*

ISBN 0-8053-4582-5 (Student)
ISBN 0-8053-4586-8 (Instructor)

Library of Congress Cataloging-in-Publication Data
Nybakken, James Willard.
   Marine biology: an ecological approach / James W. Nybakken, Mark Bertness.—6th ed.
     p. cm.
   Includes bibliographical references and index.
   ISBN 0-8053-4582-5
   1. Marine biology. 2. Marine ecology. I. Bertness, Mark D., 1949- II. Title.

QH91.N9 2004
577.7--dc22

                      2004052771

On the cover: Photograph of a Blue Chromis (*Chromis cyanea*) over Boulder Brain Coral (*Colpophyllia natans*) in the Caribbean. Copyright © 2003 Norbert Wu/www.norbertwu.com.

CHAPTER OPENING PHOTO CREDITS: • *Chapter 1:* © *Rick Doyle/CORBIS* • *Chapter 2:* © *Alfred Pasieka/Photo Researchers, Inc.* • *Chapter 3:* © *Amos Nachoum/CORBIS* • *Chapter 4:* © *James Nybakken* • *Chapter 5:* *Ralph A. Clevenger/CORBIS* • *Chapter 6:* *Art Wolfe/Getty Images/Stone* • *Chapter 7:* © *Andrew Syred/Photo Researchers, Inc.* • *Chapter 8:* *Phil Schermeister/CORBIS* • *Chapter 9:* © *CORBIS* • *Chapter 10:* *Jeff Hunter/Getty Images/The Image Bank* • *Chapter 11: Paul Hanna/ CORBIS/Reuters*

PEARSON
Benjamin
Cummings

www.aw-bc.com

3 4 5 6 7 8 9 10—DOW—08 07 06

CONTENTS

PREFACE

IN RECENT YEARS, there has been an increasing interest in the field of marine biology at the undergraduate level in our two-year community colleges, as well as in four-year colleges and universities. That interest apparently arises from the growing awareness and concern at all levels of society of the importance of the world's oceans as sources of food, reservoirs of minerals, major suppliers of oxygen, regulators of climates, and the ultimate dumping ground for the mounting burden of human waste materials. This concern has been popularized and brought into focus by the various international disputes on fishing rights, the whale problem, international law of the seas conferences, numerous television programs and books about marine life and, most recently, the worldwide decline or collapse of many marine fisheries.

As a result of this heightened public awareness, marine biology and marine ecology courses have attracted a wide spectrum of students with varying backgrounds who desire a basic understanding of the biological processes that operate in the oceans.

It is important to understand that in the oceans, just as on land, there are scientific principles that govern the organization and perpetuation of organisms and associations. Although those principles operate somewhat differently in the ocean than on land because of the physical-chemical properties of water, they can be readily explained and may be understood by those with a minimum background in other sciences and in mathematics.

The texts in marine biology available to serve these courses appeared to us to lack an ecological approach to the entire marine environment; they tended to emphasize specific areas, habitats, and organisms to the exclusion of broader concepts and processes. We were thus stimulated to write this book as an introduction to marine biology, emphasizing the ecological principles governing marine life throughout the world instead of a purely taxonomic or regional approach, so it would be useful in all parts of the world. We have, furthermore, purposely downgraded any discussion of pure oceanography, except as it bears on organization of the associations or communities, in order to increase coverage of the biology. The sequence of topics in this book closely follows the sequence in the senior author's upper division undergraduate course in marine ecology, which was taught for many years at the Moss Landing Marine Laboratories.

This text is designed for the undergraduate student in marine biology. It presumes a certain minimum background in basic concepts of chemistry, physics, and biology, but no more than would be obtained from introductory college courses in each of these fields. Familiarity with the major invertebrate phyla is helpful. Some acquaintance with the basic ecological concepts is also helpful, but if lacking, they may be obtained from Chapter 1. Although the text is generally aimed for an undergraduate majors marine biology course, we have tried to imbue this book with sufficient rigor and detail that it will also be useful in upper-division undergraduate courses in marine ecology and biological oceanography. To this latter end, we have included text citations to the primary literature.

This text stresses ecological processes and adaptations that structure marine associations and permit their persistence through time. It is not a guide to the local fauna and flora. It is presumed that familiarity with this, if necessary, will be obtained in laboratory and field trips in which the instructor will provide the taxonomic expertise.

Depending on the time available in the course, this book includes perhaps more material than can be adequately covered in a single quarter or semester. In that case, the instructor may choose to concentrate on those chapters or areas of most concern in his or her geographical region. For example, instructors in temperate zones may wish to de-emphasize coral reefs and spend less time on deep-sea biology and more on the intertidal, to which they have direct access. Suggestions for use of the book under varying course conditions are given in the Instructor's Manual. However, we hope the comprehensive nature of this book means that whatever instructional choices are made, information will be available from the book.

■ CHANGES IN THE SIXTH EDITION

The changes made in this edition maintain the rigorous treatment of all subjects established in the fifth edition and primarily reflect new knowledge and understanding about marine ecology.

New subjects in this edition include a section on thermohaline circulation or the marine conveyor belt system, a discussion of the significance of internal tides or waves and how they cause movement in deep water masses, how air-sea interactions affect regional climates, the new phototrophic organisms of the open oceans and their role in primary productivity, a whole new section on seamounts and their significance in the deep sea, the newly discovered thermal vents, successional sequences in thermal vent communities over time, the significance of positive interactions in community structure and the role of hydrodynamic processes in larval settlement, the ecology of cobble beach plant communities, and new research on the role of consumers in shaping salt-marsh plant communities.

Topics that have been substantially upgraded or changed from the fifth edition include a newly revised section on the microbial loop and carbon flux in the oceans, the role of viruses in the plankton, the significance of Bacteria and Archaea in the oceans, new information on seasonality of reproduction in the deep sea and the incidence of planktotrophy, lecithotrophy and direct development, updates on number of species, endemism and biogeography of the deep sea hydrothermal vents and seeps, additional information on the sea ice communities including food webs and comparisons of Arctic and Antarctic sea ice organisms, newer data on the general community patterns as well as the role of recruitment and postrecruitment processes on infaunal communities in soft sediments, updates on the general community structure of subtidal rocky bottoms and its variation with depth, increased information on the community structure of seagrass beds and their role as nurseries, and updated chapters on intertidal communities, tropical communities, and estuarine communities. Finally, we have reorganized the subjects in Chapter 11.

This edition continues the use of full four-color illustrations throughout the text, thus increasing the value of the illustrations as well as placing them next to the text rather than as a set of separate color inserts.

In this edition, we have again included additional information and examples that point out the controversies in marine ecology, because we believe that it is important for the student to learn that not only are there many unsolved problems in the field of marine ecology, but also that there are disagreements among scientists about how these various communities and systems are organized. In this latter connection, we think it is particularly important for the student to recognize that studies elucidating the mechanisms ordering a given community in a given geographical area likely cannot be extrapolated as a "universal" explanation for the organization of communities in similar environmental settings throughout the world.

One other comment deserves mentioning here and should be kept in mind when using this text. Considering the all-pervasive presence of humans in all parts of the planet, except possibly the deep sea, it is unlikely that there are any pristine natural habitats left and we are in fact studying and describing habitats that are in some way disturbed.

The first editions of this text were perceived by many as distinctly biased toward the Pacific coast. We think it is important to point out in this context that most of the East Coast of the United States south of New England, as well as the Gulf coast, consists of sedimentary environments, such as open sand beaches, protected sand beaches, muddy shores, and large estuaries and marshes. Rocky shores are rare. The Pacific coast, by contrast, consists mainly of rocky shores, with few estuaries and other sedimentary environments. We hope, therefore, that we will be forgiven for continuing to use Pacific coast examples in discussions of the rocky shores and commended for expanding a discussion of the dominant Atlantic and Gulf coast sedimentary environments. As part of an effort to make this book more useful globally, we have also included in this edition more examples and comparisons with communities in areas outside the North American continent.

Some who have used this book in more advanced classes have suggested that we remove the very basic introductory ecology material from Chapter 1 because their students have had it already. We have chosen not to do so in deference to those who teach more basic courses without major prerequisites and who feel that their students need it. I can suggest only that those whose classes are more advanced simply ignore it. Finally, the glossary has been updated to include all boldface terms and concepts in the text.

■ **INSTRUCTOR'S MATERIALS**

As with the fifth edition, an *Instructor's Manual with Test Bank* (0-8053-4585-X) is available to text adopters. The material has been updated with problems reflecting the updated coverage. An *Instructor's Art CD-ROM* (0-8053-4584-1) includes all of the figures from the text in Powerpoint and JPEG format. The sixth edition also includes a text specific Companion Web site (http://www.aw-bc.com/nybakken).

■ **ACKNOWLEDGMENTS**

As the author of this textbook, we are indebted to a large number of marine scientists whose research work is the foundation on which this book is based and who are too numerous to mention in their entirety here. A few of this vast number are mentioned in the text and listed in the end-of-chapter bibliographies. We repeat here our thanks to all those colleagues and friends mentioned in the prefaces to the first, second, third, and fourth editions who helped us by reviewing chapters, and who contributed illustrations for the preceding editions.

In addition to those colleagues who helped me in the first four editions of this book, we would like to thank the many who have helped with this edition by providing us with constructive criticisms of the fifth edition and pointing out errors. I am particularly indebted here to my Moss Landing colleagues James Harvey, Greg Cailliet, and William Broenkow. And thank you as well to the following colleagues who reviewed this sixth edition manuscript: Paul Dayton (Scripps Institution of Oceanography), Charles W. Holliday (Lafayette College), Kevin S. Beach (University of Tampa), Ralph J. Larson (San Francisco State University), Wayne Price (University of Tampa), Ivan Valiela (Boston University), James T. Carlton (Williams College), James McClintock (University of Alabama), John Pearse (University of California Santa Cruz), Bruce Robison (Monterey Bay Aquarium Research Institute), George Somero (Hopkins Marine Station), Brian Bingham (Western Washington University), Roger Seapy (California State University at Fullerton), Brad Peterson (Long Island University at Southampton), and Scott Shumway (Wheaton College).

Finally, we would like to thank the staff of Benjamin Cummings and particularly Beth Wilbur, Chalon Bridges, and Nora Lally-Graves for their help and encouragement in the process of this revision.

Because of the text's broad scope and necessary limitations in length, we have often been forced to treat certain complex or poorly understood processes through generalizations that some may consider seriously oversimplified. Undoubtedly, some errors and omissions may remain. Therefore, please feel free to pass on to us any potential errors, omissions, or questions of interpretation, as well as any other comments and suggestions. All observations will aid in revising future editions.

James W. Nybakken
Moss Landing Marine Laboratories
P.O. Box 450
Moss Landing, CA 95039-0450
nybakken@mlml.calstate.edu

Mark Bertness
Ecology and Evolutionary Biology
Brown University
Providence, Rhode Island
mark_bertness@brown.edu

CHAPTER 1

INTRODUCTION TO THE MARINE ENVIRONMENT

About 71% of the surface of this planet is covered by salt water. The water depth averages 3.8 kilometers (km), a volume of $1,370 \times 10^6$ km^3. Since life exists throughout this immense volume, the oceans constitute the single largest repository of organisms on the planet. These organisms include representatives of virtually all phyla and are tremendously varied. All, however, are subject to the properties of the seawater that surrounds them, and many features common to such plants and animals result from adaptations to the watery medium and its movements. Before we can consider the major associations or assemblages of organisms, we need to examine briefly the physical and chemical conditions of seawater and various aspects of its motion (oceanography), as well as some basic ecological principles and terms. These principles and terms will be central to understanding the processes discussed in the remainder of this text. Finally, we will make some comparisons between aquatic and terrestrial ecosystems to point out some fundamental differences in their organization.

Properties of Water

Water is the substance that surrounds all marine organisms. It composes the bulk of the bodies of marine plants and animals, and it is the medium in which various chemical reactions take place, both inside and outside living organisms.

■ CHEMICAL COMPOSITION

Pure water is a very simple chemical compound composed of two atoms of hydrogen (H) joined to one oxygen (O) atom. Expressed symbolically, water is H_2O. The hydrogen atoms are bonded to the oxygen asymmetrically, such that the two hydrogens are at one end of the molecule and the oxygen is at the other. The bonding between the hydrogen atoms and the oxygen is via shared electrons, each hydrogen sharing its single electron with the oxygen. In this manner, oxygen receives the two electrons needed to complete its outer electron shell, and each hydrogen the one needed for its outer shell. However, the oxygen atom tends to draw the electrons furnished by the hydrogen atom closer to its nucleus. This tendency creates a slight negative charge at the oxygen end of the molecule, while the removal of the electrons away from the hydrogens results in a slight positive charge at the hydrogen end. The resulting electrical separation creates a polar molecule.

The polar nature of the water molecule means that the hydrogen end, which is positive, will attract the negative, or oxygen, end of other water molecules. This attraction gives rise to weak bonds, called **hydrogen bonds**, between adjacent water molecules. These bonds are only about 6% as strong as the covalent bonds between the hydrogen and oxygen in the water molecule itself, and they are easily broken and re-formed (Figure 1.1). It is the hydrogen bonding between adjacent water molecules and the polarity of the water molecule that are jointly responsible for many of the unique chemical and physical properties of water. Indeed, if water were not polar and did not form such bonds with adjacent molecules, it would be a gas, not a liquid, at room temperature, and its freezing point would be lower than temperatures found over most of the earth's surface. Under such conditions, life as we know it on Earth would be impossible.

■ PHYSICAL AND CHEMICAL PROPERTIES

Because of hydrogen bonding, water tends to stick firmly to itself, resisting external forces that would break these bonds. This tendency is called **cohesion**. At air–water interfaces, the strength of cohesion forms a "skin" over the water surface strong enough to support small objects—a phenomenon called **surface tension**. The surface tension of water is the highest of all common liquids and permits the existence of associations of organisms either suspended below the surface or moving over it. Cohesion is also responsible for the

viscosity of water, a property that indicates the force necessary to separate the molecules and allow passage of an object through the liquid. Viscosity of seawater is altered by temperature and salinity. It increases with a decrease in temperature or an increase in salinity. The resistance to flow or movement that viscosity represents is important in the sinking rate of objects (pp. 56–61) and in problems of movement in the water by animals (see pp. 109–113).

Another set of properties concerns the effects of heat. In order to evaporate, water must absorb more heat in comparison with other liquids of simple molecules. This means that the **heat of vaporization** of water is high—the highest of most common substances. This is a direct result of the strength of the hydrogen bonding between molecules, which must be broken to allow the escape of a molecule. Because of its high heat of vaporization, water evaporates slowly at earth's atmospheric temperatures and absorbs considerable heat, leading to atmospheric cooling. Similarly, the high heat of vaporization means a high boiling point (100°C), with the result that most water is in a liquid phase on earth rather than a gas. Related to this is the **latent heat of fusion**, which is the amount of heat gained per unit mass when a substance changes from solid to liquid, or lost when it changes from liquid to solid. Water, again, has the highest value of most common compounds. Melting ice, therefore, takes up large quantities of heat. Correspondingly, when ice is formed, large quantities of heat are given off.

The high values for both the heat of vaporization and latent heat of fusion mean that it takes more heat to change the temperature of a given quantity of

FIGURE 1.1 Diagrammatic representation of a series of water molecules, indicating both the polar nature of the molecules and the hydrogen bonding.

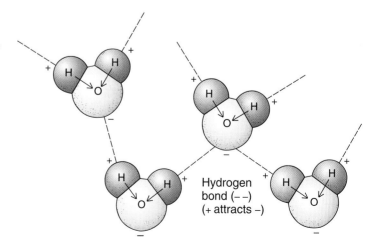

water than to change that of virtually any other common substance. Water also has a high **heat capacity**. This means that water is a strong buffer against both rising and falling temperatures and, therefore, moderates the climate. It also means that the range of temperatures exhibited within any body of water is less than that in air.

Water has a peculiar *density–temperature* relationship. Most liquids become more dense as they are cooled. When these liquids are cooled until they become solid, their solid phase is more dense than their liquid phase. This is not true for water. Like most liquids, pure water becomes more dense as it is cooled—until, however, it reaches 4°C. Cooling water below that temperature *decreases* its density, and when freezing occurs, there is a *marked* decrease in density. Consequently, ice is lighter than water and floats upon it. This property is of utmost importance to life in the oceans; otherwise, major volumes of the oceans would take the form of large blocks of ice and would be uninhabitable. Different water masses in the world's oceans have different densities due to their different temperatures and salinities. Water masses may sink or rise through other water masses, depending on their relative densities and may also spread out and flow horizontally. Such behavior is common in the world's oceans, and such density differences are responsible for deep ocean currents.

The chemical properties of water that are of importance are those concerned with its solvent capacities (Figure 1.2). Water is almost a universal solvent, with the ability to dissolve more substances than any other

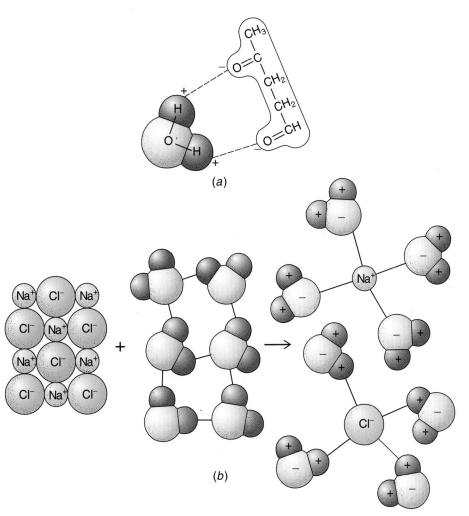

(a)

(b)

FIGURE 1.2 (a) Solvent action of water through the creation of hydrogen bonds. (b) Solvent action of water through the interaction of salt ions with charged parts of water molecules.

liquid. This is because the solvent action is of two types, one depending on the polar character of the molecule and the other on the hydrogen bonding involved. Various nonpolar organic and inorganic compounds containing oxygen atoms or hydrogen atoms bonded to either oxygen or nitrogen atoms are held in solution by hydrogen bonding. The polar dissolving action of water with respect to various salts depends on the interaction of the salt ions with the charges on the water molecule. When a salt is dissolved in water, it breaks into its component ions; thus, common table salt, NaCl, when put into water, dissociates into Na^+ and Cl^- ions. It is through the interaction of the charges on the salt ions and those on the water molecules that salts are held in solution.

Still other chemical properties of water depend on the breaking of the strong bonds between the hydrogen and the oxygen in the water molecule. In any volume of water, a few of the water molecules have separated the charges so completely that the molecule breaks into two charged parts: H^+ (hydrogen ion) and OH^- (hydroxyl ion). Here, one hydrogen atom moves away, leaving its electron with the bonding electrons on the oxygen atom, making the free hydrogen positive and leaving the OH negative. When

certain substances dissolve in water, they do so by reacting with these ions. An example is carbon dioxide (CO_2), which reacts as follows:

$$CO_2 + HOH \leftrightarrow H_2CO_3$$

These properties are all summarized, with some additional ones, in Table 1.1.

■ SEAWATER

Seawater is water in which a variety of solids and gases are dissolved. A 1,000-gram (g) sample of seawater will contain about 35 g of dissolved compounds, collectively called salts. In other words, 96.5% of seawater is water and 3.5% is dissolved substances. The total amount of dissolved material is termed the **salinity** of the water. In the past, marine biologists and oceanographers usually expressed salinity in terms of parts per thousand, abbreviated as ‰, but the new term is **practical salinity units**, or **psu**. Thus, if a typical sample of seawater has 35 g of dissolved compounds in 1,000 g, it has a salinity of 35 psu (35‰).

Dissolved substances include inorganic salts, organic compounds derived from living organisms, and dissolved gases. The greatest fraction of the dissolved material in seawater is composed of inorganic salts

TABLE 1.1

Some Properties of Water

PROPERTY	COMPARED WITH OTHER SUBSTANCES
Surface tension	The highest of all common liquids
Conduction of heat	The highest of all common liquids except mercury
Viscosity	Relatively low viscosity for a liquid (decreases with increasing temperature)
Latent heat of vaporization: The quantity of heat gained or lost per unit mass by a substance changing from a liquid to a gas or a gas to a liquid without an increase in temperature (cal/g)	The highest of all common substances
Heat capacity: The quantity of heat required to raise the temperature of 1g of a substance 1°C (cal/g/°C)	The highest of all common solids and liquids
Density: Mass per unit volume (g/cm³ or g/ml)	Density determined by (1) temperature, (2) salinity, (3) pressure, in that order; the temperature of maximum density for pure water is 4°C; for seawater the freezing point decreases with increasing salinity
Dissolving ability	Dissolves more substances in greater quantities than any other common liquid

Source: After Ingmanson and Wallace, 1973.

present as ions. Six inorganic ions make up 99.28% by weight of the solid matter: chloride, sodium, sulfur (as sulfate), magnesium, calcium, and potassium ions (Table 1.2). These can be considered the major ions. Four minor ions add an additional 0.71% by weight, so that ten ions together make up 99.99% by weight of the dissolved substances in seawater.

The salinity of various parts of the open ocean away from coastal areas varies within a narrow range, usually 34–37 psu, and averages about 35 psu. The differences in salinity are due to differences in evaporation and precipitation. Higher salinities occur in subtropical oceans, where there is a high evaporation rate, and lower values in temperate oceans, where there is less evaporation. Equatorial salinities are low because of precipitation in the intertropical convergence. In inshore areas and partially enclosed seas, the salinity is more variable, and it may be near 0 psu where large rivers discharge fresh water, rising to near 40 psu in the Red Sea and Persian Gulf.

Even though variations in salinity from near 0 to 40 psu exist, the ratios among the most abundant ions remain virtually constant (Principle of Constant Proportions). This constancy is important; it means that concentrations of nearly all ions in a given sample of water can be determined by measuring only one. This fundamental relationship is the basis for the measurement of salinity in seawater. Salinity can be established by measuring a single property, which may be the chlorinity (chloride ion concentration), for example, or the electrical conductivity or refractive index, which also depend on the salt content.

Among the remaining 0.01% of dissolved substances in seawater are several inorganic salts that are crucial to marine organisms. Included here are the inorganic nutrients phosphate and nitrate, required by plants to synthesize organic material in photosynthesis, and silicon dioxide, required by diatoms and radiolarians to construct their skeletons. In contrast to most ions, nitrate (NO_3^-) and phosphate (PO_4^{3-}) do not exist in a constant ratio with other elements or ions and tend to be in short supply in surface waters, varying in abundance as a result of biological activity. (See Chapter 2 for discussion.) Supplies of these essential nutrients may often become limiting to plant growth in some cases. (See pp. 68–72.)

Other substances essential to life processes exist as trace amounts in seawater and include such elements as iron, manganese, cobalt, and copper. Of particular interest is iron. According to Martin et al. (1994), in some parts of the world's oceans where nitrate and phosphate are abundant, plant production is limited by a lack of trace amounts of iron in the water column. (See Chapter 11.) Certain organic compounds, such as vitamins, are also present in minute amounts, but little is known of their variations.

The salt content of seawater has a definite effect on its properties. The maximum density of pure water occurs at 4°C, but in all salt water with a salinity above 24 psu, the density continues to increase to the freezing point. Because of the salt content, the freezing point is also reduced from 0°C, the amount being a function of the salinity. For seawater of 35 psu, the freezing point is −1.9°C. Upon freezing, however, the salts are excluded and the density decreases; thus, ice floats on the surface. The importance of this increase in density is that very cold and very dense surface water can be formed and sink to the bottom of the ocean basins, where it becomes the environment for deep-sea organisms and, more importantly, creates a fairly vigorous circulation that aerates the deep ocean.

Two gases dissolved in seawater are of metabolic importance: oxygen and carbon dioxide. The solubility of gases in seawater is a function of temperature: The lower the temperature, the greater the solubility.

TABLE 1.2

Major and Minor Constituents of 34.8 psu Seawater

ION	PERCENT BY WEIGHT
A. Major	
Chloride (Cl^-)	55.04
Sodium (Na^+)	30.61
Sulfate (SO_4^{2-})	7.68
Magnesium (Mg^{2+})	3.69
Calcium (Ca^{2+})	1.16
Potassium (K^+)	1.10
Subtotal	99.28
B. Minor	
Bicarbonate (HCO_3^-)	0.41
Bromide (Br^-)	0.19
Boric acid (H_3BO_3)	0.07
Strontium (Sr^{2+})	0.04
Subtotal	0.71
Total	99.99

Source: From *Deep Sea Biology: Developments and Perspectives*, by N. B. Marshall, © 1980 Garland STPM Press.

Therefore, the colder the water, the more oxygen it can hold. Even so, the solubility of gases in water is not great. At 0°C, 35 psu seawater contains about 8 ml of O_2 per liter, whereas air contains 210 ml/liter. At 20°C, 35 psu seawater contains only 5.4 ml/liter. The reason that deep ocean water does not become anoxic (devoid of oxygen) through biological activity as it flows across the ocean floor is that, when water that is to become the deep ocean water sinks from the surface, it is so cold that it has a maximum amount of oxygen, more than can be consumed by the limited populations of animals in deep water. Oxygen is not distributed uniformly with depth in the ocean. A typical vertical profile of the oxygen content of the ocean shows a maximum amount in the upper 10–20 meters (m), where photosynthetic activity by plants and diffusion from the atmosphere often lead to supersaturation (Figure 1.3). With increasing depth, the oxygen content declines, reaching a minimum value somewhere between 200 and 1,000 m in open ocean waters. This **oxygen minimum zone** may have oxygen values that approach zero in some areas. Below the zone, oxygen values increase somewhat with depth, but they usually do not reach surface values (Figure 1.3). The occurrence of the oxygen minimum zone is usually attributed to biological activity depleting the oxygen, coupled with the absence of both photosynthetic activity and contact with the atmosphere, which would otherwise permit renewal.

The increase in oxygen below the oxygen minimum zone results from the influx of the cold, oxygen-rich waters that originally sank at high latitudes.

The solubility of carbon dioxide is somewhat different from that of oxygen, since it reacts chemically in the water. Carbon dioxide is abundant in seawater, which has a considerable capacity to absorb the gas. This is because carbon dioxide, upon entering seawater, reacts with the water to produce carbonic acid via the reaction

$$CO_2 + HOH \leftrightarrow H_2CO_3$$

Carbonic acid dissociates further into a hydrogen ion and a bicarbonate ion:

$$H_2CO_3 \leftrightarrow H^+ + HCO_3^-$$

Bicarbonate may dissociate yet further into another hydrogen ion and a carbonate ion:

$$HCO_3^- \leftrightarrow H^+ + CO_3^{2-}$$

The major reservoir of carbon dioxide in the ocean is the bicarbonate ion, as can be seen from its place as one of the most abundant minor ions in seawater. (See Table 1.2.) In contrast to oxygen, carbon dioxide is more abundant in seawater than in air. Therefore, the availability of carbon dioxide is not limiting to plant growth in seawater.

The carbon dioxide–carbonic acid–bicarbonate system is a complex chemical system that tends to stay

FIGURE 1.3 Change in dissolved oxygen with depth in the eastern tropical Pacific Ocean (dashed line) and the tropical Atlantic Ocean (solid line). (Modified from *Oceanography: An Introduction*, D. E. Ingmanson & W. J. Wallace, © 1973. Reprinted with permission of Brooks/Cole, a division of Thomson Learning: www.thomsonrights.com. Fax 800-730-2215.)

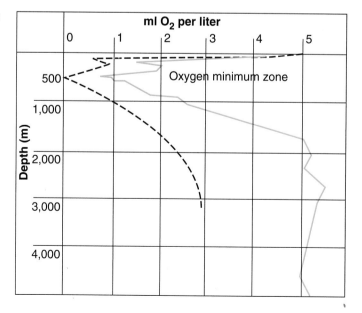

in equilibrium. If CO_2 gas is removed from seawater, the equilibrium will be disturbed, and carbonic acid and bicarbonate will shift to the left in the preceding equations, until more CO_2 is produced and a new equilibrium is established.

The reactions shown result in the production or absorption of free hydrogen ions (H^+). The abundance of hydrogen ions in solution is a measure of the solution's **acidity**. More H^+ ions mean a more acid solution, and fewer H^+ ions mean a more basic, or alkaline, solution. **Alkaline** (or basic) solutions are those that have large numbers of OH^- ions and few H^+ ions. Acidity and alkalinity are measured on a logarithmic scale of 1–14 pH units. The higher the concentration of H^+ ions on this scale, the lower is the pH value; hence, low pH values indicate acid conditions. Conversely, high pH values indicate low H^+ concentrations and high OH^- concentration. The neutral point is pH 7, where equal numbers of both ions occur (Figure 1.4).

Although pure water is neutral in pH, because dissociation of the water molecule produces equal numbers of H^+ and OH^- ions, the presence of CO_2 and the strongly alkaline ions—sodium, potassium, and calcium—in ocean waters tends to change this. Seawater is slightly alkaline, usually in the pH range of 7.5–8.4. At the pH of ocean water, the carbon dioxide–carbonic acid–bicarbonate system functions as a **buffer** to keep the pH of seawater within a narrow range. It does this by absorbing H^+ ions in the water when they are in excess and producing more when they are in short supply. This balancing act is accomplished by shifting the reactions shown previously to the right when there are too few H^+ ions, producing more bicarbonate ions and carbonate ions, and shifting the reaction to the left when there are too many H^+ ions, producing more undissociated carbonic acid, carbon dioxide, and bicarbonate ions.

Basic Oceanography

This is not a book on oceanography, which is the study of all aspects of the physics, chemistry, geology, and biology of the sea. We are concerned only with imparting a basic understanding of life in the sea and how it is organized and persists. To understand the ecology of various marine associations, however, it is necessary to know something about the structure and motion of the ocean water masses.

■ GEOGRAPHY AND GEOMORPHOLOGY OF THE OCEANS

The three major oceans are connected to each other, converging in the area around the Antarctic continent. The world ocean has been separated for convenience into four major divisions—Pacific, Atlantic, Indian, and Arctic, in order of decreasing size. Projecting from, or partially cut off from, these larger oceans are smaller marginal seas, such as the Mediterranean, Caribbean, Baltic, Bering, South China, and Okhotsk. The locations of the major oceans and seas are given in Figure 1.5.

The oceans are not equally distributed over the earth. Oceans cover more than 80% of the Southern Hemisphere, but only 61% of the Northern

FIGURE 1.4 The pH scale.

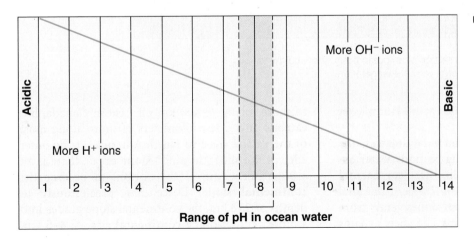

Range of pH in ocean water

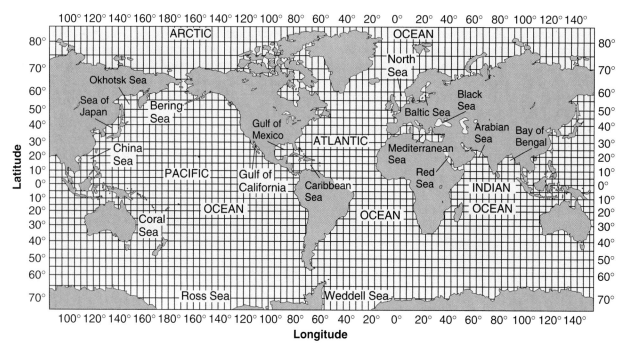

FIGURE 1.5 Major oceans and seas of the world.

FIGURE 1.6 Diagrammatic cross section of an ocean basin, showing the various geographic features. (Not to scale; large vertical exaggeration.)

Hemisphere, where most of the earth's landmasses occur.

On the margins of the major landmasses, the ocean is very shallow, overlying an underwater extension of the continent called the **continental shelf** (Figure 1.6). Forming only 7–8% of the total ocean area, the continental shelf slopes gently from shore to depths of 100–200 m. The shelf extends offshore for up to 400 km off eastern Canada, but extends only a few kilometers offshore along most of the Pacific coast of North America. At the outer edge of the shelf, the **shelf–slope** break, there is an abrupt steepening of the bottom to become the **continental slope**. Descending precipitously to depths of 2–3 km, the continental slope grades into a more gently sloping **continental rise**. At 4–5 km, the bottom becomes the flat, extensive, sediment-covered **abyssal plain.** Such plains monotonously cover the floors of vast areas of the oceans at depths between 3 and 5 km.

Abyssal plains are broken in several places by various submarine ridges. This mid-ocean ridge system is an extensive, contiguous chain that has been found in all oceans. The best known is the mid-Atlantic ridge, which bisects the Atlantic Ocean into east and west basins and runs from Iceland into the South Atlantic, where it links with a similar ridge coming from the Indian Ocean. Occasionally, the ridges break the surface to form islands. The Azores, Ascension, and Tristan da Cunha are islands formed by the mid-Atlantic ridge. These extensive ridge systems mark the boundaries of the various crustal plates of the Earth and are sites of volcanic activity.

In certain areas, the abyssal plains are cut by deep, narrow troughs called **trenches.** Most of these trenches lie in an arc bordering the islands and continents in the Pacific Ocean. The trenches have depths from 7,000 to more than 11,000 m. The deepest area known is the 11,022-m Challenger Deep in the Marianas Trench.

Finally, there may be solitary islands and submarine **seamounts,** formed by isolated volcanic action. Such mountains, as opposed to the ridges, rise individually from the abyssal plain.

■ PLATE TECTONICS

In the 1960s, when new evidence became available about the ocean floor it became apparent to scientists that the earth's crust was divided into a number of large units called **plates** and that these plates were bounded by the aforementioned ridge and trench systems. The sizes of the plates vary, but most of the planet is covered by seven major plates (Figure 1.7). The plates themselves are rigid and float on the underlying mantle. The major plates encompass both

FIGURE 1.7 Location and boundaries of the Earth's plates, midocean ridges, and trenches. Arrows indicate the direction of plate movement. (*Scientific American*, Vol. 226, No. 5, May 1972, pp. 56–57. Reproduced with permission from the Estate of Mary and Dan Todd.)

continental and oceanic crust. The plates are not static, but move slowly relative to each other. Their movement over geologic time is responsible for the differing positions of the continents through time, a phenomenon called **continental drift**.

The plates move because the oceanic ridges are centers of volcanic activity where new material is formed and added to the crust. As this volcanism takes place, the plates move laterally, in opposite directions, causing **seafloor spreading**. The opposite of this spreading phenomenon occurs in the trench systems, where the margin of one plate dives beneath the margin of the adjoining plate in a process called **subduction**. Here, the subducting crust is melted into the mantle layer again (Figure 1.8). Both the oceanic ridge system and the subduction zones of the trenches are the sites of volcanic and seismic activity.

The plate boundaries—particularly the oceanic ridge systems—are significant in marine ecology because they are weak spots in the earth's crust where the mantle pushes up, producing cracks and fissures. Water seeps into these openings and is heated to very high temperatures by the mantle material. The heated water subsequently emerges again, carrying with it many reduced chemical compounds in what are called **hydrothermal vents**. The sites of unique communities of deep-sea organisms that were first discovered in 1977, hydrothermal vents are discussed in detail in Chapter 4.

■ TEMPERATURE AND VERTICAL STRATIFICATION

Temperature is a measure of the energy of molecular motion. In the world oceans, it varies, moving north to south with changes in latitude and vertically with depth. Temperature is a singularly important factor governing the life processes and the distribution of organisms. Vital life processes, collectively termed **metabolism**, function only within a relatively narrow range of temperatures, usually between 0 and 40°C. However, a few organisms, including the cyanobacteria living in 85°C hot springs, can tolerate temperatures somewhat above and below these limits. Within the temperature range in which life processes operate, metabolism is temperature dependent. In general, for organisms that do not regulate their internal temperatures, metabolic processes are increased by a factor of 2 for each 10°C rise in temperature. With the exception of the marine birds and mammals, and some fishes, marine organisms are **poikilothermic**, or **ectothermic**, meaning that their body temperatures vary with that of the surrounding water mass.

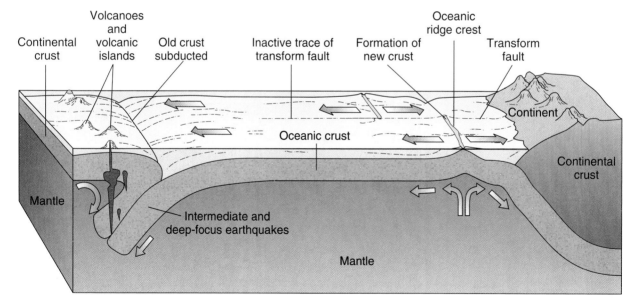

FIGURE 1.8 Diagrammatic representation of the oceanic and continental crusts, showing the formation of new crust at the oceanic ridges and the destruction of the oceanic crust in the subduction zones. (From *Marine Biology: Function, Biodiversity, Ecology,* Second Edition by Jeffrey S. Levinton, copyright © 1995, 2001 by Oxford University Press, Inc. Used by permission of Oxford University Press, Inc., p. 17, f. 2.5.)

Homeothermic, or **endothermic**, birds and mammals have the ability to regulate their own internal temperature, regardless of the temperature of the water mass. Since most marine organisms are poikilothermic and since the seawater temperatures vary latitudinally, the distribution of marine organisms follows closely the geographical differences in ocean temperatures.

On the basis of surface ocean temperatures and the overall distribution of organisms, four major biogeographical zones can be established: **polar**, **cold temperate**, **warm temperate (subtropical)**, and **tropical (equatorial)** (Figure 1.9). Transition zones between these areas also exist. Boundaries may vary somewhat with season, so the zones are not absolute.

Temperature in the oceans also has a marked variation with depth. Surface waters in the tropical regions are warm, 20–30°C, throughout the year, and temperate-zone surface waters are warm in the summer.

Below the warm surface water, the temperature begins to fall, and over a narrow depth range of 50–300 m it undergoes a very rapid decline. The depth zone of the most rapid temperature decline is the **thermocline**, below which the temperature continues to decrease with depth, but at a much slower rate, so that the water mass underneath the thermocline is nearly isother-mal all the way to the bottom. The thermocline is a persistent feature of tropical waters and occurs in temperate waters in the summer months. Thermoclines are absent in polar waters. In winter, thermoclines extend through a more narrow depth range over the relatively shallow waters of the continental shelf, compared with those which form in the open ocean. The importance of the thermocline is discussed in Chapter 2.

Temperature also has an effect on the density of seawater. Warm seawater is less dense than cold seawater of the same salinity. Density is also a function of salinity, with increasing salinity causing increasing density. However, the range of temperatures found throughout the world's oceans is greater than the range of salinities. Temperature is, therefore, more important in affecting the density. The rapid change in temperature that produces the thermocline also means that the density of seawater changes rapidly over the same depth range. Such a zone of rapid density change is called a **pycnocline**.

■ WATER MASSES AND CIRCULATION

As a result of different temperatures and salinities and their effects on density, the seawater of the world's oceans can be separated into different water

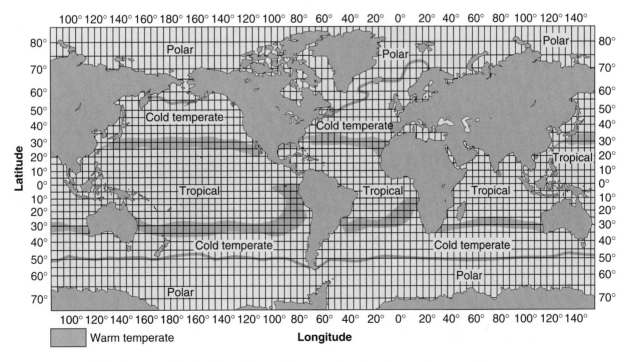

FIGURE 1.9 Major biogeographical regions of the world's oceans, based on temperature.

masses. The **upper water mass** (surface water mass) of the oceans includes all well-mixed water above the thermocline. Below the thermocline are several **deep water masses** extending to the bottom.

The upper water mass of the oceans is in constant motion, induced primarily by the action of winds blowing across the surface of the water. These winds produce two kinds of motion: **waves** and **currents**. Waves range in size from ripples only a few centime-

ters in height to storm waves, which may tower as high as 30 m. Other than height, waves are further characterized by their **wavelength**, which is the horizontal distance between the tops or crests of successive waves (Figure 1.10). The **period** of a wave is the time required for two successive wave crests to pass a fixed point. In addition to being formed from wind, waves may be generated by earthquakes, volcanic explosions, and underwater landslides, which create the

FIGURE 1.10 (a) Features of waves and change of waveform as a wave enters shallow water. (b) A cork floating in the water demonstrates the passage of a wave, indicating that the water itself does not move. As the wave passes, the cork traces out an imaginary circle (dashed line).

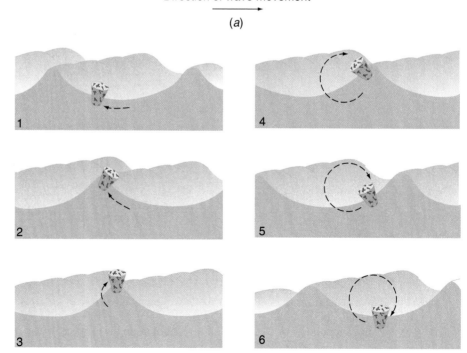

destructive waves known as **tsunamis**. The attraction of the moon and sun produces the waves known as **tides**. (See Chapter 6.)

Wind wave height in the open ocean is dependent on (1) wind **speed**, (2) the distance, or **fetch**, over which the wind blows, and (3) the **duration** that the wind blows.

All waves behave similarly. Once generated, they move outward and away from their center of origin. This repeated horizontal progression results in very little horizontal transport of the individual water molecules. Instead, the water molecules trace out a circle, moving upward and to one side as a wave crest approaches and to the other side and down to near their original position as the wave crest passes by. While the waveform and the energy are transported horizontally, the water remains nearly stationary (Figure 1.10). The passage of waves generates movement not only in the surface water molecules, but also in water down to a depth approximately equal to one-half the wavelength. With each depth interval below the surface equal to one-ninth the wavelength, the orbits followed by the water particles diminish by one-half. By the time a depth of half a wavelength is reached, the movement is almost imperceptible.

As waves enter shallow water and begin to encounter bottom effects, they slow their forward motion and the wavelength decreases. As a result, they begin to increase in height and become steeper. At a point where the water depth is 1.3 times the height of the wave, it will "break," releasing energy onto the shore. Waves are of greatest biological importance in shallow water and the surf zone, as discussed in Chapter 6.

Currents are water movements that result in the horizontal transport of water. The major shallow ocean current systems are produced by a few prominent wind belts that succeed each other latitudinally around the world and in which the winds are steady and persistent in direction. These winds are caused by differential heating of the atmospheric air masses, aided by the **Coriolis effect** (to be discussed shortly). The backbone of the system is the combination of the northeast trade winds blowing from northeast to southwest between the equator and 30°N latitude and the southeast trades in a similar position south of the equator, moving air from southeast to northwest. Between 30° and 60°N and S latitudes, the westerlies blow from the southwest to the northeast in the Northern Hemisphere and to the southeast in the Southern Hemisphere. In the

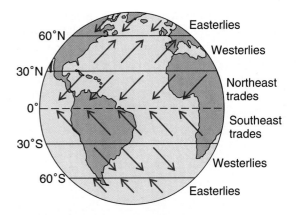

FIGURE 1.11 The main wind belts of the Earth and their prevailing direction of motion (arrows).

polar regions, cold air flows toward the equator as the polar easterlies (Figure 1.11).

These winds set the surface waters into motion, producing the slow, horizontally moving currents that can transport huge volumes of water across vast distances in the oceans. Such currents influence the distribution of marine organisms and also lead to the displacement of biogeographical zones through the transport of warm water into colder regions and vice versa. The major currents of the world's oceans are shown in Figure 1.12.

A comparison of Figures 1.11 and 1.12 shows that the ocean currents do not flow parallel to the direction of the wind. Instead, they are deflected into roughly circular **gyres** that move clockwise in the Northern Hemisphere and counterclockwise in the Southern Hemisphere. The deflections and gyres are the result of the Coriolis effect, caused by the rotation of the Earth on its axis. The spinning of the planet imparts a deflection to moving water, displacing it to the right in the Northern Hemisphere and to the left in the Southern Hemisphere. Because the rotation of the earth is from west to east and because the deflection of the currents created by the trade winds imparts a movement of water at the equator that is parallel to the equator and moving from east to west, the net water movement is from east to west, piling up water on the western side of ocean basins. As water builds up on the western sides, it meets the continental or island chain landmasses and is deflected north or south as continental boundary currents. Moving poleward, these boundary currents in turn fall under the

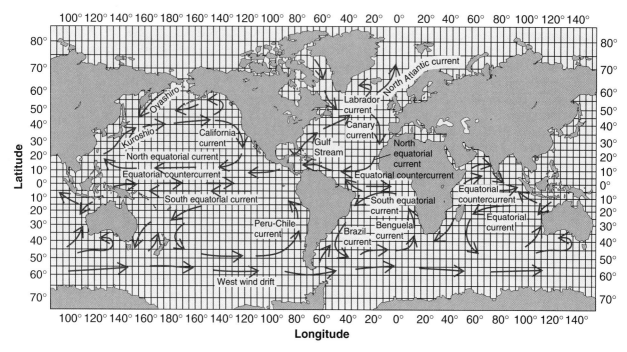

FIGURE 1.12 The major ocean surface current systems.

influence of the westerly winds. The westerlies impart yet more energy to the currents and drive them in an easterly direction, eventually to cross the ocean basins to return water to the eastern side of the basin. Continental landmasses on the eastern side deflect the moving water toward the equator, which completes the pattern. These huge circular current patterns, called gyres, are found in all major ocean basins. (See Figure 1.12.)

These global currents are surface currents. How deeply do they affect the water column? This can be estimated by noting what happens to deeper layers of water when wind stress causes the surface water to move. As previously noted, because of the Coriolis effect, the direction of the water movement is deflected from that of the initiating wind. The energy of the wind is passed down through the water column, and it sets each successively deeper water layer into motion. Each layer, however, receives a decreasing amount of energy, and therefore its velocity is lower. At the same time, due to the Coriolis effect, each layer set in motion is deflected with respect to the one immediately above it. The result is the **Ekman spiral** of current directions and velocities from the surface downward (Figure 1.13). The depth at which wind stress fails to impart motion varies, but is

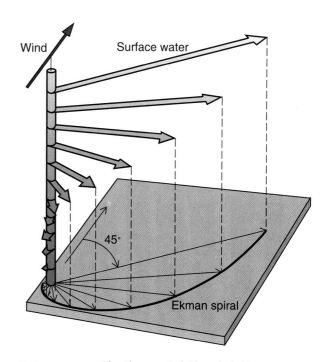

FIGURE 1.13 The Ekman spiral. The wind-driven surface current moves at an angle of 45° to the direction of the wind, to the right in the Northern Hemisphere, and to the left in the Southern Hemisphere. Successively deeper water layers are deflected even further with respect to those immediately above them and move at slower speeds. The net water movement is at 90° to the wind. (Modified from *Oceanography for Meteorologists* by Swerdrop.)

roughly limited to the upper 100 m. Such currents are called Ekman currents.

In certain areas and under certain conditions, the wind-induced lateral movements of the water may also bring about a vertical circulation, or upwelling, of water. Along the eastern margins of ocean basins, for example, the wind-driven surface currents parallel to the continental margins flow toward the equator. At the same time, the Coriolis effect tends to push these surface waters offshore. This water is then replaced by deeper water transported vertically along sloping boundaries to the surface. Similarly, along the equator, the two equatorial currents flowing west are deflected to the right north of the equator and to the

left to the south. To replace this water, subsurface water upwells to the surface (Figure 1.14). Such wind-induced upwelling is limited to the upper 200 m of the water column.

The various chemical and physical properties of the atmosphere and of the oceans interact in several ways that may be significant on different time scales. Energy is transferred from the atmosphere to the ocean in the form of winds that produce the waves and currents in the surface waters. Energy, as well as moisture, is also transferred from the ocean to the atmosphere when evaporation takes place. However, these processes do not occur on similar time scales. Energy from the atmosphere, which

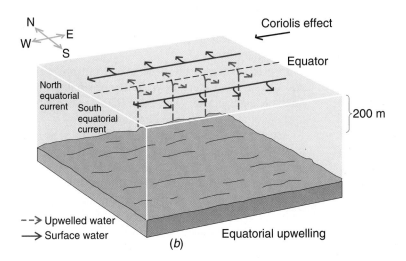

FIGURE 1.14 (a) Coastal upwelling in the Northern Hemisphere. Along the western margins of continents, the wind and Coriolis effect act to move water offshore, as indicated by the blue arrows. This water is replaced by water moving up from depths of about 200 meters (dashed arrows). (b) Equatorial upwelling. Along the equator, the Coriolis effect acts on the westward-flowing currents, pulling water north in the Northern Hemisphere and south in the Southern Hemisphere (blue arrows). This water is replaced by cool water moving up from the 200-meter depths (dashed arrows).

produces the surface waves and currents, has an almost immediate effect, whereas the heat energy from the ocean affects the atmosphere on time scales of weeks to months. Surface winds are also responsible for carrying various natural and human-made dust particles from the terrestrial environment into the oceans and thus are an important pathway for the movement of various materials into the marine environment. The **El Niño** phenomenon is an example of the sea–air interaction on a global scale. El Niño episodes occur several times each decade and begin in the tropical Pacific. Warm water builds up in the Western Pacific and moves eastward across the Pacific bringing warm water to the coasts of South America which are normally cool due to upwelling. This influx of warm water is related to the weakening of the winds at the equator. These higher sea temperatures all across the Pacific in turn alter the atmospheric circulation such that terrestrial areas which normally receive little rain are deluged and normally rainy areas undergo droughts. In the marine environment, the increased temperatures alter shallow-water marine communities and the distributions of species. They have seemed to have had particularly destructive effects on coral reefs. (See Chapter 9.)

Movement in the deep water mass is quite different from that in the surface mass. The deep water mass is isolated from the wind; therefore, its motion cannot be dependent on it. Movement in the deep water mass does, however, result from changes occurring in water at the surface. As mentioned earlier, seawater increases in density with a decrease in temperature and an increase in salinity. When seawater increases in density, it sinks. Therefore, to move water into the deep basin of the oceans, it is necessary to increase its density at the surface. This is accomplished in two ways. Warm water from the tropics or subtropics is high in salinity, due to evaporation. In the northern Atlantic Ocean, such warm saline water is transported out of the tropics by the Gulf Stream. In the region of Iceland and Greenland, it meets the very cold waters of the Labrador Current moving south. Cooling at the surface increases the density of this highly saline water, and it sinks to form North Atlantic deep water (Figure 1.15). Similarly, warm water moving south in Antarctic seas loses heat to the atmosphere, thereby causing the water masses to sink. A final, very high density water mass is

produced in the Weddell Sea in Antarctica, where very cold water becomes more saline when winter freezing occurs. This mass sinks to become the bottom water of most ocean basins.

Since these dense waters are all cold and produced at the surface, they contain large amounts of oxygen, which is then transported to the depths. Without this oxygen, deep water would be anoxic.

The currents at the surface of the world's oceans are generally horizontal currents that are wind generated and rarely disturb the deeper layers. The linkage between the surface currents and the deeper water masses is effected by a process called **thermohaline circulation**, which produces both horizontal and vertical surface and subsurface ocean currents that circulate throughout the world's oceans and establish a worldwide system of rising and sinking water masses. This circulation pattern is driven by the temperature and salinity differences that are responsible for the density gradients among adjacent water masses. A simplified model of the pattern is called the **ocean conveyor belt** (Fig. 1.16). The most prominent of the two bottom water masses is the one that sinks in the area between Iceland and Greenland. This North Atlantic bottom water flows south throughout the Atlantic, finally meeting the southern Antarctic bottom water, which was swept up as it moved northward by the strong Antarctic Circumpolar current, which moves from west to east around the Antarctic continent. At their meeting point, the two masses merge, and the resulting current continues eastward, distributing the mixture into the Indian and then Pacific Oceans. As the current progresses into the Indian and Pacific Oceans, it is continually losing water to upwelling, and the water lost in this way returns to the surface and moves via surface currents back across the Pacific, Indian, and Atlantic Oceans to the original source in the North Atlantic, thus balancing the deep water generation.

Inserted between the deep ocean bottom waters and the surface waters are a variety of smaller intermediate water masses (Fig. 1.15) that also contribute to the global thermohaline circulation.

Recently, evidence from climate records preserved in various ice cores from glaciers in the Northern Hemisphere has suggested significant swings in global climate in periods of hundreds to thousands of years. Such climate changes have in turn suggested that there could be a slowing, or even a shutdown, of the thermohaline circulation if another climate swing,

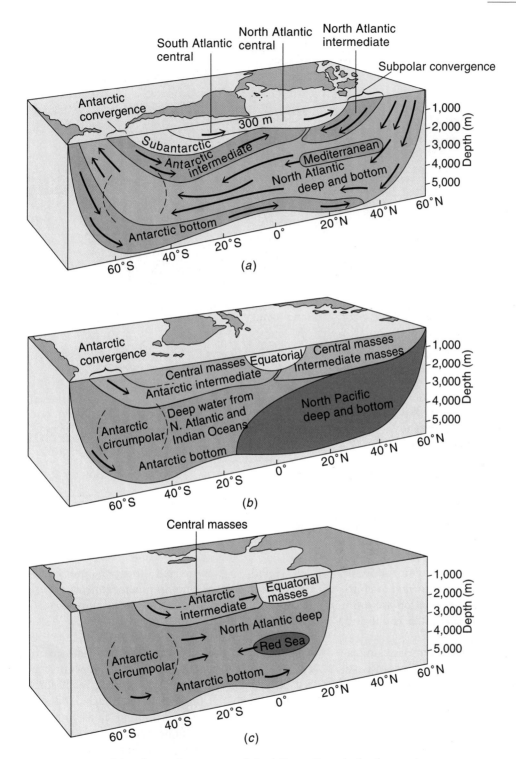

FIGURE 1.15 Subsurface water masses and circulation patterns in the three major oceans. (a) Atlantic Ocean. (b) Pacific Ocean. (c) Indian Ocean. (Modified from *Oceanography: An Introduction*, D. E. Ingmanson & W.J. Wallace, © 1973. Reprinted with permission of Brooks/Cole, a division of Thomson Learning: www.thomsonrights.com. Fax 800-730-2215.)

GREAT OCEAN CONVEYOR BELT

FIGURE 1.16 Thermohaline circulation via the global conveyor belt. The dense saline water sinks in the Norwegian Sea and the Weddell Sea and is moved into the Indian and Pacific Oceans by the Antarctic Circumpolar Current. The water returns to the surface by upwelling and is returned via warm currents. (From Broocker, 1991 in *Climate Change 1995, Impacts, Adaptations, and Mitigation of Climate Change.*)

such as global warming, changes the temperature of the planet. Although the effects of a slowing or shutting down of the thermohaline circulation are difficult to predict, one likely result is the cooling of terrestrial northern Europe.

The various deep and bottom water masses move very slowly north and south. Many hundreds of years are required for these waters to move through the ocean basins. North Atlantic deep water, for example, may have been away from the surface for hundreds of years, from the time it sank near Iceland until the time it again surfaced in the Antarctic region. It is, however, a major contributor to the productivity of Antarctic seas, because, when it surfaces, it brings with it large amounts of inorganic nutrients

(NO_3^-, PO_4^{3-}) accumulated during the many years it spent beneath the photosynthetic zone. A general outline of these deep water masses and movements is given in Figures 1.15 and 1.16.

The rise and fall of sea level caused by tides (see Chapter 6 for a discussion of surface tides in the oceans) not only generates the surface tidal currents, but is indirectly responsible for mixing in the deep ocean through the production of **internal tides** or **waves**. In general, waves are found at the interface between fluids of different densities. The most familiar waves are ocean surface waves. Internal waves occur on the pycnocline between the low-density surface water and the higher density deep waters. These waves have periods of tens of minutes to hours. The tide producing force generates not only surface tides but also internal tides. This tidal activity causes the continental slope to act as a giant wave-generating machine would by oscillating back and forth. The resulting internal waves can create mixing and turbulence. Substantial internal waves may also be generated by the oscillations of tidal currents over midocean ridges, seamounts, and other rough topography in the deep ocean (Fig. 1.17). Internal waves can also interact with wind waves at the sea surface to produce slick or glassy patches there (Fig. 1.18). Internal waves generally have higher peak-to-trough ranges than surface waves have. (Internal waves range from 10 to 100 or more meters.) When internal waves occur near the

Internal Waves

◆ Underwater waves that propagate along the pycnocline

◆ Generated by flow over tough topography

◆ Much greater amplitude than surface

FIGURE 1.17 Diagram of internal waves. (Image courtesy of Mark Benfield, © 2004.)

FIGURE 1.18 Surface manifestations of internal waves as seen from a satellite. (Photo courtesy of NASA and the Lunar and Planetary Institute.)

thermocline, they mix nutrient-rich deep water with the nutrient-poor upper water, thereby enhancing biological productivity. (See Chapter 2.)

Some Ecological Principles

Because this is a text that emphasizes habitats and the ecology of marine organisms, and because certain users of the book may lack a background in basic ecology, it is necessary at this point to cover briefly some basic ecological concepts and terms.

■ TERMS AND DEFINITIONS

Ecology is the science that treats the spectrum of interrelationships existing between organisms and their environments and among groups of organisms. It is important to realize initially that living things do not exist as isolated individuals or groups of individuals. All organisms interact with their own species, with other species, and with the physical and chemical environments that surround them. In this interactive process, the organisms have effects on each other and on the surrounding environment. Similarly, the different factors in the environment affect the activities of the organisms. The various organisms and the environmental factors may be further organized into various levels, each of which is somewhat broader than the preceding. A *species* is a natural group of actually or potentially interbreeding individuals that is reproductively isolated from other such groups. All the individuals of a given species in an area constitute a *population*. Several different populations of species that tend to occur together in a particular geographical area constitute an ecological *community*. A community or a series of communities, together with the surrounding physical and chemical environment, constitute an *ecosystem*. Ecosystems are the most complex entities and have many interacting components. They are so large and complicated that ecologists tend to study them by concentrating on their components, such as various communities or populations. Ecosystems, however, may be considered on a large or small scale, depending on the number of communities included and the dimensions of the surrounding nonliving environment. On the largest scale, it is possible to consider the Earth as a single ecosystem comprising all the various terrestrial, freshwater, and marine communities. At the other end of the scale, it is possible to consider a tide pool or freshwater pond as an ecosystem. Both have the biological and nonbiological components the definition requires.

The oceans of the world may also be considered a single ecosystem, in which a series of communities is influenced by and, in turn, influences the physical and chemical factors of the surrounding seawater. This large ecosystem may be divided into smaller sections in which the physical and chemical parameters have differential effects on the populations of organisms, changing their composition and adaptations within a given area. For example, the physical factors acting on a rocky seashore, such as waves and movements of the sea bottom, are not those which affect small organisms floating in the open ocean. Both areas, therefore, contain different organisms and different adaptations. It is the aim of this book to afford an understanding of the functioning of the total marine ecosystem by considering the functioning of its major subsections in the chapters that follow.

■ ECOSYSTEM COMPONENTS

An ecosystem is a functional unit of variable size composed of living and nonliving parts that interact. The parts and the whole system function through a sequence of operations involving energy and the transfer of energy. With few exceptions, the original energy source is the sun. Energy from the sun is captured by the **autotrophic** component: the green plants, bacteria and algae. The captured energy is stored in the chemical bonds of organic materials in the plants, which are the "food" that drives the **heterotrophic** component of the system. Heterotrophic organisms include all other life-forms that obtain their energy through the consumption of the autotrophic organisms, through the consumption of organisms that have ingested the autotrophs, or through the absorption of dissolved organic matter from the environment. Such an arrangement of autotrophs and succeeding levels of heterotrophs is called a **trophic structure**, in which each successive level is called a **trophic level**. Trophic structure is a characteristic feature of all ecosystems. The first trophic level is the autotrophic, or producer, level, at which the energy is initially captured and stored in organic compounds. As the energy is passed from level to level in such a system, the majority of it is lost through heat and metabolic use by the organisms. The amount lost at each transfer is variable, but substantial, usually ranging between 80 and 95%. The system thus becomes self-limiting in that, at a certain

point, not enough energy remains to be passed on to sustain another level. The system is visualized as either an energy pyramid or a trophic pyramid (Figure 1.19). In such a structure, the trophic level that consumes the photosynthetic organisms (autotrophs, first level) includes animals called **herbivores**. Herbivores are consumed by **carnivores**, which are eaten by still other, often larger, carnivores. All levels above the second consist of carnivores or **omnivores** (animals that eat both plant and animal material). Within each trophic level or population, the amount of living material at any instant in time is the **standing crop**.

The final component of the trophic structure of an ecosystem consists of the **decomposers**. These are organisms, chiefly bacteria, that break down the complex organic molecules of dead organisms, releasing simple molecules that are again usable by the autotrophs or by those heterotrophs which can absorb organic molecules from the environment. Decomposers exist on every trophic level.

The necessary abiotic components of the trophic structure of an ecosystem are an energy source, a nutrient source, and a water source. Photosynthetic organisms cannot fix energy and produce complex organic molecules without sunlight for energy and without a series of inorganic nutrients, of which nitrates and phosphates are the most important.

Since ecosystems are composed of from one to many communities, each in turn composed of many populations of producers, consumers, and decomposers, the transfer of energy through an ecosystem may follow several pathways. Each pathway that transfers energy from a given photosynthetic source through a given series of consumers is called a **food chain**. The combination of all food chains in a given community or ecosystem is called a **food web**. The food web is thus a summary of all the pathways by which energy moves from one level to another through a community or ecosystem.

■ BIOGEOCHEMICAL CYCLES

Among many of the chemical elements and compounds in ecosystems, there is a cycling back and forth between organisms and the physical environment. Such repeated transfers are called **biogeochemical cycles**. A few of these cycles involve chemical compounds vital to the continued maintenance of life in the ecosystem and are extremely important. Perhaps the most significant are the cycles involving carbon, nitrogen, and phosphorus (Figure 1.20).

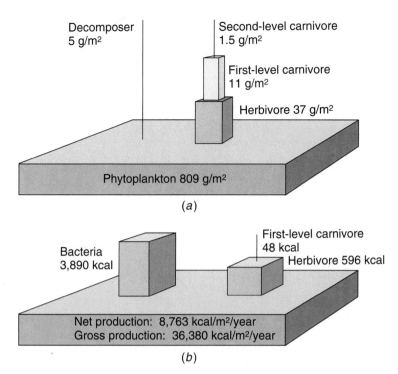

Decomposer
5 g/m²

Second-level carnivore
1.5 g/m²

First-level carnivore
11 g/m²

Herbivore 37 g/m²

Phytoplankton 809 g/m²

(a)

First-level carnivore
48 kcal

Bacteria
3,890 kcal

Herbivore 596 kcal

Net production: 8,763 kcal/m²/year
Gross production: 36,380 kcal/m²/year

(b)

FIGURE 1.19 Trophic pyramids. (a) Biomass pyramid. (b) Energy pyramid. The thickness of the bars indicates the relative amounts. (*Ecology & Field Biology*, R. L. Smith, p. 55, f3.13. Copyright © 1980 R. L. Smith. Reprinted by permission of Pearson Education, Inc.)

In all these cycles, there is a major reservoir or pool of the element, from which the element is continually moving in and out as it passes through organisms. Each cycle also contains a sink, into which a certain amount of the chemical passes and from which it is not recycled in the normal course of events. Over long periods, the loss to the sink may become limiting, unless the sink can be tapped again. Such a situation usually develops through geological action that moves the sink into an area where organisms, erosion, or other factors release the elements. Biogeochemical cycles tend, finally, to have self-regulating feedback mechanisms that keep the cycles in equilibrium.

In the carbon cycle, the reservoir is in the form of CO_2, which exists in water in the carbonic acid–bicarbonate–carbonate system (Figure 1.20a). Carbon dioxide is fixed into organic compounds by photosynthetic organisms; transferred to animals through herbivory, absorption, and predation; and returned to the reservoir via respiration and bacterial action. A nonrenewable loss occurs when heavy carbonate materials, such as shells, are deposited in deep oceans as calcareous sediments.

In the phosphorus cycle, the major reservoir is in phosphate rock. Here, erosion brings the chemical into the water, where it cycles through animals and plants and is returned to the general circulation through decay and excretion. Phosphorus is lost when it sinks to the bottom in the form of various skeletal materials that accumulate in deep sediments (Figure 1.20b).

Air is the major reservoir in the nitrogen cycle. Nitrogen gas is not used by most organisms and must be converted into a nitrogen compound before it can enter the cycle. Conversion is normally accomplished by bacteria or algae. Useful nitrogen compounds may also be produced through volcanic activity. Once in compounds, the nitrogen cycles through organisms, with some loss to deep sediments (Figure 1.20c).

■ BIOTIC STRUCTURE OF ECOSYSTEMS

Communities and ecosystems have the same trophic levels throughout the world. However, the species constituting each level differ among geographical areas, and in some areas each level may have more or fewer species than another. This arrangement implies that ecologically equivalent species replace one another geographically and that the species structure of communities is variable, even if the trophic structure is not. For example, seagrass beds in the tropics may consist of several species of grasses, whereas those in the temperate zone consist of but one. They are different species, but fulfill the same ecological role as

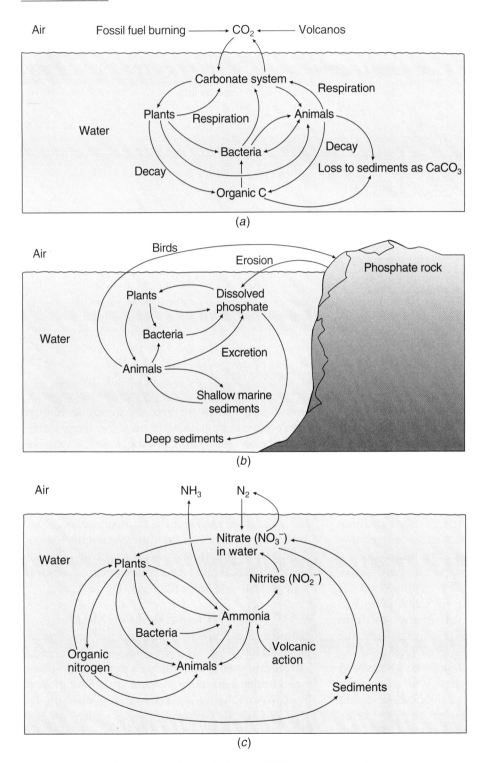

FIGURE 1.20 Biogeochemical cycles in the sea. (a) The carbon cycle. (b) The phosphorus cycle. (c) The nitrogen cycle.

autotrophs. Species that are ecological equivalents perform the same function or role in a given community. They occupy the same ecological niche.

In ecology, a **niche** is usually defined today as the role of an organism in a community. The fact that the term *niche* has had other meanings often causes confusion. A niche can be established for a species for virtually every environmental and biological variable affecting the species. For any given variable, the range within which the species can exist and reproduce is a part of its niche. The potential distribution of a species along all such niche axes or ranges would then be the **fundamental niche** for the species. This theoretical concept of a fundamental niche is difficult to comprehend, because it requires us to think about a large number of dimensions simultaneously, whereas we are used to only three. Since we cannot completely determine the fundamental niche of any organism, we tend to restrict our discussion of the niche to only a few variables. We speak then of niches with respect to food, space, breeding sites, and so forth. These are **realized niches** and represent the actual distribution of the species in the real world. They will be the niches to which the term is applied in this book.

Ecological niches may be broad or narrow. In a narrow niche, the role or function of a species in the community or ecosystem has been more finely subdivided, and the species is, therefore, more specialized. In a broad niche, the function is more generalized. At the extremes, species may be either *specialists* or *generalists* with reference to their niches. In the preceding example, the autotroph niche is broad in the temperate zone and occupied by a single generalist grass; in the tropics, several more specialized grasses subdivide the niche.

In contrast to a niche, an ecological **habitat** is the place where an organism is found. For example, the small copepod crustacean *Calanus finmarchicus* (see Figure 2.5) has the surface waters of the North Atlantic Ocean as its habitat, but its niche is that of a herbivore feeding on the microscopic algae in the area.

Most communities have a characteristic structure that consists of a few species that are abundant and a larger number of species that are uncommon. The numerically abundant species are usually called **dominants** and are often used to characterize a community. We thus can speak of a mussel bed on the rocky shore reflecting the dominance by the mussel *Mytilus edulis* or *Mytilus californianus*. This pattern of few common and many uncommon, or rare, species seems to hold at each trophic level. It also holds whether the community has many or few species in total. Species structure in ecological communities may be measured in several ways. **Species richness** is a simple listing of the total number of species in a community or trophic level. **Species diversity** is a measure that combines into a single figure both the number of species (richness) and the distribution of the total number of individuals among the species (evenness). Species diversity is expressed through various mathematical indices, which need not concern us here. High species diversity has been used in the past as an indicator of stable environments and communities, but current ecological thought recognizes many exceptions to this paradigm.

There are currently two major schools of thought in ecology that address how high species diversity is produced and maintained in communities. One school, the equilibrium school, suggests that the species composition of communities is usually in a state of equilibrium and that high diversity is due to a large number of habitats or finely divided, narrow niches maintained by various feedback mechanisms (niche diversification) and the stable nature of the physical environment that allows the equilibrium to be achieved. The other school, the nonequilibrium, or intermediate-disturbance, school of Connell (1978) and others, says that communities and species are rarely in a state of equilibrium and that high species diversity is maintained through continual or gradual environmental change and periodic disturbance. This mechanism promotes a changing species composition through the presence of many species that are not highly specialized. In Connell's model, communities which experience frequent disturbances that kill most or all the resident species would tend to be colonized by only those few species that have the ability to attain maturity and reproduce quickly before the next catastrophic event. At the other end of the spectrum, those areas which experience infrequent, small disturbances at long intervals would also tend to have low diversity. In this case, it is because the most efficient competitors have the time to eliminate most or all of the other species and take over the community. Only when the disturbances are intermediate does diversity increase. Then, enough time is available to permit the establishment of a variety of slower growing and slower reproducing species, and the interval between disturbances is short enough to avoid competitive exclusion. It is probably not possible at our current level of knowledge to decide whether one or the other or both of these theories is correct for all the various marine communities, but the evidence that has been gathered thus far seems to favor the intermediate-disturbance theory for at least some communities.

Each species in a community has certain tolerances with respect to all the environmental factors. If the limit of tolerance is exceeded in some area by a given factor (say, temperature), the species will be absent. Similarly, each species requires a certain minimum amount of various materials. If the concentrations of these necessities, such as nitrate, fall below the minimum, the species disappears. More important, if any *single factor* exceeds the tolerance level, or if any single necessary substance is reduced below the minimum, the species will be eliminated, even if all the other factors and substances are favorable. This idea is known as **Liebig's law of the minimum**, after Justus von Liebig, a nineteenth-century German chemist.

Since the *limiting factors* vary for each species, and since communities are defined on the basis of recurrent groups of species, it follows that the boundaries between communities, representing changes in various environmental factors, are also not sharp. The result is areas of transition between adjacent communities in which species gradually drop out and others come in. Such boundary zones are called **ecotones**. Compared with areas with communities that usually change rather slowly in time and space, ecotones change rapidly.

Communities are not static units: they change in structure and composition with season and over longer periods. Some communities tend to change in an orderly fashion over periods of many years until they reach a stage that perpetuates itself indefinitely, as long as the climate does not change or there is no disturbance. In such a sequence, each community modifies the environment, making it suitable for the next community. This orderly process of community change controlled through modification of the physical environment is called **ecological succession**, and the terminal, persistent community is known as the **climax**. The intermediate communities are called **seres**.

The scenario just described is the classical model, or **facilitation model**, of succession, as it was originally developed for terrestrial plant communities by F. E. Clements (1936). A second model of succession, the **inhibition model**, was proposed by Egler (1954), and it differs from the classical in that no species is competitively superior and whichever species gets to a site first holds it against later settlers. In this model, succession is not an orderly, predictable process, and there is no climatic climax. Here, succession proceeds from short-lived to long-lived species. A third model, originally proposed by Grinell (1917) and more recently discussed by Connell and Slayter (1977), called the **tolerance model**, is intermediate between the other two models. In the tolerance model, early colonizing species are not necessary and any species can start succession. Community change then occurs as more tolerant or more competitively superior species prevail.

In the marine environment, ecological succession seems to follow the second or third model. In bottom communities, following a disturbance the area is usually reoccupied by organisms that happen to have their propagules in the water and available to settle. The first occupants may change, depending on the time of year that the disturbance occurs. Furthermore, the initial occupiers often do not modify the environment to make it unsuitable for them and more suitable for subsequent organisms. In fact, they may even prevent the occupation of the area by other, later successional forms by various means and thus delay the climax community development. Some examples of these kinds of succession are given in Chapters 5 and 6. Finally, succession probably does not occur in the open waters among pelagic communities.

■ ECOLOGICAL CONTROL AND REGULATION

Populations, communities, and ecosystems are all regulated by various factors. The major controlling factors for ecosystems and communities are energy, the physical factors collectively termed climate or environment, and the interaction among various species that compose the systems. Ultimately, nearly all systems on Earth are limited by the amount of energy available from the sun. Tolerance limits of various species to such abiotic factors as temperature, light, nutrients, and salinity also limit the extent of populations and communities in the oceans. The final aspects of regulation and control considered here are those dealing with the interaction among populations which keep the populations within limits.

Biologists have long known that all species possess the reproductive potential to produce much larger populations than are observed under natural conditions and that if population explosions occur, they are quickly reduced. What are the biological factors that exert control? They are grouped under competition, predation, and parasitism and disease.

Competition is an ecological term referring to the interaction among organisms to obtain a necessary resource that exists in short supply. Competition may be intraspecific (among individuals of the same species) or interspecific (among individuals of different species). Competition may be for any number of things, but usually is limited to items such as light, food, nutrients, water, and space. In a competitive interaction, either the competitors manage to share the limited resource or one excludes the other. In the first case, both organisms are hampered, which inhibits their growth, development, and reproduction; thus, the numbers of both organisms are limited. In the second case, one organism is eliminated, again controlling population. Interspecific competition is usually between two closely related species and has led to the **competitive exclusion principle**, which says that no two species with exactly the same requirements can coexist in the same place at the same time; that is, complete competitors cannot coexist. As population numbers increase, competition usually increases, because the limited resource becomes more scarce. This increased competition then increases the stress on animals and plants and absorbs energy otherwise used for reproduction, limiting populations.

Predation can be defined as the consumption of one species by another. The animal that is the consumer is the predator; the victim, the prey. A special case concerns those animals which consume autotrophic (photosynthetic) organisms. They are called herbivores. A **grazer** is an animal that feeds on plants or sessile animals. Predators and herbivores vary considerably in their ability to regulate the numbers of organisms they consume. In some cases, the predator may be the most important factor in regulating numbers of a prey species. In other cases, a predator may have little effect on a prey population. In the first case, the removal of the predator will have a marked effect on the prey population, causing it to increase dramatically. In the latter case, removal of the predator has little effect on the prey population. We shall see examples of both of these situations in the chapters that follow. A special case of predation concerns those predators that have a profound effect, not only on their prey population, but also on the entire community of which they are a part. In this case, the removal or depletion of the numbers of the predator causes great changes in the presence and abundance of many species in the community, most of which are not the prey of the predator. As a result, the entire community structure may be changed. Such predators are called **key industry species** or **keystone species**. A simple definition of a keystone species is that it is a predator which, by preying on a competitive dominant species, limits the competitive exclusion of other species by that dominant (Paine, 1969).

Parasitism and disease are the final biological controls on populations. As with predation, they may exert strong control, or they may have little effect. **Parasites** are organisms living in or on other organisms from which they derive nourishment and shelter. Many—perhaps all—marine organisms have parasites, but we know considerably less about the role of parasites and disease in regulating populations of marine organisms than we do about their role with regard to terrestrial organisms. Perhaps the best-documented change induced in a marine community by a disease is the loss of the eelgrass beds in the Atlantic Ocean in the 1930s. (See pp. 240–241).

Larvae and Larval Ecology

A very large number of marine organisms, both fish and various invertebrate phyla, that inhabit both pelagic and benthic systems, produce larvae in their life cycles. **Larvae** are independent, morphologically different stages that develop from fertilized eggs and that must undergo a profound change before assuming adult features. They are almost always considerably smaller than the adult stage of the species. Larvae and their ecological characteristics are important in the establishment and maintenance of many marine communities and associations. The reason is that, as Chia (1988) notes, in any given marine habitat the distribution and abundance of benthic invertebrates are maintained by four factors: larval recruitment, migration, asexual reproduction, and mortality. Many benthic communities are composed of species that reproduce by producing various larval types that undergo a free-swimming, drifting, or crawling stage in the water column or on the bottom before metamorphosing into benthic adults. Even many organisms that pass their adult lives in the water column produce larvae. Therefore, a grasp of larval ecology is central to understanding how such communities persist. We discuss the basic features of larval ecology here, leaving the specific role for larvae in each larger community to be covered in the appropriate chapter.

■ LARVAL TYPES AND STRATEGIES

As noted by Vance (1973), there are three paths that development may take in a benthic invertebrate with a given amount of energy available with which to produce young. One is to produce a great many very small eggs which hatch quickly into larvae that are free swimming in the plankton. Because so little yolk is put into each egg, the larvae are dependent on food sources in the water column for their nutrition and are called **planktotrophic larvae**. A second course is to produce fewer eggs and endow each with somewhat more energy in the form of yolk. Such eggs hatch into larvae that, because of their yolk reserve, do not feed and spend less time in the water column before settling. These larvae use the plankton phase mainly for their own dispersal and are called **lecithotrophic larvae** (Figure 1.21).

A final option is to dispense with the larval phase altogether. In this course, the adult produces a few eggs, each with a very large amount of yolk. Such

eggs undergo a long development without additional energy sources. The young pass through the larval stages in the egg and hatch as juveniles. There is no free-swimming larval stage. This type of development is **nonpelagic**, or **direct, development**, and the hatchlings are called **nonpelagic larvae** or **juveniles**.

There are two major assumptions underlying each Vance model. One is that mortality in the plankton is constant over time and is not related to size of the larvae; the other is that egg size is inversely related to time spent in the plankton.

Each of the three developmental pathways noted by Vance has both advantages and disadvantages. In the case of the planktotrophic larvae, the advantages are that a very large number of young can be produced for a given amount of available energy, and wide dispersal is assured through the long time spent in the plankton. The disadvantages are that the larvae depend on the plankton for their nutrition, a notoriously unpredictable situation, and that long residence in the plankton increases the chances that a predator will consume the larvae. The chances of being consumed by a predator in the plankton are

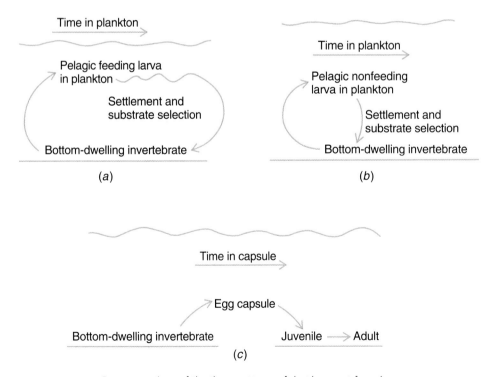

FIGURE 1.21 Representations of the three patterns of development found among benthic invertebrates. (a) Planktotrophic. (b) Lecithotrophic. (c) Nonpelagic.

extremely high, and the pelagic zone is thus a very dangerous habitat.

Lecithotrophic larvae have the advantage of spending less time in the plankton and thus have less chance of being consumed. They also are not dependent on the vagaries of the plankton for food. The disadvantage of this strategy, however, is that, because of the greater amount of energy placed into each egg, fewer eggs and larvae can be produced per unit of available energy. A shorter time in the plankton also means less ability to disperse In addition, a lecithotrophic larva presents a larger target for visual predators in the plankton.

The nonpelagic development has the advantage of reducing the planktonic mortality to zero. In this case, however, the disadvantages are that only a few eggs can be produced, because of the great amount of energy each takes, and that dispersal is nil.

Since all three types of development exist, it can be assumed that, under certain conditions, one type or another is favored. Even though we are currently ignorant of all the reasons that some areas and conditions favor one larval type over another, it is possible to suggest some that are valid.

In polar waters, nonpelagic development is common, even though productivity is reduced and confined to a narrow summer peak (p. 75) and it takes a long time for benthic animals to obtain sufficient energy to reproduce. Such development avoids the problem of having to time the reproductive cycle precisely, so as to hit the very short polar bloom period, where in failure would mean the death of most, if not all, larvae, and it ensures the highest survival rate of the few eggs that can be produced.

Lecithotrophic larvae are common under conditions somewhat similar to those in polar waters. Lecithotrophy is preferred over nonpelagic development if dispersal is of prime concern to the survival of the organism in question. Lecithotrophy is also more energy efficient than planktotrophy when planktonic mortality is high. Both lecithotrophy and nonpelagic development were thought by Thorson (1950) to be the most common mode of development in deep-sea organisms, but recent work has suggested that planktotrophy is also common and may be the dominant form of development in some taxa.

The advantages of planktotrophy are that eggs are cheap to produce, can be produced in large numbers, and have high dispersal rates. The disadvantages are primarily the great attrition rate suffered while the larvae are in the plankton and the unpredictable plankton food source. Planktotrophic larvae, therefore, should be common where the plankton food is predictable over long periods, where planktonic mortality is low, where dispersal is of importance, and where development time is short. The area that fulfills most of these criteria is the tropics, and it is there that we find most forms with planktotrophic larvae.

Indeed, it was Thorson (1950) who first suggested that there was a latitudinal gradient in the proportions of the three larval types. From his studies, primarily in cold temperate and Arctic waters, Thorson suggested that planktotrophy was most common among tropical marine organisms and that lecithotrophy and nonpelagic development became more prevalent as one approached polar latitudes. Planktotrophy also decreased, and lecithotrophy and nonpelagic development increased, in moving from shallow waters to the deep sea.

The problem with the simple Thorson models is that there is no nice fit to them if one studies a variety of marine invertebrates. Strathmann (1977, 1978) has clarified the problem by demonstrating that the predicted effect of various hypotheses about larvae and larval development is confounded when one looks at the real world. For example, species with a feeding larval form (planktotrophic) very often have low survival and recruitment most years, but occasionally they have a very successful year. One would expect that an organism that produces such larvae would be long lived as an adult and breed many times (iteroparous), simply because in most years the young would not survive and it would take several years of producing young to have a successful recruitment. However, many short-lived animals that breed only once (semelparous) have feeding planktotrophic larvae.

Similarly, it has been suggested that small animals more often are brooders, producing a few eggs, because they simply are not big enough to allocate a large amount of energy to produce numerous eggs. To ensure the survival of the few, they brood them through to a more advanced state. This hypothesis, however, seems illogical, because if brooding ensures a higher percentage of survival of young, why shouldn't all animals, large and small, do it? Certainly, the larger animals would gain by getting a higher return of juveniles per unit input of energy. In other words, there must be some other compensating advantage to sticking with the "risky" planktotrophic development other than simple survival of the young. One such advantage might be dispersal, since large animals are less likely to be moved over long distances by water currents and rafting than are small animals, particularly if the large adults are also sedentary. However, both animals that are permanent members of the plankton and those which, as adults, are sedentary or sessile on the bottom have planktotrophic larvae. Since adults in the water column are as dispersible as the larvae, this would negate the argument for dispersibility. One reason larger animals tend not to brood is that they may simply lack the physical space to brood a large set of eggs or be unable to aerate the very large egg mass properly. Strathmann (1985) has even suggested that planktotrophy may be an adaptation to prevent the suffocation of eggs in large egg masses, since eggs must have oxygen and there are physical constraints on how egg masses are constructed so that oxygen can reach interior eggs.

In sum, whereas we have several hypotheses as to why organisms may have a particular type of larval development, we can find many exceptions. Together, these exceptions suggest that there are many factors involved in selecting for a given larval type, that these factors do not act similarly for all the animals in a given habitat, and that there may be morphological or phylogenetic constraints on the development of a particular type.

Obrebski (1979) has also offered a model for the existence of different larval types. This model suggests that the type of larva—planktotrophic, lecithotrophic, or nonpelagic—should be predicted by the relative abundance of suitable substrates for settlement. In the model, those benthic organisms colonizing abundant habitat patches would have less need for long-lived planktotrophic larvae, whereas those inhabiting rare patches would benefit from having long-lived planktotrophic larvae in order to search out the rare type of habitat.

Vance's models have also been challenged by Underwood (1974), who has shown that surveys of the existing data reveal many discrepancies in the predicted modes of reproduction. Underwood has concluded that other factors are therefore of importance and that energy considerations alone are inadequate to explain the different strategies.

The above problems and exceptions to the initial simple three-division classification scheme has prompted more complicated schemes that take into account a more complete range of possibilities. For example, Levin and Bridges (1995) have proposed a classification scheme with four main divisions and several subcategories to account for various exceptions. In their scheme, the four major categories are nutritional mode, site of development, dispersal potential, and morphogenesis. Nutritional-mode subcategories include planktotrophy, facultative planktotrophy, lecithotrophy, consuming special nondevelopmental or nurse eggs, translocation of food from the parent, absorbing dissolved organic matter from seawater, and autotrophy. The subcategories of the site of development include planktonic, demersal (near the bottom), benthic, and brooding. The dispersal-potential subcategories include long-range planktonic larvae with planktonic periods of two months to a year, short-term larvae with planktonic periods of a week to several months, larvae with planktonic periods of a few hours to a few days, and, finally, those without any planktonic stage. The morphogenesis subcategories include direct development and indirect development. In this complex classification scheme, one could theoretically have any combination of the subcategories of the four major categories. For example, one could have indirect development with a planktonic, demersal, or benthic site of development and planktotrophy, lecithotrophy, or autotrophy.

We find ourselves in the somewhat awkward position of having several different hypotheses for the existence and perpetuation of the various larval types, but with the real data not seeming to fit any model completely and with several larval types coexisting in the same area. This situation must strongly suggest that other factors are also at work in determining the type of larval development. Why should this be? We can only suggest that, for each species, there is a necessity to balance all the factors, including energy allocation, dispersal, relative abundance of microhabitat space, longevity of adults, competitive ability of adults, and, undoubtedly, others, against the need for larger numbers of young. The type of reproduction that has evolved is that which balances all of these factors with survival. This idea is reinforced by the finding of certain benthic infaunal species that can alter their reproductive strategy among the three developmental patterns, depending on prevailing conditions. An example is the common polychaete worm *Capitella capitata*, which Grassle and Grassle (1974) have shown can produce either planktonic or benthic larvae, depending on the conditions.

■ LARVAL ECOLOGY AND COMMUNITY ESTABLISHMENT

Many, if not most, marine benthic communities are composed primarily of species that have a free-swimming larval stage. The exception to this rule appears to be the aforementioned polar waters. Since, over larger areas, the benthic communities seem to persist over time (see Chapter 5), but are dependent for this continued existence on the settlement of larvae from the plankton, and since the adults live only a short time, how is such stability maintained? What mechanisms are there that will act to ensure that the "right" larvae will settle and mature to keep the community going? A partial answer is found in the characteristics of larvae.

Since the pioneering work of Wilson (1952), many laboratory studies of larvae have found that larval settlement and metamorphosis are responses on the part of the larvae to complex and highly specific environmental stimuli, including light, gravity, and fluid movement. The most influential clues, however, appear to be chemical, emanating from biological sources. Most larvae have preferences as to the areas in which they will settle. Larvae do not just settle out of the water column on whatever substrate is available when the time comes for them to metamorphose into adults. Larvae have the ability to "test" the substrate, probably responding to certain physiochemical factors; if the substrate is unsuitable, they do not settle out or metamorphose. This means that certain substrates will always be attractive to certain larvae and repellent to others. Larvae also respond to the presence or absence of adults of their own species. Many larvae preferentially settle where the adults are living. They are, as Crisp and Meadows (1962) have shown, attracted to the area by a **pheromone** secreted by the adult. Since the presence of the adult means that the area is suitable for habitation, this mechanism ensures the survival of the young. It also ensures that the community will persist.

The larvae of many invertebrates also can delay their own metamorphosis for a certain period if they cannot find suitable substrates at the time they are supposed to settle. This delay of metamorphosis, however, is finite: After a certain time, metamorphosis will take place even over unsuitable substrates. The ability to delay settlement, however, is yet another factor that can be employed to ensure that the larvae will settle in the proper place.

Larvae, as Thorson (1966) has noted, also respond to other physicochemical factors, such as light, pressure, and salinity. Many free-floating larvae are positively phototactic in the early stages of their larval life. This condition ensures that they will reside in the upper, faster moving waters where dispersal is greatest. Later, as the time for settlement approaches, they become negatively phototactic and migrate toward the bottom. Some larvae are highly sensitive to light and pressure and inhabit only certain levels in the water column, areas with a definite light and pressure regime. The segregation of larvae into various layers in the water column also means that wherever a given layer impinges on the bottom, it will have only a certain complement of larvae available for settlement, a further way of ensuring a nonrandom bottom community (Figure 1.22).

These various studies designed to identify the mechanisms that control larval settlement and metamorphosis have tended to strongly favor the hypothesis of active habitat selection by the larvae. Most of the studies, however, have been done in the laboratory in still water and have not considered the alternative hypothesis that larvae may be passively deposited on the seafloor by physical processes. Butman (1987) has also suggested that various hydrodynamic processes are active as well in determining where larvae settle.

Finally, different species have different times of reproduction and, hence, different times of the year during which larvae are in the water column. As we shall see, this temporal distribution may also have an effect on the associations that develop.

Despite the considerable abilities of larvae to discriminate among substrates and select places to settle, samples taken close to each other show that there is often considerable variability from year to year in both the abundance and composition of species within bottom infaunal communities. If the larvae have such selective powers, why is there such variability? The answer to this question lies in the life histories of the various invertebrates composing the community, their interaction with each other and with the physical environment, and the effects of predators and competition.

■ LIFE-HISTORY STRATEGIES

According to MacArthur (1960), it is possible to recognize two quite different types of life-history patterns among organisms in any habitat. These patterns are introduced here and will be referred to in subsequent chapters. One can be called **opportunistic** or "*r* selected." Opportunistic marine species are characterized

FIGURE 1.22 Various factors of the environment to which larvae respond in selecting a suitable site for settling.

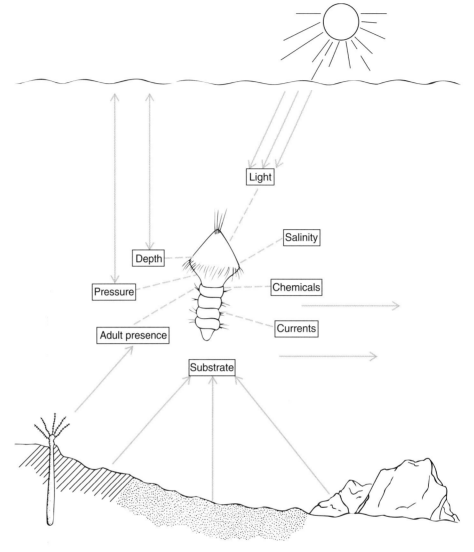

by having short life spans, rapid development to reproductive maturity, many reproductive periods per year, larvae present in the water during much or all of the year, and high death rates. Usually, they are small animals and often are sedentary or **sessile** (fixed in place). The second type of life-history pattern can be termed **equilibrium** or "*K* selected." Equilibrium species have long life spans, a relatively long development time to reach reproductive maturity, one or more reproductive periods per year, and low death rates. Usually, these species are somewhat larger than opportunists, and they are often mobile (Table 1.3).

The characteristics of opportunistic and equilibrium species define the two ends of a continuum. It

must be remembered that although some species will have all characteristics of an opportunistic species and others those of equilibrium species, still others will fall somewhere in the middle and have varying mixtures of these features. For purposes of discussion, however, it is easier to consider the ends of the spectrum.

Opportunistic species are favored where the substrate is subjected to frequent disturbance from some particular agent. In the case of shallow-water soft bottoms, the common disturbance agent is waves or some other water motion that stirs up the bottom or carries away the upper layer of sediment. Another disturbance may be the rapid deposition of sediment on

TABLE 1.3

Summary of the Characteristics of Opportunistic and Equilibrium Species

FEATURES	EQUILIBRIUM (K)	OPPORTUNISTIC (r)
1. Reproduction periods	Few per year	Many per year
2. Development	Slow	Rapid
3. Death rate	Low	High
4. Recruitment	Low	High
5. Colonizing time	Late	Early
6. Adult size	Generally large	Generally small
7. Mobility	High	Low

the bottom. In certain cases, biological activity may disturb the bottom—for instance, when large fish, such as rays, dig into it. When such a disturbance occurs, it usually kills the resident organisms and leaves an open area, which can be resettled by other organisms. Because of their attribute of having larvae in the water most of the time, opportunists quickly settle these open areas. Furthermore, if disturbances occur frequently, the opportunists will have the advantage because they can quickly mature and reproduce again before being destroyed by the next disturbance.

Equilibrium species, on the other hand, tend to inhabit areas that are not subject to frequent disturbance, allowing them to complete their life cycles. Too frequent disturbance would mean destruction before they could reach reproductive maturity; hence, no larvae would be present in the water column to settle the area again.

Whenever the bottom is subject to frequent disturbance, the benthic communities are usually composed primarily of opportunists, whereas in areas where disturbance is less frequent, more equilibrium species are found. The reason is that equilibrium species are generally better competitors than opportunistic species and, given sufficient time, will outcompete the opportunists.

Because of their different characteristics, there are differences in abundances between the two groups. Communities in which the number of opportunists is high show great fluctuations in numbers of individuals over time, whereas communities composed of equilibrium species are more constant in abundances.

In view of all these considerations, it would appear that communities would be composed of either opportunists or equilibrium species, depending on frequency of disturbance. In reality, however, samples

of the bottom usually contain representatives of both groups. How does this happen? Apparently, the equilibrium species are slower to colonize disturbed areas than are opportunists, but they are better competitors once they are there, often even pushing out (outcompeting) the opportunists. Over time, there is a sequence in which the area is first settled by opportunists immediately following a disturbance. Then, as time goes on, the equilibrium species settle in and finally outcompete the opportunists. When sampling, then, one would encounter differing numbers of both groups, depending on the time since the last disturbance. Varying numbers of both groups, therefore, can be seen in bottom samples (Figure 1.23).

Part of the reason there are different numbers of individuals and species from different areas in the same bottom is that disturbances do not affect the whole bottom equally, but act selectively. The result is a patchy array of communities, each in varying degrees of recovery from disturbance and with differing numbers of opportunists and equilibrium species.

Since wave-induced disturbance decreases with depth, it is not surprising that deeper communities tend to be inhabited by more equilibrium species than shallow-water communities do, unless another type of disturbance creates unstable conditions. Such a disturbance could be biological—for example, the reworking of sediment by fishes or by burrowing deposit feeders.

Comparison of Terrestrial and Marine Ecosystems

It is quite true that ecological principles apply equally whether one is dealing with terrestrial, freshwater, or marine ecosystems. However, the special conditions

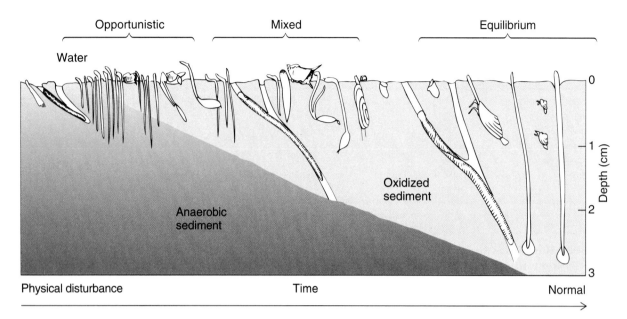

FIGURE 1.23 Changes in a soft-bottom community with time after a physical disturbance. ("Disturbance and Production on the Estuarine Sea Floor," D. C. Rhoads et al., *American Scientist,* September–October, 1978. Copyright 1978 *American Scientist.* Reprinted by permission.)

prevailing in salt water that result from the physical and chemical properties of water have, directly or indirectly, channeled the evolution and adaptation of marine organisms. The result has been some striking differences in the organization of marine communities compared with terrestrial communities. We consider these differences next.

■ PHYSICAL AND CHEMICAL DIFFERENCES

Seawater has several physical features that have a profound effect on the organization of marine communities: greater density with respect to air (seawater is more than 800 times as dense as air), a greater viscosity (about 60 times that of air), an ability to transmit sound (4 times faster than air), a low electrical resistivity (16 orders of magnitude less than that of air), and an ability to absorb light. Of these factors, the density and viscosity differences have the most widespread effects on marine organisms.

The greater density of seawater means that relatively large organisms and particles can remain suspended in it. This is not possible in air. One significant result of seawater's high density is that the marine ecosystem has evolved a whole community of small organisms that are perpetually afloat: the plankton. No such comparable terrestrial community exists floating permanently

in the air. As a result, all communities in the sea are bathed in a medium that itself contains a community! The presence of this suspended community has, in turn, led to the evolution of animals that are adapted to filter organisms and particles out of seawater. These **filter feeders** are unique to aquatic systems. There is nothing really comparable on land; the closest analog would be spiders, which capture flying insects in their webs. Furthermore, since the waters of the oceans are in perpetual motion, many of these filter feeders are sessile, extending their filtering nets up into the moving water. Sessile filtering animals are not possible in the terrestrial environment because the bathing medium flowing by contains no perpetually suspended food organisms as particles. Finally, marine organisms—especially the sessile species—have evolved motile larval forms. Such motile larvae, when put into the plankton, permit the species to disperse.

As a result of the suspension of a whole community in the bathing medium, it is more difficult to have isolated communities and isolated distributions of species in the sea.

An interesting sidelight of the difference in viscosity between air and water is that water becomes less viscous as it heats up, whereas air becomes more viscous. Tropical marine animals, therefore, encounter

COMPARISON OF TERRESTRIAL AND MARINE ECOSYSTEMS • 33

less resistance to movement than do temperate or polar animals, whereas tropical birds flying experience more drag than do temperate or polar birds.

Another physical factor indirectly responsible for differences between marine and terrestrial organisms is **gravity**. Because marine plants and animals are buoyed up by the greater density and viscosity of water, they do not need to have a significant amount of their biomass invested in structural material, such as skeletons or cellulose, in order to hold themselves erect against the force of gravity. Similarly, where movement is concerned, terrestrial animals must raise their mass against the force of gravity for each step they take. Such movement requires significantly more energy than swimming movements do in aquatic organisms; hence, terrestrial animals require more energy storage. The disadvantage of the high density of water to marine organisms is that it takes more energy to move through the water. (See Chapter 3 for a discussion.)

This absence of large amounts of structural material and energy storage in the bodies of marine organisms is reflected in the differences in the dominant biochemical compounds found in terrestrial and marine organisms. Among terrestrial plants, the dominant structural and storage compounds are carbohydrates; among marine organisms, the predominant body biochemical is protein, with smaller amounts of carbohydrates and calcium carbonate going into structure. Living organisms composed primarily of carbohydrates tend to be long lived, slow growing, and rich in stored energy. Protein-dominated organisms, on the other hand, tend to be rapid growing and with lower energy storage.

In contrast to air, water strongly absorbs light. As a result, light entering water can penetrate only to a certain depth before it is completely absorbed. Although this depth varies, it is sufficiently shallow (200–400 m) that the majority of the volume of water in the oceans is without light. This means that photosynthetic autotrophic life and primary productivity are limited to an extremely narrow band near the surface. Whereas terrestrial communities, with the exception of caves, all possess sufficient light for photosynthesis, many marine communities exist without the benefit of a photosynthetic autotrophic component.

Not only does water absorb light, it does so differentially, depending on the particular wavelength. The result is that certain wavelengths penetrate deeper than others. This is not true on land, where nearly the same spectral composition of light impinges at all levels in any community. (The importance of this difference to marine communities is discussed on pages 66–68.)

The speed of sound at 20°C in air is 346 m/sec, whereas, at the same temperature in water, it is 1,518 m/sec, more than a fourfold increase. In addition, the wavelength of sound in water increases by a factor of four. The biological consequences of these relationships concern the use of sound by animals to monitor their environment, find food, and avoid predators. The faster movement of sound in water, coupled with the more limited range of vision due to the absorption of light by water, means that hearing is more useful than sight in water and has a greater range for orienting an animal to its surroundings in water. Both terrestrial and aquatic animals emit sounds to orient to and detect prey. However, for marine animals to detect prey with sound, the sound waves must be several wavelengths smaller than they would be in air to detect the same size of prey. Since the wavelengths of sound increase fourfold in water, marine animals must use shorter wavelengths (higher frequencies) than terrestrial animals use to detect the same size of prey. (See Chapter 3 for further discussion.)

The physical factor with the greatest difference between air and water, as Denny (1990) points out, is electrical resistivity. Water has a resistivity 10^{16} times less than that of air. The chief consequence of this tiny resistivity is that some aquatic animals have developed an ability to detect electrical activity in other animals at a distance. This sense has not been found in terrestrial animals.

Because of its greater density, water exhibits greater changes in pressure with depth than does air. This steeper gradient, as we are now beginning to learn, has profound effects on the metabolism of marine organisms and their distribution. (The topic is discussed more fully in Chapter 4.)

It is also worth pointing out that, unlike air, seawater contains energy-rich organic compounds in the form of dissolved organic matter (DOM), which can itself be a potential food source.

A final physical difference between terrestrial and marine systems concerns oxygen. In air, oxygen constitutes a nearly constant 21% of the volume throughout the world. Water holds less oxygen, and its concentration can vary with temperature and salinity (pp. 5–6). Finally, oxygen diffuses through water at a rate 1/10,000 of the rate it diffuses in air. Thus, once reduced, it regenerates slowly.

These facts concerning oxygen in seawater create a situation for marine organisms in which the oxygen level can be reduced not only in the external environment, due to limited solubility or biological activity, but also in the immediate boundary layers around the organism, due to the slowness of diffusion. Such conditions have led marine organisms to develop rather extensive morphological, physiological, and behavioral adaptations to provide adequate oxygen in the face of its variable availability. Some of the more important adaptations include extensive respiratory surfaces, particularly in the form of integumental respiration, and small size, giving a greater surface-to-volume ratio.

■ BIODIVERSITY DIFFERENCES

Although the oceans are by far the largest repository of life on the planet, this fact is not reflected in all the patterns of biodiversity in the oceans, which are poorly known. At a higher taxonomic level, namely the phylum, the oceans have more diversity than the land (Table 1.4). Of the 33 animal phyla thus far recognized, only 12 occur on land and only one is endemic there, whereas 32 occur in the oceans and 14 are endemic there. This distribution is also reflected in the observation that on land 90% of the species are in the single Phylum Arthropoda, while in the oceans 90% of the diversity is spread over 8 phyla. At the species level, however, there is considerable disagreement. The estimated number of described marine species is 250,000, with perhaps another 150,000 to 200,000 (excluding microorganisms) yet to be described. However, Grassle and Maciolek (1992) have estimated the number of marine species yet unde-

scribed at 10 million, whereas May (1992) believes that the total number of marine species is unlikely to exceed 500,000. The number of terrestrial species is not known within an order of magnitude, but perhaps 1.5 million have been named, and estimates of the total number range from about 4 or 5 million to more than 50 million. This large number of terrestrial species is due primarily to the huge numbers of insect species, coupled with the large number of vascular plant species (about 350,000), which serve as habitats for many insect species. By contrast, the oceans have only 50 species of flowering plants and perhaps 12,000 to 13,000 species of algae.

The oceans and the land have similar latitudinal gradients in biodiversity, namely, an increase in species diversity in many taxa, moving from polar regions to the tropics. However, in the oceans there is also a longitudinal gradient in diversity, at least in shallow water. The Indo-Pacific—the area encompassing the Philippines, Indonesia, northern Australia, and New Guinea—has the highest diversity, which falls off both east and west from that center. In the Atlantic, the highest diversity is found in the west (Caribbean Sea) and declines to the east. The highest species diversity in the oceans occurs among the benthic organisms, not among those living in the water column. Among habitats, the highest diversity is found in coral reefs and deep-sea benthos.

■ LIFE-HISTORY DIFFERENCES

There are several life-history traits common to many marine organisms that are not seen in terrestrial organisms. For example, in contrast to terrestrial organisms, aquatic photosynthetic organisms and animals may shed both male and female gametes into the water and have external fertilization. [Terrestrial plants may shed male gametes (pollen) to be moved by air currents, but female gametes are never shed.] Furthermore, there are no marine equivalents of the many terrestrial pollinators—animals that transfer male gametes in return for a reward. Finally, for animals, parental care (brood protection) and investment of energy in offspring operate at a much lower level in aquatic systems than in terrestrial systems. Strathmann (1990) suggests that this is probably because the abundance of food in water permits feeding at a very small size, because desiccation is not a factor, and perhaps because the lower diffusion constant of oxygen in water and the higher viscosity of water may simply preclude furnishing enough oxygen to a brooded clutch of young.

TABLE 1.4

Number of Higher Animal Taxa in the Three Main World Ecosystems[a]

TAXON	MARINE	TERRESTRIAL	FRESHWATER
Phyla	32[b]	12	16
Classes[c]	73	25	33

[a] Includes parasites.
[b] Considering Mesozoa as two phyla, Pogonophora as a polychaete, and Pentastomida as a class of arthropods.
[c] Not counting those phyla which are not subdivided into classes.

■ STRUCTURAL AND FUNCTIONAL DIFFERENCES

One striking difference between terrestrial and marine communities is the insignificance of large macroscopic plants in the latter. Terrestrial communities are universally dominated by large flowering plants that are persistent and long lived. By contrast, with the exception of certain large kelp plants, the communities of the sea have no large photosynthetic organisms, and the dominant autotrophs are single-celled microscopic organisms of various groups of bacteria and algae. In turn, this means that the dominant herbivores of the sea are also small, often microscopic, animals, in contrast to the large-bodied herbivores common to terrestrial communities. There are, for example, few herbivores in the sea that compare in size with the antelopes of the African plains or the bison of the North American grasslands. The dominant marine herbivores are microscopic crustaceans called **copepods** or various protistans. (See Figure 2.5.)

Another characteristic of terrestrial plants is that significant portions of their anatomy are composed of rigid structural materials, such as wood and fiber, which themselves are relatively indigestible by most herbivores. As a result, herbivores grazing in terrestrial communities rarely remove significant amounts of the entire plant community. By contrast, the small herbivores of the seas usually consume the entire autotroph, and a major difference between terrestrial and marine ecosystems is the ability of the herbivores to remove the autotrophs completely. (See pp. 77–80.)

Terrestrial communities are characterized by a matrix of long-lived plants and a fauna that is generally shorter lived. In the seas—at least among those communities associated with the sea bottom—the autotrophs are short lived, and the matrix of the community consists of relatively large, long-lived animals. One speaks of redwood forests, oak forests, and tallgrass prairies on land, but of coral reefs, clam beds, oyster reefs, and mussel beds in the seas.

As a result of the microscopic nature of most marine autotrophs and herbivores, the majority of large animals in the sea are carnivores. Large marine animals are, therefore, on higher trophic levels than large land animals are. From the standpoint of human beings, harvestable food in the form of animal protein is available primarily from large herbivores on land (the second trophic level), such as cattle and sheep, whereas harvestable food from the sea is usually in the form of carnivores from the third or fourth trophic levels, such

as salmon, tuna, and halibut. A corollary of this observation is that most marine food chains have about five links (steps) to reach the top carnivore, whereas terrestrial food chains tend to be shorter, averaging three links to reach the top carnivore (Figure 1.24).

There is also a question about the efficiency of energy production and transfer between terrestrial and marine ecosystems. Generally, the production of organic material is higher in terrestrial than in marine ecosystems, but the efficiency of transfer from first to second trophic levels is higher in marine ecosystems. Although starting out with less energy fixed, marine systems lose less with the first transfer. Attempts at further assessments of the efficiency of energy transfer in marine food chains are complicated because of the various feedback mechanisms that tend to blur the trophic level to which to assign the animal. The primary group responsible for this blurring of the trophic level is the filter feeders. These animals feed upon suspended particles, and what is retained on their filters is based on size, not whether the particle is animal or plant, living or dead. This means that they often take in both plant and animal material and feed on several trophic levels. A second reason for the blurring of the trophic levels is the tendency of marine species to switch positions in the trophic hierarchy. Juvenile fishes, for example, may feed on herbivorous copepods, but when the fishes mature, they feed on other fishes higher in the trophic spectrum. An additional problem is the ability of various marine organisms to switch trophic levels as adults. Salmon, for example, may feed on squid at one time, but switch to lantern fish at another; thus, they shorten the food chain, since lantern fishes are themselves the food of squid. Similarly, euphausiid crustaceans (krill) may be herbivores in shallow water and carnivores in the open ocean.

The differences between terrestrial and marine ecosystems should be kept in mind as we explore the various habitats and communities of the sea.

Division of the Marine Environment

The marine ecosystem is the largest aquatic system on the planet. Its size and complexity make it difficult to deal with as a whole. As a result, it is convenient to divide it into more manageable, albeit somewhat arbitrary, subdivisions, each of which can then be discussed

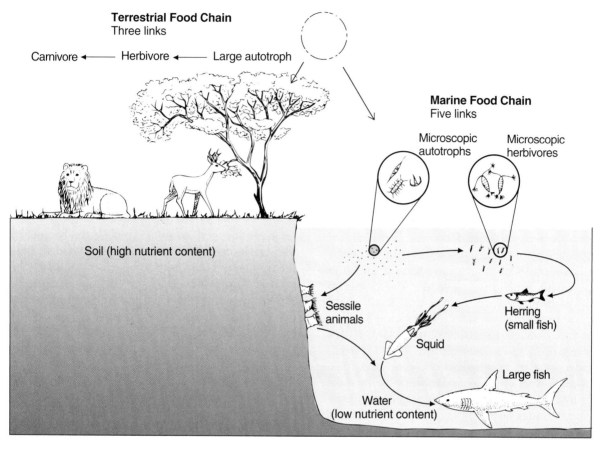

FIGURE 1.24 Diagrammatic representation of the differences between marine and terrestrial food chains.

in terms of the ecological principles that govern the adaptations of the organisms and the organization of the communities in that subdivision. No universally acceptable scheme of subdividing the marine environment has yet been proposed. The one followed here is modified from Hedgpeth (1957) and has enjoyed widespread use among biologists for more than 40 years. It probably comes as close as any to a universally acceptable division scheme.

■ MAJOR SUBDIVISIONS OF THE WORLD OCEAN

Beginning with the waters of the open ocean, subdivisions can be made both vertically and horizontally. The entire area of the open water is the **pelagic** realm; pelagic organisms live in the open sea, away from the bottom. In contrast to the pelagic realm is the **benthic** realm, comprising organisms and zones of the sea bottom. Horizontally, the pelagic realm can be divided into two zones. The **neritic** zone encompasses the water mass that overlies the continental shelves; the **oceanic** zone includes all other open waters (Figure 1.25). Progressing vertically, the pelagic realm can be further subdivided. Two schemes are possible. The first is based on light penetration. The **photic** or **euphotic** zone is that part of the pelagic realm that is lighted. Its lower boundary is the limit of light penetration and varies in depth with clarity of the water. Generally, the lower boundary is between 100 and 200 m. Because this is the zone of primary production in the ocean, it is of major importance and the subject of Chapters 2 and 3. The permanently dark water mass below the photic zone is the **aphotic** zone. Some scientists prefer to recognize a transition zone between the photic and aphotic called the **disphotic** zone. This transition area, which has enough light for vision, but not enough for photosynthesis, extends down to about 1,000 m.

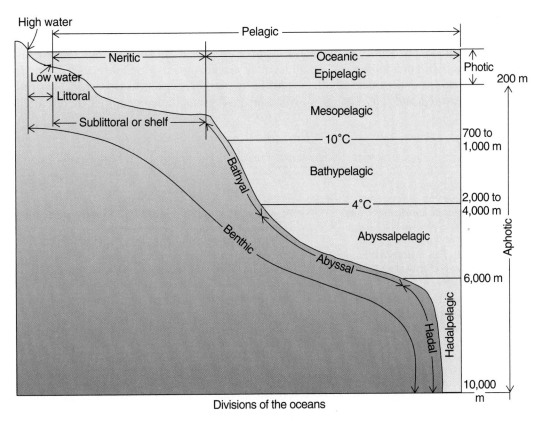

FIGURE 1.25 Divisions of the oceans (not to scale). ("The Treatise on Marine Ecology and Paleontology," J. Hedgpeth, *Ecology Memoir #67.* Copyright Geological Society of America. Used with permission.)

The pelagic part of the photic zone is also known as the **epipelagic** zone. The pelagic part of the aphotic zone can be subdivided into zones that succeed each other vertically. The **mesopelagic** is the uppermost of the aphotic areas. Its lower boundary in the tropics is the 10°C isotherm, which ranges from 700 to 1,000 m, depending on the geographical area. Next is the **bathypelagic**, lying between 10 and 4°C, or between an upper boundary of 700 to 1,000 m and a lower boundary of 2,000 and 4,000 m in depth. Overlying the plains of the major ocean basins is the **abyssalpelagic**, which has its lower boundary at about 6,000 m. The open water of the deep oceanic trenches between 6,000 and 10,000 m is called the **hadalpelagic** zone (Figure 1.25).

Corresponding to the last three pelagic zones are three bottom or benthic zones. The **bathyal** zone is that area of the ocean bottom encompassing the continental slope and rise down to about 4,000 m.

The **abyssal** zone includes the broad abyssal plains of the ocean basins between 4,000 and 6,000 m. The **hadal** is the benthic zone of the trenches between 6,000 and 10,000 m. This entire aphotic area, although encompassing the largest volume and area of the world's oceans, is little known and forms the subject of Chapter 4.

The benthic zone underlying the neritic pelagic zone on the continental shelf is termed the **sublittoral** or **shelf** zone. It is illuminated and is generally populated with an abundance of organisms constituting several different communities, including seagrass beds, kelp forests, and coral reefs. The majority of this area is covered in Chapter 5, but the specialized coral reef communities are discussed in Chapter 9.

Two transitional areas exist, one between the marine environment and the terrestrial, the other between marine and fresh water. The **intertidal** zone or **littoral** zone is that shore area lying between the extremes of high and low tide; it represents the transitional area from marine to terrestrial conditions. A well-studied region of abundant life, the intertidal zone is the subject of Chapter 6. **Estuaries** represent the transition area where fresh and salt water meet and mix. They form the subject of Chapter 8.

It is well to remember that the classification scheme discussed in this chapter is not absolute, but is used for convenience, and organisms and communities may extend into two or more zones. This is particularly true of organisms that are powerful swimmers and divers; these organisms are discussed in Chapter 3.

Summary of Key Concepts

- Water is a simple chemical compound consisting of one atom of oxygen covalently joined to two atoms of hydrogen. With weak hydrogen bonds to other water molecules, water is a polar compound. This hydrogen bonding gives rise to many of water's unique chemical and physical properties that make life possible.

- The unique chemical and physical properties of water include the highest surface tension of all common liquids, a high viscosity relative to air, a high latent heat of vaporization and fusion, high heat capacity, a peculiar density–temperature relationship, and the capability of acting as an almost universal solvent.

- Seawater is pure water in which a variety of gases and solids are dissolved. Most of these dissolved substances have ratios that are constant among themselves, thus allowing the concentrations of all substances to be identified by measuring one. Accordingly, the salinity of seawater is usually found by measuring the chlorine concentration of the water.

- The two major gases in seawater are oxygen and carbon dioxide, and their solubility is a function of temperature. Oxygen concentration also depends upon usage and production by organisms.

- Seawater is a buffer employing a carbonic acid–bicarbonate system that maintains a pH around 8.0.

- All major oceans are connected to each other. They also are connected to the continents, by shallow extensions of the latter called continental shelves. Most of the ocean basins are flat abyssal plains, cut by deep trenches in some places and with submarine ridges, volcanoes, and seamounts in other areas.

- Temperature in the oceans varies latitudinally and vertically.

- As a result of different temperatures and salinities, the oceans have water masses of different densities. The uppermost water masses are moved around by the action of the prevailing wind belts of the Earth, to form global surface currents and local waves. Under certain conditions of motion, and coupled with the Coriolis effect, these currents may diverge, causing cooler water to rise from deeper locations, a phenomenon called upwelling.

Also, some water masses may cool upon transport into cold climates or upon their formation in polar areas and sink to form deep-water masses that make up the bottom water in the world's ocean basins. These water masses move slowly and are linked back to the surface in thermohaline circulation. Internal waves or tides create turbulence and water movement in the deep ocean water masses.

- Ecology is the science that treats the spectrum of relationships that exist between organisms and their environments and among groups of organisms. The various organisms and the environmental factors may be further organized into species, populations, communities, and ecosystems, and more.

- Ecosystems are functional units of variable size composed of living and nonliving parts that interact. Ecosystems have a trophic structure, usually consisting of an autotrophic base in which energy is fixed, followed by one or more layers of heterotrophs, through which the fixed energy is passed. The number of trophic levels is limited by the loss of energy during the transfer between layers.

- Biogeochemical cycling of carbon, nitrogen, and phosphorus between marine organisms and the physical environment is vital to the continued maintenance of life in the oceans.

- Communities and ecosystems have the same trophic structure throughout the world's oceans, but the species composing the structure differ in different geographical areas.

- In ecology, a niche is defined as the role of an organism in a community. Niches may be broad or narrow and can be established for virtually any environmental or biological factor. By contrast, a habitat is the place where an organism is found.

- Most communities characteristically have a few species that are abundant or dominant and that define the community, together with many less abundant species. Species structure may be measured by species richness or species diversity. There are two schools of thought as to how high species diversity is established and maintained: the equilibrium model and the intermediate-disturbance model.

■ Communities are not static units; they change through time. Some change in an orderly fashion through a predictable sequence, whereas others progress in an unpredictable fashion. Such ecological succession is explained by the facilitation, inhibition, and tolerance models.

■ Populations, communities, and ecosystems are regulated by various chemical and physical factors existing in the environment. Populations are further regulated by four biological factors: competition, predation, parasitism, and disease.

■ Large numbers of marine animals have larvae in their life cycle. Larvae, which are morphologically different from the adult, usually undergo a profound change in morphology when they become adult.

■ There are three patterns of development in marine organisms, producing, respectively, planktotrophic larvae, lecithotrophic larvae, and juveniles or nonpelagic larvae. Several hypotheses explain under what conditions each of the patterns prevails, but none of the hypotheses account for all the nuances seen in the sea, suggesting that additional factors must be operating.

■ Many marine benthic communities are established by larvae settling from the water column. To ensure settlement into the proper environment, larvae respond to various environmental stimuli, of which certain chemicals appear to be the most influential.

■ Most organisms can be categorized into one of two life-history patterns: opportunistic and equilibrium.

■ Terrestrial and marine ecosystems exhibit some striking differences, due primarily to the physical and chemical differences in the bathing media—air and water.

■ The greater density and viscosity of water compared with air means that organisms can live permanently suspended in water, which is not the case for air. Therefore, we have plankton communities in the water. These in turn allow the existence of sessile animals that filter organisms out of the water column.

■ Water also buoys up organisms against gravity, so that marine organisms do not have significant amounts of biomass invested in structural material to hold them up.

■ Water absorbs light and does so differentially for the wavelengths of visible light. This means that light penetrates only to a few hundred meters in the ocean; therefore, the greatest volume of the oceans is without light. As a consequence, many marine communities are without an autotrophic component.

■ Because of its greater density, water also exhibits greater changes in pressure with depth—with profound effects on the metabolism of organisms.

■ Water holds less oxygen than air, and the oxygen concentration of water is not constant, as it is in air, but varies with temperature and salinity. Some parts of the ocean may approach anoxic conditions.

■ There are also differences in biodiversity between the marine and terrestrial environments. The marine environment has the greatest number of phyla, but the terrestrial environment greatly exceeds the marine in numbers of species.

■ Certain life-history traits seen in marine organisms are not found in terrestrial organisms. For example, marine organisms shed both male and female gametes into the water, but terrestrial organisms never shed female gametes. Conversely, there are no marine equivalents of terrestrial pollinators.

■ Large plants form the matrix of most terrestrial communities, but the marine autotrophs are dominated by single-celled microscopic autotrophs. This means that the herbivores in the sea are also tiny animals, in contrast to the large herbivores on land. As a result, most marine food chains have five links to reach the top carnivore, as opposed to three links on land.

■ The marine ecosystem is the largest aquatic system on the planet, and its size and complexity make it difficult to deal with as a whole. It is therefore convenient to divide it into more manageable units that can then be taken up separately.

Review Questions

ESSAY Develop complete answers to these questions:

1. What is a thermocline? How does it become established? Why is it that, once a thermocline has become established, it will not be disrupted by a summer storm? What is a pycnocline?

2. Is there a marine equivalent to the terrestrial concept of ecological succession? Is there a marine climax community? Compare and contrast the classical or facilitation model of succession with the inhibition and the tolerance models.

3. Energy is said to flow through an ecosystem, but chemicals cycle (e.g., biochemical cycles). Why is this distinction an important one to keep in mind in studying ecosystems?

4. Enumerate and discuss three life-history differences between marine and terrestrial organisms.

Bibliography

Bascom, W. 1980. *Waves and beaches, the dynamics of the ocean surface.* New York: Anchor Books.

Briggs, J. C. 1974. *Marine zoogeography.* New York: McGraw-Hill.

Butman, C. 1987. Larval settlement of soft sediment invertebrates: The spatial scales of pattern explained by active habitat selection and the emerging role of hydrodynamical processes. *Oceanog. Mar. Biol. Ann. Rev.* 25:113–165.

Caswell, H. 1981. The evolution of "mixed" life histories in marine invertebrates and elsewhere. *Amer. Nat.* 117:529–536.

Chia, Fu-shang. 1988. Differential larval settlement of benthic invertebrates. In *Reproduction, genetics and distributions of marine organisms,* edited by J. S. Ryland and P. A. Tyler. Fredensborg, Denmark: Olsen and Olsen, 3–12.

Clements, F. E. 1936. Nature and structure of the climax. *J. Ecol.* 24:252–284.

Connell, J. H. 1975. Some mechanisms producing structure in natural communities: A model and evidence from field experiments. In *Ecology and evolution of communities,* edited by M. L. Cody and J. M. Diamond. Cambridge, MA: Belknap Press, 460–490.

Connell, J. H. 1978. Diversity in tropical rain forests and coral reefs. *Science* 199(4335):1302–1310.

Connell, J. H., and R. O. Slayter. 1977. Mechanisms of succession in natural communities and their role in community stability and organization. *Amer. Nat.* 111:1119–1144.

Crisp, D. J., and P. S. Meadows. 1962. The chemical bases of gregariousness in cirripedes. *Proc. Roy. Soc. London (Biol.)* 150:500–520.

Denny, Mark. 1990. Terrestrial versus aquatic biology: The medium and the message. *Amer. Zoo.* 30(1):111–122.

Dietrich, G., K. Kalle, W. Krauss, and G. Siedler. 1980. *General oceanography: An introduction.* 2d ed. New York: Wiley.

Egler, F. E. 1954. Vegetation science concepts. I. Initial floristic composition, a factor in old field vegetation development. *Vegetation* 14:412–417.

Grassle, J. F., and J. P. Grassle. 1974. Opportunistic life histories and genetic systems in marine benthic polychaetes. *J. Mar. Res.* 32:253–284.

Grassle, J. F., and N. Maciolek. 1992. Deep sea species richness: Regional and local diversity estimates from quantitative bottom samples. *Amer. Nat.* 139: 313–341.

Grinell, J. 1917. Field tests of theories concerning distributional control. *Amer. Nat.* 51:115–128.

Gross, M. G. 1982. *Oceanography: A view of the earth.* 3d ed. Englewood Cliffs, NJ: Prentice-Hall.

Hedgpeth, J. 1957. Classification of marine environments and concepts of marine ecology. In *The treatise on marine ecology and paleoecology.* Vol. 1, *Ecology.* Memoir No. 67 of The Geological Society of America, edited by J. E. Hedgpeth. New York: Geological Society of America.

Ingmanson, D. E., and W. J. Wallace. 1973. *Oceanology: An introduction.* Belmont, CA: Wadsworth.

Krebs, C. J. 1985. *Ecology, the experimental analysis of distribution and abundance.* New York: Harper and Row.

Levin, L. A., and T. S. Bridges. 1995. Pattern and diversity in reproduction and development. In *The Ecology of marine invertebrate larvae,* edited by L. McEdward. Boca Raton, FL: CRC Press, 1–48.

MacArthur, R. 1960. On the relative abundance of species. *Amer. Nat.* 94 (874):25–34.

Martin, J., *et al.* 1994. Testing the iron hypothesis in ecosystems of the equatorial Pacific Ocean. *Nature* 371:123–129.

May, R. M. 1992. Bottoms up for the oceans. *Nature* 357:278–279.

McEdward, L. R., ed. 1995. *Ecology of marine invertebrate larvae.* Boca Raton, FL: CRC Press.

Obrebski, S. 1979. Larval colonizing strategies in marine benthic invertebrates. *Mar. Ecol. Progr. Ser.* 1:293–300.

Ormond, R. F. G., J. D. Gage, and M. V. Angel, eds. 1997. *Marine biodiversity, patterns and processes.* Cambridge, U.K.: Cambridge University Press.

Paine, R. 1969. A note on trophic complexity and community stability. *Amer. Nat.* 103:91–93.

Palmer, A. R., and R. R. Strathman. 1981. Scale of dispersal in varying environments and its implication for life histories of marine invertebrates. *Oecologia* 48:308–318.

Smith, R. L. 1980. *Ecology and field biology.* 3d ed. New York: Harper and Row.

Smith, R. L. 1986. *Elements of ecology.* New York: Harper and Row.

Steele, J. H. 1985. Comparison of marine and terrestrial ecosystems. *Nature* 313:355–358.

Steele, J. H. 1991. Ecological explanations in marine and terrestrial systems. In *Marine biology, its accomplishment and future prospect,* edited by J. Mauchline and T. Nemoto. New York: Elsevier, 101–106.

Strahler, A. 1971. *The earth sciences.* 2d ed. New York: Harper and Row.

Strathmann, R. R. 1974. The spread of sibling larvae of sedentary marine invertebrates. *Amer. Nat.* 108:29–44.

Strathmann, R. R. 1977. Egg size, larval development, and juvenile size in benthic marine invertebrates. *Amer. Nat.* 111:373–376.

Strathmann, R. R. 1978. The evolution and loss of feeding larval stages of marine invertebrates. *Evolution* 32:899–906.

Strathmann, R. R. 1985. Feeding and nonfeeding larval development and life history evolution in marine invertebrates. *Ann. Rev. Ecol. Syst.* 16:339–361.

Strathmann, R. R. 1990. Why life histories evolve differently in the sea. *Amer. Zoo.* 30(1):197–208.

Strathmann, R. R., and M. F. Strathmann. 1982. The relation between adult size and brooding in marine invertebrates. *Amer. Nat.* 119:91–101.

Thorson, G. 1950. Reproductive and larval ecology of marine bottom invertebrates. *Biol. Rev.* 25:1–45.

Thorson, G. 1966. Some factors influencing the recruitment and establishment of marine benthic communities. *Neth. J. Sea Res.* 3(2):267–293.

Underwood, A. J. 1974. On models for reproductive strategy in marine benthic invertebrates. *Amer. Nat.* 108:874–878.

Vance, R. 1973. Reproductive strategies in marine benthic invertebrates. *Amer. Nat.* 107(955): 339–352.

Whittaker, R. H. 1975. *Communities and ecosystems.* 2d ed. New York: Macmillan.

Wilson, D. P. 1952. The influence of the nature of the substratum on the metamorphosis of the larvae of marine animals, especially the larvae of *Ophelia bicornis* Savigny. *Anna. l'Insti. Oceanogr. Monaco* 27:49–156.

CHAPTER 2

PLANKTON AND PLANKTON COMMUNITIES

As we have seen in the previous chapter, certain fundamental physical differences between terrestrial and marine environments contribute to the differences we observe in the organization of the communities in the two areas. Perhaps nowhere are these differences more dramatic than in the free-floating and weakly swimming associations of organisms that we collectively call the plankton. It is appropriate to begin a consideration of the ecology of the oceans with the plankton because the chlorophyll-bearing organisms of the plankton contribute the greatest amount of photosynthesis in the oceans. In fact, some scientists believe that oceanic productivity rivals that of the terrestrial environment and may even be greater. The plankton trap most of the sun's energy that enters the oceans, which is then transferred to the many other communities of the oceans. This vital role played by the initial fixation of energy makes plankton important in the economy of the oceans. Just as there could be no life on land without the energy-fixing grasses, trees, and shrubs, there could be no life in the oceans without the energy-fixing, minute planktonic organisms.

■

■ DEFINITIONS

The term **plankton** is a general term. Planktonic organisms have such limited powers of locomotion that they are at the mercy of the prevailing water movement. This is in contrast to the **nekton**, strong-swimming animals of the open sea, capable of exercising movement against the prevailing water flow. Plankton may be further subdivided. **Phytoplankton** are free-floating organisms of the sea that are capable of photosynthesis. **Zooplankton** are the various free-floating animals. **Bacterioplankton** include the various minute heterotrophic and autotrophic bacteria. Recently, we have become aware of various **virioplankton** organisms in the oceans.

Because planktonic organisms are traditionally captured with nets of various mesh sizes, they have also been classified by size. This size classification does not distinguish between photosynthetic and nonphotosynthetic organisms. The most recent and probably most acceptable, though arbitrary, scheme

TABLE 2.1

Size Classification for Plankton Organisms

				SIZE CLASS			
CLASS NAME	Femtoplankton (0.02–.2 μm)	Picoplankton (0.2–2.0 μm)	Nanoplankton (2.0–20 μm)	Microplankton (20–200 μm)	Mesoplankton (0.2–20 mm)	Macroplankton (2–20 cm)	Megaplankton (20–200 cm)
Virioplankton	▬▬▬						
Bacterioplankton		▬▬▬					
Mycoplankton			▬▬▬				
Phytoplankton			▬▬▬▬▬▬▬▬▬▬▬▬				
Protozooplankton			▬▬▬▬▬▬▬▬▬				
Metazooplankton				▬▬▬▬▬▬▬▬▬▬▬▬▬▬▬▬▬▬			
Size (m)	10^{-7}	10^{-6}	10^{-5}	10^{-4}	10^{-3} 10^{-2}	10^{-1}	10^{-0}

Source: From *Sea Microbes*, by J. M. Sieburth, 1979.

of classification for plankton has seven size classes (Table 2.1). **Megaplankton** consist of all planktonic organisms above 20 cm. **Macroplankton** comprise those 2–20 cm in size. The **mesoplankton** fall between 0.2 and 20 mm in size. **Microplankton** encompass plankton 20–200 μm in size. These four groups are those usually captured in standard plankton nets and are called **net plankton**. The **nanoplankton** are very small organisms ranging from 2 to 20 μm in size. The **picoplankton** comprise minute organisms 0.2–2.0 μm in size and consist primarily of bacteria and cyanobacteria. The smallest class is the **femtoplankton**, 0.02–0.2 μm in size. Little is known of this last group. The nanoplankton, picoplankton, and femtoplankton cannot be captured in plankton nets because, in order to filter them out of the water column, the mesh would have to be so fine that when the nets were pulled behind a boat, the water would not pass through. As a result, these organisms can be obtained only by centrifuging samples of seawater, filtering samples of water on fine filters such as millipore filters, or allowing the organisms to settle out from water samples.

A final set of terms concerns the life-history characteristics of the zooplankton organisms. **Holoplankton** are those organisms that spend their entire lives in the plankton. **Meroplankton**, by contrast, comprise those species that spend but a part of their lives in the plankton. Meroplanktonic organisms include a large number of larvae of animals that, as adults, either live on the bottom or swim as nekton.

The Phytoplankton

Although this text is designed to cover the principles of organization of life in the sea, not to detail the various taxonomic groups, it is nonetheless worthwhile to include a brief description of the common phytoplankton. The phytoplankton include a wide range of photosynthetic and phototropic organisms that generate chemical energy from light, that are usually microscopic, and that are from the taxonomic domains Bacteria and Eukaya (and perhaps also the domain Archaea—see the Taxonomic Appendix for an explanation of the various changes in the hierarchy of classification of organisms). The larger phytoplankton—those normally captured in nets—consist of only two groups of Eukarya that dominate the net phytoplankton throughout the world: diatoms and dinoflagellates. The smaller phytoplankton, those in the nanoplankton and picoplankton size ranges that pass through nets, are more variable and seem to be dominataed by microbes from the domain Bacteria. The picoplankton group now appears to be the more important in productivity. (See pp. 89–92.)

■ DIATOMS

The **diatoms** (class Bacillariophyceae) are easily differentiated from the dinoflagellates because they are enclosed within a unique glass "pillbox" and have no visible means of locomotion. Each box is composed of two valves or frustules, with one valve fitting over another (Figure 2.1). The living part of the diatom is within the

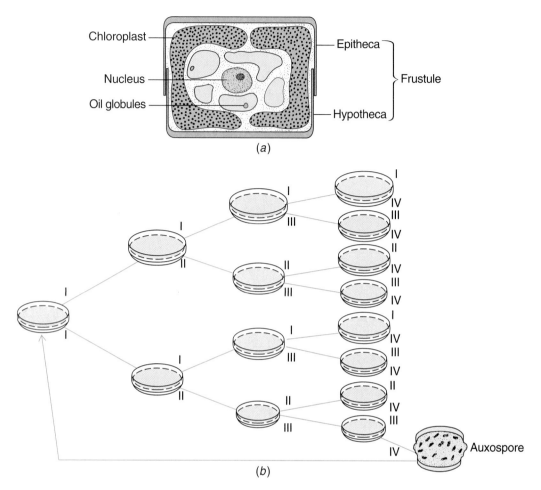

FIGURE 2.1 Diatom features. (a) Diagrammatic representation of the structure of a diatom. (b) Diagrammatic representation of the diminution of size in successive divisions and the restoration of size following auxospore formation.

box, which is constructed of silicon dioxide, the same material that is a major constituent of glass. Each box is highly ornamented with species-specific designs, pits, and perforations, and this feature has made these organisms popular with light microscopists and, more recently, scanning electron microscopists (Figure 2.2).

Diatoms may occur singly, with each individual occupying a single box, or they may occur in chains of various kinds, which themselves add to the ornamentation seen in the individual boxes (Figure 2.2; see also Figure 2.4).

In reproduction, each diatom divides into two halves, one occupying the top valve of the original box, the other the bottom. Each half then secretes a valve so that the typical box is re-created. Each new valve is secreted *within* the old valve; hence, as the process

continues through several generations, the average size of the diatom population decreases. As a result, individuals of species of diatom vary in size. Obviously, there must be a limit to the process of reduction, and it occurs after a certain number of generations, when the diatom casts off both valves and becomes a structure called an **auxospore**. Within this spore, new valves are secreted that reestablish the original size of the diatom species. (See Figure 2.1.) Although what we have just described is the general mechanism of diatom reproduction and reestablishment of size, some diatoms undergo replication without decreasing the size of their valves and thus do not require auxospore formation to reestablish that size.

Diatoms are abundant as both species and individuals. Although it is not possible to illustrate all

FIGURE 2.2 Scanning electron micrographs of diatoms. (a) *Chaetoceros* sp. (700×). (b) *Thalassiosira decipiens* (1,000×). (c) *Pseudoeunotia doliolus* (1,400×).

here, some of the more common planktonic genera are shown in Figures 2.2 and 2.4.

In recent years it has been discovered that diatoms produce a toxin called **domoic acid**, which may be potent enough to kill other organisms.

Some diatoms are not planktonic but benthic. These diatoms will be discussed in Chapter 6.

■ **DINOFLAGELLATES**

The second major group, the **dinoflagellates** (class Dinophyceae), have two flagella, which they use to propel themselves through the water. They lack an external skeleton of silicon, but are often armored with plates of the carbohydrate cellulose (Figures 2.3 and 2.4). They also have chromosomes that are condensed throughout the cell cycle. Dinoflagellates are generally small organisms and usually solitary, rarely forming chains. They reproduce by simple fission, as do the diatoms. In the case of dinoflagellates, however, each daughter cell retains half the original cellulose armor and forms a new part to replace the missing half without any diminution in size; hence, successive generations do not change in size. Some dinoflagellates also are capable of producing toxins that are released into seawater. If dinoflagellates become extremely abundant (2 to 8 million cells per liter), the cumulative effect of all the

FIGURE 2.3 Scanning electron micrographs of dinoflagellates. (a) *Dinophysis parva* (21 × 16 μm). (b) *Ceratium candelabra* (140 × 70 μm). (c) *Dinophysis caudata* (160 × 80 μm). (Biophoto Associates/Photo Researchers, Inc.)

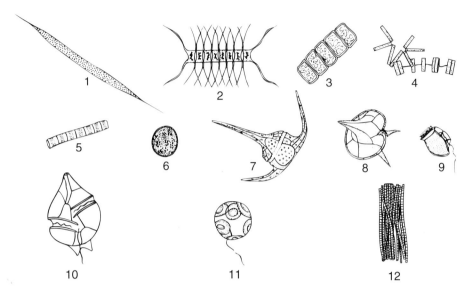

FIGURE 2.4 Diagrammatic representation of some characteristic genera of marine phytoplankton. Diatoms: (1) *Rhizosolenia*, (2) *Chaetoceros*, (3) *Navicula*, (4) *Thalassiosira*, (5) *Skeletonema*, (6) *Coscinodiscus*. Dinoflagellates: (7) *Ceratium*, (8) *Peridinium*, (9) *Dinophysis*, (10) *Gonyaulax*. Coccolithophores: (11) *Coccolithus*. Cyanobacteria: (12) *Trichodesmium*. (Not to scale.)

toxins released may affect other organisms, causing mass mortality. Such extreme concentrations, or blooms, of dinoflagellates are called **red tides** and are responsible for massive localized mortality in fish and invertebrates in various places.

Some dinoflagellates have nonmotile stages called **zooxanthellae**, which are symbionts in the tissues of many invertebrates, including corals, sea anemones, and giant clams. (See Chapter 10.) Other dinoflagellates, such as the common *Noctiluca* (see Figure 2.26a), are also highly bioluminescent (see Chapter 5) and, when present in large numbers, can actually light the wakes of boats and the breaking waves on a beach. Many dinoflagellates, such as *Noctiluca*, are not photosynthetic, but feed by engulfing food particles (phagocytosis), thus playing the role of a grazer.

■ **THE SMALLER PHYTOPLANKTON**

Haptophytes (Coccolithophoridae, Haptophyceae) are tiny (less than 20 μm) eukaryotic flagellates with one or two chromatophores, a unique threadlike appendage called the haptonema, and distinctively marked external calcareous or organic plates. They are abundant in all oceanic waters. The most widespread and abundant haptophytes are the **coccolithophores**, easily distinguished by the tiny calcareous plates (coccoliths) on their outer surface. They have complex life histories, with several morphologically different cells present in the same species and several modes of reproduction. Coccolithophores are now recognized as a major source of primary production in many ocean areas. Other, less abundant, microalgae include the silicoflagellates (Chrysophyceae), the cryptomonads (Cryptophyceae), and certain motile green algae (Chlorophyceae).

Constituents of the nanoplankton and picoplankton size classes (sometimes collectively called nanoplankton) include a number of photosynthetic organisms that are important in primary productivity and in oceanic food webs—a fact that has been realized only in recent years. (See pp. 89–94.) Most of these tiny organisms belong to the branch of life-forms known as prokaryotes, which are characterized by the absence of a nuclear envelope and of membrane-covered organelles. The prokaryotes are divided into two domains, Bacteria and Archaea, which are as different from each other as both are from the eukaryotes. The domain Bacteria includes the **cyanobacteria** (Cyanophyceae, formerly called blue-green algae) (Figure 2.4). The cyanobacteria possess chlorophyll *a*, but not in plastids. Cyanobacteria are abundant throughout the world oceans, and in the tropics they occasionally form dense mats of filaments and color the water. The Red Sea was named after the red color of the cyanobacterium *Trichodesmium* (*Oscillatoria*)

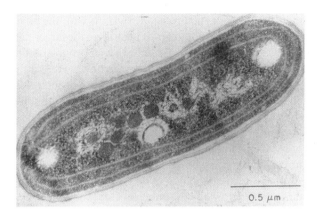

FIGURE 2.5 Transmission electron micrograph of the prochlorophyte *Synechococcus*. (Copyright Woods Hole Oceanographic Institute.)

erythraeum. Perhaps the most important are the recently discovered, very tiny (0.6–0.8 μm) **prochlorophytes**, which are probably the most numerically abundant phytoplankters in the open ocean (10^6 cells/ml). Prochlorophytes are thought to contribute from one-third to perhaps seven-tenths of all chlorophyll *a* in the midlatitude open ocean and are estimated to contribute between one-third and four-fifths of the total ocean productivity. They resemble other cyanobacteria, but are smaller and have a unique form of chlorophyll called divinyl chlorophyll. Thus far, only two genera are known: *Prochloroccus* and *Synechococcus* (Figure 2.5). They differ from each other in the type of nitrogen compound that they can use. Some cyanobacteria can fix nitrogen gas into biologically available soluble nitrogen compounds. The best known of these organisms are the very large photosynthetic colonial bacteria of the genus *Trichodesmium*. In 2001, Zehr et al. reported on their discovery of unknown single-celled cyanobacteria in the subtropical Pacific Ocean that also have been shown to fix nitrogen.

Recently, two groups of marine scientists working separately and employing different techniques found entirely new bacterial groups with previously unknown light-harvesting mechanisms (phototrophy) that likely bring energy more efficiently into the marine planktonic food chain. Beja et al. (2000) found a new protein that they called **proteorhodopsin**, which harnesses light to generate ATP. However, this type of phototrophy that uses the pigment rhodopsin had never before been seen in Bacteria, although it had been found in Archaea. The other team, using a fluorometer designed to detect wavelengths of light given off only by bacterial chlorophyll, discovered significant numbers of phototrophic

bacteria in surface waters in both the Atlantic and Pacific Oceans. These bacteria, called **alpha proteobacteria** by Kolber et al. (2000), do not produce oxygen as do photosynthetic organisms but instead use oxygen to synthesize chlorophyll. They also fix carbon so that they use light energy to break down organic matter and at the same time they somehow fix some inorganic carbon. Both of these groups of bacteria use light (through different mechanisms) to generate ATP and may account for as much as 20% of the bacterioplankton in the surface ocean waters. It is not yet known whether the two groups also use the light to fix carbon the way other photosynthetic phytoplankton do.

Pelagic heterotrophic bacteria (or bacterioplankton) are also found in all oceans. They are most abundant near the surface and are now thought to equal or exceed the total biomass of phytoplankton in the oceans. Pelagic heterotrophic bacteria are usually found in association with organic particles in the water column—collectively called **particulate organic carbon (POC)**—or on various gelatinous zooplankton pieces known as marine snow (see p. 96). They are usually motile and gram negative and occur in coccoid, rod, and spiral forms. They decrease markedly with depth, and their role in the microbial loop of the oceanic food web is now clearly recognized. (See pp. 91–94.)

Within the last decade, using molecular techniques, several scientists have found large numbers of two groups of the domain Archaea to be widely distributed in the oceans in cold surface waters as well as in cold depths below 100 meters in temperate oceans. The ecological significance of these groups is currently speculative, but Ouverney and Fuhrman (2000) have shown that both groups are able to ingest organic molecules present in seawater in extremely low concentrations.

■ THE VIRUSES

It is now becoming apparent to marine biologists that viruses may well be the most common biological agents in the oceans. They typically number about 10^{10} per liter, exceeding the bacterial abundance of 10^9 per liter, and retain this dominance throughout all habitats in all oceans. All viruses are parasitic and must infect another organism's cells in order to live and reproduce. They have a simple structure of strands of DNA or RNA surrounded by a protein or sometimes a lipid coat. They are only 20 to 200 nanometers long and are invisible to ordinary light microscopy, but visible in transmission electron microscopes (Figure 2.6). Until about 10 years ago, viruses were virtually ignored in

FIGURE 2.6 Observations of viral-like particles (VLPs) in concentrated seawater samples have revealed high abundances (up to 1 million to 100 million VLPs per ml of seawater). In this micrograph, 2% uranyl acetate negative stain reveals VLPs that were isolated from the oceans. These VLPs are a clonal virus called P-RSM2 and infect both low-light and high-light adapted strains of the marine cyanobacterium *Prochlorococcus*. These marine cyanobacteria are less than a micron in diameter. (Copyright Matthew B. Sullivan.)

marine biology and ecology, but since that time research has not only established their overwhelming abundance, but also their likely significance in controlling various photosynthetic organisms as well as various ecological processesses in the open ocean, such as nutrient cycling, particle size distributions, and sinking rates of bacterioplankton and phytoplankton. (See p. 94.)

With all the recent discoveries of microbial organisms the current challenge to marine ecologists and oceanographers is to elucidate their roles in the various marine biogeochemical processes and to understand how they might structure communities in the oceans of the world. This will not be easy because at the present time, as Azam and Worden (2004) note, these newly discovered organisms are uncultivable and prior methods for elucidating microbial diversity and significance required the organisms to be cultivated. Therefore cultivation-independent methods would appear to be necessary.

The Zooplankton

In contrast to the phytoplankton, which consist of a relatively small variety of photosynthetic organisms, the zooplankton are extremely diverse, consisting of a host of larval and adult forms representing most of the animal and many of the protistan phyla. As with the phytoplankton, two groups can be identified: the larger net zooplankton and the smaller microzooplankton (nanozooplankton and picozooplankton). We know much more about the first group, because of their size and ease of collection. Marine biologists have long considered these larger net zooplankton to be the most important grazers of the phytoplankton. In recent years, however, the importance of the smaller microzooplankton as grazers has been established. (See pp. 89–94.) Among the net zooplankton, one group is more important than the others: the subclass Copepoda (subphylum Crustacea, phylum Arthropoda) which includes small holoplanktonic crustaceans that dominate the net zooplankton throughout the world's oceans (Figure 2.7). These small animals are of vital importance in the economy of the ocean's ecosystems because they are one of the primary herbivorous animals in the sea. It is they who graze upon the aquatic pastures constituted by the larger phytoplankton, especially in coastal waters, and it is they who provide one of the links between the primary production of the autotrophs and the numerous large and small carnivores.

■ COPEPODS

Free-living planktonic copepods are generally small, between one and several millimeters in length. They swim weakly, using their jointed thoracic limbs, and have a characteristic jerky movement. They employ their large antennae to slow their rate of sinking. Most of the free-living planktonic copepods have a characteristic body shape and hence are readily recognizable (Figure 2.7a).

Copepods graze on phytoplankton either by using a filtering mechanism that removes algal cells from the water or by using the anterior appendages. As Koehl and Strickler (1981) have demonstrated, copepods produce a stream of water past their bodies by flapping several anterior appendages (second antennae, mandibular palps, first maxillae and maxillipeds). When an algal cell enters this stream of water, the anterior appendages respond by beating asymmetrically, redirecting the water so that the entrained algal cell moves toward the animal. The copepod then rapidly flings apart its second maxillae, which close over the water and the algal cell, forcing water out through the setae while retaining the algal cell (Figure 2.8). Some copepods are carnivorous, seizing prey with their appendages.

In copepods, the sexes are separate, and sperm packaged in spermatophores is transferred to the

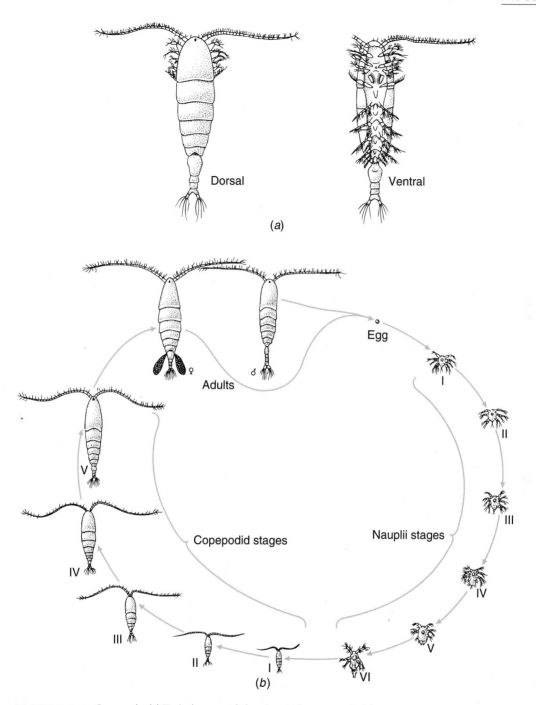

FIGURE 2.7 Copepods. (a) Typical copepod showing major anatomical features. (b) Outline of the typical life cycle of a copepod.

female. After fertilization, the eggs are enclosed in a sac attached to the female's body. The eggs hatch as **nauplius** larvae and progress through several naupliar stages and then several more **copepodid** stages before becoming adult. (See Figure 2.7.) As we shall see, the life history of the adult and larval forms affects the phytoplankton cycles observed in the sea.

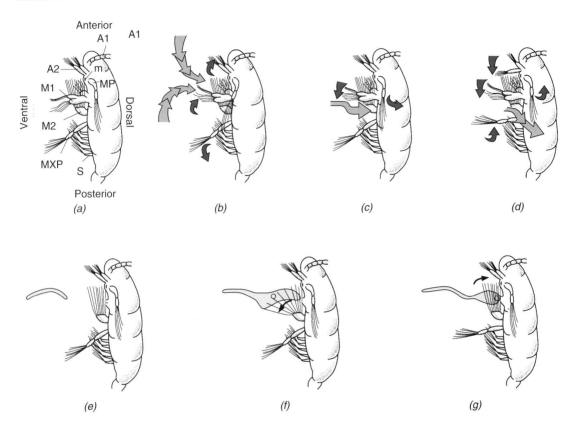

FIGURE 2.8 (a) Diagram of a copepod in feeding position as viewed from the left side. Only the left appendage of each pair is shown. Abbreviations are A1 = first antennae; A2 = second antennae; m = mouth; MP = mandibular palp; M1 = first maxillae; M2 = second maxillae; MXP = maxilliped; S = swimming appendage. (b–d) Diagrams of movement of feeding appendage (darker blue arrows) and water current (light blue arrows). Arrows with a narrow shaft and a wide head indicate movement laterally toward the reader; arrows with a wide shaft and a narrow head indicate movement away from the reader. (e–g) Diagrams indicating mechanism of food capture in calanoids. (e) shows movement of water bypassing second maxillae when no cells are present; (f) shows the second maxillae flung open when an algal cell is present (small arrow); (g) shows the final capture of algal cell by an inward sweep of the second maxillae. (From "Copepod Feeding Currents," M. A. R. Koehl & J. R. Strickler, *Limnology & Oceanography*, Vol. 26, pp. 1061–1072, 1981. Reprinted by permission of the American Society of Limnology and Oceanography.)

■ OTHER ZOOPLANKTON

Because of the great diversity of zooplankton organisms, it is not practical to cover all in detail. Thus, only the larger and more significant taxonomic groups will be considered. Within each of these taxa are often many different species.

Among the holoplankton, the kingdom Protista (see Taxonomic Appendix for changes in this taxon) (Figure 2.9) is the most important, approaching or even exceeding the copepods as grazers. Usually, however, the protists are overlooked because of the small

size of the individuals and the difficulty of capturing them. The dominant groups of the net zooplankton are in the phylum Sarcomastigophora, particularly the order Foraminiferida (forams) and the class Radiolaria (Figure 2.9a, b). Both radiolarians and foraminiferans are single-celled organisms that produce skeletons, of calcium carbonate in the case of forams and of glass (SiO_2) in the case of radiolarians. Radiolarians are exclusively marine, forams primarily so. Members of these two taxa are so abundant and widespread that their skeletons have formed thick

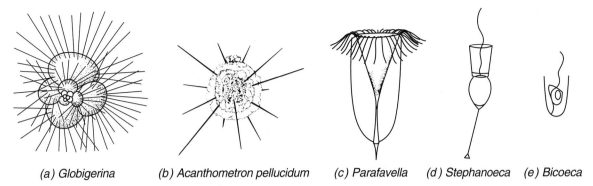

(a) Globigerina   *(b) Acanthometron pellucidum*   *(c) Parafavella*   *(d) Stephanoeca*   *(e) Bicoeca*

FIGURE 2.9 Holoplankton. Examples of some typical plankton members of the kingdom Protista. (a) Phylum Sarcomastigophora, order Foraminiferida. (b) Phylum Sarcomastigophora, order Radiolaria. (c) Phylum Ciliophora. (d) Phylum Sarcomastigophora. (e) Phylum Sarcomastigophora. (Not to scale.) (a–c after *A Guide to Marine Coastal Plankton and Marine Invertebrate Larvae* by DeBoyd L. Smith. Copyright © 1977 by Kendall/Hunt Publishing Company. Used with permission.)

layers of foraminiferan and radiolarian ooze extending over vast areas of the deep-sea floor. Among the nanoplankton, many small protistans are the most important primary consumers. The dominant groups are various ciliates of the phylum Ciliophora and small flagellates of the phylum Sarcomastigophora (Figure 2.9). Both of these groups are exceedingly abundant and are the major grazers of the nanophytoplankton (Figure 2.10).

Holoplanktonic members of the phylum Cnidaria include the various jellyfishes of the classes Hydrozoa (Figure 2.11) and Scyphozoa (Figures 2.12 and 2.13) and the curious, complex hydrozoan colonies known as siphonophores (Figure 2.14; see also Figure 2.27b). Scyphozoan jellyfishes are among the largest plank-

tonic organisms and may occasionally be found in large numbers. (See Figure 2.12.)

Closely related to the Cnidaria is the phylum Ctenophora. (See Figures 2.13 and 2.15). With few exceptions, this phylum is entirely planktonic. (See Figures 2.13c, d; 2.15, 2.26c, and 2.28a). All are voracious carnivores, capturing food with sticky tentacles

FIGURE 2.10 Epifluorescence photomicrograph of a DAPI-stained flagellate consuming a red fluorescing phytoplankter. (Photo courtesy of Nick Welschmeyer.)

FIGURE 2.11 A Hydrozoan jellyfish *Crassota alba*. (Approximate size is 15 mm.) (Photo courtesy of Dr. George Matsumoto.)

FIGURE 2.12 The large scyphozoan jellyfish *Pelagia colorata* with juvenile cancer crabs. (Approximate size is 30 cm.) (Photo courtesy of Drs. Lovell and Libby Langstroth.)

or engulfing it with an oversized mouth. Locomotion is via rows of fused, large cilia called ctenes.

The phyla Nemertea and Annelida are represented in the holoplankton by a few highly specialized forms, which are not abundant (Figure 2.16). The annelids include polychaete worms of the families Tomopteridae and Alciopidae, the latter with the largest and best-developed eyes in the Annelida. Most planktonic nemerteans live in deep water.

The phylum Mollusca is the second largest in the animal kingdom. It is usually considered to be composed of slow-moving benthic animals. However, it includes a considerable variety of specially adapted holoplanktonic forms as well. Perhaps the most highly modified planktonic mollusks are the **pteropods** and **heteropods** (Figures 2.13e, f, 2.17 and 2.26d), which are probably derived from benthic slugs and snails, respectively, and are classified in the class Gastropoda. Pteropods are of two types: shelled (order Thecosomata) and naked (order Gymnosomata)

(Figure 2.17). Shelled pteropods have fragile shells and swim using their winglike foot. They are particle feeders employing mucous nets and are preyed upon by the faster-swimming naked pteropods. Heteropods are carnivores with transparent, jellylike bodies (Figures 2.18 and 2.19). A special case is the pelagic shelled snail *Janthina*, which maintains itself at the surface by clinging to the underside of its own bubble raft. (See Figures 2.20 and 2.27c.) A final group of planktonic mollusks are the squids, many of which are powerful, fast swimmers that must be considered nektonic. There are, however, a significant number of very small squids that are not strong swimmers and must be considered planktonic. (See Figures 2.26b and 2.21.)

It is among the phylum Arthropoda that the greatest number of zooplanktonic organisms occur. In the seas of the world, virtually all belong to the subphylum Crustacea. In addition to the dominant copepods, the holoplanktonic crustaceans include members from the order Cladocera (Figure 2.22a, b), the class Ostracoda

(a) *Muggiaea*

(b) *Pelagia*

(c) *Pleurobrachia*

FIGURE 2.13
Holoplankton. Examples of some typical plankton members of the phyla Cnidaria (a, b); Ctenophora (c, d); Mollusca (e, f); and Chaetognatha (g). (Not to scale.) (a, e, f, g from *Marine Plankton, a Practical Guide* by G. E. Newell & R. C. Newell, © 1973 Century Hutchinson. Reprinted by permision of the author. b, c, d from *A Guide to Marine Coastal Plankton and Marine Invertebrate Larvae* by DeBoyd L. Smith. Copyright © 1977 by Kendall/Hunt Publishing Company. Used with permission.)

(d) *Beröe*

(e) *Clione limacina*

Side view   Ventral view

(f) *Euclio pyramidata*

(g) *Sagitta setosa*

FIGURE 2.14 The siphonophore *Nanomia bijuga*. Only the upper portion of the colony is visible in this photo. (Approximate size is 15 cm.) (Photo courtesy of Dr. George Matsumoto.)

FIGURE 2.15 A ctenophore of the genus *Pleurobrachia*. Note the long tentacles used to capture prey. (Approximate size of body is 2 cm.) (Photo courtesy of Dr. George Matsumoto.)

FIGURE 2.16 A holopelagic polychaete of the genus *Tomopteris*. (Approximate size is 8 cm.) (Photo courtesy of Dr. George Matsumoto.)

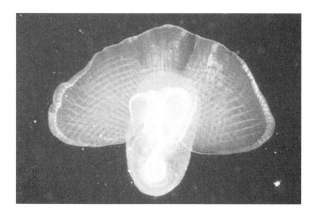

FIGURE 2.17 *Corolla*, the largest pelagic pteropod mollusk genus. These animals feed using a large mucous net, not shown here, that may exceed 2 m in diameter. (Approximate size is 8 cm.) (Photo courtesy of Dr. George Matsumoto.)

FIGURE 2.18 A small heteropod mollusk of the genus *Oxygyrus* with reduced transparent shell. (Approximate size is 20 mm.) (Photo courtesy of Dr. Roger Seapy.)

FIGURE 2.19 The heteropod *Pterosoma planum*, a visual planktonic predator. (Approximate size is 70 mm.) (Photo courtesy of Dr. Roger Seapy.)

(Figure 2.22d), and the orders Mysidacea, Amphipoda, Euphausiacea (Figures 2.22c and 2.23), and Decapoda. Most of these taxa are small filter feeders straining algae or small animals out of the water.

The phylum Chaetognatha is very small, consisting of about 100 species, which, with the exception of one genus, are planktonic. Chaetognaths, or arrow-worms, are abundant members of the plankton throughout the world. (See Figures 2.13g and 2.28d.) All are voracious predators of copepods and other planktonic organisms.

The final phylum represented in the holoplankton is the Chordata. Planktonic chordates belong to the classes Thaliacea (salps) and Larvacea. (See Figures 2.22e, f and 2.24). These gelatinous-bodied animals

are filter feeders. Larvaceans construct a "house" around themselves and pump water through a screen in the house to filter out their food. "Houses" are continually built and shed.

FIGURE 2.20 *Janthina prolongata*, a shelled gastropod living at the air–water interface suspended from a bubble raft. (Approximate size is 3 cm.) (Photo courtesy of Dr. Roger Seapy.)

FIGURE 2.21 A small planktonic squid of the genus *Chiroteuthis*. (Approximate size is 15 cm.) (Photo courtesy of Dr. George Matsumoto.)

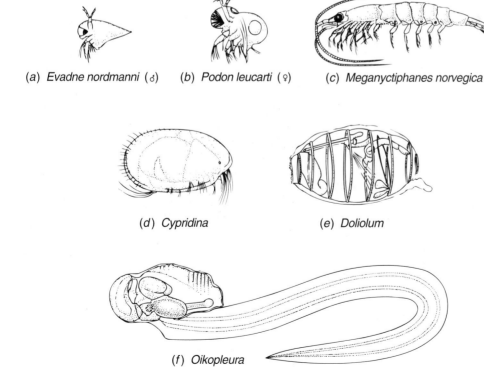

(a) *Evadne nordmanni* (♂)

(b) *Podon leucarti* (♀)

(c) *Meganyctiphanes norvegica*

(d) *Cypridina*

(e) *Doliolum*

(f) *Oikopleura*

FIGURE 2.22 Holoplankton. Typical planktonic forms of the phyla Arthropoda (a, b, c, d) and Chordata (e, f). (a, b) Order Cladocera, subphylum Crustacea. (c) Order Euphausiacea, subphylum Crustacea. (d) Class Ostracoda, subphylum Crustacea. (e) Salp of the class Thaliacea. (f) Class Larvacea. (Not to scale.) (a–c from *Marine Plankton, a Practical Guide* by G. E. Newell & R. C. Newell, © 1973 Century Hutchinson. Reprinted by permission of the author. d–f from *A Guide to Marine Coastal Plankton and Marine Invertebrate Larvae* by DeBoyd L. Smith. Copyright © 1977 by Kendall/Hunt Publishing Company. Used with permission.)

FIGURE 2.23 *Euphausia*, an individual of one of the most abundant groups of zooplankton in the oceans of the world. (Approximate size is 2 cm.) (Photo courtesy of Dr. Roger Seapy.)

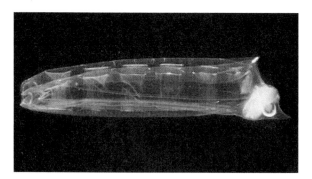

FIGURE 2.24 A salp of the genus *Iasis*. (Approximate size is 5 cm.) (Photo courtesy of Dr. Roger Seapy.)

Compared with the holoplankton, the meroplankton is an even more diverse group. A bewildering array of larval forms constitute the meroplankton. The larvae are derived from virtually all animal phyla and from all different marine habitats. The number of larval forms is greater than the number of species in the sea that produce them, because many species go through a series of larval stages before becoming adult. Each of these species may be represented in the plankton by several different stages. This is particularly true in the Crustacea, where some decapods may have as many as 18 different larval stages, and most have more than one. Phyla that are not represented in the holoplankton, such as Bryozoa, Phoronida, Echinodermata, and Porifera, have larval forms (Figure 2.25a–f). Other abundant phyla, such as Nemertea, Mollusca, and Annelida, which have but a few highly adapted holoplankton groups, have planktonic larvae (Figure 2.25g, h, and l). The Crustacea have by far the greatest number of larval types, and these also include larvae of the abundant holoplankton organisms. Even the nekton

may be represented here, since many fishes have eggs and larvae in the plankton. Although space does not allow treatment of all the types, we next discuss some adaptations of the plankton.

Flotation Mechanisms

Plankton tend to have a density (mass per unit volume) that is somewhat greater than that of seawater. This means that any planktonic organism will eventually sink in the water column. This can be deleterious for both phytoplankton and zooplankton: Phytoplankton would sink below the lighted areas of the sea and be unable to photosynthesize, and zooplankton would sink out of the area where their phytoplankton food occurs. Nonetheless, there may be some advantage to a selective sinking in the water column for phytoplankton because deeper water contains more nutrients. (See pp. 68–72.) Still, uncontrolled or continued sinking generally is undesirable for both groups. Since they are weak swimmers unable to cope with winds and currents, and since living flesh tends to sink, how do these organisms cope with the problem of staying in the upper layers of the sea?

■ PRINCIPLES

To answer the above question, we must consider some physical and chemical characteristics that will bear on the problem. The density of seawater is a function of two factors: its temperature and its salinity. Seawater becomes more dense as its salinity increases and less dense as its temperature increases. As a result, wherever there are temperature changes over the course of a year, such as in temperate seas, the density also varies. Another physical characteristic is the viscosity of seawater. Viscosity is related to temperature and salinity in that the more saline water is and the lower its temperature, the more viscous it is and the less rapidly things sink in it.

A second set of principles has to do with the effect of shape on the rate of sinking in dense liquids. Objects of different shape but similar weight fall at differing rates in both gases, such as air, and liquids, such as water. The rate of fall is proportional to the amount of resistance the body offers to the gas or liquid through which it moves. Objects with a great amount of surface for a given weight tend to fall more slowly than objects of similar weight with less surface

(a) *Echniocardium cordatum* echinopluteus

(b) *Asterias rubens* bipinnaria

(c) *Asterias rubens* brachiolaria

(d) *Labidoplax digitata* auricularia

(e) *Electra pilosa* cyphonautes

(f) *Phoronis* sp. actinotrocha

(g) Serpulid trochophore (side view)

(h) *Balcis alba* (early veliger)

(i) *Papurus puber* megalopa

(j) *Carcinus maenas* (3rd zoea)

(k) Nauplius of barnacle

(l) Pilidium larva

FIGURE 2.25 Meroplankton. Examples from several phyla. (a) Echinodermata, sea urchin. (b, c) Echinodermata, starfish. (d) Echinodermata, sea cucumber. (e) Bryozoa. (f) Phoronida. (g) Annelids, Polychaeta. (h) Mollusca, gastropod. (i, j) Arthropoda, Crustacea, Decapoda. (k) Crustacea, barnacle. (l) Nemertea. (a–j from *Marine Plankton, a Practical Guide* by G. E. Newell & R. C. Newell, © 1973 Century Hutchinson. Reprinted by permision of the author. k, l from *A Guide to Marine Coastal Plankton and Marine Invertebrate Larvae* by DeBoyd L. Smith. Copyright © 1977 by Kendall/Hunt Publishing Company. Used with permission.)

area. Operating here is the physical law that surface area increases as the square of the linear dimensions of the object, but volume increases as the cube of the same dimensions. This means that, for a given or fixed shape, the smaller the body, the greater the is surface area relative to its mass.

From the previously mentioned physical laws governing changes in density and viscosity of water and those governing resistance, it is possible to derive a very simple equation that will relate the sinking rates of organisms to these parameters. The equation is

$$SR = \frac{W_1 - W_2}{(R)(V_w)}$$

where SR = sinking rate, W_1 = density of the organism, W_2 = density of seawater, $W_1 - W_2$ = amount of overweight (that is, the amount by which the weight of the organism exceeds that of a similar volume of water), R = surface of resistance, V_w = viscosity of water.

Organisms can do nothing about the density or the viscosity of water; hence, adaptations of the organism to reduce the sinking rate must pertain to either reducing the amount of overweight or increasing the surface of resistance.

Let us consider first those adaptations designed to reduce the amount of overweight.

■ REDUCTION OF OVERWEIGHT

One mechanism that may be employed to reduce overweight is simply to alter the composition of the body fluids so that they are less dense than an equal volume of seawater. It is necessary, however, to maintain the same number of solute particles in the body fluid as in the surrounding seawater, in order to avoid osmotic problems. **Osmosis** is the passage of water across semipermeable membranes, such as cell membranes, to equalize the water content on both sides of the membrane. How is this change in composition accomplished? Perhaps the most common way is to replace heavy chemical ions in the body fluids with lighter ones. Doing this allows the organism to maintain the same osmotic condition (the same number of solute particles) while becoming lighter with respect to seawater. An example of an organism that employs this strategy is the dinoflagellate *Noctiluca* (Figure 2.26a), whose internal fluid contains ammonium chloride (NH_4Cl) and which is iso-osmotic with seawater, but is less dense, having a specific gravity of 1.01 compared with the 1.025 of seawater. Similarly, the cranchid squids, which have quite unsquid like fat, bulbous bodies, have a sac filled with NH_4Cl, making them less dense than seawater (Figure 2.26b).

In a similar fashion, certain planktonic forms, such as salps, ctenophores, and heteropods (Figure 2.26c, d) actively exclude heavy ions such as sulfate (SO_4^{2-}) from their bodies and replace them with osmotically similar, but lighter, chloride ions.

Another flotation mechanism is the development of special gas-filled floats. Since gas is much less dense than a similar volume of water, buoyancy is assured. Perhaps the most familiar examples of this approach are the floats of the Portuguese man-of-war (*Physalia*)(Figure 2.27a) and the swim bladders of fishes. If the animal is able to regulate the gas pressure in the float or bladder, it

   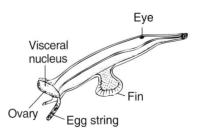

(a) *Noctiluca*   (b) Cranchid squid   (c) *Pleurobrachia*   (d) Heteropod (*Firoloida*)

FIGURE 2.26 Flotation mechanisms. Examples of organisms that exclude heavy ions, forming body fluids less dense than seawater. (Not to scale.) (From *An Introduction to the Study of Tropical Plankton*, J. H. Wickstead, 1965.)

Float —

(a) Portuguese man-of-war

— Float

— Swimming bells

(b) Siphonophore colony

FIGURE 2.27 Flotation mechanisms. Examples of organisms that use gas- or fluid-filled floats. (a) and (b) are both siphonophore colonies. (a, from *Marine Plankton: A Practical Guide*, G. E. Newell & R. C. Newell, © 1973 Century Hutchinson. Reprinted by permission of the author. b, d, from *The Invertebrates*, Vol. 1, "Protozoa to Ctenophora," by L. Hyman, 1940.)

Float —

(c) *Janthina*

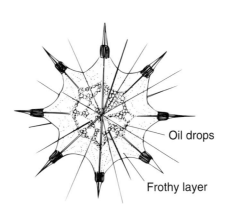

— Oil drops

Frothy layer

(d) Radiolarian with CO_2 bubbles in outer cytoplasm

can move up or down in the water column at will, as some fishes do.

Similar to the use of gas-filled floats is the employment of liquids that are less dense than water. Prominent among these liquids are oils and fats, which can also serve as food reserves for the organisms. Copepods, for example, often store excess food in the form of oil droplets under the carapace, and these droplets aid in buoying the animal. Diatoms also store food as oils.

Whereas some planktonic organisms have employed the various flotation mechanisms to alter their density and remain afloat, many others have not. If we return to our original equation, we can see that, if overweight cannot be reduced, the only other option is to increase the surface of resistance to the water and at least slow the rate of sinking.

■ CHANGES IN SURFACE OF RESISTANCE

The surface of resistance may be increased in a number of ways, and all are observed in planktonic organisms.

One of the most common observations made about plankton is that the body size is generally small. As we noted earlier, surface area increases as the square, whereas volume increases as the cube of the linear dimension. This means that the smaller the organism, the greater the surface area relative to volume. By remaining small, plankton organisms offer far more surface area of resistance to sinking per unit volume of living material than if they were large. Small size is particularly important for tropical plankton, because they live in warmer, and thus less viscous, water, where sinking would be faster.

The other way of increasing the surface of resistance is to change the shape of the body. If one were to drop a round ball and a flat coin of equal weight into water, the ball would sink more quickly because it has less surface of resistance to the fluid. In a like manner, planktonic organisms have evolved various flattened body shapes or appendages. Examples (Figure 2.28) include a host of different species from many different phyla.

Even more common than changes in body shape is the development of various spines and body projections. These structures add considerable resistance, but little to the weight. They may also deter predators by preventing them from ingesting or swallowing the organism. Such adaptations are common in the various diatoms (see Figure 2.2a), radiolarians (see Figure 2.27b), foraminiferans, and crustaceans.

■ WATER MOVEMENTS

A final mechanism of buoyancy has to do not with the organisms, but with the nature of water movements in the ocean. In the ocean, the surface water heats up during the day and cools at night. This alternating heating and cooling changes the density of the water and leads to the creation of convection cells, which are small units of water that either sink or rise according to their density. The movements are gentle, and plankton may be moved by them.

FIGURE 2.28 Flotation mechanisms. Examples of organisms that are flat or that have spines: (a) Phylum Ctenophora. (b) Phylum Mollusca, class Gastropoda. (c) Phylum Arthropoda, class Crustacea. (d) Phylum Chaetognatha. (e) Order Dinoflagellata. (f) Phylum Arthropoda, order Decapoda. (g) Phylum Echinodermata, class Ophiuroidea. (h) Phylum Protozoa, class Sarcodina. (i) Phylum Arthropoda, subphylum Crustacea. (Not to scale.) (a–g from *An Introduction to the Study of Tropical Plankton*, J. H. Wickstead, 1965. h from *Marine Plankton: A Practical Guide*, G. E. Newell & R. C. Newell, © 1973 Century Hutchinson. Reprinted by permission of the author. i from *The Cambridge Natural History*, Vol. 4, *Crustacea*, reprinted in 1968 by Wheldon and Wesley.)

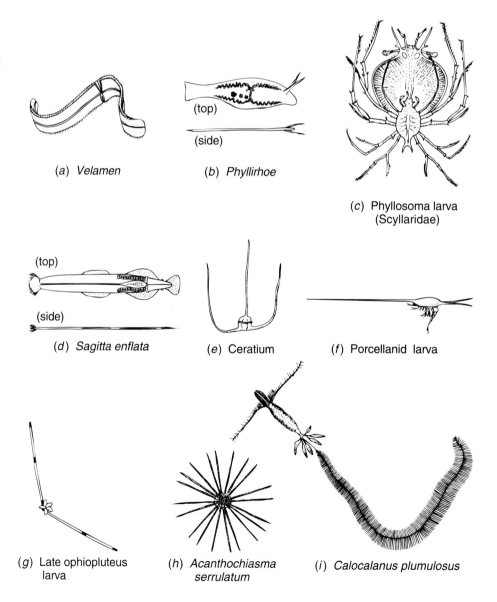

(a) *Velamen*

(b) *Phyllirhoe*   (top) (side)

(c) Phyllosoma larva (Scyllaridae)

(d) *Sagitta enflata*   (top) (side)

(e) Ceratium

(f) Porcellanid larva

(g) Late ophiopluteus larva

(h) *Acanthochiasma serrulatum*

(i) *Calocalanus plumulosus*

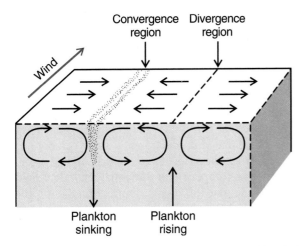

Convergence region   Divergence region

Wind

Plankton sinking   Plankton rising

FIGURE 2.29 Langmuir convection cells induced by wind.

On a small scale, the action of the wind blowing over the water can produce localized vertical movement of plankton due to the production of **Langmuir convection cells**. These cells appear when wind speeds are above 3 meters per second. Each Langmuir cell is a few meters wide and hundreds of meters long. In each, wind causes the water to move away from the center (divergence) toward the outside, where it meets the water from the adjacent cell. These two convection currents converge and move down along the line of convergence for a short distance. Then they move horizontally, where they again meet water moving in a similar direction from the adjacent cells, whereupon they rise to the surface to complete the cycle (Figure 2.29). Planktonic organisms in such cells tend to be carried downward at the margins of the cell (convergence) and upward in the center (divergence). On a larger scale, upwelling (pp. 70–72) can carry planktonic organisms toward the water surface.

Primary Production

Almost all life on the Earth depends on chlorophyll-bearing organisms being able to use the energy of sunlight to synthesize energy-rich organic molecules from inorganic materials. This is the process of photosynthesis, the general equation for which is

$$6CO_2 + 6H_2O \xrightarrow[\text{nutrients}]{\text{sunlight}} C_6H_{12}O_6 + 6O_2$$
$$\text{energy-rich organic compounds}$$

The basis for nearly all life in the sea is the photosynthetic activity of the aquatic autotrophs. There are, however, significant differences in the forms of the organisms and in the location and extent of maximum photosynthetic activity imposed by the particular physical and chemical conditions of the oceans. (See pp. 70–71.) These will be covered subsequently. Photosynthetic organisms are diverse and widespread in the world's oceans. Among these organisms are the pelagic phytoplankton and cyanobacteria, benthic microalgae, benthic macroalgae (kelps, seaweeds), seed plants (sea grasses, mangroves, salt-marsh plants), and symbiotic algal cells in corals and other animals. However, the pelagic phytoplankton, cyanobacteria, and other presently undescribed aerobic photoheterotrophs are responsible for perhaps 95% of all marine primary productivity. It is for this reason that we consider primary productivity in detail here.

■ PRIMARY PRODUCTIVITY

Primary productivity is defined as the rate of formation of energy-rich organic compounds from inorganic materials. Primary productivity is usually considered identical to photosynthesis, but this is not quite correct, since a minor amount of primary production may be produced by chemosynthetic bacteria. (See Chapter 4.) Here, however, primary productivity is confined to photosynthesis.

The total amount of organic material fixed in the primary productivity process is termed the **gross primary production**. Since some of this total production must be used by the plants themselves to operate their life processes, collectively called respiration, a lesser amount is available for transfer or use by the other organisms of the sea. **Net primary production** is the term given to the amount of total production left to support other trophic levels after losses from plant respiration.

For either gross or net primary production, the rates are usually expressed in terms of grams of carbon fixed per unit area or volume of seawater per interval of time. For example, annual net primary production may be reported in grams of carbon per square meter per year (g C/m^2/yr), a figure that represents the production integrated over all depths from the surface to the base of the euphotic zone.

A few other terms require definition at this point. The **standing crop**, as applied to autotrophs, is the total amount of the organism's biomass present in a given volume of water at a given time. Primary production and standing crop can vary considerably on a timescale of a few days to a year, and this variation is the result of a large number of factors that act both directly and indirectly on the photosynthetic process within autotrophs and on the autotrophs themselves. These factors will be discussed later in the chapter.

As we saw in Chapter 1, water has a profound effect on the light that penetrates it. Because water absorbs light, there is less light energy available as one goes deeper, and eventually, light disappears. For photosynthetic organisms, this means that, at some depth, the light energy is just sufficient for the organism to fix energy at a rate equal to the rate at which it uses energy in its own metabolic processes. In other words, there is a depth at which the respiratory usage of energy by the photosynthetic organism equals the ability of the photosynthetic mechanism to produce energy. If the organism goes deeper, the respiratory needs continue at the same rate, but the decreasing light is insufficient to allow the photosynthetic process to keep up, so there is a net loss of energy. The depth at which the rate of respiration of a photosynthetic organism is just equaled by the rate of photosynthesis is called the **compensation depth**. Above this depth, the rate of photosynthesis exceeds that of respiration, and there is a net production of carbon, or a net primary production.

The compensation depth varies over the world's oceans, because it depends on the clarity of the water. The clearer the water, the greater is the light penetration and the deeper is the compensation depth. In general, the compensation depth is deeper in clear, open ocean waters and shallower in inshore waters, where large amounts of particulate matter are found in the water.

Since the compensation depth varies with the depth of light penetration, it should also be possible to define the compensation depth in terms of light intensity. In fact, this is true: The **compensation intensity** is the light intensity at which the rate of photosynthesis equals the rate of respiration. Although this intensity is somewhat different for different phytoplankton species, it is approximately that depth at which the intensity has been reduced to 1% of the surface light intensity. The compensation depth can then be approximated as the depth to which 1% of the incident light penetrates.

It is well to remember that the compensation depth varies not only in different geographic areas of the world's oceans, as noted previously, but also from day to day and from season to season in any one geo-graphical area, as a result of changes in light with season (see pp. 66–68) and latitude, changes in water clarity, and the availability of nutrients.

■ MEASUREMENT OF PRIMARY PRODUCTIVITY

If we return to the photosynthetic equation on page 61, we can see that it should theoretically be possible to measure the rate of production of organic compounds by measuring the rate of appearance or disappearance of some component of that equation. If one could measure the rate of disappearance of CO_2 or the appearance of O_2, that would be a measure of the rate of photosynthesis. In fact, this is done, but for various reasons it is practical only for the two components mentioned.

The classical method used for many years to measure primary productivity is called the **light/dark-bottle method**. In this method, two identical bottles are employed. One bottle is completely transparent, while the other is made completely opaque by painting it black or covering it with some opaque substance, such as aluminum foil. Into each bottle is placed the same volume of seawater taken from the body of water whose productivity is being estimated. The water contains the naturally occurring phytoplankton and zooplankton organisms. The oxygen content of the water to be added to the two bottles is determined on a separate sample in order to establish the initial concentration of oxygen. The bottles are then stoppered, attached to a line, and returned to the appropriate depth in the sea. In the bottle that is dark, no photosynthesis can occur, but the phytoplankton and zooplankton within continue to respire and use up oxygen. In this bottle, then, the original oxygen content will decrease as the respiring phytoplankton and zooplankton use up the dissolved oxygen. In the light bottle, on the other hand, photosynthesis continues in excess of respiration; hence, oxygen builds up over the initial values, since it cannot escape. After the two bottles have incubated for a period determined by the investigator, they are brought up and unstoppered, and the water is analyzed for oxygen content. Usually, the oxygen content is determined by the Winkler method.

Once the oxygen content of each bottle after incubation is known, it is possible to calculate the rate of photosynthesis from that result and the initial concentration of oxygen of the water when it was put into the bottles. To understand how this is done, it is necessary to consider what has happened in each bottle. When the original seawater was added to each, it contained a certain amount of oxygen already present (initial oxygen concentration = O_1). In the light bottle, this oxygen was used for respiration by the organisms. At the same time, the phytoplankton organisms were producing oxygen through photosynthesis. The final oxygen concentration in the light bottle, then, is the sum of the oxygen added by photosynthesis and the initial oxygen concentration, less the oxygen used in respiration. Since we know the initial amount of oxygen, in order to estimate the total amount of oxygen produced by photosynthesis, we need to know how much was consumed in respiration. If there were no respiration, but only photosynthesis, in the light bottle, the increase in oxygen would be greater than that actually observed. This is where the dark bottle comes in. In the dark bottle, only respiration occurred and the oxygen concentration decreased. Since both bottles have the same volume and were incubated the same length of time, the decrease in oxygen in the dark bottle is a measure of the total respiration in the light bottle. To find the total amount of oxygen produced in photosynthesis, then, it is necessary to make a few simple additions and subtractions. First, the initial oxygen concentration of the water (before incubation) is subtracted from the final oxygen concentration in the light bottle, and the final oxygen concentration is subtracted from the initial concentration in the dark bottle. The result of the subtraction in the case of the light bottle is the **net community photosynthesis**, or the amount of photosynthesis in excess of respiration by both autotrophic and heterotrophic organisms. This is called **new production**. The result from the dark bottle is simply the respiration of the organisms. The **gross photosynthesis** is then obtained by adding the amount of oxygen respired to the net community photosynthesis, or subtracting the dark bottle from the light bottle. Gross photosynthesis is a measure of the total amount of photosynthesis that occurred during incubation. One then converts the rates of oxygen production to rates of carbon assimilated by using a conversion factor, which varies with the organic compound produced (carbohydrate, lipid, or protein), but an average conversion factor would be 1.2 to 1.4.

Of course, the values for both gross and net community photosynthesis will differ with different incubation depths, because light values change with depth and affect the photosynthetic process. This traditional method was originally applicable only in waters where there was considerable photosynthetic activity, because the sensitivity of the Winkler method is too low to pick up changes in oxygen levels under sparse phytoplankton populations and short (2- to 4-hour) incubation times. Longer incubation times would permit use of the Winkler method, but cannot be done because various "bottle effect" errors, such as an increased growth of oxygen-consuming bacteria on the walls of the container, become too large. More recently, the Winkler method has been electronically coupled to optical titration, which gives greater sensitivity.

Another method for estimating primary productivity is the ^{14}C method—the preferred technique for the past 30 to 35 years. In this method, radioactive ^{14}C is added to a bottle containing a volume of seawater populated by phytoplankton. For experimental and technical reasons, the ^{14}C is usually introduced as $H^{14}CO_3^-$ (bicarbonate). A known quantity of $H^{14}CO_3^-$ is added, and the bottle is incubated for a certain period at the appropriate depth in the sea (or in special racks on the deck of a vessel). At the end of the incubation period, the bottles are brought up and the water is filtered onto fine membrane filters or glass fibers, which catch all phytoplankton organisms. The amount of radioactivity on the filters is measured with a counter. The amount of ^{14}C that appears on the membrane (e.g., in the particles) is proportional to the rate of primary production. To correct for possible nonphotosynthetic uptake of ^{14}C, a dark bottle is often run along with the light bottle. One calculates the carbon uptake from the following equation:

$$\text{C uptake} = \frac{{}^{14}C \text{ in particles on the filter} \times \text{available inorganic carbon} \times 1.05}{\text{Total } {}^{14}C \text{ added}}$$

The available inorganic carbon in the sample is usually measured by alkalinity titration. In most marine waters with a salinity greater than 30 psu, the total available inorganic carbon is about 25 mg C per liter. The factor 1.05 is included in the equation to correct for the fact that phytoplankton discriminate between ^{14}C and ^{12}C, absorbing ^{14}C more slowly.

Both of the methods discussed have errors and problems associated with them, and both rest upon certain assumptions. In the case of the light/dark-bottle method, the assumption is that conditions in both bottles are similar, except for light. However, that may not actually be the case, and if it isn't, errors will be introduced. Bacterial growth tends to be accelerated by the introduction of surfaces; hence, the increased bacterial numbers may affect the oxygen levels in the bottles. This problem is common to both ^{14}C and light/dark-bottle methods. Another problem common to both methods is that phytoplankton organisms enclosed in a bottle do not behave normally, and their photosynthetic rate may be affected. A serious problem associated with the light/dark-bottle technique is its comparative insensitivity when concentrations of phytoplankton are low or when waters are polluted or contain high bacterial concentrations. The chief additional problems with the ^{14}C method are that there may be cell breakage on the filters, allowing some ^{14}C to leak out, and that the method does not account for ^{14}C fixed in photosynthesis and then respired out as $^{14}CO_2$, which is then redissolved in seawater. Because of the latter problem, it is not always clear whether the method measures gross or net photosynthesis or something in-between.

Recently, various independent methods used to assess carbon dioxide uptake or oxygen release have suggested that the ^{14}C method underestimates primary productivity, particularly in waters with low nutrient contents and low phytoplankton densities. Part of this problem may lie in the technique (reviewed by Peterson, 1980). Part of the problem also may result from inadvertent contamination from metals associated with bottle samples and ^{14}C stock solutions.

Because of these shortcomings, newer methods of assessing primary productivity have been, and are being, attempted. Among these are the use of large-scale optical assessment via satellite imaging and bio-optical methods of measuring photosynthesis, specifically from specially designed fluorometers. In some ways, these instruments measure the "physiological health" of the phytoplankton. The type of instrument used in such studies is called a *pulse amplitude-modulated* (PAM) fluorometer. Yet another variant, the fast-repetition-rate (FRR) fluorometer, attempts to measure the instantaneous photosynthetic rate.

Net primary productivity (NPP) for both terrestrial and oceanic biomes has now been calculated from the absorbed photosynthetically active (solar) radiation (APAR) and *e*, which is an average light utilization efficiency, in the following equation:

$$NPP = APAR \times e$$

For both the oceans and the terrestrial environment, the APAR can be derived from satellite measurements—in the case of the oceans from the measurement of chlorophyll, and in the terrestrial environment from estimates of greenness. However, *e* cannot be accurately measured from space and must be measured in the field. Figure 2.30 shows the net primary productivity for the biosphere, as determined by the APAR method.

FIGURE 2.30 Global net primary productivity calculated from satellite-derived measurements of surface chlorophyll and absorbed photosynthetically active solar radiation. (Note the different measurements for ocean and land.) (Provided by the SeaWiFS Project, NASA/Goddard Space Flight Center and ORBIMAGE.)

ocean

Chlorophyll *a* Concentration (mg / m³)

land

Normalized Difference Vegetation Index

■ STANDING CROP

Standing crop may include autotrophs, animals, or both. The **standing crop** at any point in time is the result of the difference between the factors tending to increase the numbers of individuals—reproduction and growth—and the factors tending to decrease biomass or numbers—death and sinking or lateral transport out of the area. If reproduction and growth rates are, or have been, high and death or removal low, then standing crop will be high, and vice versa.

The standing crop of phytoplankton in the oceans is a difficult factor to measure accurately, partly because of the patchy distribution of plankton organisms, partly because of the problems of sampling (see pp. 94–96 for further discussion of this), and partly because of problems inherent in the methods. The customary method of measuring standing crop is to measure some component common to all phytoplankton, usually the chlorophyll *a* content of a given volume of seawater. Since all phytoplankton must have chlorophyll to photosynthesize, the total amount of chlorophyll in a given volume of seawater should be a direct measure of the total biomass of phytoplankton present. Chlorophyll can be measured by taking advantage of its ability to fluoresce when excited by an appropriate wavelength of light, or it can be measured by directly extracting the chlorophyll from the plants with a chemical such as acetone and then measuring the intensity of color in a colorimeter. The problems with this method are numerous. The method assumes, for example, that a plant's cellular chlorophyll content is constant, but it is not; the amount of chlorophyll varies with the phytoplankton species, as well as with cells of the same species, depending on their condition. It also varies with time of day and light intensity. These variations are not insignificant and constitute a substantial problem in assessing standing crop.

Among other methods of assessing standing crop is the measurement of **adenosine triphosphate (ATP)**, which is found only in living cells and disappears rapidly on death. The method devised by Holme-Hansen and Booth (1966) depends on the use of the protein luciferin and the enzyme luciferase. In the presence of ATP, these two compounds produce light (bioluminescence), which is then measured quantitatively by a photocell. Drawbacks of the method are that (1) ATP content is variable, depending on several biological and physical factors, making any extrapolation of ATP levels to units of living cells difficult, (2) interference by microzooplankton is a

problem, and (3) biologists cannot agree on the percentage of ATP measured in a unit volume of water that is attributable to animals or bacteria. A potential advantage of the ATP method is that, once sufficient data are available to assign values accurately to all components of the plankton, the method will give biomass values for zooplankton, phytoplankton, and bacterioplankton collectively. Today ATP is not widely used for measuring standing crop, but fluorescence microscopy is used for enumerating picoplankton, bacteria, heterotrophic flagellates, and viruses.

Factors Affecting Primary Productivity

Since the entire ecosystem of the world's oceans depends for energy almost exclusively on the photosynthetic activity of the small autotrophs confined to the thin layer of lighted surface waters of the oceans, it is important to understand the conditions under which the productivity of these organisms is either enhanced or inhibited. This is in fact the first step toward understanding how our oceans function.

■ PHYSICAL AND CHEMICAL FACTORS

A series of physical and chemical factors affects the growth, survival, and productivity of terrestrial plants. Critically important factors include light, temperature, nutrient concentration, soil, and water. As regards the freely floating autotrophs, we find that the number of important factors is considerably reduced. For autotrophs already suspended in water, the water is a given and soil is of no importance. Temperature, which can vary enough in the terrestrial biosphere to influence productivity, has a considerably narrower range in the marine environment. Because this range almost never encompasses lethal limits for life, and because the changes are always gradual due to the inherent physical properties of water, temperature is of lesser importance to productivity in the ocean. Of all the physicochemical factors affecting terrestrial plant production, only two—**light** and **nutrient supply**—are significant in limiting productivity in the ocean. Since, however, the autotrophs are suspended in water and are influenced by forces that move the water, and also because both light and nutrients are affected by the water masses, a new, very important factor not seen on land enters into consideration. This factor is a composite one we may call **hydrography** and

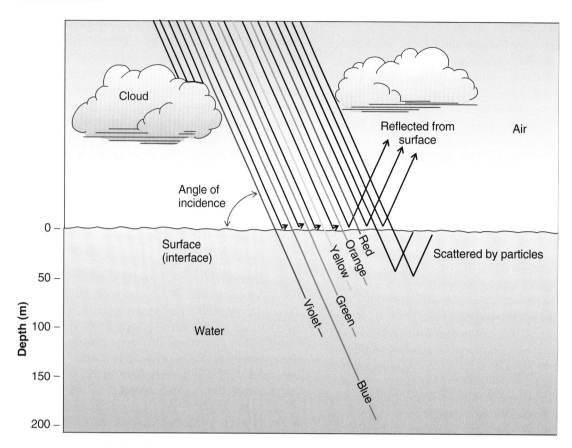

FIGURE 2.31 The fate of incident light in the ocean.

comprises all those factors which act to move water masses around in the oceans, such as currents, upwelling, and diffusion. It is the interplay among these three factors—light, nutrients, and hydrography—that provides the limits to autotrophic productivity in the oceans, as well as the geographical differences we observe. We next consider how each factor acts separately to limit or enhance production, and then we examine how all three act in concert to produce the observed productivity patterns of the oceans.

LIGHT Photosynthesis is possible only when the light reaching the autotrophic cell is above a certain intensity. This means that the phytoplankton are limited to the uppermost layers of the ocean, where light intensity is sufficient for photosynthesis. The depth to which light will penetrate into the ocean, and hence the depth at which production can occur, depends on a number of factors, including the absorption of light by the water, the wavelength of light, the transparency of the water, reflection from the surface of

the water, reflection from suspended particles, the latitude, and the season of the year.

A number of meteorological features influence light before it even touches the surface of the water (Figure 2.31). Factors such as clouds and dust interfere with light in such a way that lesser amounts survive the passage through the atmosphere to reach the water surface. This reduces the available light initially, without reference to water conditions.

When light strikes the surface of the water, a certain amount of light is reflected, depending on the angle at which the light strikes. If the angle from the horizontal is low, a large amount will be reflected. Conversely, the nearer the angle is to 90° (that is, perpendicular to the horizontal surface of the water), the greater the penetration and the less the reflection. Reflected light is lost to the system, so from the standpoint of autotroph productivity, maximum penetration is the most desirable.

The angle at which light strikes the surface of the water is directly related to the maximum height of the sun above the horizon. In the tropical regions of the Earth, the sun is directly overhead at midday, or virtually perpendicular to the sea surface, giving an angle that affords maximum penetration of light into the water column. This position changes little with the seasons, so light conditions are maximal all year long in the tropics. As one progresses toward the poles from the equator, the sun may be directly overhead during the summer months, but it may be far from that position at other times of the year. The closer one approaches the poles, the greater becomes the difference in the height of the sun above the horizon among the seasons of the year. The maximum difference is reached in the Arctic and Antarctic regions, where the sun is absent during the winter or is so low on the horizon that no light can penetrate the water. The presence of ice in these areas also reduces light penetrating into the water. This means that, as one moves away from the equator to the north or south, the amount of light penetrating the surface of the ocean and available for autotroph use changes significantly with the season. As a result, the amount of photosynthesis that can occur also changes and is maximal in the summer and minimal in the winter. Only in the tropics is the light optimal all year (Figure 2.32).

The portion of light that enters the water column is subject to further reduction from two additional processes acting on it within the water. The first is scat-

tering from various suspended particles in the water column. Suspended living or dead particles intercept the light and either absorb it or scatter it. Scattered light assumes a longer path length, which in turn results in further attenuation. Second, water itself absorbs light, making it unavailable for the phytoplankton. The amount absorbed is a function of the wavelength of the light and the path length or water depth in perfectly clear, clean water. This absorption of light by water is the reason most water masses of the ocean are dark below a certain level. Because of such absorption, photosynthesis is automatically restricted to the thin, uppermost lighted layer. In the clearest ocean water, this may correspond only to the upper 150–200 m. Water, however, does not absorb all wavelengths of light equally. Sunlight arriving at the surface of the ocean is composed of radiation in a spectrum of wavelengths measured in nanometers (1 nanometer = 10^{-9} m). This spectrum includes all the visible colors ranging from violet to red (see Figure 2.31), or wavelengths from about 400 to 700 nm. As these wavelengths enter seawater, the violet and red components are quickly absorbed by the water. The green and blue components are absorbed less rapidly, so they penetrate most deeply, but eventually, they too are absorbed. All the red and violet light is absorbed within the first few meters of even the clearest seawater, whereas 10% of the blue light may penetrate to more than 100 m under similar circumstances. It is important to realize that, for any given wavelength of light, a certain fraction of its

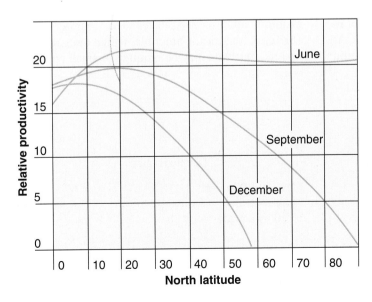

FIGURE 2.32 Relative amount of photosynthesis as a function of latitude for three seasons of the year. ("Geographical Variations in Productivity" from *The Sea*, Vol. 2, ed. M. N. Hill, © 1963 John Wiley & Sons. Reprinted by permission of John Wiley & Sons, Inc.)

remaining intensity is lost by absorption with each additional increment of water depth. Even though blue and green light penetrate deeper into the water column, the intensity decreases with depth, and it is intensity that affects autotrophs. The intensity, relative to the surface, can be estimated from the **extinction coefficient** $I_z = I_0 e^{-kz}$, where I_0 and I_z are the intensities at the surface and at depth z, respectively; k is the extinction coefficient (for pure water, $k = 0.035$), and z is the depth or path length in meters.

The depth to which a given intensity of light penetrates is a function of both the transparency of the water and the differential absorption by the water. Since differential absorption by water is constant, the changes in depth of effective light penetration are due primarily to particle concentration. Where there are large numbers of particles in the water, such as in coastal waters, the depth of light penetration may be severely reduced and the amount of light insufficient for photosynthesis below a few meters. On the other hand, in the clearest tropical waters, where few interfering particles exist, light intensity may be sufficient for photosynthesis down to 100–150 m.

Autotroph photosynthesis is light dependent. The rate of photosynthesis is high at intermediate light levels and decreases as the light intensity either decreases or increases (Figure 2.33). In contrast, the rate of respiration of the phytoplankton cells does not change much with depth. This means that, as the photosynthetic cells go deeper into the water column, the rate of photosynthesis declines as the light intensity decreases, until at some point the photosynthetic rate equals the respiration rate. At this point, there is no net production of organic material. As defined earlier, this depth is called the compensation depth, and it is approximated by the depth to which 1% of the incident radiation penetrates. The compensation depth marks the lower limit of the euphotic zone and varies geographically from a few meters in very turbid inshore waters to depths of 120–150 m or more in the open waters of tropical oceans. It also varies seasonally in temperate areas, where high turbidity during certain seasons (the phytoplankton bloom) reduces it to a few meters (see p. 69) and sparse populations of organisms increase it at other times.

The compensation depth also changes with season, due to the change in the position of the sun, and it may be virtually absent during the winter months in high latitudes.

For most autotrophs, the photosynthetic rate varies with light intensity. Near the surface of the water column, where the light intensities are the highest, most species show a leveling off or a decrease in photosynthesis (Figures 2.33 and 2.34), due either to an inhibition at high light levels or saturation of the photosynthetic apparatus such that it is not possible to increase rates. Different species have different curves of photosynthetic rate when plotted against light intensity, giving different optimal light intensities for maximum photosynthesis. (See Figure 2.33.) These differences may be of significance in seasonal succession. (See p. 87.)

NUTRIENTS   The major inorganic nutrients required by phytoplankton for growth and reproduction are nitrogen (as nitrate, NO_3^-; nitrite, NO_2^{2-}; or ammonium, NH_4^+) and phosphorus (as phosphate, PO_4^{3-}). Diatoms and silicoflagellates also require silicate (SiO_2) in significant amounts. Other inorganic and organic nutrients may be required in small amounts.

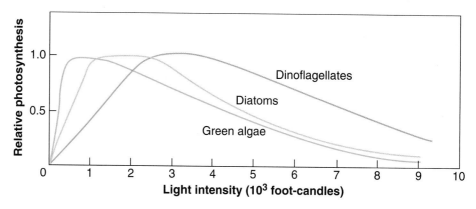

FIGURE 2.33 Curves of photosynthesis versus light intensity for three groups of phytoplankton. (From *Biological Oceanographic Processes*, 3d ed., T. R. Parsons et al. Copyright © 1984 Butterworth Heinemann. Reprinted by permission.)

FIGURE 2.34 Inhibition of photosynthesis in the surface layers of water, as measured by gross production at different levels in the water column. ("Productivity, Definition & Measurement" by E. Steeman-Nielsen, from *The Sea*, Vol. II, ed. M. N. Hill, © 1963 John Wiley & Sons. Reprinted by permission of John Wiley & Sons, Inc.)

All of these nutrients are of great importance, partly because they occur in such small amounts in seawater. They are the limiting factors for phytoplankton productivity under most conditions, and the oceans of the world can be considered nutrient-poor deserts compared with their terrestrial counterparts. For example, fairly rich agricultural land contains about 0.5% nitrogen in the upper meter of soil. In 1 cubic meter of soil, this amount of nitrogen is sufficient to permit the production of 50 kg of dry organic matter. Under ideal conditions, terrestrial plants can produce several kilograms of dry organic matter in excess of their own need per year per square meter; thus, the reservoir of nitrogen in that 1 cubic meter is sufficient to allow a plant to grow for many years. As a result, in terrestrial systems it is possible to have long-lived plants that continue to grow for many years. Forests are a good example, for they represent the accumulation of years of organic production. Of course, eventually, unless the nitrogen is renewed in the soil, nutrient exhaustion will limit growth.

On the other hand, the richest ocean water contains only about 0.00005% nitrogen, or 1/10,000 of the amount in soil. This means that a cubic meter of such water permits the production of only 5 g of dry organic matter. In contrast to soil, however, where plant roots may penetrate only a few meters, oceanic phytoplankton should, theoretically, have access to the nitrogen in a column of water that extends as deep as the plants can exist. What is this depth and does it compensate for the greatly reduced nitrogen concentration? As we learned in the previous section, planktonic autotrophs are limited in depth by light. If we assume the most ideal light conditions, we might have plant production as deep as 100–120 m. This would mean the plants could have access to a volume of 100 m^3 (1 m^2 × 100 m deep) with a potential production of 500 g of dry weight (5 × 100), or about 1/100 of the amount on dry land. In fact, however, production levels of that magnitude never occur in the ocean, and the maximum amount of organic production that can accrue under a square meter of ocean surface under ideal conditions is only about 25 g. Why the discrepancy?

Several factors act in concert to reduce the biomass yield. First, the previous example assumed that the nitrogen content of seawater was constant throughout the water column. Such is usually not the case: The upper layers of water usually have a reduced concentration compared with lower waters. Also, due only to light absorption by water, the production at 100 m would be less than at 10 m. More important, the increasing numbers of autotrophic cells have a profound effect. As the planktonic autotroph population grows in the upper 100 m of water, the cells absorb more and more of the light. As a result, less and less light penetrates to the deeper levels, so that the compensation depth begins to move upward and becomes shallower (Figure 2.35). The original 100-m reservoir of nutrients is reduced, and as plants increase in numbers, more and more of the water column and nutrient supply become inaccessible to the autotrophs; thus, the total potential productivity is reduced. Finally, the growing numbers of autotrophs absorb the nutrients, making them unavailable to other planktonic autotrophs.

The result of all these factors acting together is to reduce the theoretical primary production. The reduction in both compensation depth and absorption of nutrients is great. It has been estimated that, by the time an autotroph population reaches a density of 2 g/m^3, an original 100-m compensation depth has decreased to as little as 3.5 m, and all nitrogen has

FIGURE 2.35 Change in compensation depth with increasing numbers of phytoplankton. Dashed line indicates compensation depth.

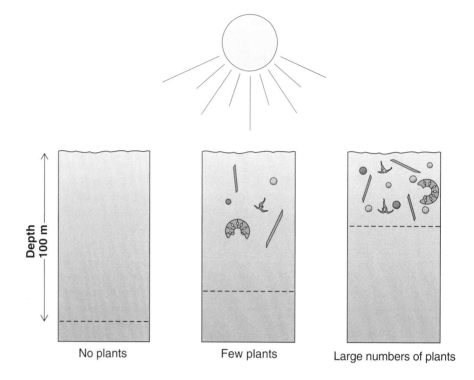

No plants   Few plants   Large numbers of plants

been transferred into autotroph bodies. In terms of primary production and of the standing crop of autotrophic material, the oceans appear to be biological deserts compared with fertile land. The rate of production between land and sea, on a global scale, may not be significantly different, but the differences in nutrient concentration preclude that rate from continuing in the sea for long periods.

If, however, the nutrients are used up in the upper, lighted zone, an untapped reservoir remains in the water mass below the photic zone. Since this mass is several orders of magnitude larger in volume than that of the lighted (photic) zone, it represents a considerable reservoir of nutrients that could enhance production greatly. If this vast reservoir could be tapped, high rates of production and large standing crops could be sustained on a long-term basis.

Unfortunately, certain physical factors prevent access to this reservoir over most of the world's oceans. Water, as we learned in Chapter 1, has different densities, depending on its temperature and salinity. Cold, saline water is more dense than warm, less saline water. Below the photic zone throughout the world's oceans, the water is cold and, therefore, dense. In the photic zone, the water varies in temperature. In the tropics it is warm, and therefore less

dense, all year. In the temperate zone, the water is warm in summer and cold in winter. In polar regions, it is cold and dense year-round. In the tropics, the difference in density between the warmer upper and colder lower layers is of such magnitude that the two do not mix; hence, the nutrients cannot reach the photic zone. In the temperate seas the same situation prevails in the summer, but in the winter the temperature of the two water masses becomes similar and mixing can occur. In polar regions, there are no significant differences in temperature, and mixing can occur year-round.

Another component is required to tap the nutrient reservoir—a mechanism to accomplish mixing. Even though the water masses approach the same density, some force must be exerted to mix them. Strong winds blowing over the surface of most of the world's oceans create the mixing force (Figure 2.36). When the water masses involved are similar in density and temperature, the wind may be of sufficient strength to cause mixing of the masses and to transport nutrient-rich water into the euphotic zone. In polar seas, this is possible at all times (when the seas are not covered by ice), in temperate seas only in the winter. The great differences in density between the photic and aphotic zones all year in the tropics and in the summer in the

Strong wind →

FIGURE 2.36 Thermal structure of temperate ocean water masses in two seasons—winter and summer—showing the relationships between nutrient concentration, temperature, density, and wind-induced mixing.

temperate zones are of such magnitude, however, that no wind (even from storms such as hurricanes) is strong enough to mix the layers. This means that the upper layer remains nutrient poor in the tropics at all times and in the temperate zone in summer, unless specific hydrographic conditions overcome the stability of the density differences.

The special hydrographic conditions that bring nutrient-rich deep water up into the photic zone are upwelling, the divergence of currents, and the slow upward trickle caused by eddy diffusion at the thermocline. Upwelling occurs where the surface water moves away from shore and is replaced by nutrient-rich deeper water brought to the surface from a few hundred meters. Persistent upwelling along the west coast of North America is responsible for high productivity in that region. Divergences are the result of transverse ocean currents flowing away from each other, bringing up deeper nutrient-rich water. Such a divergence occurs between the North Equatorial Current and the Equatorial Countercurrent in the Pacific, an area in the open tropical ocean where productivity can be high due to upwelled nutrients. (See Figure 1.14.) Note that it now seems that some fraction of the nutrients, particularly nitrogen and perhaps also some trace metals, come from dust settling out of the atmosphere. The best example of the importance of trace-metal nutrients may come from the experiments by Martin

et al. (1994) in which iron was added to an open ocean area near the Galápagos Islands. This area had sufficient nitrates and phosphates to sustain high productivity because of its position in the aforementioned tropical divergence zone, but had naturally low productivity. The addition of iron resulted in a doubling of plant biomass and a fourfold increase in plant production, indicating that iron, indeed, can limit productivity under certain conditions.

Finally, the largest reservoir of nitrogen is in the form of the nitrogen gas that enters the surface waters of the ocean from the atmosphere. Unfortunately, nitrogen gas is unusable in that form by most organisms. To be biologically useful, the nitrogen must be in the form of nitrate, nitrite, or ammonium. Whereas some microorganisms, primarily the cyanobacterium genus *Trichodesmium*, have been shown to fix nitrogen gas into soluble usable compounds, that genus has not been considered abundant enough to account for the amount of nitrogen that seems to be fixed in the oceans. Recently, Zehr et al. (2001) demonstrated that there are single-celled cyanobacteria in the 3–10-micrometer range that have the enzymes necessary to fix nitrogen and that they are abundant enough in the oceans to have a significant effect on the biogeochemistry of nitrogen. If so, we may have to modify our current understanding of the role of nitrogen in limiting oceanic production.

TURBULENCE AND CRITICAL DEPTH   Not only does vertical mixing bring up nutrients, but it also carries autotrophic cells into the depths. As long as vertical mixing is confined to the upper lighted zone, the autotrophic cells can be carried downward only a short distance and will remain where there is sufficient light for photosynthesis. When mixing includes the lower water mass, however, it is possible for the plant cells to be carried well below the compensation depth. If the mixing is especially vigorous, the autotrophic cells may spend most of their time below the compensation depth, and there will be no net production, because the time spent in the lighted zone in active photosynthesis is insufficient to fix as much organic matter as is used when the organisms are below the compensation depth. This effect of turbulence has led to the development of the concept of **critical depth** (Figure 2.37), that depth at which total gross photosynthesis of the autotrophs in the water column equals total respiration. The critical depth is calculated over a 24-hour period and is different from the compensation depth, which is the depth at which the rate of pho-

tosynthesis equals the respiration rate. The critical depth is always greater than the compensation depth, because we are dealing with a vertical mixing process in which the autotroph population is circulated above and below the compensation depth. As long as it is above the compensation depth, the autotroph population can photosynthesize in excess of respiration and build up organic matter. The decisive point is the length of time spent in each area; in turn, this depends on how deep the mixing occurs. Deeper mixing means more time spent below the compensation depth, where respiration uses up organic material more rapidly than the short time in the lighted zone can compensate for. Whenever the wind-driven vertical mixing is less than the critical depth, photosynthesis is greater than respiration and a net production occurs.

■ GEOGRAPHICAL VARIATIONS IN PRODUCTIVITY

Having considered the roles of light, transparency, and nutrients separately, we can now examine how these factors interact with each other to produce the

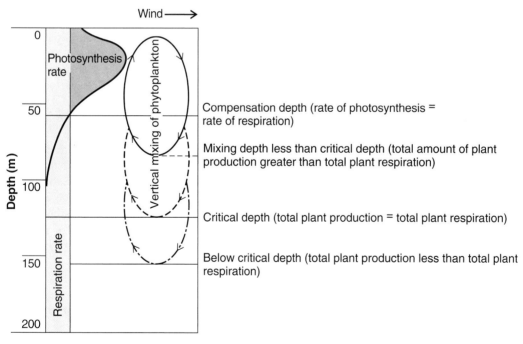

FIGURE 2.37 The relationship between the compensation depth and the critical depth—the depth to which the total phytoplankton biomass may be circulated and still spend enough time above the compensation depth to have a total production equal to its total respiration during the same period. (*Productivity of the Seas*, D. H. Cushing, Oxford Biology Reader, #78, 1975.)

geographical or latitudinal variations in productivity we observe in the oceans.

TEMPERATE SEAS   In the temperate-zone seas, the amount of light varies seasonally. As a result, the amount of solar energy entering the water varies, so that the temperature in the upper water layers is altered. The thermal structure of the water column, therefore, changes seasonally (Figure 2.38). In the summer months, the sun is high, days are long, and the upper layers heat up and become less dense than underlying layers. In other words, the water column is thermally stratified and no mixing occurs. In the fall, the amount of solar energy entering the water column decreases, days become shorter, upper layers turn cool, and thermal stratification decreases.

Finally, a point is reached where the temperature of the surface layers has been reduced to such an extent that the density of the layer is little different from that of the underlying mass. At this point, mixing can occur whenever sufficient wind is available. In winter—usually the storm season in the temperate zone—the sun is lowest on the horizon, solar energy input to the water is at a minimum, thermal stratification is at a minimum or absent, and mixing occurs. With the onset of spring, the days become longer, the solar energy increases, the temperature of the upper layers begins to rise, and the system moves toward reestablishing thermal stratification.

In contrast to the tropics, all the major factors that affect productivity change seasonally in temperate seas. This seasonal variation is reflected in

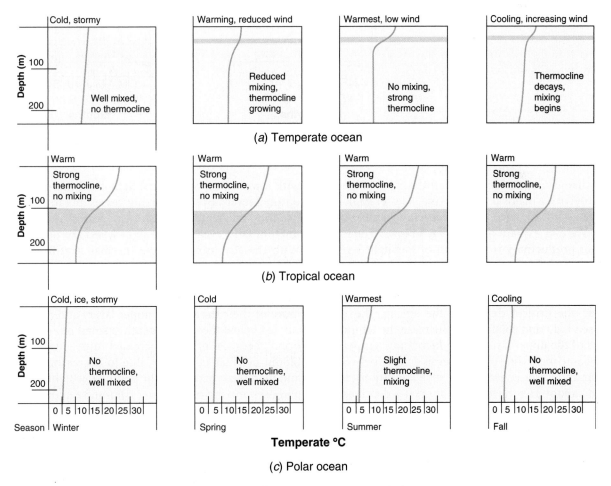

FIGURE 2.38 The thermal structure and extent of mixing in temperate, tropical, and polar seas during the four seasons of the year.

FIGURE 2.39 Summary of the seasonal cycles of net phytoplankton and zooplankton in four different geographical areas. The solid line represents phytoplankton, the dashed line zooplankton. (Modified from *Biological Oceanographic Processes*, 3d ed., T. R. Parsons et al. Copyright © 1984 Butterworth Heinemann. Reprinted by permission.)

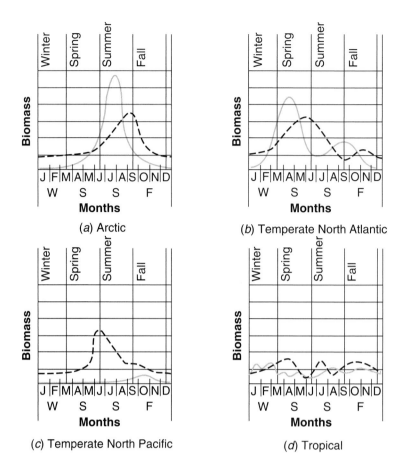

(a) Arctic

(b) Temperate North Atlantic

(c) Temperate North Pacific

(d) Tropical

the change in production over the year (Figure 2.39b), with a major peak in spring, a lesser peak in the fall, and low productivity in winter and summer. We may explain this trend as follows: The low winter productivity is the result of low light levels due to the low position of the sun on the horizon and due to the winter storms mixing the isothermal water column and carrying autotroph cells below the critical depth. In the spring, the increased light and solar energy increase the temperature of the upper layers. With increasing temperature come increasing differences in density between upper and lower layers. Under such conditions, the wind cannot mix the water to as great a depth as it can in winter; at some point, autotroph cells are no longer carried below the critical depth. Since nutrients in upper layers were replenished during the winter mixing, conditions are good for autotrophic growth, and we observe the **spring bloom** (Figure 2.39b). As spring passes into summer, the water column becomes more thermally stratified, mixing

with lower levels ceases, and light conditions reach optimal levels. Because mixing ceases due to stratification, nutrient replenishment ceases and production falls, even though light levels are optimal. With the advent of fall, the thermal stratification begins to break up and nutrients are returned to upper levels. If, in the fall, the mixing alternates with calm weather such that the autotrophs spend more of their time in the upper layers and are not carried below the critical depth, a small bloom will occur because of the increased nutrients. This bloom declines in late fall, due to decreasing light and increased mixing. In the winter, low light levels and deep mixing of the water column keep productivity low. Note that there are differences in the seasonal cycle curves of the temperate North Atlantic and temperate North Pacific in Figure 2.39b and c. These differences are the result of somewhat dissimilar hydrographic conditions, coupled with different nutrient concentrations and availability. (See pp. 82–84.)

TROPICAL SEAS   In the tropical seas, the upper waters are well lighted throughout the year because the sun does not show marked changes in height above the horizon. Light conditions are, therefore, optimal for autotroph production. At the same time, the continual input of energy from the sun maintains the surface layers of water at temperatures much higher than those in deeper waters. This means there is a great difference in density between surface and deep waters; hence, mixing does not occur. We say that such waters exhibit a **thermal stratification** that extends throughout the year. (See Figure 2.38b). Because the sun's energy creates a thermal stratification in the water column that prevents mixing and the upward transport of nutrients, productivity is low, but relatively or potentially constant all year (Figure 2.39d). Tropical seas are very clear and have the deepest compensation depths, but they are that way because there are few autotrophs in the water column due to the low nutrient content.

POLAR SEAS   In polar areas, productivity is restricted to a single short period in the polar summer, usually July or August in the Arctic (Figure 2.39a). During that time, the snow cover on the ice has disappeared, allowing sufficient light to enter the water through the ice to permit autotroph growth. In areas outside of the permanent ice pack, the breakup of the ice opens up leads, allowing sufficient light to enter the water and permitting autotroph growth. Following this single burst of growth, productivity quickly declines. Nutrients are not limiting and the water column is never strongly stratified. (See Figure 2.38c.) The reason for the lack of production at other times is primarily the absence of light. Light intensity is insufficient for a fall bloom, and during the long winter light is either absent or prevented from reaching the water column by a layer of snow over the ice pack. (See a more detailed analysis of polar seas in Chapter 5.)

■ PRODUCTIVITY IN INSHORE AND COASTAL WATERS

The latitudinal variations in phytoplankton productivity just discussed apply to open ocean areas away from the influence of landmasses. The situation in the water masses adjacent to land is somewhat different. Several factors contribute to this difference.

First, inshore waters tend to receive a considerable input of the critical nutrients PO_4^{3-} and NO_3^-, due to runoff from the adjacent land (where, as we have seen, the nutrients are far more abundant). Because of this input, inshore waters usually do not show nutrient depletion. A second factor contributing to the difference is the water depth. Most inshore waters are shallower than the critical depth; thus, the autotrophs cannot be carried below that depth in any kind of weather. Given sufficient light, production can occur at any time, even in the winter. A third factor is that shallow inshore waters rarely have a persistent thermocline; hence, no nutrients are locked up in bottom waters. A final influencing factor is the presence of large amounts of terrigenous debris in the water, which may act to restrict the depth of the photic zone and counteract the high nutrient concentration and shallow depth.

The interaction of these four factors on a latitudinal basis produces changes, both in the cycle of productivity and in total production, compared with offshore areas (Table 2.2). In temperate regions, instead of a bimodal production cycle, as is seen offshore, production remains high all through the summer. Nutrients are not limiting, due to runoff from land and the absence of a permanent thermocline. Yearly average production in inshore temperate waters is higher than in offshore waters due to the greater nutrient concentrations and the lack of critical-depth problems inshore. That production is not even higher inshore probably is due to the presence of large amounts of light-absorbing debris in shallow water and the fact that, in offshore water, production can occur to a greater depth. In other words, in shallow waters production is limited to the upper 5–10 m, whereas offshore it may go as deep as 50 m. In tropical waters, the difference between inshore and offshore waters is particularly dramatic: Inshore tropical waters have a productivity as much as ten times that of offshore waters. This must be attributed

TABLE 2.2

Production in Several Different Geographical Areas

LOCATION	PRODUCTIVITY IN C/m²/year
Long Island Sound (temperate inshore)	380
Continental shelf	100–160
Tropical oceans	18–50
Temperate oceans	70–120
Antarctic oceans	100
Arctic Ocean	<1

Source: After Raymont, 1980.

in large part to the increased nutrient concentration inshore compared with offshore areas.

Inshore production is further enhanced through contributions of benthic autotrophs, a component not usually present in offshore areas (except for the Sargasso Sea). The production of benthic algae and sea grasses is examined in Chapter 5, but is mentioned here for completeness.

Whereas inshore or neritic waters exhibit higher productivity, they also are characterized by greater variability over time and space due to local geography, river and stream discharge, storms, and tides. As a result, it is more difficult to predict the yearly productivity of these waters.

Primary Productivity of the Biosphere

Employing primarily data obtained from satellite measurements of absorbed solar radiation, and using two computer models, Field et al. (1998) have estimated the total net primary productivity of the entire biosphere and compared the percentages of that productivity produced by the oceans and the terrestrial environment. Moreover, they have subdivided these estimates into seasonal and biogeographic components (Table 2.3). Strikingly, these estimates from satellite data give an ocean productivity that is twice that of calculations made before the availability of satellite data. Thus, the annual net primary productivity of the oceans is 48.5 petagrams (1 petagram = 10^{15} grams) of carbon, while that of the terrestrial environment (exclusive of areas of permanent ice and snow) is 56.4 petagrams. How do we reconcile this roughly equal productivity with the aforementioned discussion of the reasons for low productivity in the oceans? First of all, there is the discrepancy between the amounts of land and ocean on the planet, as well as the amount of the terrestrial realm that is under permanent ice and snow. Also, the terrestrial environment has extensive areas of extreme desert, which Field and colleagues indi-

TABLE 2.3

Annual and Seasonal Net Primary Productivity (NPP) of the Major Units of the Biosphere

	OCEAN NPP		LAND NPP
Seasonal			
April to June	10.9		15.7
July to September	13.0		18.0
October to December	12.3		11.5
January to March	11.3		11.2
Biogeographic			
Oligotrophic	11.0	Tropical rain forests	17.8
Mesotrophic	27.4	Broadleaf deciduous forests	1.5
Eutrophic	9.1	Broadleaf and needleleaf forests	3.1
Macrophytes	1.0	Needleleaf evergreen forests	3.1
		Needleleaf deciduous forest	1.4
		Savannas	16.8
		Perennial grasslands	2.4
		Broadleaf shrubs with bare soil	1.0
		Tundra	0.8
		Desert	0.5
		Cultivation	8.0
Total	48.5		56.4

[a]Ocean color data are averages from 1978 to 1983. The land vegetation index is from 1982 to 1990. All values are in petagrams of carbon (1 Pg = 1015 g). Ocean NPP estimates are binned into three biogeographic categories on the basis of annual average C_{sat} for each satellite pixel, such that oligotrophic = $C_{sat} < 0.1$ mg m^{-3}, mesotrophic = $C_{sat} < 1$ mg m^{-3}, and eutrophic = $C_{sat} < 1$ mg m^{-3}.

(Reprinted with permission from C. B. Field et al., "Primary Production of the Biosphere: Integrating Terrestrial and Oceanic Components," *Science*, Vol. 281, pp. 237–240, 1998. Copyright © 1998 American Association for the Advancement of Science.)

cate are less productive than the vast tropical and subtropical oceanic gyres. (See pp. 96–97.) Given these caveats, if we calculate the net primary productivity on an areal basis, we find that the terrestrial environment has an average annual productivity of 426 g C/m^2 while the figure for the oceans is 140 g C/m^2. This lower figure stems from the aforementioned light absorption by the water and particles, the lack of abundant nutrients, and the competition between the autotrophs and the medium for the existing light. In fact, according to Field and coworkers, on average, only 7% of the photosynthetically available radiation (PAR) incident on the ocean is actually absorbed by the autotrophs while in the terrestrial environment the leaves absorb about 31% of the PAR. In addition, the autotrophs of the oceans of the world represent only 0.2% of the global photosynthetic biomass but manage to account for almost half of the biosphere's net primary productivity, primarily because the turnover time of plant organic matter in the oceans is only about 2–6 days, whereas on land it averages 19 years.

Both the oceans and the land are patchy, with areas of high and low productivity. In high-productivity areas in both locales, the maximal rate of production of carbon is similar, at about 1,000 to 1,500 g C/m^2/yr, but the oceanic areas that have such rates are much more spatially restricted. For example, 25% of the area on land can yield a net primary productivity of 500 g C/m^2/yr, while the comparable areal proportion in the oceans is 1.7%.

On a latitudinal basis, the net primary productivity of the biosphere reaches a maximum in three places (Figure 2.40). The largest centers on the equator and is due primarily to the high terrestrial productivity in the tropical forests. The next largest is in the temperate latitudes of the Northern Hemisphere and is again due primarily to the seasonal high terrestrial productivity. Finally, the smallest peak occurs in the temperate latitudes of the Southern Hemisphere and is due mainly to enhanced nutrient availability in the ocean caused by the southern subtropical convergence and the reduced land area. Figure 2.41 summarizes global oceanic annual net primary productivity.

The Ocean Ecosystem: The Classic Model

We now begin a discussion of the interactions among the various planktonic organisms, beginning with the larger or net plankton. The following sections

FIGURE 2.40 Latitudinal distribution of global net primary productivity. (A) Global total (blue line), land total (green line), and ocean total (red line). (B) Land net primary productivity April to June (blue line), July to September (green line), October to December (orange line), and January to March (red line). (C) Ocean net primary productivity, four seasons, as in B. (Reprinted with permission from Field et al., *Science*, Vol. 281, p. 239, 1998. Copyright © 1998 American Association for the Advancement of Science.)

will detail first the classical, or earlier, concepts that describe the interactions of the larger phytoplankton—diatoms and dinoflagellates—and of the larger zooplankton—primarily copepods. These ecological interactions still dominate coastal, upwelling, and polar seas. The presentation of this classical model is followed by a discussion of the importance of the microplankton.

■ GRAZING

As on land, autotroph production is transferred into the food chain of the pelagic community through the grazing activity of herbivores. Because the phytoplankton are small, the herbivores are also small. Many protistan and invertebrate planktonic species are herbivores, but by far the dominant larger herbivores, those which graze on the larger diatoms and dinoflagellates, in all oceans are various species of copepod. (See Figure 2.7.) The dominance of the herbivorous larger, or net, zooplankton by copepods

(a)

(b)

(c)

(d)

Chlorophyll *a* Concentration (mg / m³)

FIGURE 2.41 Seasonally averaged upper-ocean chlorophyll concentrations for the world's oceans, derived from satellite coastal zone scanner data. (a) Northern Hemisphere, spring, Mar. 1998–May 1998; (b) Northern Hemisphere, summer, June 1998–Aug. 1998; (c) Northern Hemisphere, fall, Sept. 1997–Nov. 1997; (d) Northern Hemisphere, winter, Dec. 1997–Feb. 1998. (Provided by the SeaWiFS Project, NASA /Goddard Space Flight Center and ORBIMAGE.)

is so great (70–90% of all of the ocean's biomass) that it is possible to consider only the copepods in discussing the effects of grazing on the larger, or net, phytoplankton populations. Furthermore, in most temperate seas, only from one to three species of copepod are numerically dominant.

Thus far, we have considered the changes in net phytoplankton production as a function only of interacting physical factors. In the oceans, however, the phytoplankton exist together with various herbivores. What effect do these herbivores have on the phytoplankton populations? Evidence suggests a great effect.

Careful inspection of the numbers of net phytoplankton cells per unit volume of water has shown that, for the temperate North Atlantic Ocean, the rate of decline of phytoplankton numbers following the spring bloom is steeper than would be predicted on the basis of the nutrient decrease. Furthermore, nutrients never disappear completely from the upper waters, as would be expected if they were limiting. It can also be shown that the curve of zooplankton abundance in the temperate North Atlantic is similar in form to that of the phytoplankton, only delayed slightly in time. (See Figure 2.39b.) All of these data have suggested to some workers (Riley, 1946; Cushing, 1959; Harvey et al., 1935) that the copepods are responsible for the regulation of the larger phytoplankton populations. This finding is reinforced by data from the North Pacific Ocean where there is no spring bloom of phytoplankton (see Figure 2.39c). This is presumably the result of the presence of feeding copepods in the North Pacific, as opposed to the North Atlantic, at that time.

The rapidity with which copepods can remove phytoplankton cells is remarkable (Table 2.4). Under experimental conditions, Fleming (1939) established a diatom population that initially had 1 million cells per liter and divided once each day. It was grazed by a copepod population adjusted to remove 1 million cells per day. Under such conditions, the diatom population would remain in a steady state of 1 million cells per liter. When the copepod density was doubled (to produce double the grazing rate), the effect was to reduce the diatom population to 27,000 cells per liter in five days. If the copepod numbers and their grazing rate were increased by a factor of five, the diatoms were eliminated in five days. Since the rate of diatom reproduction remained unchanged in these scenarios, any increase in copepod numbers in the plankton could have a significant effect on the numbers of

TABLE 2.4

Changes in the Numbers of Individuals of a Phytoplankton Population with a Constant Rate of Reproduction when Subjected to Two Different Grazing Intensities

| | PHYTOPLANKTON POPULATION DENSITY (CELLS/LITER) | | |
TIME IN DAYS	INITIAL GRAZING INTENSITY	GRAZING INTENSITY DOUBLED	GRAZING INTENSITY INCREASED FIVEFOLD
0	1,000,000	1,000,000	1,000,000
1	1,000,000	500,000	62,000
2	1,000,000	250,000	3,900
3	1,000,000	125,000	240
4	1,000,000	62,500	15
5	1,000,000	31,250	1

Source: Adapted from Fleming, 1939.

diatom individuals. It is important to understand that the decrease in diatom numbers was independent of any change in the photosynthetic rate, or the rate of primary production. It is quite possible to have a diatom population that decreases in number due to grazing, while increasing the rate of primary production. Of course, with fewer algal cells, the total amount of fixed carbon would decline with declining algal cell populations, but it would still be possible for the *rate* of fixation of carbon in each cell to increase. It is for this reason that the population density or standing crop of phytoplankton may be a poor or misleading indicator of primary productivity.

It is thus possible for the standing crop of phytoplankton to decline due to grazing, while the rate of primary production increases or remains steady. Under such circumstances, the diatom population divides rapidly because its high photosynthetic rate allows protoplasm to build up for cell division. As fast as the diatoms reproduce, however, they are consumed by the copepods and other herbivores. In this case, the major portion of the carbon fixed in photosynthesis appears not in the standing crop of phytoplankton, but in the standing crop of zooplankton. Does such a situation occur in the oceans, and if so, what is the magnitude of the fixed carbon that appears in the zooplankton compared with its magnitude in the phytoplankton? Plankton ecologists have used various mathematical equations employing the concentrations of nutrients, water transparency light availability and the photosynthetic rate to calculate what the productivity of a given body of water should be, after which they compare the result of the calculation with what is observed. For many

geographical areas, the results are remarkable. For example, Hart (1942) reported that, in the Antarctic seas, around South Georgia Island and in the English Channel, the standing crop of phytoplankton was 2% of calculated production. In oceanic areas of the Antarctic seas, where grazing was more intense, Mare (1940) found that the standing crop of phytoplankton was only 0.5% of calculated production. In other words, the majority of primary productivity resides not in the bodies of the algal cells, but in the zooplankton. These data also show that copepods are extremely efficient grazing animals and are capable of reducing the populations of phytoplankton drastically in patches of the ocean. This contrasts with the situation in natural terrestrial communities, where the grazers rarely remove all the vegetation and the standing crop of plants is high.

The ability of copepods to reduce the larger net phytoplankton populations to very low levels means that the sudden decrease in the phytoplankton curve after the spring bloom in temperate waters could be caused by the grazing of increased numbers of copepods, rather than being a result of nutrient depletion. This concept is supported by the finding that, in the Antarctic seas, where due to upwelling, nutrients are not limiting, the phytoplankton population also drops precipitously after the spring bloom. Additional evidence comes from the fact that grazing, and hence the breakup of phytoplankton cells during the passage through the copepod guts, releases the nutrients, which are then excreted back into the water. Under these circumstances, nutrients would be constantly regenerated in the upper layers and productivity never halted. Additional support for grazing as a mechanism for controlling phytoplankton is furnished by the patchiness of plankton: In the oceans, there are dense patches of phytoplankton and clear patches with copepods. One explanation is that the copepods overgraze the phytoplankton.

If the copepods are so efficient at reducing the phytoplankton population, how does the spring bloom in the North Atlantic ever get started? The answer to this question is that the copepods reproduce more slowly than the phytoplankton; thus, in early spring, following the winter low period, the phytoplankton can respond more rapidly to the improving light conditions and lowered turbulence. The phytoplankton reproduce rapidly and build up to high levels in the absence of significant grazing. The copepods, also at low levels, begin to breed as well, but their cycle is longer (weeks versus days for phytoplankton). Hence, the copepods cannot build up large enough populations until after the phytoplankton population has bloomed. (See next subsection on copepod cycles.) After that delay, however, when their populations are high, the copepods quickly graze down the plant population to the low summer level. Copepod levels also decline because there is less food and more predation, and because copepod egg production depends on food being available to the breeding adult.

The situation just described prevails in the temperate and subpolar North Atlantic. In the tropics, where nutrients limit population levels of phytoplankton to low levels at all times, the situation is different. In these areas, there are no real pulses in either the phytoplankton or copepod cycles. Instead, there is a steady but inconspicuous consumption of the small phytoplankton crop by copepods.

COPEPOD CYCLES Because the copepods dominate the larger zooplankton in all world oceans as the major grazers of the larger, bloom-forming diatoms and dinoflagellates, and because the copepods can determine the form of the phytoplankton population curve, it is worthwhile to review the major aspects of their life cycle. Such knowledge contributes to our understanding of these animals' grazing patterns. In what follows, we will concentrate on copepod cycles in temperate and polar waters.

Relatively few of the numerous copepod species have been thoroughly studied, but in most temperate and polar waters, only a few species predominate. The life cycles of some of these species—notably,

Calanus finmarchicus of the Atlantic Ocean—are well known. Since other species similar to *Calanus* also dominate Arctic and North Pacific waters and have somewhat similar life cycles, a general scheme may be established.

All copepods have a comparable pattern of development. The eggs hatch as naupliar larvae, which then pass through a series of six naupliar stages. They then enter another larval stage called the copepodid and pass through five of these stages before becoming adult. For the large, dominant Atlantic copepods of the species *Calanus finmarchicus*, the early winter months are spent in the last copepodid stage in the deeper waters of the open ocean. Here they persist without feeding, living on fat or oil reserves in their bodies. They do not migrate at this time. (See p. 83.) Those copepodids which survive the winter begin their final molt to the adult stage in late winter or early spring, at which time the population is at a minimum for the year. After molting to the adult condition, the copepods migrate to the surface waters and begin feeding. They do not begin breeding, however, until after the phytoplankton numbers have begun to increase. This delay in the onset of breeding permits the phytoplankton to escalate its numbers and produce the large populations that constitute the spring bloom. This first brood of copepod eggs then goes through the naupliar and copepodid stages and itself matures and reproduces. The resulting increase in numbers of copepods then begins to reduce the phytoplankton numbers, creating the drop in the phytoplankton curve and the increase in the zooplankton curve seen in Figure 2.39b. The cycle of generations continues through the spring and summer, and the number of generations per year depends on the environmental conditions. Off Scotland and Maine, for example, there may be three complete generations. Further north, off Norway, the number is reduced to two, and in Arctic seas, such as those off Greenland, only one generation is produced. Further south, off the Middle Atlantic states and the English Channel, four generations may appear. Each cycle, from egg production through adult to egg production again, takes about two months under favorable conditions. The life span of these large calanoid copepods is therefore about two months (Figure 2.42a), which means that adults do not live long after breeding.

The last brood produced in late summer does not complete the developmental cycle. Instead, when they reach Stage V copepodids, these animals migrate

FIGURE 2.42 (a) The life cycle of *Calanus finmarchicus* in the North Atlantic Ocean. (b) The life cycle of *Neocalanus plumchrus* in the North Pacific Ocean.

into deeper water and remain there through the winter, subsisting on stored food reserves. They return to the surface waters in the spring to feed before laying eggs. The life span of this generation is much longer than that of the others—perhaps five to seven months, depending on the area. Egg production is influenced by the abundance of phytoplankton available as food. Successive generations of copepods build up large numbers in the spring, feeding on the spring bloom of phytoplankton. The decrease in copepod numbers in the summer reflects the decrease in phytoplankton due to excessive grazing by the large numbers of copepods built up in the spring.

In sum, the life cycle of *Calanus finmarchicus* is about two months from egg through larval stages to adult, and successive generations are produced through the spring and summer in temperate and polar waters. The number of generations decreases as one moves poleward because of the increasing shortness of the season. Copepod numbers reflect phytoplankton densities.

The dominant calanoid copepods of the North Pacific Ocean, *Neocalanus plumchrus* and *Neocalanus cristatus*, have a life cycle somewhat different from that of *C. finmarchicus* (Figure 2.42b). This, coupled with certain environmental differences between the two oceans, leads to the different form of the seasonal curves for zooplankton and phytoplankton seen in Figure 2.39c. To begin with, the two *Neocalanus* species are larger than *C. finmarchicus* and feed on both phytoplankton and zooplankton. *Calanus finmarchicus*, by contrast is primarily a herbivore. The *Neocalanus* species are also less efficient in their grazing activity. Finally, the seasonal cycles are different: Both *N. plumchrus* and *N. cristatus* have only one generation per year. Both species migrate to a considerable depth (300 m) in the summer and overwinter there as adults with sufficient energy reserves to mate and lay eggs. The eggs hatch, and the larval copepods are back in the surface waters in the late winter, ready to graze before the onset of the spring phytoplankton bloom.

Besides the dominant, large *Calanus*-type copepods, other, smaller copepods in both the Atlantic and Pacific oceans, such as *Acartia*, *Pseudocalanus*, and *Microcalanus*, occur with *Neocalanus* and *Calanus* and also differ slightly in life cycle. In these forms, the generation time is shorter, and they tend to have more generations per year. Although they build up Stage V copepodids in the fall, there is no conclusive evidence of overwintering, and it may be that these small forms breed throughout the year, albeit at a reduced rate in the winter and at other periods of low phytoplankton availability. This latter type of life cycle with continuous breeding year-round is found in many tropical copepods. Because of their small size, the constancy of the conditions, and the continuous phytoplankton growth, these animals' development is rapid and many broods are produced per year, spreading out the grazing pressure more or less equally over the year to coincide with steady phytoplankton populations.

In general, then, the copepod cycles are maintained for maximum use of almost all larger or net phytoplankton production. However, a certain, but as yet undetermined, amount of energy is lost through fecal pellets and exoskeletons sinking out of the photic zone.

COPEPOD INFLUENCE ON ATLANTIC AND PACIFIC PLANKTON COMMUNITY ECOLOGY

At this juncture, it is useful to consider the differences in the life cycles of the dominant copepods in the North Atlantic and North Pacific, alongside what we know about the slight differences in environmental factors between the two oceans. That way, we can demonstrate how these interactions lead to the very different seasonal cycles of the larger phytoplankton and zooplankton in the two areas.

The environmental differences between the two oceans are slight, but significant. At the same latitude, the Atlantic is somewhat warmer, 8.5°C at depth in the winter, compared with the Pacific, at 3.8°C. The winter mixing extends to below 200 m in the Atlantic, but only to 100 m in the Pacific. Finally, the nutrient levels in the Pacific are always above 5 μm/liter, whereas in the Atlantic they are reduced to below 1 μm/liter in the summer. Other physical and chemical factors are similar between the two oceans. There are, however, some differences in organisms: Among the photosynthetic organisms, the Atlantic has larger numbers of diatoms and dinoflagellates, whereas the Pacific has larger numbers of small flagellates. The growing season is also shorter in the Atlantic than in the Pacific.

Putting all these factors together with the life-history differences among the copepod species produces surprisingly different results, as Parsons and Lalli (1988) have noted. Beginning in the winter, we find that in both oceans the dominant copepods overwinter at depth, but the difference in temperature means that the Atlantic copepods have a higher metabolic rate, so by early spring they have insufficient reserves left to lay eggs. In contrast, the larger Pacific copepods that overwinter at a lower temperature, and hence at a lower metabolic rate, retain sufficient reserves to lay eggs without resuming feeding (Figure 2.43).

In spring, the phytoplankton growth begins early in the Pacific but proceeds slowly because of reduced light and temperature. In the Atlantic, phytoplankton growth is delayed due to the depth of the mixed layer. Once it begins, however, the light is better and growth proceeds more rapidly, aided by the higher temperatures. In the Pacific, the slow phytoplankton growth and the continual presence of smaller herbivores in the surface waters, combined with the entrance into the surface waters of the recently hatched copepodids of *Neocalanus*, allows the grazers to keep pace with the plants, preventing any sudden bloom. In the Atlantic, by contrast, the zooplankton cannot keep up with plant growth and reproduction aided by the high nutrient content of the water and the absence of a dominant copepod grazer. Such a grazer is absent because *Calanus finmarchicus* must feed first in order to build up enough reserves to produce eggs, which in turn must have a finite development time before hatching. As a result, there is a delay in the grazer population increase, and a phytoplankton bloom

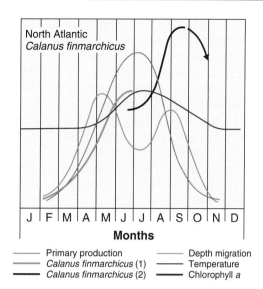

Primary production		Temperature
Neocalanus plumchrus		Chlorophyll *a*
Depth migration		

Primary production		Depth migration
Calanus finmarchicus (1)		Temperature
Calanus finmarchicus (2)		Chlorophyll *a*

Neocalanus plumchrus / *Calanus finmarchicus*

Winter

N. plumchrus: Overwinters with low metabolic loss at 3.8°C. *Result*: Copepod has sufficient reserves to lay eggs without feeding in the spring.

C. finmarchicus: Overwinters with high metabolic loss at 8.5°C. Copepod has insufficient reserves to lay eggs without feeding in the spring.

Early Spring

N. plumchrus: Phytoplankton growth starts slowly but is limited by light and temperature. *Result*: Zooplankton can track the spring bloom and suppress any increase in chlorophyll *a*; efficient phasing.

C. finmarchicus: Phytoplankton growth cannot start in mixed layer > 200 m. When growth starts, it is rapid due to higher temperatures. Zooplankton cannot keep up with phytoplankton growth and chlorophyll *a* increases; inefficient phasing.

Late Spring

N. plumchrus: Phytoplankton providing most of the primary production are all small (< 200 mµ). *Result*: Low phytoplankton cell density and small size favors alternate strategy of carnivorous feeding on microzooplankton resulting in loss of food chain efficiency. Relatively slow growth of zooplankton and slow development rate (i.e., stages) leads to large size.

C. finmarchicus: A much larger proportion of the phytoplankton is made up of large diatoms. High phytoplankton stocks and large phytoplankton size favor herbivory with rapid zooplankton growth rate and development rate to next generation, resulting in small size of adults.

Early Summer

N. plumchrus: N. plumchrus gradually reaches Stage V and migrates to depth to avoid high surface temperatures of summer. *Result*: Large size achieved of aestivating *N. plumchrus* assures winter survival at low temperature.

C. finmarchicus: C. finmarchicus grows first new generation (1). Growth rate is rapid, but phytoplankton exhaust winter nutrient supply. *C. finmarchicus* is ready to produce a second generation (2), which begins by growing faster than the first due to higher temperatures; phytoplankton production must now be based on efficient recycling of nutrients.

Late Summer

N. plumchrus: Annual cycle of *N. plumchrus* in surface layers entirely replaced by late summer opportunistic copepods and carnivores. *Result*: Phytoplankton shows small increase after *N. plumchrus* departure but this is again suppressed by summer herbivores.

C. finmarchicus: Second generation of *C. finmarchicus* grows to provide immature copepodites, which must overwinter. Other herbivores together with carnivorous zooplankton develop. Second generation is more capable of surviving winter than first generation since cooling trend favors larger body size of late development stages.

FIGURE 2.43 A comparison of the differences in the community ecology of the larger plankton organisms of the North Pacific and North Atlantic oceans. (Modified from "Comparative Oceanic Ecology of the Plankton Communities of Subarctic Atlantic & Pacific Oceans," Parsons & Lalli, *Oceanographic Marine Biology Annual Review*, Vol. 26, pp. 317–359.)

ensues. In other words, there is an efficient coupling of the phytoplankton and zooplankton populations in the Pacific, but not in the Atlantic. By late spring, the autotrophic organisms providing the bulk of the primary productivity in the Pacific are small in size and low in density. This means that the larger *Neocalanus* cannot efficiently graze on them, so *Neocalanus* turns instead to carnivorous feeding on the smaller zooplankton population, thus suppressing its numbers. In the Atlantic, the majority of the phytoplankton consists of large diatoms in high density, and they are grazed efficiently by *C. finmarchicus*. This large food supply leads to the rapid growth and maturity of *Calanus*, and they begin to produce the next generation.

In the Pacific during the early summer, the *Neocalanus* copepods have reached maturity and migrate to greater depths to avoid higher surface temperatures. Their large size and large food reserves ensure that they will survive and be able to produce eggs. Meanwhile, grazing in the surface layers is taken over by smaller zooplankton now released from the feeding pressure of the *Neocalanus*. The smaller zooplankton keep the grazing pressure on the autotrophic organisms throughout the fall and winter when *Neocalanus* are at depth. The higher level of nutrients permits some phytoplankton growth even during the low light levels of the winter. In the Atlantic, by summer, the phytoplankton have exhausted the nutrients at about the time *C. finmarchicus* begin to produce its second generation. This second generation of *Calanus* begins growth at a faster rate than the parent generation because of the higher temperatures and larger food supply that are now present. The phytoplankton numbers, however, shortly begin to decrease due to nutrient exhaustion and grazing by the high numbers of *Calanus*. Phytoplankton production now becomes dependent on nutrient regeneration in the euphotic zone. Finally, the second generation of *C. finmarchicus* grows to the last copepodid stage, but with little reserve energy, thanks to the declining phytoplankton population, and migrates to greater depths to overwinter.

A probable, but as yet unproven, result of the poor coupling of the phytoplankton and zooplankton cycles in the Atlantic is that much of the carbon fixed in the spring bloom ends up as detritus that sinks and probably sustains large numbers of benthic fishes in that ocean. By the same token, the closer coupling of the phytoplankton and zooplankton cycles in the

Pacific may be responsible for the greater numbers of pelagic fishes there.

Recent work by Miller et al. (1991) has provided additional information that has increased our understanding of how the North Pacific pelagic ecosystem functions. These researchers were also interested in the problem of why, in contrast to North Atlantic waters, the North Pacific always had substantial nutrient levels and no significant spring phytoplankton blooms. Their results showed that the autotrophs in the North Pacific are always dominated by nanoplankton-sized organisms and that the larger phytoplankton population, such as diatoms, are less common. Furthermore, the *Neocalanus* were not grazing these smaller autotrophs, indicating that there had to be another set of smaller grazers. These microzooplankton grazers were mostly protozoans and probably the main food of the *Neocalanus*. More interestingly, Miller and colleagues also discovered that these protozoans had very rapid growth rates and could always overtake and suppress any incipient blooms of the small autotrophs. Because the mixed layer is so shallow in the North Pacific, the autotrophs can overwinter in the lighted zone, albeit at a low level due to reduced light. In addition, this provide the microzooplanktonic protozoans with a continued food source. Interestingly, these protozoans give us the answer to the persistent higher nutrient levels of the North Pacific: The micrograzers produce very small fecal matter that does not sink out of the lighted zone, but is recycled in the system, mainly as NH_4^+, which itself seems to be favored over nitrate by the phytoplankton. This explanation of the situation in the North Pacific has been called the **mixing-and-micrograzer hypothesis**.

One last part of the North Pacific enigma needs to be explained: What controls the stocks of the larger phytoplankton? As noted, the diatoms and dinoflagellates are persistently present. So why do they not bloom as in the Atlantic? There is no satisfactory answer to this question at present. It could be that the *Neocalanus* species are capable of preventing blooms, but another suggestion by Martin and Fitzwater (1988) is that the area is lacking in iron, which facilitates the growth of the smaller phytoplankton at the expense of the larger ones. Without this necessary trace element (see also the section on iron limitation in the intertropical convergence), the diatoms and dinoflagellates cannot attain bloom levels.

VERTICAL MIGRATION   One of the more puzzling phenomena in the sea is the vertical migration of zooplankton. Vertical migration is the name given to the daily migration of certain zooplankton organisms down into the depths during daylight hours and up into the surface waters at night. The phenomenon is entirely separate from the seasonal migration noted above, wherein copepods descend in the fall to deeper waters to overwinter. Vertical migration may be just a special case of the more general phenomenon of the regular vertical migration of many open ocean pelagic animals from one depth zone to another and back. Movements of these other so-called deep scattering layers are discussed in Chapter 4. Here, we cover the special case of the vertical migration of epipelagic plankton.

Daily or **diel** vertical migrations have been known for more than 100 years. What makes the phenomenon so puzzling is that it is difficult to offer a satisfactory explanation as to why such small organisms should expend significant amounts of their limited energy resources for this purpose. The distances traversed in these daily movements may be 100–400 m. Since we are considering animals of only a few millimeters in size, that is the equivalent of humans walking 25 miles to work and then 25 miles back at night!

Vertical migration has been observed in all zooplankton taxa, but not all migrate. With respect to taxonomic group, size, or feeding habits, few generalities can be drawn from a comparison between those that do migrate and those that do not. This inability to find correlations makes the occurrence of vertical migration even more difficult to understand.

It is now generally agreed that the major stimulus initiating and controlling vertical diel migrations is light. Vertical migrators respond negatively to light, moving themselves deeper as the light intensity at the surface increases. Conversely, they move toward the surface as the surface light intensity decreases. In a typical pattern, the migrators are found at the surface during the night, and then, with the approach of dawn, they begin their descent. As the light intensity increases through the morning, the animals move deeper, usually keeping themselves within a certain narrow range of light intensity. At noon, or whenever the intensity is maximal, they will be at the deepest position. As the sun sinks through the afternoon and the light intensity decreases, they begin their journey toward the surface again, arriving there around sunset and remaining until just before dawn. As might be expected, reverse migrators proceed in exactly the opposite way, descending as the light decreases and rising as the light increases. Since the advent of sonar or echo-sounding gear in World War II, it has been possible to observe this phenomenon by watching the trace formed on the recorder by sound reflected from the bodies of these animals as they move up and down. Their bodies form a layer in the water column (Figure 2.44).

Since the ocean is not vertically uniform in its physical parameters, other factors may modify the light-regulation scenario. For example, temperature may also influence migration, limiting some migrators in how deep they go. In such cases, the maximum depth to which they descend is set by the temperature, which overrides the stimulus of light. Another constraint is the presence of the bottom. In open ocean areas where submarine ridges or mountains rise to within daytime

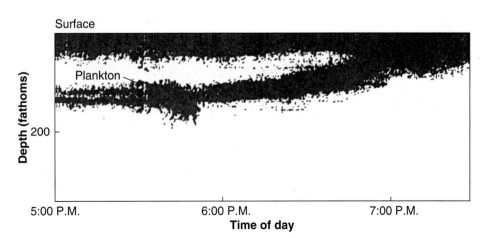

FIGURE 2.44
Sonogram record of movement of vertically migrating plankton. (After G. B. Farquhar, ed., *Proceedings of an international symposium on biological sound scattering in the ocean*, 1970, U.S. Government MC Report 005.)

depths of the vertical migrators, the migrators may congregate at the bottom. Such concentrations of animals are a prime source of food and may be why both submarine banks and seamounts are often densely populated with predators.

Although light is both the main stimulus to which the animals respond and the guide to their positioning in the water column, it does not explain why the animals make the journey in the first place. What advantage can there be for such energy-demanding movements? Over the years, a number of explanations have been offered by marine ecologists. At present, four have been presented, but there is no universal agreement as to which, if any, of the explanations is correct.

Perhaps the most popular explanation, which cannot be attributed to any single person, but has been reviewed by McLaren (1963) and Haney (1988), is that the zooplankton move from the lighted waters into dimly lit, deeper waters to avoid predation by visual, diurnal predators, such as various fishes, cephalopods, and birds that forage at the surface. Similarly, reverse diurnal migrations are a response to avoid nocturnal surface-dwelling predators or diurnally migrating nonvisual predators. The following objections, however, have been posed against this explanation:

1. The daytime depths of residence for many zooplankters still have sufficient light for visual predators to operate.
2. Other zooplankton migrate deeper than would be required by the hypothesis.
3. Many of the migrators possess and use bioluminescent organs that seem to counteract the effect of the darkness.
4. Many potential predators are also migratory.

Support for the hypothesis has come mainly from investigations in freshwater lakes. These studies have demonstrated that in systems with few or no visual predators, vertical migration is absent, whereas in those systems with such predators, vertical migration is common. Ohman (1990), however, has demonstrated that the marine copepod *Pseudocalanus newmani* may undergo vertical migration, reverse migration, or no migration depending on whether the primary predators are visually hunting planktivorous fish, nocturnally feeding nonvisual zooplankton, or are absent. Ohman's study also suggests that diel vertical migration is not a fixed trait in zooplankton species but a

dynamic behavioral trait that may be modified by prevailing conditions.

A second hypothesis, somewhat related to the first, is that vertical migrators are seeking to avoid damage from solar radiation. Since damaging ultraviolet light does penetrate several meters deep in clear oceans, this seems a reasonable hypothesis. It is also known that certain pigments, primarily carotenoids, protect organisms from damage by ultraviolet light. We have, therefore, a dilemma: The presence of pigment may protect the zooplankton from light, enabling those organisms to inhabit lighted waters, but it also makes the animal more visible to visual predators. It seems that we must assume that the greater risk of predation due to the high visibility of pigmented animals is offset by either the escape response of the zooplankton (vertical migration) or the absence of visual predators. At present, no experimental or field studies have shown a direct relationship between damage from light and vertical migration.

A third hypothesis explaining the occurrence of vertical migration, suggested by Hardy (1953), is that it allows the weakly swimming zooplankton to change their horizontal position. To understand how this is possible, it is necessary to return to the previously described principle known as the Ekman spiral (Chapter 1). As one moves downward from the surface waters, the speed and direction of a surface current change. The speed decreases and the direction of movement of the current changes so that the current moves to the right of the topmost (surface) current. (See Figure 1.13.) This means that a descending zooplankton organism will encounter currents that move not only more slowly but in slightly different directions. Once the zooplankter has descended into the deeper, slower-moving water, the surface currents will bring different water across and above the animal. If the zooplankter ascends again, it will find itself in a different water mass. What is the value of this phenomenon? Since zooplankton organisms lack the ability to swim against currents, they would automatically be confined to a given water mass in the ocean. Given their great ability to graze out the phytoplankton from an area, they would soon be without food and starve. By simply sinking into the depths and then rising again, however, they would come up in a different surface water mass, potentially ungrazed. This explanation is supported by the universal observation that phytoplankton and zooplankton are very patchy in occurrence. The mechanism might also be employed by zooplankton to avoid or depart from very dense phytoplankton patches. Hardy and Gunther (1935) have suggested that a toxic substance produced by the phytoplankton has deleterious effects on zooplankton.

A final reason for diel migration, suggested by McLaren (1963) and McAllister (1969) and refined by Enright (1977) and McLaren (1974), concerns production and energetics. This two-part hypothesis suggests that (1) the rate of phytoplankton production is greater when subjected to discontinuous nocturnal grazing as opposed to a population that is subjected to continuous grazing, and (2) the zooplankton obtain an energetic advantage by spending a portion of their time in cold, deep waters rather than maintaining themselves constantly in the warm, upper waters. Demonstrating the truth of the second part of this hypothesis rests on whether one can show that more energy is saved by the zooplankton's remaining in deeper, cold waters during the day than is expended in the animals' making the migration. It has been demonstrated that a neutrally buoyant zooplankter requires remarkably small amounts of energy to migrate, but it remains to be proven that the amount of energy saved by remaining in deeper waters exceeds the amount expended in migration. Implicit in this energetics theory is that feeding in warm water increases the efficiency of feeding and of assimilating food, while residence in deeper waters enhances the efficiency of growth.

This fourth hypothesis is supported by the observation that, in polar regions, where the temperature differential between deeper and surface waters is lowest, vertical migration is least often observed. Conversely, the strongest vertical migration patterns are observed in the tropics, where the greatest differential exists between surface water and depth. However, attempts to verify the hypothesis both in the laboratory and in the field have failed so far.

This final argument may also be used to explain the overwintering of copepods at depths where phytoplankton is scarce: Such overwintering at low temperatures conserves energy.

In summary, it can be stated that vertical migration is a worldwide phenomenon of certain zooplankton. The stimulus is primarily light, although it may be modified by other factors, such as temperature. Diel migration occurs most commonly in strongly thermally stratified seas and becomes suppressed or disappears where seas approach isothermal conditions and in temperate seas during winter months. The hypotheses that have been advanced to explain vertical migration include the avoidance of predation, the avoidance of damaging ultraviolet light, a change of the organism's position in the water column, and the use of such migration as a mechanism to increase production and conserve energy. It is probable, as Haney (1988) suggests, both that no single explanation for the occurrence of vertical migration will be found to hold for all systems and that there are multiple causes of vertical migration in different systems, as well as within a single system.

■ SEASONAL SUCCESSION IN PHYTOPLANKTON

In addition to the marked changes in abundance observed in the phytoplankton over the course of a year, there is a marked change in species composition. This change in the dominant species from season to season is called **seasonal succession**. Under seasonal succession, one or more species of diatom, dinoflagellate, or coccolithophore dominate the plankton for a shorter or longer period and then are replaced by another set of species. The pattern is repeated yearly. This "succession" is different from the typical terrestrial ecological succession in which various plant associations replace each other until finally a so-called climax community develops, which persists through time.

Figure 2.45 displays the dominant phytoplankton species at several geographic locations, to emphasize the widespread occurrence of this successional phenomenon. Some successional changes occur even in polar (East Greenland) and tropical (Bermuda) areas.

What are the factors that cause seasonal succession? Margalef (1963) and Smayda (1980) have suggested several. Considering that the seasonal succession is most often and clearly seen in temperate seas, which have a marked change in surface temperature during a year, temperature has been suggested as a cause. This may be one of the factors, but it is unlikely to be solely responsible because certain dominant species recur at different temperatures. Furthermore, temperature changes rather slowly in seawater, and the replacement of dominant species often is much more rapid.

Another suggested reason is the change in nutrient level over the year, with differing concentrations favoring different phytoplankton species. While this factor may also contribute, observations suggest that population changes are not closely correlated with it and that phytoplankton populations rise and fall much more quickly than nutrient concentrations change.

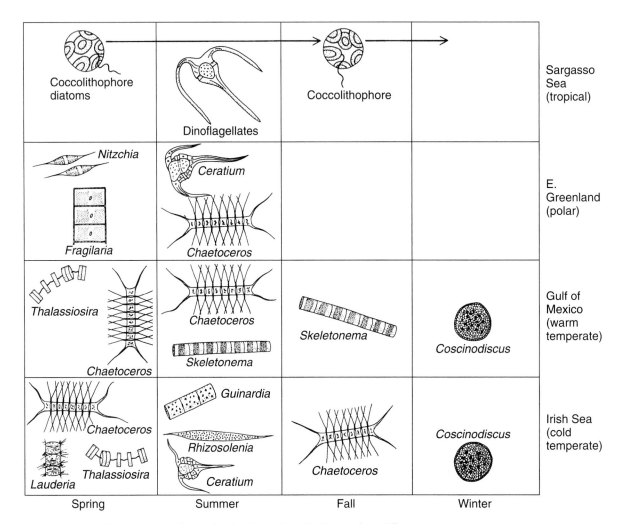

FIGURE 2.45 Seasonal succession of the dominant phytoplankton in four different geographical areas. With the exceptions of coccolithophores and *Ceratium*, which is a dinoflagellate, all genera are diatoms.

Yet another explanation for seasonal succession is that it is the result of **biological conditioning** of the seawater by the organisms in it. By biological conditioning we mean that one group of organisms, when dominant in the water column, secretes or excretes very small quantities of organic compounds or **metabolites** that have an effect on other organisms, either inhibiting or promoting their growth. At the same time, these organic compounds could affect the very organisms producing them, making the water either more or less suitable for their own existence.

Organic metabolites could, and probably do, include a number of different classes of organic compounds. Some are likely toxins, such as those released by the dinoflagellates during red tides and that inhibit the growth of other photosynthetic organisms. In these cases, the population explosion of dinoflagellates is so great that the water becomes brownish red in color from the billions of dinoflagellate cells. Although each cell secretes only a minute amount of toxin, the massive dinoflagellate numbers cause the toxin to reach concentrations that kill many creatures. The toxin can be concentrated in such filter-feeding organisms as clams and mussels, rendering them toxic to humans at certain times and places. Diatoms of some genera are also responsible for producing a toxin called domoic acid that has been linked to the deaths of various marine organisms.

Another class of metabolite is the **vitamins**. It is now known that certain phytoplankton species have requirements for certain vitamins and that there are considerable differences as to requirements among species. The B vitamins—especially B_{12}, thiamine, and biotin—seem to be the most generally required. Some species may be unable to thrive until a particular vitamin or group of vitamins is present in the water. If this is produced only by another species, then a succession of species could occur whereby first the vitamin-producing phytoplankter is present and the vitamin-requiring species follows.

Other organic compounds that may inhibit or promote various species include amino acids, carbohydrates, and fatty acids. Although it is suspected that these organic metabolites may have an important role in species succession, and it has been demonstrated in the laboratory that phytoplankton species vary both in their ability to produce necessary vitamins and in their requirements for such vitamins in order to grow, evidence is still sketchy as to the real role of these organic compounds in the sea.

There is also evidence to suggest that grazers—particularly selective grazers—can influence the phytoplankton species composition. Many copepods and invertebrate larvae pick out certain phytoplankton species from mixed groups, changing the species composition.

Finally, a growing body of evidence now suggests that all of the factors considered here are operating simultaneously to produce species succession. The importance of any one factor will vary with the particular phytoplankton species and the environmental conditions.

To summarize, there is a general seasonal pattern of succession, progressing from diatom to coccolithophorid- or dinoflagellate-dominated systems (Figure 2.45). In temperate waters, small-celled, rapidly dividing diatom species initiate the spring bloom. They are succeeded first by larger diatoms in the late spring and then by a summer warm-water dinoflagellate community. In tropical and subtropical waters, coccolithophores and dinoflagellates are generally dominant all year, although they may be interrupted by periodic diatom blooms.

The Ocean Ecosystem: A Changing Model

Up to this point, we have been considering the usual explanations for the organization and functioning of the marine plankton system. Much of this classical description, while still valid as a basic framework and still true in polar seas, coastal seas, and upwelling areas, has been considerably modified in the past 20 years, primarily in the light of new technology, including satellite oceanography, remote sensing, epifluorescence and differential interference microscopy, scanning and transmission electron microscopy, ATP assays, nucleopore filters, DNA and RNA analysis, genomic methods, and new methods for measuring minute amounts of various particles and chemicals. We now take up these new developments to build the final picture of the plankton ecosystems of the world as we understand them today.

■ MICROORGANISMS AND PRODUCTIVITY

The classical methods for studying phytoplankton productivity from net samples employing the ^{14}C method indicated that the larger net phytoplankton—diatoms and dinoflagellates—were the dominant primary producers. In an attempt to understand the relative significance of nanophytoplankton and picophytoplankton to the total phytoplankton productivity, a number of workers using various new methods demonstrated that the phytoplankton of the epipelagic zones of both temperate and tropical seas are dominated by photosynthetic organisms in the nanoplankton and picoplankton size fraction. A similar situation may prevail in polar seas, but there are currently few data to support this hypothesis. The domination by nanoplankton- and picoplankton-sized organisms is not only by numbers of individuals, but by the amount of photosynthesis as well. Malone (1980) suggests that, in most cases, the nanoplankton account for 80% or more of all photosynthetic activity and for 75% of the phytoplankton biomass in open ocean waters, especially in the tropics and subtropics (Table 2.5). Pomeroy (1974) maintains that much nanophytoplankton photosynthesis is by organisms that are less than 30 μm in size. In the past, the contribution of these tiny photosynthetic organisms was ignored because they passed through nets, were easily destroyed on filters, were sparsely distributed, and could only be studied alive.

TABLE 2.5

Summary of Nanophytoplanktons and Net Phytoplankton Photosynthesis (P) and Chlorophyll Biomass (CHL) in Different Oceanic Waters[a]

REGION	NANOPHYTOPLANKTON		NET PHYTOPLANKTON		PERCENTAGE NET PLANKTON	
	P	CHL	P	CHL	P	CHL
A. Tropical						
Eastern Equatorial Pacific	.06	20.7	0.1	3.0	14	13
North Equatorial Pacific		21.1		1.0		4
North Equatorial Atlantic					10	22
Sargasso Sea	22.5	4.3	3.6	0.4	14	9
Caribbean	0.3	12.2	0.03	1.2	9	9
Indian Ocean					7	10
B. Upwelling						
Peru Current	2.7	18.2	0.2	1.6	7	8
California Current	0.7	15.6	0.1	2.6	12	14
Costa Rica Dome	1.8	—	1.2	—	41	—
Equatorial Upwelling (91°W)	0.4	—	2.0	—	83	—
C. Temperate						
Northeast Pacific					<25	<25

[a]Numbers are means. Photosynthesis is in micrograms of carbon per liter per hour. Chlorophyll is given in milligrams per square meter of surface.
Source: Modified from Malone, 1980 in *Physiological Ecology Of Phytoplankton*, I. Morris, editor. Copyright © 1980 Blackwell Scientific. Reprinted by permission of University of California Press.

Nanoplankton and picoplankton seem to be less dominant in coastal (neritic) and upwelling areas, where net phytoplankton play a proportionately greater role, but where productivity and biomass are also more variable.

Not only are the small size classes of phytoplankton abundant, but they also show less seasonal variability in biomass than the larger net phytoplankton. Indeed, they do not exhibit the characteristic curves shown in Figure 2.39. This lack of variability is particularly noticeable in tropical and warm temperate zones, such as the central gyre of the North Pacific and the Sargasso Sea area in the Atlantic, areas that seem to be steady-state systems with very little seasonal or geographic change, according to McGowan (1974). Marine biologists attribute this steady-state situation to the stable hydrographic regime, which gives rise to the continued low nutrient concentration. The low nutrient concentration favors smaller photosynthetic organisms that have a proportionately greater surface area to absorb the nutrients but a lower relative need. Small organisms also sink more slowly.

Among the dominant groups of nanophytoplankton and picophytoplankton are the coccolithophores, very small flagellates of various kinds, and several kinds of prokaryotic and eukaryotic cells including the newly discovered prochlorophytes. According to Chisholm et al. (1988), members of the last group were discovered through the use of flow cytometry, are smaller than coccoid cyanobacteria, and reach densities of 10^5 cells per ml of water in certain deep areas of the photosynthetic zone.

A final consideration, neglected in analyses of plankton communities until now, is symbiosis. Many of the larger planktonic protozoa, particularly the Foraminiferida and Radiolaria, contain intra- and extracellular photosynthetic symbionts of different taxa. These symbionts are probably responsible for a significant, but unknown, fraction of primary productivity, according to Fenchel (1988). Even more interesting is the finding by Stoecker et al. (1987) that many planktonic ciliates retain the functioning chloroplasts of their algal food in a short-term "chloroplast symbiosis" that previously was seen only in some marine mollusks. (See Chapter 10.) The contribution of this type of symbiosis to the pelagic planktonic productivity is unknown.

■ NANOPLANKTON, RESPIRATION, AND GRAZING

Most minute photosynthetic organisms are consumed by small heterotrophic organisms. However, just as our phytoplankton nets are too large to capture nanophytoplankton and picophytoplankton, so, it appears, are the particle-capturing systems of the dominant adult copepods. What, then, are the primary consumers of these minute phytoplankters? We are not certain, but two groups of the kingdom Protista appear most significant: the nonphotosynthetic flagellates (phylum Sarcomastigophora) and various ciliates (phylum Ciliophora; see Figure 2.9). Both of these groups can ingest the tiny autotrophic cells. The flagellates act like amoebae and engulf the individual cells (phagocytosis), while the ciliates filter the cells from the water column.

How significant are these heterotrophs? Their importance can be estimated by measuring their respiration rate in the epipelagic zone of the ocean. The respiration rate of an organism is more difficult to measure than the photosynthetic rate, because the size spectrum of organisms in general is larger than that of producers, because the total respiratory rate is very low, requiring methods of concentrating the organisms, and because of inherent problems in techniques. Despite these problems, the respiration rates of net plankton and nanoplankton have been estimated by such investigators as Pomeroy and Johannes (1968) and Hobbie et al. (1972). Their mutual finding that the respiration rate of nanoplankton exceeds that of net plankton by a factor of 10 suggests that the nanozooplankton are consuming most of the energy fixed by the autotrophs. Contrary to our classical model of diatoms and dinoflagellates being grazed by copepods as the first step in the epipelagic food chains, we now believe that the initial link is from nanophytoplankton and picophytoplankton to ciliates and small flagellates. This is not to say that the classical chain is not valid—only that it is probably of less importance, at least in the open ocean.

Before we conclude our discussion of this new concept of how the food web is organized in the open ocean, we consider a few final components.

■ BACTERIA, PARTICULATES, AND DISSOLVED ORGANIC MATTER

Early studies on the epipelagic zone determined that the water had a rather stable and static amount of dissolved organic matter, which was generally considered too low to support any significant number of bacteria, except on particles where organic matter was concentrated. Recently, more sophisticated analyses performed on dissolved and particulate carbon and to detect living material have established a different picture.

The amount of dissolved organic carbon (DOC) in the waters of the world's oceans constitutes one of the largest reserves of organic carbon on our planet. The total amount has long been considered stable, with residence times of thousands of years. At the same time, we know that perhaps one-fourth of the carbon fixed by photosynthesis is lost to seawater as dissolved organic matter that "leaked" from cells. How can we reconcile this continual production with the generally stable amounts measured in the water? Where does the carbon go and why does it not build up? The answer appears to be that microorganisms, primarily bacteria, take up the dissolved organic matter directly. From field studies, Azam (1998) has stated that such direct uptake may account for up to 50% of the total oceanic primary productivity and that bacteria not only take up dissolved organic matter but attack both living and dead organic particles, thus freeing more dissolved organic matter. Fuhrman (1992) has suggested that the bacterial biomass exceeds that of the phytoplankton in oligotrophic oceans. Clearly, the microbial loop is a major biological player in the total carbon flux in the oceans.

The previous findings strongly suggest that bacteria are abundant in the water column—more abundant than we had thought. The introduction of direct plate counts with fluorescent microscopy has now demonstrated bacterial population numbers of from 10^5 cells/ml in oligotrophic waters to 5×10^6 cells/ml in coastal waters. According to Hobbie and Williams (1984), most ocean-dwelling bacteria are free living as **bacterioplankton**, are heterotrophic, and are about $0.4 \mu m$ in diameter. Larger bacteria are associated with planktonic particulate matter. These bacteria and their importance were missed by earlier scientists because of their very small size, which permitted them to pass through most filters then in use. Only the new nucleopore filters retain them. According to Sieburth (1982), bacteria passing through a $1-\mu m$ filter may account for 90% of the uptake of dissolved organic carbon in the water column.

Bacterioplankton also have a very high assimilation efficiency according to Pomeroy (1974): up to 60–65%, compared with the classic 10–20% efficiency for larger organisms. Given their large numbers, this high efficiency means that these bacteria are converting a substantial amount of DOC into particulate organic carbon (POC) in the form of

FIGURE 2.46 Diagrammatic representation of the microbial loop and its relationship to the "classical" plankton food web. Autotrophs are on the right, heterotrophs on the left. ("Marine Plankton Food Chains," T. Fenchel, *Annual Review Ecology & Systematics*, Vol. 19, pp. 19–38, 1988. With permission from the *Annual Review of Ecology and Systematics*, Volume 19, © 1988 by Annual Reviews, www.annualreviews.org.)

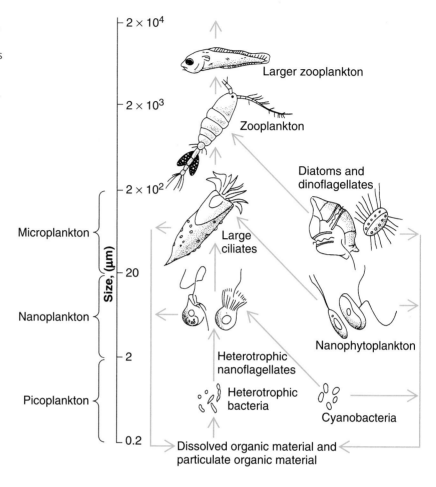

bacterial cells. It also suggests how what was formerly considered "lost" production (that is, the carbon lost through leakage from the phytoplankton) is not really lost but is recycled back into the food chains. The bacteria, in turn, are consumed by different nanozooplankton organisms and can channel a significant amount of energy into planktonic food webs. This has been termed the **microbial loop** by Pomeroy (1974) and is illustrated in Figure 2.46.

The bacteria are also partially responsible for the regeneration of nutrients in the photic zone, permitting the continued productivity of the phytoplankton even in the absence of an influx of nutrient-rich water. Such regeneration is particularly important in the highly stratified waters of the tropics and subtropics.

Johnson and Sieburth (1979) showed that, in addition to these heterotrophic bacteria, photosynthetic bacteria are ubiquitous in seawater. They are responsible for a significant fraction of the photosynthetic ac-

tivity, particularly in oligotrophic waters. Together the heterotrophic bacteria and photosynthetic bacteria are the major components of the picoplankton-size class.

An interesting twist to this story comes from recent discoveries by Kolber et al. (2001) that a group of photosynthetic bacteria called the aerobic anoxygenic photoheterotrophs, which are obligatory aerobes, metabolize carbon compounds, and respire oxygen, but also contain bacteriochlorophyll. Thus, they can and do fix carbon dioxide and are abundant in surface waters of the northeast Pacific Ocean. Currently, the group is estimated to constitute about 11% of the bacterial populations in the lighted areas. This means that in the lighted upper waters these microbes do not act exclusively as heterotrophs metabolizing carbon compounds and giving off carbon dioxide in their metabolism, but are able to fix carbon and *add* to the organic carbon pool present in the water (Fig. 2.47).

FIGURE 2.47 The flow of carbon through microorganisms in the water column of the open ocean. Components include protozoa, zooplankton, algae, cyanobacteria, viruses, heterotrophic bacteria, and aerobic (anoxygenic) photoheterotrophic bacteria. Dissolved organic carbon (DOC) is derived from the photosynthetic activities of cyanobacteria, algae, and other sources. (From Fenchet, *Science*, Vol. 292, p. 2444, 2001. Copyright © 2001 American Association for the Advancement of Science. Used with permission.)

Finally, before constructing a modern view of a planktonic food web, we need to mention one last group of potentially significant organisms that may affect our construct: the **viruses**. The existence of viruses in the marine environment has been known for more than 40 years, but it is only in the last few years that their role in plankton systems has been investigated. Viruses have no metabolic machinery of their own and must rely on a host organism to exist and reproduce. Our knowledge of the significance of viruses in planktonic systems is in its infancy. Fuhrman (1999) noted that viruses are the most common biological units in the world's oceans and may number as many as 10 billion per liter. He further suggests that viruses probably infect all marine organisms and may significantly influence many ecological processes, including nutrient cycling, bacterial and algal diversity, respiration, and algal bloom control. This causal potential has the effect of adding another semienclosed viral loop to the aforementioned microbial loop (Figures 2.47 and 2.48). All of these findings mean that we have had to change our view of the planktonic food web.

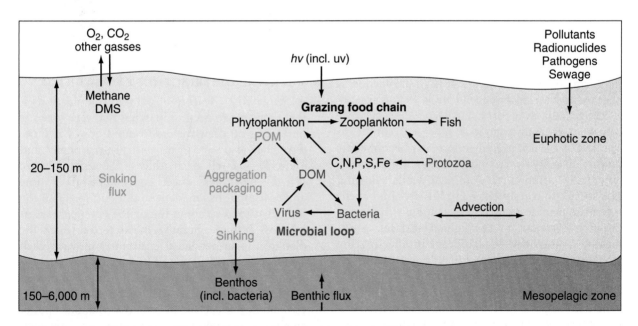

FIGURE 2.48 Modern view of the pelagic food web, emphasizing the microbial loop as a major path for organic matter and indicating its interaction with the classic grazing food chain and the sinking flux. (Reprinted with permission from Azam, *Science*, Vol. 280, p. 694, 1988. Copyright © 1998 American Association for the Advancement of Science.)

It now appears that prokaryotes are the major hosts of viruses, and viral abundance is usually more correlated with bacterial abundance than with larger chlorophyll-bearing phytoplankton. Several lines of evidence suggest that viruses are significant contributors to the mortality of bacterioplankton and hence to the structuring of open-ocean communities. These studies seem to indicate that viruses are responsible for 10–50% of bacterial mortality. There is also some indication that virus mortality may be more significant in eutrophic than oligotrophic waters. Viruses infect the larger phytoplankton cells as well, and there are indications that such infections may be significant in controlling phytoplankton blooms. If we accept the proposition that viruses infect a large proportion of bacterioplankton and even the larger phytoplankton during blooms, what are the effects on the system? The effect of viral infection of cells is lysis of the cell, which releases new viral particles, various small particulate matter, and dissolved organic molecules. The dissolved molecules can be assumed to be available to bacteria. Thus, the lysed material from the bacteria form a closed loop in which bacterial biomass is taken up by other bacteria. The loop is further fed by by the release into the water of dissolved organic matter from larger phytoplankton and grazers. By sequestering organic molecules in viruses, bacteria, and dissolved organic matter, this loop then may lead to better retention of nutrients in the virus-infected systems, since more material remains in a nonsinking form (Figures 2.47 and 2.48). Viral infection is believed to be density dependent as well as species specific. Thus, as a particular bacterial species becomes more abundant, it is more susceptible to infection. Viral infection therefore may benefit community diversity by removing too high numbers of a potentially competitive dominant while allowing rare species to continue to exist. Viruses also benefit the entire planktonic system by helping to maintain the nutrient pool in the lighted surface waters. Finally, viruses may be significant in affecting the global climate. As Fuhrman (1999) points out, when viruses infect eukaryotic phytoplankters, they induce the release of the gas dimethylsulphide (DMS). When this gas is released into the atmosphere it stimulates cloud formation, which in turn reflects solar radiation back into space. Thus, if released in large enough amounts DMS could have a cooling effect on the global climate.

■ THE MARINE PLANKTONIC FOOD WEB

In the classical marine planktonic food web, we had the net phytoplankton—the diatoms and larger dinoflagellates—as the dominant producers. They were consumed, in turn, by copepods and other crustaceans. Copepods were preyed upon by various carnivores, both fishes and other invertebrates. The final trophic level in this progression included the top carnivores, such as large fishes, cephalopods, and marine birds and mammals. With the new evidence marine biologists have uncovered, we must conclude that the dominant primary producers are various nanophytoplankton, such as coccolithophores, and such picoplankton as cyanobacteria, prochlorophytes, and aerobic photoheterotrophic bacteria, supplemented by the heterotrophic bacterioplankton, which make available and recover potentially "lost" DOC. The heterotrophic bacterioplankton are, in turn, consumed by various small protistans, mainly tiny flagellates and ciliates, which then are fed upon by larger net zooplankton. These interactions are summarized in Figure 2.46.

Where does the classical model fit into this scheme? According to Kiorboe (1993), all strongly stratified oligotrophic water masses are characterized by the dominance of a microbial type of food web, while those areas of the oceans which are weakly stratified, mixed, or turbulent are dominated by the net phytoplankton and a classic-type food chain. A comprehensive view of both the classical planktonic food web and the microbial loop is given in Figure 2.48.

■ SPATIAL DISTRIBUTION OF PLANKTON

A characteristic feature of plankton populations is that they tend to occur in **patches** that vary spatially on a scale of a few meters to hundreds of kilometers. They also vary in time—daily, seasonally, and annually—as well as vertically in the water column. It is this patchiness and its constant change in time and space that has made it so difficult for plankton biologists to learn about the ecology of plankton. At its most extreme, because the water in which the plankton is suspended is constantly moving, each sample taken by a plankton biologist removes a different volume of water, so each sample is unique and replicates do not exist. This makes it difficult to estimate species abundances and to generate applicable principles and is one reason for our unsatisfactory understanding of plankton ecology.

Patches may be generated by both physical and biological effects. Physical effects are particularly important. Large-scale patchiness can be due to changes in the stability of the water column, resulting in changes in nutrient availability. (See pp. 68–71.) Another cause of patchiness is the large-scale **advection**, or movement, of water masses in which the plankton are embedded. On the largest scales, the various ocean current systems produce convergent **gyres** in some areas, such as the North Pacific and the Sargasso Sea, and divergent systems in others, such as the upwelling areas at the equator. Areas with gyres cover large parts of the globe and define major plankton biomes.

On a smaller scale of tens to hundreds of kilometers, another source of patchiness comes from ocean **eddies** or **rings**. Rings are masses of water rotating either clockwise (anticyclonic) or counterclockwise (cyclonic) that have broken off from larger ocean current systems and have moved into different water masses. Because they have water with physical and chemical characteristics that are different from those of the waters into which they move, and because they carry plankton characteristic of the system in which they originated, they are biologically different from the surrounding water. Such rings are best known from the North Atlantic, where they are spawned from the Gulf Stream (Figure 2.49).

Another source of large-scale patchiness is the presence of **coastal fronts** in neritic waters. As water flows over the continental shelf, or other regions where the water is shallow enough, the friction with the bottom causes vertical turbulence in the water column, often transferring nutrients and changing the water chemistry. Thus, the plankton populations are also changed.

Physical processes responsible for small-scale patchiness include Langmuir cells. (See p. 61.) These cells usually give rise to patches several hundreds of meters long, because the planktonic organisms accumulate along the lines of convergence. (See Figure 2.29.) At the smallest scale, simple local turbulence is responsible for plankton patches.

Plankton may also exhibit vertical patchiness. Physical factors contributing to this type of patchiness include light intensity, nutrients, and density gradients in the water column. Phytoplankton in particular tend to be unevenly distributed vertically, leading to the existence of a **chlorophyll maximum**, wherein the chlorophyll concentration increases 2 to 20 times its concentration in other areas of the photic zone. Chlorophyll maxima are found in all temperate and tropical seas. They

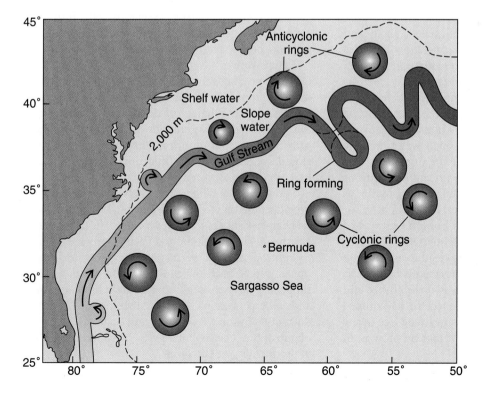

FIGURE 2.49 A schematic representation of the path of the Gulf Stream and the distribution and movement of rings. (From *Oceanus*, Vol. 17., No. 3. Copyright © 1973 Woods Hole Oceanographic Institution.)

occur at greater depths (100–150 m) in the tropics, but are a persistent feature of those waters. In temperate waters, the maximum usually occurs only during the summer and generally at depths where the light intensity is 1% of what it is at the surface (the compensation depth). Although many reasons have been suggested for the occurrence of such maxima, there is no universally acceptable explanation for its occurrence.

Biological factors causing patchiness usually create small-scale anomalies. A host of biological phenomena may be involved, but a few stand out. The ability of zooplankton to migrate vertically and graze out phytoplankton cells at a rapid rate (see pp. 77–80) can create patches where phytoplankton organisms are rare and zooplankton are abundant. Similarly, active swimming by certain zooplankton organisms can cause them to aggregate in dense groups. One particular biological feature that needs to be emphasized is the phenomenon of **marine snow**, the name given to the various kinds of aggregate particulate material derived from living organisms found in the water column. Marine snow consists of cast-off mucous nets produced by planktonic filter feeders, such as pteropod mollusks (see Figure 2.17) and larvaceans (see Figure 2.22e), and the remains of dead gelatinous larger zooplankton, such as salps (see Figure 2.24) and jellyfishes (see Figure 2.12). It may also comprise fecal pellets, algae, and microorganisms. The size of marine snow particles is usually 0.5 mm to 0.5 cm. These floating particles attract or accumulate living organisms as well as other nonliving floating and sinking particles, such as diatom frustules and fecal pellets. They become floating microcosms with populations of organisms not found in the surrounding open water. Marine snow produces a solid substrate for organisms that must attach to a surface to live. Since the gelatinous particles that make up marine snow are also a locus for bacteria and small flagellates, a certain amount of nutrients may be regenerated from the entrapped fecal pellets and other particulate matter. The regeneration of nutrients, in turn, may lead to an increase in the numbers of photosynthetic organisms, such as dinoflagellates. This microcosm is qualitatively and quantitatively different with respect to its component organisms than the organisms living free in the surrounding water. Each piece of marine snow is, in fact, a small patch.

Marine snow is widespread in the world's oceans and is one means of creating heterogeneous or patchy conditions in an otherwise homogeneous environment. Because marine snow is destroyed by plankton nets, the existence of the particles and their associated fauna and flora was recognized only when divers began looking at plankton in its natural environment.

■ THE MAJOR PLANKTON BIOMES

Given all the uncertainties so far discussed about the study and understanding of plankton, it may be surprising to learn that we now have enough data to recognize certain widespread patterns pertaining to the distribution and abundance of plankton. It is possible to establish broad associations, communities, or biotic provinces and make some reasonable statements concerning their composition and their change over time. We close this chapter with just such a summary.

As a result of the oceanic circulation that was outlined in Chapter 1, both the Atlantic and Pacific Oceans have their waters circulated in several large gyres. The two largest are the subtropical central gyres north and south of the equator in both oceans. Between these gyres lies an area of water flowing east to west: the tropical equatorial zone. Poleward from each central gyre lies a temperate subpolar gyre (Figure 2.50). Transition zones exist between the gyre systems, acting much as ecotones between terrestrial communities.

Extensive work by McGowan (1974), Reid et al. (1978), and Hayward and McGowan (1979) has established that each of the gyre systems in the Atlantic and Pacific have characteristic associations of plankton species distinct enough to qualify as biotic provinces or communities. They are also partially closed systems, showing little mixing with adjacent gyres in the transition zones.

Some of the differences between these biotic provinces are due to the different physical and chemical characteristics of the water masses. The large central North and South Atlantic and Pacific gyres are tropical to subtropical with permanent thermoclines, warm upper waters, and low nutrient concentrations. The subpolar gyres have cold water, contain high nutrients, and undergo seasonal changes. (See pp. 72–75.) Productivity is low in the central gyres, a condition termed **oligotrophic**. The subpolar gyres, by contrast, are high in productivity, or **eutrophic**.

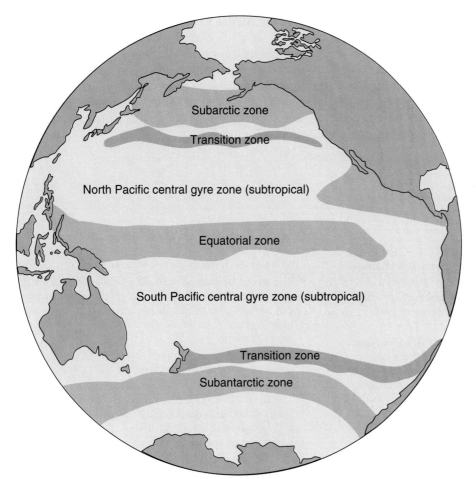

FIGURE 2.50 The main plankton biotic provinces in the Pacific Ocean. (From *The Biology of the Oceanic Pacific*, edited by Charles B. Miller, © 1974 Oregon State University Press. Used with permission.)

A number of features characterize all these biotic provinces. First, despite what we know about the general patchiness of plankton, the number and kinds of species within each province are quite uniform, on spatial scales ranging from a few meters to hundreds of kilometers. Indeed, as McGowan (1974) has said, they are "monotonous." Furthermore, at least for the very large central gyres, the species composition and relative abundances are also uniform for up to 16 years. They appear to be very stable systems.

The systems just described are also apparently old, at least in the Pacific Ocean. Evidence there suggests that the various species inhabiting them—or similar species—have been in existence for about 10 million years. With respect to the features of the plankton in the central subtropical gyres, the individual zooplankton organisms tend to be small and often transparent or blue, whereas subpolar zooplankton organisms are larger and possess other colors as well. The biomass is also smallest in the central gyres and greatest in subpolar and equatorial provinces. The central subtropical gyres and equatorial zones have the greatest numbers of zooplankton species, and the subpolar zones have the least. Carnivores are also relatively more abundant in the subtropical gyres and equatorial regions.

Phytoplankton organisms are more broadly distributed throughout the various provinces. No phytoplankton species appear to be endemic to any given province.

Perhaps the most puzzling question remaining to be answered about these large systems is "What ecological mechanisms regulate them and keep them so uniform and stable?" One of the best ways to answer this question would be to set up controlled experiments. This is not possible in the open ocean, so biologists have had to use indirect methods. At present, it appears that at least the North Pacific gyre is food limited due to its low nutrient levels, but there is no evidence of competition among the zooplankton filter feeders for the limited phytoplankton. Neither is there any evidence for niche partitioning and food specialization among more than 200 species of copepods. One might expect such partitioning in a system as stable as this. What does regulate these systems? We simply do not know.

Summary of Key Concepts

- Plankton comprise organisms that are at the mercy of the currents.

- Plankton may be subdivided by size into seven different groups, by photosynthetic capacity into three groups, or by life-history characteristics into two groups.

- Phytoplankton are photosynthetic organisms of which the larger ones are dominated by the diatoms and dinoflagellates.

- Nanophytoplankton and picophytoplankton are the most important photosynthetic organisms in the oceans and include the coccolithophores and various cyanobacteria. Recently, scientists have discovered other phototropic microorganisms in seawater that have unusual mechanisms for using light energy.

- The zooplankton are diverse and include animals from most phyla, but are dominated in the larger size classes by the copepods.

- Among the meroplankton, larval stages are more diverse than holoplankton because many species have multiple larval stages.

- Plankton have a density slightly greater than that of seawater but are buoyed up by flotation mechanisms.

- The sinking rate is a function of the overweight of the organism, the surface of resistance, and the viscosity of the water.

- To reduce sinking, plankton lower the overweight by altering the chemical composition of the body, by employing gas-filled floats, and by using compounds less dense than water.

- Plankton may reduce their sinking rate by increasing the surface of resistance of the body through different shapes and through the use of spines or other projections.

- Langmuir convection cells also serve to keep plankton afloat.

- Primary productivity in the oceans is about 95% based in the autotrophic organisms.

- Primary productivity is the rate of formation of energy-rich organic compounds from inorganic materials.

- Because water absorbs light, less energy is available for photosynthesis as one progresses deeper in the ocean. At a depth called the compensation depth, the rate of photosynthesis equals the rate of respiration. The compensation intensity is the light intensity at which the rate of photosynthesis equals the rate of respiration.

- The classic method for measuring primary productivity is the light/dark bottle, which measures changes in oxygen content. This technique has been mostly superseded today by the ^{14}C methods, which use the radioactive isotope of carbon.

- Standing crop is the amount of biomass of a particular group of plants or animals present in the water at a given time.

- The factors affecting primary productivity in the ocean are light, nutrient content, and a composite factor called hydrography that comprises all those factors which tend to move water masses around.

- The depth to which light will penetrate with enough intensity to permit photosynthesis depends on a number of factors, including wavelength, absorption by water, transparency of the water, reflection by particles, latitude, and season of the year.

- The major nutrients needed by phytoplankton for growth and reproduction are nitrates and phosphates, which together act to limit productivity because they are in short supply in most lighted areas of the ocean or are only seasonally available from the deep reservoir of the ocean depths.

- Vertical mixing of the water column replenishes nutrients in the lighted areas of the temperate oceans in winter, but the permanent substantial difference in density between the lighted surface waters and the deep waters of the tropics does not permit mixing and so leaves most of the tropics with low nutrient levels and low productivity.

- Vertical mixing also carries the phytoplankton cells below the compensation depth, circulating them between lighted and unlighted areas; this phenomenon gives rise to the concept of critical depth.

- In temperate seas, the water column is well mixed in winter and nutrients are abundant, but the scant light reduces productivity. In spring, a thermocline develops that reduces mixing while light and photosynthesis are increasing. In the summer, the thermocline is well established and no mixing occurs to replenish the nutrients, so that, although light is abundant, the nutrients are limiting. In fall, the thermocline decays and mixing begins to replenish the nutrients, but the diminishing light reduces photosynthesis and productivity.

- In tropical seas, thermal stratification extends throughout the year, but light is optimal. Thermal stratification prevents the upward transport of nutrients, so tropical seas have low productivity and hence clear water.

- In polar seas, productivity is restricted to a single short period in the polar summer. Thermal stratification is usually not present, but productivity is strongly reduced by insufficient light through most of the year.

- In inshore waters, productivity is enhanced through runoff of the critical nutrients from the land, the absence of a permanent thermocline, and the fact that the water column is shallower than the critical depth.

- The total net primary productivity of the biosphere is almost equally split between oceans and land, and the ocean productivity as estimated from satellite data is two times greater than independent calculations would suggest.

- On an areal basis, the net primary productivity of the oceans is about a quarter of that in the terrestrial environment. This reduced productivity stems from light absorption, competition for light between the autotrophs and water, and a lack of nutrients.

- Primary producers in the ocean, while responsible for nearly half the net primary productivity, make up only 0.2% of the global primary producer biomass. These producers have a turnover time averaging 2–6 days, whereas terrestrial primary producers have an average turnover time of 19 years.

- Globally, net primary productivity reaches three maxima at three latitudes. The largest centers on the equator, the next largest is in the temperate latitudes of the Northern Hemisphere, and the smallest peak occurs in the temperate latitudes of the Southern Hemisphere.

- The classic model of the interaction between the phytoplankton and zooplankton grazers has the diatoms and dinoflagellates as the dominant autotrophs and the copepods as the dominant grazers of them.

- Copepods are capable of grazing the diatom and dinoflagellate populations down to very low levels and can control the numbers of phytoplankton. Grazing may also release nutrients to the upper water mass, permitting continued productivity even in the presence of a strong or permanent thermocline.

- Differences in the life histories of the dominant copepods, the presence or absence of additional grazers, and the differing hydrographic conditions between the North Atlantic and North Pacific oceans explain the differences in ecology of the plankton communities in these two areas.

- Vertical migration is the diel movement of certain populations of zooplankton down to the ocean depths during daylight and up to the surface at night, or vice versa. Hypotheses purporting to explain the occurrence of vertical migration include the avoidance of predation, the avoidance of damaging ultraviolet light, changes in position in the water column, and the use of vertical migration as a mechanism to increase production and conserve energy.

- Seasonal succession is the change in species composition of the phytoplankton with the season of the year. Reasons for this succession include temperature changes, changes in nutrient levels, and biological conditioning of the water.

- With the use of new techniques, we now have a different understanding of the structure of the plankton communities of the open oceans. Under the microbial-loop concept, the dominant photosynthetic organisms are nanophytoplankton and picophytoplankton organisms, which are grazed by ciliate protozoa and heterotrophic nanoflagellates, in turn to be consumed by the larger net zooplankton. Part of the carbon fixed by these tiny cells "leaks out" as dissolved organic matter or is lysed by viruses. This "trove" of fixed carbon is then taken up by bacterioplankton, in turn to be consumed by the heterotrophic nanoflagellates.

- Plankton tends to be distributed in patches caused by physical factors such as advection, gyres, fronts, and eddies or by biological factors such as grazing, vertical migration, and marine snow.

- Major plankton biomes exist in the Pacific Ocean, and they are associated with the major gyre systems.

Review Questions

ESSAY Develop complete answers to these questions:

1. Are phytoplanktonic plants the only photosynthetic organisms present in the plankton? What noneukaryotic organisms may be important in primary production in the ocean? Discuss.

2. What is primary production? How does it differ from photosynthesis? How does one go about measuring primary production? How do gross, net, and total primary production differ? How is standing crop used to determine production? Discuss.

3. Name two factors that limit primary production in marine environments, and discuss their importance. (Instructors may wish to suggest other limiting factors.)

4. Is phytoplankton production nutrient limited or limited by grazing? Are phytoplankton populations limited by

"bottom-up" or by "top-down" processes? Explain these terms and the processes involved.

5. Why is there less seasonal variability in biomass in the nanoplankton of tropical and warm temperate waters, such as the central gyre of the North Pacific and the

Sargasso Sea, compared with the nanoplankton of cold temperate and polar waters? Discuss.

6. Discuss the ecological importance of the fact that plankton populations occur in patches.

Bibliography

Anonymous. 1982. *Marine pelagic protozoa and microzooplankton ecology. Ann. Inst. Oceanog. Suppl.* Tome 58.

Azam, F. 1998. Microbial control of oceanic carbon flux: The plot thickens. *Science* 280:694–696.

Azam, F., T. Fenchel, J. G. Field, J. S. Gray, L. A. Meyer-Reil, and F. Thingstad. 1983. The ecological role of water column microbes in the sea. *Mar. Ecol. Prog. Ser.* 10:257–263.

Azam, F., and A. Z. Worden. 2004. Microbes, molecules and marine ecosytems. *Science* 303: 1622–1625.

Beja, O., et. al. 2000. Bacterial rhodopsin: Evidence for a new type of phototrophy in the sea. *Science* 289: 1902–1906.

Bougis, P. 1976. *Marine plankton ecology.* New York: Elsevier.

Capriulo, G. M., and D. V. Nirivaggi. 1982. A comparison of the feeding activities of field collected tintinnids and copepods fed identical natural particle assemblages. In *Marine pelagic protozoa and microzooplankton ecology. Ann. Inst. Oceanog. Suppl.* Tome 58:325–334.

Chisholm, S. W., R. J. Olsen, E. R. Zettler, R. Goericke, J. B. Waterbury, and N. A. Welschmeyer. 1988. A novel free-living prochlorophyte abundant in the oceanic euphotic zone. *Nature* 334:340–343.

Copley, J. 2002. All at sea. *Nature* 415:572–574.

Cushing, D. H. 1959. On the nature of production in the sea. *Fish. Invest. London, 2d ser.,* 22 (6):1–40.

Cushing, D. H. 1975. *The productivity of the sea.* Oxford Biology Reader, no. 78. London: Oxford University Press.

Cushing, D. H., and J. J. Welsh, eds. 1976. *The ecology of the seas.* Philadelphia: Saunders.

Denton, E. J. 1974. Buoyancy in marine animals. *Oxford Biology Reader,* no. 54. London: Oxford University Press.

Enright, J. T. 1977. Diurnal vertical migration: Adaptive significance and timing. Part 1, Selective advantage: A metabolic model. *Limn. Oceanog.* 22:856–872.

Falkowski, P., R. T. Barber, and V. Smetacek. 1998. Biogeochemical control and feedbacks on ocean primary production. *Science* 281:200–206.

Fenchel, T. 1988. Marine planktonic food chains. *Ann. Rev. Ecol. Sys.* 19:19–38.

Fenchel, T. 2001. Marine bugs and carbon flow. *Science* 292:2444–2445.

Field, C. B., M. J. Behrenfeld, J. T. Randerson, and P. Falkowski. 1998. Primary production of the biosphere: Integrating terrestrial and oceanic components. *Science* 281:237–240.

Fleming, R. H. 1939. The control of diatom populations by grazing. *Journal du Conseil Permanent International pour l'Exploration de la Mer.* 14:210–227.

Fraser, J. 1962. *Nature adrift, the story of marine plankton.* London: G. T. Foulis and Co., Ltd.

Fuhrman, J. 1992. Bacterioplankton roles in cycling of organic matter: The microbial food web. In *Primary productivity and biogeochemical cycles in the sea,* edited by P. G. Falkowski and A. D. Woodhead. New York: Plenum Press, 361–383.

Fuhrman, J. 1999. Marine viruses and their biogeochemical and ecological effects. *Nature* 399:541–548.

Fuhrman, J. and D. G. Capone. 2001. Nifty nanoplankton. *Nature* 412:593–594.

Fuhrman, J., K. McCallum, and A. A. Davis. 1992. Novel major archaebacteria group from marine plankton. *Nature* 356:148–149.

Fuhrman, J., and C. A. Suttle. 1993. Viruses in marine planktonic systems. *Oceanography* 6(2):52–63.

Fuhrman, J. A., and C. C. Ouverney. 1998. Marine microbial diversity studied via 16s rRNA sequences: Cloning results from coastal waters and counting of native archaea with fluorescent single cell probes. *Aquatic. Ecol.* 32:3–15.

Haney, J. F. 1988. Diel patterns of zooplankton behavior. *Bull. Mar. Sci.* 43(3):583–603.

Hardy, A. C. 1953. Some problems of pelagic life. In *Essays in marine biology* (Richard Elmhurst Memorial Lectures). Edinburgh: Oliver and Boyd, 101–121.

Hardy, A. C. 1956. *The open sea: Its natural history.* Vol. 1, *The world of plankton.* Boston: Houghton Mifflin.

Hardy, A. C., and E. R. Gunther. 1935. The plankton of the south Georgia whaling grounds and adjacent waters, 1926–27. *Discov. Rep.* 11:511–538.

Hart, T. J. 1942. Phytoplankton periodicity in Antarctic surface waters. *Discov. Rep.* 21:263–348.

Harvey, H. W., L. H. N. Cooper, M. V. Lebour, and F. S. Russell. 1935. Plankton production and its control. *J. Mar. Biol. Assoc.* 20:407–441.

Hayward, T. L., and J. A. McGowan. 1979. Pattern and structure in an oceanic zooplankton community. *Amer. Zool.* 19:1045–1055.

Hill, M. N., ed. 1963. *The sea.* Vol. 2, *The composition of sea water, comparative and descriptive oceanography,* Section IV, Biological oceanography. New York: Wiley, 347–485.

Hobbie, J. E., O. Holme-Hansen, T. T. Packard, L. R. Pomeroy, R. W. Sheldon, P. J. Thomas, and W. J. Wiebe. 1972. A study of the distribution and activity of microorganisms in ocean water. *Limn. Oceanog.* 17:544–555.

Hobbie, J. E., and P. J. LeB. Williams, eds. 1984. *Heterotrophic activity in the sea.* New York: Plenum Press.

Holme-Hansen, O., and C. R. Booth. 1966. The measurement of adenosine triphosphate in the ocean and its ecological significance. *Limn. Oceanog.* 11:510–519.

Johnson, P. W., and J. McN. Sieburth. 1979. Chroococcoid cyanobacteria in the sea: A ubiquitous and diverse phototropic mass. *Limn. Oceanog.* 24:928–935.

Karner, M. B., E. F. DeLong, and D. M. Karl. 2001. Archaeal dominance in the mesopelagic zone of the Pacific Ocean. *Nature* 409:507–510.

Kiorboe, T. 1993. Turbulence, phytoplankton cell size and the structure of pelagic food webs. In *Advances in marine biology*, Vol. 29, edited by J. H. S. Blaxter and A. J. Southward. New York: Academic Press, 1–72.

Kirchman, D. L., ed. 2000. *Microbial ecology of the oceans.* New York: Wiley-Liss.

Koehl, M. A. R., and J. R. Strickler. 1981. Copepod feeding currents: Food capture at low Reynolds numbers. *Limn. Oceanog.* 26:1062–1073.

Kolber, Z. S., et. al. 2001. Contribution of aerobic protoheterotraphic bacteria to the carbon cycle in the ocean. *Science* 292:2492–2495.

Kolber, Z. S., C. L. Van Dover, R. A. Niederman, and P. G. Falkowski. 2000. Bacterial photosythesis in surface waters of the open ocean. *Nature* 407:177–179.

Malone, T. C. 1980. Algal size. In *The physiological ecology of phytoplankton*, edited by I. Morris. Berkeley: University of California Press, 433–463.

Mare, M. F. 1940. Plankton production off Plymouth and the mouth of the English Channel in 1939. *J. Mar. Biol. Assoc.* 24:461–482.

Margalef, R. 1963. Succession in marine populations. In *Advancing frontiers of plant sciences*, Vol. 2, edited by R. Vira. 137–188.

Martin, J. H., and S. Fitzwater. 1988. Iron deficiency limits phytoplankton growth in the northeast Pacific subarctic. *Nature* 331:341–343.

Martin, J. H., et al. 1994. Testing the iron hypothesis in ecosystems of the equatorial Pacific. *Nature* 371:123–129.

McAllister, C. D. 1969. Aspects of estimating zooplankton production from phytoplankton production. *J. Fish. Res. Bd. Can.* 26:199–220.

McGowan, J. A. 1974. The nature of oceanic ecosystems. In *The biology of the Pacific Ocean*, edited by C. B. Miller. Corvallis: Oregon State University Press, 9–28.

McGowan, J. A., and P. W. Walker. 1985. Dominance and diversity maintenance in an oceanic ecosystem. *Ecol. Monogr.* 55:103–118.

McLaren, I. A. 1963. Effects of temperature on growth of zooplankton and the adaptive value of vertical migration. *J. Fish. Res. Bd. Can.* 20:685–727.

McLaren, I. A. 1974. Demographic strategy of vertical migration by a marine copepod. *Amer. Nat.* 108:91–102.

Miller, C. B., ed. 1974. *The biology of the Pacific Ocean.* Corvallis: Oregon State University Press.

Miller, C. B., B. W. Frost, P. A. Wheeler, M. R. Landry, N. Welschmeyer, and T. M. Powell. 1991. Ecological dynamics in the subarctic Pacific, a possibly iron-limited ecosystem. *Limn. Oceanog.* 36(8):1600–1615.

Morris, I., ed. 1980. *Studies in ecology.* Vol. 7, *The physiological ecology of phytoplankton.* Berkeley: University of California Press.

Newell, G. E., and R. C. Newell. 1973. *Marine plankton, a practical guide.* London: Hutchinson.

Ohman, M. D. 1990. The demographic benefits of diel vertical migration by zooplankton. *Ecol. Monogr.* 60(3):257–281.

Parsons, T. R., and C. M. Lalli. 1988. Comparative oceanic ecology of the plankton communities of the subarctic Atlantic and Pacific oceans. *Oceanog. Mar. Biol. Ann. Rev.* 26:317–359.

Parsons, T. R., M. Takahashi, and B. Hargrave. 1984. *Biological oceanographic processes.* 3d ed. New York: Pergamon.

Peterson, B. J. 1980. Aquatic primary productivity and the ^{14}C-CO$_2$ method: A history of the productivity problem. *Ann. Rev. Ecol. Sys.* 11:359–385.

Platt, T., ed. 1981. Physiological bases of phytoplankton ecology. *Can. Bull. Fish. Aquatic Sci.* 210:346.

Pomeroy, L. R. 1974. The ocean's food web, a changing paradigm. *Bioscience* 24:499–504.

Pomeroy, L. R., and R. E. Johannes. 1968. Respiration of ultraplankton in the upper 500 meters of the ocean. *Deep-Sea Res.* 15:381–391.

Raymont, J. E. G. 1980. *Plankton and productivity in the oceans*, 2d ed. Vol. 1, *Phytoplankton*, Vol. 2, *Zooplankton.* New York: Macmillan.

Reid, J. L., E. R. Brinton, A. Fleminger, E. L. Venrick, and J. A. McGowan. 1978. Ocean circulation and marine life. In *Advances in oceanography*, edited by H. Charnock and G. Deacon. New York: Plenum Press, 65–130.

Riley, G. A. 1946. Factors controlling phytoplankton populations on Georges Bank. *J. Mar. Res.* 6:54–73.

Russell-Hunter, W. D. 1970. *Aquatic productivity.* New York: Macmillan.

Sherr, E. B., and B. F. Scherr. 1991. Planktonic microbes: Tiny cells at the base of the ocean's food webs. *Trends Ecol. Evol.* 6(2):50–54.

Sieburth, J. M. 1979. *Sea microbes.* New York: Oxford University Press.

Sieburth, J. M. 1982. The role of heterotrophic nanoplank-ton in the grazing and nurturing of planktonic bacteria in the Sargasso and Caribbean Seas. *Ann. Inst. Ocean. Suppl.* Tome 58:285–296.

Smayda, T. J. 1980. Phytoplankton species succession. In *The physiological ecology of phytoplankton*, edited by I. Morris. Berkeley: University of California Press, 493–570.

Smith, DeB. L., and K. B. Johnson. 1996. *A guide to marine coastal plankton and marine invertebrate larvae.* 2d ed. Dubuque, IA: Kendall/Hunt.

Spudich, J., et. al. 2000. Bacterial rhodopsin: Evidence for a new type of phototrophy in the sea. *Science* 289:1902–1906.

Stoecker, D. K., A. E. Michaels, and L. H. Davis. 1987. A large proportion of marine planktonic ciliates contain functional chloroplasts. *Nature* 326:790–792.

Waterbury, J. B., S. W. Watson, R. R. L. Guillard, and L. E. Brand. 1979. Widespread occurrence of a unicellular, marine planktonic cyanobacterium. *Nature* 277:293–294.

Wickstead, J. H. 1965. *An introduction to the study of tropical plankton.* London: Hutchinson.

Wimpenny, R. S. 1966. *The plankton of the sea.* New York: Elsevier.

Zehr, J. P., et al. 2001. Unicellular cyanobacteria fix N_2 in the subtropical North Pacific Ocean. Nature 412:635–638.

CHAPTER 3

OCEANIC NEKTON

In contrast to the plankton, the nekton comprises those organisms which have developed powers of locomotion, so that they are not at the mercy of the prevailing ocean currents or wind-induced water motion. In other words, they can move at will through the water. Most are large animals and include the largest and fastest-moving organisms in the sea. Plankton are dominated by a host of invertebrate animals, but nekton encompasses predominantly vertebrates. The fishes are the most numerous nekton, both in species and individuals, but representatives of every vertebrate class except amphibians are found here.

In the larger sense, **nekton** includes all organisms capable of sustained locomotion against the motion of water, without any consideration of their habitat. In this chapter, we will consider only nektonic animals distributed in the epipelagic zone of the open ocean—the oceanic nekton. The ecology and adaptations of nekton associated with inshore waters, such as kelp beds and coral reefs, and with the deep sea will be considered in subsequent chapters. The special adaptations of the oceanic nekton are different from those of nekton living in deeper waters and inshore waters; hence, we justify this division of the nekton.

Because these animals are fast-swimming and often wide-ranging in the immense open seas of the world, they are difficult to study at sea and are virtually impossible to keep captive under natural conditions. We know very little about most aspects of the ecology or life history of these forms. Given the dearth of such field data and the virtual absence of laboratory information, we have been forced to infer much of their ecology by studying anatomical and physiological characteristics of captured specimens.

Most students will have little opportunity for direct contact with living oceanic nekton, except for fishes and occasional seal, sea lion, or porpoise sightings or strandings. However, the immensity of the ocean area inhabited by this fauna and its significant role in the ecology of the upper layers of the seas of the world make it important to study. Some nektonic fishes, such as the tunas, sustain a major world fishery. The nekton also includes the whales, perhaps the most popular group of marine organisms in the public eye and animals over which controversy has raged in recent years with respect to their continued survival. Helping whales survive requires some knowledge of the habitat in which they live.

■

Composition of the Oceanic Nekton

The oceanic nekton is composed of a wide variety of bony fishes, sharks, and rays, as well as lesser numbers of marine mammals, reptiles, and birds. The only invertebrates that can be considered oceanic nekton are certain of the cephalopod mollusks.

Several different groups of fishes may be recognized in the nekton (Figure 3.1). First, there are fishes that spend their entire lives in the epipelagic zone. These fishes may be termed **holoepipelagic**. Included among them are certain sharks (thresher shark, mackerel shark, blue shark), most flying fishes, tunas, marlins, swordfish, saury, oarfish, and others (Figure 3.2). Holoepipelagic fishes often lay floating eggs and have epipelagic larvae. They are most abundant in the surface waters of the tropics and subtropics.

A second group of oceanic fishes is termed **meroepipelagic**. These fishes spend only part of their life cycle in epipelagic waters. This is a more diverse group and includes fishes that spend their adult lives in the epipelagic but spawn in inshore waters (herring, whale shark, dolphin, halfbeaks) or in fresh water (salmon) (Figure 3.3). Still other fishes enter the epipelagic zone only at certain times. For example, at night, certain deeper-water fishes, such as lantern fish, migrate to the surface layers to feed. This latter group, together with all other deeper-living fishes (collectively called midwater fishes), is considered in the next chapter.

Most fishes spend their early lives in epipelagic waters, but their adult lives in other areas. These juveniles form a persistent part of the epipelagic fauna, but can best be considered meroplankton, since they have a restricted ability to move. They will not be considered further here.

The second component of the oceanic nekton is the marine mammals. Nektonic marine mammals include the whales (order Cetacea) (Figure 3.4) and the seals and sea lions (order Carnivora, suborder Pinnipedia) (Figure 3.5). Other marine mammals exist, such as manatees and dugongs (order Sirenia) and sea otters (order Carnivora), but these animals are not pelagic, because they occupy inshore waters at all times. Little is known of the role of manatees and dugongs in any community. The role of sea otters in the economy of inshore waters is discussed in Chapter 5, but their new connection to the open ocean is considered at the end of the current chapter.

Nektonic reptiles are almost exclusively turtles and sea snakes. Marine iguanas exist in the Galápagos Islands, and saltwater crocodiles inhabit many island areas of the Indonesian and Australian region, but these again are littoral animals that rarely venture out of sight of land. The fossil record indicates that, during the Cretaceous period some 60 million years ago, marine reptiles were much more common and varied than today. At that time, large plesiosaurs, ichthyosaurs, and mosasaurs roamed the warm seas.

Technically, most seabirds are not nektonic, since they fly over the open ocean or swim on the surface. They, do however, feed in the upper layers, and some may dive to depths near 250 m to feed. They also occur in large numbers in some areas where it is likely that they have a significant effect on marine prey populations. Seabirds, therefore, enter the economy of these waters and are considered here. Perhaps the only group of truly nektonic birds are the flightless penguins of the southern hemisphere, but cormorants and other seabirds dive for food and spend much time swimming.

Environmental Conditions

The environmental factors acting in the epipelagic zone inhabited by the nekton are, of course, the same as those reported in the preceding chapter for the plankton (see pp. 65–72) and include light, temperature, density, and currents. However, the relative importance of various factors in selecting adaptations and life strategies of the nekton may be different. Note that the perception of this environment is very

FIGURE 3.1 Northern Bluefin Tuna (*Thunnus thynnus*). (Copyright Peter Scoones/Photo Researchers Inc.)

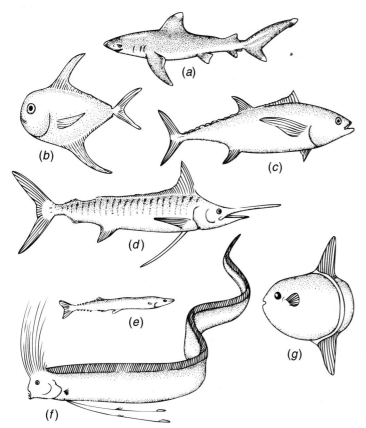

FIGURE 3.2 Some characteristic holoepipelagic fishes. (a) White tip shark (*Pterolamiops longimanus*). (b) Spiny marine bream (*Taractichthys longipinnus*). (c) Yellowfin tuna (*Thunnus albacares*). (d) Striped marlin (*Tetrapterus audax*). (e) Saury (*Cololabis saira*). (f) Oarfish (*Regalecus glesne*). (g) Ocean sunfish (*Mola mola*). (Redrawn from N. V. Parin, 1970, *Ichthyofauna of the epipelagic zone*, trans. from the Russian by the Israel Program for Scientific Translations.) (Not to scale.)

different for a large, fast-swimming fish or mammal than it is for a small copepod.

Several environmental conditions should be emphasized as being important to the nekton and for which adaptations have evolved. First, the epipelagic zone is an area with a profound three-dimensionality. Second, there is no large solid substrate anywhere; thus, the animals are always suspended in a transparent medium with no shelter from predators. Third, because of the absence of structures, there is nothing visible for animals to take bearings on when they move from place to place horizontally. Finally, the absence of a substrate also means that there is no firm support for the animals, most of which have flesh somewhat more dense than the surrounding seawater.

Three-dimensionality, combined with the absence of any obstacles, facilitates the evolution of adaptations for great mobility. In turn, great mobility, together with the ability to cover large distances, selects for the development of nervous and sensory systems that will provide and process the information neces-

sary to navigate in the area, find and capture food, and avoid predators.

Similarly, the lack of any shelter, coupled with the large size of most nektonic forms, leads to the development of faster swimming speeds as one of the few options left to escape predators and, in turn, to enable the animals to capture food. The absence of shelter also leads to the development of camouflage as another protective option. Finally, the continuous suspension of the more dense body of nektonic animals in the less dense water leads to the progressive development of various adaptations that keep the animals afloat.

Adaptations of Oceanic Nekton

■ BUOYANCY

Perhaps the most significant adaptations of nektonic animals are those which keep the animals suspended in the water and propel them through the water swiftly. It is these adaptations that have caused the

FIGURE 3.3 Some characteristic meroepipelagic fishes. (a) Ribbon halfbeak (*Euleptorhamphus viridis*). (b) Herring (*Clupea harengus*). (c) Whale shark (*Rhincodon typus*). (d) Dolphin (*Coryphaena hippurus*). (e) Salmon (*Oncorhynchus keta*). (Redrawn from N. V. Parin, 1970, *Ichthyofauna of the epipelagic zone*, trans. from the Russian by the Israel Program for Scientific Translations.) (Not to scale.)

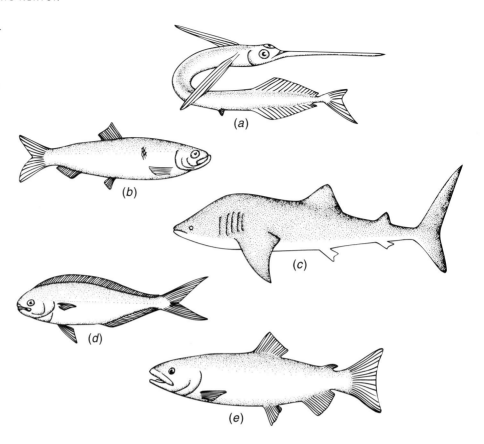

strong convergence that we observe in the morphology of otherwise very different animals.

The first order of business in living in the epipelagic zone is to stay afloat. This, as we saw in the last chapter, is also the primary concern of the plankton. Most nektonic animals have densities very close to that of seawater. Since living tissues are generally heavier than seawater, the fact that many of these large animals are nearly neutrally buoyant means that there must be areas of lower density in their bodies that counteract the higher density of most tissues.

Most fishes have a **gas** or **swim bladder** in their bodies. This structure counteracts the denser flesh of the fish and gives the fish a neutral buoyancy (Figure 3.6). Most fishes can regulate the amount of gas in the gas bladder and change their state of buoyancy. Two types of gas bladder systems are known: the **physostome**, in which there is an open duct between the gas bladder and the esophagus, and the **physoclist**, in which there is no duct. Physostomous fishes can move gases in and out of the bladder via the duct by gulping air at the surface, but they usually fill the swim bladder through the gas gland and a rete mirabile system. The **rete mirabile** is a network of small blood vessels that branch off of a large vessel. Physoclistous fishes also secrete gas into the swim bladder through the gas gland and the rete mirabile, but they must void the gas through a special absorptive section of the gas bladder called the **oval**.

In very fast moving fishes, such as bonito (*Sarda*) and mackerel (*Scomber*), which also move vertically in the water column, the gas bladder cannot adjust quickly enough to compensate for pressure changes and maintain neutral buoyancy; thus, it becomes more of a liability. Denton and Marshall (1958) noted that the fastest fishes lack gas bladders because they get enough lift from the body surface or pectoral fins during rapid locomotion.

Gas-filled cavities in the form of lungs also help sustain the neutral buoyancy of all the air-breathing nektonic animals. Some marine mammals have accessory air sacs (Figure 3.6) and can regulate their buoyancy through the amount of air they hold in their lungs. Birds also have additional air sacs, in

FIGURE 3.4 A humpback whale (*Megaptera novaeangliae*). (Copyright Mike Parry/ Minden Pictures.)

FIGURE 3.5 Antarctic fur seal (*Arctocephalus australis*). (Photo courtesy of John Heine.)

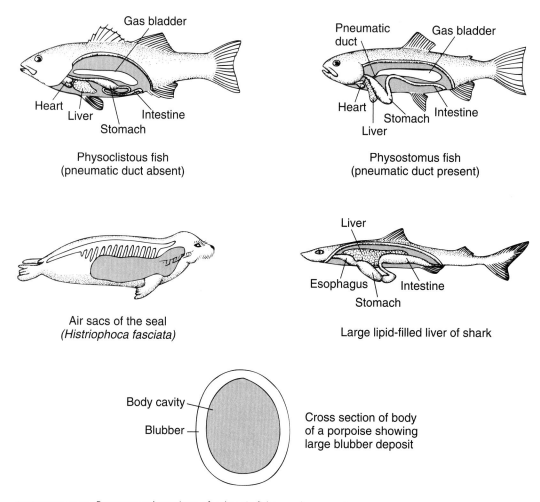

FIGURE 3.6 Buoyancy adaptations of nektonic fishes and mammals.

bones with air channels that are mostly an adaptation for flight, but that can also provide buoyancy. In most diving seabirds (excluding penguins), air trapped under the feathers provides the greatest amount of buoyancy. Among marine mammals, the sea otter and fur seals use air trapped in their dense wool undercoats for buoyancy.

Another mechanism for ensuring neutral buoyancy is the replacement of the heavy chemical ions in the body fluids with lighter ones. We observed this mechanism in plankton as well. (See pp. 58–59). The only nektonic animals in which the mechanism occurs are the squids, some of which have body cavities and tissues in which heavy sodium ions are replaced with lighter ammonium ions. As a result, an equal volume of body fluid is less dense than the same volume of seawater. Although common among plankton, this mechanism is rare among nekton because, for it to be effective, the amount of ammonium-dominated fluid must be large, and large fluid-filled spaces give the animal a rotund appearance, markedly decreasing its ability to move rapidly.

Increasing buoyancy by reducing bone or other hard parts is not an option, since strong, rigid skeletons are needed for the muscle system to propel the animals through the water. This requirement is, of course, in marked contrast to the requirements of plankton.

Still another mechanism for increasing buoyancy is to lay down lipid (fat or oil) in the body. Lipid is less dense than seawater and can be important in contributing to buoyancy. Large amounts of lipid are present in many nektonic fishes, primarily those that lack swim bladders, such as sharks, mackerels, blue-fish (*Pomatomus*), and bonito. Presumably, the lipid at least partially substitutes for the swim bladder. Lipid may be deposited in various body parts, such as muscles, bone, internal organs, and the body cavity, or may be localized in one organ. In pelagic sharks, for example, the lipid is localized in the enlarged liver, and in many shark species fat deposition in the liver is part of their development; thus, some young sharks are negatively buoyant and gradually become neutrally or positively buoyant as they mature and fat stores build up in the liver. In marine mammals, the lipid usually is deposited as a layer of fat just below the skin (blubber), where it not only aids buoyancy, but insulates against heat loss, stores energy, and adds to streamlining.

In addition to these static means of maintaining or increasing buoyancy, certain nektonic animals show some hydrodynamic mechanisms for producing additional buoyancy during movement. Perhaps the most common mechanisms are the formation of lifting surfaces in the anterior region, usually represented by pectoral fins or flippers, and the presence of a heterocercal tail—a tail with an upper lobe that is larger and better developed than the lower lobe. With this mechanism, the fins or flippers act as movable ailerons, just as on an airplane, and, when canted at the appropriate positive angle, cause the individual to rise in the water column as the tail provides the propulsive thrust (Figure 3.7). If the tail is heterocercal, its motion also provides a portion of the upward thrust. In some forms, the lift provided by the fins or flippers is aided by the whole anterior portion of the body, which can also be inclined at an angle to provide lift. The best development of these dynamic buoyancy forces occurs in negatively buoyant forms.

In general, the more primitive fishes tend to have hydrodynamic adaptations to create lift, whereas the more advanced forms seem to evolve static or passive means to achieve neutral buoyancy. The reason is that less energy is expended to obtain neutral buoyancy than to have to move constantly to achieve the lift necessary to keep a body suspended in the water. Because of air-filled lungs, air-breathing mammals tend to be nearly neutrally buoyant.

■ LOCOMOTION

A second, and related, group of adaptations in nektonic animals consists of those which enable the

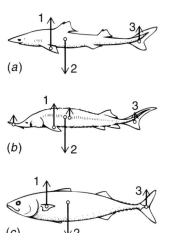

FIGURE 3.7 Various tail and fin shapes in fishes, showing the lift provided. (a) Shark (*Squalus acanthias*) with slightly heterocercal tail. (b) Sturgeon (*Acipenser stellatus*) with heterocercal tail. (c) Mackerel (*Scomber scombrus*) with homocercal tail. (1) Force created by the pectoral fins or analogs; (2) force due to residual weight; (3) vertical component of propulsive force furnished by the caudal fin. (Y. G. Aleyev, *Nekton*, © 1977 Dr. W. Junk BV, with kind permission from Kluwer Academic Publishers.)

animals to move through the water. This group can be divided into two subgroups: those necessary to create the propulsive force and those which reduce the resistance of the body to passage through the water.

The force necessary to propel a nektonic animal through the dense water is created by some part or parts of the animal's body. The most common way of moving forward is by undulating the body or the fins. Virtually all nektonic fishes show this type of motion. In the undulatory mechanism, the animal moves itself forward by sweeping the posterior part of the body and fins from side to side, an action that throws the body into a series of short curves which start at the head and move down the body. This side-to-side motion is created by alternate contractions of the body musculature, first on one side and then on the other, interacting with forces from the reacting water. Elastic properties of the body and fins and the lateral shape of the fish determine the phase relationships between the muscle contraction pattern and the undulating movements of the body. In most fishes, the wave of muscle activity along the body on each side is much faster than the wave of curvature. In the saithe (*Gadus virens*), for example, the lateral muscles behind the head and near the tail are instantaneously active.

The amplitude of lateral movements of fishes with a slender body increases gradually toward the tail. Propulsive forces are generated because the lateral movements accelerate a mass of water around the fish. The head enters undisturbed standing water. Just behind the head, water begins to accelerate and continues along the body until the end of the tail blade releases the jet of water in a direction opposite the direction in which the fish is swimming.

In short, stubby fishes, the propulsive wave is less apparent, because the amplitude on the body is low, but the wave shows a sharp increase near the caudal peduncle, which is usually narrow or even dorsoventrally flattened. The prime characteristic of the members of this group is their lunate tail blade (Figure 3.8). Propulsive forces are generated exclusively by the tail blade, acting as an undulating hydrofoil. A similar propulsive system is used by the whales, but the movements of the tail are up and down rather than from side to side. The speed of a fish is related to the speed with which the muscle contraction waves pass down the body, but it also depends on other aspects of the body shape. In general, relatively short, stubby fishes (such as tuna) are faster than very long, narrow fishes (such as eels) (Figure 3.9).

Another mode of propulsion is undulatory movements of the fins, whereby the body remains stationary and the fins are moved in one of several ways to cause forward motion. This form of locomotion is slower than the previous ones. Examples are found in the rays, such as the manta ray (*Manta hamiltoni*), certain squids (*Todarodes*), and ocean sunfish (*Mola mola*; Figure 3.10). Some elongated fishes, such as oarfish and ribbonfish, undulate the dorsal fin while holding the body rigid. In most fishes, however, the lateral fins are used for maneuvering or lift, as in sharks with a heterocercal tail.

Except in cetaceans and sea snakes, the most common form of propulsion among marine air-breathing vertebrates is paddling movements of the forelimbs, the hind limbs, or both. The limbs of marine turtles, seals, sea lions, and penguins are modified into flat, paddle-shaped appendages that the animals use to move themselves through the water as we would employ oars (Figure 3.11). The speed of movement through the water achieved with paddles depends on the frequency of the stroke. In organisms that employ few strokes, such as turtles, the speed is low. In others, such as penguins—in which up to 200 strokes per minute have been recorded by Brooks (1917) in *Pygoscelis papua*—the speed may be very fast (10 m per second with 200 strokes per minute).

When swimming, seals and sea lions cannot reach the speeds seen in whales. Eared seals (sea lions) swim by using their front flippers as paddles, but earless seals use their webbed hind feet, spread vertically like the double caudal fin of fishes. The final type of propulsion is "jet propulsion" using water. This form of propulsion is the domain of the squids, who can produce very rapid movement, up to 16 m per second.

Recently, it was discovered that swimming efficiency is improved in cetaceans by wrapping the body in an elastic sheath. The sheath acts like a rubber band to snap the tail flukes back once they have been raised, so that swimming is less energetically demanding than without the sheath.

■ SURFACE OF RESISTANCE AND BODY SHAPE

Because water is a dense medium, it is difficult to move an object through it, and it is even more difficult to move an object through it rapidly. To propel a body through water requires more energy than to move the same body at the same speed through air. It requires less energy to move an

FIGURE 3.8. Fast-swimming fishes with the characteristic lunate tail and narrow caudal peduncle. (a) Tuna (*Thunnus thynnus*). (b) Sailfish (*Istiophorus platypterus*).

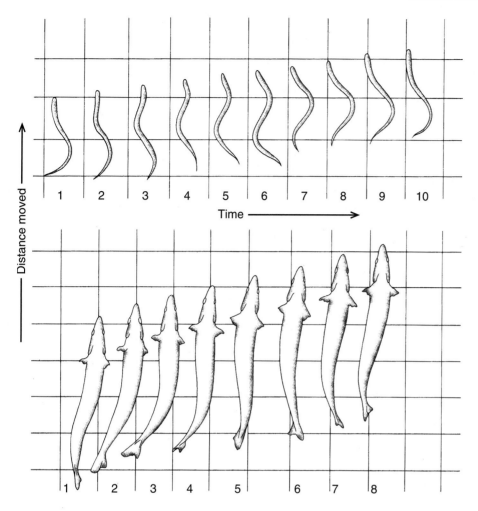

FIGURE 3.9 Propulsion in an elongated fish (top; length-to-width ratio ≅ 0.05) and a stubby fish (bottom; length-to-width ratio ≅ 0.1). (Y. G. Aleyev, *Nekton*, © 1977 Dr. W. Junk BV, with kind permission from Kluwer Academic Publishers.)

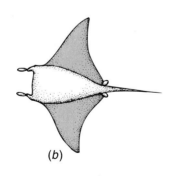

FIGURE 3.10 Fishes using fins for locomotion. (a) Ocean sunfish (*Mola mola*). (b) Manta ray (*Manta hamiltoni*). (Not to scale.) (Y. G. Aleyev, *Nekton*, © 1977 Dr. W. Junk BV. Reproduced by permission of Dr. W. Junk BV, with kind permission from Kluwer Academic Publishers.)

object through water if the body has a shape that reduces the surface of resistance to the water to a minimum. Since nektonic animals must move, and since they have finite energy resources, a major set of adaptations must be made to reduce the body's surface of resistance and to reduce the drag on the body.

There are several types of drag or resistance to movement that must be overcome. **Frictional resistance** is proportional to the amount of surface area

FIGURE 3.11 Swimming by paddle motions in three classes of marine vertebrates. (a) The green turtle (*Chelonia mydas*). (b) The fur seal (*Arctocephalus pusillus*). (c) The penguin (*Pygoscelis papua*). (Y. G. Aleyev, *Nekton*, © 1977 Dr. W. Junk BV, with kind permission from Kluwer Academic Publishers.)

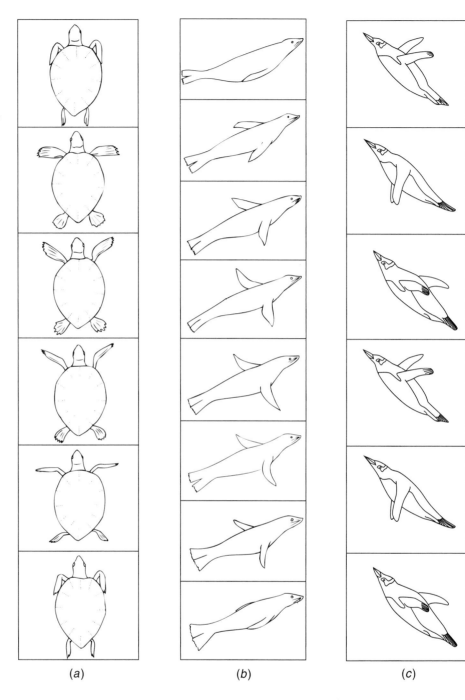

(a) (b) (c)

in contact with the water (Figure 3.12). The minimum frictional resistance is produced by a spherical object, which has the minimum surface area for a given volume of all geometric shapes. If, however, one wishes to move a nektonic object through water, another type of resistance becomes of great importance: **form resistance**, wherein the drag is proportional to the cross-sectional area of the object in contact with the water. In such a case, a spherical object presents a very large area; hence, it is an unsuitable shape for nektonic animals. To reduce form resistance to a minimum, the shape should be relatively long and thin, such as that of a thin cylinder or wire. The final type of resistance that must be

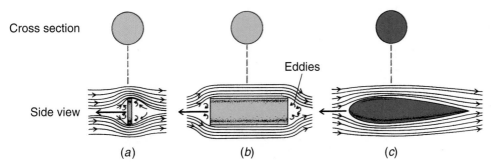

Cross section

Side view

Eddies

(a)   (b)   (c)

FIGURE 3.12 Drag forces on variously shaped objects moving through water. (a) A flat disk. (b) A cylinder. (c) A teardrop-shaped object. All have the same cross-sectional shape and area, but they differ in their three-dimensional form. The teardrop shape is best for a fast-moving fish and has about one-fourteenth the resistance of that on the disk in (a).

overcome is the so-called **induced drag** or *turbulence* due to changes in speed and direction of the flow. The flow around small animals or slow, large ones is usually smooth and is called *laminar*. With increasing speed or size, laminar flow is disrupted and thrown into vortices or eddies, greatly increasing the drag on the animal. The flow pattern along large, fast animals is not laminar, but turbulent. The water particles vibrate and have less tendency to be thrown into eddies. Turbulent flow follows the contour of a streamlined body, even at high speeds for which a laminar flow would have been disrupted. Swordfish generate turbulent flow with the rough sandpaper-like skin on their sword, thereby greatly reducing drag and making high speeds possible. Tunas have an area of scales behind the head called the corselet, which acts as a microturbulence generator. The best streamlined bodies have a teardrop shape, somewhat blunt in front and tapered to the posterior end. Such a shape gives the lowest resistance for the largest volume if the ratio between the largest diameter and the length is about 0.22. Whales, dolphins, and tunas closely approach this ideal shape (Figure 3.13).

In addition to using their body shape, nektonic animals have other adaptations for reducing drag. These adaptations generally fall under the category of streamlining the external surface of the body so that there are no protuberances that break the smooth flow of water over the body and increase drag. In this regard, species that move rapidly, such as tunas, are structured so that almost all their normally protruding body structures are recessed into depressions or grooves from which they may be

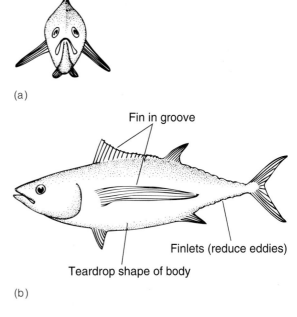

(a)

Fin in groove

Finlets (reduce eddies)

Teardrop shape of body

(b)

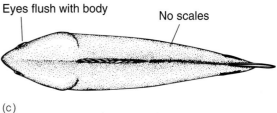

Eyes flush with body

No scales

(c)

FIGURE 3.13 Three views of a tuna, showing the adaptations necessary for fast movement. (a) Front view. (b) Side view. (c) Top view.

elevated only when needed. In fast-moving fishes, eyes, though large, do not protrude beyond the sides of the body. Pectoral and pelvic fins fit into grooves except when in use, and body scales are reduced or absent (Figure 3.13). Similarly, in marine mammals, the hair is lost or reduced in length because it produces more drag than bare skin does. Also, the mammary glands are flattened, and the males' genitalia and females' nipples do not protrude beyond the skin except when in use.

■ DEFENSE AND CAMOUFLAGE

The strongest adaptations seen in nektonic animals are those which relate to achieving rapid forward movement in the water column. These adaptations are of such importance that they take precedence over, or even preclude, any adaptations for defense against predators if such adaptations would mean decreasing rapid movement. This requirement immediately limits the number of possible defensive adaptations. At the same time, most of these nektonic animals are large as adults, which means that most have few predators; the largest (whales) often have few, if any, predators other than humans and killer whales (Figure 3.14). The need for elaborate defensive mechanisms is, therefore, somewhat reduced.

Certain defensive mechanisms remain possible within the framework of fast locomotion. Of these, the most common is *camouflage*. Since we are now dealing with an environment that has no physical hiding places—organisms are visible in all three dimensions—camouflage can take one of three directions: transparency of the body, cryptic coloring, and alteration of the body shape (cryptic body shape).

If the body of an organism is transparent and suspended in the transparent surface water of the ocean, the animal will become invisible in the water. As we saw in the plankton, such transparency is a common defensive adaptation of many of those species. Transparency, however, is not found among the nekton. The reason is that, as the size and thickness of nektonic animals increase, it becomes much more difficult to keep a body transparent, particularly when the body is highly muscular for propulsion.

Camouflage through alteration of the body shape is possible, so long as the shape does not interfere with fast locomotion. Among nektonic vertebrates, the most common manifestation is to develop a

FIGURE 3.14 A killer whale (*Orcinus orca*) surfacing through the pack ice in Antarctica. (Photo courtesy of Dr. John Oliver.)

ventral keel to the body to eliminate a conspicuous shadow on the belly of the animal when viewed from below. What is the origin of this shadow and how does body shape help reduce it? To answer this we must know something about how light appears in the surface waters of the ocean. When light enters the water, the rays penetrate in a conelike path. At the same time, some of the light is reflected, or scattered, in all directions by the particles in the water. This scattered light illuminates objects in the water from angles different from that at the surface, but its intensity is much less than the intensity of the downwelling surface light. If an animal is suspended in the water column in question, it is illuminated most intensely from above, while the scattered light illuminates from the sides and below. Because the intensity of the light from the side and below is so low, a shadow appears under the animal where its body has cut out the strongly downwelling surface light (Figure 3.15). If the body is now extended ventrally

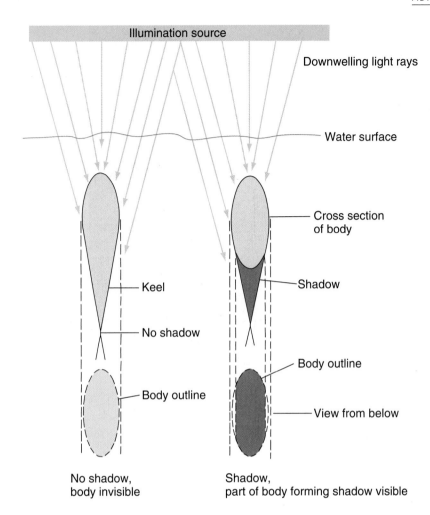

Illumination source

Downwelling light rays

Water surface

Cross section
of body

Keel

No shadow

Shadow

Body outline

Body outline

View from below

No shadow,
body invisible

Shadow,
part of body forming shadow visible

FIGURE 3.15 Diagram showing how a keel on the ventral surface of an animal eliminates the dark shadow normally cast downward by an unkeeled animal. The presence of the shadow means that an animal living deeper and looking up would see the unkeeled nektonic animal due to the shadow, but would not see the keeled animal, which would blend into the lighted background. (Y. G. Aleyev, *Nekton*, © 1977 Dr. W. Junk BV, with kind permission from Kluwer Academic Publishers.)

as a keel to form a sharp ventral edge rather than a rounded edge, the shadow will be reduced when viewed from below. This is because all surfaces on the body are now oriented so that none is illuminated exclusively by the diffuse scattered light; there is also at least some component of the intense surface light (Figure 3.15). Without the conspicuous shadow, an animal viewed from below becomes virtually invisible in the downwelling light; hence, it is camouflaged from deeper-dwelling predators. The shadow is practically eliminated if the keel has reflecting surfaces on it in the form of white pigment or scales.

Cryptic coloration is also characteristic of most nektonic animals. In the lighted upper waters of the ocean, the dominant spectral colors are blues and greens. When water is viewed from above at the surface, it appears greenish or bluish looking down into the depths. It should not be surprising then that many nektonic forms have dark blue or green on their dorsal surfaces, so that potential visual predators looking down will be hard pressed to see those forms against the general blue-green background. At the same time, viewed from below, the water looks brighter or whiter toward the surface. Any dark organism swimming in the water would thus be conspicuous when viewed from below, even if it keeled to prevent a shadow. If, however, the ventral side is colored white or silver to maximize the reflectance of light or to blend in with the downwelled light, the animal will become virtually invisible. As a result, we observe that many nekton are bicolored, or *countershaded*, dark green or blue above and white or silver below (Figure 3.16). In certain of the highly

FIGURE 3.16 Contrasting color patterns as seen on Dall's porpoise (*Phocoenoides dalli*). (Photo courtesy of Dr. Jim Harvey.)

surface-dwelling vertebrates, such as porpoises, the color pattern is more complex, with irregular bands of light and dark that actually mimic the pattern of wave-roughened surface waters (Figure 3.17). Some dolphins that live among schools of tuna are countershaded gray above and white beneath, with white spots and speckles in the gray and black spots and speckles in the white. Colored in this manner, these animals are difficult to perceive from the side, under water, amid a school of tuna.

A number of species of flying fishes have large pectoral fins. These fishes escape predators by propelling themselves out of the water and gliding for long distances on their winglike fins.

Other than all of the proceding adaptations, few specialized morphological structures have developed as a defense against predators. Most likely, the reason is that the development of various spines and armor would increase the animal's resistance in the water and interfere with the ability to move quickly. It also may be that the larger nektonic creatures are the top predators in the system and have few or no other nekton consuming them.

■ SENSORY SYSTEMS

Since nektonic animals are large, fast moving, and primarily predaceous, they might be expected to have well-developed sense organs. This is generally the case, and with few exceptions, such as lateral lines in fishes, the senses are no different from those possessed by other vertebrates in different habitats. Lateral lines are rows of small tubes, open to the water and containing sensory pits sensitive to pressure changes in the water.

Sharks and rays possess special sense organs called ampullae of Lorenzini that are sensitive to minute electric currents in the water. Kalmijin (1971) has demonstrated that predaceous cartilaginous fishes can use this electroreception to find prey. A geomagnetic sensory system in marine mammals seems to play an important role in long-distance navigation and may also explain the various mass strandings of some toothed whales.

Most sensory information received by nekton comes via light (vision), sound (hearing), or chemicals (olfaction). Eyes tend to be well developed and complex in all forms, but the size of the eye relative to the size of the body varies greatly. The eyes are usually set laterally on the head in such a way that the fields of view for each eye do not overlap, yet encompass large areas on either side. The lack of overlap of fields of view means that binocular vision is small or absent among most nekton, except for pinnipeds.

FIGURE 3.17 Cryptic coloring on the sides of a Pacific white-sided porpoise (*Lagenorhynchus obliquidens*), mimicking the wave-roughened surface of the water. (Photo courtesy of Nancy Black.)

■ ECHOLOCATION

Among nektonic mammals, the sense of hearing has generated the greatest number of specialized adaptations, a fact that suggests the great importance of hearing to these animals. Clearly, that importance resides in another fact: that sound travels 4.7 times faster in water than in air and has a much greater communication range than does sight. As a result, most nektonic animals show strong development of sound-receiving structures.

In a terrestrial environment, enhanced sound reception is usually indicated in external morphology through enlarged external ears or pinna of mammals. Such structures, however, would create an excessive drag on aquatic vertebrates; hence, they are reduced or absent. To make up for their absence, aquatic mammals tend to develop other structures of the head that receive sound waves.

Sound reception and production are most highly developed in those cetaceans which use them for **echolocation**, in much the same way we use **sonar** to determine depth. In echolocation or sonar, sound waves are sent out from a source in a particular direction. The sound waves pass uninterruptedly through the water until they encounter an object with a refractive index different from that of water. When the waves strike such an object, they are reflected and return to the source. The time interval between the production of the sound and its movement to a target and back after reflection, is a measure of the distance between the source and the object (Figure 3.18). As the distance changes, so will the time necessary for the "echo" to return. Continual production of sound waves and sensory evaluation of the reflected waves during swimming give a nektonic animal a constant check on all objects in its path. Knowing the distances to objects makes it possible for the echolocating animal either to avoid them (if they are predators) or to close in on them (if they are a source of food).

Low-frequency sound is used by some animals to communicate among themselves in the water column. Examples are the "singing" of humpback whales

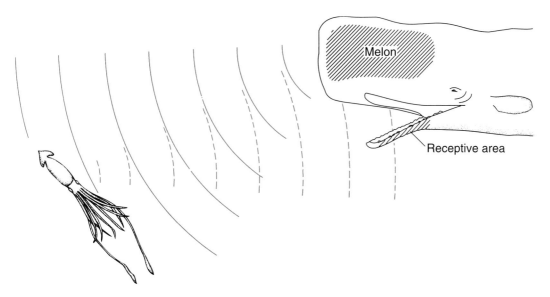

FIGURE 3.18 Echolocation in a sperm whale. Sounds are made in the nasal passages, focused by the melon, and sent out (solid lines). The returning echoes (broken lines) are received by the lower jaw and transmitted to the ear by bone conduction.

and the low frequencies (20–80 Hz) produced by blue whales. Low-frequency sound, however, does not produce information about the fine structure of objects. To obtain that information, high-frequency sound waves must be produced and reflected from the object in order for the animal to discriminate detail. Most nektonic animals that have highly developed echolocation can vary the frequency of the sound produced.

It is in the toothed whales that echolocation reaches its zenith. These animals possess elaborate morphological modifications of the head and respiratory systems that permit them to send and receive sound waves varying over a wide range of frequencies. Discriminatory ability of high-frequency sound is quite remarkable. Dolphins, for example, are reported by Kellogg (1958) to be able to distinguish between two fish species of similar size and shape. Norris et al. (1961) report porpoises able to distinguish even between objects differing only in thickness.

Toothed whales have a peculiarly bulging, rounded forehead and, associated with it, a dorsally placed external nasal opening or blowhole. Internally, a complex series of air sacs is associated with the nasal passages leading from the blowhole to the lungs. The rounded forehead is caused by the presence underneath of a large, fat-filled structure called the **melon**. This fatty organ reaches its greatest development in

the sperm whales, in which it is called the spermaceti organ and extends as much as 40% of the animal's length. The relationship of all these structures is diagrammed in Figure 3.19.

Although we do not yet completely understand how the elaborate echolocation system operates in sound production and reception, enough is understood to permit a description of how we believe that this apparatus functions. In the toothed whales, sounds are produced by the movement of air through the nasal passage and associated air sacs. During diving, sounds are produced by the recirculation of internal air. Special muscles in and around the nasal passages and air sacs allow these channels to change shape and volume and to change the frequency of the sound produced. The fatty melon is apparently used as an acoustical lens to focus and direct the train of outgoing sound waves, and this directionality increases with increasing frequency of the sound, so the animals can pinpoint objects by means of high-frequency sound. Directionality is further enhanced by the bones of the parabolically shaped skull of these toothed whales. Reception of the reflected waves is centered in the bone and fat deposits of the lower jaw and in the middle ear. In contrast to most mammals, in which the middle ear resides in bone attached to the skull, the toothed whale's middle ear is loosely

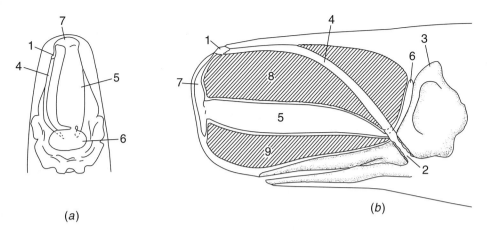

FIGURE 3.19 The anatomy of the head of a sperm whale (used for sound production, sound reception, and buoyancy control). (a) Dorsal view of the head of a sperm whale (*Physeter macrocephalus*). (b) Side view of the head of a sperm whale. (1) Nasal opening or nostril; (2) bony nasal duct; (3) sound-focusing surface of cranium; (4) left nasal passage; (5) right nasal passage; (6) frontal air sac; (7) distal air cavity; (8) and (9) upper and lower spermaceti sac (acoustic lens). (Y. G. Aleyev, *Nekton*, © 1977 Dr. W. Junk BV, with kind permission from Kluwer Academic Publishers.)

attached to the skull by ligaments and surrounded by special air- and fat-filled spaces.

The elaborate morphological modifications of the head region of toothed whales are directed primarily at producing and receiving a wide range of sound frequencies, which allow the animals both to navigate without bumping into objects and to pinpoint food organisms for capture. Toothed whales also possess very large brains relative to their body size—brains that are second only to those of humans in the development of the cerebral hemispheres. These large brains may enable the animals to process acoustical information rapidly.

■ **REPRODUCTION AND LIFE CYCLE**

Among epipelagic fishes, no special reproduction mechanisms are apparent that would sharply set them apart from their benthic or shallow-water relatives. Characteristically, however, holonektonic bony fishes, such as tuna and marlin, spawn eggs that float and undergo development in the open ocean waters. Some species' eggs even have threadlike structures that permit them to adhere to various kinds of floating debris. Because these floating eggs are planktonic, they are subject to tremendous losses to predation. As a result, the fishes produce huge numbers of eggs to offset the equally large losses. Parin

(1970) reports that skipjack and albacore tuna, for example, produce 2 and 2.6 million eggs, respectively, while striped marlin spawn over 13 million, and ocean sunfish produce an incredible 300 million eggs. Spawning is often intermittent and is often extended over a period of many months.

Among the pelagic sharks, a different reproductive strategy is observed. These fishes produce only a few large eggs or embryos. Parin (1970) notes that the thresher shark (*Alopias*), for example, has but 2 embryos and the blue shark (*Prionace glauca*) up to 54. If so few eggs were to start out as small as those of bony fishes and have to undergo development for a substantial time in the plankton, the chances any would survive are very poor. Sharks enhance the chances for the survival of their few offspring by producing very large eggs and by retaining the eggs in the female for a much longer period of time so that, at birth or hatching, the young are large and immune to most potential predators.

We know relatively little about the growth of pelagic fishes, but available knowledge suggests that their growth rates are very high. Tunas, for example, seem to increase their weights by 2–6 kg per year and length by 20–40 cm (Table 3.1). Correlated with this rapid growth, most nektonic fish seem to be short lived; even the large tunas appear

TABLE 3.1

Age and Length in Three Species of Tuna[a]

SPECIES	Age, Years							
	1	2	3	4	5	6	7	8
	Length, cm							
T. alalunga	18–66	32–84	54–89	73–94	70–95	82–100	98–105	—
T. obesus	45–50	62–70	74–94	85–116	97–138	115–155	—	—
T. albacares	54–103	70–136	85–155	99–154	109–134	110–130	127–145	135–160

[a]Albacore (*Thunnus alalunga*), bigeye tuna (*Thunnus obesus*), and yellowfin tuna (*Thunnus albacares*). *Source:* After Parin, 1970.

to live only 5 to 10 years. By contrast, the pelagic sharks may live 20 to 30 years.

Marine birds and turtles retain the reproductive characteristics of their terrestrial relatives. All produce shelled eggs, which are laid on land. Marine birds often congregate in large groups to nest on islands or cliffs that are inaccessible to terrestrial predators (Figure 3.20). This ensures that the usually helpless young (**altricial**) bird will survive until it is old enough to fly. However, it also makes such birds extremely vulnerable to human predation or pollution, since a large fraction of the existing population of a species may be present in one small area and can be wiped out easily. For example, as a result of the massive *Exxon Valdez* oil spill in Alaska, about 300,000 murres were killed, representing probably 70% of the breeding population in that area. Most marine birds have definite breeding seasons and may migrate thousands of miles from their feeding grounds to their breeding grounds.

FIGURE 3.20 Sea birds (black-legged kittiwakes) nesting on a cliff in the Pribilof Islands. (Photo by James Nybakken.)

Marine turtles lay eggs in holes that they excavate in the sand above the high-tide level on beaches in various parts of the tropics. This is the only time these animals normally return to land. Immediately upon hatching, the young turtles instinctively head for the ocean, where their continued development occurs, but about which we know virtually nothing. As with birds, turtles migrate thousands of miles and congregate off certain beaches for breeding. Females haul out to lay eggs only on certain beaches, and since both eggs and adults are considered excellent food by humans, most marine turtles have been drastically reduced in numbers in recent years in all areas of the world (Figure 3.21). Some sea snakes give birth to living young in the water; others lay eggs on beaches.

With respect to reproduction in marine mammals, there are two groups: those which give birth on land and those which give birth in the water. We know considerably more about reproduction in the group that breeds on land because they may be observed by humans at that time. Knowledge of the reproductive patterns of those giving birth in the water is generally limited to observations made on captive animals in aquaria.

Seals, sea lions, and walruses give birth to live young on land or on floating ice (Figure 3.22). Their young often cannot swim at birth and may require some time before they are capable of venturing into the water. Harbor seal (*Phoca vitulina*) pups, however, can swim immediately after birth. During their terrestrial period, the seals, sea lion, and walrus pups grow very rapidly on their mother's rich milk and gain the strength and insulating layers of fat and fur they need to survive in the cold, open water. Many sea lions and seals, such as fur seals, Steller's sea lions and elephant seals are polygamous, and sea lions and fur seals are territorial on their breeding grounds (Figure 3.23). The largest and most aggressive males (territorial bulls) defend a small area of beach, which they protect from other bulls (Figure 3.24). Should another bull encroach upon the territory occupied by the bull and his females or attempt to mate with a female, the resident territorial bull will fight. Although such battles may be mostly bluff and noise, fights do occur in which one or the other bull may be seriously injured or killed. Technically, elephant seals are not territorial since they defend groups of females, not space. In those seals and sea lions in which such behaviors occur, the males are usually much larger than the females.

The territorial behavior and aggressiveness that these animals exhibit on their breeding grounds does not extend to their life away from that domain. Their behavior suggests that breeding space is limiting, whereas pelagic food and space are not. This

FIGURE 3.21 Sea turtle (*Lepidochelys olivacea*) laying eggs on a beach. Dislodged egg in foreground. (Photo courtesy of Gregory Dimijian.)

FIGURE 3.22 Females (brown) and pups (black) of the northern fur seal (*Callorhinus ursinus*) on the breeding grounds in the Pribilof Islands. (Photo by James Nybakken.)

FIGURE 3.23 A Steller's sea lion rookery on Año Nuevo Island in California. (Photo by James Nybakken.)

FIGURE 3.24 Northern fur seal (*Callorhinus ursinus*): breeding bulls and females. (Photo © Roy Corral/Corbis.)

method of breeding also means that relatively few males—the territorial bulls—do the breeding, and most others are excluded (Figure 3.24). The exclusion is, however, not complete, and through genetics we are finding that even nonterritorial males sometimes impregnate females. As with the marine birds, many of these pinnipeds migrate long distances to their breeding grounds; for example, the northern fur seal (*Callorhinus ursinus*) is pelagic all over the northern Pacific Ocean, but most migrate back each summer to breed on two small islands making up the Pribilofs in the Bering Sea.

In contrast to the pinnipeds, the cetaceans give birth to their offspring in the water. The young of whales, therefore, must be able to swim at birth and instinctively know how to surface for air. They also remain closely associated with the mother (Figure 3.25). Whereas young pups of sea lions may be left on the breeding grounds for days while the female forages for food in the open ocean, the juvenile whales always remain close to their mothers, where they are protected from predators. As with the

pinnipeds, certain cetaceans may also migrate for breeding purposes. Often, this is a migration of thousands of miles, from the feeding areas in cold water to the calving grounds in warmer waters (Figure 3.26). The animals move to warm waters for calving because the newborn do not have the insulating blubber of the adults. They have a better chance of survival in warm water until the additional insulating layers are formed.

The young of both pinnipeds and whales grow rapidly, adding many kilograms per day. Blue whales, for example, can grow from 3 tons at birth to 23 tons at weaning, seven months later. The reason for such rapid growth is that the milk of pinnipeds and cetaceans is extraordinarily rich in fats (as much as ten times as much fat as in cow's milk) and is produced in large quantities.

Pinniped pups are nursed throughout the time they are in the rookeries. At the end of the season, in some species, they are abandoned by the females and must fend for themselves in the open ocean. Cetacean calves are nursed for varying times, depending on the

FIGURE 3.25 A humpback whale (*Megaptera novaeangliae*) with calf. (Photo courtesy of Tony Stone Images.)

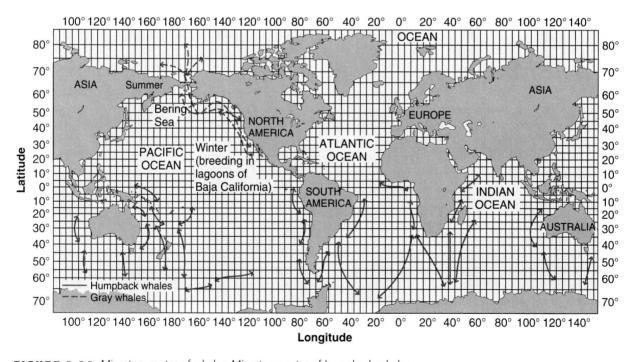

FIGURE 3.26 Migratory routes of whales. Migratory routes of humpback whales (*Megaptera novaeangliae*) in the Southern Hemisphere and migratory route of the gray whale (*Eschrichtius robustus*) in the North Pacific. (a from *Whales and Dolphins*, E. J. Slijper. Copyright © 1976 Springer-Verlag. Reprinted by permission. b reprinted with the permission of Scribner, an imprint of Simon & Schuster Adult Publishing Group, from *A Natural History Of Marine Mammals* by Victor B. Scheffer. Copyright © 1976 Victor B. Scheffer.)

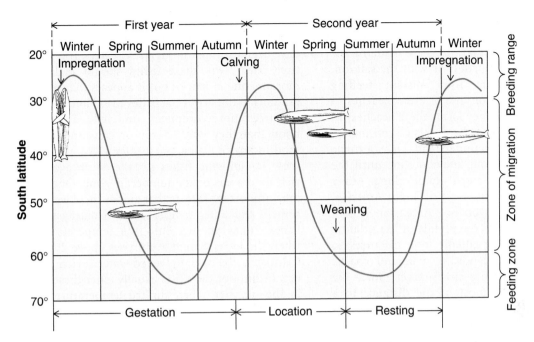

FIGURE 3.27 The reproductive cycle of the fin whale (*Balaenoptera physalis*) in the Southern Hemisphere. (From *The Stock Of Whales*, N. A. Mackintosh, 1965, Fishing News Books. Reprinted by permission of Blackwell Science.)

species. Nursing lasts 15 months in sperm whales and as much as several years in pilot whales.

Because of the enormous amount of energy put into milk production to sustain a single offspring, usually one young at a time is produced, and young are produced annually (in most pinnipeds) or at even longer intervals (in walrus and certain whales; Figure 3.27). Stocks of these animals can be easily reduced, and it may take a long time to recover their numbers.

Most marine mammals are long lived. Gray seals have been recorded to live 46 years and harbor seals 36 years, while small whales, such as the bottlenose porpoise, may live 32 years, and large whales, such as the sperm whale and fin whale, live up to 77 and 80 years, respectively. Correlated with the long life span is the delayed onset of sexual maturity and reproduction. Male fur seals, for example, do not become harem masters until 9 or 10 years of age, and sperm whales apparently do not breed until about age 20.

■ **MIGRATIONS**

As noted in the previous section, a large number of marine mammals, birds, and reptiles undertake extensive migrations for breeding or other purposes.

Migration for breeding is a common characteristic of air-breathing marine vertebrates.

Nektonic fishes also undertake extensive horizontal migrations equivalent in distance to those taken by the air breathers. These migrations are of great importance, but, unfortunately, have been little analyzed. Migrating holonektonic fishes include the various tunas and the saury.

Salmon, which spend most of their lives dispersed in the open ocean, migrate back to the freshwater streams in which they were spawned to reproduce. In this respect, they are similar to the marine air-breathing vertebrates. The ability to return to breed in the same stream where they hatched requires considerable navigational abilities on the part of the fish. From recent studies, it appears that part of the key to this navigational ability resides in the fishes' sense of "smell"; that is, these fish navigate to find the coast but then use olfaction to follow various waterborne "scents" to locate their home streams. The suggested mechanism is that, when the young salmon migrate down the streams to enter the sea, they are imprinted with the "odors" of the various streams in succession. This sequential memory is then reversed on their return, allowing them to find their way back to the same stream.

Equally remarkable migrations are made by the various species of sea turtles. As with the salmon, these animals migrate from distant feeding grounds to congregate on one or a few beaches to lay their eggs. How do they locate these beaches? In the case of beaches along the shores of continents, this is not as difficult to understand. Here, a turtle could simply follow along the shoreline until the beach with the right "smell" came along. Many turtles along the Atlantic coast of South America, however, regularly nest on tiny Ascension Island, some 1,400 miles out in the middle of the Atlantic Ocean. Others migrate equally long distances to specific beaches on islands or mainland coasts. How do they find such a small target? Since the turtles must navigate most of the distance to a mere pinpoint of an island or beach out of sight of land and without any landmarks in the ocean, they must have some ability first to know where they are in order to set a course to find the nesting ground. This is an ability that Lohman (1992) has called "map sense."

Although we still do not know exactly what cues are used by the turtles to fix their positions and navigate, recent studies have suggested some likely possibilities. Perhaps the most significant is the ability to detect magnetic fields. This is theoretically a good choice since several geomagnetic parameters vary in a predictable manner according to latitude. This ability to detect and follow magnetic fields is used by certain marine mammals, and there is now evidence that turtles may also have it. A second mechanism is to orient to wave propagation or direction. Many areas of the open ocean have consistent wave and swell directions that would permit their use in navigation. A final mechanism would be to follow some chemical cue, such as that used by salmon. Initial experiments with hatchling sea turtles by Lohman (1992) suggest that some combination of cues from the Earth's magnetic field and a seasonal consistent pattern of wave or swell may be the most likely mechanism.

Other extensive and complex migrations are undertaken by tunas and their relatives. Tunas are primarily tropical fishes that make extensive migrations across oceans within the tropical zone and that also move into temperate waters during the warmer seasons. Bluefin tunas tagged in Florida have been recovered across the Atlantic in the Bay of Biscay, and Pacific albacore tagged off California have been captured in Japanese waters. It is not entirely clear why these fishes should make such monumental journeys, but it appears that such migrations both permit the fishes to exploit their food resources more completely and reduce the likelihood that they will destroy or consume all their food in any one area. Such migrations also enable these fast-moving fishes to take advantage of the rich food areas of the temperate zone. One of the most important factors that guides these we migrations is the water temperature. Primarily tropical fishes, tunas venture into the temperate waters when the water temperature rises above 20°C. Off California, the cold, upwelled water is rich in food organisms, but the tunas usually enter these waters in the summer, when surface temperatures reach 20–21°C. If the year is cold and the temperatures of the surface waters do not rise to that point, the tunas are absent. Atlantic bluefin tuna, however, regularly summer off Newfoundland in water below 20°C, so not all tunas are restricted to warm water. Apparently, tunas always return to tropical waters to spawn and to spend the early parts of their lives (Figure 3.28).

Recent studies of albatross tagged so that their movements can be followed by satellites has shown that birds nesting in Hawaii regularly make foraging trips as far east as the west coast of North America to procure food for their chicks. These trips entail a round-trip distance of about 6,000 miles!

There may also be migratory differences between the sexes. Stewart and DeLong (1995) found that male northern elephant seals (*Mirounga angustirostris*) in the North Pacific Ocean migrate to the Gulf of Alaska, whereas the females migrate offshore between 35 and 45 degress north latitude (Figure 3.29). This divergence of migration between the sexes is apparently due to different food habits. LeBoeuf et al. (2000) report that the females feed on prey in the open ocean, the males on inshore quarry.

The ability of migratory animals to make accurate journeys across what appear to be featureless seas has generated a profound interest in marine scientists about how these animals manage to navigate accurately. Over the years, a number of hypotheses implicating a wide variety of sensory

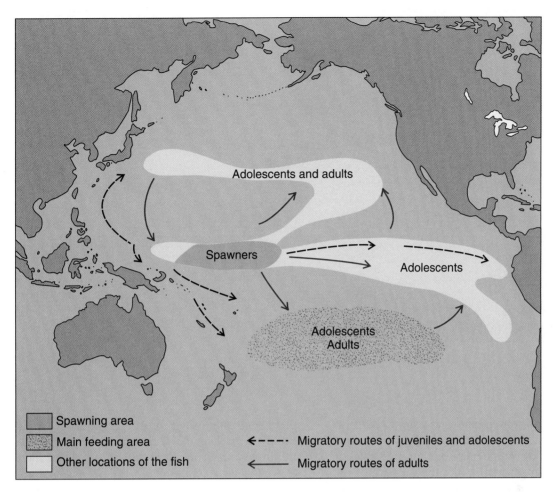

Spawning area
Main feeding area
Other locations of the fish

← - - - - Migratory routes of juveniles and adolescents
←——— Migratory routes of adults

FIGURE 3.28 Centers of distribution and migratory routes of the bigeye tuna (*Thunnus obesus*) in the Pacific Ocean. (Modified from *Rand McNally Atlas of the Oceans.* Reprinted by permission of Mitchell Beazley.)

systems have been suggested to explain migration. Proposed mechanisms have included the use of some sort of celestial compass, odor or chemical imprinting (suggested for salmon), wave direction (in turtles), cueing on underwater topography, echolocation, the detection of thermal structure, and, most recently, the detection of the flux density of the earth's magnetic field.

The ability of animals to use magnetic fields to navigate was not considered possible until recently, because of the absence of this sense in humans and the apparent lack of any known biological mechanism capable of transmitting the weak geomagnetic field into an animal's nervous system. The discovery of magnetotactic bacteria with linear chains of membrane-bound crystals of magnetite (Fe_3O_4) provided the biological mechanism for transducing the geomagnetic field to the nervous system. Subsequently, the finding by Walker et al. (1984) of membrane-bound magnetite with abundant nervous connections in tuna demonstrated the existence of an anatomical basis for magnetotaxis in a migrating nektonic animal.

Magnetic reception has now been suggested as the system used for long-distance navigation by turtles and marine mammals, but experimental proof has been difficult to come by due to the paucity of good small-scale geographic maps of magnetic anomalies and the difficulty of setting up laboratory tests.

FIGURE 3.29 Two young male elephant seals (*Mirounga angustirostris*). (Photo courtesy of Dr. Jim Harvey.)

The most interesting data that seem to confirm the presence of this magnetic sense in whales come from the well-known phenomenon of mass strandings. If cetaceans are using the flux density of the earth's magnetic field to navigate by, they can do so in two ways. First, the topography of the local field can be used the way a road map is read, with the animals following certain magnetic contours. Second, a timer based on regular changes or fluctuations in the magnetic field permits the animals to know their position and progress. Problems arise in this system when the geomagnetic contour lines cross land or when the pattern of the timer is disrupted by irregular fluctuations in the magnetic field. Whenever these two conditions occur, magnetically navigating whales may beach themselves. Analyses of mass strandings, by Kirschvink (1990) for the east coast of the United States and by Klinowska (1990) for England, have demonstrated that all sites of mass strandings have only one physical feature in common: The geomagnetic contour lines cross from the sea onto land. If the whales were following a strategy of traveling parallel to the geomagnetic lines, which would work well in the open ocean, but which would give no clue as to where land and water meet, the animals would strand themselves. Furthermore, the dates of the strandings were consistent with times when there were irregular fluctuations in the normal patterns of geomagnetic fields.

There are also short-term movement patterns related to the availability of food and the avoidance of predators. Many dolphins, such as dusky dolphins, Hawaiian spinners, and Bay of Fundy harbor porpoises, show offshore nighttime and onshore daytime movements.

Although many migrations are undertaken for breeding purposes, such as those of certain of the large baleen whales and the northern fur seal, there may be other reasons for migration as well. Predation is one possibility, and it may be that some baleen whales migrate to warmer locations to calve because killer whales are less abundant at low latitudes. Such a ploy would protect the young.

■ SPECIAL ADAPTATIONS OF MARINE BIRDS AND MAMMALS

The warm-blooded marine mammals and birds require some special adaptations to permit them to live in ocean waters. These adaptations are concerned primarily with maintaining the animal's temperature, diving, and osmotic regulation.

Water has a higher thermal conductivity than air, which means that water is quick to extract heat from a warm body. Humans experience this property when they become chilled after a short time in water of even 80°F; in air of the same temperature, they would be comfortable indefinitely. Marine mammals, which maintain elevated body temperatures with respect to the surrounding water, must adapt to prevent their body heat from being drained away.

One means of slowing the rate of heat loss is to have a large body. As you may recall from the plankton section (see p. 59), that the ratio of the surface area to the volume of any body is lower for a large organism than for a small organism. The larger the body, the smaller the surface area relative to the volume of the body. Since heat production is a function of the volume of a body, whereas heat loss is a function of a wetted surface area, this means that large animals, which have a low ratio of surface area to volume, can thermoregulate better than small animals can. Thermoregulation is aided by the smaller surface area that is in contact with the environment through which heat may escape. All nektonic marine mammals are large, and it may be that the reason there are no mouse-sized marine mammals is that they would simply die from chilling. There are small marine birds (petrels, auklets), but these animals are completely immersed in the water only for a short time, during intermittent dives. Others, such as gulls, rarely dive, and only a fraction of their body is in contact with the water at any time.

A second adaptation that prevents or reduces heat loss is a thick insulating layer of blubber or fat just beneath the skin. This layer reaches its greatest thickness in whales, where it may be 2 feet thick. In pinnipeds such as the walrus and elephant seal, subcutaneous fat may constitute as much as 33% of the weight. Blubber and fat are poor conductors of heat; hence, they protect the animal from losing internal heat. The thicker the blubber or fat layer, the less the heat loss. Marine mammals inhabiting polar waters, therefore, have thicker layers of fat than temperate and tropical species.

A final adaptation concerns the circulatory system. The areas of a marine mammal that offer the greatest surface area to the water (and, therefore,

the greatest opportunity for heat loss), and also that lack much of the protective layer of fat are the fins, flippers, and flukes. What adaptations prevent massive heat loss through these extremities? In cetaceans, the answer is that the arteries that bring the warm blood out to these extremities are surrounded by a number of smaller veins that bring blood back to the central core of the mammal. Because of this arrangement, the heat of the blood in the arteries can be absorbed by the cooler blood returning in the veins before it is lost to the external water through the thin flesh of the outer extremities (Figure 3.30). In effect, these animals have a countercurrent system of circulation designed to save heat.

Because most of their adaptations conserve body heat, marine mammals (especially pinnipeds) may, on occasion, become too warm. Hot, still days may mean considerable stress from overheating. On these uncommon occasions, the animals must act to dissipate heat. They do so by waving their flippers in the air while increasing the blood flow to the extremities and restricting the flow back to the core through the veins. The result is greater heat loss and subsequent cooling. Seals and sea lions may also open their mouths and pant like a

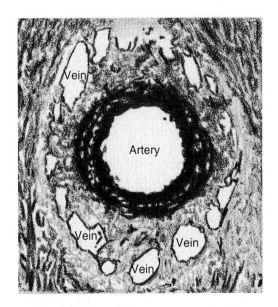

FIGURE 3.30 Section of the tail fluke of a bottle-nosed dolphin (*Tursiops truncatus*), showing the arrangement of veins surrounding an artery (25×). (From P. F. Scholander and W. E. Schevill, *Journal of Applied Physiology*, Vol. 8, p. 281, 1955. Reprinted by permission.)

dog. If they are resting on land, they may go into the water to cool off.

Most nektonic marine mammals (particularly the pinnipeds and the whales) regularly dive, or are capable of diving, to depths far greater than that to which humans can dive. Whereas exceptional humans may be capable of free dives as deep as 60 m while holding their breath for 6 minutes, various seals and porpoises are known to dive to depths of 160–1,800 m and to hold their breaths for 6–48 minutes. The champion is the sperm whale, which can dive as deep as 2,250 m and hold its breath for 80 minutes. How is it possible for the animal to do this? The answer rests with a number of anatomical and physiological adaptations.

One of the first questions to be asked regarding these prolonged deep dives is "How do the animals avoid the bends?" The **bends** is a serious affliction of human divers in which the nitrogen they breathe under pressure in deep water as part of their air supply bubbles out of solution in the blood when the diver ascends too quickly. The bubbles produced then lodge in the joints, brain, and lungs, causing paralysis and even death. The key here is that humans get the bends due to breathing gas *under pressure*, such as air from a scuba tank. Since marine mammals do not inhale pressurized gas, but depend on gas inhaled at the surface at regular atmospheric pressure, this condition has less chance of occurring in them. Even more, the chance of its occurrence is further reduced by certain other adaptations, the most important of which is the collapse of the lungs. In a deep dive, the pressure of the outside water causes the gas exchange tissues of the lungs of the diving mammal to collapse, thereby forcing the residual gas in the absorptive area of the lung into the nonabsorptive cartilaginous air passageways; thus, nitrogen stops diffusing into the blood. With no nitrogen entering the blood, there can be no nitrogen to bubble out when the animal surfaces. The collapse of the lungs is aided by the fact that diving marine mammals have few ribs that are attached to the sternum, and the sternum is shortened. The rib cage, therefore, is easily pushed in. Finally, it has also been suggested that the peculiar "foam" of emulsified fat droplets and mucus that many whales have in their respiratory passages absorbs nitrogen gas, so that it cannot enter the bloodstream where it could cause the bends. Recently, Williams et al. (2000) have docu-mented the fact that the collapse of the lungs during a dive reduces the animals' volume, and hence their buoyancy, making them negatively buoyant and allowing them to descend rapidly without expending any energy.

A second question pertaining to deep diving is "How can the animals survive so long without access to a supply of oxygen?" A consideration of this problem will suggest that the only mechanisms enabling the animals to survive are for the animals to store more oxygen than nondiving forms do and to conserve carefully the amount they have. Since the major organ system involved in oxygen transport is the circulatory system, it is there that we may look first for adaptations.

Many diving birds and mammals have a larger blood volume than their terrestrial relatives have. The elephant seal, for example, has a blood volume of 12% of its body weight, according to Elsner (1969), whereas humans have a blood volume of only 7% of their body weight and domestic dogs only 9%. The larger volume allows more oxygen to be held in the body. This oxygen capacity is increased further, as Scholander (1940) notes, because marine mammals also possess a higher oxygen capacity per unit volume of blood than terrestrial mammals do (40 ml O_2/100ml blood in elephant seals, versus 16–24 ml O_2/100 ml in humans). But it is not as great as might be expected: Neither the increased blood volume nor the increased oxygen capacity is sufficient to account for the ability of pinnipeds and cetaceans to remain underwater for extended periods. For instance, harbor seals have about the same weight as humans and possess about twice the amount of oxygen in the form of increased blood volume and oxygen capacity. Thus, one might expect them to be able to remain submerged twice as long as humans. Yet they can stay under five to ten times as long (Figure 3.31). Hence, other adaptations must also be at work.

Marked slowing of the heartbeat during the period of submersion is one of these additional adaptations. This **bradycardia** is common to all diving, air-breathing vertebrates. The decrease in heartbeat is quite dramatic. For example, under experimental conditions, the heartbeat of the Pacific bottle-nosed dolphin (*Tursiops truncatus*), drops from about 90 beats per minute at the surface to about 20 beats per minute during a 5-minute dive (Figure 3.32).

FIGURE 3.31 A harbor seal (*Phoca vitulina*) resting underwater on a rock. (Photo courtesy of Drs. Lovell and Libby Langstroth.)

Of more importance than slowing the heartbeat are two other adaptations. First, the circulatory system cuts off the blood supply to various organs and organ systems during a dive, including the muscles, digestive system, and kidneys. This cutoff conserves the limited oxygen supply in the blood for the more sensitive and vital tissues, such as the brain, central nervous system, and heart. Thus, a limited supply of oxygen lasts longer, as it is used only by selected organs. In this way, the animal may remain submerged much longer than if it had to supply oxygen to all its tissues. The second adaptation is related to the first: The muscular system and other organs are extremely tolerant of anaerobic conditions and continue to function when the blood flow is cut off. This results in the buildup of large amounts of lactic acid in the muscles during a dive.

Lactic acid is the end product of anaerobic metabolism. The muscular system can continue to function in the absence of a blood supply not only because it is very tolerant of lactic acid and anaerobic conditions, but because of yet another adaptation. The muscular system of marine mammals is rich in an oxygen-containing compound called **myoglobin**, which has a structure similar to that of hemoglobin, but which is better at storing oxygen. When the animal dives, the blood supply to the muscles is shut off. Initially, the muscles have a large supply of oxygen in the myoglobin, but as the dive continues, the oxygen of the myoglobin is depleted, and the muscles continue to operate anaerobically, building up lactic acid. It now appears, however, that few dives are of a duration that exceeds the aerobic capacity of the animals.

When a dive has been completed and the animal returns to the surface and begins breathing again, the blood is recharged with oxygen, the blood supply is restored to the muscles and other organs, and oxygen replenishes the myoglobin. The blood picks up the accumulated lactic acid, where it is eliminated by oxidation in the liver. At this point, the animal is ready to dive again.

Marine mammals and birds have salt concentrations in their blood and body fluids lower than the concentration of the surrounding seawater. This means that they face a potential osmoregulatory problem in that water has a tendency to pass from their bodies to the outside in order to equalize the salt concentrations between the inside and outside of the animal. Marine mammals and birds must compensate for this water loss either by drinking

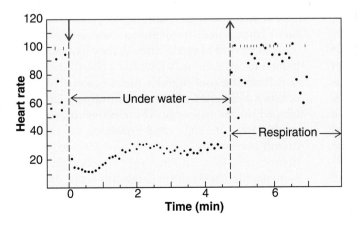

FIGURE 3.32 Change in heart rate in the bottle-nosed dolphin (*Tursiops truncatus*) during diving. Arrows indicate the beginning and end of the dive. (Reprinted by permission of *Nature*, Vol. 212, No. 407, p. 123. Copyright © 1966 Macmillan Magazines, Ltd.)

seawater or by obtaining water from their food. Marine mammals derive most of their water from the metabolic breakdown of their prey. If they drink seawater, they gain an unwanted quantity of salt, which they then must get rid of in some way. Marine birds and turtles eliminate this excess salt through special salt-secreting glands near the orbit of the eye. Cetaceans, however, do not have such glands, nor do they have sweat glands, as do terrestrial mammals. The only organ left to them for the elimination of salt is the kidney. Excess salt must be removed via the kidneys and washed out of the body with large amounts of water. The urine is highly concentrated, and the large multilobular kidneys are efficient at removing salts without losing much water.

As noted previously, we now know from satellite tracking of certain albatrosses that these birds regularly travel immense distances to forage for food for their chicks. (See page 126.) This knowledge has raised the question of how the albatrosses, and related procellariiform seabirds such as petrels and shearwaters, find their patchily distributed food sources—mainly krill, squid, and fish—in a vast and featureless ocean. The first clue to the answer is that these birds possess an anatomically elaborate sense of smell, which, in contrast, is poorly developed in most birds. Included in this mechanism is an elaborate olfactory epithelium housed within a tubular-shaped nose, as well as large olfactory bulbs in the brain that relay information to higher centers. Because of the tubular nature of the nostrils, these birds are also called "tube-nosed" seabirds (Figure 3.33).

Nevitt (1999) has demonstrated that at least some of these birds respond quickly to the presence of the gas dimethyl sulfide (DMS) in the air. It is known that when krill feed upon phytoplankton, their prey release dimethyl sulfide, which can persist for hours or even days. The presence of krill in turn tends to attract their predators, which may in turn be preyed upon by the birds and other predators. Since krill occur in patches whose location is unpredictable in the vast reaches of the oceans, a mechanism that would allow them to be detected from large distances would make foraging more certain and less wasteful of energy. Nevitt (1999) suggests that some of the tube-nosed birds find these patches by detecting the presence of DMS, or perhaps other specific olfactory cues, in the air.

FIGURE 3.33 Giant petrel, showing the tubular nostrils. (Copyright Fritz Polking; Frank Lane Picture Agency/Corbis.)

Ecology of Nekton

As noted in the introduction to this chapter, because of the difficulties of observing and performing experiments on natural populations of nekton, we know very little about the ecology of these animals. Most of our information concerns their feeding relationships and probable trophic links.

■ FEEDING ECOLOGY AND FOOD WEBS

Basically, all adult nekton are carnivores preying on either the smaller plankton or other nekton. The largest number of nekton are predators on other nekton. The plankton feeders consume the larger zooplankton and include such fishes as flying fish and sardines (Clupeidae), but the best-known plankton feeders are the baleen whales, a group that includes all of the larger whales, such as the blue whale, the largest animal that has existed on this planet. These tremendous animals have large mouths devoid of teeth; instead, they have sheets of **baleen** or whalebone, which hang down like a curtain from the roof of their mouth cavities. Baleen consists of a series of closely spaced parallel plates fringed on the free edge. These plates form a sieve mechanism. In right and bowhead whales, the mouth is very large, with long, finely fringed baleen that permits them to sieve out small copepods and euphausiids (krill) from large amounts of water while slowly moving forward. In the more streamlined rorqual whales, which range in size from the

FIGURE 3.34 Humpback whales (*Megaptera novaeangliae*) lunge-feeding at the surface. (Copyright Ralph Lee Hopkins, Wilderland Images.)

9-m minke to the 25-m blue whale, the feeding is somewhat different. These animals have relatively smaller mouths with short baleen. They can, however, open their mouths to an angle greater than 90° between the upper and lower jaws. Rorqual whales feed by lunging with wide open mouths into masses of krill or schools of small fishes, taking in large volumes (up to 70 tons) of water and prey (Figure 3.34). The large volume distends the accordionlike furrows or pleats in the throat. The animals then contract the throat muscles and push the water out through the baleen, using their large tongue while keeping the mouth partially closed. Prey are trapped against the baleen (Figure 3.35). Baleen whales are not the only marine mammals to feed on krill: The crab-eater seal of the Antarctic does so also, but it filters out the krill on the cusps of its peculiarly shaped teeth (Figure 3.36). Penguins are also important consumers of krill.

One exception to plankton feeding among baleen whales is the gray whale (*Eschrichtius robustus*), which is a bottom feeder. Gray whales dive to the bottom and take in large quantities of sediment, with whatever invertebrates reside there. They filter the sediment through their baleen and retain small invertebrates, such as amphipods. Gray whale feeding leaves large, irregular depressions on the shallow seafloor, and the whales are important sources of disturbance in these areas.

Plankton-feeding fishes are varied and include such common forms as flying fishes, salmon, and the largest fishes—the whale sharks. As with the baleen whales, the dominant food organisms are various crustaceans of the zooplankton, including euphausiids, copepods, and amphipods. The type of zooplankton taken varies among different areas of the ocean, as well as with different seasons. In the case of the Pacific Ocean salmon, the diet varies considerably, not only with the season, but also in different years. There is some evidence that this change in diet with time is related to the competition for food among the several salmon species. During years of high abundance of certain salmon, the intensified competition leads to an expanded range of food items taken and to changes in the composition of the fish's diet. For example, Parin (1970) notes that, in 1956, when pink salmon (*Oncorhynchus gorbuscha*) were low in numbers, 73.2% of the diet of chum salmon (*O. keta*) consisted of euphausiids. In 1957, however, when pink salmon were abundant, chum consumed ctenophores and jellyfish (72.6% of their diet). It does not appear that planktivores specialize on any one kind of plankton; rather, they are size selective.

FIGURE 3.35 (a) Cross section of the head of a baleen whale, indicating the functional anatomy of the feeding apparatus. (b) The feeding process in a baleen whale: (1) mouth open and tongue depressed, water and plankton flow in; (2) mouth closed, tongue raised, water passes out through the baleen, which strains out the plankton. (c) Photograph of the baleen of a gray whale (*Eschrichtius robustus*). (a, b modified from *Whales*, E. J. Slijper, © 1969 Routledge. Reprinted by permission. c © Copyright Harry Taylor/Dorling Kindersley Media Library.)

(a)

(b)

(c)

Whereas few of the vertebrates of the epipelagic zone of the oceans are plankton feeders, a large number of small fishes that live in the mesopelagic zone (see Chapter 4) are planktivores. These fishes migrate into the epipelagic zone at night to feed. Thus, the competition for plankton may be greater than first anticipated, but our understanding of these relationships is poor at present. These same small fishes also form a food source for some of the carnivorous fishes of the epipelagic.

The nekton-feeding fishes, birds, and mammals dominate the open-ocean nekton. Their diet generally includes fishes, squids, or large crustaceans. The size of prey taken usually depends on the size

FIGURE 3.36 The teeth of the Antarctic crab-eater seal (*Lobodon carcinophagus*), used to filter euphausiids. (Photo courtesy of AIBS from September 1964 cover of *BioScience*. Used by permission of Dr. V. B. Scheffer.)

of the predator, with the larger species taking progressively larger prey species. The largest carnivore in the oceans is the sperm whale (*Physeter macrocephalus*), which preys upon the largest squid, diving deeply to obtain these cephalopods. Apparently, sperm whales are the only animals that prey upon the giant squid, *Architeuthis*. (The stomach of one sperm whale yielded a giant squid 10.5 m long and weighing 184 kg.) The second largest carnivore of the nekton is the killer whale (*Orcinus orca*), which preys upon fishes, penguins, porpoises, seals, and sea lions. (See Figure 3.14.) Killer whales have also been known to attack the much larger baleen whales, which they do in packs of 3 to 40. In similar attacks on gray whales (*Eschrichtius robustus*), the killer whales usually go after the tongue of the whale. The capacity of killer whales is truly enormous; one specimen, reported by Slijper (1976), was captured with parts of 13 porpoises and 14 seals in its stomach.

The smaller-toothed whales, seals, and sea lions tend to feed on various squids and fishes, which they capture and swallow whole. Both groups have been attacked by humans because of the belief that they feed heavily on commercially important fishes such as salmon and, therefore, compete with humans. Although they do take commercially important fishes, it remains to be proven that such fishes constitute a significant part of their diet. The truly fish-eating whales include dolphins of the genera *Delphinus, Stenella, Tursiops,* and *Lagenorhynchus*. Examination of the stomachs of five species of delphinids revealed a diet composed of 90% of mesopelagic lantern fish. These dolphins have a short, slender rostrum with many teeth. Those dolphins feeding on squid tend to have reduced numbers of teeth, but increased ridges in the palate to hold the slippery squid. Included among squid eaters are the rare beaked whales (Ziphiidae) and the pilot whales. The types and size ranges of food for most whales are summarized in Figure 3.37.

Pelagic birds are generally more restricted in their feeding areas, because they cannot dive as deeply as the marine mammals. Hence, they feed mainly in the surface waters, consuming various small fishes and squids. Since certain species often occur in tremendous numbers, it may be that they have a significant effect on the fish and squid populations of the upper layers, but we are ignorant of the details of this relationship at present.

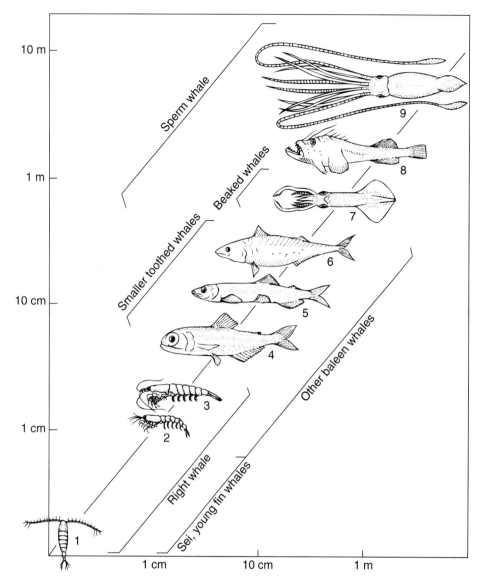

FIGURE 3.37 Size ranges of preferred food items of cetacean groups on a logarithmic scale: (1) calanoid copepods; (2, 3) various euphausiids; (4) myctophid (lantern) fish; (5) capelin; (6) mackerel; (7) onychoteuthid squid; (8) bathypelagic angler fish; (9) architeuthid squid, sporadically eaten by sperm whales. (From *The Ecology of Whales and Dolphins*, by D. Gaskin. Reprinted by permission of Heinemann Educational Publishers, a division of Reed Educational & Professional Publishing, Ltd.)

Pelagic birds may also be competing with various fishes and smaller marine mammals for the same food resource.

We know most about the food of various nektonic fishes. Of these fishes, the commercially important species, such as tunas, albacore, and marlins, are best known. They consume fishes, squids, and crustaceans.

Tunas have been studied intensively, and their food is extremely varied; as many as 180 different food items have been reported for skipjack tuna (*Katsuwonus pelamis*). Part of this varied diet is due to the great areas over which tuna move, encountering many different kinds of prey. Small tunas tend to feed on species in the surface layers of the epipelagic zone, whereas larger ones capture mesopelagic organisms in deeper areas. As with planktivores, the composition of the diet varies with season and location. Marlins and swordfishes have diets similar to that of tunas.

Nektonic sharks have a significant role in the feeding relationships of nekton, feeding on several different species of fishes, including tunas, and on marine birds, mammals, and squids. The largest carnivorous shark, the white shark, has the most varied diet, which includes not only fishes, birds, and mammals, but also other sharks. One specimen was even found with the remains of a basking shark in its stomach.

One of the most consistent features of feeding in nektonic fishes is their general lack of selectivity or specialization. Most fishes feed on any food of the appropriate size that is present.

With the limited information now available, it is possible to outline the food webs in the open epipelagic zones in the oceans of the world, but only in a very simplified and general way, as much more study is needed to define all of the links and their relative importance. These food webs are different among the tropical, temperate and polar oceans (Figure 3.38). In such cold temperate oceans as the North Pacific, the main net phytoplankton are diatoms, which are consumed by various zooplankton. These serve as food for plankton-feeding fish, mainly salmon and migrating mesopelagic forms. Top carnivores include the sharks and marine birds and mammals.

In the Antarctic, primary production is also centered on the diatoms, and the foremost herbivores are copepods and euphausiids. These are consumed by baleen whales, penguins, and certain fishes. The top-level carnivores are various marine mammals.

In the tropical waters, the primary producers are dinoflagellates and coccolithophores. The herbivores include a wide variety of zooplankton organisms. These, in turn, provide food for a number of planktivorous fishes, such as flying fishes at the surface and lantern fishes from the mesopelagic zone. All are food for the larger, first-level predatory fishes and squids, which in turn are preyed upon by larger predators, such as marlins, sword-

fish, and sharks. Finally, at the top are the largest sharks—the white and the mako—which attack tuna, swordfish, and marlins.

Although the food webs just described are highly simplified and do not include all the newly discovered important nanoplankton and picoplankton producers and consumers (see pp. 89–94), they are included here to illustrate the differences between areas. Note that there are marked differences between the tropics and the temperate–polar areas. First, the food webs are more complex in the tropics, with more links and more trophic levels, due partly to the greater number of species present in the tropics and the general absence of the larger, swifter predaceous fishes (tuna, marlin) in colder waters. The second observation is that marine mammals and birds play a larger role in the food webs of the polar regions than in those of the tropics. Indeed, larger numbers of marine mammals occur in colder, more productive seas than in warmer, less productive seas.

ECOLOGICAL SIGNIFICANCE OF MARINE MAMMALS

Although both the public and the media have exhibited interest in the plight of the world's whales and dolphins and in their past and future decimation by humans, there is little scientific information about their ecological importance in either pelagic or inshore ecosystems. As indicated earlier in this chapter, the importance of many, if not most, nektonic organisms in the dynamics of the open ocean is hard to determine. Preliminary investigations, as reviewed by Katona and Whitehead (1988), suggest that, although the productivity of cetaceans is low, their biomass is large enough to be significant on an oceanwide basis. The significance of these animals can be appreciated by noting that Kanwisher and Ridgway (1983) have suggested that cetaceans probably consume more prey than the entire world's fisheries. If true, this remark suggests that the energy flow through the cetacean component of the open ocean food webs may be extremely important, particularly in such areas as the Antarctic seas, about which the two researchers estimated that the unexploited baleen whale populations there consumed 190 million metric tons of krill (*Euphausia superba*) per year, approximately two times the current world fisheries catch. (See Chapter 11.) Now that the

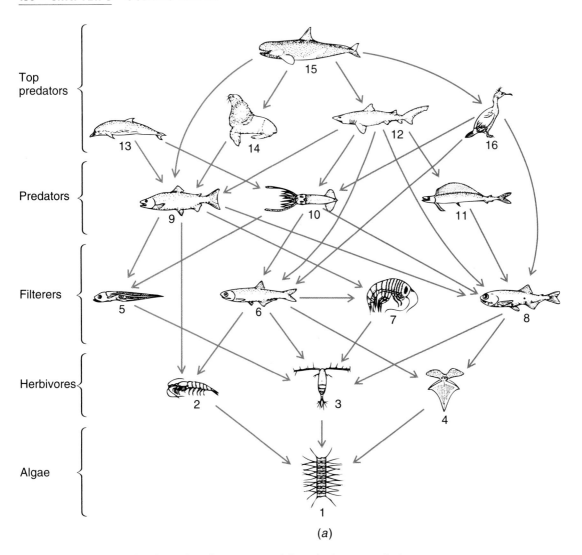

FIGURE 3.38A Food webs and trophic structure of the pelagic community in three different geographical areas. (a) Cold waters. Algae: (1) diatoms. Herbivores: (2) euphausiids; (3) copepods; (4) pteropods. Filterers: (5) juvenile fishes; (6) anchovy; (7) hyperiid amphipods; (8) lantern fishes. Predators: (9) salmon; (10) squid; (11) lancet fish. Top predators: (12) mackerel shark; (13) porpoise; (14) seals and sea lions; (15) killer whales; (16) marine birds. (continues on pages 139–140)

numbers of these Antarctic whales have been severely reduced (by perhaps 90%) by hunting, evidence pointing up their importance has arisen from the results of this unparalleled overexploitation. Populations of Antarctic birds and pinnipeds tripled following the destruction of the whales. The reason for this population explosion, some say, is the availability of krill not consumed by the whales.

Whales also are important because their carcasses sink rapidly and provide a significant food resource for deep-sea benthic creatures. (See Chapter 4.) Gray whales disturb large areas of the subarctic Bering and Chukchi seas on a scale equivalent to large geological or meteorological forces. (See Chapter 5.) Finally, whales produce a considerable amount of the sound in the world's oceans,

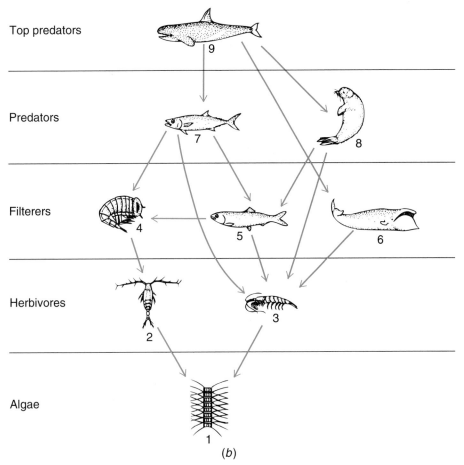

Top predators

Predators

Filterers

Herbivores

Algae

(b)

FIGURE 3.38B
(b) Antarctic seas. Algae:
(1) diatoms. Herbivores:
(2) copepods; (3) euphausiids.
Filterers: (4) hyperiid amphipods; (5) planktivorous fish;
(6) baleen whales. Predators:
(7) predatory fishes; (8) seals
and sea lions. Top predators:
(9) killer whales.

and it is possible that such signals may be used as cues by other species. Clearly, although removing whales from the ocean ecosystem would probably have little effect on the primary productivity or the total biomass of organisms, it would certainly alter some systems, perhaps more than we realize. Certainly, the richness of our human lives would be much diminished.

The ecological significance of other marine mammals in various communities is less well understood, but suggestive. Preen (1995) has hypothesized that dugong grazing on seagrass beds in Australia reduced the seagrass biomass by up to 96% and prevented the expansion of the dominant species. As mentioned earlier, it has also been suggested that the severe reduction in the populations of whales and several species of fishes in the Bering Sea that were major predators of krill and other zooplankton led

to the release of krill for other species. This situation, coupled with other environmental changes, may have led to the fish assemblage in the area becoming dominated by the less nutritious pollock (*Theragra chalcogramma*). It has accordingly been hypothesized that, because of the diminution in the numbers of the more nutritious forage species of fishes, the populations of harbor seals (*Phoca vitulina*) and Steller's sea lions (*Eumetopias jubatus*) have suffered sharp declines recently from lack of food. Estes et al. (1998) have extended this hypothesis a step further by linking the recent declines in sea otter (*Enhydra lutris*) populations in the western Aleutian Islands to predation by killer whales (*Orcinus orca*)—animals never before known to feed on otters—because seals and sea lions, their main food source, have had their populations reduced so dramatically (Figure 3.39).

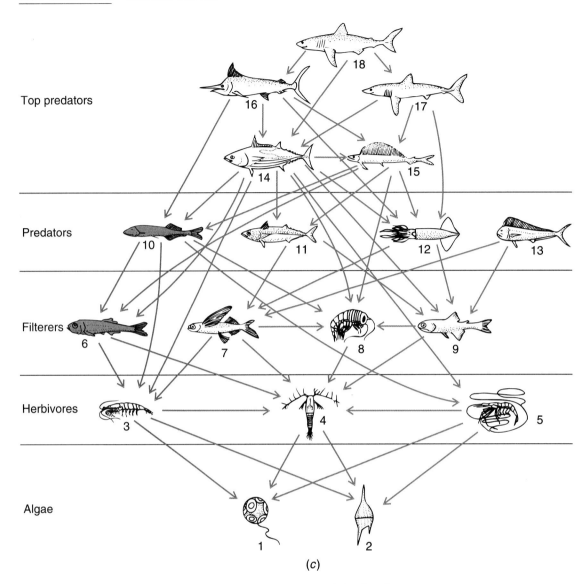

Top predators

Predators

Filterers

Herbivores

Algae

(c)

FIGURE 3.38C (c) Tropical seas. Algae: (1) coccolithophores; (2) dinoflagellates. Herbivores: (3) euphausiids; (4) copepods; (5) shrimp. Filterers: (6) vertically migrating mesopelagic fishes; (7) flying fishes; (8) hyperiid amphipods; (9) lantern fishes. Predators: (10) mesopelagic fishes; (11) snake mackerel; (12) squid; (13) dolphin (*Coryphaena*). Top predators: (14) tuna; (15) lancetfishes; (16) marlins; (17) medium-sized sharks; (18) large sharks. (a–c, from *Icthyofauna of the Epipelagic Zone*, N. V. Parin, translated from Russian by the Israel Program for Scientific Translations.)

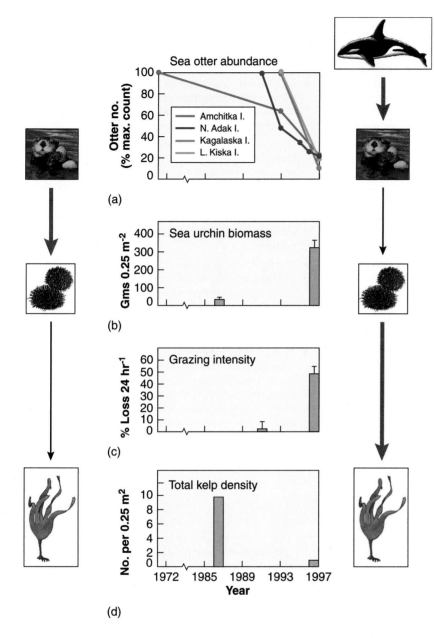

FIGURE 3.39 (a) Changes in sea otter abundance over time in the Aleutian Islands with concurrent changes in (b) sea urchin biomass, (c) grazing intensity, and (d) kelp density. The diagrammatic representations in the margin show the proposed mechanisms of change. The one on the left shows the organization of the kelp forest system before the decline of the sea otter, and the one on the right indicates how the system changed with the advent of the killer whale as the terminal predator. Heavy red arrows represent strong interactions, light black arrows weak interactions. (Reprinted with permission from "Killer Whale Predation on Sea Otters Linking Oceanic and Nearshore Systems" J. A. Estes et al., *Science,* Vol. 282, pp. 473–476, 1998. Copyright © 1998 American Association for the Advancement of Science.)

Summary of Key Concepts

- Nekton includes all organisms capable of sustained locomotion against the motion of water.
- Nektonic organisms are difficult to study in nature and difficult or impossible to keep in captivity.
- The oceanic nekton comprises a wide range of fishes and a lesser number of mammals, reptiles, birds, and cephalopod mollusks.
- Environmental conditions significant to nekton include the three-dimensionality of the medium, the transparency

of the medium, the lack of shelter from predators, and the absence of physical boundaries or structures.

- Buoyancy adaptations of nekton include the presence of a gas bladder in most fishes, air sacs and blubber in marine mammals, air channels in the bones of marine birds, lipid-filled livers in sharks, and hydrodynamic adaptations to provide lift.
- Adaptations to increase locomotory ability fall into two groups: those serving to create the propulsive force and

those reducing the resistance of the body to passage through the water.

■ Most fishes, whales, and sea snakes propel themselves by undulating movements of the body, whereas pinnipeds, turtles, and birds paddle with their limbs.

■ To reduce drag and therefore increase locomotory speed, most nektonic organisms have teardrop-shaped bodies and reduced forms of those projections from the body which would break the smooth flow of water over the body.

■ Defense against predators in nekton includes camouflage through color patterning and through alteration of the body shape to eliminate or reduce shadows.

■ Nektonic animals tend to have well-developed sensory systems, and most information comes to them via light (vision), sound (hearing), or chemicals (olfaction).

■ Sound reception has generated the greatest number of specialized adaptations and is most developed in cetaceans, who use echolocation or sonar to navigate and to locate and identify food.

■ Most pelagic bony fishes produce immense numbers of eggs that float and that are subject to equally large losses; in contrast, pelagic sharks produce a few eggs which hatch into large juveniles that are much less subject to predation.

■ Marine birds and reptiles retain the reproductive characteristics of their terrestrial relatives, producing shelled eggs laid on land.

■ Whales give birth in the water; pinnipeds give birth on land or ice. The young of whales and pinnipeds both grow rapidly, sustained by milk that is very high in fat content. All marine mammals appear to be long lived.

■ Many nektonic animals undertake extensive migrations for breeding or feeding. The means of navigation in marine mammals and turtles appears to be linked to an ability to detect the earth's magnetic field.

■ Marine birds and mammals have special adaptive features that enable them to survive in cold water. Layers of fat to reduce heat loss and circulatory systems that conserve heat are chief among these features.

■ Many marine mammals are deep divers. Adaptations that permit such diving include collapse of the lung, large amounts of myoglobin, a large blood volume, bradycardia, a cutoff of blood to certain tissues and organs, and a great tolerance of anaerobic conditions.

■ Most nektonic organisms are carnivores preying upon either the smaller plankton or, in most cases, other nekton.

■ The food webs and trophic structure of the pelagic community differ in different geographical areas of the oceans. In cold temperate waters, the top carnivores are sharks and marine birds and mammals. In Antarctic waters, the top carnivores are various marine mammals. In the tropics, the top carnivores are various fishes.

■ The ecological significance of marine mammals in the open-ocean systems is hard to determine, but the demise of the baleen whales in the Antarctic has led to large increases in Antarctic birds and pinnipeds due to the availability of krill not consumed by the whales. This situation suggests that whales were important in the open-ocean ecosystem. Other significant mammals are dugongs and various pinnipeds.

Review Questions

ESSAY Develop complete answers to the following questions:

1. Compare and contrast feeding in the blue whale with that in the gray whale.

2. Discuss the ecological significance of the Cetacea in the world's oceans.

3. Why do marine mammals usually produce only one offspring at a time? How can a species survive with such a low fecundity?

4. Discuss why nektonic animals, although possessing very different evolutionary histories, show a strong convergence in morphology.

5. Compare and contrast nektonic organisms with planktonic ones. Which of the two groups possesses greater diversity? Discuss.

6. Nektonic organisms live in a profoundly three-dimensional world, but it is not uniform. Discuss how this world is structured vertically.

Bibliography

Aleyev, Y. G. 1977. *Nekton.* The Hague: Dr. W. Junk BV.

Andersen, H. T., ed. 1969. *The biology of marine mammals.* New York: Academic Press.

Brooks, W. S. 1917. Notes on some Falkland Island birds. *Bull. Mus. Comp. Zool.* 61:135–160.

Carey, F. G. 1973. Fishes with warm bodies. *Sci. Amer.* 228(2):36–44.

Carr, A. 1965. The navigation of the green turtle. *Sci. Amer.* 212(5):79–86.

Clarke, M. R. 1979. The head of the sperm whale. *Sci. Amer.* 240(1):128–141.

Denton, E., and N. Marshall. 1958. The buoyancy of bathypelagic fishes without a gas-filled swim bladder. *J. Mar. Biol. Assoc.* 37(3):753–767.

Dunson, W. A., ed. 1975. *The biology of sea snakes.* Baltimore: University Park Press.

Elsner, R. 1969. Cardiovascular adjustments to diving. In *The biology of marine mammals*, edited by H. T. Andersen. New York: Academic Press, 117–145.

Estes, J. A., M. T. Tinker, T. M. Williams, and D. F. Doak. 1998. Killer whale predation of sea otters linking oceanic and nearshore ecosystems. *Science* 282:473–476.

Gaskin, D. E. 1982. *The ecology of whales and dolphins.* London: Heinemann.

Hardy, A. 1965. *The open sea: Its natural history.* Vol. II, *Fish and fisheries.* Boston: Houghton Mifflin.

Kalmijin, A. J. 1971. The electric sense of sharks and rays. *J. Exp. Biol.* 55:371–383.

Kanwisher, J. W., and S. H. Ridgway. 1983. The physiological ecology of whales and porpoises. *Sci. Amer.* 248(6):110–121.

Katona, S., and H. Whitehead. 1988. Are cetaceans ecologically important? *Oceanog. Mar. Biol. Ann. Rev.* 26:553–568.

Kellogg, W. N. 1958. Echo ranging in the porpoise. *Science* 128:982–988.

Kirschvink, J. L. 1990. Geomagnetic sensitivity in cetaceans: An update with live stranding records in the United States. In *Sensory abilities of cetaceans, laboratory and field evidence*, edited by J. A. Thomas and R. A. Kastelein. New York: Plenum Press, 639–649.

Klinowska, M. 1990. Geomagnetic orientation in cetaceans: Behavioural evidence. In *Sensory abilities of cetaceans, laboratory and field evidence*, edited by J. A. Thomas and R. A. Kastelein. New York: Plenum Press, 651–663.

Kooyman, G. L., and R. J. Harrison. 1971. *Diving in marine mammals.* Oxford Biology Reader, no. 6.

LeBoeuf, B. et al. 2000. Foraging ecology of northern elephant seals. *Ecological Monographs* 70(3): 353–382.

Lohman, K. J. 1992. How sea turtles navigate. *Sci. Amer.* 266(1):100–106.

Nevitt, G. 1999. Foraging by seabirds on an olfactory landscape. *Amer. Sci.* 87:46–53.

Norris, K. S. 1969. The echolocation of marine mammals. In *The biology of marine mammals*, edited by H. T. Andersen. New York: Academic Press, 391–423.

Norris, K. S., J. H. Prescott, P. V. Asa-Dorian, and P. Perkins. 1961. An experimental demonstration of echolocating behavior in the porpoise, *Tursiops truncatus* (Montagu). *Biol. Bull.* 120:163–176.

Parin, N. V. 1970. *Ichthyofauna of the epipelagic zone.* Jerusalem: Israel Program for Scientific Translations.

Preen, A. 1995. Impacts of dugong foraging on seagrass habitats: Observational and experimental evidence for cultivation grazing. *Mar. Ecol. Prog. Ser.* 124:201–213.

Scheffer, V. B. 1976. *A natural history of marine mammals.* New York: Scribner.

Scholander, P. F. 1940. Experimental investigations on the respiratory function in diving mammals and birds. *Hvalradets Skrifter Norske Videnskaps Akad., Oslo* 22:1.

Seachrist, L. 1994. Sea turtles master migration with magnetic memories. *Science* 264:661–662.

Slijper, E. J. 1976. *Whales and dolphins.* Ann Arbor: University of Michigan Press.

Stewart, B. S., and R. L. DeLong. 1995. Double migrations of the northern elephant seal *Mirounga angustirostris. J. Mammalogy* 76(1):196–205.

Walker, M. M., J. L. Kirschvink, S-B. R. Chang, and A. E. Dizon. 1984. A candidate magnetic sense organ in the yellowfin tuna, *Thunnus albacares. Science* 224:751–753.

Williams, T. M. et al., 2000. Sink or swim: Strategies for cost-efficient diving by marine mammals. *Science* 288:133–136.

Würsig, B. 1979. Dolphins. *Sci. Amer.* 240(3):136–148.

Würsig, B. 1989. Cetaceans. *Science* 244:1550–1557.

CHAPTER 4

DEEP-SEA BIOLOGY

The most extensive habitat on the planet inhabited by living organisms consists of the permanently cold, dark waters and bottoms of the deep oceans. The shallow waters fringing the continents and islands of the world make up less than 10% of the total area of the world's oceans. The lighted upper waters of all oceans are an even smaller fraction of the total volume of ocean living space. Of the 70% of the planet's surface covered with water, perhaps 85% of the area and 90% of the volume constitute the dark, cold environment we call the deep sea. Consider that 79% of the entire volume of the biosphere is composed of marine water with depths greater than 1,000 m.

Although this area is the largest habitat on Earth, its biology is the least known and explored, due primarily to the difficulty of sampling such a remote area and to the fact that, until recently, it was virtually inaccessible to humans. Although the recent advent of deep manned and remotely operated submersible vehicles would now allow much of this vast area to be observed by scientists, very little of it has been visited by these vehicles because of their high cost and relatively small numbers.

It is important to consider the deep sea, even though it is unlikely to be sampled or observed by most users of this book, because it constitutes such a huge amount of the biosphere and because it is a frontier where new and exciting discoveries are being made. It is also vital to study the deep sea because it may become increasingly important as a source of materials and as a dumping site for human wastes.

This chapter outlines our present understanding of the basic principles of adaptations and the organization of life-forms in this vast and intriguing area.

■

Zonation

When we refer to the **deep sea** in this chapter, we mean that part of the marine environment which lies below the level of effective light penetration for phytoplankton photosynthesis in the open ocean and deeper than the depth of the continental shelves (>200 m). The upper part of the deep sea, a transitional area that receives some light, has been called the *dysphotic zone* (Figure 4.1). The entire area below it is permanently dark and is known as the **aphotic** zone, as opposed to the lighted, or **photic**, zone, where all primary photosynthetic production occurs. In tropical waters, the start of the aphotic zone is deeper (~600 m), whereas in temperate waters, it is shallower (~100 m).

Several schemes of zonation for the deep sea that have been proposed over the years are reviewed by Menzies, George, and Rowe (1973), but none has been universally acceptable to all scientists researching the area. There are a number of reasons for this, but the primary one is the lack of sufficient ecological information. Most schemes of zonation have been produced by simply dividing the water column into discrete zones based on depth, temperature changes, or both. Another approach has been to divide the deep ocean on the basis of species abundance, distribution, or associations. A problem with this approach is that certain species are distributed at different depths in differ-

ent ocean basins, depending on special hydrological or ecological characteristics of the region. Similarly, the changes in temperature occur at different depths in different areas. Still, kinds of species may be better for indicating zones than the associated environmental parameters. We shall follow the zonation scheme outlined in Chapter 1 and illustrated in Figure 1.25. The relevant deep-sea zones are listed in Table 4.1 and are briefly reviewed here.

The deep sea can be divided into two primary areas: the benthic zone (associated with the bottom) and the pelagic zone (associated with the open water). Since the physical environment differs between them, these two areas are inhabited by different associations of organisms. We know more about benthic deep-sea fauna than about pelagic fauna. The benthic fauna, as noted in Chapter 1, can be subdivided into those which occupy a **bathyal** zone on the continental slope and those living in a large **abyssal** zone constituting most of the deep-sea floor. The animals of the trenches live in a separate **hadal** (ultra-abyssal) zone, but little is known of their biology.

In the pelagic division, there is an upper zone immediately below the photic zone, which contains many animals that migrate vertically into the photic zone at night. This is usually called the **mesopelagic** zone (dysphotic, twilight). This zone is inhabited by many species of animals with well-developed eyes and a variety of light organs. The dominant fishes here are silvery black, and the crustaceans are red. Because this

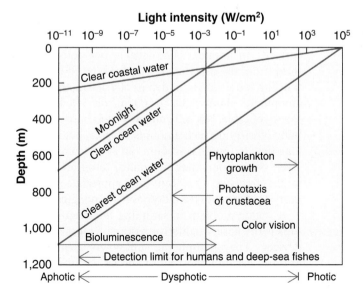

FIGURE 4.1 Schematic diagram showing the penetration of sunlight into the clearest ocean water and into clear coastal water in relation to minimum intensity values for some biological light receptors. (From *Biological Oceanographic Processes*, 3d ed., T. R. Parsons et al., 1984. Copyright © 1984 Butterworth Heinemann. Reprinted by permission of Elsevier.)

TABLE 4.1

Deep-Sea Faunal Zones

LIGHT	PELAGIC ZONES	DEPTH RANGE	BENTHIC ZONES	DEPTH RANGE
Present (photic)	Epipelagic or photic	0–200 m	Continental shelf or sublittoral	0–200 m
Present (disphotic)	Mesopelagic	200–1,000 m (?)	Bathyal	200–4,000 m (?)
Absent (aphotic)	Bathypelagic	(?) 1,000–4,000 m (?)		
	Abyssalpelagic	(?) 4,000–6,000 m (?)	Abyssal	4,000–6,000 m (?)
	Hadalpelagic	6,000–10,000 m (?)	Hadal	6,000–10,000 m

Note: Question mark indicates boundary uncertain.
Source: After Hedgpeth, 1957.

is the most accessible of the deep-sea pelagic zones, we know the most about it. The numbers of organisms in the mesopelagic zone appear to be the highest of all the deep-sea pelagic zones. The zone extends down to about 700–1,000 m, with the depth varying with location, clarity of water, and other factors.

Below the mesopelagic zone, it is much more difficult to establish any pelagic zones that are universally acceptable. The region between the lower limit of the mesopelagic zone and the upper limit of the deep trenches (6,000 m) has been divided into two zones by Hedgpeth (1957): an upper **bathypelagic** zone and a lower **abyssalpelagic** zone; we continue Hedgpath's division here. The boundaries of these zones are uncertain and vary from one authority to another. Both of these areas are characterized by low numbers of species and individuals—considerably fewer than are observed in the mesopelagic zone. The organisms tend to be white or colorless, with reduced eyes and few bioluminescent organs.

The open water of the trenches is often called the **hadalpelagic** zone. It appears set off from the other areas, and we are so ignorant of it that we do not know whether a separate pelagic community exists in these areas.

Sampling the Deep Sea

In the preceding section, we alluded to the problems associated with sampling and obtaining information about the deep sea. Before continuing this discussion, it is useful to indicate briefly how our information is obtained from the deep sea and why it is difficult to obtain and is often biased. An analogy has been made that characterizing the deep sea and its communities with the gear we have traditionally used is like characterizing terrestrial communities by taking samples

with a butterfly net towed behind an airplane! This analogy may be exaggerated, but it is close enough to have a ring of truth.

The traditional types of gear used to sample the deep-sea benthos include various types of nets, grabs, and dredges of similar design to those used in shallow water. Representative types are illustrated in Figure 4.2. Pelagic animals are sampled with several types of midwater trawls (Figure 4.3). Because the density of organisms is so low in pelagic areas below the mesopelagic zone, these midwater nets must have a large gape and must sieve a lot of water to obtain any reasonable number of animals and to mitigate the problem of animals simply avoiding the nets.

The main problem in sampling the deep sea, pelagic or benthic, is that the deeper the sampling, the more cable is necessary to get to the desired depth and the longer is the time needed to make a single haul. Furthermore, as more and more cable is payed out in a deep haul, the greater are the weight and the chance of snarling the cable or otherwise fouling the gear. In order for most nets towed on the bottom to fish correctly, an amount of line equaling one-and-a-half to three times the depth must be payed out. If one were trawling at only medium depths of 4,000 m, this would amount to about 5,000–6,000 m, or 5–6 km (3–4 miles), of cable. Not only do few vessels have the capability of carrying such amounts of cable, but it takes a long time to feed that amount out and then retrieve it. Twenty-four hours may be required to make one trawl in the trenches. One can thus appreciate why the number of samples from deep water is not great.

A major problem in sampling the deep-sea benthos is the basic qualitative nature of most dredging devices. They sample an unknown area of the bottom. Because densities of species and individuals are difficult or impossible to estimate from these

(a)   (b)   (c)

FIGURE 4.2 Two types of deep-sea sampling gear. (a) The epibenthic sled. (b) The box core. (c) Sequence showing the box core taking a sample. (a, from *The Abyssal Environment and Ecology of The World's Oceans*, R. J. Menzies et al., © 1973 John Wiley & Sons. b, c from *Deep Sea Biology*, John Gage and Paul Tyler. Copyright © 1991 Cambridge University Press. Reprinted by permission.)

FIGURE 4.3 Rigging a large midwater trawl for fishing. (Courtesy of Dr. Roger Seapy.)

samples, comparative studies of deep and shallow water have been difficult. Fortunately, recent investigations in the deep sea have employed large box corers or multicorers, which overcome this problem by sampling a known bottom area (Figure 4.2b).

Finally, in sampling the midwater animals, there is the problem of animals avoiding the net. Trawling in deep water usually means that the ship must travel relatively slowly because of the great weight of the cable it drags. This slowness in turn means that many fast-swimming pelagic animals can avoid the net simply by swimming out of its path. If this happens to any great extent, then characterizing the communities from such net hauls gives a misleading idea of what is in fact there.

Since 1970, new instrumentation has permitted deep-sea biologists to consider aspects of the ecology of deep-sea animals that were not possible in earlier years. Deep-sea cameras of various designs have done much to increase our understanding of the real nature of deep-sea communities, particularly bottom communities. The advent of deep manned and remotely operated submersibles with various devices for capturing organisms, including pressurized retrieval systems, has further advanced our understanding of these areas by permitting the observation and retrieval of living organisms for experimental studies (Figures 4.4, 4.5, and 4.6).

(a)

(b)

FIGURE 4.4 (a) The deep submersible *Alvin*. (b) The deep submersible *Alvin* in the water. (Photos courtesy of a, Norbert Wu, b, Dale Stokes/Mo Yung Productions/Norbert Wu.)

FIGURE 4.5 A remote-operated vehicle being deployed. (Photo by James Nybakken.)

Environmental Characteristics

Before considering the life in the deep ocean, it is necessary to establish the physical and chemical conditions under which the organisms must function. The major point to emphasize in considering these factors is that, at any given level or position in the deep ocean, they remain constant over long periods.

■ LIGHT

In the deep ocean, the entire area is without sunlight, except at the very upper parts of the mesopelagic zone,

FIGURE 4.6 Some deep-water benthic animals from the bathyal zone of the eastern Pacific. (a) *Tromikosoma panamense* (echinoid). (b) *Colessendeis* sp. (pycnogonid). (c) *Umbellula lindahli* (pennatulid). (d) *Freyella microplax* (brisingid sea star). (e) *Granelodone pacifica* (octopod). (f) *Scotoplanes globosa* (sea cucumber).

where some light penetrates. Since the area is either dark at all times or has such extremely low light levels, photosynthesis is not possible, precluding any photosynthetically based primary productivity. What light is present is usually produced by the animals themselves. (See pp. 163–165.) The absence of light means that the organisms have to rely on senses other than vision to find food and mates and to maintain various interspecific and intraspecific associations. The absence of sunlight may also have had a selective effect on the locomotory habits of the animals and on their propulsive systems. For example, life in darkness may reduce the intensity of predator–prey interactions, thus permitting the existence of a more sluggish locomotory habit.

■ PRESSURE

Of all the environmental factors acting on deep-sea organisms, pressure shows the greatest range. Pressure increases 1 atmosphere (atm) (14.7 lb/in.^{2} or 1 kg/cm^2) for each 10 m in depth. Since depths, including trenches, vary from a few hundred meters to more than 10,000 m, the range of pressure is from 20 to more than 1,000 atm. Most of the deep sea is under pressures between 200 and 600 atm. In no other marine environment does pressure exhibit such a great range or play such a potential role in the distribution of organisms.

Until recently, scientists lacked detailed information on the direct effects of pressure on most deep-sea organisms. The reason was primarily that virtually all organisms trawled from the deep sea arrived at the surface dead or dying. Only in recent years, with the advent of traps that incorporate a special pressure-maintaining chamber, have undamaged larger metazoan animals been retrieved from the deep sea in good condition. Some of these have been maintained for experimental purposes, and we are obtaining more knowledge of the biological effects of pressure.

That pressure must have a significant effect on deep-sea organisms can be extrapolated from experiments done on the first group of organisms successfully retrieved alive from the deep sea: bacteria. Figure 4.7 shows the results of culturing a deep-sea bacterium under different pressure regimes. The fact that bacteria from the deeper areas virtually cease growth and reproduction at lower pressures, but grow and reproduce actively at higher pressures corresponding to their natural environment, suggests that there are special adaptations to pressure that can profoundly affect organisms. Since metazoan animals are anatomically and physiologically more complex than bacteria, it is probable that multicellular animals will evince an even stricter pressure-dependent physiology.

There is now considerable evidence that both proteins and biological membranes are strongly affected by pressure and must be modified to work in animals living in the deep sea. For proteins, Siebenaller and Somero (1989) have shown that the direction of adaptation has been to reduce the pressure sensitivity of the structure at the cost of other aspects of the enzyme systems. For example, the enzyme systems of deep-sea fishes have higher activation energies but lower catalytic efficiencies than those of their relatives in shallower water. Recently, Yancey et al. (2002) have shown that adaptations to high pressures also appear to involve the accumulation within cells of deep sea organisms of small organic molecules that stabilize protein function and structure in the presence of elevated hydrostatic pressure.

Studies were done by Siebenaller and Somero (1978) on the enzyme systems of two closely related fishes living at different depths. The studies demonstrated that differences in hydrostatic pressure as small as 100 atm or less are sufficient to alter markedly the functional properties of enzymes—namely, their ability to bind to the appropriate substrate—and the rate at which enzyme reactions proceed. Such changes are comparable to the changes in enzyme activity caused by temperature differences. These results strongly suggest that hydrostatic pressure plays a major role in organisms' adaptation to the deep-sea environment. Somero (1992) has noted that proteins of deep-sea metazoan species are especially resistant to the disruption of their three-dimensional (tertiary and quaternary) structure under conditions of high pressure and that this sensitivity varies among closely related species living at different depths. Together, the sensitivity and the variation suggest that biochemical adaptations are significant in establishing depth-distributional patterns among species.

FIGURE 4.7 Graph of the amount of growth of a deep-sea bacterium at different pressures. (Reprinted with permission from "Isolation of a Deep Sea Barophilic Bacterium and Some of Its Growth Characteristics," A. A. Yayanos et al., *Science*, Vol. 205 (4408), pp. 808–809, 1979. Copyright © 1979 American Association for the Advancement of Science.)

In a review of physiological and biochemical adaptations of deep-sea fishes, Somero (1982) has pointed out other differences in muscle enzymes between shallow- and deep-dwelling fishes. Deep-dwelling fishes tend to have muscle enzymes that are less efficient than those of shallow-dwelling fishes and that are present in their cells in low concentrations. The combination of these two characteristics probably results in the low metabolic rates, as well as the sluggish movements, of some deep-sea fishes. Most of these fishes are "float-and-wait" predators. Interestingly, unlike the enzymatic activity in locomotory muscle, there appears to be no reduction in enzymatic activity in the brain with increasing depth of occurrence. This finding supports the conjecture that depth-related decreases in metabolism are specifically a result of selection for reduced swimming ability. Needs for neural activity in the central nervous system appear not to decrease with depth.

With respect to biological membranes, the major effect of pressure is on the lipids that make up the membranes. Because lipids are more compressible than water and can undergo phase changes under pressure, there is the potential for pressure effects on membranes. Higher pressures reduce the fluidity of membranes. To adapt to high pressures, bacteria incorporate more fluid lipids into their membranes. This adaptation is termed the **homeoviscous adaptation**. Cossins and Marshall (1989) have demonstrated that the ratio of saturated to unsaturated fatty acids in mitochondrial membranes in the livers of deep-sea fishes declines with increasing depth, suggesting a similar adaptation.

Many laboratory studies of various kinds of cells from protozoan to mammalian have demonstrated that pressure affects the morphology of the cells, including their ability to form mitotic spindles and to undergo mitosis. Zimmerman and Zimmerman (1972) reported that amoebas lose their pseudopods and ball up, and *Tetrahymena* loses its ability to locomote with cilia. Pressure also has a significant effect on certain physiological and biochemical processes, such as typical muscle physiology. Perhaps the most dramatic effect of pressure is its effect on proteins. Protein synthesis and function are markedly affected by pressure in laboratory situations, usually adversely. Since proteins are vital in living systems as enzymes and as major building blocks, it would seem that deep-water animals must have some special adaptation with respect to protein structure and function in order to survive.

The body of laboratory studies of the observed effects of pressure on shallow-water animals, coupled with the known effects of pressure on cells and macromolecules (proteins) and the recent experimental evidence from living deep-sea animals, are strong indications that the physiology of deep-sea animals is profoundly affected by pressure and that pressure may be of considerable significance in zonal patterns of distribution.

There is yet another effect of pressure—one which affects organisms that secrete calcium carbonate shells. As pressure increases and temperature decreases, both the solubility of calcium carbonate and its tendency to dissociate increase. This means that, as one goes deeper into the ocean, it becomes more and more difficult for organisms to lay down shells. In the deep sea, the number of shell-forming species decreases, and the remaining ones tend to have thin shells.

■ SALINITY

Below the first few hundred meters in the world's oceans, the salinity is remarkably constant throughout the depths (Figure 4.8). There are some minor differences in salinity, but none that are ecologically significant, except perhaps in the Mediterranean Sea and the Red Sea, where the bottom water is higher in temperature and salinity.

■ TEMPERATURE

The two areas of greatest and most rapid temperature change in the oceans are the transition zone between the surface waters and the deep waters—the area known as the **thermocline**—and the transition between the deep-sea floor and the hot-water flows at the hydrothermal vents. Thermoclines vary in thickness from a few hundred meters to nearly a thousand meters (Figure 4.9). Below the thermocline, the water mass of the deep ocean is cold and far more homogeneous. Thermoclines are strongest in the tropics, where the temperature of the epipelagic zone is usually above 20°C. From the base of the epipelagic zone, the temperature drops over several hundred meters to 5 or 6°C at 1,000 meters. The temperature continues to decrease toward the bottom, but the rate of change is much slower (Figure 4.9). Below 3,000–4,000 m, the water is nearly isothermal and has a temperature between 1 and 4°C. What is ecologically significant is that, at any given depth, the temperature is practically unvarying over long periods of time. There are no

FIGURE 4.8 Salinity and temperature changes across the North Atlantic Ocean with respect to depth. (From *Introduction to Oceanography*, D. A. Ross, p. 305, © 1982 Prentice Hall, Upper Saddle River, NJ.)

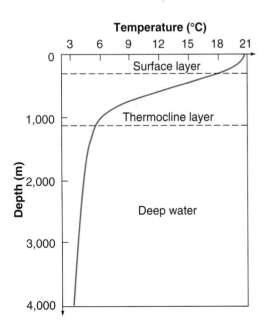

FIGURE 4.9 The thermocline layer in the ocean. (From *Oceanology: An Introduction*, D. E. Ingmanson & W. J. Wallace, © 1973. Reprinted with permission of Brooks/Cole, a division of Thomson Learning.)

seasonal temperature changes, nor are there any annual changes. Perhaps nowhere on Earth is there another habitat with such a constant temperature. It should be pointed out, however, that a few deep-sea environments have somewhat higher temperatures. For example, the Mediterranean Sea has bottom temperatures of about 13°C and the Red Sea 20°C, both due to their separation from the main deep ocean basins and the sinking of very warm saline water from the surface. Such temperatures, however, are in direct contrast to temperatures in the hydrothermal vents. In these systems the temperature of the water as it emerges from the "black smoker" chimneys may be as high as 400°C (it is kept from boiling by the high hydrostatic pressure), while a few meters away it may be back down to 2–4°C. (See pp. 175–187.)

■ OXYGEN

Even though the deep ocean waters are far from any source of oxygen replenishment, either by interaction with the atmosphere or through production by autotrophs, essentially no abyssal or hadal areas lack sufficient oxygen to support life. The few exceptions include the Cariaco trench off Venezuela in the Caribbean, which is anoxic at the bottom, and the Santa Barbara Basin off California, which is seasonally anoxic at the bottom. The oxygen present in deep water masses entered the water when those masses were at the surface. Virtually all the water of the deep sea has its origin at the surface in the Arctic or Antarctic areas (see Chapter 1, pp. 16–17), where the oxygen-rich cold water sinks and flows north or south to make up the deep water of the world's oceans. Since that water no longer gains oxygen after sinking, the reason these deep masses are not depleted of oxygen by the respiration of deep-water organisms is that the density of such organisms is so low and they have such low metabolic rates that, together, the two factors are insufficient to deplete the oxygen supply. This observation is supported by the fact that the oxygen concentration declines in the 20 m or so just above the bottom in the deep sea. It is also near the bottom that the densest concentrations of organisms occur in the deep sea.

Another peculiar feature of oxygen concentration in the deep sea is the presence of the **oxygen minimum zone**, usually at 500–1,000 m. (See Figure 1.3.) Oxygen values are higher both above and below this zone. In the oxygen minimum zone, oxygen values may fall to less than 0.5 ml/liter. The occurrence of the zone is mainly the result of the respiration of organisms, coupled with the lack of interchange of the water mass with more oxygen-rich water. The reason that an oxygen minimum zone occurs at 500–1,000 m and not deeper is that the numbers of animals in deeper waters are so low that they never deplete the available oxygen, whereas at 500–1,000 m, organisms—particularly bacteria consuming sinking organic matter—are abundant. Above 500 m, oxygen depletion does not occur, even though the animal biomass is high, because there is constant replenishment from air and by autotrophs.

Animals living in the oxygen minimum zone often have special adaptations to extract oxygen from the low-oxygen environment, or else they can subsist on minimal amounts of oxygen.

■ FOOD

The deep sea is removed from the photosynthetic zone and has no primary production except for certain areas where chemosynthetic bacteria are found.

(See pp. 175–187.) All organisms living in the deep sea outside the hydrothermal vents and cold seeps depend on food that ultimately is produced elsewhere, where photosynthesis is possible, and subsequently transported into the deep sea. The deep sea, along with caves in the terrestrial environment, are, therefore, unique among the world ecosystems in that they have virtually no indigenous primary productivity.

Food thus must come into the deep sea by sinking from the surface waters. Since the populations of organisms are greatest in the upper layers of the oceans, the chance that any food particle will be able to sink through all these voracious animals without being consumed or without decaying is small. As a result, relatively few food particles are available to the animals of the deep sea, as opposed to those in the surface waters. Exceptions do occur, however, such as in the eastern equatorial Pacific, where, within 2° of the equator, high surface productivity provides a high carbon input in the form of phytoplankton detritus. The paucity of food is probably the reason the density of deep-sea animals is relatively low and the size of most of them is small. Without sufficient energy in the form of food, large numbers of organisms cannot be sustained.

The probability that a particle of food will decay or be consumed during sinking increases with increasing depth, since there is more time and there are more organisms to consume it. Organic particles sinking through the water column are continuously degraded by bacteria associated with the particles, and the deeper a particle occurs, the more likely it is to be depleted of its more nutritious organic substances, such as proteins and nucleic acids. Thus, the deeper the organisms, the less food is available to them and the lower is its quality.

A portion of the particles sinking from the photic zone are particles of material that are not directly suitable as food. An example is the shed chitinous exoskeletons of crustaceans. Most organisms cannot digest chitin. These materials, however, are acted upon by bacteria during their descent into the deep sea and, after reaching the bottom, are converted into suitable food in the form of bacterial protoplasm. Because the residence time for a food particle is longer on the bottom than in the water column, there is an increase in bacteria in the bottom oozes of the deep sea; thus, this allows increased numbers of larger organisms to survive. Densities of the benthic

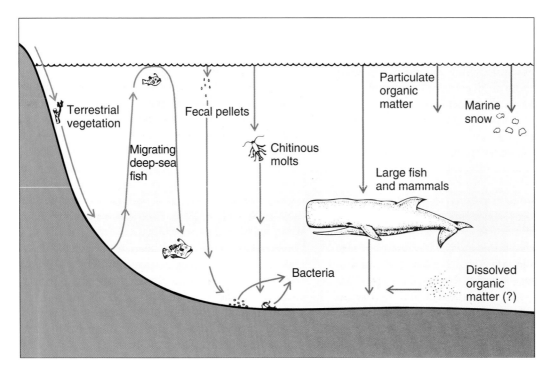

FIGURE 4.10 Food sources for the deep sea.

infaunal organisms may exceed those of pelagic organisms at the same depth. These increased densities also explain the reduction in oxygen in the near-bottom water.

The amount of food available to the deep sea is correlated with the amount of primary productivity in the surface waters above, or else with the proximity to a secondary source, such as organic debris from terrestrial habitats. Deep-sea areas under productive surface waters and near islands or continents have more food than deep waters under areas of low surface productivity or far from land and hence have higher numbers of organisms. Because primary productivity is highly seasonal at many latitudes, the amount of organic matter reaching the sea floor varies considerably during the year. (See Chapter 2.)

Several types of food sources are available in the deep sea. Directly available foods include those deep-sea organisms which spend their early or larval stages in the lighted upper waters and then migrate into the depths, to become food for the predators. An example is the group of deep-sea angler fishes whose young live in the upper waters of the ocean while the adults are restricted to the deep sea. Another directly

available food source is the large dead bodies of marine mammals and fishes that may sink quickly before being completely consumed (Figure 4.10). They would seem an unpredictable type of food, but, by placing baited cans with cameras attached into the deep sea, Isaacs and Hessler of Scripps Institution of Oceanography have shown that food is quickly detected and brings in consumers from great distances (Figure 4.11). That this unlikely food source must be reliable is also supported by the fact that several groups of deep-sea gastropod mollusks have evolved to live on whale and shark remains. (See p. 182.)

Indirect food sources include indigestible animal and plant remains, which first must be acted on by bacteria. Examples are chitin and the wood and cellulose of land plants. Other potential sources are colloidal or dissolved organic material in the water and gelatinous plankton matter (marine snow). At present, the relative importance of these sources cannot be evaluated.

We can summarize by saying that food is scarce in the deep sea, compared with other areas of the oceans, and that it decreases with depth in both quantity and quality and with distance from land. This food

(a)   (b)

FIGURE 4.11 Two photographs showing the rapid attraction of deep-sea ani- mals to bait and its subsequent consumption. (a) At the time of reaching the bottom. (b) Six hours and 40 minutes after reaching the bottom, showing the bait being con- sumed by the amphipod *Hirondella gigas* (at 9,605 m in the Philippine trench). (Photos courtesy of Dr. Robert Hessler, Scripps Institution of Oceanography.)

scarcity is responsible for the low densities and low per-unit-mass energy content of most deep-sea organisms.

■ BIOMASS

Because the deep sea is dependent primarily upon organic matter sinking from the surface waters, and this organic matter is progressively degraded or con- sumed as it sinks, there is obviously less to sustain the deep-sea organisms as one goes deeper. This trend has an effect on the biomass that can be sustained at any given depth. Angel (2003) reports that, between the surface photosythetic layer and 1,000 meters, the total surface biomass decreases by 90%. Between 1,000 and 4,000 meters, it declines to 1% of the bio- mass in the surface waters. Hence, the deep sea at all depths displays a paucity of individual organisms, which are more widely separated the deeper you go.

Adaptations of Deep-Sea Organisms

The organisms that live in the deep sea possess a number of adaptations correlated with the condi- tions of the environment. Because experimentation with living metazoan animals from the deep sea has been limited until recently, we have little direct evi- dence of how many adaptations function. Accord- ingly, we can offer only educated guesses based on

the conditions under which we know the animals live. We also know little about physiological and bio- chemical adaptations at present.

One adaptation that can be observed, even in or- ganisms from the upper mesopelagic zone, is that of *color*. Mesopelagic fishes tend to be either silvery gray or a deep black (Figures 4.12 and 4.13). They are not countershaded, as are epipelagic fishes. Mesopelagic invertebrates, by contrast, may be either purple, bright red, or orange. Mesopelagic jellyfish are often a dark purple or red, whereas the crustaceans, such as copepods, mysids, and shrimps, are brilliant red

FIGURE 4.12 *Anoplogaster cornuta*, a "daggertooth" fish from the mesopelagic zone. Notice the very large mouth and large teeth for seizing prey. (Photo courtesy of Michael Kelly.)

FIGURE 4.13 *Argyropelecus lychnus*, a mesopelagic hatchetfish. Note the lines of black photophores along the ventral portion of the body, the tubular eyes, and the silvery reflecting scales. (Photo courtesy of Michael Kelly.)

FIGURE 4.16 The giant deep-sea mysid *Gnathophausia*, an example of deep-sea gigantism. (Photo © E. Widder/HBOI/ Visuals Unlimited.)

FIGURE 4.14 A mysid crustacean of the genus *Boreomysis*. (Photo courtesy of Gary McDonald.)

(Figures 4.14–4.19). Since these organisms live in virtually unlighted waters, black organisms will be invisible, but why the red color? Young (1983) has suggested that, because most bioluminescent light is blue, the red pigmentation would protect these animals from the revealing rays of the bioluminescent flashes used by predators.

Organisms living even deeper, in the abyssal and bathyal zones, lack pigment and are often colorless or a dirty white. This is particularly true of animals dwelling on the bottom, although some taxa, such as

FIGURE 4.15 A deep-sea prawn of the genus *Notostomus*. Note the deep orange-red color, characteristic of many deep-sea crustaceans. (Photo courtesy of Dr. Roger Seapy.)

FIGURE 4.17 A deep-sea jellyfish, *Benthocodon hyalinus*. Note the characteristic deep purple color. (Photo courtesy of Dr. George Matsumoto.)

FIGURE 4.19 A deep-sea ctenophore, *Lampocteis sanguineventer*. (Photo courtesy of Dr. George Matsumoto.)

FIGURE 4.18 The deep-water jellyfish *Arctopodema amplum*. (Photo courtesy of Dr. George Matsumoto.)

anemones, may be colorful. Fishes, however, may be black at all depths.

Another adaptation, seen particularly in mesopelagic and upper bathypelagic fishes, is the presence of large eyes. The eyes in these fishes are much larger relative to their body size than those in fishes of the lighted epipelagic zone (Figure 4.20). Often, the presence of large eyes is correlated with that of light organs. (See pp. 163–166.) These fishes dwell in the uppermost parts of the deep sea, where some small amounts of light may penetrate. Many also migrate vertically into the epipelagic zone at night. The large eyes give them maximum light-collecting abilities in low-light areas. The large eyes are also presumably needed to detect the low light intensities of the light organs. Increased surface area is only one visual adaptation; these fishes also have enhanced "twilight vision" derived from the visual proteins (opsins) in their eyes that have the right spectral sensitivities for the type of light that is present in the ocean depths. Fishes dwelling in the deepest parts of the ocean (abyssalpelagic, hadalpelagic) show a different trend. Many of these animals have small eyes or lack eyes entirely (Figure 4.21). Because they dwell in permanent darkness, eyes are not necessary. Generally, fishes in depths of 2,000 m and above have eyes, often large; below 2,000 m, eyes are small, degenerate, or lost.

FIGURE 4.20 Large eyes in three mesopelagic fishes of the family Myctophidae. (a) *Myctophum punctayum*. (b) *Lampanyctus elongatus*. (c) *Diaphus metopoclampus*. (After *Aspects of Deep Sea Biology*, N. B. Marshall. Copyright 1954 Philosophical Library, New York. Reprinted by permission.)

Yet another adaptation in fishes is **tubular eyes**. (See Figure 4.13.) Fishes with peculiar tubular eyes are found in several families. Such eyes give the fishes an extremely bizarre appearance. In these fishes, the eye is a short black or silver cylinder topped with a spherical, transparent lens. In a few species, each eye has two retinas, one at the base of the cylinder and one on the wall of the cylinder. Tubular eyes tend to give the animal wide binocular vision so that a light flash will be seen by both eyes, making it more likely that the brain will receive adequate stimulation. Therefore, the tubular eye appears to be an adaptation to increase sensitivity to light. It also affords depth perception for eating small prey because the fields overlap.

Among the invertebrates, certain squids of the family Histioteuthidae exhibit another peculiar adaptation: They have one eye much larger than the other

FIGURE 4.21 Fish with reduced eyes, a gulper eel of the genus *Saccopharynx*. (Photo courtesy of Michael Kelly.)

FIGURE 4.22 The dimorphic eyes of the squid *Histioteuthis*. (Photo © Peter David/Getty Images.)

(Figure 4.22). Such squids also display numerous light-producing organs called photophores. Work by Young (1975) has demonstrated that these squids, which live in the mesopelagic zone at depths between 500 and 700 m, usually orient themselves with the arms and small eye downward and the large eye upward. In this position, the large eye gathers the faint light downwelling from the surface, while the smaller eye responds to light from other animals. It has been further demonstrated that another midwater squid, *Abraliopsis*, responds to downwelling light by turning on its ventral photophores just enough to match the downwelling light; hence, the animal becomes invisible when viewed from below (that is, the squid is prevented from casting a silhouette).

Food scarcity in the deep sea seems to be the reason for another series of adaptations. Most deep-sea fishes have large mouths—larger relative to their size than fishes of other areas (Figures 4.23 and 4.24; see also Figure 4.12). Furthermore, the mouth is often filled with long teeth that are curved toward the throat, an adaptation to ensure that whatever is caught does not escape. Even more bizarre is that, in some fishes, the mouth and skull are so hinged that the animals can open their mouths much wider than their own bodies, enabling them to swallow prey larger than themselves (Figures 4.24–4.27). Thus, food is not passed up simply because of size! Still other fishes, such as the angler fish (Ceratoidea), have responded to the scarce food by establishing themselves as traps, using a highly modified part of the dorsal fin as a lure (Figure 4.27). In this case, the lure is a luminescent organ. Other types of lures are found in many other fishes, such as the stomiatoids, and occur on barbels attached to the chin (Figure 4.28). One curious adaptation reported by McAllister (1961) is that many carnivorous fishes have

FIGURE 4.24 The deep-sea fish *Aristostomias scintillans*. Note the greatly expanded mouth cavity that allows the animal to swallow very large prey. The long barbel from the jaw region has a luminescent organ at the end. (Photo courtesy of Dr. Roger Seapy.)

black linings on their coeloms, so bioluminescence from ingested prey will not shine out and attract predators to them!

Because food is scarce and the density of organisms is very low, there is a potential problem of finding a mate in this vast, dark area. One adaptation to this problem is found in angler fish, *Ceratias*, in which the large individuals are all female. The males are tiny and are parasitic, attached to the body of the female (Figure 4.29). Since the males are always present to provide the sperm, a search for a mate is not necessary. Of course, the male first has to find the female, and this is presumably accomplished via olfaction.

The body size of deep-sea organisms is a peculiar paradox. Since food is scarce, it might be anticipated that most organisms would be small. Indeed, as regards fishes, this appears to be the case. Most fishes captured in the deep sea are smaller than their shallow-water relatives. As far as we now know, the deep sea contains few large fishes, although some have been photographed by the aforementioned baited "monster camera." It is also possible that the larger animals are present, but are able to avoid our capture methods. For example, no specimen of the giant squid genus *Architeuthis* has ever been captured in a net or observed by a camera or video in situ, but we know that they exist because moribund specimens have washed ashore and parts of the squid have been found in the guts of sperm whales. In contrast to the

FIGURE 4.23 *Linophryne,* a genus of fish with a very large mouth. (Photo courtesy of Michael Kelly.)

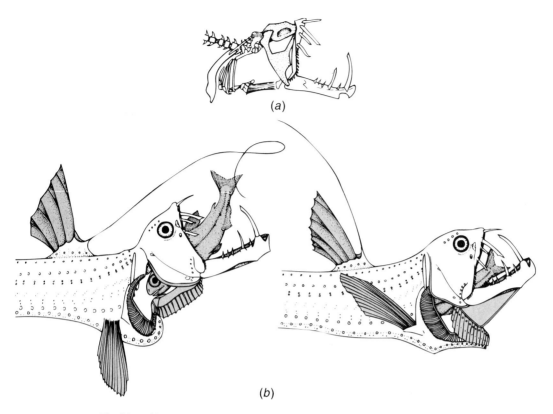

(a)

(b)

FIGURE 4.25 The hinged jaw in *Chauloidus*. (a) Anatomical construction of the skull. (b) Operation during swallowing. (a, Illustration from *The Open Sea* by Sir Alister Hardy. Reprinted by permission of Houghton Mifflin Company. Reprinted by permission. b, after *Aspects of Deep Sea Biology*, N. B. Marshall. Copyright 1954 Philosophical Library, New York. All rights reserved.)

FIGURE 4.26 Deep-water angler fish, *Melanocetus johnsoni*, before and after swallowing a larger fish. (Illustration from *The Open Sea*, by Sir Alister Hardy, Copyright © 1959 and renewed 1987. Reprinted by permission of Houghton Mifflin Company. All rights reserved.)

fishes, certain species of some invertebrate groups, particularly amphipods, isopods, ostracods, mysids, and copepods, attain a much greater size than their shallow-water relatives. This phenomenon of larger size with increasing depth is called **abyssal gigantism.** (See Figure 4.16.) The size that certain of these abyssal giants attain is remarkable. The isopod *Bathynornus giganteus* reaches 15 in. (42 cm), and the amphipod *Alicella gigantea* reaches 6 in. (15 cm).

Gigantocypris, the giant mesopelagic ostracod, has a diameter of several centimeters. The copepod *Gausia princeps* reaches a size of 10 mm, nearly ten times the size of most calanoid copepods.

Scientists currently disagree about the reasons for abyssal gigantism, but there are two main theories. The first attributes the condition to the peculiarities of metabolism under conditions of high pressure. The second suggests that the combination of low

FIGURE 4.27 A deep-sea angler fish, *Melanocetus johnsoni*. Note the extremely large mouth and the luminescent "lure" above the head. (Photo courtesy of Dr. Roger Seapy.)

FIGURE 4.28 Luminescent barbel of the stomiatoid fish *Idiacanthus* sp. (Photo courtesy of Norbert Wu/Peter Arnold, Inc.)

temperature and scarce food reduces the growth rates in these crustaceans but increases their longevity and the time it takes them to reach sexual maturity, such that a larger size is obtained.

The large size may also be the result of natural selection. Large size, long life, and delayed sexual maturity confer certain advantages on organisms in the deep sea, including the production of larger eggs and subsequently larger young, which can then feed on a wider range of food sizes. Thus, large size circumvents the need for special food in this food-short area. Large animals are also more mobile and can cover more area in search of both food and potential mates. Further, increased longevity means a longer period of sexual maturity; hence, these animals have more time to find mates.

Although these abyssal giants have captured much attention, they are rare. Most of the deep-sea benthic infauna (polychaete worms, crustaceans, and mollusks) are much smaller than their shallow-water counterparts. In fact, the small size of benthic invertebrates, and not gigantism, is the major characteristic of deep-sea fauna. Given the vast area of the deep sea, some giants should be expected, and similar examples of giant taxa can also be found in the less extensive shallow-water areas—for example, the giant clams found on tropical reefs.

The bottom substrate of most deep-ocean areas is a soft ooze. Benthic organisms inhabiting this soft material tend to have delicate bodies, long legs, or, in

FIGURE 4.29 Large female angler fish, *Linophryne argyresca*, with attached small parasitic male. (From "Deep Sea Angler-Fishes (Ceratoidea)," C. T. Regan & E. Trewavas, *Dana Report*, No. 2, 1932. Reprinted by permission of Oxford University Press, Ltd., and the Carlsberg Foundation.)

FIGURE 4.30 Adaptations to raise animals above the ooze. (a) Cirrate octopod. (b) A stalked crinoid. (c) A sea cucumber of the genus *Peniagone*. (a, courtesy of Dr. Clyde Roper. b, courtesy of Dr. H. Mullins. c, redrawn from Hansen, *Galathea Report*, Vol. 2.)

(a)

(b)

(c)

the case of sessile animals, long stalks to raise them above the ooze (Figure 4.30). Some fishes have long, narrow fins that serve the same function.

In general, as one goes deeper in the oceans, the anatomical organization of the animals becomes more simplified. This relationship is best seen in a comparison of the features of mesopelagic and bathypelagic fishes as outlined in Table 4.2. It is also reflected in the generally slow mode of life of most deep-sea creatures. Measurements of oxygen uptake on the deep-sea floor at several localities by Smith and Hinga (1983) gave results from 0.02 to 0.1 ml of oxygen per m^2 per hour, or more than 100 times less than that measured in shallow water. Part of this result is due to the reduced density of organisms, but part also reflects the low metabolic activity of the organisms that are present.

Biochemical adaptations to the deep sea have been difficult to ascertain in the past because of the lack of live animals to study. However, newer data on chemical composition suggest that profound changes may occur. Childress and Nygaard (1973) report that the water content of the body tissues of fishes increases with increasing depth, while the lipid and protein concentrations decrease. In other words, the animals become more like jellyfishes. In addition, Childress and Nygaard (1974) found that the protein content of crustaceans also decreased with depth. Presum-

ably, this decrease in both groups is related to the scarcity of food from which to produce the protein. Caloric content also decreases with depth.

A final set of adaptations in deep-sea organisms has to do with bioluminescent organs and will be discussed separately.

■ BIOLUMINESCENCE IN THE DEEP SEA

Marine bioluminescence is not confined to deep-sea organisms. Indeed, it is a widespread phenomenon in the sea. Many people have observed it in the form of "phosphorescent seas," which result from light production in the surface waters by untold millions of dinoflagellates. Whereas bioluminescence is widespread, it is in the deep-sea mesopelagic organisms that bioluminescence reaches its highest and most complex development. In the mesopelagic zone, as many as 90% of the organisms have the ability to produce light. The prevalence of bioluminescence in the deep sea and the large numbers of organisms that can produce light have been documented by photos and videos taken by remotely operated vehicles (ROV) and manned submersibles.

Bioluminescence is the chemical production of light by living organisms. The mechanism of light production is well known from studies of such terrestrial organisms as fireflies, and the same mechanism is

TABLE 4.2

Anatomical Differences in Mesopelagic and Bathypelagic Fishes

FEATURES	MESOPELAGIC, PLANKTON-CONSUMING SPECIES	BATHYPELAGIC SPECIES
Color	Many with silvery sides	Black
Photophores	Numerous and well developed in most species	Small or regressed in most species; a single luminous lure on the female of most angler fish
Jaws	Relatively short	Relatively long
Eyes	Fairly large to very large with sensitive pure-rod retinas	Small or regressed, except in the males of some angler fishes
Olfactory organs	Moderately developed in both sexes of most species	Regressed in females, but large in males of *Cyclothone* species and angler fish
Central nervous system	Well developed in all parts	Usually weakly developed system
Myotomes	Well developed	Weakly developed
Skeleton	Well ossified, including scales	Weakly ossified; scales usually absent
Swim bladder	Usually present, highly developed	Absent or regressed
Gill system	Gill filaments numerous, bearing very many lamellae	Gill filaments relatively few, with a reduced lamellar surface
Kidneys	Relatively large, with numerous tubules	Relatively small, with few tubules
Heart	Large	Small

Source: From *Deep Sea Biology: Developments and Perspectives* by N. B. Marshall, © 1980 Garland STPM Press.

FIGURE 4.31 Light organ patterns of some deep-sea squids. (a) *Nematolampus*. (b) *Pterygioteuthis*. (c) *Histioteuthis*. (d) *Ctenopteryx* male. (From *The Biology of Cephalopods*, eds. M. Nixon & J. B. Messenger, © 1977 Academic Press, Ltd.)

(a)   (b)   (c)   (d)

used by aquatic organisms. In the reaction, a substrate called luciferin is oxidized in the presence of the enzyme luciferase to produce a molecule that emits light. Luciferin and luciferase may differ from organism to organism, but the reaction is similar. The spectrum of color that is produced varies from species to species, but, in general, in the mesopelagic the wavelength ranges between 460 and 480 nanometers.

Not only are large numbers of deep-sea organisms capable of producing light, but they also have developed the most elaborate organs for doing so. These light-producing organs are called **photophores** and are particularly abundant in fishes and squids (Figure 4.31), but are present in other invertebrates as well (Figure 4.32; see also Figures 4.13, 4.24, 4.27, and 4.28).

The largest number of animals with photophores is found in the upper areas of the deep sea—the mesopelagic and upper bathypelagic zones. The incidence of organisms with bioluminescence decreases in the deepest parts of the sea.

Photophores in deep-sea organisms are of several types and range in structure from simple to complex. The simplest photophores consist of a series of glandularlike cells that produce the light or a simple glandular cup holding a bacterial culture that produces the light. (See also p. 498.) In either case, the source is surrounded by a screen of black pigment cells. The more elaborate photophores have one or more of the following in addition: lenses to focus the light, a color filter, and an adjustable diaphragm of pigment cells. Still other photophores are able to move by

FIGURE 4.32 A mesopelagic amphipod of the genus *Cystisoma* showing the transparent body and huge eyes. (Photo courtesy of Dr. Roger Seapy.)

muscular action. Some squids combine their photophores with overlying layers of skin that contain chromatophores, allowing them to alter the color or intensity of the light produced.

That the production of light is of considerable adaptive significance to its possessor is demonstrated by (1) its prevalence in the deep sea and (2) the complex anatomical organs present to produce it. It is also apparent that bioluminescence serves different functions in different animals. Several alternative hypotheses for the use of bioluminescence in the deep sea have been reviewed by Young (1983), who suggests that most are used primarily as a means of food capture or defense against predators.

Many organisms, particularly fishes and squids, have their photophores placed ventrally. For those animals living in the mesopelagic zone, where some

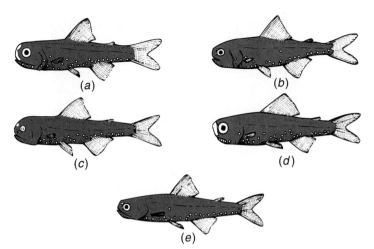

FIGURE 4.33 Species-specific light-organ patterns (open circles) in the lantern-fish genus *Diaphus*. (a) *D. macrophus*. (b) *D. lucidus*. (c) *D. splendidus*. (d) *D. garmani*. (e) *D. effnlgeus*. (From *Aspects of Deep Sea Biology*, N. B. Marshall. Copyright © 1954 Philosophical Library, New York. Reprinted by permission.)

light downwells from the surface, a visual predator looking up at them could see a silhouette. The photophores, producing a similar intensity as the downwelled light, would cause the silhouette to disappear and make the animal less vulnerable to predation. This is, in effect, camouflage. Since some organisms can adjust their downward-pointing photophores to the incident radiation level from above, such camouflage is a viable defense mechanism. Hatchetfish (Sternoptychidae) have even taken things a step further: On their sides, they have highly reflective surfaces of the white pigment guanine that function as mirrors. When a predator approaches from the side it will see light reflected from the hatchetfish that is of the same color and intensity as the background, leaving the fish virtually invisible.

Photophores might also be used to produce a "blinding" flash of light, which would momentarily startle a potential predator and allow disengagement and escape. Since such light production may also have the opposite effect of attracting more distant predators, however, this use of bioluminescence contains an element of risk. Some deep-sea squids (*Heteroteuthis dispar*) distract potential predators by shooting out a luminescent cloud rather than a black "ink" cloud as is done by squids in lighted waters. The cloud masks the escape of the squid.

As noted before, photophores might also be employed as lures to attract prey within range of the predator. This is undoubtedly the function of the elaborate light organ (esca) on the modified dorsal fin (illicium) of the angler fish and the "lures" of other midwater fishes. (See Figures 4.24, 4.27, and 4.28.)

Photophores may also be used for simple illumination of an area so that a predator may pick out its prey.

A final set of uses of bioluminescence involves recognition. In a perpetually dark area, such as the deep sea, the particular pattern of distribution of photophores over the body of the animal would be visible at a distance. Elaborate patterns of photophores occur on fishes in particular. The pattern is different for different species, thus allowing individuals to recognize their own species. Such recognition may be important in maintaining schools and in finding mates (Figure 4.33). In the latter case, certain fishes also have photophore patterns that differ between males and females, thereby assisting mate selection. It is interesting to note that one of the first scientists to observe living mesopelagic fishes in their natural habitat, William Beebe, remarked while looking out from a bathysphere that he could tell the different myctophid (lantern fish) species in the dark from their photophore pattern. If he could, it seems likely that the fish can also.

Other uses of photophores may occur, and more may become recognized with increased knowledge of this area, but the previously presented hypotheses represent the current most plausible ones.

In sum, paraphrasing Young (1983), we can say that in this homogeneous twilight area without physical barriers or structures for concealment and with abundant predators, life becomes a peculiar battle in which the judicious use of bioluminescence is of major importance in concealing one from predators, finding and attracting prey, and ensuring survival. No other area on Earth is quite like it.

Community Ecology of the Benthos

Although we have samples of organisms from the pelagic environment of the deep sea, the problems associated with capturing these organisms and with their ability to avoid nets make the available information on the community composition of the pelagic fauna incomplete at best and suspect at worst. The exception to this is the mesopelagic zone, from which many samples and observations have been taken. Considerably more is known about the deep-sea benthic communities because the sedentary and sessile nature of the animals makes their avoidance of nets less problematic. As a result, the samples taken and the observations of this area by deep submersibles and cameras have given us a reasonable idea of its composition. Ecologically, all benthic organisms throughout the world's oceans can be divided into two groups on the basis of position. The **epifauna** are benthic organisms that live on or are otherwise associated with the surface of the bottom. **Infauna** are organisms that live within the bottom substrate.

■ FAUNAL COMPOSITION

Nearly all the major animal groups have representatives in the deep sea, but the relative abundances of the groups vary. Crustaceans—particularly isopods, amphipods, tanaids, and cumaceans—are common in the deep sea. In the Atlantic abyssal area, they constitute 30–50% of the fauna. Polychaete worms are also abundant, making up 40–80% of the fauna in the Atlantic. Quite common in abyssal areas are sea cucumbers (Holothuroidea), usually of large size, and brittle stars (Ophiuroidea). Holothurians are often the most common organisms in deep-sea photographs (Figure 4.34). This, plus the fact that they form 30–80% of the biomass of living organisms taken in some trawls, suggests that they are a major component of the abyssal community. Since holothurians are deposit feeders, the abyssal oozes are likely an excellent food source for these animals—hence their abundance.

Starfish (Asteroidea), sea lilies (Crinoidea), and sea urchins (Echinoidea) are also present (Figures 4.35 and 4.36). Among the sponges, the deep sea is populated by the glass sponges (Hexactinellida), a group rarely found in shallow water. Sea anemones (Anthozoa), sea pens (Pennatulacea), sea fans (Gorgonacea),

FIGURE 4.34 A group of deep-sea holothurians of the genus *Scotoplanes* on the ocean bottom in 3,000 m. Holothurians are one of the major groups of large megafaunal organisms characteristic of sedimentary areas of the deep sea. (Photo courtesy of Dr. Barbara Hecker.)

hydroids (Hydrozoa), and zoanthids (Anthozoa) are the common representatives of the phylum Cnidaria (Figures 4.37 and 4.38). A number of fishes are also present, but their mobility makes it difficult to estimate their relative abundance. Figure 4.39 diagrammatically summarizes zonation changes and typical abyssal animals.

In the trenches (in the hadal zone), the available data suggest that there is a higher percentage of peracaridean crustaceans, polychaetes, and holothurians and a lower percentage of sea stars, sea urchins, and brittle stars in comparison with their percentage in the abyssal zone.

The major fishes of the bottom waters of the deep sea are rattails (Macrouridae), cusk eels, bythidids (brotulas), liparids (snailfishes), and certain eels (*Synaphobranchus*) (Figure 4.40). A marked characteristic of the deep sea is the relative lack of the more derived acanthopterygian fishes and the domination by the evolutionarily older or primitive fishes. In pelagic zones below the bathypelagic, fishes tend to be rare, while bottom-dwelling fishes are generally slow moving and sedentary.

One of the dominant features of the deep-sea benthos is the very small size and fragility of most individuals there. For example, in the exploration of the shallow waters of the continental shelf, a 1-mm sieve is often used to screen out infaunal organisms. Such a screen used in the deep sea would retain practically no organisms. Furthermore, the deep-sea animals

(a)   (b)

FIGURE 4.35 Two views of the sea bottom at 1,300 m on the Gay Head–Bermuda transect showing variation in the abundance of organisms. (a) Area barren of large benthic animals. The fish is a rattail. (b) Concentration of the echinoid *Phormosoma placenta*. (Photos courtesy of Dr. Fred Grassle; reprinted with permission from J. F Grassle et al., *Deep-Sea Res.*, Vol. 22, Pergamon Journals, Ltd., 1975.)

FIGURE 4.36 A typical deep-sea brisingid starfish. (Photo courtesy of Dr. Barbara Hecker.)

FIGURE 4.38 A sea anemone, *Liponema brevicornis*. (Photo courtesy of Dr. Barbara Hecker.)

FIGURE 4.37 Sea pens of the genus *Kophoblemnon* in 3,000 m of water. (Photo courtesy of Dr. Barbara Hecker.)

would fall apart unless they were carefully handled. The fragility of these animals is due to several circumstances, including the fact that the environment is virtually motionless and that calcium carbonate is difficult to deposit.

■ **COMMUNITY STRUCTURE**

In contrast with the shallow-water benthos or the epipelagic zone, the deep-sea sediment-inhabiting benthos contains few suspension feeders. According to Jumars and Gallagher (1982), no more than 7% of the species are suspension feeders because of the very slow water movement and the scarcity of food

FIGURE 4.39 Zonation of benthic animals with depth on the Atlantic continental margin of Europe. (From *The Face of the Deep*, B. Heezen and C. Hollister, © 1971 Woods Hole Oceanographic Institution.)

particles in the water. As a result, there is relatively little food to take out of the water, and a filter feeder would have to expend more energy creating a current than it would receive in food—a losing situation.

Carnivores also are less abundant among the deep-sea invertebrates, although their numbers are uncertain due to the difficulty of separating omnivores and scavengers from carnivores. Herbivores are present in low numbers, feeding mainly on plant remains, such as wood. The majority of animals, however—some 80% by number—are deposit feeders ingesting the organic-rich sediment.

Most attempts at analyzing the age structure of these benthic organisms through size–frequency graphs have given the same results no matter what time of the year they were sampled. Whereas such monotonous similarity in the graphs suggests that these animals reproduce throughout the year and have no seasonal breeding time, we do know that there is seasonal breeding for some deep-sea animals. As with their numbers, the life spans of most of the deep-sea animals are not known.

In an attempt to understand colonization and recruitment rates in the deep-sea benthos, Grassle and Morse-Porteous (1987) set down trays of defaunated sediments at two sites in the North Atlantic (1,800 and 3,600 m). Results indicated similar rates of colonization at the two depths, but at much slower rates than those observed in shallow waters. Furthermore, even after five years, the species composition and

FIGURE 4.40 A rattail fish (*Coryphaenoides filifer*) at 3,000 m. A brisingid starfish is in the foreground. (Photo courtesy of Dr. Barbara Hecker.)

density of the experimental trays were different from those of the surrounding natural area. The most common colonizers were polychaete worms of the families Spionidae, Capitellidae, and Sigalionidae. Experiments with organically enriched but azoic sediments demonstrated that organisms would preferentially settle in large numbers in the organically enriched areas or patches. Screening out predators resulted in high densities of colonizers. It appears that normally heavy predation, coupled with occasional escapes from predation, may explain the patchiness of the fauna.

■ DIVERSITY

Because of the low density of benthic populations, the small size of most invertebrates, and the paucity of adequate samples, the deep sea was considered a biological desert for many years. In the early 1960s, this notion was abandoned as sampling efforts intensified and methods of dredging and of processing samples improved. As a result, the desert idea was replaced by the opposite belief, first postulated by Sanders (1968), that the deep sea is a highly diverse community. The concept of species diversity is founded on the intuition that large groups of co-occurring species are characterized by a greater complexity of biological interactions. As noted in Chapter 1, species diversity has two important components: the number of species and their patterns of relative abundance. For example, if two areas have the same number of species, one area has a greater

diversity if it has a more even distribution of individuals among the resident species. The idea of a highly diverse deep-sea fauna is based entirely on relative abundance patterns. In the deep sea, not only is the total number of individuals sparse, but also each deep-sea species is commonly represented by only a few individuals. In other words, most deep-sea species are rare. This pattern had led to the popular contention that deep-sea diversity is as high as that of coral reefs, or among the highest in the world. In contrast, shallow-water assemblages are often dominated by one or several abundant animals, but there are also many rare species.

The suggestion that the deep sea is a highly diverse area had the salutory effect of generating even more interest in, and sampling of, the deep sea by other scientists, whose efforts led to a more critical evaluation of Sanders's work and the formulation of alternative hypotheses.

If one accepts the idea of a highly diverse deep sea, one has the choice of several conflicting hypotheses attempting to explain that diversity. The first is termed the **stability–time hypothesis**, originally suggested by Sanders (1968). This hypothesis says that high diversity occurs because highly stable environmental conditions have persisted over long periods and have allowed species to evolve that are highly specialized for a particular microhabitat or food source. Since most of the benthic animals are detritus feeders in the deep sea, this means they have become specialized for a particular narrow range of particle sizes. Such specialization is virtually unknown among deposit feeders elsewhere in the oceans and has not been demonstrated in deep-sea benthos. Moreover, the stability–time hypothesis has been falsified as an explanation of diversity patterns among the insects found in sugarcane fields and among the marine crustaceans inhabiting coral heads.

A second hypothesis is the **cropper** or **disturbance theory** suggested by Dayton and Hessler (1972). The basic premise of the theory is that organisms generally increase in numbers until they reach the limit of some resource that is in least abundance. At that point, competition occurs, and species are eliminated (in accordance with the principle of competitive exclusion). Since, in the deep sea, animals are scarce compared with the total amount of space available to inhabit, competition for space does not occur (except, perhaps, for animals living on hard substrates, which are rarer). The only other resource likely to be subject to competition is food, which as noted earlier, is in very short supply. The cropper theory suggests that none of the deep-sea animals are food specialists and that they are generalists feeding on anything they can engulf, not even distinguishing whether it is living or dead—hence the term cropper, as one that feeds indiscriminately on anything smaller than itself. High diversity is maintained because the intense "cropping" by all levels and sizes of animals prevents any species from building up its population to the point where it would be competing with another for food. High diversity is due, then, to intense predation at all levels, which allows a large number of species to persist, eating the same food, but never becoming abundant enough to compete with each other. A central argument against this theory is that heavy predation should result in the evolution of definite life-history features, including the production of large numbers of young, a short life span, and fast growth. As far as is known, however (Table 4.3), these are the exact opposite of the life-history characteristics found in the deep-sea organisms, which grow slowly, live long, and produce few offspring. However, if cropping results in differential mortality of age groups, either set of life history traits can, in theory, evolve.

A final hypothesis, which does adequately explain the original reports that diversity increases with increasing depth, is the **area hypothesis**. The idea is simple: Since species number is positively

TABLE 4.3

Summary of General Biological Characteristics of Deep-Sea Organisms

Reproduction and Development
Few eggs, large, rich in yolk
Slow gametogenesis
Late reproductive maturity
Reduced gonadal volume
Slow embryological development
Breed usually once (semelparous)

Physiological
Low metabolic rate
Low activity level
Low enzyme concentration
High water content
Low protein content
Small size

Ecological
Slow, indeterminate growth
High longevity
Slow colonization rate
Low population densities
Low mortality due to low predation pressure

correlated with area (i.e., the larger the area, the more species inhabit it), diversity is highest in the deep sea because it covers the greatest area. Both diversity and the area of the sea bottom decrease with decreasing depth. This correlation is significant because, although the actual numbers of species per unit bottom area have not been established, trends in relative abundance patterns can be related to the size of large biogeographical areas. The correlation, however, applies only to samples taken in the Atlantic Ocean. Workers in the Pacific Ocean have not found a general increase in diversity with increasing depth. On the contrary, both components of diversity (species density and relative abundance) are highest at *intermediate* bathyal depths rather than in the abyss.

Additional data and analyses of various data sets have produced a picture that, though still not completely resolving the diversity issue, has helped to clarify some aspects of it and put it more in perspective. Rex (1981) has demonstrated that, for some benthic organisms in certain areas, diversity does not continue to increase as one goes deeper

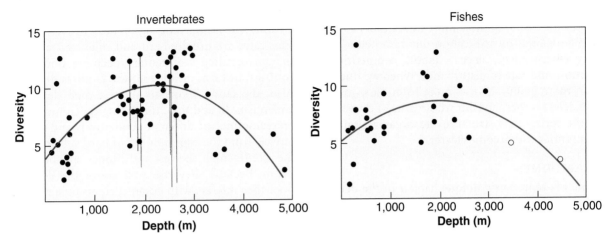

FIGURE 4.41 Patterns of species diversity in the deep-sea benthos based on samples from the northwest Atlantic Ocean. (With permission from "Community Structure in the Deep Sea Benthos," M. Rex. *Annual Review of Ecology and Systematics*, Vol. 12, © 1981 by Annual Reviews. www.annualreviews.org.)

into the ocean. In fact, he has shown that diversity is higher at intermediate depths and decreases both downward and upward to the continental shelf, giving a curve of parabolic shape (Figure 4.41).

The reasons for this parabolic diversity pattern and for deep-sea diversity in general remain unanswered. Data now available, however, suggest that there are various patterns of distribution of deep-sea benthos which change with depth and that, at different depths, the patterns result from differing sets of interacting factors. For example, it is now apparent from photographs and observations that the deep ocean floor is not a homogeneous, featureless plain, but has various mounds, holes, fecal deposits, and tracks. Furthermore, because of the stability of the water, these features persist for long periods and are therefore suitable "microhabitats" for various organisms. This heterogeneity can lead to increased diversity in the deep sea just as it does in the topographically diverse rocky intertidal areas (Chapter 6) and coral reefs (Chapter 9). In other words, the deep sea is "patchy," and organisms are distributed "patchily." If one samples several patches, one naturally gets higher diversity. Moreover, the patches exist on different scales of size and so do the organisms. This means that diversity also changes on different size scales; thus, Rex (1981) showed that for small animals maximum diversities occurred at greater

depths. This correlation is perhaps due to the longer persistence of microhabitat features or to the absence of competition with larger organisms. Predation and competition may also act differentially in the deeper waters, where the density of animals is low due to decreased energy input from above. Thus, predators are rare because their prey is rare or absent. In such cases, it may be that competition or microhabitat specialization is the dominant structuring force. In shallower waters, predators are more abundant due to an increased density of prey and may act strongly to structure the community, relegating competition or microhabitat specialization to a minor role.

Grassle and Morse-Porteous (1987) have suggested that high diversity in the deep sea is maintained by a combination of three factors: (1) the patchy distribution of organically enriched areas in a background of low productivity; (2) the occurrence of discrete small-scale disturbances (mainly from biogenic agents) in an area of otherwise great constancy; and (3) the absence of barriers to dispersal among species distributed over a very large area.

In sum, it appears that none of the early hypotheses are by themselves adequate to explain all deep-sea diversity patterns and that the observed patterns are the result of many factors interacting simultaneously, with some factors having more influence in certain areas, at certain depths, or at certain times. Among these more influential factors are microhabitat diversification due to a heterogeneous environment, energy input, competition, predation, and rate of disturbance. When we finally obtain more quantitative samples from more areas of the deep sea, perhaps this whole issue of diversity and the forces that structure the organization of communities will become clearer.

■ SEAMOUNTS

Seamounts represent a unique habitat in the deep-sea environment. They are typically extinct volcanos that rise abruptly, and usually individually (though often in chains or groups), from the deep ocean floor but do not reach the surface. They are found in all ocean basins. Seamounts are defined by oceanographers as independent features that rise at least 1,000 meters above the surrounding seafloor. Their peaks are usually hundreds to 1,500 or more meters below the surface. They therefore must be considered as within the deep sea. There are an estimated 30,000 seamounts in the world's oceans, but only a few have been studied (Fig. 4.42). Their interest to marine biologists and ecologists is due to several features. Because they are isolated from each other by deep ocean areas, they are like islands creating the same biogeographical interest. Their substrates are usually hard and thus contrast with the surrounding sedimentary deep-sea floor. This contrast, in turn, means that the fauna is different. Also, the fauna seems to show a high degree of endemism. In addition, the ecological communities on seamounts are different qualitatively and quantitatively from the communities of hard substrates on the continental shelves and slopes at the same depth. Because they rise well above the deep-sea floor they interact with ocean currents on a range of scales that can have effects on both the pelagic ecosystem above the seamount and the benthic organisms and communities living on the seamount. Finally, because they have high densities of fish, seamounts have been targeted in recent years by fishermen, and several populations of fishes have been overexploited.

Seamounts have effects on the ocean circulation that vary with the size, shape, and depth of the summit of the seamount, as well as with its position with respect to currents and neighboring seamounts. These effects are complex and are currently poorly understood. One such effect, examined by Boehlert and Genin (1987), is that the local

FIGURE 4.42 Map showing the locations of Seamounts in the world's oceans.
(From K. Stocks, 2003. Map of Sampled Seamounts. Seamounts Online.)

hydrographic conditions produced by seamounts somehow enhance primary productivity in the epipelagic waters above the submerged peak, which in turn leads to increased densities of zooplankton and thus explains the high concentrations of fishes on and around seamounts. Another suggestion for the high concentrations of fish on seamounts is that they are supported by feeding on diurnally migrating zooplankton trapped by the seamount as they descend. As Rogers (1994) suggests, it is also possible that the high densities of fishes on seamounts have more to do with the fishes' life histories, reproduction, and interaction with the benthic fauna of the seamount than with enhanced primary productivity or the trapping of zooplankton. Clearly, we need more information to assess this situation. What is apparent from current studies is that the planktonic communities over seamounts are often quite different from those in the surrounding deep water.

The benthic fauna of the hard substrates of seamounts is dominated by suspension feeders, such as various species of gorgonians, antipatharians, sponges, and true corals (Fig. 4.43). In some seamounts that come within 200 to 300 meters of the surface there are benthic macroalgae. Wherever

sediment accumulates, the epifauna seems dominated by the unusually large protozoans called xenophyophores, which apparently enhance species diversity. The sedimentary infauna is dominated by various polychaetes. Recruitment to seamount populations appears to be variable, with some evidence of input from outside. Figure 4.44 summarizes our current knowledge of seamounts.

Certainly, part of the recent interest in seamounts stems from the fact that they are known to support stocks of commercially important fishes and invertebrates. In their search for new fishing grounds in the 1960s, the fishermen of Russia, Australia, and New Zealand began to trawl on the seamounts, bringing up large numbers of fish of which the most significant were orange roughy (*Hoplostethus atlanticus*) and pelagic armourhead (*Pseudopentaceros wheeleri*). The most significant invertebrates collected were the precious corals (*Corallium*) and the black corals (*Antipathes*), both significant in the jewelry trade. The two fish species were quickly overexploited for a number of reasons, including a lack of understanding of the recruitment to the fishery, the longevity of the fish and its late age at maturity, low fecundity, and small geographical range. Not only were the fishes

FIGURE 4.43 Photo of suspension feeding octocorals from the Davidson Seamount. (Photo © NOAA/MBARI 2002.)

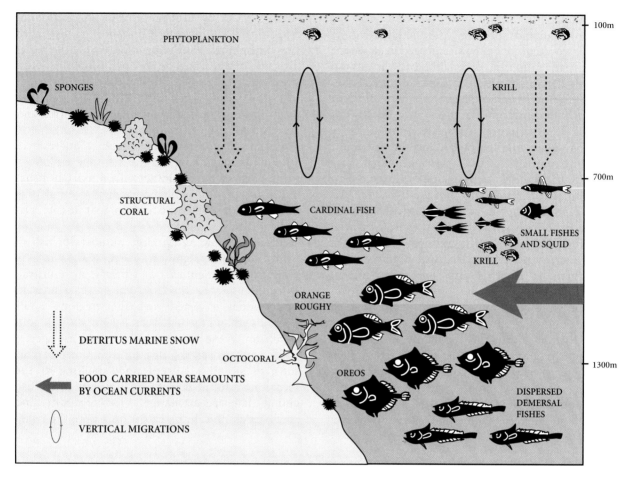

FIGURE 4.44 Diagrammatic representation of the organisms associated with seamounts.

overexploited, but the benthic communities of the seamounts were destroyed by the trawling gear. As a result of the devastation of a number of seamounts by fishing, there is now considerable interest in conserving them, particularly the ones that have not as yet been exploited.

■ LIFE-HISTORY PATTERNS

Life-history patterns with respect to the reproduction and development of deep-sea animals are difficult to generalize, primarily because we still lack data about those aspects of the life history of most these animals. Still, in accordance with the low metabolic rates and low activity rates of deep-sea animals, we find that perhaps most of them have slow or delayed sexual maturity, smaller gonads, and

slow embryological development. Some, in fact, may reproduce only once during their life span (semelparous). Mead et al. (1964) report hermaphroditism to be common among fishes.

In shallow-water areas and terrestrial areas, there are usually cyclic changes in some environmental parameter that initiate and control reproduction. As a result, reproduction is generally confined to a particular time of year. In the deep sea, where light is absent and nearly all environmental parameters are monotonously constant, it would appear that a cyclic pattern of reproductive activity should be absent. What little evidence we have so far, however, is mixed. Reproductive data from benthic organisms in the deep sea off San Diego, studied by Rokop (1974, 1979), showed that

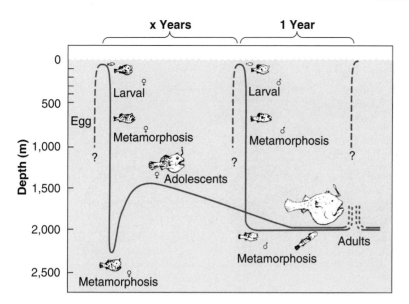

FIGURE 4.45 Life cycle of an angler fish. The fish spends its larval life at the surface and descends upon metamorphosis. (From *Aspects of Deep Sea Biology*, N. B. Marshall. Copyright 1954 Philosophical Library, New York. Reprinted by permission.)

reproduction is constant and continuous throughout the year and has no seasonal peaks in four bivalve mollusks. However, Lightfoot et al. (1979) found a distinctly seasonal cycle of reproduction in two other bivalve species from Rockall Trough in the Atlantic. Similar findings of both seasonality and its absence exist for various Crustacea, while continuous reproduction appears to be the pattern in deep-sea echinoderms. In some cases, seasonal reproduction has been linked to a pulse of particulate organic matter sinking from the euphotic zone after a bloom in the surface waters.

Since food is scarce, it might be suspected that most deep-sea organisms would produce few offspring, which would not spend time as larvae in the open water. (See Chapter 1, pp. 26–28, on larval strategies.) However, what information we have seems to suggest a mixture of strategies. In gastropod mollusks for which planktotrophy, lecithotrophy, and direct development can be readily inferred from the shape, sculpture, and number of whorls in the larval shell, the data suggest that planktotrophic development increases with depth. Bouchet and Waren (1979) have reported that, in the Atlantic at depths to 1,000 m, less than 25% of the gastropods had planktotrophic development, while below 4,000 m over 50% had planktotrophic development. Most of the remainder had direct development. Most fishes also have pelagic eggs and larvae.

Two general reproductive patterns exist in the deep sea. In the first, the early stages of the life history are spent in the lighted surface waters, and the juveniles migrate down to adult depths as they mature or metamorphose. Such is the pattern exhibited by certain deep-sea angler fishes (Figure 4.45), myctophids, stomiatids, many gastropod mollusks, and certain squids. In the other pattern, no migration occurs and the young spend their various stages in the same area as the adult. This pattern is exhibited by the unique cephalopod *Vampyroteuthis infernalis* and certain prawns. Table 4.3 summarizes the general biological features of deep-sea animals.

■ HYDROTHERMAL-VENT AND COLD-SEEP COMMUNITIES

Perhaps one of the most significant and exciting discoveries in deep-sea biology—indeed, in marine biology in general—occurred in 1977. At that time, a series of dives with the deep submersible *Alvin* was made in the Galápagos rift zone some 200 miles northeast of the islands. As reported by Ballard (1977), these dives to 2,700 m revealed a spectacular abundance of hitherto unknown marine animals living in and around four hot-water geysers in an otherwise barren ocean floor. The water temperature in these vent areas was 8–16°C, considerably above the normal temperature prevalent at the bottom at these depths (about 2°C). This warm water was very high in reduced sulfur compounds,

FIGURE 4.46 Chemosynthesis in hydrothermal-vent communities. (a) Sulfur bacteria in the vents use the energy in hydrogen sulfide gas to build complex organic molecules and grow. The source of carbon is carbon dioxide gas, which is also dissolved in the water. The bacteria either are eaten by heterotrophs, live symbiotically with the larger animals (to which they furnish organic material), or die and form particulate organic matter (detritus), which in turn is used as a source of food by heterotrophs. (b) Comparison of photosynthesis and chemosynthesis. (a, after "Analysis of Marine Ecosystems" A. R. Longhurst, © 1981 Academic Press, Ltd.)

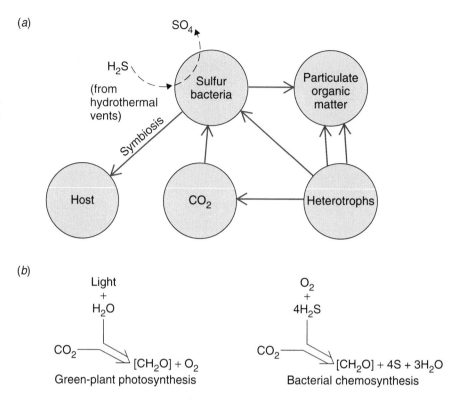

primarily hydrogen sulfide (H_2S), which was found to be used as an energy source by bacteria at the base of the food chain, often as symbionts in the larger organisms (Figure 4.46; see also Chapter 6 on sulfide bacteria in mud shores).

Since that time, further exploration with submersibles, remotely operated vehicles, and towed camera systems has revealed similar vent communities, at depths ranging from 1,500 to 3,200 m, at several additional locations in the eastern Pacific Ocean (Figure 4.47). In addition, other hydrothermal-vent communities have been reported from the western Pacific, in the Marianas Back Arc Basin (3,590–3,660 m), in the North Fiji and Lau basins (1,990 m), at several sites on the mid-Atlantic Ridge and in a subduction zone off eastern Japan. Tunnicliffe et al. (1998) report that about 30 hydrothermal sites have now been the subject of study by biologists (Figure 4.47). That we have not discovered all vent systems on the planet is demonstrated by the discovery by Van Dover et al. (2001) of a vent field in the Indian Ocean on the Central Indian Ridge and a similar discovery of vents on the Gakkel Ridge system in the Arctic Ocean by

Edmonds et al. (2003). Up until recently, all the hydrothermal vents that were discovered had been found on the ridge systems marking the boundaries between tectonic plates. However, Van Damm (2001) has reported a new vent system in the Atlantic that is off of the Mid-Atlantic Ridge and that is very different from all the others, at least partly because it vents cooler fluids (40–75°C) and the waters are alkaline, not acidic.

Similar communities of organisms have been found that depend, at least in part, upon chemoautotrophic production associated with the emission of reducing chemicals that are not always associated with the edges of tectonic plates. These communities are called **cold-seep** communities and have been found in 24 deep-sea locations in the Atlantic, the eastern and western Pacific, and the Mediterranean Sea at depths between 400 and 6,000 m (Figure 4.48). The largest number of sites, 13, are found off North America. According to Sibuet and Olu (1998), seeps are found on both active and passive plate margins. Known American seep sites include the Gulf of Mexico at the base of the Florida Escarpment at 3,200 m, on the continental slope off

FIGURE 4.47 Distribution of major vent fields around the world. Each field may include several sites. IND = southeast Indian Ridge; LFM = Lau, Fiji, Manus, Woodlark, and Lihir sites; MBJ = Marianas, Bonin, and Okinawa sites; NEP = Explorer, Juan de Fuca, and Gorda ridges; nEPR + 9° through 21 ° N East Pacific Rise; GAL = Galápagos Rift; sEPR = 17° through 25 ° S; ATL = mid-Atlantic Ridge sites; IO = Indian Ocean Ridge site. (From "A Biogeographical Perspective of the Deep-Sea Hydrothermal Vent Fauna," Tunnicliffe et al., *Adv. in Marine Biology*, vol. 34, p. 360, 1998, AP London.)

of Louisiana at 600–700 m, in Monterey Bay at 900–3,500 m, and two sites in the Atlantic, one off of eastern Canada and the other off of North Carolina. Undoubtedly, more will come to light as deep-sea exploration continues. Some of these cold seeps are associated with "cold" hypersaline brines or other hydrocarbon seeps.

All these vent areas and cold seeps display an astonishing assemblage of large (megafaunal) animals forming a unique association in the deep sea. Truly they are "oases" in an otherwise low-density, low-productivity area. Each hydrothermal-vent system seems to have an ecologically similar set of dominant species, though not always a similar set taxonomically. Tunnicliffe (1992) has reported a total of one new class, three new orders, and 22 new families of animals from the vent systems. Eighty-two percent of the 443 species recovered

from the vents since their first exploration in 1977 appear to be endemic to the vents, and most were previously unknown (Figure 4.49). The dominant animals, by size, at these vents usually include one or more of the following: large clams, mussels, and vestimentiferan worms (Figures 4.50, 4.51, and 4.52). The last of these belong to a special deep-sea taxon formerly known as the Pogonophora and are characterized by the complete lack of a digestive tract as adults. (Pogonophora used to be classified as a phylum, but has now been reduced to a family of polychaete worms.) Vestimentiferans appear absent from the West Pacific vent systems, where the dominant symbiont-containing animal is a newly discovered species of gastropod mollusk, *Alvinoconcha hessleri*. Other large animals associated with vents include crabs, a large number of snails, a variety of polychaete worms, an odd

FIGURE 4.48 Locations of cold seeps (black circles) and whale carcasses (black whale symbols) in the world's oceans. Only the better known ones are illustrated. Jps = Shallow Japan; Jpd = Japan Trench; Ale = Aleutian Trench; Ore = Oregon Margin; Mon = Monterey Canyon; Per = Peru Margin; Lsp = Louisiana Slope; Fla = Florida Margin; Bar = Barbados; Lfn = Laurentian Fan. (From "A Biogeographical Perspective of the Deep-Sea Hydrothermal Vent Fauna," Tunnicliffe et al., *Adv. in Marine Biology*, vol. 34, p. 383, 1998, AP London.)

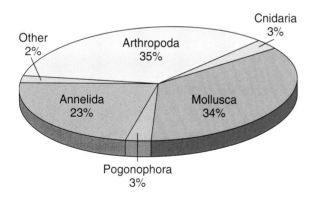

FIGURE 4.49 Overall composition of the total invertebrate fauna, as recorded from the vents. (From "A Biogeographical Perspective of the Deep-Sea Hydrothermal Vent Fauna," Tunnicliffe et al., *Adv. in Marine Biology*, vol. 34, p. 381, 1998, AP London.)

benthic siphonophore, enteropneust worms (called "spaghetti worms" from their appearance in photos; Figure 4.53), and five new species of fishes. The faunal components of the vent systems are highly adapted to live in these areas with highly toxic chemicals—a fact which suggests that the suite of adaptations necessary for exploiting the vents represents a considerable barrier to evolving other groups to colonize such habitats.

The vent megafauna differs among the explored sites. In the Atlantic there are no vestimentiferan worms and many of the sites are dominated by swarming shrimps (*Rimicaris*) and mussels (*Bathymodiolus*). In the Northeast Pacfic, the vent sites are dominated by vestimentiferans of the genus *Ridgea*, polychaetes, and small gastropods. To the south, the East Pacific rise vents are dominated by vestimentiferan worms of the genus *Riftia*, mussels (*Bathymodiolus*), polychaetes, and clams. At the Galápagos—the first vent system discovered and the most studied—the dominants are the tube worms (*Riftia*), large clams

FIGURE 4.50 A vent clam field dominated by the giant white clam *Calyptogena magnifica*. (Photo © Woods Hole Oceanographic Institution.)

FIGURE 4.51 Large mussels (*Bathymodiolus*) around a hydrothermal vent. (Photo © Woods Hole Oceanographic Institution.)

(*Calyptogena*), mussels (*Bathymodiolus*), and large numbers of polychaetes and gastropods. In the western Pacific, the vents of the Marianas Back Basin have a fauna distinct from that of the East Pacific. Gastropods with chemoautotrophic symbionts are abundant, along with stalked barnacles, shrimp, and certain crabs. Vestimentiferans are absent. The vents in the Fiji area are also different and are dominated by the large gastropod *Ifremeria nautilei*, which contains chemoautotrophic bacteria. These vents also have mussels, as well as tube worms distinct from those in the East Pacific. The Indian Ocean vents are dominated by swarming shrimp (*Rimicaris*), anemones (*Marianactis*), and a strange scaly-footed gastropod. The newly discovered vents in the Arctic Ocean have not yet been characterized.

The cold-seep and hypersaline brine communities are less well known than the hypothermal-vent communities. According to Sibuet and Olu (1998), the dominant seep species are bivalve mollusks of several families, pogonophoran worms, and certain sponges. Of the currently known 211 species from seeps, a large majority are endemic, and only 13 species are shared between vents and seeps. Diversity at some of the seep sites also appears to be higher than at vents. Also, whereas the vent systems seem to depend primarily on the utilization of sulfides for energy, the cold-seep areas may depend on sulfide or methane (Figure 4.54).

FIGURE 4.52 Giant vestimentiferan tube worms, *Riftia pachyptila*, near a Galápagos thermal vent. (Photo © Woods Hole Oceanographic Institution.)

The final, and probably the least known, of these deep-sea sulfide communities was discovered in 1987 when the submersible *Alvin* chanced upon a 21-m whale skeleton at 1,240 m in the Santa Catalina Basin off southern California. Subsequent study of this skeleton revealed chemoautotrophic bacteria and certain other animals associated with vent systems on the outside of the bones, and vesicomyid clams (like the *Calyptogena* at the vents) in the sediments around the bone. In this case, the sulfide source was produced by anaerobic bacteria inside the whale bones decomposing the large lipid content of the bone (Figure 4.55). As the sulfide diffused out of the bone, it provided the energy source for the bacteria. Although whales have been much reduced throughout the world's oceans, Smith (1992) estimates that such sulfide-rich whale falls may have an average spacing of one per 25 km in the North Pacific and may give credence to the hypothesis that these falls may be the stepping-stones that permit the sulfide-based communities to disperse over the vast distances between the vent systems.

All of the organisms of the sulfide or methane communities associated with seeps or vents are confined to an area within only a few meters of the sulfide or methane source. Both vents and seeps may be separated from other such associations by hundreds or thousands of kilometers. The latter fact has aroused the interest of marine biologists who are curious not only about how such communities persist, but also about how they disperse to find the vents; they are virtually looking for a needle in a haystack. Although we still have much to learn about these communities, we have made much progress since their discovery.

All vent systems and cold seeps depend on the primary productivity of **chemoautotrophic bacteria** to survive. Potential energy in the form of reduced sulfur compounds, such as hydrogen sulfide, is spewed from the hot vents. Cold seeps ooze either sulfides or, in some cases, methane or other hydrocarbons. (See

FIGURE 4.53 Mats of enteropneust worms, *Saxipendium coronaturn*, called "spaghetti worms," draped over rocks around hydrothermal vents. (Photo © Woods Hole Oceanographic Institution.)

Utilization of Sulfides

Utilization of Methane

FIGURE 4.54 Diagrammatic representation of the utilization of sulfides by clams and tube worms and of methane by mussels and sponges. The activity of both methanotrophic bacteria and sulfate-reducing bacteria in the sediments results, respectively, in the oxidation of methane from pore fluid to carbon dioxide and in the reduction of seawater sulfate to sulfide. (From *Deep Sea Research II*, Vol 45, p. 532, 1998. Copyright 1998 with permission from Elsevier Science.)

Figure 4.54.) These reduced compounds are oxidized by the chemoautotrophic bacteria to release energy, which they use to form organic matter from carbon dioxide. (See Figure 4.46.) Since the vent water is anoxic, the bacteria depend on the surrounding cold ambient water for the oxygen necessary to produce the organic material. Because sulfide is rapidly oxidized by oxygen in the ambient water and vent waters are rapidly diluted, sulfide does not persist in the water column for any distance from the vents. This is why the organisms that require both oxygen and sulfide must live in a very narrow volume where vent water and ambient water begin to interact. Productivity is very high. Karl et al. (1984) measured production rates of 19 μg C/g/h at one vent. The primary consumers are the various animals that filter

FIGURE 4.55 Diagrammatic representation of the chemical changes occurring in whale bone. (After Whale Falls, C. Smith, 1992, *Oceanus*, Vol. 35, No. 3, p. 77. Copyright © 1992 Woods Hole Oceanographic Institution.)

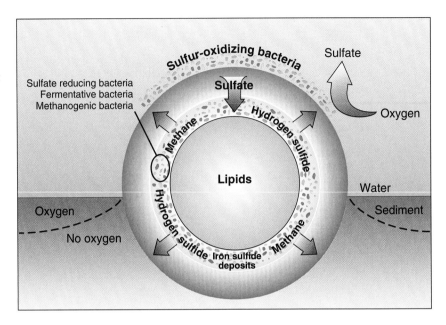

the bacteria from the water, graze the bacterial film from the rocks, or are locked in a symbiotic association with the bacteria. (See Figure 4.54.) In addition to the chemoautotrophic bacteria, there are other free-living microorganisms in the vent areas, including bacteria suspended in the vent plumes, suspended in the warm water outflows, and clinging to the rock surfaces as mats.

The animals that form a symbiotic association with the chemoautotrophic bacteria in the vent systems are of particular interest. The large vestimentiferan worm *Riftia pachyptila*, and the largest clam, *Calyptogena magnifica*, both contain symbiotic chemosynthetic bacteria in their tissues. *Riftia* has no digestive tract as an adult, and there is compelling evidence that it obtains all its nutrition from the bacteria. Although the clam retains its ability to filter feed, it, too, apparently derives most of its nutrition from the symbiotic bacteria. What is curious about this kind of symbiotic system is that the bacteria must have a reduced sulfide source in order to function. According to Arp and Childress (1983), this is brought to them by the blood of *Riftia* and *Calyptogena*. (See Figures 4.50 and 4.52.) Under normal circumstances, the reduced sulfides would poison or kill a typical aerobic animal, since sulfides bind to the cytochrome system, interfering with oxygen uptake and, therefore, respiration. The bloods of *Riftia* and *Calyptogena*, however, contain a protein factor that binds to the sulfide to transport it, freeing

it again in the presence of the bacteria that take it up. This mechanism is assumed, since there is no inhibition of oxygen uptake in the presence of the sulfide. How other vent animals deal with sulfide is not yet known. In the cold seeps and brine pools, by contrast, the dominant mussels (*Bathymodiolus*) have a methane-based symbiosis. The seeps and brine pools are rich in methane, which is taken up by the gills of the mussels, where intracellular bacteria oxidize it and provide energy for both the mussels and the bacteria. (See Figure 4.54.)

Hessler and Smithey (1983) have noted that vent animals are distributed with respect to water temperature in the vent systems (Figure 4.56). Water temperature in these systems is quite variable. Water that rises directly to the surface following contact with the hot subsurface rock can exit at 400°C, forming "black smoker" chimneys. No large animals are associated with these extreme temperatures. Other vent water may mix with cooler water on its way to the surface, giving rise to temperatures that vary from near ambient to 50°C. Most of the vent animals are associated with water that is below 30°C and with hydrogen sulfide levels of 400 μm or less. It is curious that, rather than being exposed to water that is at the average of the mixing of the vent and ambient temperatures, most vent animals appear to live where they encounter discrete parcels of the two types of water; thus, they must be adapted to considerable temperature change that occurs rapidly. As noted

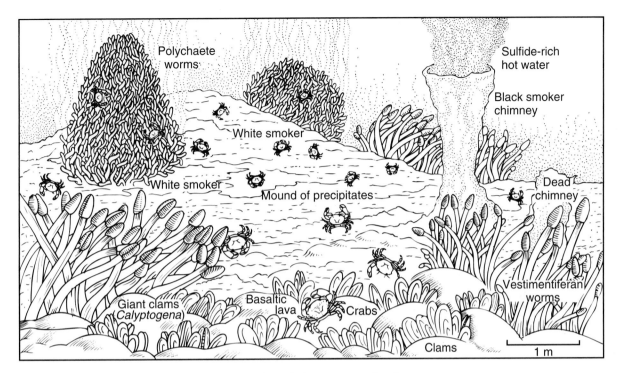

FIGURE 4.56 A diagrammatic representation of a typical vent community, showing the presence of the larger macrofaunal organisms. (Modified from "The Geology of Deep Sea Hot Springs," K. M. Haymon and K. C. McDonald, *American Scientist*, Vol. 73, No. 5, pp. 441–449, 1985. Copyright © 1985 American Scientist. Reprinted by permission.)

above, most vent animals appear to be restricted to the vent areas, and we do not know what keeps them from invading the adjacent deep-sea area, nor do we know what prevents the normal deep-sea fauna from invading the vent systems. However, it seems reasonable to speculate that the vent animals—notably those with symbionts—are restricted to the vents by the need for sulfide to fuel their metabolism. Temperature may also be a factor.

Childress and Fisher (1992) suggest that the vent systems are highly unstable environments that can vary in structure and in water flow over time periods as short as weeks or months. There are also long-term changes brought on by a reduction in the flow of vent water and the eventual cessation of the flow. In addition, vent communities may be destroyed by the outpouring of fresh lava. Just such an event was reported by Tunnicliffe (1992) for a site on the East Pacific Rise. When that happens, the vent community must be reestablished on the new lava. (See page 185.)

According to Grassle (1985), the life span of individual vents is only in the tens of years, and scientists have observed inactive vents with the remains of vent animals. This observation suggests that these vent species have life spans also measured in decades. Recent estimates of the life span of *Calyptogena* by Grassle (1985) gave the ages of the oldest at 30 years. However, the growth rate and size reached by *Calyptogena* varies with the vent site, suggesting that the growth rates among vent animals may be more variable than we think. Given these circumstances, it would seem that the vent animals would experience strong selective pressure to grow rapidly to sexual maturity, produce many young, and have an efficient means of dispersal. What we now know suggests that the dominant animals do grow much faster than surrounding deep-sea animals that are away from the vents, but we are uninformed about the reproductive potential of most. In the case of dispersal, we are similarly without much data, except as regards the Mollusca. Among the gastropod mollusks at the vents, a strange situation seems to prevail: Most species seem to lack planktotrophic larvae and have nonplanktotrophic development; hence, these animals have low dispersal. It is an enigma how such development could facilitate the location of widely separated vent systems by the larvae, but somehow it must, because Grassle (1985) reports that close relatives of present-day vent animals have existed for 200 million years.

Certain areas of the hydrothermal vents in the Galápagos have been the object of follow-up studies, reported by Childress (1988), that have demonstrated how quickly these areas can change. The particular area called "the Rose Garden," when discovered in 1979, was dominated by large numbers of vestimentiferan worms and contained significant, but smaller, numbers of large clams and mussels. When revisited in 1985, a little over five years later, the number of worms had declined markedly, while the number of mussels had increased dramatically, to the point where these animals now dominated the area. The clams were also abundant. Among the associated fauna, filter feeders had generally declined in number, siphonophores and enteropneusts were virtually absent, and mobile scavengers and carnivores, such as galatheid crabs and gastropod mollusks, were more abundant. According to Hessler et al. (1988), these striking community changes seemed to be a result of two factors. The first was a change in physical conditions that was manifested mainly by a decline in the vent water flux, which led to a decline in the vestimentiferans and probably also the filter feeders, but favored the mussels. The second was continuing recruitment into the area, which favored the crabs, mussels, and clams. It is also possible, however, that the vestimentiferan population declined as that of the mussels grew, effectively cutting off the vestimentiferans from their sulfide source by absorbing the sulfide before it could get to the more distant absorbing crowns of the worms. The latter scenario does not require any change in the water circulation. Interestingly, at the present time (2003) this "Rose Garden" community has vanished and a new one called "Rose Bud" is starting up.

It would appear that mortality, at least among the vestimentiferan worms, is very high. Tunnicliffe et al. (1990) have shown that, for the vestimentiferan *Ridgeia pescesae* on the Juan de Fuca vent system, 44% of the animals were removed over a 26-day period due to collapsing sulfide/sulfate chimneys on which they lived or to predation by fish.

As a result of observations at the Rose Garden, a general pattern of community change over time

has been suggested by Hessler et al. (1988) for the Galápagos hydrothermal vents. Beginning with an established vent system dominated by vestimentiferans, clams, and mussels, one sees the vestimentiferans gradually diminish, while the clam and mussel populations grow. This trend may be the result of a decrease in water flux or sulfide concentration or the result of competition with the more efficient mussels. As flow diminishes in the vents, suspension feeders tend to decline, as do the clams, leaving the mussels as the most persistent organisms. Since the same suite of dominant organisms is not present at all hydrothermal vents, the generalizability of this successional scheme is unknown.

Recently, it has been possible to undertake manipulative field experiments at the hydrothermal vents in order to gain a more thorough understanding of how the communities are structured. Working at the hydrothermal vents on the East Pacific Rise, Micheli et al. (2002) employed caged and uncaged basalt blocks to assess the role of predation in structuring the vent communities along a gradient of decreasing vent fluid flux. On the caged blocks that excluded predatory fishes, crabs, and cephalopod mollusks for a period of 8 months, the researchers found an increased abundance of small gastropods and amphipod crustaceans and a decreased abundance of juvenile vestimentiferan worms, tube-dwelling polychaetes, and mussels. Predation effects were most significant nearest the vents themselves. Stomach content analyses of the most abundant predator, a zoarcid fish *Thermarces cerberus*, revealed that it fed primarily on vent snails and amphipods, thus substantiating the predator exclusion caging experiments. The results suggest that large mobile predators can influence the structure of vent communities, both directly, by reducing the populations of small grazing gastropods, and indirectly, by reducing gastropod grazing which in turn reduces the recruits of the vestimentiferans, polychaetes, and mussels. Since the different vent systems have different arrays of species these kinds of manipulative experiments will also have to be done at more of them in order to assess the relative importance of various biotic and abiotic factors in structuring the vent systems around the world.

In 1991 the discovery of newly formed hydrothermal vents in an area of recent volcanic activity on the East Pacific Rise area offered a unique opportunity to scientists to follow the changes in the establishment of a vent community from bare volcanic rock. In 1992 an in situ observatory was emplaced to monitor the megafaunal succession and the chemical changes in the vent fluids. Chemical sampling and video and photo documentation were carried out once each year from 1992 to 1995. According to Shank et al. (1998), immediately after the 1991 eruption the area had extremely high hydrogen sulfide concentrations and the surface of the new lava was blanketed with microbial mats over which a fauna of mobile amphipods, copepods, octopods, crabs, and shrimp swarmed. Within a year the bacterial mat was reduced by more than 50%

and a vestimentiferan worm *Tevnia jerichonana* began to settle in the area. In 1993 the hydrogen sulfide concentrations were cut nearly in half and the large vestimentiferan worm *Riftia pachyptila* colonized the site, overgrowing the *Tevnia* population. In 1994 the sulfide concentrations decreased, the *Riftia* population continued to grow, and mussels (*Bathymodiolus thermophilus*) appeared. By 1995, galatheid crabs and serpulid worms became abundant, the hydrogen sulfide continued a slow decrease in concentration, and the mussels colonized the *Riftia* tubes (Figure 4.57). Between 1994 and 1996 there was a severalfold increase in the number of species in the new vent community.

(a)　　　　(b)　　　　(c)

(d)　　　　(e)

FIGURE 4.57 Temporal sequence of faunal community development at the East Pacific Rise. Arrow in c, d, and e indicates comparable corresponding location on the seafloor in successive years. (a) No vent megafaunal organisms were present when vigorous hydrothermal activity was initiated, and profuse microbial material expelled from a fissure within 15 m of the subsequent location of Biomarker #9, view is to the southeast. (b) By March 1992, the fissure was colonized by an extensive population of *Tevnia jerichonana* next to Biomarker #9, with *Riftia pachyptila* noticeably absent (view is to the northeast). (c) By December 1993, numerous *Riftia pachyptila* had settled and rapidly grown to form a dense thicket, engulfing the existing *T. jerichonana*. (d) Between December 1993 and October 1994, continued rapid colonization and growth by large numbers of *Riftia pachyptila* dramatically increased the density, lateral extent, and vertical growth within this colony. (e) By November 1995 the colony was now occupied by over ~2000 *R. pachyptila*. About half of the *T. jerichonana* assemblage documented in 1992 was alive in 1995 at the base of the *R. pachyptila* tubes (*T. jerichonana* not visible in this image). The staining of worm tubes with a rust-colored, ferrous-oxide precipitate was coincident with increased concentrations of iron in the diffuse vent fluids supporting this colony. (Photos and caption from "Temporal and Spatial Patterns of Biological Community Development at Nascent Deep-Sea Hydrothermal Vents (9°50' N, East Pacific Rise)," T. M. Shank, et al., *Deep-Sea Research II* 45, 1998, pp. 465–515.)

Shank et al. (1998) hypothesize that if there is no more volcanic disturbance, then, in 5 to 10 years, the vestimentiferans will be replaced by mussels and vent clams, the number of suspension feeders will decline, and that of carnivores will increase. They also suggest that the successional sequence described here is a generalized pattern to be expected at new vents along the East Pacific Rise.

Although most attention has been directed at the vestimentiferan worms and the large clams and mussels because of their size and symbiotic relationship with the chemoautotrophic bacteria, note that most vent species do not have symbiotic bacteria and feed by grazing on bacteria or by filtering. At our current state of knowledge, it is not possible to quantify a food web for the vents, but our knowledge is sufficient to outline a trophic structure (Figure 4.58). Such a structure is characterized by the simplicity of the relationships and limited by the supply of chemicals required for the chemoau-

totrophic reactions which supply most of the energy that fuels the system. There are four pathways for carbon and energy flow within the food web. The first is the export of production in the form of bacteria and organic particulate matter through the vent plume that is found at levels of 100–500 m above the benthic vent community and is carried away by currents. This production is fed upon by various planktonic organisms, which occur in abundance in the plume. The second path is a suspension-feeding loop in which the bacteria, particulate organic matter, and perhaps some indigenous plankton are removed from the water by resident vent filter feeders. A third trophic pathway is that of grazing. The chemoautotrophic bacteria and other bacteria form mats over the bottom and are grazed upon by numerous benthic animals, among which the gastropod mollusks are the most abundant. The final pathway is through the symbiotic chemoautotrophic bacteria that reside in the tissues of the

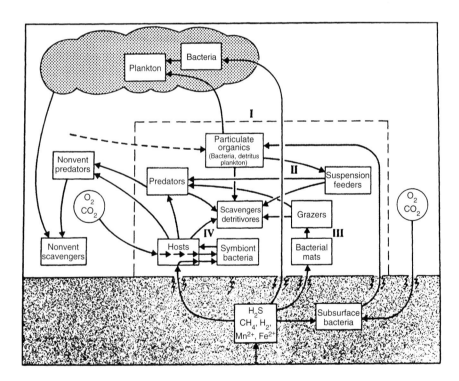

FIGURE 4.58 The food web of a hydrothermal vent community. The four distinct carbon-energy pathways are indicated by roman numerals. I, Export production of bacteria and particulate organic material in the vent plume. II, Suspension-feeding pathway, involving indigenous vent filter feeders. III, Grazer–scavenger pathway, involving bacterial mats. IV, Symbiotic chemolithoautotrophic bacterial pathway, involving large invertebrate hosts. (After "Biology of Hydrothermal Vents" V. Tunnicliffe, 1992, *Oceanographic Marine Biology Annual Reviews*, Vol. 29, p. 341.)

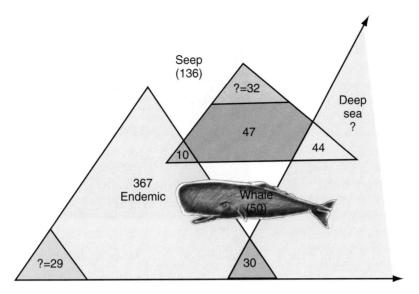

FIGURE 4.59 Overlap of species among sulfide-rich habitats and the deep sea. Vent and seep habitats are represented by triangles overlapping with each other and with the deep sea. Species numbers in the triangles represent the numbers of endemics; the numbers of overlap species are also indicated. (A question mark means that the species is not yet identified). The whale represents the species found on decaying whale bones. Of the total whale bone species, two are known from whale bones, seeps, and vents; one is found at bones, vents, and the deep sea; four are known from vents and bones; and one is known only from bone and seeps; the rest are endemic to bones. (From "A Biogeographical Perspective of the Deep-Sea Hydrothermal Vent Fauna," Tunnicliffe et al., *Adv. in Marine Biology*, vol. 34, p. 393, 1998, AP London.)

dominant vestimentiferan worms, clams, and mussels. This pathway has the greatest biomass in the vent systems. In turn, these dominant animals are eaten by various predators both in and out of the vent community. Scavengers receive input from all the pathways, and some of the vent animals may be involved in more than a single pathway.

Biogeographically, the three chemosynthetic-based communities display a very high degree of endemism and very low overlap with each other or with the surrounding deep-sea benthos (Figure 4.59). Seeps have a greater depth distribution than vents and also show a decrease in species richness with depth. Single-site species richness is higher at seeps then at vents and may be due to the longer-term stability of the seeps. Seeps tend to have more nonseep species than vents and hence more links with the surrounding communities. At the same time, seeps have more symbiont-containing species than the vents have.

Midwater Community Ecology

Because of the increased difficulty in sampling pelagic deep-sea organisms due to net avoidance and migration, the lack of observations on these animals from deep submersibles, their avoidance of light, and their small size, we know less about the ecology of midwater animals than we do about the deeper but sedentary and sessile benthos.

■ DEEP SCATTERING LAYERS AND VERTICAL MIGRATION

Early in World War II, when the use of echo sounders was being tested within the oceans to detect submarines and the bottom, the devices (now known as sonar) regularly began to pick up echoes. They came, not from the bottom, but from sources in the water column itself, hundreds or even thousands of meters above the known bottom. These areas of sound reflection in midwater were termed

FIGURE 4.60 A sonogram recording showing the presence of deep scattering layers rising toward the surface. (Photo Courtesy of Dr. Roger Seapy.)

Surface

DSL

Ocean floor

deep scattering layers (DSLs). The depths of the DSLs varied. During the day, there were often two or three layers varying in depth from 200 to 700 m (Figure 4.60). At night, these different layers migrated toward the surface, where they often merged into a single broad band. At dawn, the layers returned to deeper waters. Although first interpreted as the result of some physical discontinuity in the water, DSLs are now known to be concentrations of midwater animals, and the movement of the layers toward the surface at night is the result of a vertical migration similar to that undertaken by surface zooplankton. (See pp. 85–87). Presumably, these midwater animals are migrating into the surface waters at night to feed on the abundant plankton there.

The major animals forming the DSLs include the lantern fish (Myctophidae) and other, similar mesopelagic fishes (see Figure 4.20), shrimplike crustaceans such as euphausiids (krill), sergestid shrimp, various squids, and certain deep-floating hydrozoan cnidarians called siphonophores (see Figures 4.14, 4.15, 4.16, and 4.19). The composition of the scattering layer varies with geographical area.

Deep scattering layers have been found in all oceans except the Arctic. They are best developed in areas with high surface productivity and are faint in areas of low productivity.

■ **SPECIES COMPOSITION AND DISTRIBUTION**

The mesopelagic zone of the deep sea is dominated by small fishes of the families Myctophidae, Gonostomatidae, and Sternoptychidae and by euphausiid and decapod crustaceans (Figure 4.61). Blackburn (1977) has noted that these combined groups constitute more than 80% of the biomass of the

mesopelagic zone. One estimate for the Pacific Ocean has fishes making up 20–45% of the biomass, shrimps 15–25%, and euphausiids 35–50% in the upper 1,000 m of water. Other components include other crustaceans, such as mysids, amphipods, and copepods; chaetognaths; scyphozoan jellyfishes; and cephalopods. This dominance of the pelagic deep-sea fauna by small fishes, euphausiids, and shrimp seems to be widespread in all oceans, but as Omori (1974) has noted, the relative abundance of each group changes with latitude and depth (Figure 4.62). Seasonal changes in the composition of the assemblage appear minor, but have received little study.

There are two components of the pelagic species assemblages in the mesopelagic zone. One includes those animals which do not migrate vertically: certain fishes (*Cyclothone, Sternoptyx*), mysids, some shrimps (penaeids, carideans), and a few euphausiids, amphipods, cephalopods, and cnidarians. The other component is called the **interzonal fauna** and undergoes a diel vertical migration. These animals live at variable depths by day and migrate into surface waters by night. This component includes most of the mesopelagic fishes (Myctophidae, Gonostomatidae) and most euphausiids and decapods; hence, the composition of the community changes depending on whether it is sampled at night or during the day (Figure 4.62).

■ **FOOD AND FEEDING**

Information on the food of midwater organisms is sparse, but is most adequate for two dominant groups: the fishes and the decapod crustaceans. Studies of the diet of midwater fishes have concentrated on the numerically abundant families Myctophidae,

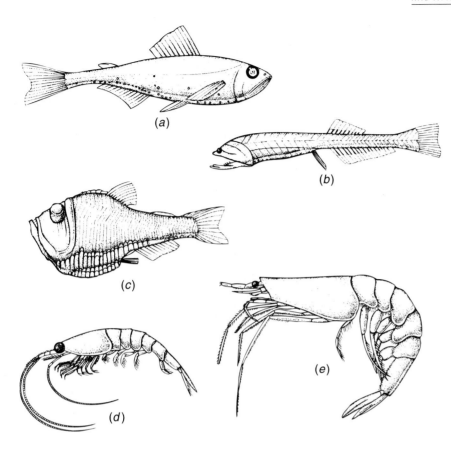

FIGURE 4.61 Representatives of the major groups of organisms of the pelagic zones of the deep sea. (a) Lantern fish, *Tarletonbeania crenularis* (Myctophidae). (b) Gonostomatid fish, *Cyclothone pallida* (Gonostomatidae). (c) Hatchet fish, *Argyropelecus affinis* (Sternoptychidae). (d) Euphausiid, *Euphausia pacifica*. (e) Sergestid shrimp, *Sergestes similis*.

Gonostomatidae, and Sternoptychidae. Whereas there is often considerable variation in species composition in the diets of these fishes, the principal components in all cases, as noted by Hopkins and Baird (1977), are smaller crustaceans, such as copepods, euphausiids, ostracods, amphipods, and small decapod shrimps (Table 4.4). Of these groups, note Gartner et al. (1997), copepods are consumed the most, but there may be considerable selectivity.

Changes in the diet of these fishes occur with age. In general, the larger and older individuals take a greater proportion of the larger crustaceans than do the smaller fishes (Table 4.4). Seasonal changes in diet have also been observed in some fishes, perhaps reflecting the availability of prey. Geographical variation in diet also occurs in widely distributed species. Although it is generally thought that fishes in these energy-poor deep-water environments should be opportunists and generalists, Gartner et al. (1997) report that true generalists are rare.

Omori (1974) reports that most decapod crustaceans are also predators or omnivores that make use of various available food items. Analyses of the stomachs of midwater shrimp suggest that copepods, euphausiids, and other small crustaceans constitute the majority of the food. There is also evidence of a certain amount of cannibalism. The shrimp are not specialists, but opportunists, taking what is available within a certain size range.

In spite of the paucity of food in the pelagic areas of the deep sea—a condition that causes the organisms to have the varied diets observed among fishes and crustaceans—there is a certain amount of selectivity and partitioning of food resources among animals. The most common form of partitioning occurs when species feed in different depth zones in the water column. Different kinds or abundances of prey occur in different depth zones; thus, predators feeding in these zones encounter and consume prey populations of differing composition. Even within a given zone, the prey composition of the resident predators varies, indicating a certain degree of resource partitioning, probably due to anatomical differences in the prey-capture mechanisms of the predator. Such structural modifications allow the

FIGURE 4.62 Differences in biomass composition of the midwater community in the tropical Pacific with depth and between day and night. (Based on *Oceanic Sound-Scattering Prediction*, N. R. Andersen & B. J. Zahuranec, eds., 1977, © Plenum Publishers.)

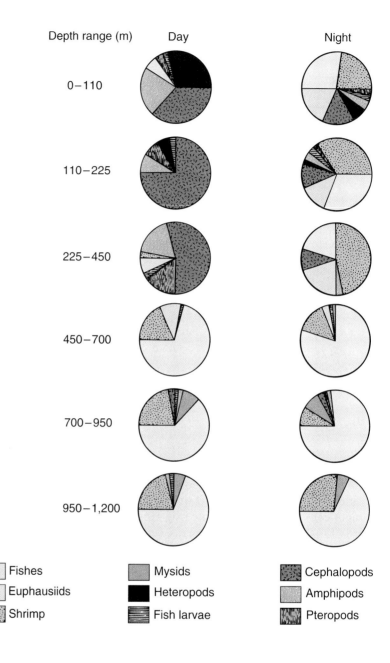

Fishes	Mysids	Cephalopods
Euphausiids	Heteropods	Amphipods
Shrimp	Fish larvae	Pteropods

predator to prey only on organisms of certain sizes or shapes. Since many organisms migrate through depth zones on a diel basis, the available prey composition varies over a 24-hour period at any given depth. Therefore, a predator that feeds during the day at a given depth may encounter prey different from that which a predator feeding at the same depth at night would meet up with.

In a study of the potential competition among mesopelagic fishes and shrimp in the eastern Gulf of Mexico, Hopkins and Sutton (1998) found considerable niche overlap among the species if single niche parameters were considered. However, if the entire spectrum of niches, including food size, food composition, and vertical distribution, was examined, then considerably less niche overlap and resource partitioning were found at the species level. The researchers attributed the high species diversity of the fishes and shrimp in the Gulf of Mexico to such partitioning.

TABLE 4.4

Changes in Diet with Size and Age in Six Midwater Fishes

	SIZE RANGE (MM)	PERCENT OF FOOD <5 MM	PERCENT OF FOOD >5 MM	TREND IN DIET COMPOSITION (INCREASING PERCENT)
Argyropelecus aculeatus	10–20	99	1	Ostracods, copepods
(WN central Atlantic)	50–60	12	88	→ fish
Argyropelecus affinis	21–34	94	6	Copepods, ostracods
(off southern California)	52–68	68	33	→ euphausiids, salps, chaetognaths
Argyropelecus sladeni	10–20	66	34	Ostracods, copepods,
(off southern California)	30–50	24	76	chaetognaths → euphausiids
Sternoptyx diaphana	10–24	100	0	Copepods, ostracods
(Pacific subantarctic)	40–80	60	40	→ euphausiids
Sternoptyx obscura	10–20	86	14	Copepods → ostracods,
(off southern California)	30–50	63	37	amphipods, chaetognaths
Valenciennellus tripunctulatus	30–55	85	15	Small copepods →
(E. Gulf of Mexico)	30–35	58	42	larger copepods

Source: *Oceanic Sound-Scattering Prediction*, N. R. Andersen and B. J. Zahuranec, eds. 1977.

■ LIFE-HISTORY PATTERNS OF MIDWATER ORGANISMS

Most fishes and decapod crustaceans of the mesopelagic zone are short-lived creatures. According to Mauchline (1977), myctopid and gonostomatid fishes tend to become sexually mature at from one to three years and probably live no more than two to four years.

The reproductive cycle among midwater fishes is not yet well known. Data from Clarke (1973, 1974) on species of myctophids and stomiatoid fishes from the central North Pacific Ocean suggest that the fishes there have a seasonal reproductive cycle and spawn in the spring and summer. Juveniles do not appear to undergo vertical migration as do adults, and they are distributed in shallower water than are the adults.

Decapod shrimps of the dominant mesopelagic genera *Sergestes*, *Pasiphaea*, and *Gennadas* mature in their first or second year and have life spans of one to three years. Euphausiids of midwater zones generally mature in their first year and live perhaps two years. On the other hand, the bathypelagic and abyssalpelagic crustaceans are estimated to live as much as two to seven times longer.

The lower temperatures and the greater scarcity of food in the bathypelagic and abyssalpelagic zones probably decrease the rate of growth, delay the onset of sexual maturity, and therefore, increase the longevity of these animals. The extreme expression of this trend is abyssal gigantism (pp. 160–161).

Mesopelagic decapod shrimp also show seasonal reproductive activity, usually spawning in late winter or spring in temperate-zone seas. In bathypelagic shrimp, it appears that spawning is prolonged over more of the year. Little is known about the movements and distribution of larval shrimp.

Summary of Key Concepts

- The deep sea is that part of the marine environment which lies below the level of light penetration and beyond the continental shelf.
- The deep-sea pelagic faunal zones are the mesopelagic (dysphotic), bathypelagic, abyssalpelagic, and hadalpelagic zones.
- The deep-sea benthic faunal zones are the bathyal, abyssal, and hadal zones.

- It is difficult to sample the deep sea because of the need for large amounts of cable, the uncertainty that the gear is sampling correctly, the qualitative nature of most gear, and the time involved in obtaining a sample.
- Newer instruments to sample the deep sea include manned and unmanned submersibles.
- In the deep sea, the major environmental factors that affect animals and their interactions are temperature,

pressure, the absence of light, food, and oxygen content.

- The range of pressures is greatest in the deep sea and has a profound effect on the metabolism of the animals there, including enzyme systems, membranes, and ability to secrete calcium carbonate.

- Oxygen is adequate for life in most of the deep sea except in certain areas, called oxygen minimum zones, where very low levels may preclude some organisms.

- Food is scarce in the deep sea and decreases with depth and distance from land, leading to low densities and small sizes of deep-sea organisms.

- Adaptations observed in mesopelagic animals include red color in crustaceans and, in fishes, black color, tubular eyes or large eyes, large mouths with recurved teeth to enable the animal to feed on large prey, and lures to attract prey.

- Most deep-sea animals are small, but some crustacean groups show abyssal gigantism.

- As one goes deeper, the anatomical organization of the animals becomes simplified and they show increased water content.

- Bioluminescence is widespread, particularly in the mesopelagic zone, where many animals produce light, often in elaborate photophores.

- Bioluminescence is of major importance in concealing animals from predators, in finding and attacking prey, in finding mates, and in ensuring the animal's survival.

- The faunal composition of the deep-sea benthos includes both epifaunal and infaunal organisms, and virtually all major animal phyla are represented.

- The community structure of the deep-sea benthos shows a paucity of suspension feeders, due to the low density of food in the water, low density of carnivores and herbivores, and dominance by detritus feeders.

- The age structure, life span, colonization times, and recruitment patterns of the deep-sea benthos are poorly known.

- Deep-sea diversity, particularly in the benthos, is very high. Several theories, having to do with stability time, croppers and the relation between numbers of species and the sizes of the areas they inhabit, have been suggested to explain it, but none appear to be individually adequate to explain all the patterns, which are probably the result of many factors acting simultaneously. Among these factors are microhabitat diversification, energy input, competition, predation, and rate of disturbance.

- Seamounts are submerged volcanic mountains that rise to within hundreds or thousands of meters of the surface. The fauna is often highly endemic and the communities are dominated by suspension-feeding organisms. Many seamounts have been overexploited for fish and coral.

- Two general reproductive patterns exist in the deep sea: Either the young spend their life stages in the surface waters and then the animals migrate to the deep sea as they mature or metamorphose, or no migration occurs and the young spend their life stages in the same area as the adults.

- One of the most significant discoveries in deep-sea biology was the finding of a previously unknown set of organisms living around hot water springs called hydrothermal vents. These vents have subsequently been discovered at plate boundaries in a number of areas in the Pacific, Indian, Arctic, and Atlantic oceans.

- Similar communities, called cold seeps have been discovered where reducing chemicals are seeping out from the seafloor.

- Both seeps and vents are areas where the reducing chemicals sulfides and methane are produced and the existence of the communities inhabiting the area depends on the primary productivity of the chemoautotrophic bacteria that oxidize the sulfides and methane and use the energy to fix carbon dioxide into organic compounds.

- The communities at vents and seeps are unique, and they have a high diversity and biomass, with the dominant animals displaying a special symbiotic relationship with the sulfide-oxidizing bacteria. The animals are confined to an area within a few meters of the sulfide source, and seeps and vents may be separated from each other by hundreds or thousands of kilometers.

- Individual vents have a life span only in the tens of years, whereas seep systems may have longer life spans.

- Observations of the vents at the Galápagos over time suggest that initially they are dominated by vestimentiferan worms, clams, and mussels, but that over time the numbers of worms decrease while those of the clams and mussels increase. As the flow of the vents diminishes over time, many animals disappear. The mussels appear to be the most persistent animals around vents.

- More recently, a sulfide-driven community similar to that inhabiting seeps has been discovered on whale bone.

- Midwater community ecology is not very well known because of the lack of observations, the avoidance of nets and of light by the organisms in question and the small size of the organisms.

- Deep scattering layers are concentrations of sound-reflecting midwater animals found in all oceans except the Arctic.

- Midwater organisms appear to be dominated by various small fishes, crustaceans, and jellyfishes which seem to have partitioned niches in such a manner that permits them to avoid competition for food; this behavior allows very diverse numbers of species to coexist.

- Most mesopelagic animals are short lived, living just a few years, whereas the bathypelagic and abyssal pelagic animals are longer lived.

Review Questions

ESSAY Develop complete answers to these questions:

1. Choose one of the following questions to discuss: How are deep-sea fishes like the Venus flytrap? Why are many deep-sea fishes float-and-wait predators? Compare and contrast sit-and-wait predators in terrestrial systems and float-and-wait predators in marine systems.

2. Why do we believe that bioluminescence in the deep sea is of great importance to the organisms possessing it? Discuss.

3. At what oxygen concentrations are the deep-sea waters? Why are they not zero? Where is the oxygen added to the deep-sea waters? Why are the deep-sea waters not depleted by organisms found in them?

4. Consider the three hypotheses that attempt to explain the relatively high species diversity of the deep-sea benthos. Discuss the relative merits of each. (Instructors may wish to ask students for ways to develop tests of these hypotheses.)

5. Why are planktonic larval forms not as advantageous for benthic animals of the abyss as they are for animals of other benthic environments?

Bibliography

Andersen, N. R., and B. J. Zahuranec, eds. 1977. *Oceanic sound scattering prediction*. New York: Plenum Press.

Angel, M. V. 2003. The pelagic environment of the open ocean. In *Ecosystems of the Deep Oceans, Ecosystems of the World 28*, edited by P. A. Tyler. Amsterdam, Elsevier, pp. 39–79.

Angel, M. V., and M. Sibuet, eds. 1990. Deep sea biology. *Prog. Oceanog.* 24 (1–4):1–343.

Anonymous. 1984. Deep-sea hot springs and cold seeps. *Oceanus* 27:3–78.

Arp, A. J., and J. J. Childress. 1983. Sulfide binding by the blood of the hydrothermal vent tube worm *Riftia pachyptila*. *Science* 219:295–297.

Ballard, R. D. 1977. Notes on a major oceanographic find. *Oceanus* 20(3):35–44.

Blackburn, M. 1977. Studies on pelagic animal biomasses. In *Oceanic sound scattering prediction*, edited by N. R. Andersen and B. J. Zahuranec. New York: Plenum Press, 283–299.

Boehlert, G. W., and A. Genin. 1987. A review of the effects of sea mounts on biological processes. Pp 319-334 In *Seamounts, islands and atolls. Geophysical Monograph 43*, edited by B. H. Keating, P. Fryer, R. Batiza, and G. W. Boehlert.

Bouchet, P., and A. Waren. 1979. Planktotrophic larval development in deep water gastropods. *Sarsia* 64:37–40.

Brauer, R., ed. 1972. *Barobiology and the experimental biology of the deep sea*. North Carolina Sea Grant Program, University of North Carolina, Chapel Hill.

Bruun, A. 1957. Deep-sea and abyssal depths. In *The treatise on marine ecology and paleoecology*, edited by J. E. Hedgpeth. Vol. I, *Ecology*. Memoir 67, Geol. Soc. Amer., 641–672.

Childress, J., ed. 1988. Hydrothermal vents: A case study of the biology and chemistry of a deep sea hydrothermal vent of the Galápagos Rift. The Rose Garden in 1985. *Deep-Sea Res.* 35 (10–11A):1677–1849.

Childress, J. J., and C. R. Fisher. 1992. The biology of thermal vent animals: Physiology, biochemistry and autotrophic symbioses. *Oceanog. Mar. Biol. Ann. Rev.* 30: 337–441.

Childress, J., and M. Nygaard. 1973. The chemical composition of midwater fishes as a function of depth of occurrence off southern California. *Deep-Sea Res.* 20:1093–1111.

Childress, J., and M. Nygaard. 1974. Chemical composition and buoyancy of midwater crustaceans as a function of depth of occurrence off southern California. *Mar. Biol.* 27:225–238.

Clarke, T. S. 1973. Some aspects of the ecology of lantern fishes (Myctophidae) in the Pacific Ocean near Hawaii. *Fish. Bull. U.S.* 71:401–434.

Clarke, T. S. 1974. Some aspects of the ecology of stomiatoid fishes in the Pacific Ocean near Hawaii. *Fish. Bull. U.S.* 72:337–351.

Cossins, A. R., and A. G. Marshall. 1989. The adaptation of biological membranes to temperature and pressure: Fish from deep and cold. *J. Bioenerg. Biomembr.* 21:115–135.

Dayton, P., and R. H. Hessler. 1972. The role of biological disturbance in maintaining diversity in the deep sea. *Deep-Sea Res.* 19:199–208.

Dietz, R. S. 1962. The sea's deep scattering layers. *Sci. Amer.* 207(2):44–50.

Edmonds, H. N. et al. 2003. Discovery of abundant hydrothermal venting on the ultraslow-spreading Gakkel Ridge in the Arctic Ocean. *Nature* 421: 252–256.

Ernst, W. G., and J. G. Morin, eds. 1982. *The environment of the deep sea*. Englewood Cliffs, NJ: Prentice-Hall.

Gage, J. D., and P. A. Tyler. 1991. *Deep sea biology*. New York: Cambridge University Press.

Gartner, J. V., R. E. Crabtree, and K. J. Sulak. 1997. Feeding at depth. In *Deep sea fishes*, edited by D. J. Randall and A. P. Farrell. San Diego: Academic Press, 115–182.

Grassle, J. F. 1985. Hydrothermal vent animals: Distribution and biology. *Science* 229:713–717.

Grassle, J. F., and L. Morse-Porteous. 1987. Macrofaunal colonization of disturbed deep sea environments and the structure of deep sea benthic communities. *Deep-Sea Res.* 34:1911–1950.

Hardy, A. 1956. *The open sea.* Vol. 1, *The world of plankton.* New York: Houghton Mifflin.

Haymon, K. M., and K. C. McDonald. 1985. The geology of deep sea hot springs. *Amer. Sci.* 73(5):441–449.

Hedgpeth, J. 1957. Classification of marine environments. In *The treatise on marine ecology and paleoecology.* Vol. 1, *Ecology.* Geol. Soc. of Amer. Memoir 67, 18–27.

Heezen, B. C., and C. D. Hollister. 1971. *The face of the deep.* New York: Oxford University Press.

Hessler, R. R., and W. M. Smithey, Jr. 1983. The community structure of megafauna. In *Hydrothermal processes at seafloor spreading centers,* edited by P. A. Rona et al. New York: Plenum Press, 735–770.

Hessler, R. R., W. M. Smithey, M. A. Boudrias, C. H. Keller, R. A. Lutz, and J. Childress. 1988. Temporal changes in the megafauna at the Rose Garden hydrothermal vent (Galápagos rift; eastern tropical Pacific). *Deep-Sea Res.* 35 (10–11):1681–1709.

Hopkins, T. L., and R. C. Baird. 1977. Aspects of the feeding ecology of oceanic midwater fishes. In *Oceanic sound scattering prediction,* edited by N. R. Andersen and B. J. Zahuranec. New York: Plenum Press.

Hopkins, T. L., and T. T. Sutton. 1998. Midwater fishes and shrimps as competitors and resource partitioning in low latitude oligotrophic ecosystems. *Mar. Ecol. Prog. Ser.* 164:37–45.

Idyll, C. P. 1976. *Abyss.* New York: Crowell.

Ingmanson, D. E., and W. J. Wallace. 1973. *Oceanology: An introduction.* Belmont, CA: Wadsworth.

Jones, M. E., ed. 1985. Hydrothermal vents of the eastern Pacific, an overview. *Bull. Biol. Soc. Washington,* no. 6:1–547.

Jumars, P. A., and E. D. Gallagher. 1982. Deep-sea community structure: Three plays on the benthic proscenium. In *The environment of the deep sea,* edited by W. G Ernst and J. G. Morin. Englewood Cliffs, NJ: Prentice-Hall.

Karl, D. M. (editor). 1995. *The Microbiology of Deep sea Hydrothermal Vents.* Boca Raton, FL: CRC Press.

Karl, D. M., D. J. Burns, K. Orrett, and H. W. Jannasch. 1984. Thermophilic microbial activity in samples from deep sea hydrothermal vents. *Mar. Biol. Lett.* 5:227–231.

Lightfoot, R., P. A. Tyler, and J. D. Gage. 1979. Seasonal reproduction in deep sea bivalves and brittlestars. *Deep-Sea Res.* 26A:967–973.

Longhurst, A. R. 1981. *Analysis of marine ecosystems.* New York: Academic Press.

MacDonald, A. G. 1975. *Physiological aspects of deep sea biology.* New York: Cambridge University Press.

Marshall, N. B. 1954. *Aspects of deep-sea biology.* New York: Philosophical Library.

Marshall, N. B. 1980. *Deep sea biology: Developments and perspectives.* New York: Garland STPM Press.

Mauchline, J. 1972. The biology of bathypelagic organisms, especially crustacea. *Deep-Sea Res.* 19:753–780.

Mauchline, J. 1977. Estimating production of midwater organisms. In *Oceanic sound scattering prediction,* edited by N. R. Andersen and B. J. Zahuranec. New York: Plenum Press.

McAllister, D. E. 1961. A collection of oceanic fishes from off British Columbia with a discussion of the evolution of the black peritoneum. *Bull. Nat. Mus. Canada* 172:39–43.

Mead, G. W., E. Bertelsen, and D. M. Cohen. 1964. Reproduction among deep sea fishes. *Deep-Sea Res.* 11:569–596.

Menzies, R. J. 1964. Improved techniques for benthic trawling at depths greater than 2000 meters. *Biol. Antarctic Seas. Antarctic Res. Set., Amer. Geophys. Union* 1:93–109.

Menzies, R. J., R. Y. George, and G. T. Rowe. 1973. *The abyssal environment and ecology of the world oceans.* New York: Wiley.

Micheli, F. et al. 2002. Predation structures communities at deep sea hydrothermal vents. *Ecological Monographs* 72(3):365–382.

Nixon, M., and J. B. Messenger, eds. 1977. *The biology of cephalopods.* New York: Academic Press.

Omori, M. 1974. The biology of pelagic shrimp in the ocean. *Adv. Mar. Biol.* 12:233–324.

Parsons, T. R., M. Takahashi, and B. Hargrave. 1984. *Biological oceanographic processes,* 3d ed. New York: Pergamon.

Regan, C. T., and E. Trewavas. 1932. *Deep sea angler-fishes [Ceratoidea].* Dana report no. 2. The Carlsberg Foundation and Oxford University Press.

Rex, M. A. 1981. Community structure in the deep sea benthos. *Ann. Rev. Ecol. Sys.* 12:331–353.

Robison, B. H. 1995. Light in the oceans midwaters. *Sci. Amer.* 273(1):60–64.

Rogers, A. D. 1994. The biology of seamounts. *Advances in Marine Biology* 30:305–350.

Rokop, F. 1974. Reproductive patterns in the deep-sea benthos. *Science* 186:743–745.

Rokop, F. 1979. Year around reproduction in deep sea bivalve molluscs. In *Reproductive ecology of marine invertebrates,* edited by S. E. Stancyk. Columbia: University of South Carolina Press, 189–198.

Rona, R. P., K. Bostrum, L. Laubier, and K. L. Smith, eds. 1983. *Hydrothermal processes at seafloor spreading centers.* New York: Plenum Press.

Ross, D. A. 1982. *Introduction to oceanography.* Englewood Cliffs, NJ: Prentice-Hall.

Rowe, G. T., ed. 1983. *The sea.* Vol. 8, *Deep sea biology.* New York: Wiley.

Sanders, H. L. 1968. Marine benthic diversity, a comparative study. *Amer. Nat.* 102:243–282.

Sanders, H. L., and R. R. Hessler. 1969. Ecology of the deep-sea benthos. *Science* 163:1419–1424.

Shank, T. M. et al. 1998. Temporal and spatial patterns of biological community development at nascent deep sea hydrothermal vents (9°50′ N, East Pacific Rise. *Deep Sea Research II* 45:465–515.

Siebenaller, J., and G. N. Somero. 1978. Pressure adaptive differences in lactate dehydrogenases of congeneric fishes living at different depths. *Science* 201(4352):255–257.

Siebenaller, J., and G. N. Somero. 1989. Biochemical adaptation to the deep sea. *Crit. Rev. Aquat. Sci.* 1:1–25.

Sibuet, M., and K. Olu. 1998. Biogeography, biodiversity and fluid dependence of deep cold-seep communities at active and passive margins. *Deep-Sea Res.* II 45:517–567.

Smith, C. R. 1992. Whale falls. *Oceanus* 35(3):74–78.

Smith, K. L., and K. R. Hinga. 1983. Sediment community respiration in the deep sea. In *The sea*, Vol. 8, *Deep sea biology*, edited by G. T. Rowe. New York: John Wiley and Sons, 331–370.

Somero, G. N. 1982. Physiological and biochemical adaptations of deep sea fishes: Adaptative responses to the physical and biological characteristics of the abyss. In *The environment of the deep sea*, edited by W. G. Ernst and J. G. Morin. Englewood Cliffs, NJ: Prentice Hall, 256–278.

Somero, G. N. 1992. Biochemical ecology of deep-sea animals. *Experientia* 48:537–543.

Tunnicliffe, V. J. 1991. The biology of hydrothermal vents: Ecology and evolution. *Oceanog. Mar. Biol. Ann. Rev.* 29:319–407.

Tunnicliffe, V. J. 1992. Hydrothermal vent communities of the deep sea. *Amer. Sci.* 80(4):336–349.

Tunnicliffe, V. J., F. Garrett, and H. P. Johnson. 1990. Physical and biological factors affecting the behaviour and mortality of hydrothermal vent tubeworms (vestimentiferans). *Deep-Sea Res.* 37(1):103–125.

Tunnicliffe, V. J., A. G. McArthur, and D. McHugh. 1998. A biogeographical perspective of the deep-sea hydrothermal vent fauna. *Adv. Mar. Biol.* 34:353–442.

Van Damm, K. L. 2001. Lost city found. *Nature* 412:127–128.

Van Dover, C. L. et al. 2001. Biogeography and ecological setting of Indian Ocean hydrothermal vents. *Science* 294:818–823.

Yancey, P. H. et. al. 2002. Unusual organic osmolytes in deep sea animals: Adaptation to hydrostatic pressure and other perturbants. *Comp. Biochem. Phys. A* 133: 667–676.

Yayanos, A. A., A. S. Dietz, and R. van Boxtel. 1979. Isolation of a deep sea barophilic bacterium and some of its growth characteristics. *Science* 205(4408):808–809.

Young, R. E. 1975. Function of the dimorphic eyes in the midwater squid, *Histioteuthis dofleini. Pac. Sci.* 29(2):211–218.

Young, R. E. 1983. Oceanic bioluminescence: An overview of general functions. *Bull. Mar. Sci.* 33:829–845.

Zimmerman, A. M., and S. B. Zimmerman. 1972. Commentary on high pressure effects on cellular systems. In *Barobiology and the experimental biology of the deep sea*, edited by R. W. Brauer. Chapel Hill: North Carolina Sea Grant Program, University of North Carolina.

CHAPTER 5

SHALLOW-WATER SUBTIDAL BENTHIC ASSOCIATIONS

In this chapter, we deal with the bottom communities and conditions for life in the shallow waters fringing the landmasses of the world: the continental shelf region of the world's oceans. We have considerably more knowledge about this region than we have about the deep sea. It is more accessible and more studied because the continental shelf areas are the principal fishing grounds of the world. This chapter covers major subtidal associations, except for coral reefs. Because of their many unique properties, coral reefs are covered separately in Chapter 9. At its upper end, the shallow subtidal area is contiguous with the intertidal regions of the world's oceans, and at least some of the principles we discuss in this chapter also will be applicable to the intertidal zone.

■

■ COVERAGE AND DEFINITIONS

This chapter covers that area of the oceans which lies between the level of the lowest low water on the shore to the edge of the continental shelf at a depth of about 200 m. In our scheme of classification, this area is known as the **sublittoral** zone. Overlying it are the waters of the **neritic** zone. (See Figure 1.22 on the scheme of zonation.) Most of this zone is composed of soft sediments, sand and mud, and a much lesser area of hard substrate. On an areal and numerical basis, the bottom is dominated by infaunal organisms.

Infaunal organisms are usually divided into categories based on size. **Macrofauna** are organisms that are greater than 0.5 mm. The term **meiofauna** is applied to organisms that lie within the size range of 0.5–0.062 mm. The final size class of infaunal organisms is **microfauna**, which includes organisms below 0.062 mm. The latter group is composed primarily of

protozoa and bacteria. This chapter will deal primarily with macrofauna. Meiofauna and microfauna are covered in Chapter 7.

In the temperate regions the sublittoral zone can be divided into two sectors: a shallow zone that extends as deep as the available light for photosynthesis and a zone in which the light is below the level of that required for photosynthesis. The first zone is dominated by macroalgae such as kelps living on hard substrates (see p. 221) and by various seagrass communities persisting on soft substrates (see p. 233).

Environmental Conditions

The continental shelf waters show more variability in environmental conditions than do either the epipelagic zone of the open ocean or the deep sea. Perhaps the most important physical factor that acts

196

on the bottom communities is turbulence, or wave action. In these shallower waters, the interaction among waves, currents, internal waves, and upwelling creates turbulence that keeps inshore waters from becoming thermally stratified except for brief periods, at least in the temperate zone. As a result, nutrients are rarely limiting or locked in a bottom reservoir. Also, productivity is generally higher than in similar waters offshore because of the abundance of nutrients, from both runoff from land and recycling. (See Table 2.2.) This high productivity sustains large populations of both zooplankton and benthic organisms.

Wave action is an important factor in this area. Long-period ocean swell and storm waves have an effect that can extend to the bottom in these shallow waters. In soft bottoms, the passage of such waves may cause large surging motions in the bottom water, greatly affecting the stability of the substrate. Substrate particles may be moved around and resuspended, profoundly affecting the infaunal animals in the substrate. Wave action also determines the type of particles present. Heavy wave action removes fine particles by keeping them in suspension, leaving mainly sand; thus, fine silt sediments can occur only in areas that have low wave action or that are too deep to be affected by wave action.

Salinity in this region is more variable than in the open ocean or deep ocean. Except for areas where large rivers discharge massive amounts of fresh water, however, the salinity does not change enough to be of ecological significance.

Temperature is also more variable in inshore waters and shows seasonal change in the temperate zone. The temperature changes may be used by organisms as cues to begin or end various activities, such as reproduction.

Light penetration in these turbulent waters is reduced, compared with open ocean areas. The combination of large amounts of debris, both from land and from the breakup of kelp and seagrasses, and the high plankton densities due to the abundant nutrients, can reduce light penetration to just a few meters.

The food supply is abundant in this area, partly because of the increased productivity by the plankton, but also because of the production of attached plants, such as kelp and seagrasses. This is one of the few areas in the sea where macroscopic plants have any significance in production. A final food source is runoff from land. Although there are large

autotrophs in the sublittoral shelf, there are relatively few large grazing animals. The major use of kelps and seagrasses as food occurs only after they have been broken up into detrital particles.

The soft bottoms in the sublittoral zone are essentially without a diversity of large topographic features, and the vast expanses extend monotonously for long distances. Small-feature diversity exists in the form of ripple marks, worm tubes, fecal mounds, and so on (Figure 5.1). Without large

FIGURE 5.1 A shallow-water soft bottom. Note the ripple marks. (Photo courtesy of Tony Stone Images.)

topographic relief, the only apparent difference from one place to another is in the grain size and composition (silica vs. calcium carbonate) of the substrate. Subtidal hard substrates, on the other hand, may have considerable relief with many potential habitats. The lack of major topographic relief in the infaunal areas means fewer habitats for animals to occupy and fewer ways for making a living. The number of infaunal species is generally less than the number of epifaunal species, because there are fewer niches available. The most abundant infaunal animals are deposit feeders, ingesting the abundant detritus that rains down, or are suspension feeders, filtering the abundant plankton or floating detritus out of the water column. A lesser number, mainly worms, crustaceans, mollusks, and echinoderms, are predators. Bottom fishes, on the other hand, are predominantly carnivores.

It is convenient to divide the subtidal continental shelf region into four major habitats, which are readily recognizable and topographically or physiognomically different: (1) open, unvegetated sedimentary environments (the most common in terms of area), (2) hard substrates dominated by low-encrusting plants and animals, (3) kelp beds and forests, and (4) seagrass beds. We discuss each of these separately.

Unvegetated Sedimentary Environments

■ GENERAL PATTERNS

Unvegetated marine sediments are the most abundant marine benthic habitat in the world. Counting the deep-sea floor, these habitats comprise probably 80% or more of the ocean bottom. On the continental shelf areas of the world's oceans there are general latitudinal differences in the composition of the open sedimentary environments. The temperate zones tend to have more sand in the sediments, the tropical zones more mud, and the polar zones more gravel, except for the Arctic which has large amounts of riverine-derived mud. (See p. 243 on polar communities.)

■ SAMPLING

From the times of Petersen in the early 1900s (see next section) until about the 1960s, the common method for sampling the organisms of the soft sediments of the continental shelf was to drop a **grab** (Figure 5.2). Various types of benthic grabs have been developed over the years, and each has certain advantages and disadvantages. Most have the advantage of being quantitative; that is, they take a sample of known areal extent from the bottom. Standard treatment of these grab samples has been to screen out the organisms on

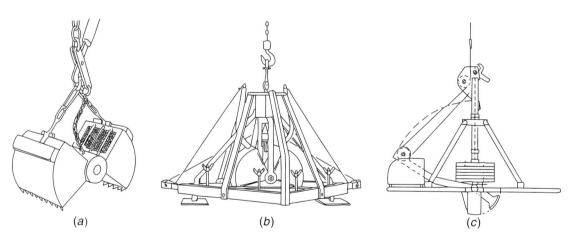

(a)   *(b)*   *(c)*

FIGURE 5.2 Various types of grab samplers. (a) Peterson grab. The jaws close as the cable is pulled upward, thus securing the sample. (b) Smith–McIntyre grab. This is a spring-loaded grab, in open position here, that will close when it hits the bottom. (c) Reineck box corer. The "box" is closed by a knife-edge actuated by pulling up on the cable. (a and c from *Methods for the Study of Marine Benthos*, 2d ed., N. A. Holmes & A. D. McIntyre, eds., 1984, Blackwell Science. Reprinted by permission. b from Hardy, A. C., 1959, *The Open Sea: It's Natural History, Part II, Fish and Fisheries*, New Naturalist Series, 37. Collins, London.)

one or a series of screens. Mesh sizes of 1.0 mm and 0.5 mm have been the most commonly employed.

Although such sampling was quantitative and gave a good indication of the kinds of organisms present and their relative abundance, it was a remote method in which the benthic biologist never actually saw the community. It was also not possible to perform any manipulative experiments with the community. This was changed in the 1960s with the advent of the self-contained underwater breathing apparatus (popularly known as scuba), which allowed the biologist to investigate an underwater community firsthand and to undertake manipulative experiments, such as using cages to keep various organisms in or out of the community. Thus, changes over time could be followed. Experimental ecology arrived with scuba, but because the device was not used until later in sedimentary environments, as opposed to the hard substrates, we have fewer data for the soft sediments.

Grabs are still used today, particularly in areas below safe diving depths (about 30 m). Within safe diving depths, firsthand work by divers is more popular and productive.

Other sources of data include remote, towed cameras or video systems and side-scan sonar. The latter is capable of recording small-scale detail of the bottom topography very accurately.

■ INFAUNAL ASSOCIATIONS

Four taxonomic groups of dominant macrofauna are present in sublittoral soft bottoms: class Polychaeta, subphylum Crustacea, phylum Echinodermata, and phylum Mollusca. Polychaete worms are the most abundant and are represented by numerous tube-building and burrowing species. The dominant crustaceans are the larger ostracods, amphipods, isopods, tanaids, and mysids, and a few of the smaller decapods. These animals inhabit mainly the surface of the sand and mud. Mollusks are represented primarily by various burrowing bivalve species and some scaphopods, with a few gastropods at the surface. Echinoderms particularly common in the subtidal benthos include brittle stars, heart urchins, sand dollars, sea cucumbers, and a few predatory sea stars. Although all these taxa are represented at all latitudes, the relative abundances and dominance vary with latitude.

The first quantitative work in marine ecology was done on the sublittoral benthic infauna. During the early years of the twentieth century, C. G. Joh. Petersen, a Danish biologist, began his investigations into the benthos of the Danish seas. Petersen was interested in evaluating the role of bottom organisms in supporting the fish populations on which the Danish fishing industry depended. To carry out his research on a quantitative basis, he invented a grab, the now-famous Petersen grab, which picked up a definite area of the bottom, usually 0.5 or 0.1 m^2 (Figure 5.2a). Since he had a sample that covered a known area of bottom, he could then count the numbers of organisms in that sample and extrapolate to obtain the numbers in the whole bottom area. Once this was done, he would have some estimate of the amount of food available to the fishes and could guess the fish biomass that could be supported.

Analyzing many grab samples over the years, Petersen (1918, 1924) observed that extensive bottom areas were occupied by recurrent groups of species and that other areas were inhabited by different associations of species. In all cases, relatively few species made up most of the individuals and biomass. These findings contrasted with those of nonquantitative dredge samples taken in the same waters, which indicated that the fauna might be nearly identical in the two kinds of area. The value of the quantitative study was that, for the first time, it gave a way to evaluate differences between communities based on relative abundance, not just the presence or absence of species.

After several years of evaluating these quantitative samples, Petersen observed that the different areas and their dominant organisms remained relatively constant and uniform over time. He then proposed that these associations were communities and named them on the basis of the dominant animals. For example, he had a *Macoma balthica* community from inner Danish waters in 8–10 m of water, which was dominated by the bivalve *M. balthica*, but was also characterized by the polychaete *Arenicola marina* and the bivalves *Mya arenaria* and *Cardium edule* (Figure 5.3). In this same manner, he named and characterized a whole series of communities from the sublittoral zone in Scandinavian waters.

Following Petersen's work, various biologists began to investigate other shallow-water benthic areas around the world. The leader of this research was another Dane, G. Thorson. What Thorson (1955) discovered, particularly as regards the temperate-zone

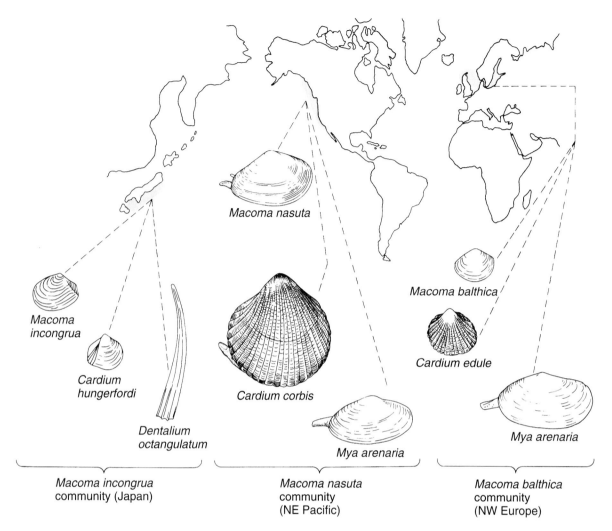

FIGURE 5.3 Diagram illustrating the parallel bottom communities dominated by the bivalve *Macoma* in three different areas in the North Temperate Zone. (Modified from "Bottom Communities" in *The Treatise on Ecology and Paleontology*, Vol. 1, G. Thorson, ed. J. Hedgpeth, 1957. © 1957 Geological Society of America. Used by permission.)

waters, was that communities of organisms similar to those found by Petersen were found in similar habitats around the world. The concept of **parallel bottom communities** arose from this discovery. According to that ecological concept, similar sediment types at the same depths around the world contain similar communities. Granted, the species are not the same, but they are closely similar ecologically and taxonomically (Figure 5.3). Additional work in tropical waters by Thorson (1966) revealed that this

concept did not extend to such areas—at least not in the same way.

That certain recurrent groups occupy the same types of substrate over large areas of the world's oceans seems established. This suggests that such associations are not random but represent real interacting systems in which some combination of factors ensures the persistence of the community.

Petersen described the communities he observed, and others later validated their continued existence in various parts of the world. What was lacking, however, was an explanation of *how* such communities persisted and *what kinds* of interactions ensured continued success. Since Petersen's time, answers to these questions have come to light. We discussed them in what follows.

■ COMMUNITY PATTERN AND STRUCTURE

Most marine organisms have patchy distributions. We noted this observation with respect to the plankton in Chapter 2. (See pp. 94–96.) It is also the situation one encounters when analyzing benthic organisms. Even in the early days of Petersen, not all quantitative samples had either the same species or the same relative abundances. They also varied with the season. Petersen was able to characterize his communities primarily on a statistical basis that was dependent on the presence of a large number of samples. Even in these seemingly monotonous open sedimentary environments removed from the more physically variable surface or intertidal areas, organisms exist in a patchy fashion, in both time and space. The patchiness is the result of the various physical factors and interactions among organisms and gives rise to the diverse patterns we observe in the distribution of benthic associations. Certain cycles that may occur on a daily, seasonal, or annual basis or over a period of years also originate in the interactions that produce patchiness. A further complicating factor is that variability in benthic associations occurs not only on a horizontal scale, but on a vertical scale as well, since many of these infaunal organisms burrow to different depths in the substrate.

Within any sedimentary benthic community there are usually hundreds of macrofaunal (greater than 0.5 mm in size) organisms, such that it is difficult to work with individual species in any attempt to understand the various ecological relationships. Fortunately, as Woodin (1983) has pointed out, the organisms in these environments can be classified into a limited number of ecological categories, called *functional groups*. One can then use the functional groups to compare assemblages of species and to arrive at generalizations regarding how the communities operate. A functional group includes all the species of different animal taxa that use and affect the environment in similar ways. For example, in Chapter 1 we noted the division of animals into opportunistic and equilibrium species. That is one functional classification.

Another approach would be to classify various species as **sediment destabilizers** and **sediment stabilizers**. Destabilizers, or **bioturbators**, include both motile and sedentary organisms whose activities—movement in the case of the motile ones and biological activities in the case of the sedentary ones—cause the sediment to move, become resuspended, erode, or otherwise change. Examples of this kind of organism are the mobile deposit-feeding clams, such as *Nucula*, *Yoldia*, and *Macoma*, studied by Rhoads and Young (1970) and the sedentary deposit-feeding sea cucumber *Molpadia oolitica*, which Rhoads and Young (1971) found destabilized the sediment by its feeding and defecation (Figure 5.4). Sediment stabilizers include various seagrasses, whose roots bind

FIGURE 5.4 Representatives of the two functional groups of sedimentary organisms: sediment stabilizers, and sediment destabilizers or bioturbators.

Phoronopsis   Owenia   Zostera   Yoldia   Macoma   Molpadia

Sediment stabilizers   Sediment destabilizers

FIGURE 5.5 Subtidal tube-dwelling anemonelike cnidarian *Pachycerianthus fimbriatus*. (Photo courtesy of Drs. Lovell and Libby Langstroth.)

sediments and whose upper parts change the local hydrographic regime (see pp. 233–236), as well as various invertebrate tube builders, which, in high density, form mats or areas in which sediment is stable. Examples of tube dwellers that stabilize sediments include amphipod crustaceans, phoronid worms, and tube-building anemones and polychaetes (Figures 5.4 and 5.5). This classification of sediment stabilizers and destabilizers is an improvement on Woodin's (1976) earlier classification of burrowing deposit feeders, tube builders, and suspension feeders.

Another functional division is the more traditional trophic classification of animals into herbivores, omnivores, and carnivores (Figure 5.6). Whereas it is usually easy to classify most infaunal organisms into such functional divisions as the ones just mentioned, some of the organisms are interme-

diate between categories. For example, maldanid polychaetes are tube builders, but they are also deposit feeders. To place such an organism in a stabilizing or destabilizing category will require information regarding its effect on the substrate. Note also that any functional group includes a number of different taxa of invertebrates and may further include different nutritional types. For example, we may have mobile burrowers that are predators, deposit feeders, or suspension feeders, and they may also be divided into those which are opportunistic and those which have reached equilibrium in their life-history features. In other words, categorization by functional group can be done several ways and does not necessarily exclude certain other features of the life history of the organism. The only trophic mode absent from open sedimentary communities is macroautotrophy: There are no large autotrophs, and any herbivores in this system must feed either on plant material that is in the plankton (suspension feeders), on benthic microalgae (diatoms) or on material that is brought in from outside as debris (deposit feeders or scavengers).

■ COMMUNITY STRUCTURE AND CHANGE

Basically, changes in sedimentary community structure occur through physical factors, biological factors, or some combination of the two. The simplest cases involve the dominance of physical factors. For example, Oliver et al. (1979) have demonstrated that, on the Pacific coast of California, subtidal benthic communities are zoned along a gradient of increasing wave disturbance to the bottom. In shallow water (less than 14 m), the bottom sediments are regularly disrupted by waves, and the dominant inhabitants are various small, mobile crustaceans. In the deeper zones (greater than 14 m), the fauna is dominated by tube-building polychaete worms. Evidence showed that wave action prevented the establishment of worm tubes in the shallow water. Wherever wave action was diminished, the polychaete zone became shallower, and wherever wave action was stronger, the worm zone was driven deeper.

A more complicated example in which both biological and physical factors act is seen in an East Coast situation. Mills (1969) studied a sandy area in Barnstable Harbor, Massachusetts, that had been dominated by the snail *Ilyanassa obsoleta* for some time. This community had low biomass and low

Temperate latitudes

Polar latitudes

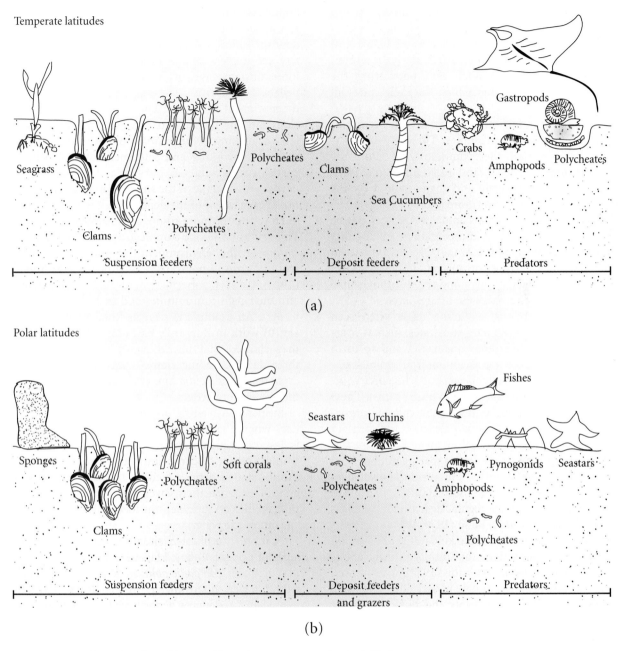

FIGURE 5.6 Examples of trophic assemblages in temperate- and polar-latitude unvegetated sediment habitants. (From M. Bertness et al., *Marine Community Ecology*, p. 254, figure 10,1. Copyright © 2001. Used with permission of Sinauer Associates, Inc.)

numbers of individuals of associated organisms, probably due to the mechanical disturbance of the sediments by the snails. Because of the low density of infaunal animals, this sandy area was colonized in spring by a tube-building amphipod, *Ampelisca abdita*. The *Ampelisca* quickly built a dense mat of

tubes that excluded *Ilyanassa*. Selective deposit feeding by the *Ampelisca* allowed fine particles of sediment to accumulate, and the increased topographic diversity created by the tubes also changed the physical environment of the area. This in turn led to a change in the biological environment and the colonization of the mats and sediments by an abundance of polychaete worms favoring fine sediments. The *Ampelisca* tube mats, however, are not stable, and stormy weather, particularly in winter, causes large areas to wash out, permitting the return of *Ilyanassa*, after which the cycle starts all over again.

The role of biological factors in deciding community structure is more complex. The dominant biological factors that alter community structure are competition, predation, and recruitment. We shall consider each separately and then examines some test cases in which two or more interact.

Competitive interactions among organisms can be direct or indirect. Direct competition involves either contact between organisms or the direct interference in some aspect of the other organism's life cycle. Indirect competition usually involves one organism interrupting or preventing another's use of a resource.

Direct interactions, including those often seen between organisms on hard substrates, such as overgrowth, pushing off other organisms, and so forth (see Chapter 6), are not as common in sedimentary environments, probably because such activity involves having a solid substrate to push against. These interactions also may be inherent in colonial organisms, which are conspicuously absent in sedimentary substrates. Direct competition involving aggressive activities leading to exclusion or spacing out has been reported by several authors in various polychaetes and small crustaceans, but most studies were done in the laboratory. Still, the regular distribution patterns found by Levin (1981), Reise (1979), Wilson (1981), and others suggest that these kinds of actions occur. Some direct interactions that displace other species have been reported by Levinton (1977), who showed that movement of the mobile bivalve *Yoldia limatula* within the substrate disrupts the burrows of the more sedentary bivalve *Solemya velum*, leading to its replacement. Myers (1977) reported a similar situation in which a burrowing sea cucumber does the same thing to small bivalves.

Indirect interactions leading to changes in community structure and organization appear to be more common in sedimentary communities and usually involve some sort of change in the condition of the sediment.

Burrowing deposit feeders are most abundant in soft, muddy sediments—areas with a high concentration of organic matter. It has been observed in several areas that, where burrowing deposit feeders are prevalent, suspension feeders are rare or absent. The deposit-feeding organisms burrow through the top few centimeters of the bottom and create a very loose, unstable layer of fine particles, including fecal pellets. This layer is easily resuspended by the slightest motion of the water. The resuspended sediment clogs the fine filtering structures of the suspension feeders, making feeding impossible. Furthermore, the constant reworking and settlement of resuspended particles often winds up burying and suffocating the newly settled larvae of the suspension feeders. It does not kill the larvae of the deposit feeders, because they burrow into the more compact substrate underneath. This exclusion of one trophic group through modification of the environment by another is called **trophic group amensalism** by Rhoads and Young (1970). Via such a mechanism, deposit feeders maintain their own communities and exclude suspension feeders. An example is seen in Rhoads and Young's (1970) work in Buzzard's Bay, Massachusetts, where they showed that reworking of the sediments by three deposit-feeding clams—*Nucula proxima, Macoma tenta*, and *Yoldia limatula*—effectively excluded the suspension feeders.

Suspension-feeding organisms seem more abundant where the substrate is more sandy, where there is less organic matter, and where deposit feeders would find food scarce and burrowing difficult. In the absence of deposit feeders, the suspension feeders can establish themselves. Once established, they may further exclude deposit feeders by filtering larvae out of the water.

A more complicated situation occurs over large areas of Cape Cod Bay, Massachusetts, where a large sea cucumber, *Molpadia oolitica*, occurs in high densities (2–6 per square meter) in silt–clay muds. This animal is a sedentary deposit feeder that lives head down vertically in the substrate. It ingests sediment at depths and deposits loose fecal matter at the surface, forming a mound around the anal opening. The reworking of the sediment produces a loose, easily suspended layer of sediment with high water content on the surface of the substrate, except in the immediate vicinity of the fecal cone. The areas between the cones are unstable and exclude suspension feeders (amensalism). The stable cone areas attract the tube-building, suspension-feeding polychaete *Euchone incolor* and other tube-dwelling polychaetes, which in turn help further stabilize the fecal cone area and

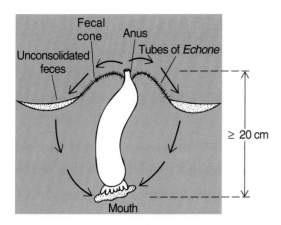

FIGURE 5.7 *Molpadia oolitica*. Diagrammatic representation of feeding position, showing surface cone and depression and subsurface excavation produced by intensive feeding. (From "Animal–sediment relations in Cape Cod Bay, Massachusetts, II." D. C. Rhoads & D. K. Young, 1971, vol 3. Copyright © 1971 Springer-Verlag.)

attract other suspension feeders (Figure 5.7). Note in this example that *Molpadia* is a relatively large animal. To effect changes that will influence other animals, an animal must be large enough or abundant enough to change large areas of the sediment.

Since the formulation and advancement of trophic group amensalism as a general mechanism to explain interactions among feeding guilds and community structure, additional studies have produced mixed results, with some supporting the paradigm and others refuting it. In a review of animal–sediment relationships, Snelgrove and Butman (1994) suggest that trophic group amensalism is not a universal process structuring soft-sediment communities, for several reasons. One is that the two groups of feeders often do co-occur, and neither is always restricted to a given type of sediment. They also point out that some clam and polychaete species are able to switch between suspension and deposit feeding, depending upon hydrodynamic conditions in the water layer nearest the bottom. For example, Levinton (1991) found that three species of the bivalve genus *Macoma* switched from deposit feeding to suspension feeding when the flow rate of water was high enough to suspend organic particles in the water column. Finally, there seems to be little evidence to support the tenet of the trophic group amensalism paradigm that muddy sediments are generally detrimental to either larvae or adults of

suspension feeders when deposit feeders are present. Snelgrove and Butman conclude that associations between sediments and various suspension and deposit feeders are caused primarily by physical processes that produce a given set of sedimentary conditions of which the most significant are the boundary layer hydrodynamics.

Tube-building organisms can stabilize the substrate with their tubes. (See Figure 5.5). They also extend down into the substrate. By stabilizing the substrate, they prevent fine particles from being resuspended. At the same time, the presence of the tubes in the substrate in sufficiently high densities restricts the available space for burrowing animals. The animals cannot burrow through the rigid tubes; thus, the tube builders indirectly restrict burrowing animals. Such exclusion by interference with normal activities is an example of **competitive interference**. Although burrowing deposit feeders rarely co-occur with suspension feeders or with tube dwellers, and vice versa, it is possible to find these different functional groups in the same geographical area, usually with sharp boundaries between them. These boundaries generally do not coincide with any physical discontinuity and are presumed to result from either predation or some other physical or biological disturbance that has removed the resident organisms. Since most animals dwelling in sedimentary environments are solitary and do not reproduce asexually, any patch created by disturbance must be reoccupied either by the immigration of adults or the settlement of larvae. (See pp. 25–29 in Chapter 1.)

The second biological factor that is important in determining the species structure of infaunal communities is predation. Both invertebrate and vertebrate predators take a toll on the infaunal organisms. Predator activity may be responsible for clearing certain small areas of macrofauna and creating a disturbance that will be followed by recolonization. This is one way that patchy distributions occur on the bottom. Such clearing of areas is known to result from the feeding activities of certain flatfishes, rays, whales, and walruses. (See Chapter 3.)

Sediments, as we have noted, are three-dimensional substrates, and it may be possible for some prey organisms to use this three-dimensionality as a refuge from predators. If, for example, the predator cannot dig into the substrate to go after its prey and the prey can burrow deeply, then the depth becomes a refuge for the prey. If a predator can dig as far down into the substrate as the prey can, however, then the substrate provides no refuge for the prey. Moreover, in its digging activities, the predator may have a widespread effect on other members of the infaunal community by destroying or disrupting the surrounding sediments. For example, VanBlaricom (1982) showed that, in a subtidal area off southern California, large rays created pits in the sand in the process of removing resident clams. In digging the pits, the rays destroyed most other infauna, and these areas of substrate became places where resettlement could take place.

Depending on their position and effect on the sediment, predators were classified into three different types by Woodin (1983). **Surface predators** are exposed at the surface and take organisms at or near the surface without disrupting the structure of the sediment (Figure 5.8). Most fishes and various crabs are surface predators, as are certain invertebrates that may be infaunal themselves, that but capture and ingest other infaunal adults, juveniles, or larvae. A peculiar group that needs to be included here consists of browsers that take off only parts of the prey. An example would be the nipping

of siphons from clams by various fish species, as reported by deVlas (1979) and Peterson and Quammen (1982). An interesting "cannibalistic" relationship exists between various suspension feeders and their own and other larvae. These larvae may be captured and ingested as they settle to the substrate and are intercepted by the feeding nets of the suspension feeders. A special case includes such predators as worms of the genus *Glycera*, which form burrows in the substrate, but feed mainly at the surface, emerging from their burrows to seize prey (Ockelmann and Vahl, 1970).

Burrowing predators include those which Woodin (1983) terms "weasel" predators, meaning that they move down various tubes or channels provided by the deeper-dwelling prey and attack them. An example would be various nemertean worms and perhaps some gastropods which have a long proboscis that can be extended down into the tubes. An interesting case is that of the sea star *Pisaster brevispinus*, which feeds on deep-dwelling clams on the Pacific coast of North America. According to Van Veldhuizen and Phillips (1978), *Pisaster* somehow senses the buried clams as it moves across the surface. It then extends the tube feet

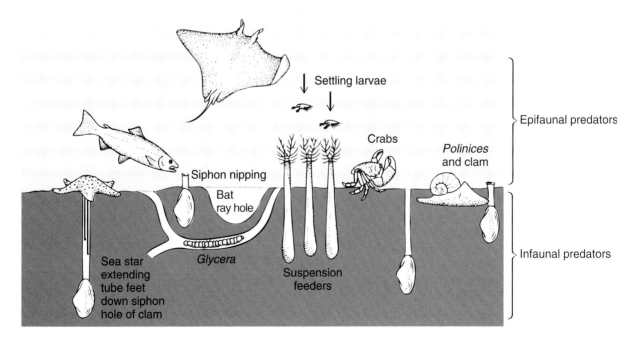

FIGURE 5.8 Various types of predators and their effects on shallow-water sedimentary communities.

from around the mouth area deep into the sediment (it can extend them to a length of 20 cm), attaches them to the clam, and pulls the clam to the surface without greatly disturbing the sediment. *Pisaster* can also extend its stomach down to the clam to digest it (Figure 5.8).

Digging predators excavate holes to get to their food and include such different animals as the blue crab *Callinectes sapidus*, the gastropod *Buccinum undatum*, and the rays. Digging predators have a greater effect on the general infaunal organisms than do weasel predators, because they also disturb the sediment.

A final category includes predators that burrow through the substrate and live within it at all times. These are the **infaunal predators** and include such organisms as nemertine worms and a number of carnivorous polychaete worms.

Different types of predators have different effects on their environment and create diverse patterns in sedimentary communities. In general, the digging predators have the most widespread and recognizable effects, because they create a change in the sediment. Predation by surface predators most likely affects the size classes of prey, particularly if there is a deep refuge for the adults, which leaves the juveniles most vulnerable. The siphon nippers would have their greatest effect on the growth of the prey. The effects of other predators are difficult to predict without knowledge of the prey they prefer, relative predator densities, and other factors.

It is difficult to determine the relative roles of predator-induced and wave-induced disturbances on the structure of soft-bottom communities, mainly because it is difficult to devise good experiments that will exclude predators from areas of the bottom while allowing wave action to proceed. Virnstein (1977) and other ecologists have devised various "cages" that, when placed over areas of the bottom, exclude or include predaceous fishes and invertebrates. Carefully done, such studies have produced increased numbers of infaunal organisms within the cage compared with outside (Figure 5.9). This suggests that predation is important in determining at least the numbers of infaunal organisms that are present and perhaps also the species composition.

It has been particularly difficult to estimate the importance of infaunal predators in determining benthic community structure, because we cannot observe those predators and we have difficulty devising ways of excluding them in manipulative experiments. Indeed, most of the ecological studies of the effects of predation on sedimentary communities have been based on studies of epifaunal predators. In a review, Ambrose (1991) noted that infaunal predators are often abundant and, with few exceptions, seem able to significantly reduce the infaunal prey populations. Interestingly enough, in 63% of the cases the prey was another predator, suggesting that multiple trophic levels may be a characteristic of these soft-bottom communities. Another characteristic of infaunal predators is that they confine their activities mainly to the layers at or near the surface, affecting shallow-dwelling taxa more than deep-dwelling ones.

FIGURE 5.9 Effects of a cage that excludes predaceous starfishes and fishes from the infauna, increasing the infaunal density inside the cage compared with that outside.

Although studies quantifying the ability of infaunal predators to reduce numbers of prey are rare, data from long-term studies in the Wadden Sea in northern Europe have shown that the nemertine and polychaete predators can consume from 20 to 50% of the standing stock of prey. The percentages suggest that predation by these infaunal predators may help explain the absence of an established competitive dominant in manipulative experiments in which the epifaunal predators have been excluded while the infaunal predators remain. Perhaps the exclusion of epifaunal predators does not always result in increased infaunal prey densities because the infaunal predators take the place of the excluded epifaunal predators.

■

WE CAN SUMMARIZE *our understanding of the relative roles of competition and predation on the structure of sedimentary communities by paraphrasing from a review by Wilson (1991), who concludes that there is no evidence of either predation or competition affecting community structure, and there is also no evidence of any competitive dominant in any soft-sediment system. Although we have made much progress in understanding the organization of infaunal communities in the last 25 years, we still lack the ability to predict the effects of either competition or predation. We lack knowledge of growth rates, life spans, and the nature of the food and population dynamics of the dominant soft-bodied invertebrates, particularly deposit feeders. Without these data, it is too early to generate any unified theory.*

■

Most benthic infaunal and epifaunal invertebrates are recruited from larvae settling out from the plankton. Therefore, one primary process that could produce changes in infaunal adult abundances and community structure is recruitment. Despite this fundamental relationship, few studies have analyzed the role of recruitment in determining adult abundances and community structure, probably because it is difficult to isolate the effect of recruitment from various postrecruitment mortality sources. In a review of the role of recruitment, Menge and Farrell (1989) suggest that it may be an influence in structuring adult communities

when one or more of three conditions prevail. First, the recruiting organisms generally must be free of other sources of mortality, such as predation and competition. Second, the recruiting organisms must settle in such numbers that they either swamp the predators with sheer numbers or keep pace with the predation rate. Finally, the recruitment rates of all species in the community must be low, and none of the organisms that settle must be competitive dominants. It is probably safe to conclude that recruitment has the potential to influence community structure in these sedimentary communities or indeed in any benthic community, but the lack of definitive studies permits us only to speculate.

The role of recruitment in determining the structure of soft-sediment communities was reviewed by Olafsson et al. (1994), who concluded that, despite the now-extensive studies on the population dynamics of soft-bottom marine invertebrate communities, there is not yet compelling evidence demonstrating that recruitment shapes the adult communities. However, Olafsson and colleagues also point out the difficulties in separating settlement intensity, mortality, and variability from various postsettlement processes that may also affect mortality, making any conclusions tentative at present. They also note that several postsettlement processes—namely predation by large epifaunal predators, inhibitory adult–juvenile interactions, disturbances of the sediment, and food limitations—could also regulate adult community structure, particularly if several operate jointly.

Another factor that complicates community structure in unvegetated sediments is evidence that there may be significant postsettlement dispersal of juvenile and adult animals, thus giving rise to a constant flux of animals in and out of the community. This transiency is likely to lead to substantial variability in a given community over time.

Soft-sediment animals are subject to various diseases and parasitic infestations that affect communities by causing mortality, influencing reproduction and growth, and altering the behavior of organisms. However, our knowledge of the significance of the effects of diseases and parasitism on species or communities is virtually unknown, except in the case of commercially important species. For example, the protozoan *Perkinsus marinus* infects the oyster *Crassostrea virginica* on the east coast of the United States, where it has decimated oyster populations.

Studies have shown that the rate of infection and the impact on the resident populations vary with environmental conditions.

■ POSITIVE INTERACTIONS

Most ecological studies have tended to focus on what might be called the negative impacts of biological and abiotic factors on community structure. Such studies emphasize the importance of predation, competition, and the interaction of physical factors with organisms as the most important in structuring marine communities. Recently, however, ecologists such as Bertness and Calloway (1994) have suggested that *positive* interactions, or facilitation, may also have significant effects on community structure. Facilitation is defined as the set of interactions among species that positively affect at least one species involved. A common form of facilitation is the modification of the habitat by species that some ecologists term "ecosystem engineers." In unvegetated sedimentary habitats, such engineers include tube-building and mound-building species, as well as reef builders such as polychaetes and oysters. These species facilitate the presence of other species by modifying the basic habitat so that it is suitable for those species. They may also provide refuges from predation. The importance of facilitation with respect to negative interactions is presently under debate among marine ecologists.

■ VERTICAL DISTRIBUTION AND COMPETITION

Competition among benthic marine invertebrates is generally restricted to competition for food and space. In the subtidal soft-bottom areas of the oceans, where the dominant species are suspension feeders and deposit feeders, competition is usually for space. Food, in the form of either detritus or plankton, is abundant. Suspension feeders, of course, must have their feeding structures in the water column. It is possible, however, to get more suspension feeders in a given area of bottom if, instead of placing them all on the surface, they are distributed vertically in the substrate, with only the feeding organ at the surface (Figure 5.10). Depending on the orientation of the animal in the substrate, the feeding structure may take up less room than the whole animal would. Deposit feeders may feed not only at the surface but at any level in the substrate that contains organic material. Such animals do, however, require a connection to the surface to obtain oxygen from the overlying water. In a similar manner, more deposit feeders could be accommodated and the direct competition for space reduced if the different species were distributed vertically in the substrate.

Such vertical distribution of species is a well-known feature of soft bottoms, occurring in both mud and sand (Figure 5.11) and in both suspension- and deposit-feeding organisms. Experiments by Peterson (1977) in which certain species were removed from given depth strata resulted in an increase in the abundance of other species occupying the same depth level. Still other experiments, in which the density of given species at given levels was increased above natural levels, resulted in rapid emigration out of the high-density areas by those species. These experiments suggest that there is competition for space among species, and it results not only in a vertical spacing of species but also in the maintenance of certain densities.

Competition for space may be the reason for the occurrence of **commensalism** among certain species.

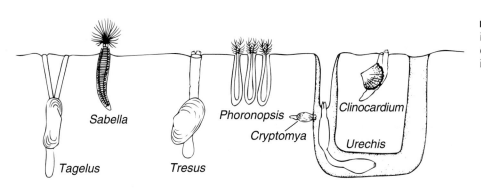

FIGURE 5.10 Various infaunal suspension feeders distributed at different depths in the substrate.

Sabella

Tagelus

Tresus

Phoronopsis

Cryptomya

Clinocardium

Urechis

FIGURE 5.11 Vertical distribution of benthic infauna in (a) an Atlantic coast bay, Quisset Harbor, Massachusetts, and (b) a Pacific coast lagoon, Mugu Lagoon, California. (a from "Ecology of Shallow Water Deposit Feeding Communities" in *Ecology of Marine Benthos*, edited by B. Coull. Copyright © 1977 University of South Carolina Press. Reprinted by permission. b from "Competitive Organizations of the Soft-Bottom Macrobenthic Communities of Southern California Lagoons," C. H. Peterson, *Marine Biology* Vol. 43, 1977, pp. 343–359. Copyright © 1977 Springer-Verlag.)

(a) East Coast

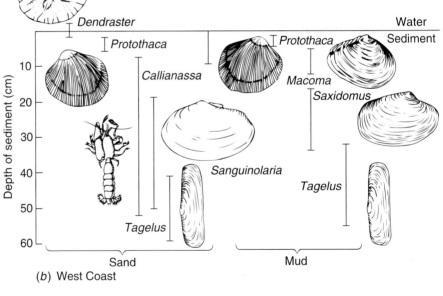

(b) West Coast

(See Chapter 10.) Competition is so severe that it selects for organisms that share space. An example is the clam *Cryptomya californica*, which lives with its siphons in burrows of the shrimp *Callianassa californiensis* (see Figure 5.11) or the worm *Urechis caupo* (see Figure 5.10).

■ COMMUNITY PATTERNS, STABILITY, AND VARIABILITY

Community structure in open sedimentary environments changes over differing time periods and produces certain patterns or patches that recur on different areal and time scales. We shall now take what we know of physical and biological factors and their roles in structuring communities and see how these act together over time to create varying community patterns. It is our purpose here to document the types of patterns and explain how they might have come about and why they change, using some examples from the scientific literature. We note, as Woodin (1983) does, that there is a definite geographical bias to our fund of knowledge: Most well-studied cases come either from the temperate Pacific coast of North America or from the temperate Atlantic coasts of North America and Europe. Thus, the coasts of other continents and those of islands effectively receive short shrift. Still, the two well-studied areas have definite differences in infaunal and predator abundances, the Pacific coast being most dense and, in general, relatively more abundant in large predators than the Atlantic coast. As a result, the relative importance of competition and predation in structuring these communities differs between the two areas.

In some instances, communities are stable over periods of several years, due likely to the dominance of one or a few physical factors or to a dominant animal that controls its own environment and is long lived. A good example is the previously discussed sea cucumber *Molpadia oolitica*. As we noted earlier, *Molpadia*, by its feeding activities, creates a bottom topography of stable cones or mounds interspersed with unstable depressions. By creating these unstable areas, *Molpadia* prevents other organisms—mainly suspension feeders—from settling in the area. The exception is the more stable area of the fecal cones, where the suspension feeders can and do settle and add to the stability of the cones. Such areas, once established, remain for years, perhaps because of the life span of *Molpadia*. Fager (1968) found that, on the southern California coast, a shallow subtidal sand community dominated by nine large epifaunal invertebrates (two sea anemones, one sea pansy, one brittle star, three snails, one starfish, and one hermit crab) remained constant for a six-year period. He also found that this community lacked substantial interspecific interaction, suggesting that the community was structured by physical factors alone.

In certain situations, communities show regular cyclical oscillations between states. For example, in the earlier-mentioned sedimentary system in Barn-

stable Harbor, Massachusetts, Mills (1969) demonstrated that the area was dominated at certain times by the gastropod *Ilyanassa obsoleta*, but at other times by the tube-dwelling amphipod *Ampelisca abdita*. In this case, both physical and biological factors act, but alternately. *Ampelisca*, apparently a dominant space competitor, interferes with *Ilyanassa* feeding, so it outcompetes *Ilyanassa* in seasons with low storm activity, leading to the mat development. In stormy winter weather, however, the tube mats are rolled up by the heavy wave action, freeing the substrate for *Ilyanassa*; thus, this community alternates between two states. Such regular patterns may also result from pulses of reproductive activity and subsequent settlement of the young. Typically, the pattern is to have high numbers of juveniles in spring and summer followed by more or less massive mortality produced by competitive interactions or predation; hence, there are smaller numbers in winter.

The importance of predation in sedimentary environments is demonstrated by a study by Virnstein (1977) in subtidal sand communities in Chesapeake Bay. Virnstein used cages to exclude several large predators, including the blue crab (*Callinectes sapidus*) and two bottom-feeding fish: the spot (*Leiostomus xanthurus*) and the hogchoker (*Trinectes maculatus*). He found that, inside the cages, where predators (at least the epifaunal ones) were absent, the infaunal invertebrates increased in density and diversity. The species that had the largest population increases were the shallow-living opportunistic ones, those most likely to be preyed upon by epifaunal predators. The deep-dwelling species showed little change, primarily because they probably avoid epifaunal predation by their habits.

Although data regarding multiyear changes in sedimentary communities are sparse, we do know that in some cases changes occur at long intervals. One way of obtaining such a pattern would be through the variable recruitment success of the dominants. For example, if a dominant, long-lived species produces many juveniles each year, but in most years all or nearly all die, and then in one year all or most live to grow up, the resulting adult population and the community will show oscillations spaced in years. Such a situation seems to prevail for the amphipod *Pontoporeia affinis* in the Baltic Sea (Figure 5.12). Such a long-term cycle may affect other dominants as

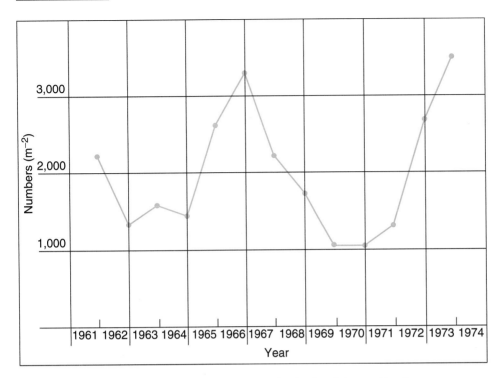

FIGURE 5.12 Graph of the population changes in the density of the amphipod *Pontoporeia affinis* in the Baltic Sea. (From *The Ecology of Marine Sediments*, John Gray. Copyright © 1981 Cambridge University Press. Reprinted by permission.)

well. For example, also in the Baltic Sea, *Macoma balthica* oscillates in dominance with *Pontoporeia* because at high densities of *Pontoporeia* the juveniles of *Macoma* are outcompeted. The oscillatory period is about six or seven years.

The previous discussion centered on the capability of a single factor—or of two or more factors acting alternately—to impose a pattern of stability. In most benthic sedimentary environments, however, there is a mosaic of patches, each with a different structure of species. Usually, the pattern is the result of several factors acting simultaneously or in some sequence. These are the most complicated situations and can be understood best by again resorting to some well-studied examples.

One of the least complicated situations was that studied by Kastendiek (1976) at depths of 2–13 m on sand at Zuma Beach in southern California. Here the dominant species are the sea pansy *Renilla kollikeri*, the sand dollar *Dendraster excentricus*, the nudibranch *Armina californica*, and the starfish *Astropecten armatus*. Off Zuma Beach, *D. excentricus*

forms a dense band (up to 1,500 individuals per square meter) that lies parallel to the shore and runs the entire length of the beach (Figure 5.13). The band moves into shallower water during calm

FIGURE 5.13 Sand dollars, *Dendraster excentricus*, in a characteristic upright feeding position in a subtidal bed in California. (Photo courtesy of Drs. Lovell and Libby Langstroth.)

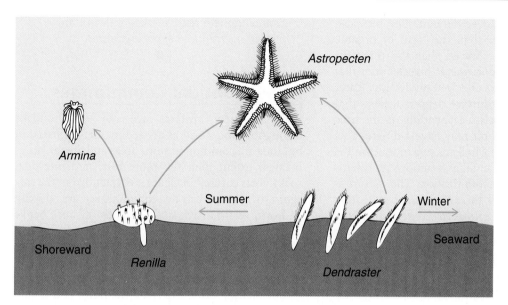

FIGURE 5.14 The food web of a southern California subtidal sand flat.

periods of the year and offshore during stormy times. *Renilla kollikeri* occupies an area inshore of the sand-dollar bed. *Armina californica* is a specialist predator of *R. kollikeri*, and *A. armatus* feeds on both *R. kollikeri* and *D. excentricus* (Figure 5.14). In this system, *Renilla kollikeri* is prevented from moving further shoreward because the increased wave motion in shallower water stops the sea pansy from anchoring itself. The animal cannot extend further seaward because of the sand-dollar bed. When *R. kollikeri* enter the sand-dollar bed, *Dendraster excentricus* uproot and eliminate them. During the time of year when the sand dollars move shoreward, that portion of the *R. kollikeri* bed it occupies is destroyed.

Armina californica and *Astropecten armatus* have a synergistic effect. *Astropecten armatus* is repelled by the expanded autozooids of *R. kollikeri*, which, however, has a characteristic behavior when attacked by *A. californica*: The animal closes up all of its polyps and raises itself up off the substrate. The colony thus acts like a sail. The turbulent water then uproots *R. kollikeri* and tumbles it away from *A. californica*; thus, *A. californica* gets only one or two bites before the escape response. However, a closed-up *R. kollikeri* is readily eaten by *A. armatus*, so escape from one predator is not escape from the other. *Astropecten armatus*, however, is capable of ingesting only *R. kollikeri* up to 40 mm in rachis diameter. *Armina californica*, in turn, is able to consume

R. kollikeri colonies if they become lodged against objects, such as sand dollars.

The sand-dollar bed forms a barrier to *Astropecten armatus*, which normally is distributed seaward of the bed and is not able to penetrate shoreward of it. *Astropecten armatus* also consumes sand dollars. What controls the edges of the sand-dollar beds is not known.

In this system, there are both physical and biological controls on the community. *Renilla kollikeri* is controlled seaward by *Dendraster excentricus* and shoreward by waves, and the population is further regulated by predators, of which one is unable to attack it due to the sand-dollar bed. Kastendiek predicts that, should that bed disappear, *Astropecten armatus* and *Armina californica* will destroy the *R. kollikeri* population.

A final example is one given by VanBlaricom (1982) that demonstrates how a temporal mosaic of patches of different groups of infauna may come to exist in sedimentary environments. VanBlaricom investigated the community structure of an open sedimentary environment at 17-m depth off La Jolla in southern California. In this area, the infaunal community is disturbed by two large rays—the round stingray *Urolophus halleri* and the bat ray *Myliobatis californica*—that dig pits to expose their prey. In so doing, they also destroy most of the other infauna in the area of the pit or expose them to other predators. As a result, the pits represent open areas available for settlement. The pits also accumulate food material in the form of organic detritus. VanBlaricom showed that the pit areas attract first a set of opportunistic invertebrates that are nocturnal swimmers and feed on detritus. Later animals tend to be subsurface deposit feeders that displace the early colonists. Depending on the time since its creation, each pit may have a different set of organisms. This scenario leads to a mosaic on the bottom of differing infaunal associations representing various stages in recolonization. The picture is further complicated by the activities of still other predators. The starfish *Astropecten verrilli* consumes crab larvae soon after they settle; this reduces the abundances of two crabs: *Cancer gracilis* and *Portunus xantusii*. The two crabs, in turn, feed on infaunal organisms, so sea star predation is important to the maintenance of the infaunal community. Finally, the speckled sand dab (*Citharichthys stigmaea*), a fish that is a visual predator, ingests invertebrates stirred up by the rays' digging, further depleting the infauna in the pits.

∎

IN SUMMARY, *marine subtidal infaunal communities of unvegetated sedimentary environments are characterized by patchy distributions of organisms and a certain variability in species abundances and composition over time. The communities are maintained by the settlement of larvae from the plankton. (See Chapter 1.) The ability of larvae to choose areas for settlement, coupled with the ability to delay their metamorphosis, ensures that settlement is not random. The patchiness and variability result from continual random disturbances created by the motion of the water or by biological activity such as predation, which eliminates the population in a small area (a patch). This is followed by recolonization and change with time due to a complex series of biological interactions involving different life-history strategies, (opportunist versus equilibrium), competitive interference, and exclusion and predation. The result is that different areas have different species abundances and composition depending on the time from the most recent disturbance and the relative stability of the substrate.*

Competition for space is ameliorated through vertical stratification of the organisms.

∎

Rocky Subtidal Communities

Rocky subtidal surfaces are not as abundant throughout the world as are sedimentary substrates, but they are common in some areas and are far larger in areal extent than the more restricted intertidal rocky areas covered in Chapter 6. Included in the category of hard-substrate communities are sloping bedrock, vertical rock walls, various offshore banks and pinnacles, and cobble fields, as well as kelp forests and coral reefs. We discuss only the first here; kelp forests are covered in the next section, and coral reefs are dealt with in detail in Chapter 9.

On rocky subtidal surfaces, the floral and faunal composition of the communities is determined in large part by the slope and type of rock. Within the lighted area of the upper sublittoral region in the temperate zone, macroalgae dominate the horizontal and gently sloping substrates, while the vertical rock faces are dominated by various epifaunal invertebrates (Figure 5.15). Several factors contribute to these differences, according to Witman and Cooper (1983). One is that the horizontal surfaces have higher light levels than the vertical faces, thus favoring the algae. Another is that photonegative invertebrate larvae are attracted to the more shaded vertical faces. There is also some evidence that the vertical habitats are more sheltered from physical and biological disturbances than the open horizontal and gently sloping rock surfaces are. In addition, vertical faces may be less susceptible to sea urchin grazing because the urchins have difficulty attaching to such surfaces. (See pp. 218–219.) This generalized pattern of different associations of organisms on different rock surfaces is not absolute, and various invertebrates occur on horizontal and gentle slopes as well. Moreover, the local distribution and abundance of various species in the communities are modified by various biological interactions such as predation, competition, and recruitment. Other physical factors such as water flow, wave action, and sedimentation are also of significance.

The differences in community structure between the horizontal gently sloping areas and the vertical

(A) Gulf of Maine

Key:

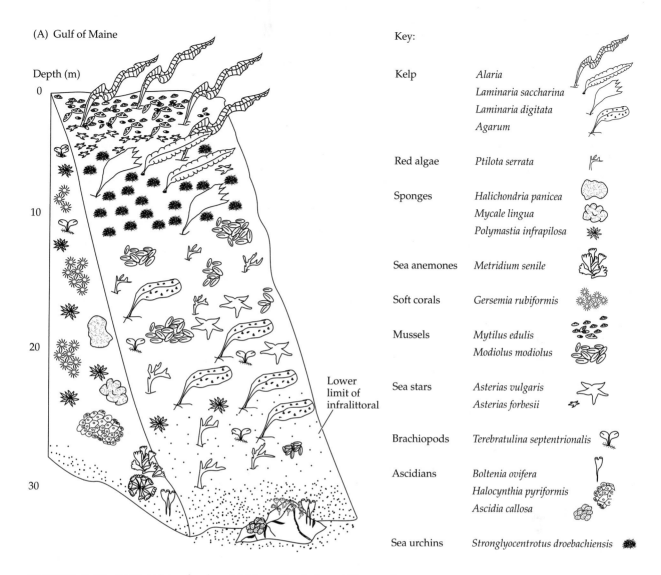

Kelp	*Alaria*
	Laminaria saccharina
	Laminaria digitata
	Agarum
Red algae	*Ptilota serrata*
Sponges	*Halichondria panicea*
	Mycale lingua
	Polymastia infrapilosa
Sea anemones	*Metridium senile*
Soft corals	*Gersemia rubiformis*
Mussels	*Mytilus edulis*
	Modiolus modiolus
Sea stars	*Asterias vulgaris*
	Asterias forbesii
Brachiopods	*Terebratulina septentrionalis*
Ascidians	*Boltenia ovifera*
	Halocynthia pyriformis
	Ascidia callosa
Sea urchins	*Stronglyocentrotus droebachiensis*

Depth (m): 0, 10, 20, 30

Lower limit of infralittoral

FIGURE 5.15 A comparison of rocky subtidal zonation in the cold temperate zone versus the tropics. (a) A typical wave-exposed site in the Gulf of Maine, showing communities on a sloping rock shelf (front slope) versus a vertical rock wall (left-side diagram) down to a depth of 33 m. The lower limit of the algal-dominated sublittoral zone extends to approximately 25 m, as indicated by shotgun kelp, *Agarum cribosum*, on the slope. A shallow bed of laminarian kelp is truncated at about a 10- to 15-m depth by sea urchin grazing. Dense carpets of blue-mussel recruits can cover the substrate down to about 5 m in depth. Not shown in the upper 10 m of the slope community is an understory of red algal turf. Beds of horse mussels, *Modiolus modiolus*, are a conspicuous feature of slopes at a 10- to 20-m depth. Note that macroalgae are absent from vertical rock walls, which are covered by soft corals, sponges, ascidians, and aggregations of the sea anemone *Metridium senile* in the deeper zone. (b) Subtidal zonation at a wave-exposed side of an island in the Galápagos. The sublittoral zone is much shallower in the tropics, here ending at a depth of approximately 5 m. Note that sea fans (gorgonians) and black corals have replaced the temperate subtidal zone's kelp—the major species forming the vertical structure of the community there. Boring sea urchins (*Eucidaris*) increase the spatial complexity of the substrate by excavating holes on sloping, on shelves, and on rock walls. Small aggregations of the ahermatypic coral *Tubastrea* sp. are more abundant on vertical than on horizontal-sloping rock faces. In general, differences in community structure between horizontal-sloping rock surfaces and vertical rock walls tend to be less conspicuous in the tropical rocky subtidal zone. Barnacles are an important component of rocky subtidal communities in the Galápagos, where they are preyed on by fishes, crabs, and lobsters (not to scale). (From M. Bertness et al., *Marine Community Ecology*, pp. 342–343, figure 13.2a &b. Copyright © 2001. Used with the permission of Sinauer Associates, Inc.)

(B) Galapagos

Key:

Algae	Red algal turf	
Sponges	Red encrusting sponge	
Sea anemones	*Anthopleura sp.*	
Hydroids	*Lytocarpus sp.*	
Gorgonians	*Pacifigorgia sp.*	
	Muricea californica	
	Lophogorgia sp.	
Black corals	*Antipathes panamaensis*	
	Antipathes galapagensis	
Corals	*Tubastrea sp.*	
Barnacles	*Megabalanus galapaganus*	
	Balanus trigonus	
Sea urchins	*Eucidaris thouarsii*	
	Tripneustes depressus	
Sea star	*Nidorellia armata*	
Ascidians	*Pyura sp.*	
	Leptoclinides sp.	

Lower
limit of
infralittoral

FIGURE 5.15 continued

faces also occur in the subtropical and tropical subtidal zones, but are less obvious because of the lack of large fleshy macroalgae in both areas and the dominance of the lighted zone by hermatypic corals. Figure 5.15 is a diagrammatic comparison between temperate and subtropical rocky zonation.

The differences in community structure between the horizontal and vertical rock faces decrease as the depth increases; the algae drop out of the picture due to a lack of sufficient light. Below the level of sufficient light for photosynthesis (about 30 meters), various invertebrates dominate all the rock surfaces.

In general, the range in variation in physical factors decreases with depth. Temperature is most variable in the depth range of the thermocline, which is usually between 3 and 12 m in temperate waters. Below that depth, thermal stress is unlikely to affect organisms unless the communities occur where there are large internal waves or periods of intense upwelling. Similarly, physical disturbance from wave action decreases with depth because the wave energy decreases. This effect is substantiated by noting that subtidal communities below a certain depth tend to have few disturbance patches, unless they are affected by severe episodic storms or hurricanes that create substantially larger waves which cause catastrophic mortality. The velocity of flowing water also decreases with depth but is modified by topographical features such as pinnacles and ridges. As with seamounts, the flow speed increases across pinnacles, banks, and ridges. This enhanced flow increases the flux of food particles in the water and enhances the abundance of suspension feeders in the benthos. Sedimentation increases with the decrease in water movement; hence, in the rocky subtidal zone the deeper areas are subject to greater sedimentation. According to Witman and Sebens (1988), this sedimentation may be the cause of the decline in suspension-feeding organisms at depths below 50 m in Maine.

There are also biological gradients with depth in the subtidal rocky areas. Grazing pressure tends to be lowest in the first few meters of the subtidal zone, due to the effects of wave action and extreme water movement. For example, sea urchins and certain snail grazers are often missing from the shallowest subtidal areas that are affected by waves, but become abundant at depths where the motion of the water has diminished. This phenomenon has been noted in the Gulf of Maine, in southern California waters, in the Galápagos Islands, and in the Seychelles. The most probable reason for it is that the urchins and snails cannot maintain their attachment in such turbulent areas. By contrast, certain predators, such as various sea stars, seem to be able to feed in these high hydrodynamic environments, likely due to their higher attachment strength and low profile. Highly mobile predators such as lobsters and crabs can forage in the shallowest subtidal zone only during the time the hydrodynamic forces are below a threshold level that would sweep them away. Thus there is a relationship among the hydrodynamic forces and the biology of the consumers. Competition among various algal species also seems to decrease with depth as the density of macroalgae organisms thins out and certain species are eliminated due to the decreased availability of light. Similarly, the competiton among the large algae and various invertebrates decreases with depth as the algae dwindle or finally drop out. Within the shallow competition zone the algae have a deleterious effect on the sessile invertebrates by overgrowth and by a sweeping action of their fronds, which removes or damages the adjacent invertebrates. Finally, the decreased flow velocity of water with increasing depth implies that various invertebrates with planktonic larvae have larger dispersal ranges in the shallow, turbulent subtidal area than those living in deeper zones have, provided that the larvae remain in the same depth zone.

As noted earlier, the subtidal, hard substrates are densely covered with low-growing plants and sessile invertebrates (Figure 5.16). The prominent motile animals are usually sea urchins, crabs, and gastropod mollusks. There is little, if any, open primary space (Figure 5.16). Most of these surfaces are without

FIGURE 5.16 Anemones of the genus *Tealia* and the hydrocoral *Stylantheca* (*Allopora*) *porphyra* with surrounding tunicates, sponges, and coralline algae on subtidal rocks of central California. (Photo courtesy of Drs. Lovell and Libby Langstroth.)

significant topographic relief, giving a homogeneous two-dimensional surface. Nevertheless, they have a high species diversity or species richness, leading biologists to ask how so many species can coexist in such a relatively unstructured flat area. In contrast to the sedimentary environment, there is no third dimension (depth) to be occupied by additional species. Furthermore, there does not appear to be a fine partitioning of the ecological niches, as in other environments. Most of the animals of this area are either colonial encrusting species that spread over surfaces by asexual means or single individuals that settle gregariously, forming large, space-occupying clumps. Both groups are suspension feeders. Included in the first group as dominants are sponges, tunicates, bryozoans, and various cnidarians, such as anemones and hydroids. The second group is composed of bivalve mollusks, algae, sea urchins, and gastropod mollusks. These animals do not appear to be food limited.

It now appears from studies done primarily in Massachusetts by Sebens (1985, 1986), in Maine by Witman (1987), and in Australia by Fletcher (1987) that the organization and persistence of communities in the rocky subtidal areas, as well as the co-occurrence of many species, are explained by a combination of biological and physical factors that include disturbances from storms, competition for space, grazing, recruitment, and mutualism. The disturbances generate patches of free space that are subsequently colonized by larvae or juveniles of nearby adults or by the spread of the adjacent colonial organisms. Competition among these organisms for the limited space, coupled with the seasonality of certain organisms and mutualistic interactions among a number of species, completes the cycle. This means that, in any area at any time, either there is a mosaic of patches of different ages with different sets or abundances of species, resulting in the high species diversity, or there is a different zonation pattern. Now let us examine how each of these comes about using a few examples. In the Massachusetts subtidal zone, Sebens (1985, 1986) found that the most significant biological disturbance was the sea urchin *Strongylocentrotus droebachiensis*. Where these urchins have been abundant for a period of years on flat or sloping surfaces, they have grazed off most sessile and sedentary organisms, leaving only a thin layer of encrusting coralline red algae (*Lithothamnion, Phymatolithon, Clathromorphum*). Although other animals also graze the bottom areas, none appear to have the effect of the urchin. Within this gene rally depauperate grazed area, there is one refuge from the

urchin grazing and other predation: the beds of the mussel *Modiolus modiolus*. Experiments by Witman (1985) demonstrated that the consumption of several prey items was much greater outside than inside the mussel beds. Two groups of predators patrol these areas: a diurnal set consisting of several fishes, and a nocturnal set of crabs (*Cancer irroratus* and *C. borealis*) and the lobster (*Homarus americanus*).

Where the urchins are rare or absent, the hard bottom consists of a 10- to 20-cm-thick canopy of fleshy algae (Figure 5.17). Below this canopy are encrusting coralline algae and patches of various colonial animals, such as ascidians, plus small, motile crustaceans and the tubes of polychaete worms and amphipod crustaceans.

Vertical rock walls in this area usually are inaccessible to urchins and have a different association of organisms. On these walls are a variety of encrusting, mostly colonial invertebrates dominated by a colonial compound ascidian, *Aplidium pallidum*, the sea anemone *Metridium senile*, the octocoral *Alcyonium siderium*, and several sponges (Figure 5.17). This community appears to be structured by competition among the organisms for space. Since most of these organisms are colonial, competition involves the direct overgrowth of one by another whenever their edges meet. The species on the vertical faces have been shown by Sebens (1985, 1986) to exist in a competitive hierarchy that ranges from those able to overgrow all others to those which can be overgrown by all others. In general, the colonial species with the larger or more massive growth forms are competitively dominant. Thin, flat colonies are competitively inferior.

The experimental addition of urchins to vertical faces results in the removal of all the encrusting forms except the coralline algae. Caging urchins out of horizontal surfaces results in a dense growth of foliose algae.

The situation described would seem to lead to monotonous landscapes of sea urchins and encrusting coralline algae on horizontal surfaces and to massive aggregations of the dominant space competitor, *Aplidium*, on vertical walls. This, however, does not happen, and there are horizontal areas of foliose algae and vertical walls with various numbers of other, inferior space competitors. How does such a distribution come about? It has to do with a number of life-history and growth features of some of these invertebrates. The sea urchins, for example, are preyed on by fishes, several crabs, and (uncommonly) lobsters. They are also destroyed by an undetermined disease and winter

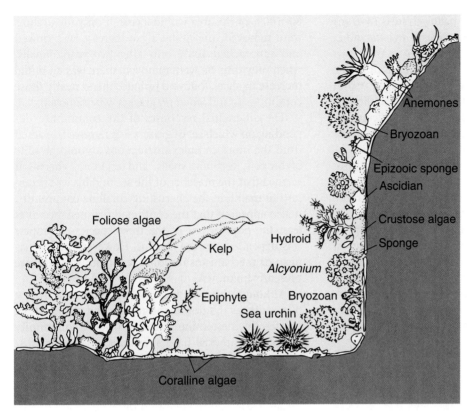

FIGURE 5.17 A diagrammatic representation of the characteristic flora and fauna inhabiting the vertical and horizontal surfaces of some Massachusetts subtidal rocks. Horizontal surfaces are dominated by a canopy of foliose algae and other epiphytic species. If the surfaces are being actively grazed by sea urchins, coralline algae predominate. Vertical surfaces support diverse assemblages of invertebrates and algae that compete for space; some of these species are seasonal, such as the erect bryozoans and hydroids that form a canopy layer during the warm months. (Modified from "The Ecology of the Rocky Subtidal Zone," K. Sebens, *American Scientist*, Vol. 73, No. 5, pp. 548–557, 1985. Copyright © 1985 American Scientist. Reprinted by permission.)

storms. Whenever massive urchin mortality occurs in an area, the community changes and will remain in that alternate state until urchin grazing again reduces the area to crustose corallines. This may take some time, so we get patches. Also, urchins are not able to gain access to vertical walls, because they cannot attach themselves to the colonial animals. If some disturbance does clear the vertical surface of *Aplidium*, the urchins may gain access to that patch of vertical surface and keep it grazed down to corallines for a long period. Such an area may exist next to a vertical face with a well-developed association of colonial animals dominated by *Aplidium*.

Inferior colonial competitors may coexist with *Aplidium* for several reasons. In one example, the erect octocoral *Alcyonium* tends to grow upward as a juvenile. At the same time, *Aplidium* refrains from reproducing for its first two years. On an open surface, therefore, if *Alcyonium* and *Aplidium* settle at the same time, the time lag in *Aplidium* growth will allow *Alcyonium* to grow up high enough to avoid overgrowth. Although *Aplidium* will eventually grow over the surface, it will not displace the *Alcyonium*, which expands above it, maintaining only a narrow attach-

ment to the rock under the *Aplidium*. *Alcyonium* also inhibits urchin grazing, either physically, by presenting a vertical obstruction, or through the action of its nematocysts on the urchins. This, in turn, may aid the other members of the wall community. *Alcyonium* has crawling larvae that will not mature unless they settle more than 2 cm away from the parent colony. They also need a certain type of substrate—either rock or coralline algae. Since neither is usually present where the adults are, there is usually no recruitment. Because the dispersal is so space limited, there may be walls that lack any of the species only meters away from adult *Alcyonium*. The anemone *Metridium*, also an inferior space competitor, is found in patches but because of a different set of factors. Once established, *Metridium* holds its space by using its special "sweeper tentacles," heavily armed with nematocysts, to actively deter any other competitor. Meanwhile, *Metridium* reproduces asexually by fragmenting and forms a dense clone that can persist for years.

Further north, in the Gulf of Maine, Witman (1987) investigated subtidal rocky surfaces on extremely exposed sites at depths of 4–20 m. Here, the deeper surfaces (11–18 m) were dominated by the mussel

Modiolus modiolus, while the shallower areas (4–8 m) were dominated by the kelps *Laminaria digitata* and *L. saccharina*. Experimental manipulations in these two communities demonstrated the causes of this dramatic zonation pattern. The key species was again the urchin *Strongylocentrotus droebachiensis*. The shallower-dwelling kelps were prevented from moving deeper by the sea urchins grazing on them, thus maintaining the integrity of the mussel bed. Storm-generated dislodgment was the most significant factor in the mortality of the mussels. In the absence of urchins, the kelps would attach to the mussels, making them 30 times more subject to dislodgment and mortality by storms. Therefore, the urchins increase *Modiolus* survivorship by removing the kelp and decreasing the probability of dislodgment. This *Modiolus*–*Strongylocentrotus* interaction is mutualistic in that it takes significantly more force to dislodge the urchins from the *Modiolus* surfaces than from surrounding bare rock surfaces or kelp beds. The *Modiolus* beds also provide a refuge from predation for the recently recruited small urchins.

Finally, the ability of both kelps and mussels to recover from dislodgment due to storm disturbance indicated that the kelp recovery was much faster, the kelp occupying all open patches. This fast recovery and spreading suggested the kelps were competitively superior. Indeed, in the absence of sea urchin grazing, they would probably dominate the entire depth range. Similar subtidal zonation patterns, with a shallow zone dominated by kelps and a deeper zone dominated by urchins, have been reported in other temperate regions in Nova Scotia and New Zealand and suggest a more universal pattern, perhaps maintained by a similar mechanism.

A somewhat similar mechanism was found by Fletcher (1987) to maintain subtidal rocky areas on the east coast of Australia. In this area, the subtidal rock is covered with encrusting coralline algae on which are found high densities of a suite of grazers that includes a sea urchin, *Centrostephanus rodgersii*; three limpets, *Patelloida alticostata*, *P. mufria*, and *Cellana tramoserica*; and two turbinid gastropods, *Australium tentiforme* and *Turbo torquata*. Manipulative experiments here, however, gave different results depending on which group of grazers was removed. The removal of all grazers resulted in the rapid colonization of the crustose algae by fleshy, foliose algae. If only the turbinid snails were removed, there was no significant change in the community. If limpets and urchins were removed, the fleshy algae covered 80–100% of the area within a year. If only the urchins were removed, the fleshy algae increase was slowed and approached 100% only after two years. Finally, when only limpets were removed, there was an initial increase in algae, followed by a decline to nearly those conditions found when no grazers were removed.

The differential responses of the community, depending on which set of grazers was removed, resulted from the more complex interactions among the suite of limpets, turbinid snails, and urchins. Whereas it seemed that the presence of the sea urchins was necessary to maintain the encrusting coralline community, it also appeared that the continued maintenance of at least the populations of the three species of limpets was dependent on the urchins. However, the mechanism or mechanisms responsible for this situation are presently unknown.

Although these studies emphasize the importance of urchins as the dominant grazers and the keystone species in determining the type of community formed on subtidal hard substrates, one cannot discount the importance of other grazing animals. This has been demonstrated by Ayling (1981), who investigated three communities in the subtidal zone of New Zealand, each dominated by a different group of animals: (1) encrusting coralline algae, (2) sponges, and (3) ascidians. There were three groups of grazers: an urchin, *Evechinus chloroticus*; a suite of herbivorous gastropods; and a grazing fish, *Parika scaber*. At high densities, urchin grazing was responsible for the perpetuation of a low-diversity community consisting primarily of crustose coralline algae. However, the removal of the suite of grazing gastropods resulted in a change in the community to one with more fleshy ephemeral algae, even in the presence of the urchins. Removal of the grazing fish, which fed primarily on sponges and ascidians, caused a major shift in the community structure from one dominated by crustose coralline algae to one dominated by ascidians and sponges. Finally, disease and parasites also played a role: Ayling reported episodic outbreaks of a bacterial or fungal disease that destroyed large numbers of the otherwise long-lived sponges.

■

IN SUMMARY, *on hard substrates we have a two-dimensional surface. In areas so far studied, all in cold or warm temperate seas, the communities are dominated by (1) an encrusting community of coralline algae, ascidians, or sponges that compete for space among themselves by overgrowth; (2) beds of mussels forming large clumps; or (3) kelps. In all cases, sea urchins appear to be the keystone species, for without them, the community changes. Where urchin densities are high, the communities are dominated by crustose coralline algae or mussels, both highly resistant to urchin grazing. Where urchins are absent or in lesser numbers, communities of fleshy algae or kelp, or both, dominate. Physical disturbance, primarily by storms but also by disease and predation, causes mortality among the urchins, freeing the more ephemeral algae or kelps to occupy more space. This "typical" picture may be modified by suites of smaller gastropods and fish grazers, both of which may alter the community structure to favor certain encrusting organisms. Finally, the differences in the life histories of various competitively inferior organisms may lead them to persist even in the face of their superior competitors.*

■

The existence of different groups of organisms or patches of organisms in the same area is what Sutherland (1974) has called **alternate stable states**. These states result when some factor or set of factors disturbs the dominant species, causing a community to switch to a new composition, which in turn remains stable until another significant disturbance occurs.

Although there has been some argument among ecologists as to whether alternate stable states is a real phenomenon, the fact remains that very different community structures can and do exist near one another under virtually identical environmental conditions. This sort of coexistence suggests that some perturbing factor has caused a community to change from one configuration to another and that there is some mechanism which allows the two configurations to persist. One of the most intriguing examples of this phenomenon is that reported by Barkai and McQuaid (1988) with respect to two subtidal rocky communities in South Africa. Off the west coast of South Africa are two islands: Malgas and Marcus. They are separated by 4 km, so they share similar environmental conditions. However, the rocky subtidal communities are quite different in each. At Malgas Island, the rock lobster *Jasus lalandii* and algae dominate the community, and the lobsters make up 70% of the biomass. Few other organisms, save for a few gastropods of the species *Burnupena papyracea*,

are found. At Marcus Island, by contrast, rock lobsters are rare, and the community is dominated by extensive beds of the mussel *Choromytilus meridionalis*, large populations of the carnivorous gastropod *B. papyracea*, and numerous holothurians and sea urchins. Why are there such extreme differences, and how are they maintained? Through experimental manipulations, Barkai and McQuaid demonstrated that on Malgas Island the rock lobsters prey on the settling mussels, preventing their establishment. They also prey on the gastropod *Burnupena* but are deterred from preying on those gastropods which have a protective bryozoan (*Alcyonidium nodosum*) on their shell. As a result, some gastropods continue to persist in the face of heavy lobster predation. At Marcus Island, when rock lobsters were transferred into the subtidal zone, all were attacked and killed immediately by the carnivorous gastropod *Burnupena*, reversing the normal predator–prey relationship established at Malgas Island. Since there are normally no lobsters at Marcus Island, what do the *Burnupena* prey on? The answer seems to be that they scavenge on dead or dying mussels, but are unable to successfully attack healthy mussels. Therefore, it appears that predation maintains the current communities on both islands, but instead of differences in predators at each, the change is maintained by a reversal of predator–prey roles (Figure 5.18). These two different communities persisted through the four years of the study as alternate stable states maintained by the very unusual mechanism of predator–prey reversal. Since the local fishermen report that the two islands had similar lobster populations 20 years ago, the unanswered question is how the change came about.

The concept of alternate stable states is not limited to subtidal communities, but is applicable to many marine communities, and we shall see it again in our considerations of intertidal and coral reef systems.

Kelp Beds and Kelp Forests

Throughout a large part of the cold temperate regions of the world, hard subtidal substrates are inhabited by a community dominated by very large brown algae known collectively as **kelps** (Figure 5.19). These associations are technically known as **kelp beds** if the algae do not form a surface canopy and **kelp forests** if there is a floating surface canopy (Figures 5.20 and 5.21). The terms, however, are often used interchangeably.

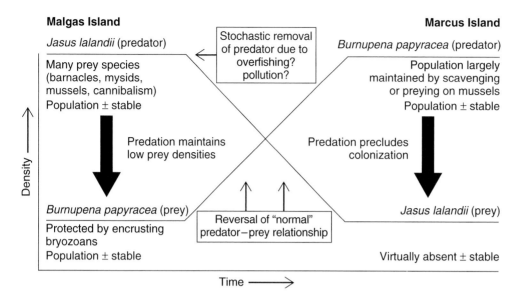

FIGURE 5.18 Schematic summary of rock lobster and whelk predator–prey role reversal at Marcus and Malgas islands. (Reprinted with permission from "Predator–Prey Reversal in a Marine Benthic Ecoysystem," A. Barkai & C. McQuaid, *Science*, Vol. 242, pp. 62–64, 1988. Copyright © 1988 American Association for the Advancement of Science.)

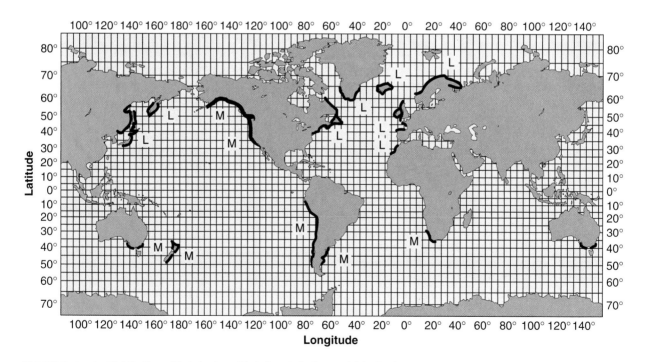

FIGURE 5.19 Distribution of kelp beds and kelp forests in the world, indicating the dominant genera. L: *Laminaria*; M: *Macrocystis*. (Reprinted with permission from "Seaweeds: Their Productivity and Strategy for Growth," K. H. Mann, *Science*, Vol. 182, pp. 975–983. Copyright © 1973 American Association for the Advancement of Science.)

FIGURE 5.20 A kelp bed composed of an overstory of *Laminaria dentigera* and an understory of the surfgrass *Phyllospadix*. (Photo courtesy of Drs. Lovell and Libby Langstroth.)

■ STRUCTURE AND DISTRIBUTION

Kelps are attached to the substrate by a structure called a **holdfast** rather than by true roots (Figure 5.22). From the holdfast arises a stemlike or trunklike **stipe**, which ends in one or more broad, flat **blades**. In some kelps, at the base of the blade is a **pneumatocyst**, or *float*, which keeps the blade upright, sometimes at the surface. Kelps obtain their nutrients directly from the seawater, as do the organisms of the phytoplankton. Kelps depend on the constant movement of water past them to avoid exhausting their nutrients. Since these shallow waters are constantly moved by wave action and currents, and the nutrients are replenished by turbulence, upwelling, and runoff from land, nutrients are not depleted over most of the range. However, at the southern end of the range in North America, from San Diego to the southern limit of kelps in Baja California, nutrients can be depleted in late summer and fall and during El Niño events. Similarly, at the

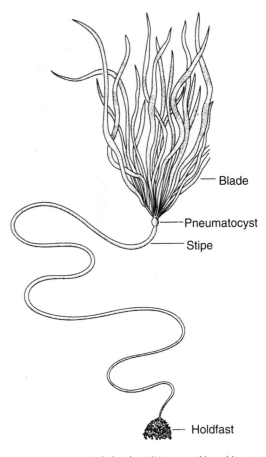

FIGURE 5.21 A view inside a giant kelp forest dominated by the kelp genus *Macrocystis* on the California coast. (Photo courtesy of Drs. Lovell and Libby Langstroth.)

Blade

Pneumatocyst

Stipe

Holdfast

FIGURE 5.22 A kelp plant (*Nereocystis*) and its structure.

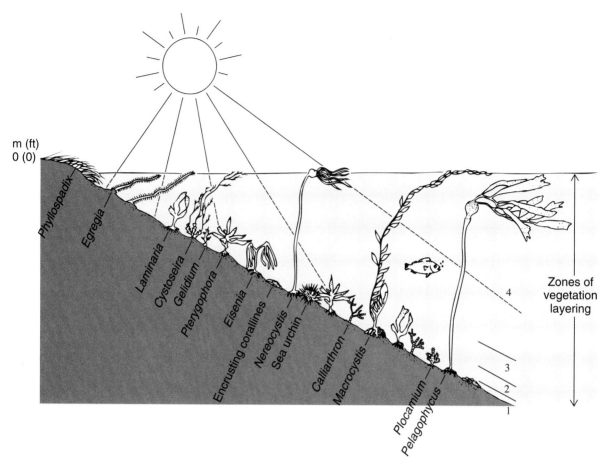

FIGURE 5.23 The distribution of some common seaweeds within giant kelp forests on the Pacific coast of North America. Plants in the four zones of vegetation layering include (1) small filamentous species and encrusting coralline algae; (2) bottom-canopy plants, such as *Gelidium, Calliarthron,* and *Plocamium*; (3) understory-canopy kelps, such as *Pterygophora, Eisenia,* and *Laminaria*; and (4) midwater and surface-canopy plants, such as *Egregia, Macrocystis,* and *Nereocystis*. This is a generalized diagram and some species do not necessarily occur at the site shown. (Modified from M. S. Foster and D. R. Schiel, 1985, *The ecology of giant kelp forests in California: A community profile,* U.S. Fish and Wildlife Service Biological Report 85 (7.2).)

southern end of the kelp range in eastern North America, nutrients can be depleted but due to summer warming not El Niño.

In contrast to the monotonous, level landscape of the soft-bottom or the short-growing colonial organisms of other subtidal rock surfaces, the kelps can form an extensive three-dimensional habitat composed of several vertical layers or strata (Figure 5.23). As a result, a large number of potential habitats are available, and the variety of life is greater.

The major kelps that dominate the structure of the kelp forests are the genera *Macrocystis* and *Nereocystis*. Kelp beds are dominated by the genera *Laminaria, Pterygophora,* and *Ecklonia*. The Pacific coast of both North and South America is dominated by *Macrocystis,* but *Lessonia* is common in Chile. *Laminaria* is dominant in Atlantic waters, in the Sea of Japan and in the Sea of Okhotsk. (See Figure 5.19.) In contrast to most algae, which are small and generally do not exceed a few centimeters or a decimeter in length, the major kelps are giants, with lengths the equivalent of the height of trees on land. *Macrocystis* and *Nereocystis* on the Pacific coast of North America may reach 20–30 m in length. Such massive autotrophs grow upward from the bottom and spread their blades at the surface of the water, where they obtain the maximum amount of light. The blades form a canopy similar to that of terrestrial forests, cutting off light to the substrate below

FIGURE 5.24 (a) The surface of a kelp forest, showing the thick canopy of blades. (b) The interior of a California kelp bed. (a, photo courtesy of Peter Arnold; b, photo courtesy of Norbert Wu.)

(Figure 5.24). Below the canopy, extending up from the bottom, is a set of "understory" algae forming another layer (Figure 5.25). Finally, the ground-cover algae and turf community form the last layer.

The extent of the kelp beds and forests on various coasts depends on several factors. First, for most species, a hard substrate must be available for attachment. Second, the kelps must have light and thus can establish themselves only in water depths where the young, small autotrophs receive enough light to grow. The amount of light received at any depth is a function of the clarity of the water, and where clear water prevails, kelp beds and forests extend out from shore to depths of 20–30 m. Where such shallow water is extensive, the beds may extend as much as several kilometers from shore. In more turbulent, murky water, the depth of the beds may be more restricted. Finally, kelps seem limited by high temperature and associated low nutrients, as well as by grazing. Kelp beds and forests occur throughout the world wherever the water is cool, but they are absent from most warm temperate and tropical areas. Their considerable extent along the Pacific coast of North and South America is due to the cold upwelling on the North American coast and the cold water of the Humboldt Current along the South American coast.

■ ECOLOGY AND LIFE CYCLE

Kelps have a phenomenal growth rate. Scagel (1947) reported that *Nereocystis luetkeana* grew by 6 cm/day, and, according to North (1971), *Macrocystis pyrifera* commonly grows 20 cm/day on the California coast. Such growth rates mean that plants grow from the bottom to the surface in a short time. Kelp forests and beds are also extremely productive. Mann (1973) reports that their net annual productivity ranges from 800 g C/m^2 in California kelp beds to perhaps 2,000 g C/m^2 in Indian Ocean kelp beds. These figures are several times the production of the phytoplankton on the same per-area basis.

Some kelps are annuals while others are perennials. Perennial kelps often lose their stipes and blades, but

FIGURE 5.25 Cross section showing the inhabitants of a generalized giant kelp forest. The numbers to the right indicate vegetation layers. (See legend for Figure 5.22.) Three broad zonal associations along the depth gradient are shown: (Z1) in-shore of the giant kelp community; (Z2) within the giant kelp community; and (Z3) offshore from the giant kelp community. Various subcommunities or associations are indicated by the circular diagrams: (a) animals associated with the surface of *Macrocystis* and other seaweeds (polychaetes, isopods, bryozoans); (b) animals found in giant kelp holdfasts (small sea urchins, brittle stars, crustaceans, polychaetes; although shown on the outside, these organisms occupy the spaces within the holdfast); (c) plants and animals characteristic of horizontal surfaces (various sea stars, urchins, benthic fishes, understory algae); (d) organisms most common on vertical surfaces (primarily sessile animals, such as sponges, ascidians, bryozoans, sea anemones). Some of the organisms shown do not necessarily co-occur at any one site. (From M. S. Foster and D. K. Schiel, 1985, *The ecology of giant kelp forests in California: A community profile*, U.S. Fish and Wildlife Service Biological Report 85 (7.2).)

regrow new ones from the holdfast. The life span of *Macrocystis pyrifera* plants in California usually ranges from one to seven years. The life cycle of kelp alternates between an asexual macroscopic stage termed the **sporophyte** and a sexual microscopic stage termed the **gametophyte** (Figure 5.26). Only the sporophyte stage is considered here, because almost all of the ecological work has been done with this stage and little is known about the ecology of the gametophyte in the natural environment. However, studies by Reed et al. (1988) and Deysher and Dean (1986) have suggested that patterns of recruitment of sporophytes, at least in southern California, are determined in part by episodic, infrequent occurrences of "recruitment window" conditions that affect the gametophytes.

The kelps provide the framework for the kelp forest or kelp bed community. Associated with these dominants are many other species of algae, invertebrates, fishes, and occasionally, marine mammals such as the sea otter.

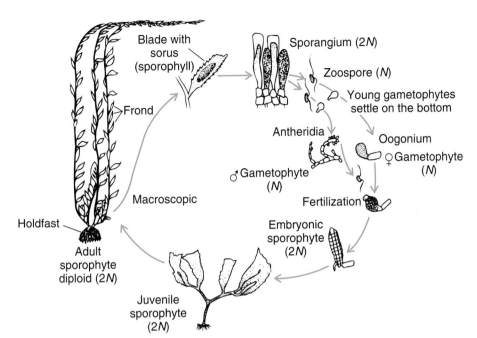

FIGURE 5.26 Life cycle of a typical kelp plant (*Macrocystis*). (From M. S. Foster and D. K. Schiel, 1985, *The ecology of giant kelp forests in California: A community profile*, U.S. Fish and Wildlife Service Biological Report. 85 (7.9).)

Despite the enormous productivity of the kelps, relatively few herbivores graze directly on them. It has been estimated that only 10% of the net production enters the food webs of the kelp bed through grazing. The remaining 90% enters the food chains in the form of detritus or dissolved organic matter.

Since the general economy of kelp beds and forests depends on the continued existence of the giant kelps, it is of interest to know the causes of kelp mortality. Adult plants are only occasionally destroyed by grazing herbivores, but they are vulnerable to destruction by mechanical forces—mainly wave action—and to nutrient depletion—mainly nitrogen. Since kelps occur in shallow water, often on open coasts, storm waves can have a devastating effect. The waves rip the plants from the substrate and cast them up on the beach. Furthermore, once a plant has been uprooted, its movement through a bed, as Rosenthal and associates (1974) have demonstrated, often entangles the plant among the stipes and blades of other plants. The result is that even more plants are pulled away from the substrate and destroyed. Severe storms can virtually denude kelp beds of their stipes and blades, but if the holdfast and some meristem tissue from the kelps remain, new stipes and blades can be produced.

The large El Niño events that produced devastating effects on the coral reef communities in the Pacific Ocean (see p. 452) also had catastrophic effects on the extratropical kelp forests. Those events around the California coast have been best documented. The storm seasons of 1982–83 and 1997–98 were perhaps the worst of the last century, according to Tegner and Dayton (1987) and Dayton et al. (1999). These storms caused massive mechanical damage to the kelp forests. In 1982 at San Diego, the Point Loma kelp forest was reduced from 600 hectares to 40 hectares. Off the Palos Verdes Peninsula in Los Angeles, the cover was reduced from 196 hectares to 18. In addition to destroying much of the canopy, the severity of the waves ripped up holdfasts, leading to further destruction. Finally, recovery in many places, particularly in southern California, was delayed because plants not directly killed by the storms were retarded in growth by the high water temperatures and the reduction in nitrogen that accompanied the storms. Devastation from El Niño was not universal, and some areas seemed to escape damage for unknown reasons. Among these areas were the sandy ones around Santa Barbara, where old holdfast material served as recruitment and growth centers. Most of these forests recovered in a few years. Dayton et al. (1999) describe a similar cycle of crash and recovery in southern California kelp forests following the 1997–98 El Niño.

In this case the recovery of the Point Loma kelp forest was delayed, according to Graham (2002), by an episodic explosion of grazing amphipods that denuded the new kelp plants.

Whereas storms—especially severe ones—are usually devastating to existing kelp forests, they may have vastly different effects on the subsequent community that develops, which is dependent on the state of the community before the arrival of the storms. Ebeling et al. (1985) made regular observations on a southern California kelp forest over a five-year period. During that time, the forest was subjected to two severe storms. The first removed the canopy of *Macrocystis*, but left the understory kelps, such as *Pterygophora californica*, intact. This resulted in a drastic reduction of the drift kelp used as food by the urchins *Strongylocentrotus purpuratus* and *S. franciscanus*. The urchins then emerged from their sheltered cracks and crevices and, in the absence of effective predators, consumed most of the remaining living kelp. This, in turn, weakened the other parts of the food web and led to further declines in kelp forest associates, such as fishes. The second storm in the area caught the urchins in the open, destroyed them, and cleared many surfaces. The newly opened surfaces were then colonized by kelp plants that produced a fully developed kelp forest a year later. Thus, two storms led to dramatically different communities, which were dependent on the type of community existing immediately prior to the storm.

Another potential cause of mortality is grazing, particularly by various sea urchins and fishes. Here the evidence is more equivocal. There is little doubt that sea urchins of many species can have significant effects on kelp beds and forests, ranging from wholesale removal of plants to selective feeding and even to the provision of suitably cleared space in which new kelp plants may settle. The creation by urchins of so-called barren grounds has been reported in many regions in which kelps occur. In recent years, greatly increased numbers of urchins in certain areas, such as New England and southern California, have coincided with a diminished abundance of kelps—undoubtedly because of the urchins' grazing pressure. However, the causes of the increased urchin abundance are not always clear. In southern California, for example, the urchins do not have to undergo a population explosion; they can simply change their behavior from feeding on drift algae to actively foraging on live plants, or vice versa. Such a change in behavior

has been demonstrated by Harrold and Reed (1985), who studied the grazing activity of red urchins (*Strongylocentrotus franciscanus*) in adjacent barren and kelp-dominated areas. In barren areas, they found drift algae sparse; hence, the urchins were poorly nourished and were actively grazing all surfaces. In the kelp areas the urchins were well nourished, lived mainly in cracks, moved little, and fed on abundant drift kelp.

During Harrold and Reed's study, a substantial recruitment of kelp occurred on both the barren site and the kelp site. On the barren site the large recruitment transformed the site into a kelp-dominated area, increasing the amount of drift algae. The urchin population accordingly underwent a behavioral transformation, moving into cracks and crevices, feeding on the drift algae, and reducing active surface grazing to levels found in the kelp-dominated areas. In effect, there was a change in the community structure facilitated by a behavioral change in the dominant urchin, which was triggered by a switch in the mode of feeding. The switch was dependent on the availability of drift algae, which was in turn dependent on the abundance of new kelp. In this case, the grazing intensity was independent of the urchin density, and the ability of the kelp to colonize a barren area successfully seemed regulated by a favorable set of hydrographic conditions (Figure 5.27).

The previous studies demonstrate that it is not necessary to alter the number of urchins to effect a change in the kelp forest community structure and that storms can have a dramatic effect on both urchins and kelps. But what about predators? Until recently, it had been assumed that urchin populations are kept in check by various predators and that many, if not all, urchin population explosions are the result of dramatic decreases in numbers of the major predator. In other words, urchins are capable of destroying kelp beds, but under normal conditions predation checks prevent such destruction. In most cases, however, the evidence to support this contention is poor or the effects due to predation cannot be separated from other effects. For instance, it was originally thought that increases in urchin densities in eastern Canada and New England wiping out the kelp beds there and leading to "urchin barrens" of coralline algae were due to the reduced numbers of lobsters, which were believed to prey on the urchins. Later, as Elner and Vadas (1990) report, it was found that lobsters fed mainly on crabs and rarely on urchins and, further, that the urchins were subject to

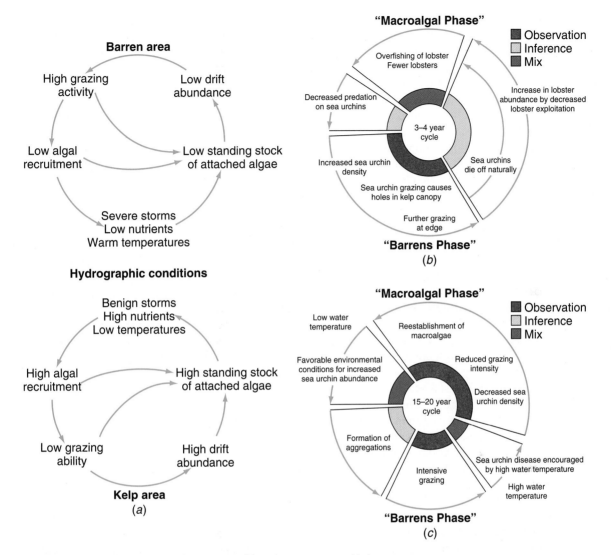

FIGURE 5.27 (a) Diagrammatic summary of how barren areas and kelp areas are maintained at San Nicolas Island, California. (b) Proposed sequence of events causing the establishment of urchin barrens in the sublittoral zone off Nova Scotia according to a paradigm that has the lobster as the keystone predator of urchins. (c) Proposed sequence of events causing a 10- to 15-year temperature- and disease-related cycle between urchin barrens and macroalgal dominance. (a, after C. Harrold and D. Reed, 1985, Food availability, sea urchin grazing and kelp forest community structure. *Ecology* 66(4): 1160–1169. b, c, from "Inference in Ecology: The Sea Urchin Phenomenon in the Northwestern Atlantic" R. W. Elner and R. L. Vadas, *American Naturalist*, Vol. 13, pp. 108–125, 1990. Copyright © 1990 University of Chicago Press. Reprinted by permission.)

massive die-offs from disease brought about by high water temperatures with a periodicity of 10–15 years, thus allowing the reestablishment of the kelp beds (Figure 5.27). Similarly, in southern California, the sheephead fish (*Semicossyphus pulcher*) was thought to be the major predator on urchins, but more recent studies have suggested that it is not. It is likely that storms or changes in the inshore waters in the area, due to sewage outfalls and perhaps to hydrographic effects caused by El Niño, have led to changes in the numbers of both kelp and urchins or to a modification in the urchins' behavior that has manifested itself in alterations to the community structure.

The only well-documented case in which urchins have been shown to be controlled by a predator is in Alaska. Estes and Palmisano (1974) demonstrated that the sea otter (*Enhydra lutris*) is the dominant predator on urchins. In the Aleutian Islands, these investigators found that islands with large otter populations had low urchin densities and a lush growth of kelps. Islands on which otters were absent were virtually without kelps and had enormous urchin densities.

In a later study, Estes and Duggins (1995) demonstrated that, over periods of 3 to 15 years, at sites where kelp or urchins had been either continuously present or continuously absent, the abundances of each had remained unchanged. However, at sites that sea otters were recolonizing, the sea urchin biomass declined dramatically, by 50% in the Aleutian Islands and by almost 100% in southeastern Alaska. At colonizing sites in southeastern Alaska, the increase in kelp was fast and dramatic. In the Aleutian Islands, however, kelp recovery at the recolonization sites was much slower. This difference was because the urchins had significant annual recruitment and growth in the Aleutians, whereas urchin recruitment was episodic, with years separating recruitment pulses, in southeastern Alaska. The rapid return of urchins in the Aleutians tended to slow the recovery of the kelps, whereas the long absence of urchins in southeastern Alaska permitted the quick development of the kelp forest.

The almost complete destruction of the sea otter during the nineteenth century along most of the North American Pacific coast probably allowed the increase in sea urchin numbers and could have permitted the gradual decrease in the kelp beds. The recovery of the sea otter in California and Alaska has been followed by decreased sea urchin numbers and increased kelp forests in a few areas studied.

The importance of urchins to the integrity of kelp beds in California is suggested by comparing the recovery of the beds from the two most devastating El Niño events in the twentieth century, those of 1957–59 and 1982–83. In the former case, according to Tegner and Dayton (1987), the kelp forests had dense populations of the sea urchins *Strongylocentrotus purpuratus* and *S. franciscanus*, which normally feed on drift kelp. After the 1957–59 event, these urchins survived and, because of the lack of drift, moved out to feed on the adult plants, reducing them even more. Persistence of these urchins precluded the recovery of the kelp forest until well into the 1960s. By contrast, the 1982–83 El Niño was a stronger event with a more devastating effect on the kelp forests. (See p. 227.) Yet most of these forests recovered in a shorter time, probably because, by 1982–83, the fishery for sea urchins had reduced the populations so much that urchin grazing could not prevent the recruitment of new kelp or the regrowth of surviving plants. It is also possible that the anomalous environmental conditions associated with the 1982–83 El Niño may have had deleterious effects on urchin recruitment.

Other predators of sea urchins exist in kelp beds, notably several starfish species. In some areas, these animals have been suggested as important in controlling urchin populations (Figure 5.28). However, we now know that sea urchin recruitment is highly variable and that they can also be destroyed by disease or large ocean swells, so that control by predation is not the whole story. The role of other organisms in structuring the kelp bed community is not as well known, except for recent evidence of killer whales as predators of sea otters in Alaska. (See Chapter 3, p. 139.)

A generalized food web for invertebrates for a California kelp forest is given in Figure 5.28. Note that, in this qualitative food web, the primary producers have only the two species of urchins as major grazers. The other grazer, *Astraea (Megastrea) undosa*, is a herbivorous gastropod that is unlikely to exert any considerable effect on the kelps. Sea otters are not present in this area, and urchin control is presumably through the activities of the four predaceous species of starfish, the octopus, and the one fish, coupled with the aforementioned variability in recruitment, disease, and wave action.

Kelp forests also show various patterns of zonation, but the zonation is not consistent from one area to another (Figure 5.29). Zonation is determined by changes in physical factors with depth, such as light and wave action; by differences in topographic relief; and by competition among species. Although kelps may be found in shallow near-shore waters in some areas, a general zonation pattern would have a shallow zone nearest the shore dominated by surfgrass and small coralline algae. Large kelps are absent because the wave surge removes them. At about 5 m, depth conditions are right for kelps, and the forest develops and extends to 20–30 m. As greater depths

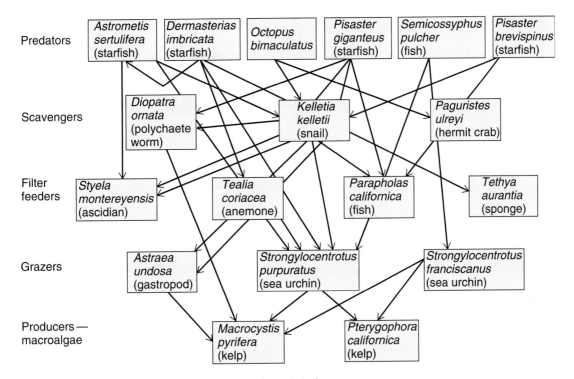

FIGURE 5.28 The food web of a southern California kelp forest. (From R. J. Rosenthal, W. D. Clarke, and P. K. Dayton, 1974, *U.S. National Marine Fisheries Service Fisheries Bulletin*, Vol. 72.)

are approached, the light intensity becomes limiting, and the edge of the kelp forest seems to be set at that point where there is insufficient light for kelp growth or where rocky substrates are covered with sand.

The above studies have elucidated a number of factors that can influence kelp forests and beds, but they tend to be short-term studies and do not integrate abiotic with biotic factors that act over a longer period. Such an integrative approach covering a multiyear time frame was undertaken by Dayton et al. (1998, 1999) in a kelp forest off San Diego and gives probably the best understanding of how these communities react to the interplay of changes in oceanic physical factors with biotic factors such as competition, predation, and grazing and their variation over time. The species of kelp that were examined in this multiyear study were the dominant species *Macrocystis pyrifera*, with a floating surface canopy; the subordinate species *Pterygophora californica* and *Eisenia arborea*, which have an intermediate-level canopy; and *Laminaria farlowii*, with a prostrate benthic canopy. The period of years over which the study was carried out encom-

passed a nutrient-rich cold-water event (La Niña) and a nutrient-depleted warm-water event (El Niño) and a depth range of 8–23 m that included various physical factors critical to kelp growth. Interactions among the kelps was examined by clearing patches and by manipulating the *Macrocystis* recruiting densities across the depth gradient. The results of this study suggest that the strongest competitive interactions occurred during the nutrient-rich La Niña, when there was strong canopy growth by *Macrocystis* resulting in large competitive impacts on the subordinate species, but little effect on *Macrocystis* recruitment and little intraspecific competition. *Pterygophora* and *Eisenia* appeared to have their recruitment limited by light, and competition with *Macrocystis* resulted in reduced growth and reproduction. *Macrocystis* competition also had significant effects on *Laminaria* growth and reproduction and there was evidence of significant effects of *Pterygophora* on *Laminaria* growth. All of these patterns were dependent upon depth because the density of the canopy is greater in shallower water. These patterns were quite different in the nutrient-poor years

FIGURE 5.29 Different zonation patterns of the dominant organisms of five southern California kelp forests. (From M. S. Foster and D. R. Schiel, 1985, *The ecology of giant kelp forests in California: A community profile*, U.S. Fish and Wildlife Service Biological Report 85 (7.2).)

when the *Macrocystis* canopy was more severely affected than the subordinate algal canopy. Recovery from the low-nutrient, high-temperature and the high-nutrient, low-temperature events was also found to be different. All kelps recovered quickly from the El Niño, but recovery from the nutrient-rich La Niña which induced strong competitive interactions among the kelps, was much slower, and some patches seemed to undergo long-term changes. For example, Dayton et al. (1999) noted that some understory patches which had persisted for 25 years through the largest El Niño and one of the most destructive storms in a century were destroyed by light limitations induced by the La Niña event. They also report that the dominant kelp *Macrocystis* is highly sensitive to nutrient availability, which in turn influences the amount and direction of competition, with cascading effects on the recruitment, survivorship, and reproduction of the subordinate understory kelps. This chain of events suggests that the bottom-up forcing of nutrient availability could be one of the most important features of such a system. One of the most significant conclusions of the Dayton et al study is that scaling in time should include interdecadal oceanographic regime changes that have pervasive long-term effects on the kelp communities.

Finally, we should not leave the kelp forests without mentioning a disconcerting fact that may also play a significant role in the kelp forest and other marine communities: As Dayton et al. (1998) point out, most of the large animals have been removed from kelp forests and have been so for decades. Species long gone include sea otters, most large fishes, and abalone, while others, such as urchins, lobsters, and rockfish, are still present but in much reduced numbers. Most of these species have been removed by human action with the result that we are unable to know what a "natural" kelp community really is.

Seagrass Communities

Many areas of the shallow sea bottom are covered with a lush growth of aquatic flowering plants adapted to live submerged in seawater. These plants are collectively called seagrasses.

■ COMPOSITION AND DISTRIBUTION

Worldwide, seagrasses encompass only about 50 species, according to den Hartog (1977)—a small number in comparison with their ecological importance. Seagrass beds form dense carpets of as many as 4,000 blades per square meter over extensive areas of the bottom, making them one of the most conspicuous communities of the shallow waters of temperate and tropical seas (Figure 5.30). Seagrass beds or meadows may have a standing biomass of 2 kg/m^2 and are usually well defined, by a visible boundary, from surrounding unvegetated areas. Seagrass beds vary in size from isolated patches to a continuous carpet that covers many square kilometers. They also vary in the density of the plants.

Most seagrass species are similar in appearance, with long, thin, straplike leaves that have air channels and a monopodial growth form. The plants arise from a creeping rhizome (Figure 5.31). Compared with freshwater aquatic plants, the seagrasses have a lesser number of species and are not as diverse in morphology.

Seagrasses occur from the midintertidal region to depths of 50 or 60 m. They seem most abundant, however, in the immediate sublittoral area. The number of species is greater in the tropics than in the temperate zones. All types of substrates are inhabited by these grasses, from soupy mud to granitic rock, but the most extensive beds occur on soft substrates. The dense, intertwined rhizomes and roots of the grasses form a strong mat that penetrates the substrate and secures the plant against motion of the water, thereby stabilizing the bottom sediments.

■ ENVIRONMENTAL CONDITIONS

Seagrass beds are strongly influenced by several physical factors. The most significant is water motion: currents and waves. Since seagrass systems exist in both sheltered and relatively open areas, they are subject to differing amounts of water motion. For any given seagrass system, however, the water motion is relatively constant. Seagrass meadows in relatively turbulent waters tend to form a mosaic of individual mounds, whereas meadows in relatively calm waters tend to form flat, extensive carpets. The seagrass beds, in turn, dampen wave action, particularly if the blades reach the water surface. This damping effect can be significant to the point where just 1 m into a seagrass bed the wave motion can be reduced to zero. Currents are also slowed as they move into the bed.

The slowing of wave action and currents means that seagrass beds tend to accumulate sediment. However, this is not universal and depends on the current regime under which the bed exists. Seagrass beds under the influence of strong currents tend to have much of the

FIGURE 5.30 A *Posidonia* seagrass bed in the Mediterranean Sea. (Photo courtesy of John Heine.)

FIGURE 5.31
Diagrammatic three-dimensional illustration of an eelgrass meadow. (Redrawn from G. W. Thayer, W. J. Kenworthy, and M. S. Fonseca, 1984, *The ecology of eelgrass meadows of the Atlantic coast: A community profile*, U.S. Fish and Wildlife Service Program, FWS/OBS-84/02.)

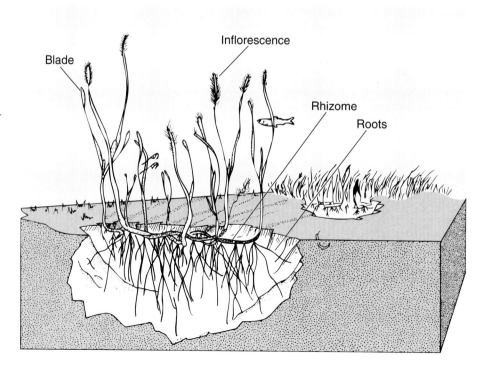

lighter particles, including seagrass debris, moved out, whereas beds in weak current areas accumulate lighter detrital material. It is interesting that temperate seagrass beds (*Zostera*) accumulate sediments from sources outside the beds, whereas tropical seagrass beds derive most of their sediments from within.

Since most seagrass systems are depositional environments, they eventually accumulate organic material that leads to the creation of fine-grained sediments with a much higher organic content than that of the surrounding unvegetated areas. This accumulation, in turn, reduces the interstitial water movement and the oxygen supply. The high rate of metabolism of the microorganisms in the sediments causes the sediments to be anaerobic below the first few millimeters. According to Kenworthy et al. (1982), anaerobic processes of the microorganisms in the sediment are an important mechanism for regenerating and recycling nutrients and carbon, ensuring the high rates of productivity measured in those beds (Figure 5.32).

Other physical factors that have an effect on seagrass beds include light, temperature, and desiccation. For example, water depth and turbidity together or separately control the amount of light available to the plants and the depth to which the seagrasses may extend. Although early on it was suggested by Setchell (1929) that temperature was critical to the growth and reproduction of eelgrass, *Zostera marina*, it has since been shown that this particularly widespread seagrass grows and reproduces at temperatures between 2 and 4°C in the Arctic (Hudson's Bay) and at temperatures up to 28°C in Chesapeake Bay. Still, extreme temperatures, in combination with other factors, may have dramatic deleterious effects. For example, in areas of the subarctic and cold temperate North Atlantic, ice may form in winter. Robertson and Mann (1984) note that when the ice begins to break up, the wind and tides may move the ice around, scouring the bottom and uprooting the eelgrass. In contrast, at the southern end of the eelgrass range, in North Carolina, temperatures over 30°C in summer cause excessive mortality. Seagrass beds also decline if they are subjected to too much exposure to the air. The effect of desiccation is often difficult to separate from the effect of temperature. Most seagrass beds seem tolerant of considerable changes in salinity

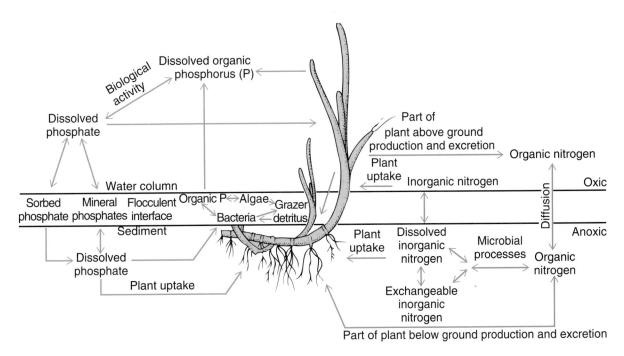

FIGURE 5.32 Diagrammatic representation of the phosphorus and nitrogen cycles in a seagrass bed. (Modified from G. W. Thayer, W. J. Kenworthy, and M. S. Fonseca, 1984, *The ecology of eelgrass meadows of the Atlantic coast: A community profile*, U.S. Fish and Wildlife Service Program, FWS/OBS-84/02.)

and can be found in brackish waters as well as in full-strength seawater.

■ PRODUCTIVITY

Because seagrass beds are densely covered with plants (the result of extensive nutrient cycling and regeneration) and because they cover such extensive areas in continental shelf waters, they have a very high productivity rate and contribute significantly to the total production of inshore waters. McRoy and McMillan (1977) have estimated the production of temperate beds to be 500–1,000 g C/m^2/year, and McRoy and Helfferich (1977) report values of over 4,000 g C/m^2/day for *Thalassia* beds in the tropics. Seagrass blades can grow as much as 1 cm per day and are continually being shed and replaced.

Seagrass beds are among the most productive areas of the oceans. (cf. with plankton productivity, Table 2.2, and with that of kelp, pp. 225–226.) Most of the data on productivity, however, are for the two most-studied species: *Zostera marina* (eelgrass) of the North Temperate Zone and *Thalassia testudinum* (turtle grass) of the tropics. Little is known about the other species.

In contrast to other productivity in the ocean, which is confined to various species of algae and bacteria dependent on nutrient concentrations in the water column, seagrasses are rooted plants that absorb nutrients from the sediment or substrate. They are, therefore, capable of recycling nutrients into the ecosystem that would otherwise be trapped in the bottom and rendered unavailable.

■ STRUCTURE AND BIOLOGICAL INTERACTIONS

Most seagrass beds have only one or, at most, a few dominant species of seagrasses. Communities in which two or more grasses are present have been little studied and may be different in biological interactions. In general, the number of seagrass species increases from the temperate zone to the tropics. On the Atlantic coast of the United States, the only seagrass present in temperate waters is eelgrass, *Zostera marina*. In the Caribbean Sea, three main species are found—*Thalassia testudinum, Halodule wrighti*, and *Syringodinium filiforme*—occuring in various mixed stands. In Japan, five species of *Zostera* occur and as many as three can occur in the same bed.

Seagrass beds expand at different rates, depending on environmental conditions, but their expansion can be rapid. Thayer et al. (1975) found that a bed in North Carolina covering 30% of a bay in 1969 had increased to cover 55% by 1973. Unfortunately, we do not have data on how long these beds remain in any given environment.

Seagrass beds are complex systems whose physical structure is dominated by the leaves, roots, and detritus of the seagrasses themselves. Furthermore, they contain large numbers of epiphytic and epizoic organisms, burrowers, and motile animals, all interacting in ways that are only partly known at present (Figure 5.33). The roots and leaves provide horizontal and vertical complexity, in addition to abundant food. The diversity of organisms within seagrass beds is much higher than in the unvegetated surrounding areas.

The various trophic levels represented in the seagrass beds undergo shifts in abundance in response to various changes in the seagrasses, as well as to seasonal changes in environmental factors. Therefore, there may be considerable temporal variation in the composition and abundance of the associated organisms.

On the seagrass blades, there is usually a community of epiphytic and epizoic organisms which are so abundant that the biomass of these organisms exceeds that of the blade on which they reside. This community may also reduce the productivity of the seagrass by reducing the light intensity or even eliminating the light altogether at the surface of the blade. The epiphytic algae can represent a significant biomass and a substantial portion of the primary production in seagrass beds. They also seem to have higher nutritional value to grazers than the seagrass blades themselves have. Little is known of the importance of this component of the community to the general economy of the beds.

Despite the obvious position of seagrass beds as primary production units in inshore waters (and in direct contrast to the situation in the terrestrial environment, where grasses are heavily grazed by a variety of vertebrate and invertebrate herbivores), surprisingly few animals consume seagrasses directly. This may in part be due to the fact that seagrasses have large amounts of chemical compounds that reduce their nutritional value to grazers, as well as large amounts of relatively indigestible structural carbohydrates. It appears that the major grazers are certain birds, a few fishes, sea urchins, and the green sea turtle (*Chelonia mydas*). Despite the small numbers of grazers, their effect may be significant in certain areas. Among the birds, the black brant (*Branta nigricans*)

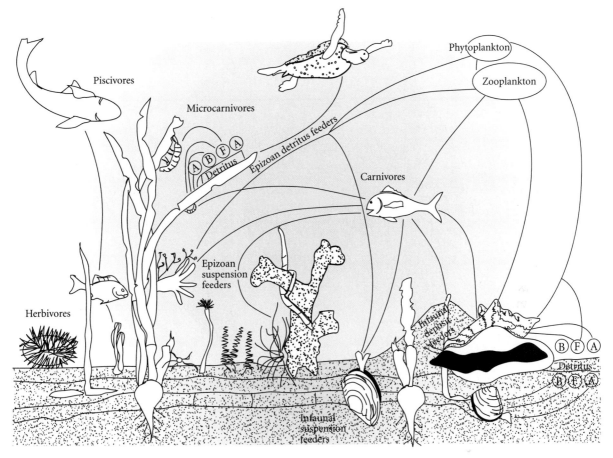

FIGURE 5.33 Diagrammatic representation of pathways of energy flow among consumer groups in Caribbean seagrass meadows. (From M. Bertness, et al., *Marine Community Ecology*, p. 319, fig. 12.2. Copyright © 2001. Used with permission of Sinauer Associates, Inc.)

and American brant (*Branta bernicla*) geese are major grazers of seagrasses (Figure 5.34). McRoy (1966) reported that the black brant consumes 4% of the standing crop of eelgrass in Izembeck Lagoon in Alaska, and Cottam (1934) indicated that on the U.S. east coast 80% of the diet of American brant consisted of eelgrass. Feeding by these and other herbivorous birds can have a dramatic effect on seagrass beds. For example, Wilkins (1982) estimated that in one year Canada geese (*Branta canadensis*) consumed 20% of the eelgrass standing crop in lower Chesapeake Bay, while in Australia black swans (*Cygnus atratus*) uprooted 94% and consumed 82% of net production of *Zostera muelleri*. In many of these cases, the beds are almost cleared of parts of the seagrasses that are above the sediments. Similarly, in the tropics, urchins such as *Lytechinus variegatus* have

been reported by Camp et al. (1973) to denude large areas of seagrass. The widespread tropical urchin genus *Diadema* has also been implicated in the destruction of tropical seagrass beds, and green turtles are known to feed on seagrasses, but because of their current low numbers and the absence of historical data, we know little of what effect this grazing may have had. Ogden (1980) reports, however, that in the few areas where these turtles are still abundant they create very visible grazing scars in seagrass beds. This finding suggests that, at high population densities in the past, they may have had significant effects, especially since, in the Caribbean Sea, the diet of the turtle seems to be almost exclusively seagrasses. Among fishes, the pinfish (*Lagodon rhomboides*) of the U.S. east coast appears to feed heavily on eelgrass and can reduce small beds to a short stubble.

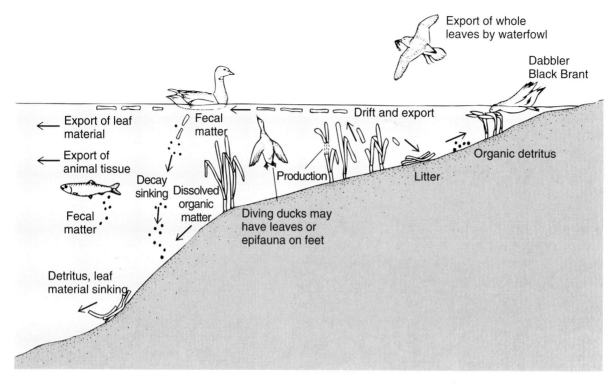

FIGURE 5.34 Diagrammatic representation of the principal pathways of movement of organic material out of a seagrass bed. (Modified from J. C. Zieman, 1982, *The ecology of the seagrasses of South Florida: A community profile*, U.S. Fish and Wildlife Service Program, FWS/OBS-82/25, 124.)

Seagrasses' response to grazing varies with the intensity of the grazing, according to Williams and Heck (2001). At low to intermediate grazing intensity there are some physiological adjustments, and the plants increase their rate of photosynthesis and their production of new nutrient-rich leaves, tending to make the plants more nutritious to grazers. However, this increased production cannot be sustained at high grazing intensity, and heavy grazing by large herbivores such as urchins, turtles, and dugongs can lead to a reduced standing crop or even to the local disappearance of the aboveground parts of the plants.

A wide variety of predators, both resident and migratory, is associated with seagrass beds, and they have diverse feeding strategies. Despite their abundance, there is little experimental evidence that they regulate either the structure of the seagrass beds or the abundance of prey. Various experiments in which predators have been caged out of areas usually have not demonstrated significant increases in prey inside the cages. There is, however, evidence that seagrass beds may serve as refuges from predation, compared with unvegetated areas around them. For example, Peterson (1982) found that the densities of two clams—*Mercenaria mercenaria* and *Chione cancellata*—were reduced in areas where seagrasses were removed, relative to their densities in control areas where there was seagrass cover. Peterson attributed the difference to greater predation by three species of whelks of the genus *Busycon*. In this case, the seagrass bed provided a refuge, probably by binding the sediment with its roots, thereby preventing the whelks from gaining access to the clams.

Fishes are abundant in seagrass beds, and studies of food habits on the U.S. east coast by Thayer et al. (1975) indicate that many fishes feed within the bed, removing considerable biomass (Figure 5.34). In North Carolina, Thayer et al. (1975) also found that the total mass of food in the guts of fishes leaving a meadow was three times that of the fish entering. By contrast, Ogden (1980) reports that in the tropics most herbivorous fishes are not resident in the turtlegrass beds, but migrate in during the night from surrounding reefs. Grazing by such fishes, as well as by urchins, is responsible for the formation of "halos" of grazed areas around reefs.

Certain larger fishes, such as sharks and rays, are actually or potentially of importance in structuring seagrass communities. For example, Orth (1975) reports that the feeding of the cownose ray (*Rhinoptera bonasus*) destroys large areas of seagrass beds on the U.S. east coast. These fish dig into the bottom to pull out clams and in so doing destroy seagrasses and alter the sediments. Not only do the rays drastically reduce clam densities (from 60 to 100 *Mya arenaria* per square meter to zero!), but they also reduce densities of the associated fauna.

The conventional wisdom has been that seagrass beds serve as nursery grounds for various commercially important fishes and shellfishes, but more recent studies by Heck et al. (1995) and others suggest that this is not the case, at least in temperate waters. Evidence for the importance of seagrass beds as nursery grounds is more compelling in tropical regions and may be related to the increased predation pressure characteristic of such areas.

■ RELATIONSHIPS TO OTHER SYSTEMS

With the exceptions noted, relatively little of the living seagrass tissue is consumed directly by grazers. Instead, most seagrass organic production either is transformed into detritus and consumed or further broken down in situ, or else it is transported out of the system to be used by other shallow-water communities. How is this large food source channeled to other shallow-water systems? There are at least four ways: Organic material can be transported as whole plant parts with associated organisms, as detritus, as dissolved organic matter, or as living tissue or fecal material of various motile vertebrates (Figures 5.34 and 5.35). The importance of seagrass beds to adjacent communities varies, depending on geography, hydrographic conditions, and depth. Shallow-water beds tend to export more material, because they are more often disturbed by water movement and by ice scour and because they can be reached more easily by various birds. Deep-water beds, by contrast, are usually inaccessible to birds and also are less disturbed by water motion. In these beds, however, movement by fishes may be more significant. The

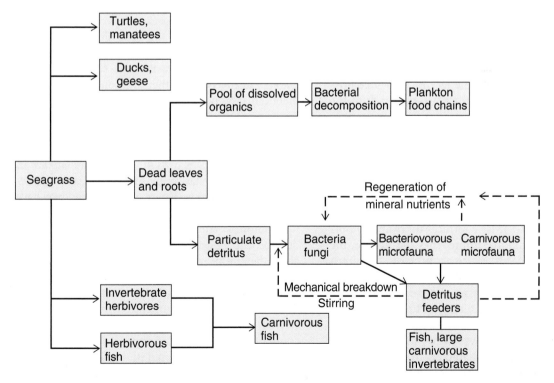

FIGURE 5.35 The pathway for the channeling of eelgrass into the food web. (Modified from *Seagrass Ecosystems*, C. P. McRoy & C. Helfferich, eds., 1977, Marcel Dekker.)

detrital material produced in seagrass beds may be transported to communities at considerable distances from the beds, enriching those communities. Seagrass has been suggested as a source of energy for organisms as far removed as the abyssal benthos.

■ ECOLOGY

Ecologically, seagrass beds serve a number of important functions in inshore areas. They are a major source of primary productivity in shallow waters around the world and thus an important source of food for many organisms (in the form of detritus). In addition, they stabilize the soft bottoms, on which most species grow, primarily through their dense, matted root system. This stabilization of the bottom by roots is sufficiently durable to withstand storms as severe as hurricanes. The roots in turn, shelter many organisms. As a result, numerous animals that do not have a direct trophic relationship to the grasses are commonly found in seagrass beds. The beds are nursery grounds for some species that spend their adult lives in other areas. Included among these species are several of commercial importance, such as the shrimp *Penaeus duorarum* of southern Florida. These beds may also harbor adults of commercially important species, such as the bay scallop, *Argopecten irradians*, off the U.S. east coast. Indeed, Pohle et al. (1991) have shown through field experiments that the juveniles of the bay scallop have a dramatically increased survival rate in the face of crab predators when they are tethered by their byssal threads to the blades of seagrasses at distances of 20–35 cm above the substrate. The destruction of seagrass beds, therefore, may have a serious effect on certain commercial fisheries.

Seagrass beds may also trap sediment and, therefore, build up the bottom. Such a situation has been recorded for the *Posidonia* beds in the Mediterranean Sea, which build extensive terraces up from the bottom. When the buildup approaches the surface, the floating leaves break the force of the waves, forming a calm water habitat over the bed.

The leaves of the seagrasses also create a protective canopy, shielding the inhabitants of the bed from the effects of strong sunlight. Where the beds become intertidal, the leaves may cover the bottom substrate at low tide, protecting the inhabitants from desiccation. In Florida, seagrasses are the primary substrate for most attached macroalgae. Very few algae colonize soft sediments, so with limited hard substrate in

the sublittoral zone, the Florida seagrasses fulfill an important function.

The importance of seagrass beds and what might happen when they are destroyed was dramatically pointed up by a natural destructive event that affected the dominant seagrass *Zostera marina* in the North Atlantic Ocean. In the early 1930s, the *Zostera* beds on both sides of the Atlantic were destroyed by a mysterious "wasting disease." There was no universal agreement about what caused this destruction until Short et al. (1987) confirmed that a pathogenic strain of the protistan *Labyrinthula* caused the symptoms. During the epidemic of the 1930s, the *Zostera* beds in the Pacific Ocean were not affected.

The effects of the *Zostera* disappearance were different in different areas. In Europe, the level of the beach fell; with the loss of sediment, many bared rocks sprouted a new growth of algae in place of *Zostera*, and new sandbars formed (Figure 5.36). The organisms also changed, but less than was anticipated. The eelgrass beds had been dominated by detritus-feeding animals. After the loss, the same bottom area had more suspension feeders and, in general, more species. In Europe, very few annual species associated with *Zostera* disappeared completely. In the United States, by contrast, the destruction of the *Zostera* beds meant that the animals associated with the beds almost completely disappeared. The reason for this difference is probably that, in European waters, the alga *Fucus* quickly replaced *Zostera*, offering the same protected environment. This was not the case in the northeastern United States. Since the 1930s, however, the *Zostera* beds have slowly returned to reoccupy their former habitat. More recently Robblee et al. (1991) reported the widespread demise of the eelgrass beds of *Thalassia testudinum* in Florida Bay. This catastrophic loss has the same characteristics as the wasting disease of *Zostera* and may have a similar cause. The extent and ecological results of the new outbreak are not known. During the last few decades, the *Zostera* beds have once again declined along the European and American Atlantic coasts, but this decline seems to be due, not to another episode of the wasting disease, but to eutrophication of the shallow waters through human activity.

Den Hartog (1977) reports that, in some seagrass communities, a successional pattern similar to that found in terrestrial communities has been identified (Figure 5.37). An example is the seagrass beds of the tropical Caribbean. Here, the climax community is dominated by *Thalassia testudinum*, but there is a

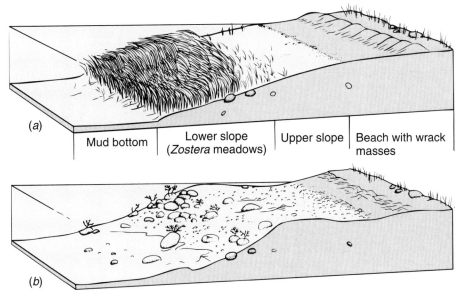

(a)

| Mud bottom | Lower slope (*Zostera* meadows) | Upper slope | Beach with wrack masses |

(b)

FIGURE 5.36 Representation of (a) the conditions of the eelgrass beds of Scandinavia before the onset of the wasting disease and (b) the extreme conditions following the complete destruction of the beds. (From *Seagrass Ecosystems*, C. P. McRoy & C. Helfferich, eds., 1977, Marcel Dekker.)

series of stages leading up to *Thalassia* that are different, depending on the substrate. In the temperate north Atlantic where the dominant seagrass is *Zostera marina*, no series of successional stages has been uncovered, and *Z. marina* colonizes directly; that is, it is both pioneer and terminal stage. In the temperate north Pacific Ocean, the seagrass *Phyllospadix scouleri* occurs on hard substrates on waveswept shores. (*Zostera* occurs on soft bottoms in protected areas.) Turner (1983) has shown that there is an obligate succession leading to the development of *P. scouleri* beds. Open areas in surfgrass beds are settled first by *Ulva*, a green ephemeral alga, which is then replaced by several red algae. Finally, *Phyllospadix* colonizes either by the rhizomes growing into the space or by seed recruitment. Rhizome invasion does not depend on any preexisting species and may occur directly, but for seeds to establish the surfgrass, macroalgae must be present for the seeds to attach to. Furthermore, the seeds attach preferentially, or at least have a better likelihood of survival, if the algae are turf-forming algae with bushy branches and a narrow central axis; thus, a true facilitated succession occurs.

Some seagrass beds—particularly those on the Pacific coast of North America—have been invaded by alien species. The Japanese eelgrass *Zostera japonica* invaded the Pacific northwest coast in the 1950s, according to Williams and Heck (2001). However, instead of displacing the native *Zostera marina*, *Z. japonica* occupied primarily the mudflat areas above the distributional limits of *Z. marina* and increased the total eelgrass area, thus providing the native geese with an increased food source. On the other hand the invasion of the *Z. marina* beds by the alien mussel *Musculista senhousia* in the 1920s was not as beneficial: According to Williams and Heck (2001), the mussel inhibits the growth of the eelgrass and tends to take over any habitat where the eelgrass has been disturbed, suggesting that *M. senhousia* might eventually eliminate the *Zostera* beds. (See also Chapter 11).

Some Special Communities

In addition to the widespread communities described in this chapter there are a few of more restricted distribution that have received considerable study; hence, they deserve mention here as examples of how communities are organized, how they may vary temporally, and how they are maintained.

■ A SEA-PEN COMMUNITY

In Puget Sound, Washington, at depths of 10–50 m on soft bottoms, is a community dominated by dense stands of the long-lived cnidarian sea pen, *Ptilosarcus gurneyi* (Figures 5.38 and 5.39). In an extensive study, Birkeland (1974) found that *Ptilosarcus gurneyi* provides the major food source for seven predators: four starfishes (*Hippasteria spinosa*, *Dermasterias imbricata*, *Crossaster papposus*, and *Mediaster aequalis*) and

FIGURE 5.37 Successional sequences in seagrass beds. (a) Tropical areas. (b) Temperate areas.

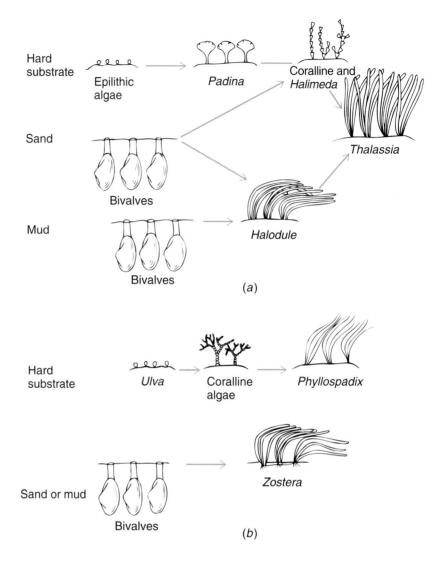

three opisthobranch mollusks (*Armina californica, Tritonia festiva,* and *Hermissenda crassicornis*). *Ptilosarcus* has a life span of about 15 years; takes several years to reach sexual maturity; has irregular, unpredictable recruitment, resulting in stands of differing age structure; and never reaches sizes at which it is immune to predation.

Adult *Ptilosarcus* are readily consumed by three starfishes—*Hippasteria, Dermasterias,* and *Mediaster*—plus the opisthobranch *Armina*, which seems to prefer the largest sea pens (Figure 5.40). Of these four predators, *Hippasteria* has the most restricted diet, feeding only on *Ptilosarcus* (Figure 5.41); the others are capable of consuming other prey. In Puget Sound at least, *Armina* seems to feed exclusively on *Ptilosarcus*. Juvenile *Ptilosarcus* are preyed upon

by *Hermissenda, Tritonia,* and *Crossaster,* but this is uncommon.

In the Puget Sound community, one might expect that, because they do not completely depend on *Ptilosarcus*, the starfish generalists (*Dermasterias, Mediaster,* and *Crossaster*) would maintain high population numbers and reduce the *Ptilosarcus* population, so that the specialists (*Armina* and *Hippasteria*) would have nothing left. In other words, the generalists can outcompete the specialists. Yet they do not. Why? The answer lies with still another starfish, the top predator *Solaster dawsoni*, which preys on *Mediaster, Dermasterias,* and *Crossaster,* reducing their numbers sufficiently so that competitive exclusion does not occur (see Figure 5.39), thus allowing *Hippasteria* to persist.

FIGURE 5.38 A North Pacific sea pen (*Ptilosarcus*). (Photo courtesy of Randy Morse.)

In sum, this community is thus maintained by a complex food web that has as its key industry species the sea pen *Ptilosarcus*, which is preyed on by several specialist and generalist predators. The generalists (*Mediaster, Dermasterias,* and *Crossaster*) are unable to dominate and outcompete the specialists (*Hippasteria* and *Armina*) because a top predator, *Solaster*, keeps their numbers in check.

■ A SEA-PANSY COMMUNITY

Off San Diego in 5–10 m of water on a sand bottom exists quite a different community. As noted on page 211, this community was first studied by Fager (1968) between 1957 and 1963, when it was dominated by cnidarians, primarily the sea pansy *Renilla kollikeri* and the anemones *Harenactis attenuata* and *Zaolutus actius*. Associated with these organisms were three gastropods (*Nassarius fossatus, Nassarius perpinguis,* and *Polinices recluzianus*) and two echinoderms (*Amphiodia occidentalis* and *Astropecten armatus*). In this community, Fager found that the pattern of distribution of the major organisms was primarily aggregated, and the populations remained constant over the six-year period. Furthermore, there was little evidence of any interaction among the species and no evidence that these large macroinvertebrates were either preyed on by mobile fishes or themselves preyed on the smaller infaunal organisms.

Eleven years later, in 1974, Davis and VanBlaricom (1978) studied the same site. Whereas Fager had observed essentially constant densities of the major species over six years, Davis and VanBlaricom found both long- and short-term changes in the abundances and densities of seven of the dominant species. The new studies suggest that shallow-water benthic communities are not completely stable over long periods, even though they may appear so for several years. It also seems that a much longer time interval is necessary to evaluate what is happening in such areas.

Biology of Polar Seas

The Arctic and Antarctic regions of the Earth appear outwardly similar in that both are characterized by the dominating presence of ice and snow, year-round cold temperatures, and drastic changes in photoperiod that prevent photosynthesis during a significant part of the year. Most of these outward similarities, however, mask fundamental differences that make the two polar regions very dissimilar both physically and biologically.

In this section we discuss, first, the differences between the two polar regions that give rise to the different biological associations. We then discuss the biological disparities and conclude with a brief look at some examples.

Perhaps because of the profound interest and support of the U.S. government through its Antarctic research programs over the last 30 years, as well as a similar commitment from other world powers to Antarctic research, we have a better understanding of shallow-water Antarctic communities than we have

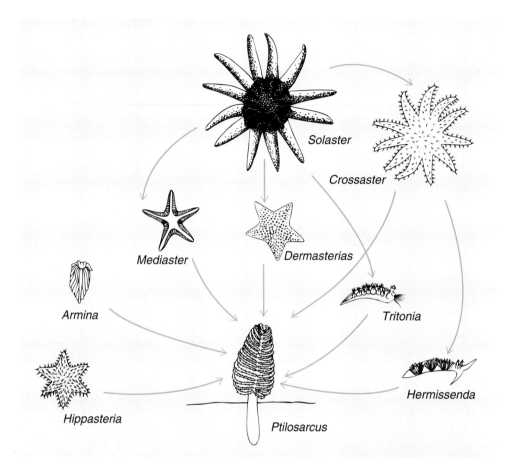

Solaster

Crossaster

Mediaster

Dermasterias

Armina

Tritonia

Hippasteria

Ptilosarcus

Hermissenda

FIGURE 5.39 The food web of a North Pacific sea pen community.

FIGURE 5.40 The nudibranch *Armina californica* eating the sea pen *Ptilosarcus gurneyi*. (Photo courtesy of Dr. Ron Shimek.)

FIGURE 5.41 The sea star *Hippasteria* feeding on the sea pen *Ptilosarcus gurneyi*. (Photo courtesy of Dr. Ron Shimek.)

of those of the Arctic. This gap is reflected here in the coverage of each.

■ PHYSICAL CONDITIONS OF THE ARCTIC AND ANTARCTIC REGIONS

The Arctic Ocean system is an isolated sea surrounded by landmasses that leave only two outlets to other oceans: the Bering Strait to the Pacific and the Fram Strait to the Atlantic. Both of these outlets have relatively shallow sill depths, 70 m in the Bering Strait and 400 m in the Fram Strait (Figure 5.42). The Arctic Ocean Basin is surrounded by extensive shallow shelves such that the deep water of the central basin is cut off from other oceans.

The circulation pattern consists of two major anticyclonic currents: the Beaufort gyre over the Canadian Basin and a transpolar current across the Eurasian part of the basin that exits through the Fram Strait (Figure 5.42).

The landmasses surrounding the Arctic Ocean have several large rivers that discharge significant amounts of sediment into the basin such that the basic substrate overlying the extensive continental shelves is particulate matter. The large freshwater discharges also lead to the development of a low-saline, stratified surface layer.

The Arctic Ocean pack ice is a persistent multiyear system that is harder and thicker than that of the Antarctic, which is usually seasonal and only a year old or younger. The whole central Arctic Ocean is permanently covered by ice (Figure 5.42).

By contrast, the Antarctic marine ecosystem is open to all oceans and is a circumpolar ring of water surrounding a central landmass. The Antarctic continental shelf is narrow and falls off rapidly to deep water. The northernmost marine boundary is defined by the circular Antarctic Convergence, which is situated over deep water in a zone of persistent and strong westerly winds (Figure 5.43). Immediately south of the Antarctic Convergence, nutrient-rich circumpolar water rises toward the surface and fertilizes the Antarctic surface waters (Figure 5.43). Near the continent and beneath the ice shelves, seawater is cooled to freezing or below and flows downslope to form the bottom water of the Earth's deep ocean basins. Since the Antarctic continent is permanently covered by ice and lacks rivers, little or no sediment accumulates on the continent and no stratified surface layers form.

Compared with the Arctic, the Antarctic undergoes only small seasonal variations in temperature, both horizontally and vertically. In winter, pack ice covers more than half of the Antarctic marine area, but it is reduced to only 10% in late summer. (See Figure 5.42). Antarctic pack ice is only 1–2 m thick, in contrast to the permanent 3–4 m-thick Arctic pack ice. Both the Arctic and the Antarctic have pronounced year-to-year variations in ice cover.

The Antarctic Ocean occupies a ring around the Earth between 50–60° and 65–70° south latitude, while the Arctic Ocean is an enclosed Mediterranean-type sea beyond 70° north latitude. Thus, the Antarctic Ocean endures strong, but less pronounced, seasonal light conditions than does the Arctic Ocean, and the plankton experience strong diurnal, but weak seasonal, vertical migrations. Conversely, the Arctic plankton show weak diurnal vertical migrations, but pronounced seasonal ones.

These physical differences between the two polar regions are summarized in Table 5.1.

Because the polar inshore waters are covered with ice most of the year, benthic organisms and communities are rarely disturbed by wave action. However, polar shallow-water areas are subject to a stress factor unknown in other areas: ice. During the winter months, ice forms a thick layer over the continental shelf areas, extending down several meters. Wherever the ice layer makes contact with the bottom, the fauna is destroyed, either by mechanical action of the ice grinding against the bottom or by freezing the animals. This means that, in polar areas, the shallowest subtidal areas are devoid of permanent communities down to the level to which the ice extends. Those areas may, however, be inhabited by mobile or transient organisms during ice-free times of the year. Indeed, Conlan et al. (1998) studied ice scour disturbance in Barrow Strait in the high Arctic of Canada and found a gradient of high ice scour by shallow drift ice in inshore waters to less scour by less frequent icebergs and ice shelves at 30 m. These researchers also found that ice scours at all depths were dominated by the same set of disturbance-associated fauna, which was quite different from the undisturbed benthic community outside the scours (Figure 5.44).

In Antarctic seas, in addition to the action of the thick layer of annual sea ice affecting the bottom, there is an area below that layer which is subject to a different kind of ice action. In this area, which extends down to approximately 30 m, ice platelets begin to form on the bottom around any convenient nucleus. As the platelets increase in size, they may

Occurrence of Ice

☐ Permanent during entire year   ▨ Occasional in winter and spring

▨ Always or frequently during year   ▨ Occasional icebergs

FIGURE 5.42 Geography of the Arctic and Antarctic regions, showing the seasonality of the ice cover and the major currents (black arrows). (From "On the Biology of Polar Seas," by G. Hempel, in *Marine Biology of Polar Regions and Effects of Stress on Marine Organisms*, Gray & Christiansen, eds., © 1985 John Wiley & Sons Limited. Reproduced with permission.)

FIGURE 5.43
Diagrammatic represen-
tation of the currents off
the Antarctic continent.
(After "On the Biology of
Polar Seas," by G. Hempel,
in *Marine Biology of Polar
Regions and Effects of
Stress on Marine Organisms*,
Gray & Christiansen, eds., ©
1985 John Wiley & Sons
Limited. Reproduced with
permission.)

TABLE 5.1

Oceanographic Features of Arctic and Antarctic Oceans

	ANTARCTIC	ARCTIC
Area ($\times 10^6$ km)	35–38	14.6
Pack ice		
Mean maximum	22	12–13
Mean minimum	2.6	6–7
General features		
Form	Circumpolar ring between 50–60°S and 65–70°S	Mediterranean Sea around North Pole, almost enclosed by land at 70–80°N
Shelf	Narrow, open to all oceans, with large exchange in deep layers	Broad, two narrow openings (Bering Strait, Fram Strait)
Current systems	Circular currents (west wind drift, east wind drift), large eddies (Weddell Sea, Ross Sea)	Transpolar current, Beaufort gyre
Pack ice	High seasonality	Little seasonality
Cover in winter	c.50%	90%
Cover in summer	c.10%	80%
Age, thickness	Mainly one year, 1.5 m	Mainly multiyear, 3.5 m
Icebergs	Large, tabular icebergs abundant and carrying coarse material	Small, irregular icebergs, only in Greenland and Bering Sea, not in Arctic Basin
River input	None	Much
Stratification	Low vertical stability	Surface layer (20–50 m) stable all year
Nutrients in euphotic zone	High all year	Seasonal depletion

Source: Modified from "On the Biology of Polar Seas" by G. Hempel, in *Marine Biology of Polar Regions and Effects of Stress on Marine Organisms*, Gray & Christiansen, eds., © 1985 John Wiley & Sons Limited. Reproduced with permission.

(a)   1 cm   1 cm   (b)

FIGURE 5.44 Dominant fauna (a) outside and (b) inside a one-year-old ice scour in Barrow Strait, Canadian Arctic, at 20 m depth. The undisturbed benthos (a) is dominated by the large bivalves *Mya truncata*, *Serripes groenlandicus*, and *Macoma calcarea*; the sea urchin *Strongylocentrotus pallidus*; the sea cucumber *Cucumaria frondosa*; the anemone *Tealia* sp.; the polychaete *Phyllodoce groenlandicus*; and the kelp *Agarum cribosum*. Dominant occupants of a one-year-old scour (b) are tubicolous spionid polychaetes (vertical tubes) and *Capitella capitata* (horizontal tube); the mobile polychaetes *Ophryotrocha spatula* (spatulate posterior) and *Nereimyra punctata*; the amphipod *Monoculodes vibei*; and various cumaceans. Diatoms also colonize in abundance (the tufts). (From "Ice Scour Disturbances to Benthic Communities in the Canadian Arctic," by K. E. Conlan et al., *Marine Ecology Progress Series*, Vol. 166, pp. 1–16, 1998. Reprinted by permission of Inter-Research.)

surround various sessile and sedentary invertebrates living on the bottom. Since ice is generally less dense than the surrounding water, as the platelets grow they lift off the bottom and rise toward the thick sea ice above, carrying with them the trapped invertebrates. These organisms are permanently lost from the community. Such an ice formation is termed **anchor ice** (Figure 5.45). As a result of anchor ice formation and its tendency to remove organisms, a zonation of the benthic fauna with depth is created.

■ BIOLOGICAL COMPARISONS OF ARCTIC AND ANTARCTIC SEAS

As might be anticipated, the differences in physical characteristics between the Arctic and the Antarctic Oceans are reflected in differences in biology. Perhaps the most significant difference between the two polar regions is that the Antarctic is far richer in species of benthic organisms (Table 5.2). It also has an extremely high degree of endemism (Table 5.3). The high diver-

sity of Antarctic species is matched by a high biomass. By contrast, the Arctic fauna is impoverished and consists mainly of organisms derived from the Atlantic Ocean. In general, the biomass figures for any given depth in the Antarctic are one to two orders of magnitude higher than those of the Arctic. The high biomass of the Antarctic benthos is probably related to the nutrient-rich water and high offshore productivity, while the high species richness and endemism, according to Hempel (1985), are related to a long evolutionary history and isolation, as well as to the presence of both hard and soft substrates.

Arctic benthic communities are often dominated by one or only a few species, but Antarctic communities usually have several species that are common or dominant. Because of the extensive areas of sediments, the Arctic benthic fauna is mainly an infauna, whereas the Antarctic benthos is characterized by large numbers of epifaunal species. According to Hempel (1985), fishes are less

FIGURE 5.45 Anchor ice area in the Antarctic. (Reprinted with permission from "Anchor Ice Formation in McMurdo Sound, Antarctica, and its Biological Effect," P. K. Dayton et al., *Science*, Vol.163, p. 273, 1969. Copyright © 1969 American Association for the Advancement of Science.)

TABLE 5.2

Comparison of the Number of Species of Macrobenthic Organisms from the Arctic and the Antarctic

	ARCTIC	ANTARCTIC
Sponges	200	300
Polychaetes	c. 300	650
Amphipods	262 (Barents Sea)	470
Pycnogonids	29 (Barents Sea)	100
Ascidians	47	129
Bryozoans	200 (Barents Sea)	310
Mollusks	224 (Barents Sea)	875
Isopods	49 (Kara Sea)	299

Source: After "On the Biology of Polar Seas," by G. Hempel, in *Marine Biology of Polar Regions and Effects of Stress on Marine Organisms*, Gray & Christiansen, eds., © 1985 John Wiley & Sons Limited. Reproduced with permission.

TABLE 5.3

Degree of Endemism in Selected Groups of Antarctic Marine Organisms

	GENERA (%)	SPECIES (%)
Isopoda and Tanaidacea	10	66
Fishes	70	95
Pycnogonida	14	90
Echinodermata	27	73
Echinoidea	25	77
Holothuroidea	5	58
Bryozoa	—	58
Polychaeta	5	57
Amphipoda	39	90

Source: "On the Biology of Polar Seas," by G. Hempel, in *Marine Biology of Polar Regions and Effects of Stress on Marine Organisms*, Gray & Christiansen, eds., © 1985 John Wiley & Sons Limited. Reproduced with permission.

important in Antarctic benthic communities and are restricted mainly to the near bottom.

Some of the differences between the Arctic and Antarctic benthic communities may be due to differences in the disturbance regimes in the two regions. Arctic benthic organisms are subjected to a larger number of disturbances, such as increased ice scour from land fast ice, large variations in salinity due to the sizable river runoff in the summer and the sinking of water with very high salinity (80–180 psu) during the winter freezing.

Whereas both regions are disturbed by ice bergs running aground, the Arctic has more biotic disturbances due to the presence of crabs, large fishes, and bottom-feeding marine mammals, all of which are absent from the Antarctic.

Most polar benthic invertebrates appear to be long-lived, slow-reproducing animals, but it is unclear whether this is due to some inherent response to low temperatures or to some other feature of the polar environment, such as low food resources. Among the benthic invertebrates in the Antarctic, there is little evidence for any temperature adaptation with respect to reproduction, development, or growth. Pearse et al. (1991) note that, contrary to "Thorson's rule," which holds that cold-water benthic organisms should show more nonpelagic development, pelagic development is common. The only likely difference from temperate and tropical organisms is that most Antarctic pelagic larvae are lecithotrophic and take a very long time to develop to metamorphosis. How the planktonic larvae, which are produced by some of the most abundant species, survive the poor food conditions during the Antarctic winter is still not known.

While the shallow Antarctic seas have relatively few fish species—120, according to Knox (1994)—most are endemic, probably because they are isolated by deep water from all other continents. These fishes have generated considerable interest among physiologists, because they have developed special mechanisms to keep their body fluids from freezing in the below-freezing water and because some of them (Channichthyidae) lack respiratory pigments in their blood. Such specialized endemic fish are not present in the Arctic.

There are also differences between the polar oceans with respect to the ecology of marine birds. In Antarctic seas, the distributions tend to be latitudinal, with the bird communities similar in all ocean basins at a given latitude. In the Arctic Ocean, distribution is longitudinal with strong differences between communities at similar latitudes in different parts of the ocean basin. Another difference is that, in the Antarctic seas, a whole community of birds is adapted to forage near the ice edge, and the open water near the ice margin is also a significant foraging area. In the Arctic, Hunt (1991) reports few birds forage in the marginal ice area or at the edge of the ice.

■ ECOLOGICAL COMMUNITIES OF POLAR SEAS

SEA-ICE COMMUNITIES A marine environment unique to polar seas is that occupied by organisms inhabiting the sea ice. Sea ice provides an extensive habitat for a set of organisms ranging in size from bacteria to marine birds and mammals. Here, we are interested mainly in the organisms that inhabit the ice on a long-term basis and that interact with each other.

A large number of microbes inhabit sea ice in both the Antarctic and the Arctic, but the research on microbial diversity has, according to Lizotte (2003), been directed chiefly at the protistans—principally diatoms—because of their dominance in sea ice (Figure 5.46). Arctic sea-ice diatoms comprise about 200 species while the Antarctic diversity is less—on the order of 100 species. The most

FIGURE 5.46 Sea-ice diatoms can reach such densities that their photosynthetic pigments color the underside of ice floes brown. The sea-ice organisms grow throughout the ice in brine channels, the brown vertical lines in the image on the right, as well as in porous ice layers, which can be seen as dark brown horizontal layers. (Photos © David Thomas.)

FIGURE 5.47 Psychrophilic bacteria. The small fluorescing dots and rods in the photomicrograph grow to high numbers in sea ice. Their productivity, in terms of carbon biomass, generally lags behind the primary production of sea-ice algae (large fluorescing circles), except in winter, when bacterial activity may exceed that of the primary producers. (Photos © Soenkke Grossman.)

diverse nonphotosynthetic protistan group is the ciliates. The prokaryotic microbes are represented by both Bacteria and Archaea, and the Archaea of sea ice appear most closely related to those from hydrothermal vents (Figure 5.47). Little is known at present about the significance of these two domains to sea-ice communities. Viruses have been found in only a few sea-ice studies and little is known about them.

In Antarctica, the circumpolar ring of sea ice around the continent ranges in size seasonally from 4×10^6 to 20×10^6 km^2. The biological communities that develop in the sea ice are quite variable, reflecting the different physical, chemical, and biological features that characterize the physically distinct types of ice and the seasonal differences that occur in the sea ice.

More than 200 species of marine organisms have been reported to live in association with sea ice, according to Garrison (1991). The major groups, outside of the aforementioned microbes, include heterotroph protistans, small metazoans, and fishes (Figure 5.48). There is still disagreement as to whether these ice associations are true communities, since many of the species are also found in adjacent open waters. It has been suggested that the ice may be a temporary habitat for planktonic species and may be important in providing a "seed" population for plankton blooms at the sea edge.

The algal species inhabiting the ice depend on light for photosynthesis and are, therefore, less abundant under the thick ice layers of the Arctic Ocean and most abundant at the edges of the ice. Stromberg

(1991) reports that, in the Antarctic, where the ice is thinner and more seasonal, these algae are more widespread (Figure 5.46). The algae are grazed by various microbial organisms, but also by various metazoan animals.

As with plankton, the sea-ice communities are dominated by microbes, whose preeminent position is enhanced by the small spaces and extreme conditions that exclude most large marine organisms. The microbes' dominance also has the effect of simplifying the food web. According to Lizotte (2003), there is a well-developed microbial food web in the sea ice, but our ability to quantify it has been constrained by sampling difficulties; the variety of sea-ice microhabitats; the fact that the sea-ice systems are open, with unknown inputs and outputs, to the sea; and the short-term existence of the sea ice in many areas. This microbial community or association may be significant not only to plankton, but to polar benthic communities as well, because there is evidence that benthic organisms feed on the sea-ice biota. A conceptual food web for the microbial ice community is illustrated in Figure 5.49.

The metazoan organisms found in sea ice include both pelagic forms, such as calanoid copepods and rotifers, and benthic forms, such as turbellarians, nematodes, and harpacticoid copepods. There is a significant difference in the relative abundances of these taxa in Arctic and Antarctic sea-ice assemblages: Arctic sea ice is dominated by nematodes and rotifers, while Antarctic sea ice is dominated by copepods and turbellarians (Figure 5.50). The highest numbers of these metazoans are in the bottom layers

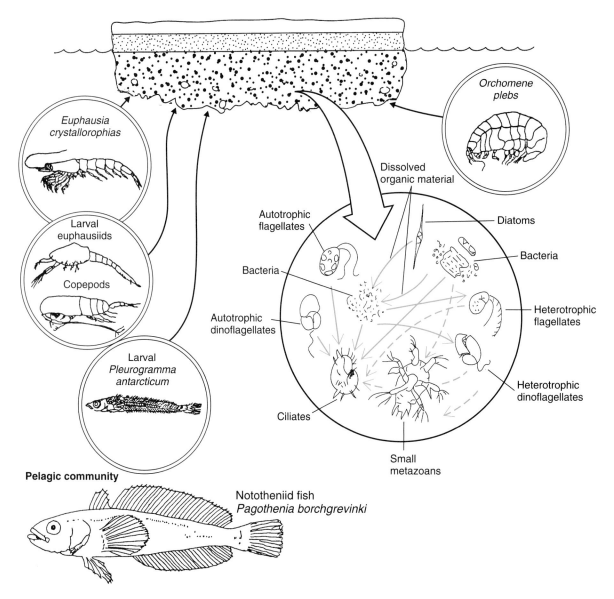

FIGURE 5.48 A conceptual food web for the pack-ice microbial community. Solid arrows show feeding relationships documented. Dashed arrows show predator–prey relationships known in other systems, but not yet confirmed in sea ice. None of the feeding rates is known. (After "Antarctic Sea Ice Biota," D. L. Garrison, *American Zoologists*, Vol. 31, pp. 17–33, 1991. © 1991 American Society of Zoologists.)

of the ice, where a variety of ice structures provides a wide range of habitats. The ice–water interface has the most complex array of larger nooks and crannies and is occupied by the larger metazoan animals, of which the gammarid amphipods and euphausiids are the most abundant (Tables 5.4 and 5.5). As with sea ice, there is a significant difference in the relative abundances of these taxa in the Arctic and Antarctic ice–water interface. In the Arctic, amphipods are the most conspicuous metazoans and euphausiids are rare, while in the Antarctic, euphausiids are dominant. Some fishes are also found in this region. In the

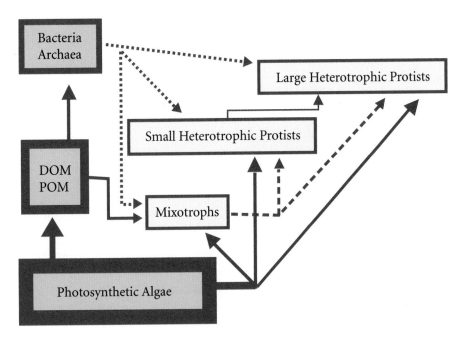

FIGURE 5.49 Conceptual model of the main feeding relationships in sea-ice microbial food webs. The loop returning dissolved organic matter (DOM) and particulate organic matter (POM) from all organismal reservoirs is not shown. The wall thickness for each reservoir represents relative biomass for typical summer conditions. Each of these reservoirs has been observed to dominate sea-ice microbial biomass at some time and place. (From Thomas, D. N., and G. S. Dieckmann (editors). 2003. *Sea Ice, an Introduction to Its Physics, Chemistry, Biology, and Geology*, p. 201, Figure 6.1. Reprinted by permission of Blackwell Publishing.)

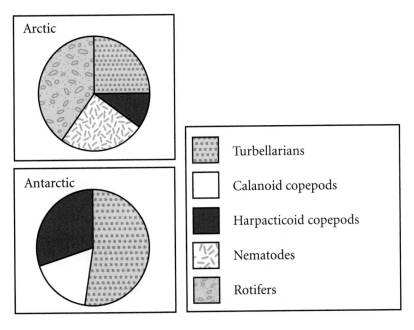

FIGURE 5.50 Differences in metazoan abundances in sea ice in the Arctic and Antarctic Oceans. (From Thomas, D. N., and G. S. Dieckmann (editors). 2003. *Sea Ice, an Introduction to Its Physics, Chemistry, Biology, and Geology*, p. 213, Figure 7.1. Reprinted by permission of Blackwell Publishing.)

Arctic, the Arctic cod (*Boeogadus saida*) and the glacial cod (*Arctogadus glacialis*) live in close association with the ice–water interface, and in the Antarctic, the adult and young of the broadhead fish (*Pagothenia borchgrevinki*) and the young of the Antarctic toothfish (*Dissostichus mawsoni*) are found just under the ice.

A final difference between the Arctic and Antarctic communities concerns the zooplankton. Despite the similarity of the zooplankton of the two oceans with respect to the major taxa represented, there is one major difference that has profound implications for the organization of perhaps all Antarctic marine communities: South of the Antarctic Convergence

TABLE 5.4

Dominant Metazoans Found in Sea Ice [a]

Nematodes
Arctic | *Theristus melnikovii, Cryonema tenue*

Rotifers
Arctic | *Synchaeta hyperborea, S. tamara*

Harpacticoid copepods
Arctic | *Halectinosoma* sp., *Harpacticus* sp., *Tisbe furcata*
Antarctic | *Drescheriella glacialis*

Calanoid copepods
Antarctic | *Peralabidocera antarctica, Stephos longipes*

Cyclopoid copepods
Arctic | *Cyclopine gracills, C. schneideri*

[a] From Table 7.1, p. 212, of Thomas, D. N., and G. S. Diekmann (editors) 2003. *Sea Ice, An Introduction to its Physics, Chemistry, Biology, and Geology*, Oxford, Blackwell.

TABLE 5.5

Dominant Metazoan Species Found at the Ice–Water Interface[a]

Calanoid copepods
Antarctic | *Stephos longipes, Peralabidocera antarctica*
Arctic | *Pseudocalanus* sp., *Calanus glacialis*

Euphausilds
Antarctic | *Euphausia superba, E. crystallorophias*

Gammeriid amphipods
Antarctic | *Paramoera walkeri, Cheirimedon fougneri, Pontogeneia Antarctica, Eusirus antarcticus, Orchomene plebs*
Arctic | *Apherusa glacialis, Gammarus wilkitzkii, Onismus nanseni, O. glacialis, O. litoralis, Ischyrocerus anguipes, Weyprechtia pinguis*

Fishes
Antarctic | *Pagothenia (= Trematomus) borchgrevinki*
Arctic | *Boreogadus saida, Archogadus glacialis*

[a] From Table 7.2, p. 217, of Thomas, D. N., and G. S Dieckmann (editors), 2003. *Sea Ice, an Introduction to its Physics, Chemistry, Biology, and Geology*. Oxford, Blackwell.

the euphausiid *Euphausia superba* is the dominant herbivore in the pelagic system (Figure 5.51). No such dominating species occurs in the Arctic Ocean. The Antarctic krill, *E. superba*, is the largest of the euphausiids, reaching 6 cm. It occurs in enormous numbers, but is also subject to great interannual population fluctuations. Because *E. superba* is the staple food for a large number of Antarctic species, including such disparate forms as whales, seals, penguins, and fishes, any change in the populations of krill will affect the populations of all those vertebrate species as well (Figure 5.52). This is the reason for the central position of this species in the Antarctic seas. Recently, Loeb et al. (1997) have suggested that population fluctuations of krill are related to the ice cover and the abundance of the salp *Salpa*

FIGURE 5.51 The dominant Antarctic euphausiid, *Euphasia superba*. (Photo © Robin Rowley.)

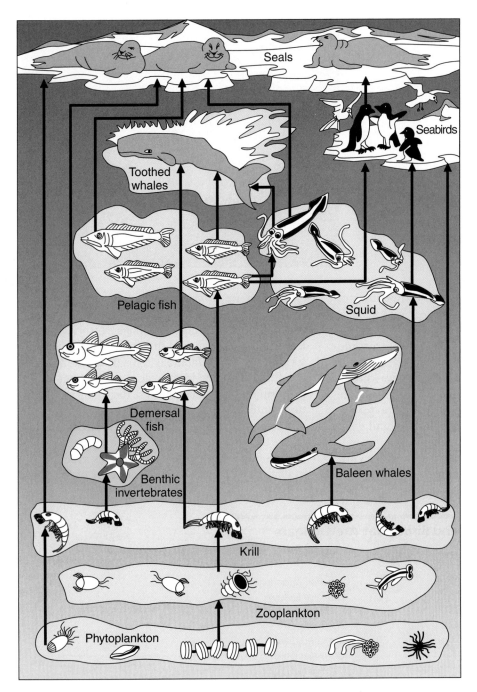

FIGURE 5.52 Diagrammatic representation of the central position of the krill *Euphausia superba* in the Antarctic oceanic ecosystem. (After "Ecosystem Management and the Atlantic Krill," S. Nicol and W. de la Mare, *American Scientist*, Vol. 81, No. 5, pp. 36–47, 1993. Copyright © 1993 American Scientist/Elyse Carter. Reprinted by permission.)

FIGURE 5.53 The salp *Salpa Thompsoni*. (Photo © Valerie Loeb, Moss Landing Marine Laboratories.)

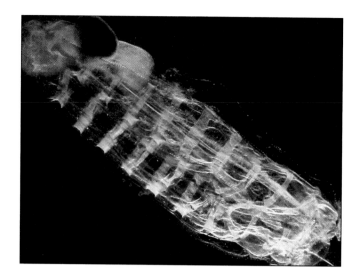

thompsoni (Figure 5.53). Salps are competitors of krill for phytoplankton but cannot feed on ice algae, as can the krill. In years of extensive ice cover, the krill are favored, leading to good recruitment and high standing stocks, while salps decline. The reverse is true in years of low ice cover. Although we know a fair amount about *E. superba*, we are still unsure about many aspects of its biology. We do not, for example, know what the standing stocks are and what a safe level of harvest might be. These are important questions because, in the last 25 years, a fishery for the krill has developed. (See Chapter 11.)

The sea ice is also a critical habitat for certain marine birds and mammals. Indeed, certain of the marine birds and mammals are obligate associates of sea ice in both the Arctic and the Antarctic. Such obligate species are listed in Table 5.6. A major use of the ice for the obligate marine birds, seals, and sea lions is as a platform for reproduction where the young can be nutured until they are self-supporting and where they can also avoid predators. Another use for the ice is as a place where the birds and seals can moult and renew their feathers and pelage. Sea ice also serves as a food source in that the undersurface ice community of fishes and

TABLE 5.6

A List of Polar Mammals and Birds That Are Obligate Associates of Sea Ice [a]

Mammals

Antarctic	* Crabeater seal	*Lobodon carcinophagus*
	* Leopard seal	*Hydrurga leptonyx*
	* Ross seal	*Ommatophoca rossii*
Arctic	* Ringed seal	*Phoca hispida*
	* Bearded seal	*Erignathus barbatus*
	* Polar bear	*Ursus maritimus*
	* Bowhead whale	*Balaena mysticetus*
	* Narwhal	*Monodon monoceros*

Birds

Antarctic	* Emperor penguin	*Aptenodytes forsteri*
	* Adélie penguin	*Pygoscelis adeliae*
	* Snow petrel	*Pagodroma nivea*
Arctic	* Ivory gull	*Pagophila eburnea*
	* Ross gull	*Rhodostethia rosea*

[a] modified from Table 8.1 of Thomas, D. n., and G. S., Dieckmann (editiors). 2003. *Sea Ice An Introduction to its Physics, Chemistry, Biology, and Geology.* Oxford, Blackwell.

invertebrates is available to the predators. For example, the crabeater seals feed almost exclusively on euphausiids, which are abundant under the ice and hence available without deep diving. A final note that needs to be mentioned is that of global warming. Reducing the extent of the ice will curtail the primary productivity of the polar seas. Together with the decrease in ice cover, a diminution of primary productivity will lead to reductions in populations of critical herbivores such as *Euphausia superba*, with cascading effects on various predators (Figure 5.52). The loss of ice for reproduction, moulting, and resting by various obligate seals and birds will in turn have an immediate effect, reducing the populations of these essential animals.

AN ANTARCTIC SOFT-SEDIMENT COMMUNITY   Oliver and Slattery (1985) studied a rare soft-bottom community in McMurdo Sound, Antarctica. The community was composed primarily of burrowing and tube-building polychaete worms and small crustaceans, occurring in densities of over 100,000 individuals per square meter. This dense assemblage was divided into three faunal groups of species that had similar characteristics: canopy, understory, and subsurface organisms (Figure 5.54). The canopy was dominated by the anemone *Edwardsia* and consisted primarily of suspension feeders. The understory included mainly mobile crustaceans that were deposit feeders and predators. The subsurface species group was composed primarily of tube-dwelling polychaetes that were mainly deposit feeders.

Experimental manipulations of this community suggested that its structure was regulated by the understory predatory small crustaceans. These animals consumed the juveniles of the dominant polychaete species. The size of the adults of the dominant polychaetes was a refuge from the predators. Therefore, the dense assemblage was dominated by large and long-lived species.

This community probably persists for long periods because of several co-occurring factors. First, there are no large organisms or predators, such as walruses, gray whales, skates, or rays, that could introduce biological disturbances. As we have seen earlier (p. 133), these predators remove large areas of infauna in other shallow-water areas. In this community, such a biological disturbance would destroy the assemblage, because the adults would be removed while the motile crustacean predators would remain and devour the settling juveniles. Second, the adults are long lived and become immune to predation from the small crustaceans when they reach a certain size. This means that there is a long time over which they reproduce—indeed, enough time to allow a few juveniles to escape the predation of the crustaceans and replace the few adults that may die.

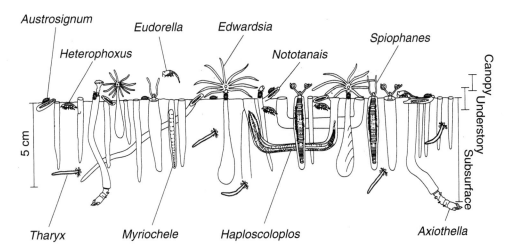

FIGURE 5.54 Schematic of a dense soft-bottom Antarctic assemblage, showing the three faunal groups: canopy, understory, and subsurface species. (From "Effects of Crustacean Predators on Species Composition and Population Structure of Soft-bodied Infauna from McMurdo Sound, Antarctica," J. S. Oliver & P. N. Slattery, *Ophelia*, Vol. 24, No. 3, pp. 155–175, 1985. Copyright © 1985 Apollo Books. Reprinted by permission.)

FIGURE 5.55 Vertical zonation of invertebrates in Antarctic shelf areas. A few motile
animals forage in Zone I, which
is otherwise barren of sessile
animals. Cnidarians dominate
the sessile animals of Zone II,
sponges Zone III. (From *Antarctic
Biology*, Vol. 1, M. W. Holdgate,
1970. Reprinted by permission of
Academic Press, Ltd.)

Zone I: 0–15 m

Zone II: 15–30 m

Zone III: below 33 m

In contrast to other soft-bottom infaunal communities that seem to be structured by competition (see
pp. 210–214), in this one there is no evidence of a
competitive dominant.

**AN ANTARCTIC HARD-BOTTOM COMMU-
NITY**   Antarctic hard-bottom communities appear
more common than soft-bottom communities. They

have also had a longer period of investigation, beginning in the late 1960s. Indeed, the more than 30 years of
marine ecological work in the McMurdo Sound area of
Antarctica has given us a better understanding of community interaction in this remote area than we have of
many temperate or tropical areas closer to home.

Dayton, Robilliard, and Paine (1970) recognize
three zones in the shallow inshore waters of

McMurdo Sound (Figure 5.55). Zone I extends 0–15 m and is essentially devoid of life, due to scouring by sea ice and the almost universal coverage by anchor ice. Only in ice-free periods can mobile organisms enter this zone.

Zone II extends from about 15 m down to the limit of anchor-ice formation at approximately 30 m. This zone has less anchor ice and exhibits a fauna of numerous sessile animals, mainly anemones and other cnidarians. The anemones here often feed upon rafted-in jellyfishes or sea urchins, while the other cnidarians are suspension feeders. The anemones appear to be territorial.

Zone III begins abruptly where the anchor ice ceases. Zone III continues down to an undetermined depth, with its substrate changing from a cobbly–rocky bottom to one that is a thick mat of sponge spicules. Sponges dominate this area, but the area has a rich and diverse fauna of other invertebrates as well, of which starfish are the most conspicuous motile animals. Zone III is noteworthy because of its extreme physical stability, which has resulted in a complex, highly diverse community that, according to Dayton et al. (1974), seems to be regulated by

FIGURE 5.56 Abundant starfish of the genus *Odontaster* on the bottom of McMurdo Sound, Antarctica. (Photo courtesy of Hunter Lenihan.)

FIGURE 5.57 Sequence showing the consumption of the large Antarctic starfish *Acodonaster conspicuus* by the small starfish *Odontaster validus* and the nemertean *Lineus corrugatus*. (From P. K. Dayton, G. A. Robilliard, R. T. Paine, and L. B. Dayton, 1974, "Biological accommodation in the benthic community at McMurdo Sound, Antarctica." *Ecol. Monogr.* 44(1):105–128.)

biological factors (predation), rather than physical factors as in Zones I and II.

The dominant organisms in Zone III are very slow growing—so slow, in fact, that in many cases the growth rates are too low to measure in one year. One exception is the sponge *Mycale acerata*, which has a very rapid growth rate and seems capable of outcompeting all other sessile invertebrates and taking over the substrate space. *Mycale* is prevented from dominating the substrate, however, because it is preyed on by two sea stars *Perknaster fuscus* and *Acodonaster conspicuus*—which keep it in check. *Acodonaster conspicuus* also preys on several other dominant sponges, but is prevented from reaching population densities that would allow it to destroy the sponge community by yet another asteroid, *Odontaster validus*, which consumes the larvae, young, and even adults of *A. conspicuus* (Figures 5.56 and 5.57). Thus, we have a community that is regulated through finely tuned biological interactions, primarily predation.

Summary of Key Concepts

- Continental shelf waters show more variability in environmental conditions than does the open sea or the deep sea. Temperature and salinity are more variable, light penetration is reduced, and food supply is abundant.

- Unvegetated sedimentary environments were first sampled remotely by grabs. More recently, scuba, towed cameras, and video systems have been the instruments of choice.

- Unvegetated sedimentary systems are dominated by polychaetes, crustaceans, mollusks, and echinoderms.

- The first quantitative work in marine ecology was done by Petersen, who invented the quantitative grab. He was succeeded by Thorson, who established the concept of parallel bottom communities—that certain recurrent groups occupy the same types of substrate over large areas of the continental shelves.

- Marine organisms have patchy distributions in time and space and may be categorized into functional groups such as sediment stabilizers, sediment destabilizers, detritivores, carnivores, omnivores, and herbivores.

- Changes in community structure occur through physical factors, biological factors, or some combination of the two. The simplest cases involve the dominance of physical factors. Biological factors are more complex and include predation, competition, and recruitment.

- Competition may be direct, involving either contact between organisms or interference in some aspect of the organisms life cycle, or it may be indirect, with one organism interrupting or preventing the use of one or more resources by another.

- Burrowing deposit feeders are most abundant in mud, whereas suspension feeders are most abundant in sand.

- Trophic group amensalism is the exclusion of one trophic group via modification of the environment by another group.

- Competitive interference is the exclusion of one group of organisms via interference with their normal activities.

- Three categories of predators are surface, burrowing, and digging predators.

- Caging out epifaunal predators in soft sediments usually results in increased numbers of infaunal organisms, suggesting that predation is an important process. However, caging usually does not exclude infaunal predators, which may well be important.

- Recruitment may have the potential to determine community structure, but the difficulty of separating the effect of recruitment from various sources of postrecruitment mortality means that we can only speculate at present.

- Competition for space may occur in benthic infauna both vertically and horizontally and may affect the density of the organisms over a given area or volume or may lead to more occurrences of commensalism.

- Community structure in open sedimentary environments changes over time and produces certain patterns or patches that recur on differing areal and time scales. The changes result from a complex interplay of physical and biological factors.

- Rocky subtidal communities are greatly influenced by the slope and composition of the rock.

- Rocky subtidal communities are usually covered with low-growing plants and sessile or sedentary invertebrates, occupying a two-dimensional space and leaving little open space.

- The communities on horizontal or gentle rock slopes are different from those on vertical rock slopes.

- The upper part of the rocky subtidal zone is usually dominated by various algae.

- Most animals of subtidal rock are either colonial encrusting species able to spread over surfaces by asexual means or single individuals that settle gregariously, forming large, space-occupying clumps. Most are suspension feeders.

- The organization and persistence of organisms in the rocky subtidal zone and the high species diversity there are explained by a combination of biological and physical factors, including disturbances from storms, competition for space, grazing, recruitment, and mutualism.

- Disturbance generates patches of open space that are colonized by larvae or juveniles of nearby adults or by the spread of adjacent colonial organisms. Competition for space among these organisms, coupled with the seasonality of certain organisms and mutualistic interactions among certain species, completes the cycle.

- In any area at any time, either there is a mosaic of patches of organisms of different ages, different species, or different abundances of species, resulting in high species diversity, or there is a varying zonation pattern.

- The existence of different groups of organisms or patches of organisms in the same area has been called *alternate stable states.*

- In cold temperate parts of the world, the subtidal hard substrates are occupied by large brown algae, giving rise to kelp forests and kelp beds.

- Kelps form extensive three-dimensional habitats composed of several vertical layers.

- The extent of kelp beds and kelp forests is dependent upon a hard substrate for attachment, sufficient light, low temperatures, available nutrients, and grazing.

- Kelp plants have a life cycle that alternates between an asexual macroscopic stage termed the *sporophyte* and a sexual microscopic stage called the *gametophyte.* Kelps have very high productivity.

- The causes of kelp mortality are wave action, depletion of nutrients, warm temperatures, and grazing.

- Multidecadal studies of kelp forests substantiate the importance of changes in the oceanic regime to the composition and persistence of the kelp forests.

- Seagrasses have a worldwide distribution and form dense carpets of blades in shallow waters of temperate and tropical seas.

- All seagrasses have a similar appearance, with thin straplike blades arising from a creeping rhizome that stabilizes the sediment.

- Seagrass beds are most strongly influenced by water currents and wave action, which shape how the beds form.

- Seagrass beds are depositional environments, often with high organic content that can make the sediment under the bed anaerobic.

- Other physical factors affecting seagrass beds include light, temperature, and desiccation.

- The productivity of seagrass beds is among the highest in the marine environment. Because seagrass is a flowering plant, it extracts nutrients from the sediment and can recycle otherwise trapped nutrients back into the ecosystem.

- Seagrass beds generally have only one, or at most a few, dominant species of grasses.

- Seagrass beds are complex systems with a narrow vertical three-dimensionality and complexity. They contain large numbers of epiphytic and epizoic organisms, burrows, and motile animals.

- Seagrasses are grazed by relatively few animals—notably, various birds, a few fish, sea urchins, and the green turtle—but these may consume a large fraction of the annual seagrass production.

- A wide variety of predators is associated with seagrass beds, but there is little evidence that they regulate the structure of the beds or the abundance of prey.

- Seagrass beds may offer some refuge from predators to some prey organisms.

- Much seagrass productivity may be exported in the form of detritus to other systems.

- Seagrass beds serve a number of functions in shallow inshore waters, including being a major source of primary productivity, stabilizing the bottom, and serving as a nursery grounds for many species, as a trap for sediment, as a protective canopy shielding the inhabitants underneath from strong sunlight, and as a substrate for various epizoic and epiphytic organisms.

- Seagrass beds are subject to catastrophic, episodic destruction by disease.

- Some seagrass beds show a successional pattern, whereas others do not.

- Examples of some special communities are the sea-pen community in Puget Sound and the sea-pansy community in southern California.

- Polar communities in shallow water differ from other communities due to the drastic seasonal changes in light, the ice and snow, and the constant extremely low temperatures.

- The Arctic and Antarctic oceans have very different physical conditions, which lead to different communities. The Arctic Ocean is an isolated sea surrounded by landmasses with extensive continental shelf areas and receiving the influx from many rivers, leading to mainly sedimentary benthic environments. In contrast, the Antarctic ocean is a circumpolar ring of water surrounding a central landmass with no rivers that produce sediment and with a narrow continental shelf that drops off quickly into deep water.

- A major disturbance factor in both the Arctic and Antarctic is ice, which can destroy benthos by freezing the organisms and lifting them or by gouging them out.

- Biologically, the Antarctic organisms are more diverse and show more endemism than those of the Arctic.

- Associated with the sea ice, particularly in Antarctica, is a whole set of organisms, from bacteria to krill, and a whole well-developed microbial food web that is significant to both the water column community and the benthos.

- Antarctic seas are characterized by the central position and importance of the krill *Euphausia superba.*

- The Antarctic soft-bottom community is composed primarily of burrowing and tube-building polychaete worms and small crustaceans dominated by large and long-lived species. There is no evidence of a competitive dominant.

- Antarctic hard-bottom communities are dominated by sponges with a rich and diverse associated fauna, of which various starfishes are conspicuous and are regulated primarily by predation.

Review Questions

ESSAY Develop complete answers to these questions:

1. Why are there fewer niches available in the soft-bottom, sublittoral region than around rocky coasts? Discuss.

2. Some ecological wag has said, "Nichy nachy, the environment is patchy." Discuss three ways (one physical, one chemical, and one biological) in which the shallow-water subtidal benthos is (or becomes) patchy.

3. Compare a kelp forest with a terrestrial forest. How are they similar? How are they different? Consider, for example, light, nutrient uptake, vertical zonation, and general structure.

4. "Sediments are profoundly three dimensional, but they are not uniform." Present evidence that supports this statement and discuss its significance to the infauna.

5. Discuss examples whereby abiotic and biotic factors interact to structure the marine sublittoral zone into patches of organisms.

6. Explain the phenomenon called alternate stable states; how do these states develop and why do they change?

7. What important roles (functions) do seagrass beds serve in inshore regions? Discuss at least _____ of these. (Instructors should decide the number that their students are to discuss.)

8. Present an argument that would explain why the fauna of the Antarctic region has a high degree of diversity, a high endemic rate, and a high biomass compared with Arctic Ocean fauna.

Bibliography

Ackley, F. F., and C. W. Sullivan. 1994. Physical controls on the development and characteristics of Antarctic sea ice biological communities—a review and synthesis. *Deep-Sea Res.* I 41:1583–1604.

Ambrose, W. 1991. The importance of infaunal predators in structuring marine soft bottom communities. *Amer. Zoo.* 3:849–860.

Ayling, A. M. 1981. The role of biological disturbance in temperate subtidal encrusting communities. *Ecology* 62(3):830–847.

Barkai, A., and C. McQuaid. 1988. Predator–prey role reversal in a marine benthic ecosystem. *Science* 242:62–64.

Bertness, M., and R. Calloway. 1994. Positive interactions in communities. *Trends Ecol. Evol.* 9:191–193.

Birkeland, C. 1974. Interactions between a sea pen and seven of its predators. *Ecol. Monogr.* 44:211–232.

Camp, D. K., S. P. Cobb, and J. F. Van Breedveld. 1973. Overgrazing of seagrasses by a regular urchin, *Lytechinus variegatus. Bioscience* 23:37–38.

Conlan, K. E., H. S. Lenihan, R. G. Kvitek, and J. S. Oliver. 1998. Ice scour disturbance to benthic communities in the Canadian High Arctic. *Mar. Ecol. Prog. Ser.* 166:1–16.

Cottam, C. 1934. Past periods of eelgrass scarcity. *Rhodora* 36:261–264.

Coull, B. C., ed. 1977. *Ecology of marine benthos.* Columbia: University of South Carolina Press.

Crisp, D. J., and P. S. Meadows. 1962. The chemical bases of gregariousness in cirripedes. *Proc. Roy. Soc. London (Biol.)* 150:500–520.

Davis, N., and G. VanBlaricom. 1978. Spatial and temporal heterogeneity in a sand bottom epifaunal community of invertebrates in shallow water. *Limn. Oceanog.* 23(3):417–427.

Dayton, P. K., G. A. Robilliard, and A. L. DeVries. 1969. Anchor ice formation in McMurdo Sound, Antarctica, and its biological effects. *Science* 163.

Dayton, P. K., G. A. Robilliard, and R. T. Paine. 1970. Benthic faunal zonation as a result of anchor ice at McMurdo Sound, Antarctica. In *Antarctic ecology*, Vol. 1. New York: Academic Press, 244–258.

Dayton, P. K., G. A. Robilliard, R. T. Paine, and L. B. Dayton. 1974. Biological accommodation in the benthic community at McMurdo Sound, Antarctica. *Ecol. Monogr.* 44(1):105–128.

Dayton, P. K., M. J. Tegner, P. B. Edwards, and K. L. Riser. 1998. Sliding baselines, ghosts, and reduced expectations in kelp forest communities. *Ecol. Appl.* 8:309–322.

Dayton, P. K., M. J. Tegner, P. B. Edwards, and K. L. Riser. 1999. Temporal and spatial scales of kelp forest demography: The role of oceanographic climate. *Ecological Monographs* 69: 219–250.

Den Hartog, C. 1977. Structure, function and classification in sea grass communities. In *Seagrass ecosystems*, edited by C. P McRoy and C. Helfferich. New York: Dekker, 89–121.

DeVlas, J. 1979. Annual food intake by plaice and flounder in a tidal flat area in the Dutch Wadden Sea with special reference to consumption of regenerating parts of macrobenthic prey. *Neth. J. Sea Res.* 13:117–153.

Deysher, L. E., and T. A. Dean. 1986. In situ recruitment of sporophytes of the giant kelp, *Macrocystis pyrifera* (L) C. A. Agardh: Effects of physical factors. *J. Exp. Mar. Biol. Ecol.* 103:41–63.

Ebeling, A. W., D. R. Laur, and R. J. Rowley. 1985. Severe storm disturbance and reversal of community structure in a southern California kelp forest. *Mar. Biol.* 84:287–294.

Elner, R. W., and R. L. Vadas. 1990. Inference in ecology: The sea urchin phenomenon in the Northwest Atlantic. *Amer. Nat.* 136:108–125.

Estes, J., and D. O. Duggins. 1995. Sea otters and kelp forests in Alaska: Generality and variation in a community ecological paradigm. *Ecol. Monogr.* 65(1):75–100.

Estes, J., and J. Palmisano. 1974. Sea otters, their role in structuring near-shore communities. *Science* 185:1058–1060.

Fager, E. W. 1968. A sand bottom epifaunal community of invertebrates in shallow water. *Limn. Oceanog.* 13(3):448–464.

Fenchel, T. 1977. Aspects of the decomposition of seagrasses. In *Seagrass ecosystems*, edited by C. P. McRoy and C. Helfferich. New York: Dekker, 123–146.

Fletcher, W. J. 1987. Interactions among subtidal Australian sea urchins, gastropods and algae: Effects of experimental removals. *Ecol. Monogr.* 57(1):89–109.

Foster, M. S., and D. R. Schiel. 1985. *The ecology of giant kelp forests in California: A community profile*. U.S. Fish and Wildlife Service Biological Report 85 (7.2).

Garrison, D. L. 1991. Antarctic sea ice biota. *Amer. Zoo.* 31:17–33.

Graham, M. H. 2002. Prolonged reproductive consequences of short term biomass loss in seaweeds. *Marine Biology* 140: 901–911.

Gray, J. S. 1981. *The ecology of marine sediments*. New York: Cambridge University Press.

Harrold, C., and D. Reed. 1985. Food availability, sea urchin grazing and kelp forest community structure. *Ecology* 66(4):1160–1169.

Heck, K. L., K. W. Able, C. T. Roman, and M. P. Fahey. 1995. Composition, abundance, biomass and production of macrofauna in a New England estuary: Comparison among eelgrass meadows and other nursery habitats. *Estuaries* 18: 379–389.

Hempel, G. 1985. On the biology of polar seas, particularly the southern ocean. In *Marine biology of polar regions and effects of stress on marine organisms*, edited by J. S. Gray and M. E. Christiansen. New York: Wiley, 3–33.

Holme, N. A., and A. D. McIntyre, eds. 1984. *Methods for the study of marine benthos*, 2d ed. Oxford: Blackwell Scientific Publications.

Hunt, G. L. 1991. Marine ecology of seabirds in polar oceans. *Amer. Zoo.* 31:131–142.

Jones, N. S. 1950. Marine bottom communities. *Biol. Rev.* 25:283–313.

Kastendiek, J. 1976. Behavior of the sea pansy, *Renilla kollikeri* Pfeffer (Coelenterata, Pennatulacea), and its influence on the distribution and biological interaction of the species. *Biol. Bull.* 151:518–537.

Keegan, B. F., P. O. Ceidigh, and P. J. S. Boaden, eds. 1977. *Biology of benthic organisms*. New York: Pergamon.

Kenworthy, W. J., J. C. Zieman, and G. W. Thayer. 1982. Evidence for the influence of seagrass on the benthic nitrogen cycle in a coastal plain estuary near Beaufort, North Carolina. *Oecologia* 54:152–158.

Knox, G. A. 1994. *The Biology of the southern ocean*. Cambridge, UK: Cambridge University Press.

Lenihan, H. S., and F. Micheli. 2001. Soft sediment communities. In *Marine community ecology*, edited by Bertness, M., S. Gaines, and M. Hay. Sinauer Associates, Sunderland, MA, 253–287.

Levin, L. A. 1981. Dispersion, feeding behavior and competition in two spionid polychaetes. *J. Mar. Res.* 39:99–117.

Levinton, J. S. 1977. Ecology of shallow water deposit feeding communities, Quisset Harbor, Massachusetts. In *Ecology of marine benthos*, edited by B. Coull. Columbia: University of South Carolina Press.

Levinton, J. S. 1991. Variable feeding behavior in three species of *Macoma* (Bivalvia: Tellinacea) as a response to water flow and sediment transport. *Mar. Biol.* 110: 375–383.

Lizotte, M. P. 2003. The microbiology of sea ice. In *Sea ice, an introduction to its physics, chemistry, biology and geology*, edited by D. N. Thomas and G. S. Dieckmann. Oxford: Blackwell, 184–210.

Loeb, V., V. Siegel, O. Holm-Hansen, R. Hewitt, W. Fraser, W. Trivelpiece, and S. Trivelpiece. 1997. Effects of sea-ice extent and krill or salp dominance on the Antarctic food web. *Nature* 387:897–900.

Mann, K. H. 1973. Seaweeds: Their productivity and strategy for growth. *Science* 182:975–983.

McRoy, C. P. 1966. The standing stock and ecology of eelgrass in Izembeck Lagoon, Alaska. Master's thesis, University of Washington.

McRoy, C. P., and C. Helfferich, eds. 1977. *Seagrass ecosystems*. New York: Dekker.

McRoy, C. P., and C. McMillan. 1977. Production ecology and physiology of sea grasses. In *Seagrass ecosystems*, edited by C. P. McRoy and C. Helfferich. New York: Dekker, 53–88.

Menge, B. A., and T. M. Farrell. 1989. Community structure and interaction webs in shallow marine hard bottom communities: Tests of an environmental model. *Adv. Ecol. Res.* 19:189–262.

Mills, E. L. 1969. The community concept in marine zoology, with comments on continua and instability in some marine communities: A review. *J. Fish. Res. Bd. Canada* 26:1415–1428.

Myers, A. C. 1977. Sediment processing in a marine subtidal sandy bottom community. II. Biological consequences. *J. Mar. Res.* 35:633–647.

Nicol, S., and W. de la Mare. 1993. Ecosystem management and the Antarctic krill. *Am. Sci.* 81(1):36–47.

North, W. 1971. Growth of individual fronds of the mature giant kelp, *Macrocystis*. In *The biology of giant kelp beds* (*Macrocystis*) *in California*, edited by W. North. *Beihefte zur Nova Hedwigia*, vol. 32, Warsaw: J. Cramer, 123–168.

Ockelmann, K. W., and O. Vahl. 1970. On the biology of the polychaete *Glycera alba*, especially its burrowing and feeding. *Ophelia* 8:275–294.

Ogden, J. C. 1980. Faunal relationships in Caribbean seagrass beds. In *Handbook of seagrass biology: An ecosystem perspective*, edited by R. C. Phillips and C. P. McRoy. New York: Garland STPM Press, 173–198.

Olafsson, E. B., C. H. Peterson, and W. G. Ambrose. 1994. Does recruitment limitation structure populations and communities of macro-invertebrates in marine soft sediments: The relative significance of pre- and post-settlement processes. *Oceanogr. Mar. Bio., Ann. Rev.* 32:65–109.

Oliver, J. S., and P. N. Slattery. 1985. Effects of crustacean predators on species composition and population structure of soft bodied infauna from McMurdo Sound, Antarctica. *Ophelia* 24(3):155–175.

Oliver, J. S., P. N. Slattery, L. W. Hulberg, and J. W. Nybakken. 1979. Relationships between wave disturbance and zonation of benthic invertebrate communities along a subtidal high energy beach in Monterey Bay, California. *Fish. Bull.* 78:437–454.

Orth, R. J. 1975. Destruction of eelgrass *Zostera marina* by the cownose ray *Rhinoptera bonasus* in the Chesapeake Bay. *Chesapeake Sci.* 16:205–208.

Pearse, J. S., J. B. McClintock, and I. Bosch. 1991. Reproduction of Antarctic benthic marine invertebrates: Tempos, modes and timing. *Amer. Zoo.* 31:65–80.

Petersen, C. G. Joh. 1918. The sea bottom and its production of fish food. A survey of the work done in connection with the valuation of the Danish waters from 1883–1917. *Rap. Danish Biol. Stat.* Vol. 25.

Petersen, C. G. Joh. 1924. A brief survey of the animal communities in Danish waters. *Amer. J. Sci., Ser. 5* 7(41):343–354.

Peterson, C. H. 1977. Competitive organization of the soft bottom macrobenthic communities of southern California lagoons. *Mar. Biol.* 43:343–359.

Peterson, C. H. 1982. Clam predation by whelks (*Busycon* spp.): Experimental tests of the importance of prey size, prey density and seagrass cover. *Mar. Biol.* 66:159–170.

Peterson, C. H., and M. L. Quammen. 1982. Siphon nipping: Its importance to small fishes and its impact on growth of the bivalve *Protothaca staminea*(Conrad). *J. Exp. Mar. Biol. Ecol.* 63:249–268.

Phillips, R. C., and C. P. McRoy, eds. 1980. *Handbook of seagrass biology*. New York: Garland STPM Press.

Pohle, D. G., V. M. Brieelj, and Z. Garcia-Esquiral. 1991. The eelgrass canopy: An above bottom refuge from benthic predators for juvenile bay scallops *Argopecten irradians*. *Mar. Ecol. Progr. Ser.* 74(1):47–59.

Rasmussen, E. 1973. Systematics and ecology of the Isefjord marine fauna (Denmark). *Ophelia* 11(12):1–495.

Reed, D. C., D. R. Laur, and A. W. Ebeling. 1988. Variation in algal dispersal and recruitment: The importance of episodic events. *Ecol. Monogr.* 58:321–335.

Reise, K. 1979. Spatial configurations generated by motile benthic polychaetes. *Helgol. Wiss. Meeresunters.* 32:55–72.

Rhoads, D. C., and D. K. Young. 1970. The influence of deposit feeding organisms on sediment stability and community structure. *J. Mar. Res.* 28(2):150–178.

Rhoads, D. C., and D. K. Young. 1971. Animal–sediment relations in Cape Cod Bay, Massachusetts. II. Reworking by *Molpadia oolitica*. *Mar. Biol.* 11(3):255–261.

Robblee, M. B., T. R. Barber, P. R. Carlson, Jr., M. J. Durako, J. W. Fourgurean, L. K. Muehlstein, D. Porter, L. A. Yarbro, R. T. Zieman, and J. C. Zieman. 1991. Mass mortality of the tropical seagrass *Thalassia testudinum* in Florida Bay (U.S.A.). *Mar. Ecol. Progr. Ser.* 71:297–299.

Robertson, A. I., and K. H. Mann. 1984. Disturbance by ice and life history adaptations of the seagrass *Zostera marina*. *Mar. Biol.* 80:131–142.

Rosenthal, R. J., W. D. Clarke, and P. K. Dayton. 1974. Ecology and natural history of a stand of giant kelp, *Macrocystis pyrifera*, off Del Mar, California. *Fish. Bull.* 72(3):670–684.

Scagel, R. F. 1947. *An investigation on marine plants near Hardy Bay, B.C.* Provincial Dept. Fisheries, Victoria, B.C., Canada, 1–70.

Sebens, K. 1985. The ecology of the rocky subtidal zone. *Amer. Sci.* 73:548–557.

Sebens, K. 1986. Spatial relationships among encrusting marine organisms in the New England subtidal zone. *Ecol. Monogr.* 56(1):73–96.

Setchell, W. A. 1929. Morphological and phenological notes on *Zostera marina*. *Univ. Calif. Pub. Botany* 14:389–452.

Short, F. T., L. K. Muehlstein, and D. Porter. 1987. Eelgrass wasting disease: Cause and recurrence of a marine epidemic. *Biol. Bull.* 173(3):557–562.

Snelgrove, P. V. R., and C. A. Butman. 1994. Animal–sediment relationships revisited: Cause versus effect. *Oceanogr. Mar. Biol. Ann. Rev.* 32:111–177.

Stromberg, J. O. 1991. Marine ecology of Polar seas: A comparison Arctic/Antarctic. In *Marine biology: Its accomplishment and future prospect*, edited by J. Mauchline and T. Nemoto. New York: Elsevier, 247–261.

Sutherland, J. P. 1974. Multiple stable points in natural communities. *Amer. Nat.* 108:859–873.

Tegner, M. J., and P. K. Dayton. 1987. El Niño effects on southern California kelp forest communities. *Adv. Ecol. Res.* 17:243–279.

Tevesz, M. J. S., and P. L. McCall, eds. 1983. *Biotic interactions in recent and fossil benthic communities*. New York: Plenum Press.

Thayer, G. W., S. M. Adams, and M. W. LaCroix. 1975. Structural and functional aspects of a recently established *Zostera marina* community. In *Estuarine research*, Vol. 1., edited by L. E. Cronin. New York: Academic Press, 517–540.

Thayer, G. W., W. J. Kenworthy, and M. S. Fonseca. 1984. *The ecology of eelgrass meadows of the Atlantic coast: A community profile*. U.S. Fish and Wildlife Service. FWS/OBS-84/02.

Thomas, D. N., and G. S. Dieckmann. 2002. Antarctic sea ice, A habitat for extremophiles. *Science* 295: 641–644.

Thomas, D. N., and G. S. Dieckmann (editors). 2003. *Sea Ice, an Introduction to Its Physics, Chemistry, Biology and Geology*. Oxford: Blackwell.

Thorson, G. 1950. Reproductive and larval ecology of marine bottom invertebrates. *Biol. Rev.* 25:1–45.

Thorson, G. 1955. Modern aspects of marine level bottom animal communities. *J. Mar. Res.* 14:387–397.

Thorson, G. 1957. Bottom communities (sublittoral or shallow shelf). In *The treatise on marine ecology and paleoecology*, Vol. 1, *Ecology*. Geol. Soc. Amer. Memoir 67, 461–534.

Thorson, G. 1966. Some factors influencing the recruitment and establishment of marine benthic communities. *Neth. J. Sea. Res.* 3(2):267–293.

Turner, T. 1983. Facilitation as a successional mechanism in a rocky intertidal community. *Amer. Nat.* 121:729–738.

VanBlaricom, G. R. 1982. Experimental analyses of structural regulation in a marine sand community exposed to ocean swell. *Ecol. Monogr.* 52:283–305.

Van Veldhuizen, H., and D. Phillips. 1978. Prey capture by *Pisaster brevispinus* (Asteroidea, Echinodermata) on soft substrate. *Mar. Biol.* 48:89–97.

Virnstein, R. W. 1977. The importance of predation by crabs and fishes on benthic infauna in Chesapeake Bay. *Ecology* 58:1199–1217.

Warher, G. F. 1984. *Diving and marine biology, the ecology of the sublittoral*. New York: Cambridge University Press.

Wilkins, E. W. 1982. Waterfowl utilization of a submerged vegetation (*Zostera marina* and *Ruppia maritima*) bed in the lower Chesapeake Bay. Master's thesis, College of William and Mary.

Williams, S. 1987. Competition between seagrasses *Thalassia testudinum* and *Syringodinium filiforme* in a Caribbean lagoon. *Mar. Ecol. Progr. Ser.* 35:91–98.

Williams, S., and K. L. Heck. 2001. Seagrass community ecology. In *Marine community ecology*, edited by Bertness, M., S. Gaines, and M. Hay. Sinauer Associates: Sunderland, MA, 317–337.

Wilson, W. H., Jr. 1981. Sediment mediated interactions in a densely populated infaunal assemblage: The effects of the polychaete *Abarenicola pacifica*. *J. Mar. Res.* 39:735–748.

Wilson, W. H., Jr. 1991. Competition and predation in marine soft sediment communities. *Ann. Rev. Ecol. Syst.* 21:221–241.

Witman, J. D. 1985. Refuges, biological disturbance, and rocky subtidal community structure in New England. *Ecol. Monogr.* 55:421–445.

Witman, J. D. 1987. Subtidal coexistence: Storms, grazing, mutualism and the zonation of kelps and mussels. *Ecol. Monogr.* 57(2):167–187.

Witman, J. D., and R. A. Cooper. 1983. Disturbance and contrasting patterns of population structure in the brachiopod *Terebratulina septentrionalis* from two subtidal habitats. *J. Exp. Mar. Biol. Ecol.* 73: 57–79.

Witman, J. D., and P. K. Dayton. 2001. Rocky subtidal communities. In *Marine community ecology*, edited by Bertness, M., S. Gaines, and M. Hay. Sinauer Associates, Sunderland, MA, 339–366.

Witman, J. D., and K. P. Sebens. 1988. Benthic community structure at a subtidal rock pinnacle in the central Gulf of Maine. Pp. 67–104 In *Benthic productivity and marine resources of the Gulf of Maine*, edited by I. Babb and M. DeLuca. National Undersea Research Program Research Report 88–3.

Woodin, S. A. 1976. Adult–larval interactions in dense infaunal assemblages: Patterns of abundance. *J. Mar. Res.* 34:25–41.

Woodin, S. A. 1983. Biotic interactions in recent marine sedimentary environments. In *Biotic interactions in recent and fossil benthic communities*, edited by M. J. S. Tevesz and P. L. McCall. New York: Plenum Press, 3–38.

Zieman, J. C. 1982. *The ecology of the seagrasses of South Florida: A community profile*. U.S. Fish and Wildlife Service Program FWS/OBS-82/25. 12.

CHAPTER 6

INTERTIDAL ECOLOGY

Although it constitutes by far the smallest area of all in the world's oceans, the intertidal zone is perhaps the best known. It is an extremely narrow fringe area a few meters in extent between high and low water. Because it is the most accessible of the occan's zones to humans, the intertidal zone has been studied more intensively than any other marine habitat. Only here can aquatic organisms be directly observed during low tide and without special equipment. The intertidal zone has been of interest and use to humans since prehistoric times.

Despite being restricted in size, the intertidal zone has the greatest variation in environmental factors of any marine area. Stemming from this variation is a tremendous diversity of life, which may be as great as or greater than that found in the more extensive subtidal habitats. The intertidal zone is an extension of the marine environment and is inhabited almost exclusively by marine organisms. Although it may be exposed to air as much as half the time, terrestrial fauna and flora have not invaded this zone to any extent, except at the uppermost fringes.

The richness, the diversity of environmental factors, and the ease of access to this area attract a disproportionate amount of scientific attention. More knowledge is available for the relatively small intertidal zone and its organisms, and the interactions among those organisms are better known, than in any other area. Thanks to such interest and knowledge, the intertidal zone has produced more unifying concepts with respect to the organization of marine communities than any other oceanic zone. As a result, a disproportionate amount of space in this book is devoted to a discussion of this zone.

■

Environmental Conditions

In part, such a tremendous range of environmental factors is found in the intertidal zone because it is exposed to the air for a certain amount of time during the day and most physical factors show a wider range in air than in water. (See Chapter 1, pp. 1–4.)

■ TIDES

The periodic, predictable rise and fall of the level of the sea over a given time interval is called a **tide**. This is the most important environmental factor influencing life in the intertidal zone. Without the tide, the zone would not exist. With very few exceptions, most shore areas of the world are subject to tides. The only major seas that virtually lack tidal action are the Mediterranean, Black, and Baltic. In these bodies, the water level on the shoreline fluctuates primarily because of winds. This does not mean, however, that all such seashores have the same tidal range or even the same type of tide. The reasons for different tides and different tidal ranges are complex and have to do with the interactions of the tide-generating forces: the sun and moon, the rotation of the Earth, and the geomorphology of the ocean basins. A complete explanation of tides is beyond the scope of this text, but a simplified description will be given next.

Tides occur due to the interaction between the gravitational attraction of the sun and the moon on the Earth and the centrifugal force generated by the rotating Earth–moon system. As a result of these forces, the water in the ocean basins is pulled into bulges. The gravitational attraction of one body for another is a function of the mass of each body and their distance apart. In the case of the sun and the moon, the gravitational force of the moon on the Earth is about twice that of the sun on the Earth, even though the sun is many times more massive than the moon. The reason for the moon's stronger pull is the much greater distance of the sun from the Earth.

The Earth and moon form an orbiting system that revolves around their common center of mass. Because of the large size of the Earth relative to the moon, the center of mass is located inside the Earth. The revolution of the Earth–moon system creates a centrifugal force (acting outward) that is balanced by the gravitational force between the two bodies. The gravitational force, however, is much stronger than centrifugal force on the side of the Earth facing the moon and much weaker on the side opposite. As a result, the side facing the moon has the water pulled into a bulge (high tide). On the opposite side of the Earth, the gravitational force of the moon will be the least, and the stronger centrifugal force will pull water into a bulge *away* from earth (another high tide; Figure 6.1); thus, we have the two high tides. These high tides circle the Earth, following the position of the moon as the Earth revolves on its axis once every 24 hours. Low tides are about halfway between the high tides.

As just described, the Earth–moon system would produce two high tides and two low tides of equal magnitude each day. As we know, however, tides are not equal, nor are there always four tides per day. How do these discrepancies arise? In part, it is because the Earth's axis is not vertical with respect to its plane of orbit around the sun. Instead, the axis is inclined 23.45° from the vertical, and as a result, during the Earth's rotation on its axis, a given point on the Earth's surface is subjected to different tidal heights, as Figure 6.1a shows. Note also from the figure that, at the highest latitudes (in polar seas), only a single high tide would occur. Further changes in the height of the tides result from changes in the moon relative to the Earth as the moon moves in its orbit around the Earth. Since the orbit of the moon is not circular, but elliptical, there are times when the moon is closer to the earth (the point where it is closest is called the **perigee**) and others when it is farther away (the point where it is most distant is called the **apogee**). Tides are greater at perigee and diminished at apogee.

Figure 6.1b shows the effect of the sun during spring and neap tides (Figure 6.1b). Spring (extreme) tides show the greatest range (both high and low) and result when the moon and sun are directly aligned and their gravitational forces augment each other. Neap tides, by contrast, are tides showing a minimum range and result when sun and moon are at right angles to each other and so do not augment each other.

Differences both in number of high and low tides per day and in height in various parts of the world are due to peculiarities of the various ocean basins in which the tides occur. The extremely high tides of the Bay of Fundy in Nova Scotia and in Cook Inlet in Alaska result from the geometry of the basin in which the basic tidal forces are acting.

Locations having a single low and a single high tide per day are said to exhibit **diurnal tides** (Figure 6.2). Locations with two highs and two lows per lunar day display **semidiurnal tides**. Those having a mixture of

FIGURE 6.1 Origin of tides. (a) The moon raises a bulge on the side of the Earth nearest it, due to a gravitational force greater than the counteracting centrifugal force. On the opposite side, the centrifugal force is stronger and throws another bulge outward. Because of the inclination of the Earth on its rotational axis, as point A rotates, it undergoes two high tides of different height within a 24-hour period. (b) Position of the moon and sun at neap and spring tides.

(a)

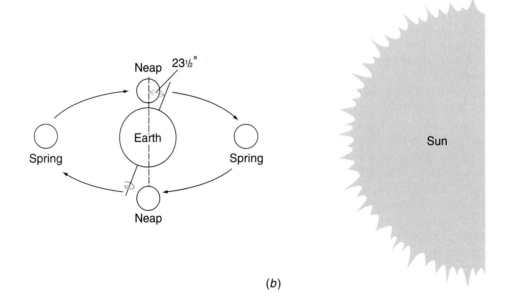

(b)

diurnal and semidiurnal tides show **mixed tides.** The heights of the highs and lows vary from day to day as the positions of the sun and moon change relative to each other, each being alternately fully aligned and 90° out of alignment every 14 days.

Changes in tidal level over time can have two direct effects on the presence and organization of intertidal communities. The first effect results from the time that a given area of the intertidal zone is exposed to the air relative to the time it is submerged in water. It is the *duration of exposure* to air that is most important, because during that time is when marine organisms will be subjected to the greatest temperature ranges and the possibility of desiccation (drying out due to loss of water). The longer the duration of exposure, the greater the chance of encountering a lethal temperature or of desiccating beyond tolerable limits. Since most of these animals

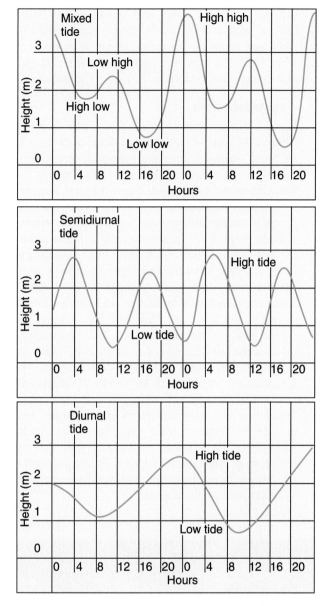

FIGURE 6.2 Mixed, semidiurnal, and diurnal tide curves. (Modified from *Introduction to Oceanography*, 4th ed., D. A. Ross, 1988, p. 287, © 1988 Prentice Hall, Upper Saddle River, NJ.)

also must wait until they are covered with water in order to feed, the longer the duration of exposure, the less opportunity the animal has to feed and obtain sufficient energy. Animals and plants of the tidal zone vary in their ability to tolerate exposure to air, and this differential is part of the reason for the different pattern of distribution of organisms observed on many rocky shores. (See pp. 278–280.)

In locations with semidiurnal tides, when the two highs or two lows are not equal, the duration of immersion is not smooth going from low to high tide

levels, and there are breakpoints where the duration changes greatly over a short distance. An organism living just a few centimeters above one of these points would be subjected to a greatly increased exposure to air relative to one just a few centimeters below. The existence of these **critical tide levels** was suggested by Doty (1946) as the reason for the sharp changes in the distribution of fauna and flora observed on rocky shores (see Figure 6.12b), since such shores were thought to delineate an important environmental breakpoint.

The second effect of tide is the result of the time of day during which the intertidal zone is exposed to the air. As mentioned before, the lethal ranges for physical parameters are encountered by marine organisms only when they are exposed to the air. In any daily cycle, these ranges vary greatly at different times of the day. If, for example, the low tides in the tropics occurred only during the hours of darkness, the intertidal fauna and flora would be exposed to much lower temperatures and less desiccation than if they were exposed at midday. As a result, we might expect a greater diversity of organisms in the intertidal zone of a tropical area where low tides regularly occur during the night or during the early morning or evening than in an area where they occur at midday. In the cold temperate zones, the reverse would be true, because in winter the lowest temperatures would occur in early morning or at night and might freeze the animals, whereas at midday the temperatures might be high enough to prevent freezing.

A final effect of the tide is that, because of its great predictability, it induces certain rhythms in the activities of shore organisms. These rhythms affect processes such as the time of spawning. For example, the grunion, a fish of the Pacific coast of the United States, spawns on the beach only on certain nights of the highest spring tides (Figure 6.3), and so does the horseshoe crab (*Limulus polyphemus*) of the Atlantic and Gulf coasts. Tidal rhythms may also affect feeding or other activities. Many intertidal organisms are quiescent when the tide is out and resume normal activity, such as feeding, only when the tide is in. Others, such as snails and chitons, which might be dislodged when the rocks are swept by waves, feed when the tide is out, especially at night when stresses due to temperature and desiccation are minimal.

■ TEMPERATURE

Because of its inherent physical characteristics (see Chapter 1), water—particularly a large body such as an ocean—shows a minimum range of temperature change. Furthermore, this range rarely exceeds the lethal limits of the organisms. Intertidal areas, however, are regularly subjected to aerial temperatures that may exceed a number of organisms' lethal limits for varying periods. Because of exposure to aerial conditions during low tide, intertidal organisms can be subjected to more dramatic variations in temperature than can organisms in any other marine habitat. In New England, for example, intertidal organisms can be encased in ice during winter low tides, while rock temperatures can exceed 40°C on hot summer days. Daily exposure to temperatures fluctuations of as much as 20°C are not uncommon in many rocky intertidal habitats. The temperatures experienced by intertidal organisms, however, differ dramatically among habitat types. Organisms on open rock surfaces face the largest temperature fluctuation, while organisms living under seaweeds or marsh grasses are buffered from extreme temperature fluctuations (Figure 6.4). Organisms that live buried in mud or sand are even more buffered from temperature fluctuations.

FIGURE 6.3
Grunion spawning at high tide. (Photo courtesy of Jeff Foott.)

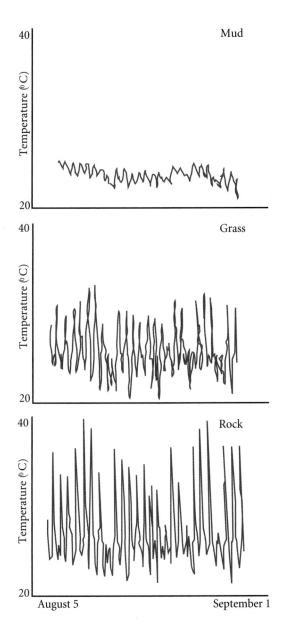

FIGURE 6.4 The influence of intertidal habitats on the temperature experienced in the intertidal zone. (Modified from *The Ecology of Atlantic Shorelines*, M. D. Bertness, p. 220. Copyright © 1999 Sinauer Associates, Inc. Reprinted by permission.)

Ice can have significant effects on intertidal organisms on far northern or far southern shores. Wherever sea ice forms on the shore, or wherever large icebergs impinge on the intertidal zone, the result is catastrophic mortality by freezing or mechanical abrasion. Wethey (1985) has suggested that the presence of this disturbance might explain the absence from New England rocky shores of long-lived

organisms otherwise common on intertidal shores that are unaffected by ice.

■ **WAVE ACTION**

In the intertidal zone, wave action exerts more influence on organisms and communities than in any other area of the sea. This influence is manifested both directly and indirectly. Wave action affects shore life directly in two major ways. First, it has a mechanical effect that smashes and tears away objects. The many pictures and stories of the destruction of homes, harbors, and ships by waves from storms are proof. The action is similar in the intertidal zone; hence, any creatures that live there must be adapted to this persistent force in some way. (See pp. 275–276.) On shores of loose sand or gravel, the wave action is even more profound. There, waves move the entire substrate around (see pp. 309–313), influencing the shape and slope of the beach. Exposure to waves may be limiting for organisms that are unable to withstand their force, and necessary to others that depend on the waves to bring them food, oxygen, and other essentials of life and that cannot exist outside heavy-wave areas.

Second, wave action extends the limits of the intertidal zone, by throwing water higher on the shore than it would normally reach as a result of the tide alone. This continual splashing allows marine organisms to live higher in exposed wave-swept areas than in sheltered areas within the same tidal range and often mediates temperature extremes and effects of desiccation (Figure 6.5).

■ **SALINITY**

Salinity may change in the intertidal zone in two situations that can affect organisms. First, the intertidal zone may be exposed at low tide and subsequently flooded by heavy rains or by runoff from heavy rains. This means that organisms will be subjected to greatly reduced salinities. Since most intertidal organisms show a limited tolerance to decreased salinity, extreme reductions of salinity cause them to swell by osmosis and die. The second situation concerns tide pools—areas that retain seawater at low tide. Tide pools may be flooded with freshwater runoff from heavy rains, thus reducing their salinity, or they may show *increased* salinity due to evaporation during the day (see pp. 302–305).

FIGURE 6.5 Changes in the extent of vertical zonation with change in exposure to wave action.

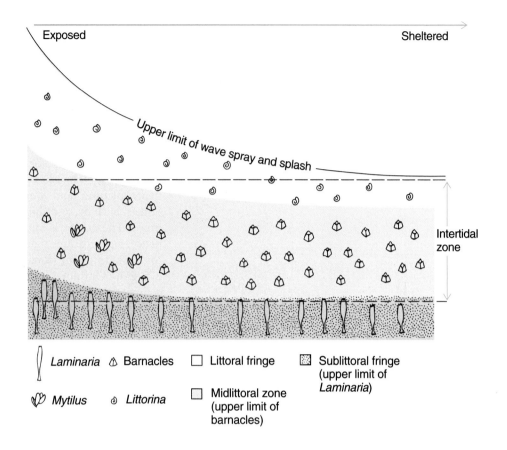

Exposed

Sheltered

Upper limit of wave spray and splash

Intertidal zone

Laminaria △ Barnacles ☐ Littoral fringe ▦ Sublittoral fringe (upper limit of *Laminaria*)

🐚 *Mytilus* ◎ *Littorina* ☐ Midlittoral zone (upper limit of barnacles)

Adaptations of Intertidal Organisms

Because intertidal organisms are primarily marine in origin, the adaptations we observe are generally for avoiding or minimizing the stresses of daily exposure to air. The major stress from the marine environment is wave action.

■ RESISTANCE TO WATER LOSS

When intertidal marine organisms are exposed to air during low tides, they begin to lose water by evaporation. If they are to survive, they must either have body systems that will tolerate considerable water loss during the hours of exposure to the air or have some mechanism that reduces their water loss until an external supply of water becomes available again.

The simplest mechanism for avoiding water loss is seen in mobile animals, such as crabs. These animals simply move from the exposed surface areas of the intertidal zone into moist cracks, crevices, or bur-

rows where water is available. Alternatively, they may seek refuge under a covering of moist algae. Such animals thus avoid the adverse environmental condition of the shore by actively selecting suitable micro-habitats. A similar situation occurs with some species of anemones, such as *Anthopleura xanthogrammica* of the Pacific coast of North America (Figure 6.6). This animal is soft bodied with no way of preventing water loss. However, since it normally is found only among mussels or in crevices where water loss is reduced, physiological adaptation is not needed. On the Atlantic coast, the anemone *Anthopleura varioarmatus* covers itself with shell fragments to shield itself from drying when exposed to air. According to Bertness and Leonard (1997), many organisms that live at high tidal heights on rocky shores also live in dense groups or in association with seaweed canopies that buffer individuals from water loss. Living in dense groups decreases water loss in high intertidal barnacles, mussels, sea anemones, and some sea-weeds. Garrity (1984) found that, on tropical rocky

FIGURE 6.6 The large green Pacific coast anemone *Anthopleura xanthogrammica*. (Photo © Brandon D. Cole/ Corbis.)

shores, some high intertidal snails also live in dense aggregations to minimize water loss.

Another simple mechanism is that seen in several of the high intertidal algae genera, such as *Porphyra*, *Fucus*, and *Enteromorpha*. These plants cannot move and have no mechanism for avoiding water loss. They simply are adapted to withstand a severe loss of water from their tissues. Specimens of these genera are dry and brittle after a long exposure at low tide, yet will quickly take up water and resume normal body processes when the tide returns. Kanwisher (1957) found that they can tolerate as much as a 60–90% loss of water and still recover. A similar tolerance to great water loss is seen in some intertidal sedentary animals. For example, Boyle (1969) reported that chitons can tolerate a 75% water loss (Figure 6.7). According to Davis (1969), limpets can lose 30–70% of their water, depending on the species.

Other intertidal organisms have mechanisms for preventing water loss. These mechanisms may be structural, behavioral, or both. Acorn barnacles of many species are predominant members of the

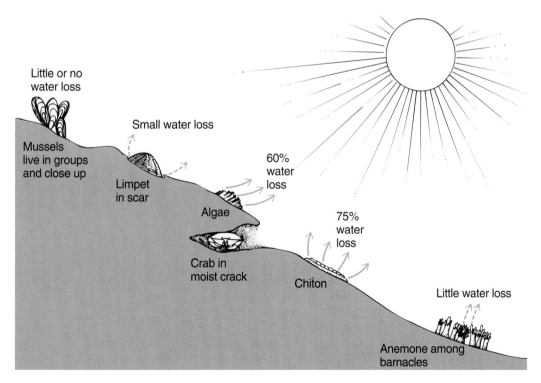

FIGURE 6.7 Diagrammatic representation of adaptations to water loss in intertidal organisms.

intertidal zone all over the world. These animals are sessile and avoid water loss by closing their shells at low tide. Here, then, the presence of an impermeable shell cuts down on evaporative water loss. Limpets of the genera *Patella, Acmaea,* and *Lottia* are also dominant animals of the rocky intertidal zone. Certain species of limpets have a "home scar" into which their shells fit exactly and to which they return at low tide. By fitting their shells into these grooves, they greatly reduce water loss (Figure 6.7). Other limpets that have no scar simply clamp down tightly against the rock so that none of their tissues are exposed. Other gastropods, such as periwinkle snails (*Littorina*), have opercula that completely seal off the aperture to their shells. At low tide, they may pull into the shell, closing the aperture with the operculum and thus reducing water loss. Certain bivalves, such as *Mytilus edulis,* and oysters, such as *Crassostrea virginica,* survive in the intertidal zone by closing their valves tightly to prevent water loss. Still other organisms, such as the anemone *Actinia* and the hydroid *Clava squamata,* produce mucus that reduces water loss. Finally, many inhabitants of sand or mud simply burrow into the substrate to prevent desiccation.

■ MAINTENANCE OF HEAT BALANCE

Intertidal organisms that are exposed to extremes of heat and cold show behavioral and structural adaptations to maintain their internal heat balance. Many intertidal organisms avoid high temperatures by reducing heat gain from the environment or by increasing heat loss from their bodies (or both). Heat gain from the environment is reduced in several ways. One way is to have a relatively large body size compared with similar species either lower in the intertidal zone or in the subtidal zone. A large body size means less surface area relative to volume and less area for gaining heat. At the same time, a larger body takes longer to heat up than a smaller one. In a study of tropical rocky shores by Vermeij (1971), large snails of many species commonly live at higher shoreline elevations than smaller individuals that are more vulnerable to heat gain. Yet another mechanism for reducing heat gain is to reduce the area of body tissue in contact with the substrate. This is, however, difficult to achieve for most intertidal animals, because they need the extra purchase provided by a large area of tissue in contact with the substrate to avoid being swept off by waves. Some shore animals, however, do have a small area of attachment or none at all. These organisms inhabit areas where wave action is not severe. For example, many high-shore tropical littorine snails attach to an overhang or a ledge by means of a thread of mucus, thereby eliminating all body contact with the substratum. In those animals which lose attachment, the ability to reattach as the tide comes in is well developed.

Heat loss is manifest in a number of adaptations. One mechanism, found in such hard-shelled organisms as mollusks, is greater elaboration of ridges and other sculpturing on the shell (Figure 6.8). Ridges and sculpturing act as radiators and facilitate heat loss. Examples of such sculptured mollusks are commonly found in the tropics and include *Tectarius muricata*

FIGURE 6.8 Differences in heat absorption between smooth, dark shells and sculptured, light shells.

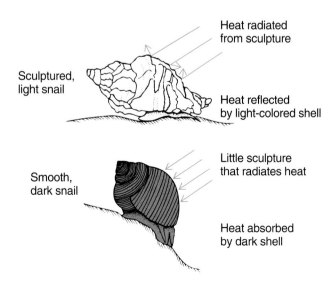

Sculptured, light snail

Heat radiated from sculpture

Heat reflected by light-colored shell

Smooth, dark snail

Little sculpture that radiates heat

Heat absorbed by dark shell

and *Nodolittorina tuberculata*. Also, dark-colored bodies gain and lose heat by radiation more rapidly than light-colored ones. Therefore, light-colored shelled gastropods lose heat more slowly than dark ones, but they also heat up less as well, keeping cooler for a longer time when the sun shines on them. Many tropical and subtropical snails of the high intertidal zone, such as *Nerita peloronta* (Caribbean) and *Littorina unifasciata* (New Zealand), are much lighter in color than their lower-level relatives. Presumably, their coloring slows the heat gain.

Finally, and most importantly, heat loss can occur through the evaporation of water. Unfortunately, intertidal organisms must face the problem of desiccation and may not be able to afford to cool themselves through evaporation and thereby risk desiccation. According to Segal and Dehnel (1962), many intertidal organisms have adaptations that allow them to cool themselves by evaporating water while avoiding excessive desiccation. To facilitate this balance, these animals have an extra water supply for cooling. The extra water is held in the mantle cavity of barnacles and limpets, and it exceeds the amount the animal needs to survive desiccation. For example, members of the barnacle genus *Tetraclita* have a thick porous shell that traps water at high tide and slowly allows evaporation at low tide, providing evaporative cooling. Moreover, most high intertidal animals can withstand a remarkable amount of desiccation. (See previous section.) In addition, many animals can reduce the rate of water loss from their tissues when necessary. Finally, those animals living in intertidal sedimentary environments escape most of the problems of water loss by living in a substrate that retains water even at low tide.

At higher latitudes, freezing can be a problem for intertidal organisms that are exposed to aerial conditions at low tide during winter. Intertidal organisms deal with this potential problem in a number of ways. Kanwisher (1960) found that some organisms produce antifreeze compounds to limit freezing of their tissues, while others have mechanisms for preventing cell damage when their tissues do freeze.

■ MECHANICAL STRESS

Wave action reaches a zenith in the intertidal zone. As a result, it is necessary for any organism that lives intertidally to adapt to resist the smashing and tearing effects of waves. Wave action has different effects on rocky shores than on sandy shores and requires

different adaptations, discussed later. Here, we discuss some adaptations of dwellers on rocky substrates.

Perhaps the most common adaptation to wave stress among animals is the limitation of size and shape (Denny et al., 1985). Most common organisms that frequent the splash zone have small, squat bodies with streamlined shapes that minimize their exposure to the lift and drag of wave forces. This adaptation can be seen in animals such as limpets, barnacles, and chitons that are common in the rocky midintertidal zone. As animals increase in size, they become more vulnerable to destruction or removal by waves, unless they compensate with additional means of attachment. Unlike intertidal animals, intertidal algae are often large and cope with wave stress by being flexible and bending toward the substrate as the waves pass over, thereby minimizing their surface area that is in contact with the waves (Figure 6.9).

Another adaptation, observed in barnacles, oysters, and serpulid polychaete worms, is permanent attachment to the substrate. Similarly, most algae in the intertidal zone are attached to the bottom via a holdfast. Other organisms develop strong, but temporary, attachments that give them limited movement. An example of such an attachment is the byssal threads of *Mytilus*, which anchor the animals securely, but can be broken and remade to allow limited, slow movement. Living in dense groups also buffers mussels and seaweeds from wave stress.

Most intertidal mollusks are adapted to wave shock by possessing a strong, enlarged foot that clamps them to the substrate and a shell that is low and flat, offering little resistance. Limpets and chitons are good examples of streamlined mollusks with an enlarged foot. In addition, many intertidal snails, including dog whelks and periwinkles, develop a larger foot for attachment in response to being reared in wave-exposed habitats (Trussell et al., 1993). This type of developmental plasticity is also found in other organisms. A study by Palumbi (1984) found that sponges develop denser body walls in wave-exposed than in wave-protected habitats. Most organisms on wave-swept shores also lack delicate sculpturing, which is easily broken in the waves. Motile organisms, such as crabs, have no structural mechanism to resist being swept away and survive only because they seek shelter from the waves in crevices or under rocks.

FIGURE 6.9 A large seaweed coping with high wave energy by bending into the flow, thus reducing its surface area and drag to the wave. (Modified from *The Ecology of Atlantic Shorelines*, M. D. Bertness, p. 220. Copyright © 1999 Sinauer Associates, Inc. Reprinted by permission.)

Low flow   High flow

■ RESPIRATION

Since most animals inhabiting the intertidal zone are marine, they have respiratory surfaces or gills to extract O_2 from water. Usually, these structures are thin-walled extensions from the body surface that are highly susceptible to desiccation in air and that usually do not function well unless immersed in water. Such organs would be at a disadvantage during aerial exposure in the intertidal zone. Among intertidal animals, there is a tendency to enclose the respiratory surfaces in a protective cavity to prevent them from drying. This is seen particularly in various mollusks, whose gills are in a mantle cavity that is itself protected by the shell. Another adaptation, especially of high intertidal gastropods, is a reduced-size gill and the formation of a vascularized mantle cavity, which then serves as a lung for aerial respiration. A similar situation prevails in barnacles, whose mantle tissue acts as a respiratory organ.

Because animals with protected respiratory organs must also conserve water at low tide, they often close up (by means of an operculum) or clamp down (chitons, limpets) such that gaseous exchange is reduced. To conserve O_2 and water, most animals

are quiescent during low tide. Intertidal fishes often are specialized for cutaneous respiration through the reduction of gills and the proliferation of blood vessels in the skin. Bridges (1988) suggests that many fishes may satisfy over half their O_2 needs by gas exchange through the skin.

■ FEEDING

All intertidal animals must expose the fleshy parts of their bodies in order to feed. For those on hard substrates, this means exposing those parts most susceptible to desiccation. As a result, most diurnal rocky intertidal animals are active only during the time the tide is in and they are covered with water. This is true whether the animals are grazers, filter feeders, detritus feeders, or predators. By contrast, organisms living in soft substrates are often protected by the water-saturated substrate and may be active feeders at low tide. Nocturnal animals may also be active during less stressful evening, night, or early morning low tides.

■ SALINITY STRESS

The intertidal zone may be flooded by fresh water after heavy rains, leading to low salinities. In the opposite situation, the evaporation of tide pools during low tides can lead to elevated salinities. Either of these situations create an osmotic stress on intertidal organisms—particularly those lacking adaptations to tolerate salinity changes. Intertidal organisms without mechanisms for controlling the salt content of their body fluids are **osmoconformers**. Many osmoconformers, such as starfishes and sea urchins, that inhabit intertidal habitats cannot tolerate large fluctuations in salinity and die en masse after heavy rains.

■ REPRODUCTION

Because so many intertidal organisms are sedentary or sessile, many rely on external fertilization and have planktonic larvae or free-floating eggs that disperse and live in the water column. These larvae are of many types. (See Figure 2.23 on meroplankton.) To increase their chances of successful fertilization, slow-moving intertidal organisms such as chitons, starfishes, and many segmented worms that reproduce by external fertilization form aggregations before releasing their gametes. Since gametes are rapidly diluted in seawater once they are released, aggregation, or the clumping of organisms, increases the chances that the gametes will be fertilized. Most mobile intertidal organisms, such as crabs and snails, reproduce by internal fertilization, are protected early in their development, and have a dispersive planktonic larval stage. There are notable exceptions to all these generalities, of course. For instance, some intertidal snails attach their eggs under rocks or in protected cracks and crevices, and the eggs develop directly into crawl-away juveniles with no planktonic stage. In another case, in spite of being sessile, intertidal acorn barnacles reproduce by internal fertilization that is facilitated by an enormous penis three to five times the length of their body.

Other reproductive adaptations of intertidal organisms include breeding cycles that are synchronized with the tides. An example is *Mytilus edulis*, in which gonads mature during periods of spring tides and spawning occurs on subsequent neap tides. In this case, synchronized spawning of all individuals ensures the maximum number of gametes in the water at the same time to facilitate fertilization. In *Littorina neritoides*, fertilized eggs are spawned during spring tides. This maximizes larval dispersal. Semiterrestrial shore animals, such as the Caribbean land hermit crab *Cenobita*, migrate from the shore to the sea to deposit their eggs.

Rocky Shores

Of all the intertidal shores, those composed of hard material—the rocky shores, particularly those of the temperate zones—are the most densely inhabited by macroorganisms and have the greatest diversity of animals and seaweeds. Rocky shores contrast sharply with the almost barren appearance of the surface of sand and mud shores. It is these densely populated, topographically diverse, and species-rich rocky areas that have fascinated marine biologists and ecologists for many years. In the last 30 years, these areas have been the subject of many classic studies that have enhanced our understanding of how the associations of species that inhabit such regions interact to maintain or change the community.

■ ATLANTIC AND PACIFIC ROCKY SHORES

There are significant differences in the geographical, physical, and biological factors shaping intertidal associations and communities among rocky intertidal areas in different parts of the world. These differences are illustrated by comparing temperate Atlantic and Pacific rocky intertidal habitats of North America. The temperate Atlantic coast is geologically old and dominated by eroding sediments, so it generally lacks rocky intertidal areas south of Cape Cod, where sedimentary intertidal areas dominate. In contrast, the Pacific coast intertidal zone is geologically young and dominated by rock from California up to Alaska, and sedimentary intertidal areas are less common. Thus, rocky shores occur over an extensive range of latitude on the Pacific coast, but are confined to northern latitudes on the Atlantic coast. Also, the range of aerial temperatures over the Pacific coast is much narrower than over the Atlantic, on both a seasonal and a daily scale, due largely to ocean circulation patterns. Thus, intertidal organisms in New England and eastern Canada can be exposed to winter temperatures as low as −20°C and summer temperatures of nearly 40°C, whereas, on the Pacific coast, even in Canada and southern Alaska, the temperatures rarely go much below freezing in winter, and persistent cloud and fog cover in summer lower what would otherwise be high temperatures. As a result, Atlantic organisms are killed regularly by seasonal freezing or ice action in winter, neither of which occurs in the Pacific.

The species diversity in the intertidal zone of the Atlantic is also markedly less than that of the Pacific, because, during the ice ages, the continental glaciers scoured the Atlantic shoreline, causing widespread extinction of the rocky intertidal organisms when the lack of suitable substrate prevented them from retreating southward. Since the glaciers retreated only 10,000 to 20,000 years ago, there has been insufficient time for reinvasion. The Pacific rocky intertidal community, by contrast, did not experience these extinctions, because rocky-shore organisms simply migrated south during recent glaciations. As a consequence, rocky intertidal organisms in the Pacific are much older geologically. Neither chitons nor limpets are common herbivores in the temperate Atlantic, whereas they are dominant organisms in the Pacific. Indeed, the Pacific coast has over 10 common intertidal limpet species, while the Atlantic has but two, and in Puget Sound there are five common intertidal chitons of large size that play an important ecological role, whereas in New England there are but three chitons, all small, uncommon, and primarily subtidal.

■ ZONATION

One of the most striking features of rocky shores anywhere in the world is the prominent horizontal banding, or **zonation**, of the organisms that live there. Each zone or band is set off from those adjacent to it by differences in color, morphology of the dominant organism, or some combination of color and morphology. These horizontal bands or zones succeed each other vertically as one progresses up from the level of the lowest low tides to true terrestrial conditions (Figures 6.10 and 6.11). This pattern of zonation on intertidal rocky shores is similar to that which one observes with increasing elevation on a mountain, where the different horizontal zones of trees and shrubs succeed each other vertically until, if one progresses far enough, permanent snow cover is reached. The major difference between these two areas is the scale: Mountain zones are perhaps kilometers in extent, as opposed to intertidal zones extending at most a few meters vertically.

Rocky intertidal zones vary in vertical extent, depending on the slope of the rocky surface, the tidal range, and the exposure to wave action. Where there is a gradual slope to the rock, individual zones may be broad. Under similar tidal and exposure condi-

FIGURE 6.10 An intertidal area at low tide, illustrating the conspicuous banding or zonation of organisms. (Photo courtesy Norbert Wu.)

tions on a vertical face, the identical zones would be narrow. In the same manner, exposed areas have broader zones than protected shores, and shores with greater tidal ranges have broader vertical zones. (See Figures 6.5 and 6.11.)

The fact that these prominent zones can be observed on rocky shores throughout the world under

(a)

(b)

FIGURE 6.11 Zonation in the rocky intertidal. (a) An intertidal area in Scotland. (b) An intertidal area in British Columbia. Note in b the difference in the extent of the zones. Line A–J indicates the lower limit of lichens, line C–J the upper limit of the "black" zone, line D–K the upper limit of barnacles, line E–L the upper limit of *Fucus*, and line G–N the lower limit of barnacles. (a, from J. R. Lewis, 1964, *The Ecology of Rocky Shores*, English Universities Press, London. b, from *Life Between the Tidemarks on Rocky Shores*, T. A. Stephenson and A. Stephenson, © 1972 W. H. Freeman and Company. Used with permission.)

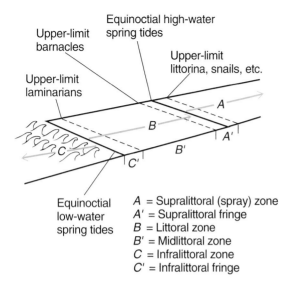

FIGURE 6.12 The Stephensons' universal scheme of zonation for rocky shores. (From *Ecology and Field Biology*, 4th ed., Robert L. Smith, p. 900. Copyright © 2000 Addison Wesley. Reprinted by permission of Addison Wesley Educational Publishers, Inc.)

many different tidal regimes led Stephenson and Stephenson (1949) to propose, after some 30 years of study, a "universal" scheme of zonation for rocky shores (Figure 6.12). Their scheme, which was really a framework using common terms that would allow diverse areas to be compared, established zones based on the distributional limits of certain common groups of organisms and not on the tides. The interconnections hypothesized reflect the knowledge of the Stephensons, as well as other intertidal ecologists, that distribution patterns of the organisms and zones vary not only with the tides, but with slopes and exposure. Therefore, under similar tidal conditions, there could be different bandwidths, due to different exposures or slopes of rocks. It was this universal scheme that established a standard format for describing shore zonation, replacing a bewildering host of schemes and names fashioned by earlier biologists.

The Stephensons' scheme has three main divisions of the intertidal (littoral) zone. The uppermost is termed the **supralittoral fringe**. Its lower limit is the upper limit of barnacles, and the zone extends to the upper limit of snails of the genus *Littorina* (periwinkles). The dominant organisms are the littorine snails and black encrusting lichens (of the *Verrucaria* type). The extreme high water of spring tides reaches part of this division, but most of its water comes from

splashing of the waves. Above the supralittoral fringe is the maritime terrestrial, or supralittoral, zone.

The middle part of the intertidal zone is termed the **midlittoral zone** and is the broadest in extent. Its upper limit coincides with the upper limit of barnacles, while its lower limit is the point where large kelps (*Laminaria*, etc.) reach their uppermost distribution. This zone is often subdivided and contains a host of different organisms. Perhaps the only universally present dominant group is the barnacles.

The lowest zone of the Stephensons' scheme is the **infralittoral fringe**, which extends from the lowest low tide up to the upper limit of the large kelps. This is an extremely rich zone composed of organisms that can tolerate only limited exposure to air. It is really an intertidal extension of the **infralittoral zone** (the Stephensons' term), or what we know as the **sublittoral** area.

The Stephensons' scheme *described* zonation on rocky shores, but did not *explain* why zonation occurs. It is the explanation of zonal patterns that intrigues many marine biologists.

■ CAUSES OF ZONATION

Whereas it is fairly easy to recognize and measure the extent of the zones on a rocky shore, it is more difficult to explain why organisms are distributed so predictably in such zones. Both physical and biological factors explain the phenomenon.

PHYSICAL FACTORS The most obvious explanation for zonation is that it is the result of the tidal action on the shore and reflects the different tolerances of the organisms to increasing exposure to the air and to the resultant desiccation and temperature extremes. One difficulty with this explanation is that the rise and fall of a tide tends to follow a smooth curve, with no obvious sharp breaks corresponding to the often sharp boundaries observed in the intertidal zones. If, however, one observes a whole series of tidal curves, such a series of breaks does become apparent. For example, Figure 6.13a shows a typical tidal curve for a mixed-tide area on the California coast. Figure 6.13b gives the maximum time of continuous submergence for various tide levels. The graph shows that there are certain points on this curve that reflect sharp increases in exposure to air. The existence of these points can also be deduced from Figure 6.13a, which represents a single typical lunar tide cycle. Consider an organism at point X on

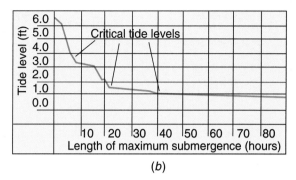

FIGURE 6.13 (a) Typical tidal curve for a mixed-tide area on the Pacific coast of North America. (b) Maximum time of continuous submergence for various tide levels on the Pacific coast of North America. (b from "Critical Tide Factors That Are Correlated with the Vertical Distribution of Marine Algae and Other Organisms along the Pacific Coast," Maxwell S. Doty, *Ecology,* Vol. 27, No. 4, October 1946, p. 320, Figure 4. Reprinted by permission of the Ecological Society of America.)

the graph. At this point, the tide will cover the organism at least once every 6 hours. If, however, the organism moves up only a few inches, the tide will not cover it for 12 hours. Thus, is a great change in exposure time with a very short vertical movement. Points exhibiting sharp increases in exposure time to terrestrial stresses over short distances have been termed *critical tide levels* by Doty (1946) and were offered as one of the early explanations for the zonation patterns described previously.

The critical-tide hypothesis has been tested in various places by several scientists since its original formulation by Doty in 1946. In general, it is difficult to find good correlations, particularly at the lower tide levels. The absence of good correlations can be attributed partly to the diverse topography of the different shores and to the variation in exposure. Thus, a species may be able to exist above the critical tide level if the rocky shore is exposed to persistent violent wave action, which would throw water up higher and decrease desiccation. Similarly, caves, overhangs, and crevices remain moist when exposed areas are dried out; thus, organisms are able to persist above the critical tide level. The effects of differences in topography and exposure suggest, however, that the ultimate upper limits are typically set by two physical factors: desiccation and temperature.

In the Gulf of Maine, on the most wave-exposed headlands, the zonation is simple. The uppermost zone is dominated by the barnacle *Semibalanus balanoides*, because it is the only primary space holder able to survive there (Figure 6.14a). At intermediate heights, *Semibalanus* is replaced by *Mytilus*, because the latter is competitively dominant. The intermediate zone grades into an Irish-moss zone and then a kelp zone at the lowest level, where the wave action precludes the presence of urchins (Figure 6.14a). This simple pattern is due primarily to the heavy wave action, which largely excludes the crabs, fishes, urchins, and snails that are the primary consumers of the dominant organisms. On nearby wave-protected shores, the situation is quite different (Figure 6.14b). Below the barnacle zone are a profusion of tough algae, a number of mobile consumers, and few mussels. The lack of wave action allows consumers, snails, crabs, sea stars, and sea urchins to control the communities. On these shores, the large, but unpalatable, seaweeds *Ascophyllum* and *Fucus* conspicuously dominate, forming canopies that protect a sizable assemblage of understory organisms from desiccation. This assemblage includes herbivores such as *Littorina littorea* and carnivores such as the snail *Nucella*, the starfish *Asterias*, and crabs. These consumers limit the abundance of most sessile

FIGURE 6.14 (a) Typical zonation on a wave-exposed open coast in New England. (b) Typical zonation on a wave-protected coast in New England. (Modified from *The Ecology of Atlantic Shorelines*, M. D. Bertness, p. 225. Copyright © 1999 Sinauer Associates, Inc. Reprinted by permission.)

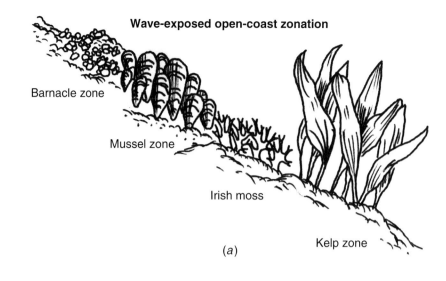

Wave-exposed open-coast zonation

Barnacle zone

Mussel zone

Irish moss

Kelp zone

(*a*)

Protected-coast zonation

Barnacle zone

Fucoid canopy understory with mussels, barnacles, snail grazers, and predators

Irish moss

Algal crusts
Urchin barrens

(*b*)

organisms, and substrate in the understory is dominated by open space (Figure 6.14b).

Previously, we noted that desiccation is a serious problem. Several natural observations, as well as field experiments, suggest that desiccation can set the upper limits to organisms and zones. In the North Temperate Zone, for example, rocks that have a north-facing slope often have the same organism occurring higher than do adjacent south-facing slopes. Since such rocks experience no difference in wave exposure and tides, the only reason for the difference in height is that north faces dry out more slowly than south faces. Similarly, field experiments by Frank (1965) and others in which seawater was slowly dripped down slopes have resulted in the organisms extending higher up (Figure 6.15). In other experiments, organisms were transplanted above their normal positions. Notable in this respect are barnacles, which cannot move. Transplanted by Foster (1971) above their normal tidal height, they died,

FIGURE 6.15 Change in zonation in the presence of a water seep.

younger barnacles more quickly than older ones. Again, desiccation is suspected.

Often acting in concert with desiccation is temperature. Aerial temperatures can exceed the lethal limits of intertidal organisms; hence, those organisms may die from either freezing or "cooking." Upper limits of zones may be attributed in part to the temperature tolerances of the intertidal organisms. In addition, high temperature promotes desiccation, and the synergistic effect of these factors may be even more devastating than each acting alone. A good example of the importance of temperature comes from the New England intertidal zone, inhabited by, among other organisms, two barnacles: *Chthamalus fragilis* and *Semibalanus balanoides. Chthamalus* occupies the high intertidal zone, because it can tolerate higher temperatures than *Semibalanus,* but *Semibalanus* is competitively dominant over *Chthamalus* and outcompetes it throughout the rest of the intertidal zone. Wethey (1984) demonstrated that if *Semibalanus* was shaded to reduce the thermal stress in the uppermost intertidal zone, it would outcompete *Chthamalus* and take over there (Figure 6.16). Indeed, north of Cape Cod, the summer temperatures never become high

FIGURE 6.16 Zonation of the two acorn barnacles *Chthamalus* and *Semibalanus* in New England under different conditions. (Modified from *The Ecology of Atlantic Shorelines,* M. D. Bertness, p. 216. Copyright © 1999 Sinauer Associates, Inc. Reprinted by permission.)

enough to exclude *Semibalanus*, and it displaces *Chthamalus* throughout the high intertidal zone. At the other end of the temperature spectrum, differences in tolerance to freezing may also play a role in zonation. In New England rocky habitats, the alga *Chondrus crispus* dominates many subtidal areas, but is killed by freezing temperatures, which appear to limit it to subtidal areas. In contrast, the alga *Fucus* is common in the intertidal zone, but suffers little mortality from freezing temperatures or encasement in ice. Of course, ice can also scour organisms from the rock, leaving many far northern Atlantic shores in a constant state of recovery.

Finally, sunlight may act adversely to limit organisms on the shore. Sunlight includes wavelengths in the ultraviolet (UV) region that can harm living tissue. Water absorbs these wavelengths and protects most marine animals. However, intertidal animals have direct exposure to such rays at low tide. The higher an organism is in the intertidal zone, the greater is its exposure to these rays. At present, there is little information concerning whether UV radiation controls the distribution of organisms, but the recent finding of UV-absorbing compounds in shallow-water organisms suggests the importance of this form of radiation.

Although a large body of evidence supports the hypothesis that physical factors are the strongest determinants of the upper limits of organisms, there are many exceptions. Underwood and Denley (1984) point out that other processes in the high intertidal zone may be confused with physical effects. For example, some animals in the zone—particularly sessile filter feeders—may be killed by starvation during periods when they are not covered with water long enough to feed. Larvae may also not choose to settle in the the high intertidal zone, so that upper limits may be set by larval settlement. Underwood (1980) found that the upper limit of several algal species in Australia was set by snail grazing. In each of these cases, the upper limits of the organisms are the result of biological and physical factors acting either independently or together. In addition, the upper limit of many intertidal mussels, barnacles, and seaweeds has been shown to be driven by positive interactions among individuals in groups. Bertness and Leonard (1997) found that, in these organisms, solitary individuals die at high intertidal heights, while members of dense aggregations are protected from thermal and desiccation stress by

neighbors. Thus, the benefit of the group—a biological interaction—rather than simple physical stress sets the high intertidal border for these species, and its upper limit can gradually shift upward as the aggregation grows.

BIOLOGICAL FACTORS   Although early work on the intertidal zone focused on the importance of physical factors in setting zonal patterns, later work clearly established the great importance of a number of biological factors in setting various observed distribution patterns. In general, these biological factors are more complex than physical factors, often subtle, and closely linked to other factors. These features are probably why it took more time to understand how biological factors act. The major biological factors are **competition, predation, grazing** (herbivory), **positive interactions**, and **larval settlement (recruitment)**. We shall discuss them in order.

Competition. Competition for resources does not occur if the resource is so plentiful that adequate supplies are available for all individuals. In the rocky intertidal zone, one resource that is commonly in limited supply is space. As a result, there is intense competition for space that can contribute to zonation patterns.

On the intertidal shores of Scotland is a distinct zonation of barnacles, with the small *Chthamalus stellatus* living in the highest zone and the larger *Semibalanus balanoides* occupying the major portion of the midintertidal zone. Studies by Connell (1961) showed that *Chthamalus* larvae settled out throughout the zones occupied by both species of barnacle, but survived to adulthood only in the upper zones. Their disappearance in the midlittoral region was due to competition from *Semibalanus balanoides*, which either overgrew, uplifted, or crushed young *Chthamalus*. *Semibalanus* was prevented from eliminating *Chthamalus* from the entire midintertidal zone because, unlike *Chthamalus*, *Semibalanus* did not have the tolerance to drying and high temperatures that prevailed at the higher tidal levels; hence, it could not survive there. Here, then, is a case where zonation is at least partially a function of biological competition (Figure 6.17).

Intertidal algae—particularly those on temperate shores—also often show abrupt limits to their upper and lower distribution. Earlier, these limits

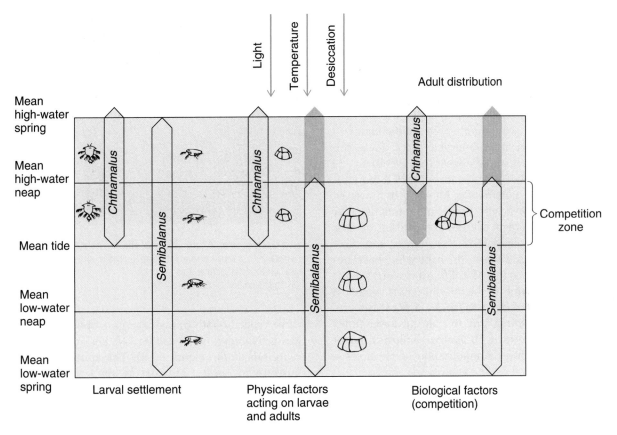

FIGURE 6.17 Intertidal zonation as a result of the interaction of physical and biological factors. The larvae of two barnacles, *Chthamalus stellatus* and *Semibalanus balanoides*, settle over a broad area. Physical factors—mainly desiccation—then limit the survival of *S. balanoides* above the mean high water level of neap tides. Competition between *S. balanoides* and *C. stellatus* in the zone between mean tide and the mean high water level of neap tides eliminates *C. stellatus*.

were ascribed to critical tide levels, but it has been suggested as well that the limits could be due to competition for space or access to light. In experimental studies done on the Pacific and Atlantic coasts of North America, evidence of competition appeared. Dayton (1975) found that the dominant kelps—*Hedophyllum sessile*, *Laminaria setchelli*, and *Lessionopsis littoralis*—all outgrew and outcompeted certain smaller species in the lower intertidal zone. These smaller species were generally fast-growing ones that quickly colonized open areas. As noted in Chapter 1, they are **opportunistic**, or **fugitive, species.** On the New England coast, studies of tide pool algae by Lubchenco (1978) indicated that *Enteromorpha intestinalis* was a dominant competitor for space, as opposed to *Chondrus crispus*. In the absence of grazers, *E. intestinalis* would quickly outcompete *Chondrus* and take over the space.

Competition among the mussel *Mytilus californianus* and several species of barnacle on the Pacific coast of Washington is a more complex example. In this case, studies by Dayton (1971) and Paine (1966, 1974) showed that *M. californianus* was the dominant space competitor on open-coast shores. Given enough time and freedom from predators, *M. californianus* eventually overgrew and outcompeted all other macroorganisms living on primary substrate and came to visually dominate the entire midintertidal zone. The takeover, however, was slow. Theorizing from the two studies, we can say that wherever open space occurs, it may be rapidly colonized by other organisms, including three species of barnacles: *Balanus glandula, B. cariosus,* and *Pollicipes polymerus.* These, in turn, displace any rapidly growing algal species. The barnacles persist on the rocks only until the mussels enter, whereupon the barnacles then recruit on the mussels themselves. The mussels outcompete and destroy the barnacles by settling on top of them and smothering them. Since nothing appears to be large enough to settle and smother *M. californianus,* it remains in control of the intertidal space. Given this competitive edge, it would appear that eventually the rocky Pacific coast of Washington could be a monotonous band of *M. californianus.* Mussel beds, however, only appear superficially to be monotonous; they actually support a high diversity of organisms, such as segmented worms, crustaceans, anemones, and other small invertebrates living in the habitat the beds create. It is also curious that *M. californianus* forms dense clumps or bands only in the intertidal zone when it is perfectly capable of living subtidally. Since it is a premier space competitor, why this abrupt lower limit? The reason has to do with another biological factor that prevents such a monopolization of resources—namely, predation.

Predation. The role of predators in determining the distribution of intertidal organisms is best documented on the Pacific coast of Washington and is discussed here as an example of how complex biological interactions create distributions.

Other than the space-dominating *Mytilus californianus,* the dominant abundant intertidal animals on the Pacific coast are the barnacles *Balanus cariosus* and *B. glandula.* These two species occur abundantly on rocks in the intertidal region, even though they are competitively inferior to *M. californianus,* because a predatory starfish, *Pisaster ochraceus,* preferentially preys on *M. californianus,* preventing it from completely overgrowing the barnacles (Figure 6.18). *Pisaster ochraceus* is a voracious predator of mussels, consuming them at a rate that prevents them from occupying all the space.

At the same time, *Balanus glandula* is found primarily as a band of adults in the high intertidal zone, while *B. cariosus* occurs as scattered, large individuals or clumps in the midintertidal zone. This pattern, as Connell (1970) has shown, is also due to predation. Like *Chthamalus* in Scotland, *Balanus glandula,* is capable of living anywhere in the intertidal zone and, indeed, settles throughout the zone. The same is true for *B. cariosus.* That both show a restricted distribution is due to predation by

FIGURE 6.18 A large mass of the mussel *Mytilus californianus* with a large aggregation of its major predator, the starfish *Pisaster ochraceus.*
(Photo © Brandon D. Cole/Corbis.)

three species of predatory gastropods of the genus *Nucella: N. lamellosa, N. emarginata,* and *N. canaliculata* (Figure 6.19). The abundance and motility of these predators is such that, within 12 to 15 months, they consume all the young *B. glandula* that settle out in the midintertidal zone. *Balanus glandula* survives only in a narrow band at the top of the intertidal zone, where the *Nucella* species are prevented from entering because of the likelihood of excessive desiccation. In the case of *B. cariosus,* however, the situation is somewhat different. There is no high-level refuge for this barnacle, and, as with *B. glandula,* the young are consumed by the *Nucella.* Size is the defense for *B. cariosus.* Once it reaches two years of age, it is too large for *Nucella* to attack. The only mystery is how *B. cariosus* survives *Nucella* for two years. It may have to do with periodic events such as big freezes at low tide that kill the whelks, or it may be due to just plain random luck. As a result, the pattern of distribution for *B. cariosus* is random clumps of large, older barnacles (Figure 6.20). It is not known what regulates *Nucella* populations. These whelks are preyed on by large *Pisaster* and are vulnerable to periodic stresses, such as freezing during winter low tides. Such periodic events may reduce the whelk population enough for *Balanus cariosus* to establish itself successfully.

Although it prefers mussels, *Pisaster ochraceus* can consume *Nucella* and barnacles of any size and is the primary predator of small- and medium-sized

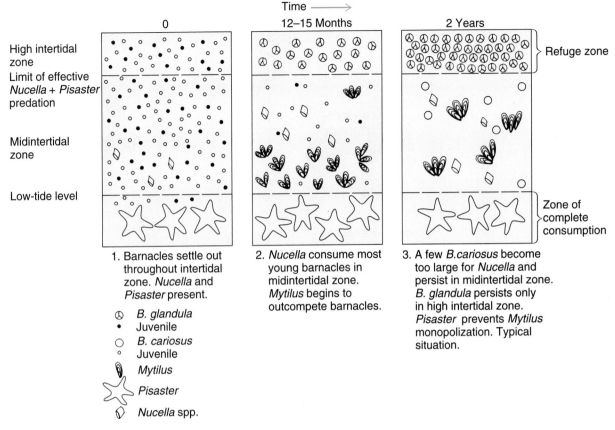

Time ⟶

| 0 | 12–15 Months | 2 Years |

High intertidal zone

Limit of effective *Nucella* + *Pisaster* predation

Midintertidal zone

Low-tide level

Refuge zone

Zone of complete consumption

1. Barnacles settle out throughout intertidal zone. *Nucella* and *Pisaster* present.

    ⊗ *B. glandula*
    • Juvenile
    ○ *B. cariosus*
    ° Juvenile
    Mytilus
    ☆ *Pisaster*
    Nucella spp.

2. *Nucella* consume most young barnacles in midintertidal zone. *Mytilus* begins to outcompete barnacles.

3. A few *B. cariosus* become too large for *Nucella* and persist in midintertidal zone. *B. glandula* persists only in high intertidal zone. *Pisaster* prevents *Mytilus* monopolization. Typical situation.

FIGURE 6.19 Interaction of predation and physical factors in establishing the zonation of the dominant intertidal organisms on the rocky shores of the northwest Pacific coast of North America.

Mytilus. Indeed, *P. ochraceus* is the top predator in the system. Because of its ability to influence the structure of the entire intertidal community by consuming *Mytilus* and preventing monopolization of the space, Paine (1966) has called *P. ochraceus* a **keystone species**, as defined in Chapter 1.

The reason the upper intertidal zone is a refuge from predators is that *Pisaster* and *Nucella* can feed only when the tide is in, and they require a long period to attack their prey successfully. The short period of immersion of the upper intertidal zone does not allow sufficient time for them to make successful forays into that area. In the subtidal zone, unlimited time is available; hence, starfishes have sufficient time to attack and consume their prey. This is the reason *Mytilus* usually does not live in the subtidal zone, even though it can live there in the absence of predators (Figure 6.21).

To explore the generality of his early studies in Washington, Paine and his colleagues (1985) have

undertaken similar studies of starfish removal from intertidal shores in Chile and New Zealand. In Chile, the dominant space occupier was the mussel *Perumytilus purpuratus*, and its predator was the sun star *Heliaster helianthus*. In New Zealand, the mussel was *Perna canaliculus*, and the predator was the sea star *Stichaster australis*. When predator removal studies were done in these areas and compared with the Washington studies, a notable difference appeared. In Washington and New Zealand, when the starfishes were permitted to return to the areas from which they were formerly excluded, the mussel community persisted in monopolizing the space it had occupied when the starfishes were excluded for as long as 14 to 17 years. In other words, there was no quick return to the conditions seen in control areas where predators were always present. By contrast, in Chile, when starfishes were returned to the excluded areas, the community rapidly returned to the condition of the

FIGURE 6.20 Interactions among mussels (*Mytilus*), barnacles, and their predators on the northwest Pacific coast of North America; these interactions allow barnacles to persist in the intertidal zone.

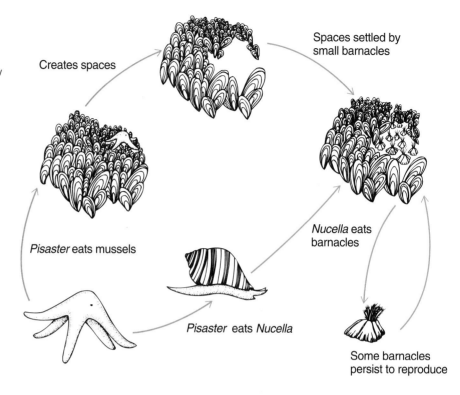

Creates spaces

Spaces settled by small barnacles

Pisaster eats mussels

Nucella eats barnacles

Pisaster eats *Nucella*

Some barnacles persist to reproduce

FIGURE 6.21 Composition of the low intertidal zone in the New England area with and without mussel predators.

Without predators (exposed)

With predators (protected)

 Mytilus edulis

⊗ *Balanus balanoides*

Chondrus crispus

 Asterias

◇ *Nucella*

undisturbed control. What are the reasons for this difference? Paine et al. (1985) suggest that the main reason is that in Washington and New Zealand, during the absence of their main predator, the mussels were able to attain a size that rendered them immune to predation when the predators were reintroduced. In the case of Chile, *Perumytilus* never reaches a size at which it is immune to predation—hence the quick return to a "normal" pattern. These findings suggest that in Washington and New Zealand the community will not return to the preexclusion state until the current large mussels either die or are removed by some other destructive event, such as wave action or crushing by waveborne objects (see p. 297), either of which may take years. Thus, whereas we may expect all three of these communities eventually to converge to their "normal" condition, the time involved may be markedly different and, in Washington and New Zealand, may lead to an intertidal zone with a mosaic of patches—some dominated by mussels and others with scattered clumps, depending on the conditions and length of time the dominant predator was excluded.

Menge (1976) and Lubchenco and Menge (1978) have also demonstrated that predation is important in setting zonal patterns in the intertidal zone of the North Atlantic Ocean. In the low intertidal zone of New England, the competitive dominant species is *Mytilus edulis*. Although this mussel is able to outcompete and eliminate the barnacle *Semibalanus balanoides* and the alga *Chondrus crispus* from the primary rocky substrate, it is prevented from doing so by the starfishes *Asterias forbesi* and *A. vulgaris* and the snail *Nucella lapillus*, all of which prey on *M. edulis*. Where these predators are absent—namely, in the most exposed, wave-beaten areas—*M. edulis* eliminates *Semibalanus* and *Chondrus* from the rocks in the mid- to low-intertidal zone. The exposed, wave-beaten areas exhibit a pattern in which the low intertidal zone is an *M. edulis* band, while in protected areas a diverse lower zone exists with *Mytilus*, *Semibalanus*, and *Chondrus* present (Figure 6.21).

Clearly, the role of predators in intertidal systems is more complex than that implied by the original keystone species concept, and some researchers have questioned the usefulness of the concept in general. Originally, Paine (1969) defined a keystone species as a single carnivore that preferentially preyed on and controlled the abundance of a prey species which, in turn, could competitively exclude other species and

so dominate the community. Subsequently, the term was used by other workers to refer to such noncarnivorous species as herbivores, mutualists, prey, and so forth; hence, confusion arose. Menge et al. (1994) have redefined the keystone species concept as originally stated by Paine and have clarified it by defining two other predator–prey relationships that do not involve keystone species. **Diffuse predation** occurs when total predation is strong and capable of controlling the abundance of a competitively dominant species, but when the predation is spread over several predators, not just one. If the total effect of predation on the competitive dominant prey is low, such that predation alone does not control the abundance of the prey, the condition is called **weak predation**.

In a more recent test of the generality of the keystone concept, Menge et al. (1994) investigated the high temporal and spatial constancy of the lower limits of *Mytilus* distribution at several sites in Washington and Oregon in which the abundance of *Pisaster* showed great variation. The authors compared sites that varied in size from meters to kilometers, adding a question of scale that had not been posed earlier. They were able to conclude that, while *Pisaster* predation varied dramatically, *Pisaster* behaved as a keystone species, eliminating *Mytilus* at exposed sites. At more protected sites, however, *Pisaster* predation was much weaker and variable, and *Mytilus* was controlled by other factors, such as recruitment and burial by sand.

Recently, Connolly and Roughgarden (1998) found evidence of a physical explanation for the differences in the *Pisaster–Mytilus* community structure observed along the extensive geographical range of the two species. They point out that, especially in the spring and summer along the Pacific coast, strong winds blow southward, causing a surface layer of water to move offshore and a cold-water upwelling to replace the offshore-moving surface water (a phenomenon known as Ekman transport). As a result, a near-shore circulation pattern is established, with water moving shoreward at depths, upwelling and surfacing near the shore, and then moving offshore at the surface (Figure 6.22). Various larvae of intertidal organisms become entrained in this offshore-moving "Ekman layer" and accumulate at frontal zones where the offshore-moving water meets the south-moving California Current. To replenish the intertidal zone, these larvae must end up on the shore, to settle and metamorphose. However, this can happen only when

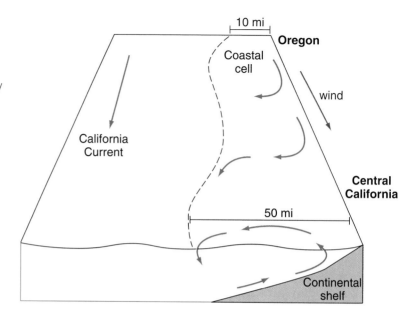

FIGURE 6.22 Diagrammatic representation of the California Current system. The dashed line represents the front between the upwelled water and the offshore California Current. (After "A Latitudinal Gradient in Northeast Pacific Intertidal Community Structure," S. R. Connolly & J. Roughgarden, *American Naturalist,* Vol. 151, p. 313, 1998. Copyright © 1998 University of Chicago Press. Reprinted by permission.)

the strong southward-blowing winds diminish and the front moves onshore. Although it in fact happens periodically, the strength of the offshore currents is greater, the fronts are further offshore, the upwelling season is longer, and the relaxing of the winds is less frequent in California, as opposed to Oregon and Washington (Figure 6.22). As a result of these differences in circulation pattern, there is lower recruitment to the intertidal zone in California than in Washington, which in turn means lower adult abundances of the intertidal invertebrates, more free space, and, hence, weaker competitive and predatory interactions among the dominant intertidal invertebrates in California. The significant high recruitment in the Pacific Northwest means that competition, predation, and disturbance are the dominant factors structuring the *Pisaster–Mytilus* community in the region, while the infrequent recruitment in California reduces such biological interaction and leads to a community composition determined primarily by the infrequent recruitment events. Recent work by Sanford et al. (2003) has also suggested that these differences in community composition have led to evolutionary differences in the feeding behavior of dog whelks in California and Washington.

Finally, changes in temperature have been demonstrated to affect keystone species predation and hence perhaps community structure. In laboratory and field experiments, Sanford (1999) showed that slight decreases in water temperature dramatically reduced predation by *Pisaster ochraceus* on *Mytilus californianus* and *M. trossulus.* Such results suggest that El Niño/La Niña events and global warming may have profound effects on intertidal community structure.

It seems safe to conclude that the keystone species concept remains valid, at least as applied to some temperate rocky shores. At the same time, it is not a universal concept, and its occurrence is variable within the geographic range of the species or communities in question. Under certain environmental conditions, the concept may be replaced by other community regulatory mechanisms, such as diffuse or weak predation, recruitment, or physical factors. This scenario illustrates an important principle: The relative importance of biological factors and physical factors on rocky shores varies over comparatively short distances. The variation is both vertical and horizontal. As a consequence, one cannot extrapolate from single experimental studies done on small areas to form models for the organization of large geographic areas of the intertidal zone.

Grazers. The role of **grazers**, or herbivores, in regulating upper and lower limits of algal species is well studied and documented, and evidence suggests that grazing is important in most systems. In the intertidal zone, the dominant grazers are various gastropod mollusks, certain crustaceans, sea urchins, and fishes, with their relative importance varying latitudinally and vertically (Figure 6.23). Their relative abilities to keep algae in check may also vary.

FIGURE 6.23 The main algal grazers at different intertidal zones in temperate and tropical systems. Numbers of symbols indicate relative importance.

In the past 25 years, considerable evidence has accumulated by various methods—the experimental removals of grazers, caging experiments, and natural disasters, among others—so our understanding of the role of grazing is now fairly well advanced. Grazing affects algal zonation, species diversity, patchiness, and succession. Before experimental methods showed otherwise, it was assumed that the upper limits of algal species on rocky shores were a function of the species' ability to tolerate physical factors. The lower limits of algal species distributions were not thought to be set by physical factors, because most algae grow faster when experimentally transplanted to lower levels or when grown constantly submerged.

Since most algae do grow better in the intertidal zone when submerged, why don't we find them there naturally? The answer coming from many researchers is that they are grazed out. For example, in various experiments that have removed grazing animals, the result has been the appearance lower on the shore of a number of algae otherwise restricted to higher levels, where grazing is less intense. This effect has been shown by Lubchenco (1978, 1980) for *Chondrus* in New England, where the main grazer is the urchin *Strongylocentrotus droebachiensis*. Himmelman and Steele (1971) found the same thing for the alga *Alaria esculenta* in Great Britain, where *S. droebachiensis* is also responsible (Figure 6.24).

Grazing can also determine which species of algae dominate particular communities. For example, Dethier and Duggins (1988) removed the chiton

Katharina tunicata from the low intertidal zone in Washington State, and the site developed from an area with few macroscopic plants into a kelp bed dominated by the kelps *Hedophyllum sessile*, *Alaria marginata*, and *Nereocystis luetkeana*. However, a similar removal of *Katharina* from a low intertidal area in Alaska did not result in the development of a kelp bed, suggesting again that it is unwise to generalize from local experiments to broad geographical areas.

Grazing may also set the upper limits of algal distribution. One of the more interesting pieces of evidence supporting this idea comes from the *Torrey Canyon* oil spill disaster of 1968. This spill killed the main intertidal molluscan grazers on the shores of southern England. Southward and Southward (1978) documented a subsequent rise in the upper limits of a number of intertidal algae. Following the return of the limpets and other grazing gastropods, the algae were once again grazed down.

The mechanisms by which the removal of grazers facilitates the upward or downward extension of algae vary. In the simplest case, the algae are directly limited by the grazer. In other cases, the absence of the grazers may permit a rapid establishment of ephemeral algae, trapping moisture and permitting the newly settled sporelings of the dominant algae to survive. In still another scenario, the upward extension of algae may be seasonal due to changes in grazing pressure. For example, Castenholz (1961) showed that a summer decrease in high intertidal diatom

FIGURE 6.24
Effect of sea urchin removal on kelp growth on the Isle of Man, Great Britain. (Modified from *Pacific Seashores: A Guide to Intertidal Ecology*, T. Carefoot, 1977, Douglas & MyIntyre, Ltd.)

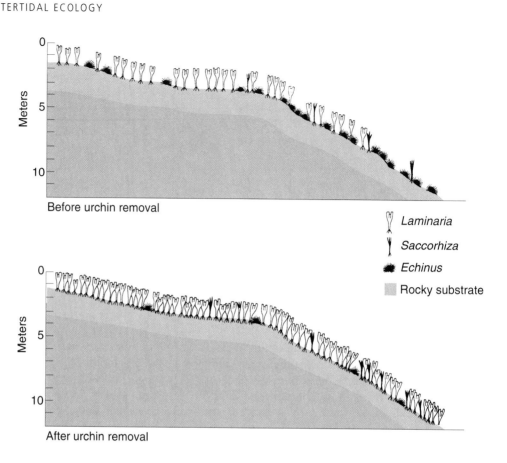

Before urchin removal

After urchin removal

Laminaria
Saccorhiza
Echinus
Rocky substrate

cover was due to a summer increase in *Littorina* grazing activity, in contrast to the less active winter condition. In a more complicated situation, grazing and physical factors interact. For example, Hay (1979) found that in New Zealand the alga *Durvillea* extended its range upshore in the winter, when grazers were absent, but was cut back in the summer by a combination of increased grazing and desiccation at the higher levels.

Surprisingly enough, some of the limits of algae are set by the effect of the algae themselves on potential grazers. In New Zealand, for instance, Underwood and Jernakoff (1981) showed that the downshore extensions of the main gazing limpets, *Cellana* and *Patelloida*, are limited by certain algae. In this case, the algae settle and grow faster than the limpet grazers can remove them. As a result, the algae form a substrate that prevents the limpets from attaching, so they cannot enter the area and graze. Obviously, in this kind of situation there is a continuum: At some level, a balance point defines one area where grazers are more efficient at grazing than the algae are at

growing, thus eliminating the algae; at the other extreme, the algae grow faster and exclude the limpets by creating an unsuitable surface for attachment.

Another consideration in the zonation of algae species is the ability of some algae to grow fast enough to escape grazing by becoming too large for the grazers to feed on. In these cases, the algae will mature only in areas where the grazers are low in density or are restricted from grazing during some seasons; thus, the algae can reach a size at which they are immune to grazing.

So far, we have considered grazing and its effects on vertical zonation patterns. Grazing may, however, also affect the horizontal patterns of zonation—particularly along wave-stress gradients. The best-studied examples come from the Atlantic coasts of Europe and North America. In Europe, exposed coastlines have an intertidal zone dominated by mussels, barnacles, and limpets, whereas sheltered shores are covered with dense stands of fucoid brown algae. Grazer removal experiments and natural experiments, such as the *Torrey Canyon* oil spill, led Southward and Southward

(1978) to produce a conceptual model to explain this pattern. Very simply, this model says that, on sheltered shores, limpets and other grazers are rare, likely due to crab predation. This permits the algae to grow and flourish and the algal canopy reduces the ability of limpets to settle on the shore. On exposed shores, crab predation is reduced, limpets are abundant, and heavy limpet grazing leaves the shore dominated by mussels and barnacles. On the opposite side of the Atlantic, in New England, limpets are rare and the dominant grazer is the snail *Littorina littorea*. On wave-protected rocky shores in New England, snail grazing limits the distribution of all but the most unpalatable seaweeds, while on wave-exposed shores, mussels outcompete most seaweeds for space in the intertidal zone.

Algal grazing has led to the evolution of a wide variety of algal defenses against the grazers. The most obvious adaptations to grazing are morphological, and one of the most common is the laying down of calcium carbonate in the tissues. A hard calcium carbonate exterior is difficult for most grazers to penetrate. Furthermove, calcification reduces the palatability of algae to grazers and makes the grazer exert more energy in feeding. Both crustose and erect coralline algal species occur. Steneck (1982) has even shown that, for one crustose coralline alga, *Clathromorphum*, the meristem tissue and reproductive sites are all on the underside, protected from grazing. *Clathromorphum's* crust has been shown to be dependent on the presence of grazers for local persistence. The growth form of an organism may also discourage grazing. Hay (1981) has shown, for example, that the formation of short turfs in algae reduces grazing mortality. Still other algae, such as the Pacific coast *Egregia*, have considerable structural or "woody" tissue that deters grazers.

Chemical defenses are also common. Many algae contain noxious or toxic compounds. For example, on the Pacific coast of North America, the genus *Desmarestia* contains sulfuric acid in a form concentrated enough to erode the calcium carbonate teeth of the grazing urchin *Strongylocentrotus franciscanus*. Other algae contain various alkaloids, phenolic compounds, and halogenated metabolites that inhibit grazing.

Finally, some algae evolve defenses at certain stages of their life histories. A number of species have two growth forms: a slow-growing, well-defended crust and a fast-growing upright form. In areas with heavy grazing pressure, the slow-growing, but grazer-resistant, crusts predominate, but in areas of low grazing pressure or in areas from which grazers have been excluded, the erect form dominates. Littler and Littler (1980) have suggested that having these two alternative growth forms in a single species is a strategy of **bet hedging**. The erect form has fast growth, high reproductive output, and high competitive ability, but low grazing resistance. The crust form, by contrast, has slow growth rates and low competitive ability, but high grazing resistance. Another life-history feature that may have evolved in algae to cope with grazing has to do with their reproduction. Many algae—particularly those favored by grazers because of their high fraction of digestible material and low fraction of structural tissue—have evolved the opportunistic, or r-selected, habit of high reproductive output. This adaptation also makes them successful in colonizing newly available space. Quick maturity and high reproductive output mean that these species should be able to colonize an open area and reach maturity before the grazers find and decimate them. Other species have evolved the opposite, or K-selected, features. These species are relatively slower growing, perennial, and long lived. To survive pressure from grazers, they either have evolved chemicals that make them unpalatable or grow large enough to be immune from grazers. Large size usually also means an enhanced competitive ability, at least for light. A final interesting example of a life history designed to cope with grazers is the seaweed *Porphyra*, a delicacy in Japan used to wrap sushi. *Porphyra* exhibits two growth forms. In the winter, when grazing pressure is low, it has a fast-growing, highly palatable sheetlike morphology, whereas, during the summer months, when grazing pressure is intense, it has a stage that bores into calcium carbonate shells to avoid grazers.

Grazing pressure in intertidal habitats varies dramatically with geography. In the temperate zone, grazing is one of the main structuring agents of the rocky intertidal communities. On the Pacific coast of North America, a large number of species of limpets and chitons, plus a few species of crabs, snails, and sea urchins, are responsible for controlling community structure. On European shores, limpets and snails also are efficient at structuring the communities. On the Atlantic coast of North America, limpets are rare, but the snail *Littorina littorea* plays important role in shaping these communities (Figure 6.25). Grazing becomes less important as one moves poleward. Generally, as one moves toward the equator, the importance

FIGURE 6.25 Littorina removal cage on a Narragansett Bay cobble beach. The picture was taken approximately four weeks after grazing snails were removed. (Photo © Mark Bertness.)

of grazing increases, and so does the variety of grazing organisms. In particular, crabs and fishes become more important as grazers in the tropics. The rocky intertidal zone in the tropics is relatively barren, due to extremes of the physical factors and the large number of mobile grazers.

Grazing pressure also varies markedly at different intertidal heights. In the supralittoral fringe, grazers either are uncommon or have little time to graze. The rigorous and stressful physical environment is thus the most important agency causing changes in algal composition. The midlittoral zone is dominated by barnacles and mussels, but is strongly influenced by gastropod grazers—primarily limpets—and may vary in algal cover, depending on exposure. The variation in cover, in turn, may limit the numbers of gastropods. Barnacles, however, may also be present, due to the ability of limpets and other grazers to keep the competing algae in check. In the sublittoral fringe, algal growth can exceed the capabilities of many intertidal grazers to control it, but sea urchins can control algal production at low intertidal elevations. Therefore, both competition for space and light among the various algae and urchin grazing are the dominant interactions that structure communities at this level. The various factors and their relative importance are summarized in Figure 6.26.

Positive interactions. While most studies of the role played by biological interactions in rocky intertidal communities have focused on negative processes such as competition and consumer effects, recent work by Bertness and Leonard (1997) has shown that positive interactions can also play an important role in generating patterns in such communities. In particular, the high intertidal borders of many sessile rocky intertidal

species (for example, mussels, barnacles, and seaweeds) can be set by the group benefits of temperature and amelioration of desiccation that accrue to dense aggregations of these organisms. According to Bertness and Leonard (1997), solitary mussels, barnacles, and seaweeds typically cannot survive as high in the intertidal zone without neighbors buffering them from the physical stresses of the high region of the zone. Similarly, intertidal seaweeds can buffer both sessile and mobile organisms, such as snails, barnacles, mussels, and other seaweeds, from high intertidal stresses during low-tide exposures and can extend the high intertidal distributions to higher levels. The high diversity of organisms that live in mussel beds is also largely a consequence of the beds providing refuge from intertidal physical stresses. Many of the organisms found in such beds would be incapable of living in the intertidal zone without the habitat modification effected by the beds. Another type of positive interactions influencing the distribution of intertidal organisms is more common under physically mild environmental conditions at lower tidal heights. According to Hay (1986), this set of interactions consists of the associational defenses that arise when a palatable prey organism survives being consumed through living in association with well-defended or impalatable prey. A good example on the Atlantic Coast of New England is the palatable seaweeds that live attached to the intertidal canopy-forming seaweed *Ascophyllum*. The vulnerable seaweeds are not consumed because grazers avoid grazing on *Ascophyllum*.

Recruitment. Since most of the organisms that live in intertidal habitats have a planktonic larval stage, and since larval recruitment can be highly variable in time and space, larval recruitment can play a leading

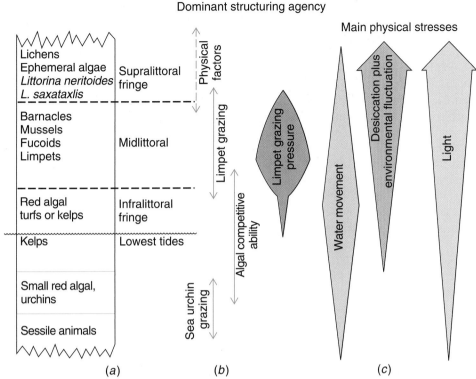

Dominant structuring agency

Main physical stresses

(a) Supralittoral fringe — Lichens, Ephemeral algae, *Littorina neritoides*, *L. saxataxlis*

Midlittoral — Barnacles, Mussels, Fucoids, Limpets

Infralittoral fringe — Red algal turfs or kelps

Lowest tides — Kelps

Small red algal, urchins

Sessile animals

(b) Physical factors; Limpet grazing; Algal competitive ability; Sea urchin grazing; Limpet grazing pressure

(c) Water movement; Desiccation plus environmental fluctuation; Light

FIGURE 6.26
Structuring forces affecting the intertidal zone of the North Atlantic Ocean. (a) The main intertidal zones and their dominant organisms (after the Stephensons' universal scheme). (b) Dominant structuring agencies; overlap suggests a balance between factors. (c) Main physical stresses on marine algae and animals. (Modified from "Grazing of Intertidal Algae by Marine Invertebrates," S. J. Hawkins & K. G. Hartnoll, 1983, *Oceanography and Marine Biology: An Annual Review*, Vol. 21, pp. 195–282.)

role in generating the structure of intertidal communities. Elucidating the factors that lead to variation in larval recruitment among intertidal organisms has been an important area of research over the past two decades. Most larvae are microscopic, so that their dispersal in the water column is largely at the mercy of local and regional oceanographic conditions. The supply of larvae can be controlled by forces operating at both large and small spatial scales. As described earlier, Connolly and Roughgarden (1998) showed that regional oceanographic conditions along the Pacific Coast of North America lead to predictably heavy larval recruitment of intertidal organisms on the coasts of Washington and Oregon and predictably low levels of recruitment on the coast of California. Similarly, on the Atlantic coast in Narragansett Bay, the larval recruitment of barnacles can vary from year to year from total failure to hundreds of recruits per square centimeter of substrate space in the intertidal zone. Gaines and Bertness (1994) have also shown that this large variation in recruitment is driven by regional oceanographic conditions. In years when rainfall in the region increases the rate at which water is flushed out of the bay, most barnacle recruits get washed out of the bay and recruitment is low. Conversely, when rainfall is low, larvae remain in the bay and larval recruitment is high. Both of these examples lead to large-scale spatial and temporal recruitment patterns that strongly influence intertidal community structure and dynamics. What control most larvae have in their large-scale dispersal is generally through behavior regulating their position in the water column and their exposure to different oceanographic patterns. The larvae of many intertidal invertebrates, for example, are highly buoyant and are released during extreme spring tides, which transport the larvae offshore to develop. As the larvae grow, they lose their buoyancy and sink to deeper waters, where they are moved back to shore and exposed to suitable adult habitats. At intermediate spatial scales measured in tens of meters, local oceanographic patterns also often drive larval distributions. Habitats exposed to high flows, for instance, typically have heavier recruitment than habitats with low water movement, for the simple reason that high flows deliver more larvae to shorelines (Gaines and Bertness, 1993). Since these mechanisms generate spatial and temporal variation in the densities of intertidal organisms, they directly mediate the importance of competition, predation, and grazing in determining the structure of intertidal communities.

At small spatial scales measured in centimeters, both hydrodynamics and larval habitat selection can influence recruitment patterns. Butman (1992) suggests that local hydrodynamic conditions can dictate where larvae have the most contact with potential settlement surfaces. Small features on surfaces can create fluid flows that deposit larvae in some micro-habitats and limit the exposure of larvae to other habitats. The larvae of marine invertebrates have been known for over 50 years to be capable of considerable habitat selection. Larvae have been shown to settle at specific tidal heights, with other members of their species, and with favorable food, and to avoid potential competitors and predators (Morgan, 2001). Larval settlement preferences, however, come into play only in habitats where larvae have a chance, and in fact get, to choose among places to settle. A good way to think about the role of larval supply to intertidal habitats is that larval production and large-scale oceanographic patterns typically set the concentration of larvae in a region, more local oceanographic conditions determine the flux of larvae to a particular habitat, and small-scale features of a habitat and larval habitat selection together generate local larval recruitment patterns (Figure 6.27).

Variation in larval recruitment can play a leading role in shaping intertidal communities. Underwood and Denley (1984) found that shores in New South Wales that are protected from wave shock have grazing gastropods as dominants, but where there is moderate wave action, the midshore is dominated by the barnacle *Chamaesipho columna*. On the most exposed areas, a second barnacle, *Tesseropora rosea*, occupies midtide levels, and *Chamaesipho* in habits the upper reaches. Neither barnacle settles into areas dominated by grazing gastropods, even if the grazers are removed. On exposed shores, neither settles into areas occupied by the adults of the other species. In other words, the whole intertidal zonation pattern in this system is structured by the patterns of larval choice and not by predation, competition, or grazing. Gaines and Roughgarden (1987) have demonstrated that variation in shoreline larval recruitment can be coupled with biological interactions in the water column. These researchers showed that juvenile rockfish in kelp forests immediately offshore of the intertidal zone in central California are significant predators of the larvae of intertidal barnacles. The rockfish can reduce the number of recruits to one-fiftieth of the number that would settle in the absence of the fish. The number of barnacle larvae reaching shore depends on the stock of rockfish, which, in turn, is a function of the extent of the kelp

FIGURE 6.27 Conceptual model of the forces that control the supply of larvae to intertidal habitats. (From Pinead, J., "Spatial and Temporal Patterns in Barnacle Settlement Along a southern California Rocky Shore," *Marine Ecology Progress Series,* Vol. 107, pp. 125–138, 1994. Copyright 1994 Inter-Research Science Center. Used with permission.)

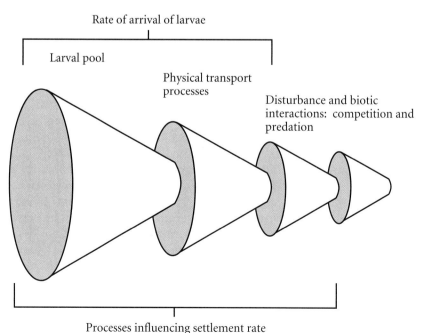

Rate of arrival of larvae

Larval pool

Physical transport processes

Disturbance and biotic interactions: competition and predation

Processes influencing settlement rate

forest. Whenever the kelp forest is reduced, as in El Niño years, the number of rockfish also is reduced, leading to good barnacle recruitment, and vice versa. Thus, in this system the dynamics of the intertidal community are directly coupled to these of the kelp forest community. (See Chapter 5 for factors influencing kelp forest dynamics.)

Variation in the larval settlement of the dominant intertidal organisms that is driven by the mechanisms just outlined leads to spatial and temporal variation in the importance of competition, predation, and grazing in intertidal communities. Understanding this coupling is difficult because it requires studying intertidal systems simultaneously at small and large spatial scales, but it is an area of considerable current research. In an attempt to understand the relative importance of competition, predation, and larval recruitment in structuring rocky intertidal communities, Menge (1991) used multiple regression analysis on these factors to reexamine his studies of New England and Panama shores. His results indicated that in New England recruitment explained 11%, and predation and competition 50–78%, of the structure of intertidal communities, whereas in Panama recruitment explained 39–87%, and predation and competition only 8–10%, of the structure. These findings suggest that the importance of recruitment in shaping intertidal communities is highly variable geographically.

■ PATCHINESS

The conspicuous distribution patterns of organisms in the rocky intertidal zone—the zonation—can be explained by one or more of the following factors acting separately or together: wave exposure, desiccation, temperature, competition, predation, larval settlement, and grazing. However, vertical zonation bands are never perfect: Often, smaller-scale local variation alters zonation and affects local distribution patterns. Such small-scale patchiness, or local distribution patterns, can result from either biological or physical factors. Patchiness is a basic feature of the structure of rocky intertidal areas just as it is of plankton (Chapter 2) and subtidal communities (Chapter 5). Usually temporally and spatially variable, patchiness is generated by any physical or biological activity that creates an open space for recolonization, produces a different substrate, or alters the prevailing physical conditions of an area.

Perhaps the easiest of these local distribution patterns to explain are those caused by physical factors, primarily variation in the nature, slope, and exposure of rock surfaces. The presence of different types of rock in a local area almost always means the presence of different organisms, or at least different abundances of the same organism. An outcrop of sandstone among an otherwise granite rock shore usually means a different complement of organisms, due to differences in surface texture and thermal characteristics.

Similarly, slopes facing south and north often have different organisms, or else the densities and levels at which the same organism is found differ. Either of these phenomena reflects differences in temperature and desiccation. On the Pacific coast of North America, for example, two limpet species, *Lottia digitalis* and *L. scabra*, live in the high intertidal zone, but, as Haven (1971) has shown, their fine-scale distributions are due to changes in the substrate. *Lottia scabra* is found primarily on horizontal surfaces, where it has a home scar or depression in which it remains during low tide. *Lottia digitalis* has no scar and is found on the cooler, moist, shaded vertical slopes. *Lottia scabra* can survive on the drier horizontal rocks because of its scar, which is a tight enough fit to prevent desiccation at low tide. *Lottia digitalis*, by contrast, without the desiccation protection of a scar, must live where desiccation is not as intense (that is, north-facing slopes and crevices). Similarly, certain species do not prefer to settle on vertical faces or overhangs. These areas are then settled by other species that are unable to compete on the flat surfaces. The presence of cracks or crevices in the rock may help algal spores "escape" from grazers and create a patch of algae where others are absent due to grazing (Figure 6.28). Cracks and crevices also may shelter predators, in turn affecting the community. Moran (1980) filled the cracks in an intertidal area with concrete. This reduced the density of the predatory gastropod *Morula marginalba*, and led to increased barnacle numbers and an altered community structure.

Much small-scale patchiness probably results from the destructive effects of wave action and such waveborne objects as logs that may be hurled against the shore, smashing into and scouring away various organisms. This kind of battering was studied by Dayton (1971) in Washington State. He found that any point on an open shore had a 5–30% chance of being hit by a log over a three-year period. On exposed shores, many mobile species, such as snails and crabs, find refuge in crevices or other protected areas.

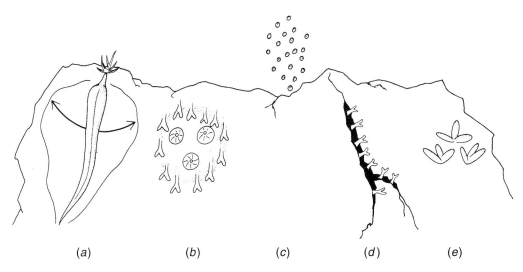

(a)   (b)   (c)   (d)   (e)

FIGURE 6.28 Causes of patchiness in algae on rocky shores. (a) Sweeping action of algal fronds. (b) Irregular spatial and temporal distribution of grazers. (c) Fluctuations in recruitment. (d) Refuge from grazing provided by pits and cracks in rock. (e) Escape of sporelings from grazers.

FIGURE 6.29
Buildup and removal of mussel clumps by wave action.

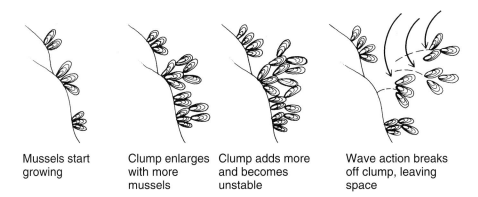

Mussels start growing

Clump enlarges with more mussels

Clump adds more and becomes unstable

Wave action breaks off clump, leaving space

Heavy wave action from storms, even when unaccompanied by battering, also is responsible for the removal of organisms from the intertidal zone. Menge (1976) found that winter storms could clear as much as 90% of the exposed intertidal zone in New England, whereas less than 10% of the same space was free of organisms during summer and fall.

The premier space competitors, *Mytilus edulis* in New England and *M. californianus* on the Pacific coast of Washington, form large clumps that grow out from the substrate. As the clumps get larger, fewer mussels are attached directly to the substrate by their byssal threads and more are attached to each other. This kind of attachment creates an unstable situation in which waves are more likely to tear off a clump; hence, older and larger clumps become more vulnerable (Figure 6.29). Small-scale distribution anomalies may also result from biological interactions of various types. Some of these interactions are passive, such as animals inhabiting moist areas under seaweeds to survive exposure to air. On wave-exposed shores, however, sessile organisms near larger seaweeds, such as *Ascophyllum* in New England, can be scoured away by algal whiplashing, creating a conspicuous halo or patch of bare space around many larger seaweeds.

FIGURE 6.30 Clones of the anemone *Anthopleura elegantissima*, showing the separation caused by fighting between adjacent clones. (Photo © Mark Bertness.)

Other biological effects leading to patchiness are more direct. For example, the intertidal anemone *Anthopleura elegantissima* of the Pacific coast of North America tends to occur in large clumps, uniformly separated from each other. Each clump is produced by the continued asexual fission from individuals formed by the original anemone, and each clump consists of genetically identical individuals, or **clones**. Francis (1973) discovered that the sharp boundaries between adjacent clones are the result of active fighting between the outer individuals of each clone (Figure 6.30). In these fights, special tentacles called acrorhagi are used to inflict tissue damage to other anemones. As a result of such clonal "border wars," the clones become separated and anemone-free areas wide enough to prevent two anemones from reaching each other arise (Figure 6.31).

Another example of a biological interaction that produces patchiness is **territoriality**—a behavior of an individual or a group that actively keeps other members of the same or different species out of an area. This phenomenon is well known among various birds and mammals, but it also occurs among certain invertebrates. On the north Pacific coast of America, the largest limpet in the intertidal zone is *Lottia gigantea*. Stimson (1970) has shown that individual female *Lottia* create a territory in the zone and actively defend it against other individuals of *Lottia* and against other limpets. Invading limpets are aggressively pushed away, leaving territorial *Lottia* with their own algal gardens,

FIGURE 6.31 Closeup of the aggregating anemone *Anthopleura elegantissima*, showing the anemone-free areas (brown) between two genetically different clones. (Photo courtesy of Drs. Lovell and Libby Langstroth.)

providing them with more food to produce a high number of eggs.

Another more complicated combination of factors producing small intertidal patches is the distribution of the sea palm *Postelsia palmaeformis*, which occurs in isolated clumps on wave-swept Pacific coast rocky shores and was studied by Dayton (1973) and Paine (1988). (See Figure 6.32.) *Postelsia* is an annual alga that lives in disturbance-generated bare patches in mussel (*Mytilus californianus*) beds. The adult plants are immune to overgrowth by the mussels. *Postelsia* has very limited dispersal of its propagules, with most settling within a meter of the adult. These propagules may

FIGURE 6.32 The sea palm *Postelsia palmaeformis*, characteristic of areas of direct wave exposure and high wave shock on the Pacific coast of North America. (Photo courtesy of Dr. Richard Mariscal.)

settle and begin to grow on almost any hard surface, but they are much more likely to survive to reproductive size if they settle on bare rock surfaces. Although capable of settling, *Postelsia* does not persist to adulthood in the presence of the common turflike algae *Halosaccion* and *Corallina*. Therefore, the continued persistence and successful reproduction of *Postelsia* are dependent on a cyclical set of relationships. The heavy wave action is necessary to clear away patches of *Mytilus* back to the bare rock and to permit the settlement of the *Postelsia* propagules. The *Mytilus* are, in turn, necessary in eliminating the algal turf (*Halosaccion* and *Corallina*) that can outcompete *Postelsia*. The mussels do that by invading and overgrowing the lower-growing algae. If the mussels are not removed, they will cover the intertidal zone. The removal of small patches of mussels by violent wave action opens an area for *Postelsia* to occupy, provided that its propagules colonize before the low-growing *Halosaccion* and *Corallina* take over the patches. The wave-swept area, therefore, becomes a mosaic of patches of *Halosaccion* or *Corallina*, and *Postelsia*. Patches of *Postelsia* result from its settling earlier than the two turflike algae, and *Halosaccion* or *Corallina* results from its first settlement of a patch. *Postelsia* may then persist in the patch it occupies only if the adult plants are ripped out after reproducing, leaving bare rock, or if another patch opens up within the dispersal range of the *Postelsia* propagules.

Predation and grazing can also create small patches or patchy distributions. Predation is the reason for the occurrence of *Balanus cariosus* (see p. 286) in small clumps in the intertidal zone. Grazing by sea urchins may result in tide pools that are free of algae, while others without urchins have an abundant growth. On the New England coast, tide pools without *Littorina littorea* are dominated by *Enteromorpha intestinalis*, whereas those with the snail have *Chondrus crispus*.

■ SUCCESSION

The creation of open spaces in any intertidal zone typically triggers the colonization of **opportunistic**, or **fugitive, species**. These species quickly settle in open areas, mature, and reproduce before the slower-growing dominant species retake the space and force them out through competition. In rocky intertidal habitats the most common opportunists are the smaller, filamentous green and red algae and such smaller animals as hydroids. These weedy, opportunistic species are replaced over time by competitively dominant species (Figure 6.33). Successional sequences in the rocky intertidal zone can depend on the time of the initial disturbance, since timing will usually determine the availability of propagules. Generally, the first organisms to settle on freed space are those with the life-history characteristics of opportunistic species. (See Table 1.2.) The disturbance-and-recovery process can cause rocky intertidal habitats to resemble mosaics of patches in different stages of recovery. Depending on the time since their clearance, patches will be passing through stages of succession toward the dominant community characterizing the zone. This temporal succession accounts for the different organisms in different patches in the same zone in the same area.

The ecological concept of succession, discussed in Chapter 1, can follow different models or hypotheses explaining the sequence of events following a disturbance. In the rocky intertidal zone, succession typically follows the inhibition and facilitation models. (See Chapter 1, p. 24).

The sequence of events characterizing succession is variable and dependent on several combinations of events in the rocky intertidal zone. Sousa (1980) found that, on the Pacific coast of North America, succession in boulder fields followed the inhibition model. Once the early algal colonizer *Ulva* settled, it inhibited colonization by the perennial red algae, such as *Chondracanthus*. As long as *Ulva* remained healthy, it would prevent colonization by any other

Time ⟶

First stage: bare rock with diatom/bacterial film | Small algal species | Barnacles replace algae | Mussels smother barnacles and eventually dominate

⊛ Barnacles   🦪 Mussels   🌿 Algae   ⸰° Diatoms   ⌇ Bacteria

FIGURE 6.33 Succession in a northwest Pacific coast intertidal mussel bed in the absence of *Pisaster*.

alga. However, this inhibition is broken by the grazing of the crab *Pachygrapsus crassipes*, which permits the succession to proceed to a community of perennial red algae. These middle successional red algae, in turn, slow the invasion by the late successional dominant, *Chondracanthus canaliculata*, but eventually *Chondracanthus* colonizes the area because it is more resistant to desiccation and overgrowth.

In other instances, there appears to be a successional sequence in which pioneer species are necessary before the later species can colonize an area (the facilitation model of succession). For example, Hawkins (1981), working on English rocky shores, showed that successional sequences were quite different on barnacle-covered rocks that were kept free of limpets than on areas that were scraped free of barnacles and also kept free of limpets. In the case of the barnacle-covered rocks, colonization was determined by the availability of propagules; but in scraped areas, larger green algae and *Fucus* could not settle until surfaces had been "prepared" by the growth of diatoms and the green alga *Ulothrix*. The reasons for these differences in sequence are not clear, but Hawkins suggests that the barnacle shells may have provided spatial refuges free from grazers, thereby facilitating "escapes" of the later successional algae,

which could then recruit directly. Another suggestion is that, although a mat of green algae enhances *Fucus* settlement, early *Fucus* growth is inhibited by the mat. As a result, *Fucus* does not begin to take over until either the mat dies or the slow-growing sporelings of *Fucus* finally reach a height or size such that they are superior to the green algae mat and take over. In this example, succession is a mixture of both the inhibition and facilitation models.

A final topic is the **seasonal succession** of algal communities. In places such as south Texas and Hong Kong, the algal flora is completely different in the winter, from that in the summer. In Hong Kong, which has a monsoonal climate, Kennish et al. (1996) found that foliose algae such as *Ulva, Porphyra*, and *Dermonema* dominated the shore during the winter, but died off in the hot summer and were replaced by encrusting algae such as *Ralfsia*, coralline crusts, and encrusting cyanobacteria. In south Texas, Lowe and Cox (1978) described a winter algal community on Galveston Island dominated by *Enteromorpha, Bangia*, and *Gelidium* that shifted in the summer to a community dominated by *Cladophora, Bryocladia*, and *Ceramium*.

We conclude this section by noting that continued physical disturbance of differing degrees of severity and extent, coupled with seasonal variation in the types of propagules available, along with the fluctuations in the spatial distribution of dominant animals, ensures that succession is always occurring in the intertidal zone. Disturbance and succession constitutes one of the major sources of patchiness in the zone.

■ HORIZONTAL DISTRIBUTION PATTERNS

As noted in the preceding sections, in all geographical areas numerous changes in community structure appear as one moves from areas of high wave exposure to protected and quiet waters. These changes are caused by wave exposure restricting consumer impacts and increasing the supply of larvae and particulate food to benthic invertebrates. Other organisms cannot tolerate heavy wave action and are absent from wave-swept shores. Usually, these are delicate species or species with poor means of holding onto the substrate. Similarly, certain species cannot tolerate quiet water areas, due perhaps to their inability to compete, a shortage of (the right kind of) food, or the presence of predators that are absent in wave-swept areas. In New England, for example, *Mytilus edulis* outcompetes both *Semibalanus balanoides* and *Chondrus crispus* on exposed shores, because mussel predators cannot tolerate the associated wave action. In protected areas, predators reduce mussel densities, allowing barnacles and algae to exist and altering the appearance of the zones. In Australia, protected shores have larger numbers of grazers, whereas exposed shores have two barnacle zones and few grazers. On European shores, limpets dominate the exposed areas and large algae are uncommon, while on sheltered shores, limpets are rare and large algae flourish.

Striking horizontal distribution patterns are also conspicuous along estuarine salinity gradients, but have not received much attention. Superficially, it appears that only organisms capable of withstanding osmotic stress—particularly, variations in osmotic conditions—are able to live in lower-salinity areas of estuaries. While this is undoubtedly true, it is probably an overly simple explanation. Since all organisms with marine origins grow and reproduce best in full-strength seawater, and since there must be a high metabolic cost of living in habitats where organisms must metabolically cope with osmotic stresses, it is likely that the organisms living in physically harsh estuarine habitats do so at least partially to escape their competitors and predators.

Latitudinal patterns in the structure and organization of rocky intertidal communities are also pronounced. At lower latitudes, consumer pressure is much more intense than at higher latitudes, due to algae-consuming and shell-crushing fishes and crabs that are restricted to tropical and semitropical latitudes. Intense solar radiation at low latitudes causes severe physical stresses on intertidal communities at these latitudes. As a consequence of both of these agents, organisms in low-latitude intertidal communities tend to be highly dependent on living in cracks and crevices as refuges from consumers and physical stresses, whereas open space in these systems is often bare or dominated by algal crusts that are resistant to both consumers and physical stresses. At higher latitudes, intertidal organisms are typically not as dependent on refuge habitats and open surfaces are often covered with organisms. At extremely high arctic latitudes, however, ice scour becomes extreme, and virtually all sessile organisms on rocky shores are again restricted to living in cracks and crevices.

■ TIDE POOLS

A characteristic feature of many rocky shores is the presence of tide pools of various sizes, depths, and locations. Certain conditions affecting life in tide pools differ markedly from conditions in the surrounding intertidal zone and necessitate a separate discussion. Our remarks in this section will be restricted to pools that undergo a complete interchange with ocean water during the tidal cycle (Figure 6.34). Note that the biotic communities of tide pools and the factors influencing their structure are less well understood and studied than those of the emergent rocky surfaces. It has even been suggested that tide pools do not represent an intertidal habitat, because they are never exposed to the air during a tidal cycle. However, the fluctuations in physical and chemical factors in tide pools are a function of the tidal cycle; hence, it seems logical to consider tide pools here. Furthermore, many of the organisms found in tide pools are similar to those found on the adjacent exposed rock.

At first glance, tide pools appear more or less ideal places for aquatic organisms seeking to escape the harshness of the intertidal zone during its exposure to air. In reality, however, escape from such physical factors as desiccation may mean exposure to others that operate more severely in tide pools.

Tide pools vary in size and in the volume of water they contain. Since water is a great moderator of harsh

FIGURE 6.34 An intertidal rock pool at low tide. (Photo © William Smithey Jr./Getty Images/Taxi.)

physical conditions, larger pools with greater volume fluctuate less in regard to physical factors. Other factors that influence the physical and chemical conditions of the water in tide pools include the surface area, the depth of the pool, its height in the intertidal zone, exposure to wave action, the degree of shading available, and the drainage pattern. In addition, the physical and chemical environment of tide pools can vary vertically with depth and horizontally across the pool. Finally, all physical factors may fluctuate diurnally and seasonally. Given all these variables, Metaxas and Scheibling (1993) suggest that it is unlikely that any two tide pools will be similar in all characteristics; therefore, individual pools are physically unique. It then follows that ecologists cannot replicate experimental manipulations in different tide pools, and this may be a significant reason that there has been so little experimental ecological work done on pools. It may also be the reason that most studies have been primarily descriptive.

Three major physical factors are subject to variation in tide pools. The first is temperature. Whereas the ocean itself is a vast reservoir that heats and cools very slowly and usually within very narrow limits, the same is not true of tide pools. These relatively small bodies of water are subject to more rapid changes. Shallow tide pools exposed to the sun on warm days may quickly reach lethal or near-lethal temperatures. Similarly, tide pools in cold temperate or subpolar regions may have temperatures in the freezing range in winter. An additional problem is temperature fluctuation. A pool may heat or cool over the several-hour period during which it is exposed to the air, but when the

tide returns, the pool will be flooded with ocean water. This flooding will suddenly change the temperature of the whole pool. The daily variation in temperature of a tide pool can be as much as 15°C, depending on the pool's height in the intertidal zone and its volume, degree of shading, and wave exposure. Finally, a change in temperature during exposure at low tide may cause temporary thermal stratification of the pool. Organisms inhabiting such pools must still be adapted to considerable temperature fluctuation.

The second factor that varies in tide pools is salinity. During low tide, the water in tide pools can evaporate, increasing the pool's salinity. Under hot tropical conditions or windy conditions, the increase in salinity can be dramatic enough to reach the point of precipitating out salt. In contrast, when heavy rains flood pools with fresh water during low tides, the pool's salinity can drop dramatically. Fluctuations in tide-pool salinity vary with the position of the pool on the shore and the other factors mentioned, but values of 5 and 50 psu have been reported. Salinity stratification also may develop in tide pools, due to freshwater runoff during heavy rains or as a result of freezing in winter and evaporation in summer. When the tide returns, pools are flooded with seawater and there is an abrupt return to normal conditions.

The final physical factor undergoing change in tide pools is oxygen concentration. Since the amount of oxygen that can be held in seawater is a function of temperature, tide pools that heat up during low tide will lose oxygen. Under normal conditions, the loss may not be serious enough to produce oxygen stress, but if the pool is crowded with organisms, it may become considerably stressful. For example, a pool filled with algae that was exposed at night could produce a situation in which the lack of photosynthesis coupled with high respiration could reduce the oxygen level significantly. Indeed, in some tropical tide pools, oxygen levels have been recorded as having fallen to only 18% of saturation. It is also possible to develop oxygen stratification in deeper pools.

Tide pools are areas of refuge from desiccation for intertidal organisms, but in turn, these organisms suffer from rapid changes in temperature, salinity, and, occasionally, oxygen; thus, the fauna and flora in these pools are restricted to organisms that are able to tolerate such ranges. The organisms that inhabit tide pools are similar to those on the adjacent emergent substrates, but often with differences in abundances between the pools and exposed surfaces. For example, several genera of algae, such as *Spongomorpha* and *Corallina* in the intertidal zone of Maine, *Prionitis* in Washington, and *Fucus* in Nova Scotia, are more abundant in pools. Other species are either absent or occur in lower numbers in pools than on exposed surfaces. An example is the alga *Ascophyllum nodosum* in New England. Finally, the occurrence of tide pools permits some organisms, including a great many algae and invertebrates—and, particularly, numerous fishes—to extend their range upward in the intertidal zone beyond the levels to which they would be limited on the adjacent emergent surface.

Zonation patterns have been described for tide-pool organisms, both within the intertidal zone and vertically within tide pools. Along the intertidal gradient, such green algae as *Enteromorpha, Cladophora,* and *Chaetomorpha* dominate the upper tide pools, while the brown algae *Fucus* and *Laminaria* and the red corallines, *Lithothamnion* and *Corallina*, are abundant in the lower intertidal pools. Benthic invertebrates also show zonation. High-tide pools are inhabited largely by littorine snails and harpacticoid copepods, whereas low-tide pools have a greater diversity of snails and other invertebrates. According to Metaxas and Scheibling (1993), fish zonation in tide pools has not been quantitatively documented, but qualitatively, there is a decrease in the number of species with increasing height of the pool. In general, both in number of species and in biomass, high-tide pools are depauperate compared with lower-level pools.

The role of biological interactions in determining the community structure of tide pools is poorly doc-

umented compared with studies of the exposed rocky intertidal zone. This paucity of information is due not only to a relative lack of studies, but also to the variability among adjacent tide pools at the same intertidal level when the pools are subjected to experimental manipulation. Such variability has precluded obtaining statistically significant results in many cases, although correlative data often indicate a trend. We discuss some of these data here.

Herbivory has similar effects in tide pools and emergent rock surfaces, namely, altering the abundance of macroalgae. For example, Paine and Vadas (1969) demonstrated that the removal of sea-urchin grazers from tide pools in Washington resulted in increases in macroalgae. Similarly, when Lubchenco (1978) added littorine snails to a pool, the density of the dominant alga, *Enteromorpha*, was reduced, whereas when she removed littorines from a pool dominated by the alga *Chondrus crispus*, the density of *Chondrus* decreased (Figure 6.35). In Nova Scotia, Chapman and Johnson (1990) found that littorinid snails had a negative effect on the abundance of several species of the brown algal genus *Fucus*, a positive effect on the ephemeral algae, and no effect on the crustose alga *Hildenbrandia*.

Very few studies have investigated the role of predation in structuring tide pool communities. Lubchenco (1978) suggested that crab and bird predators controlled the distribution of littorine snails in New England tide pools, and the snail distribution in turn controlled the abundance and distribution of algae. Recent experimental work has confirmed this relationship, but has further shown that the simple presence of green crabs in tide pools causes the littorine snails to leave the pools, triggering an increase in algal biomass without actual predation of the snails (Trussell et al., 2004).

Evidence for the importance of interspecific competition in regulating community structure in tide pools is also sparse. Both Lubchenco (1982) and Chapman (1990) have demonstrated decreases in canopy cover in *Fucus* due to competition with ephemeral algae. Kooistra et al. (1989) have shown competitive dominance, as indicated by overgrowth by the sponge *Halichondria panicea* in tide pools in Brittany, France.

Although recruitment may be a significant factor structuring tide pool communities, no studies have considered this factor directly. Similarly, there is little or no information about the role of disturbances, particularly large-scale episodic events, in the structuring of tide pool communities.

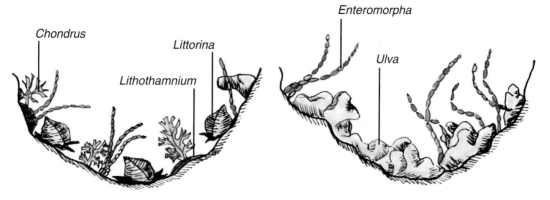

High-diversity pools with snails

Dominated by a mixture of palatable green seaweeds (*Ulva, Enteromorpha*) and slower-growing and well-defended species such as algal crusts (*Lithothamnium*) and Irish moss (*Chondrus*).

Low-diversity pools without snails

Dominated by fast-growing, competitively dominant green seaweeds such as *Ulva* and *Enteromorpha*. Green crabs colonize these pools and can limit the recruitment of herbivorous snails.

FIGURE 6.35 Diagrammatic representation of New England tide pools with and without grazers. (Modified from *The Ecology of Atlantic Shorelines*, M. D. Bertness, p. 234. Copyright © 1999, Sinauer Associates, Inc. Reprinted by permission.)

■ TROPICAL INTERTIDAL SHORES

Thus far in our discussion of the biological factors that determine community structure, we have employed examples primarily from temperate regions of the world. What about tropical rocky shores? Are the communities and organisms there similar to those on rocky shores in temperate regions? Unfortunately, there are few studies of the tropical intertidal zone from which to draw generalities. The most extensive are from the rocky intertidal zone of the Pacific coast of Panama and were carried out by Menge et al. (1986) and Lubchenco et al. (1984), but less extensive studies from the Caribbean and Hong Kong suggest that there are marked differences among tropical areas. On the Pacific coast of Panama, the general composition of rocky-shore communities is similar to that of temperate rocky shores. The major differences are the absence of fleshy algae and an abundance of herbivorous and predaceous fishes and crabs. As a consequence of the intense consumer pressure on these rocky shores, the intertidal zone is dominated by bare space and grazer-resistant encrusting algae. Other intertidal organisms are restricted to cracks and crevices where they are protected from both consumers and extreme heat stress during low-tide exposures.

The foraging behavior of intertidal gastropods in the Bay of Panama illustrates the formidable constraints on these organisms. Bertness et al. (1981) showed that, in contrast to intertidal snails in most temperate-zone habitats, snails in the Bay of Panama have extremely restricted foraging time. Whereas temperate-zone snails forage, on the one hand, whenever they are covered by the tide, and on the other, whenever the tide is low and the temperature is not too high, in the Bay of Panama intertidal snails generally do not forage during high tide or during daytime low tides. Instead, Panamanian snails are restricted to foraging only as the tide is changing and they are intermittently splashed with the incoming or outgoing tide and during evening low tides. They cannot forage during high tide, because shell-crushing fishes such as puffers and triggerfishes patrol the shore, so that snails remaining on exposed surfaces experience extremely high mortality. The only exception to this situation is that very large snails, such as adult *Murex* that have achieved a size refuge from fish predation, forage freely during high tides. During daytime low tides, harsh tropical physical conditions restrict snails to crack and crevice refuge habitats. An interesting exception to the avoidance of low-tide foraging in these tropical snails is the abundant grazer *Nerita scabra*, which, according to Garrity (1984), forages during overcast daytime low tides, but only in dense groups that retain moisture and are evaporatively cool.

Working in the mid- and low intertidal zone in the Bay of Panama, Menge and colleagues (1986) found that space is dominated by crustose algae and that community structure is determined by complex interactions among a diverse assemblage of herbivores and predators that the authors divided into four consumer groups: large fishes, small fishes and crabs, herbivorous mollusks, and predaceous gastropods (Figure 6.36). Experimental removal of various combinations of the four consumer groups suggested that the effects of each group on its prey species or prey group were often indirect and were either positive or negative. Both the small fishes and crabs and the large fishes reduced the abundance and cover of sessile colonial and solitary invertebrates, as well as fleshy algae, but were incapable of totally removing them. The molluscan herbivores, in turn, grazed the fleshy algae until only the very resistant algal crusts, a group otherwise competitively inferior to the fleshy algae, survived. The molluscan herbivores also inhibited the recruitment of sessile invertebrates. Finally, the predaceous gastropods preyed on the sessile invertebrates and reduced their numbers as well.

The results of these studies indicate that there was no keystone species in the Bay of Panama that prevented overgrowth by a superior space competitor, as we see in many temperate intertidal habitats. Instead, the dominance of the tropical habitats by crustose algae is maintained through persistent intense predation and herbivory by a diverse assemblage of fishes and invertebrates on potentially dominant sessile invertebrates and fleshy algae. The dominance of open substrate by crustose algae and the failure of any organism or group of organisms to occupy a majority of the available open space are maintained by a combination of intense consumer pressure, extreme heat, and desiccation that together prevent much of the area from being occupied by marine organisms. In contrast to temperate rocky intertidal systems, in tropical Panama extreme physical factors and consumer pressure largely limit most intertidal invertebrates to living in crack and crevice refuges.

In the Caribbean and other low-latitude areas, tidal amplitudes are small—often less than 25 cm—and intertidal rocky substrate is dominated by limestone. The high tidal amplitude in the Bay of Panama is really an anomaly for tropical shores that is due to the funnel shape of the Gulf of Panama locally amplifying the tides. In these habitats, littorine snails can be abundant, but little else other than algal crusts are common. Like rocky shores in Panama, these areas appear to be physically stressful and heavily grazed by fish. More studies are needed on tropical rocky shores before we can confidently make generalizations about their structure and organization.

Cobble Beaches

Cobble beaches typically occur on shorelines where recent ice ages have deposited a large supply of rocks that have been scoured from the landscape by advancing glaciers. In North America, cobble beaches are characteristic of Long Island Sound, Narragansett Bay, and the shores of some Maine estuaries on the Atlantic coast and of Puget Sound on the Pacific coast. Cobble beaches are not found in shoreline habitats with low water movement, because in those habitats heavy sedimentation will typically cover the underlying substrate with sand or mud. Cobble beaches therefore occur in habitats with a supply of cobble substrate and at least moderate water movement.

The biological communities that develop on cobble beaches are relatively simple, for a couple of reasons. First, the cobbles in these habitats are prone to disturbance. As Sousa (1980), found, particularly during storms, or seasonally in heavy fall and winter weather, cobbles can be overturned, whereupon they flip over and roll, crushing both sessile and mobile organisms living on or under the stones. Second, as relatively small objects, cobbles are not buffered from thermal changes, as are massive rocky benches or rocks embedded in mud or sand. As a consequence, cobbles can heat up dramatically during summer low-tide exposures (Bertness, 1989). In Narragansett Bay, intertidal cobbles can get as hot as 45°C during summer low tides. Because of the high disturbance and the thermal instability of cobbles, cobble beach communities typically include a very simple assemblage of organisms. In Narragansett Bay, cobble beaches typically are dominated by bare rock surfaces, barnacles, encrusting algae, and fucoid algae-covered surfaces, with the grazing snail *Littorina littorea* and shore crabs the most conspicuous mobile organisms. Bertness (1984) showed that grazing by *Littorina* strongly limited algal biomass in these communities and led to the dominance of algal crusts.

Recent work on New England cobble beaches has also demonstrated the importance of habitat

Open surface

Crevice

FIGURE 6.36 Diagrammatic illustrations of the organisms and physical heterogeneity at Taboguilla Island, Panama. The distant views are shown for the high (a), mid (d), and low (g) zones; each of these is less than 2 m across the foreground. Close-up views are illustrated for open surfaces (b) and crevices or holes (c) for the high, mid (e, f), and low (h, i), zones. Species identifications are (1) Chthamaloid barnacles (live and dead); (2) *Littorina* spp.; (3) *Nerita scabbricosta*; (4) *Tetraclita*; (5) *Siphonaria gigas*; (6) *Ostrea palmula*; (7) mixed algal crusts: *Schizothrix*, *Ralfsia*, and *Hildenbrandia*; (8) *Fissurella* spp.; (9) *Thais reelones*; (10) *Acanthina*; (11) *Ceratozona*; (12) anemones; (13) corallina curst; (14) *Balanus inexpectatus* (live and dead); (15) *Pachygrapsus*; (16) *Siphonaria maura*; (17) encrusting bryozoan; (18) colonial tunicate; (19) mixture of *Abietinaria* and filamentous algae; (20) *Chama*; (21) *Echinometra*; (22) sponges; (23) *Ascidia*; (24) rasp marks of grazing fishes. (After "Structure, Persistence, and Role of Consumers in a Tropical Intertidal Community," J. Lubchenco et al., 1984, *J. Exp. Marine Biology & Ecology*, vol. 78, pp. 23–73, 1984. Copyright © 1984 with permission from Elsevier Science.)

modification and facilitation in the biological communities that develop in those habitats. On moderately wave exposed cobble beaches in Narragansett Bay, disturbance and thermal labiality prevent almost all organisms from colonizing or persisting in the area. As a result, these beaches are almost entirely devoid of plants and animals. The situation changes dramatically, however, if the cordgrass *Spartina alterniflora* successfully colonizes the areas either through ice rafting or the establishment of seedlings. Once cordgrass becomes established and spreads laterally parallel to the beach, it both stabilizes and shades the substrate, limiting cobble mobility and impeding the cobbles' ability to heat up. The most dramatic consequence of this habitat modification is that a cobble beach plant community of halophytic plants develops behind larger cordgrass stands. As Bruno (2000) found, these plants include *Salicornia, Sueda, Atriplex,* and *Limonium,* and they are entirely dependent on cordgrass beds buffering wave disturbances that would otherwise prevent the plants from establishing and maintaining populations on cobble beaches (Figure 6.37). Cordgrass beds on these cobble beaches also facilitate the presence of a mussel reef community composed of ribbed mussels, blue mussels, barnacles, and green and brown seaweeds. The mussel reef community is facilitated by both substrate stabilization and shading and, like the cobble beach plant community, is entirely dependent on cordgrass for its existence in these habitats. Both the cobble beach and mussel reef communities are exam-ples of whole community facilitation by a dominant ecosystem engineer. Similar facilitation of assemblages of organisms dependent on substrate stabilization by vascular plants is probably also common on sandy and muddy shores, but has not been examined.

Sandy Shores

Intertidal exposed sand beaches and protected sand flats are common throughout the world and, because they are used for various recreational activities, are certainly better known than rocky shores to most humans. At the same time, they present a marked change from the previously described temperate rocky shores, because, in contrast to the crowded life on the latter, exposed sand beaches appear devoid of macroscopic life. The environmental factors acting on these shores create conditions such that virtually all organisms inhabiting the area bury themselves in the substrate. Intertidal protected sand flats by contrast are more often populated, with large numbers of macroorganisms visible on the surface.

Exposed sand beaches are subject to more wave action and usually face the open sea, whereas sand flats commonly have less exposure to wave action and usually face a bay or lagoon that is protected from the open sea by a barrier island. Exposed sand beaches usually have a pronounced slope toward the sea, whereas sand flats have little or no or slope.

FIGURE 6.37 A cobble beach plant community in Narragansett Bay, Rhode Island. (Photo © Mark Bertness.)

In this section, we first explore the action of physical factors on sand beaches and intertidal protected sand flats and see how these factors act to impose certain adaptations on the larger organisms found in these areas. We then investigate the operative biological factors and how they act together with the physical factors to structure the communities. The specialized conditions and organisms existing in the tiny spaces between the sand grains are the subject of Chapter 7.

■ ENVIRONMENTAL CONDITIONS

Sand beaches and flats are subject to an array of physical factors similar to those characterizing rocky intertidal shores, but the relative importance of those factors in structuring the community and their effect on the substrate differ.

Perhaps the most important physical factor governing life on exposed sand beaches is wave action and its attendant correlation with particle size and slope. Anyone who has visited such marine beaches is aware of the fact that the particle size and slope of the sand can differ among beaches and also among seasons on the same beach. Particularly in the temperate zone, the profile of the beach also changes between winter and summer. Beaches are defined by three factors: *particle size, wave action,* and *slope*. All three are interrelated such that defining two fixes the third. The importance of particle size to the distribution and abundance of organisms rests with the effect of a particle's size on *water retention* and on the suitability of that size of sand for burrowing. Through its capillary action, fine sand holds much water in its interstices after the tide has retreated. Coarse sand and gravel, by contrast, allow water to drain away quickly as the tide retreats. Since the organisms inhabiting the intertidal zone are aquatic, they are well protected against desiccation on a fine-sand beach, but subject to desiccation on a coarse-gravel beach. The latter environment is the less hospitable to those organisms. Fine sand is also more amenable to burrowing than coarse gravel is (Figure 6.38).

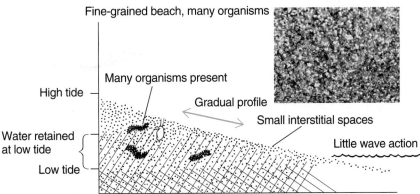

FIGURE 6.38 Comparison of the physical conditions found on fine-grained and coarse-grained beaches.

The slope of a beach is the result of the interaction between the particle size, wave action, and the relative importance of swash and backwash water. **Swash** is water running up a beach after a wave breaks; this action carries particles with it, and the particles may accrete on the beach if they remain there. **Backwash** is water flowing back down the beach; this action removes particles from the beach, at a rate depending on their size.

Protected sand flats are much less seasonally variable than exposed sand beaches and usually consist of finer-grained sand at all times. In protected sand flats, both waves and water currents affect the grain size of the sand. The magnitude of the wind is largely a function of the size of the body of water over which the wind blows, and the large embayments on the east coast of the United States, such as Chesapeake Bay and Pamlico Sound, are big enough to generate waves that can have an effect on grain size. In New England, because the embayments are all small, wind and wave action are less important.

A second important physical factor in sand beaches and tidal flats—one that was not of concern on rocky shores—is also the product of wave action: wave-induced *substrate movement*. The particles on sand or gravel beaches are not large enough to be stable when waves strike the shore. As a result, with each passing wave, the substrate particles are picked up, churned in the water, and redeposited (Figure 6.39). Particles are being continually moved and sorted. The reason fine sand occurs only where wave action is light and coarse sand where it is heavy is that, in heavy wave action, the smaller particles remain in suspension so long that they are carried away from the beach, leaving only those heavy (coarse) enough to settle immediately. Light wave action also means that only particles relatively close to the surface are affected by the passage of a wave, whereas heavy wave action disturbs the substrate to a greater depth. Because substrate mobility can severely limit the organisms that live on sandy or muddy beaches, submerged aquatic vegetation (either salt marsh grass or sea grass) or benthic animals (such as mussel and oyster beds) that can stabilize substrate material can play a major role in these habitats as ecosystem engineers. These habitat-stabilizing organisms can play the same community-facilitating roles in

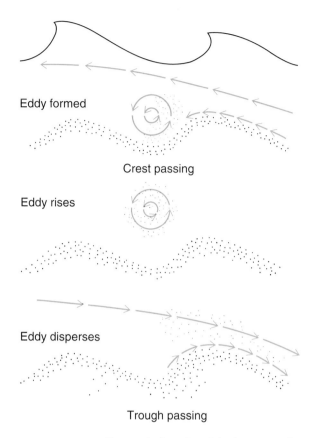

FIGURE 6.39 Transport of sand particles by wave action. (After *Treatise on Marine Ecology and Paleoecology*, J. Hedgpeth, 1957, Vol. 1, *Ecology*, © 1957 Geological Society of America. Used with permission.)

their habitats as that of cordgrass stands on cobble beaches.

Within this continuum of grain size, slope, and wave action, a number of different types of beach have been described. The extremes of this classification are the dissipative and reflective beaches (Figure 6.40). On a **dissipative beach**, wave action is strong, but the wave energy is dissipated over a broad, flat surf zone located some distance from the face of the beach. This action leads to a gentle swash and fine sediments. The beach face accordingly has a gentle slope and is maximally eroded, with much of the sediment occurring in the broad surf zone and in berms or bars parallel to the beach. A **reflective beach** is most fully developed where wave action impinges directly on the beach face. On these beaches, the sediment is coarse, there is no offshore surf zone, and waves produce large swashes

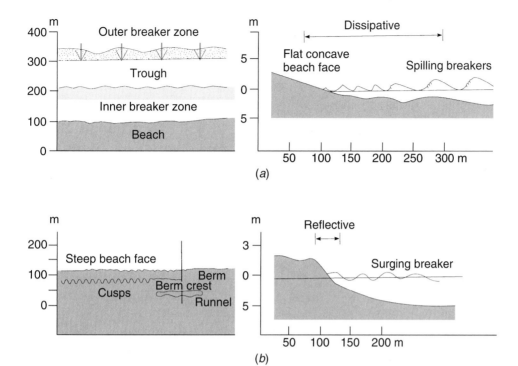

FIGURE 6.40 (a) Dissipative beach. The left side of the figure shows the beach from above; the right side is a cross-sectional view. (b) Reflective beach. The left side of the figure shows the beach from above; the right side is a cross-sectional view. (After *Ecology of the Sandy Shores*, A. C. Brown and A. McLachlan. Copyright © 1990 with permission from Elsevier Science.)

up the beach face. The beach slope is steep, and the backwash and swash collide to deposit sediment. Wave energy is directed against, the face of the shore and reflected off.

Studies by McLachlan (1983) and others have demonstrated that the abundance and diversity of macrofaunal organisms are more strongly correlated with the particle size and slope of open beaches than with wave action. Highly exposed beaches often have more abundant fauna than less exposed beaches if the latter have coarse grains and steep slopes. Fine-sand beaches usually have flatter slopes, and wave action is dissipated in the surf zone rather than on the beach itself. This means that a highly exposed beach can support a rich macrofauna if it is dissipative and not reflective. Indeed, a high biomass of filter feeders is strongly correlated with dissipative conditions. Intertidal protected sand flats show much less local variability in grain size.

Since the substrate—or at least its surface layers—is in constant motion and may be picked up and held in suspension for greater or lesser amounts of time, depending on the strength of the waves, it follows that any change in wave intensity on a given beach will mean a change not only in grain size, but also in the profile or shape of the beach. Such a change happens when gentle wave action is replaced by heavier waves, which dig deeper into the beach, pick up more and heavier grains, and hold them in suspension longer, so they are deposited some distance away, usually offshore. This changed profile is a common seasonal occurrence on many temperate-zone beaches, where a gentle slope of fine sand during the summer months is replaced by a steep, coarse beach during winter storms (Figure 6.41). It is significant for organisms that the depth of substrate moved can be a meter or more!

FIGURE 6.41 Beach profiles showing the difference between winter and summer.

Because of this ceaseless movement of surface layers of the sediment, few large organisms can permanently occupy the surface of open sand or gravel beaches. This is the reason for the barren appearance of such beaches.

In contrast to rocky shores, sand beaches and sand flats usually have a smooth, uniform profile; thus, they lack the great topographic diversity of rocky shores: There are no crevices, overhangs, permanent tide pools, or slopes facing different directions offering different moisture conditions. As a result, such environmental factors as temperature, desiccation, wave action, and insolation act uniformly at each tidal level on the beach or flat.

Because of its significant effects on particle size and on the slope of a beach, wave action is a dominant environmental factor acting on the exposed sand beaches, creating the special conditions that make it difficult or even impossible for many organisms to inhabit these areas. In contrast, the much-reduced wave action in protected sand flats, makes them suitable for a host of epifaunal and infaunal invertebrates. Both sand flats and sand beaches offer some advantages to marine organisms with respect to certain physical factors. Sand is an excellent buffer against large changes in temperature and salinity. Measurements of temperature with depth on sand beaches at low tide have demonstrated that, below the first few centimeters, the temperature is very nearly that of the surrounding seawater. This mitigation of temperature is due partly to the insulating properties of sand and partly to the water held in the interstices of the deeper layers. In a similar manner, changes in salinity are also minimal below 10–15 cm on beaches, even if the upper layers have fresh water flowing over or falling on them. Again, this is because the water held in the interstices is salt water, which has a higher density than fresh water. As a result, the fresh water remains perched on the surface. Furthermore, sand is a barrier to direct sunlight (insolation) impinging on any organism living in the sand and thereby causing harmful effects, because the sand is opaque to light and reflects or absorbs it in the surface layers. Nor is desiccation a problem, as long as the beach sand is fine enough to hold water by capillary action during periods of low tide. Any organism burrowing into the sand will be constantly moist.

A final physical factor acting on beaches is oxygen content. Oxygen is never limiting in the water bathing the beach, because the turbulence of wave action ensures constant saturation. Where it might become limiting is in the substrate itself. The water held in the beach or sand flat mitigates changes in temperature and salinity and also contains oxygen, which is available to the organisms inhabiting the area. This supply of oxygen, however, is used up by the respiration of the organisms and must be replenished through an interchange of the water with the sea above. Replenishment depends on the fineness of the sediments: Fine sediments have a slow rate of exchange, coarse sediments a rapid rate. On fine-grained beaches—particularly protected sand flats—water interchange is slow and may result in reduced oxygen or anoxic conditions below the uppermost layers. Counteracting this effect are the various organisms that live in the sediment, such as crabs and worms, which may deepen the oxygenated layers through building and maintaining burrows.

■ ADAPTATIONS OF ORGANISMS

The dominant environmental factor acting on open sand beaches is wave action, which creates the unstable, constantly moving substrate. To inhabit this area, organisms must be adapted to tolerate the instability. There are two routes to adaptation. The first is for the organism to burrow deeply enough into the substrate such that the organism is deeper than the depth of sediment affected by the passing wave. This strategy is employed by many large clams, such as *Tivela stultorum*, the Pismo clam. Deep burrowers are usually also aided by developing a heavy shell, which helps to keep the animal situated in the substrate. They often have long siphons that enable them to burrow deeply. The only difficulty with this type of adaptation is that a severe storm may generate waves large enough to pull these animals out of the sand and deposit them on the beach. That is the reason for the catastrophic destruction that occasionally occurs in these forms.

The second route of adaptation is for the animal to burrow very quickly as soon as the passing wave has removed it from the substrate. This is the more common mechanism and is employed by many annelid worms, small clams, and crustaceans. A good example of this type of adaptation can be found in the various sand crabs of the family Hippidae, which populate many open beaches around the world. These animals have a short body with limbs highly modified to dig quickly into wet sand. As soon as they are freed from the substrate by a passing wave, or as soon as they leave the substrate by moving to the surface, they are able to burrow quickly before

(a)

(b)

(c)

FIGURE 6.42 Some common sandy beach animals. (a) Sand dollars (Dendraster), (b) a moon snail (Polinices), and (c) a ghost crab (Ocypode). (a, © Marty Snyderman / Visuals Unlimited. b, Photo by Constantinos Petrinos/Nature Picture Library. c, Copyright © De la Harpe, Roger / Animals.)

the water motion can carry them offshore. Small clams of the genus *Donax* also do the same thing, and razor clams of the genera *Siliqua* and *Ensis* are extraordinarily fast at burrowing.

Other adaptations are correlated with the burrowing of these organisms. For example, some mollusks, such as *Olivella biplicata*, have very smooth shells that reduce the animal's resistance in burrowing, while others, as Stanley (1969) has noted, have developed special ridges on the shell that grip the sediment and aid in penetrating the substrate. Similarly, some shore echinoderms, such as sand dollars, have much-reduced spines that allow them to burrow into the sand.

A special adaptation observed by Chia (1973) in small sand dollars (*Dendraster excentricus*) (Figure 6.42a) is that they accumulate iron compounds in a special area of their digestive tracts. The iron serves as a "weight belt" that keeps the animals securely fastened to the substrate in the presence of wave action.

Reproductive problems facing sandy-beach organisms include when to reproduce and whether to reproduce repeatedly (iteroparously) or only once (semelparously). A critical factor here is to coordinate gamete production or the release of young with the tidal cycle so that the young are not stranded by the tide or consumed by predators. To this end, most sandy-beach animals display lunar rhythms of reproductive activity. Many are also iteroparous, although at present we cannot demonstrate that the semelparous forms are at any particular disadvantage in this kind of environment. The mode of development has a latitudinal gradient, with the tropical species having a majority of planktotrophic larvae and the temperate-zone species having more lecithotrophic larvae.

Sandy beach macrofaunal animals are vulnerable to predation, and many show adaptations to avoid being preyed upon. Chief among these is the ability to burrow deeply, as is exhibited in the bivalves *Donax* and *Tellina* and the gastropod *Bullia*. Also, the tidal migrations of *Bullia* and the hippid crabs, tending to keep the animals in the swash zone, may reduce predation by birds or fishes, which usually cannot feed in that area. Crustaceans of the upper beach areas often avoid diurnal predators by residing in deep burrows during the day or by hiding under piles of wrack (Figure 6.43).

A final set of adaptations has to do with preventing clogging of respiratory surfaces by the resuspended sand: The intake siphons of sandy-beach clams are often fitted with various screens that prevent the entrance of sand, but allow the passage of water. Similarly, in sand crabs, the antennae, when held together, form a tube to the surface through which water enters

FIGURE 6.43 Endogenous activity rhythms of the semiterrestrial isopod *Tylos*. The animal is active only during low tide at night, when it emerges from its burrow to roam the shore in search of the kelp on which it feeds. (After "Sandy Beach Ecology," by A. McLachlan, in *Sandy Beaches as Ecoystems*, edited by A. McLachlan & T. Erasmus, pp. 321–360. Copyright © 1983 Dr. W. Junk, with kind permission from Kluwer Academic Publishers.)

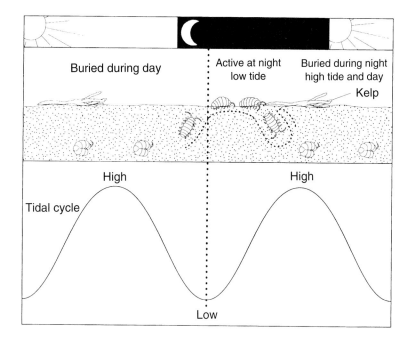

the branchial chamber. The antennae are densely clothed with closely spaced hairs that prevent the entrance of sand.

■ TYPES OF ORGANISMS

The organisms most conspicuously absent from sand beaches are large plants. No macroscopic flowering plants or large algae occur on open sand beaches, presumably because there is no way for them to attach and maintain themselves in the face of wave action. Perennial macroscopic plants, primarily seagrasses, are common in the lowest intertidal areas of many protected sand flats of the North American Atlantic coast (see the discussion of seagrass beds in Chapter 5), as are certain ephemeral algae, such as *Ulva* or *Enteromorpha*, which may be seasonally abundant. On open sand beaches the only primary producers are certain benthic diatoms and various phytoplankton living in the surf. These microfloral organisms often exhibit vertical migration either within the sediments or between the sediment and the water column. Migrations may lead to concentrations of these autotrophs into highly productive patches. Protected sand flats, however, support a large and diverse microflora of benthic diatoms, dinoflagellates, and cyanobacteria, all of which together typically appear as a brownish or greenish film on the sediment surface.

The second major group of organisms that are conspicuous by their absence is the sessile animals, such as barnacles and mussels, so dominant on rocky shores. Again, there is no place for them to attach. The sand beaches and intertidal flats are dominated by representatives from three invertebrate groups: polychaete worms, bivalve mollusks, and crustaceans. Various combinations of species from these three groups dominate sand beaches and protected sand flats throughout the world. There is a tendency for crustaceans to be more abundant on exposed and tropical beaches and bivalves to be more abundant on protected and temperate shores. The total number of macrofaunal species of all groups increases with decreasing wave exposure, but the highest biomass comes from exposed beaches.

■ FEEDING BIOLOGY

The absence of large multicellular plants and algal mats means that there is very little primary productivity on open sand beaches. Although diatoms are present, the opaqueness of the sand ensures that the diatom population is restricted to the surface layers. As a result, there are no macroscopic herbivores on a sand beach. Since there is virtually no primary productivity, the animals living on the beach must depend for food upon the phytoplankton carried in seawater above the beach, the organic debris brought in by the waves, or other beach animals. These food sources dictate a benthic invertebrate association on open sand beaches dominated by filter feeders, detritus feeders, and scavengers. On protected sand flats, productivity is confined to the microfloral films, seagrasses, and ephemeral macroalgae, but even the last do not appear to be grazed extensively, leaving 90–95% of these macrophytes to be broken down into detritus. There are relatively few resident carnivores among open-beach animals, because to be a carnivore would require active movement across the substrate in search of prey, a strategy that is probably not feasible in the face of the heavy wave action. In addition, the erratic nature and the unpredictability of large food items, either living or dead, mean that specialized feeders are rare, and most carnivores and scavengers are opportunists.

Suspension-feeding animals filter particles out of the water column. These particles may be plankton, or, particularly on the sand beach, they may be various organic particles resuspended from the bottom by the passing waves. The dominant group of suspension feeders on sand beaches is the bivalve mollusks, such as razor clams (*Siliqua*, *Ensis*), surf clams (*Tivela*, *Spisula*), and coquinas (*Donax*). On protected sand flats, suspension feeders include small bi-valves (*Gemma*) and amphipods (Haustoriidae; Figure 6.44). Filter-feeding bivalves in sand beaches and flats not only transfer a significant portion of the primary productivity in the water column to the benthic organisms wherever they are abundant, but also affect the clarity of the water. The importance of this latter function to benthic communities was demonstrated by Ulanowicz and Tuttle (1992), who reported that the loss of the extensive oyster beds and their water filtering in Chesapeake Bay may have led to decreased water clarity and decreased productivity by benthic algae and seagrasses.

Open sand beaches tend to have less organic detritus than protected sand flats have, but enough debris from various sources finds its way to the beach to be a reliable source of food for certain organisms. This detrital material is often carried up and down the beach, suspended in the wave wash, rather than being

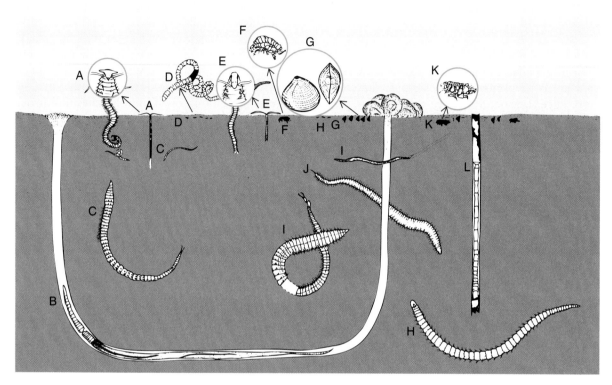

FIGURE 6.44 Some representative New England sand-flat benthic invertebrates, indicating general life habits. (Not to scale.) Surface deposit feeders: (A) *Spiophanes bombyx* (spionid polychaete); (B) *Saccoglossus kowalewskyi* (protochordate); (E) *Pygospio elegans* (spionid polychaete). Burrowing deposit feeders: (C) *Aricidea* sp. (paraonid polychaete); (D) oligochaete; (H) *Exogone hebes* (syllid polychaete); (I) *Scoloplos* spp. (orbiniid polychaete); (J) *Nephtys* spp. (nephtyid polychaete). Suspension feeders: (G) *Gemma gemma* (venerid bivalve); (F) *Protohaustorius deichmannae* (haustorid amphipod); (K) *Acanthohaustorius millsi* (haustorid amphipod). Deep deposit feeder: (L) *Clymenella torquata* (malkanid polychaete). (From R. Whitlatch, 1982, *The Ecology of New England Tidal Flats: A Community Profile*, U.S. Fish and Wildlife Service, Biological Services Program.)

deposited on the bottom. Therefore, the mechanisms employed by detritus feeders on the open beach frequently are different from those used by detritus feeders on protected shores.

The sand crabs of the family Hippidae, such as *Emerita analoga* and *Blepharipoda occidentalis* of the Pacific coast of America, employ a unique mechanism to trap debris from the wave wash. (See Figure 6.42.) MacGinitie and MacGinitie (1949) observed that *Emerita* orient themselves so that their heads point shoreward. They remain completely buried as the incoming wave passes over them shoreward. After the wave's passage, as the water begins to run

down the beach, *Emerita* stick out their very large second antennae, the thickly clothed hairs of which form a net that, when spread into the backwash, intercepts all particles within a certain size range. The particles are then wiped off by the animal's mouth and are ingested. As the tide rises and falls, the position of the wave-wash front also moves. These crabs take advantage of the moving front to maximize their feeding time by being shifted up and down the beach with the tide. The animals surface as a wave rolls in, are carried shoreward some distance by the wave, and then quickly reburrow into the sand. Then, as the tide recedes, the crabs emerge and roll down the beach before reburrowing and commencing feeding.

A similar method of feeding is employed by the gastropod mollusk *Olivella columellaris* on the Pacific coast of Central America. The animal produces a mucous net that it holds across the wave backwash. The net filters and captures small particles, and the mollusk feeds by consuming the entire net with the particles it has caught. Thus, it must produce a new net at frequent intervals.

Most sand beaches also include one or more polychaetes. Much like earthworms on land, they burrow through the sand, ingesting it and digesting the organic particles. However, these animals are rarely abundant on open sand beaches, because the amount of organic material in the sediment is much less than in protected shores.

Sand dollars, one of the echinoderms that are common on sandy beaches, are generally detritus-feeding animals. These animals usually burrow through the sand, and as they do so, the finer organic particles (but not the sand grains) fall down between the short spines, are trapped in mucus, and are carried to the mouth. Some species, such as the Pacific coast *Dendraster excentricus*, according to Timko (1976), have even modified this basic feeding process to remove suspended particles by projecting part of the body up into the water column.

On protected sand flats, detritus feeders are more abundant in terms of numbers of both species and individuals. These deposit feeders play an important role by converting detrital organic material into macrofaunal biomass, which is subsequently available to higher trophic levels. Because of their abundance, size, and feeding mechanisms, deposit feeders are also significant bioturbators, disturbing the upper layers of the substrate and thus influencing the nature of the sediments. By far the dominant group is the polychaete worms. Two types of deposit feeders are recognized: surface deposit feeders and burrowing deposit feeders. (See Figure 6.44.) Most deposit feeders are generalized feeders whose diet is limited by the size of the particle they can ingest. Surface deposit feeders may live in vertical tubes (spionid and terebellid polychaetes) or burrow into the surface layer (amphipods). Still others, such as the mud snails *Ilyanassa* and *Hydrobia* and the fiddler crabs, collect particles from the surface and, according to Levinton and Lopez (1977), may affect primary productivity by consuming the small surface-dwelling microalgae. Burrowing deposit feeders consist primarily of polychaete worms.

Resident carnivores are rare, particularly on open sand beaches, but the few that are found usually are polychaete worms, large crabs, or gastropod mollusks. The polychaetes, such as species of the genera *Nephtys* and *Glycera*, burrow through the sand and seize small prey with their eversible proboscis. The carnivorous gastropods are represented most commonly by burrowing snails of the families Naticidae (moon snails) and Nassariidae (dog whelks). Moon snails burrow slowly just below the surface layers of the sand in search of bivalve mollusks, which they usually consume by boring holes through the shell and draining out and digesting the contents. Nassariids of the genus *Bullia* are probably the best known of the larger open-sand gastropods. They are common on beaches all around Africa and as far east as India. Extreme opportunists, Nassariids consume virtually all animal material available, live or dead. They detect food with their chemical sense organ, the osphradium, and move quickly to it by crawling or surfing. Once on the food, they attach via their proboscis or foot, to avoid being separated from it by the continual movement of the food caused by the swash. On protected sand flats, epifaunal carnivorous crabs such as *Callinectes* and *Carcinus*, various snails such as *Urosalpinx* and *Busycon*, and certain polychaetes are common predators of various infaunal invertebrates. On the Atlantic coast of North America, the horseshoe crab, *Limulus*, may be a seasonally abundant predator. A number of transient predators also occur: At low tide, various shorebirds actively feed on beach organisms, and at high tide, fishes invade the area. The rays, skates, and flatfishes are the most significant predators of the benthic invertebrate communities (Figure 6.45).

■ COMMUNITY ORGANIZATION ON EXPOSED SAND BEACHES

Since few organisms are visible on the surface of the intertidal sand beaches, these areas do not display the surface zonation patterns that are so obvious on rocky shores. Furthermore, since such physical factors as desiccation and temperature limit organisms on rocky shores, but are far less important to organisms living in the sand, it might be suspected that the zonation of organisms on exposed sand beaches would be

FIGURE 6.45
Generalized food web for a
sand beach in California.

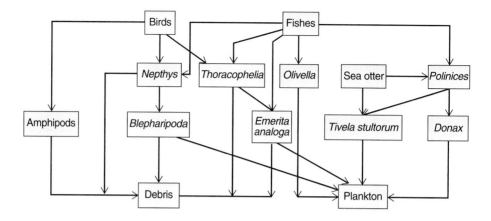

reduced or absent. It turns out, however, that there is a zonation of sandy shore organisms, but it is neither as clearly defined nor as well understood as that of rocky shores. The lack of clear zonation patterns is due partly to the habit of some dominant organisms of moving up and down the beach for feeding and partly to the dearth of studies on what determines distribution patterns in this kind of environment. One of the major contributors to the lack of understanding is researchers' inability to set up field experiments that exclude certain organisms from open sand beaches. Wave action precludes such work.

From the few data available, it is possible to establish a three-zone division of most sandy beaches that closely follows the universal scheme proposed by the Stephensons for rocky shores. The highest parts of the sandy beach, corresponding to the supralittoral fringe, are usually inhabited by talitrid amphipod crustaceans (beach hoppers) in the temperate zone and by the fast-moving ghost crabs (*Ocypode*) in the warm temperate and tropical zones. Both groups excavate burrows and are scavengers. The broad midlittoral area, which corresponds to the area inhabited by barnacles and mussels on rocky shores, is much more variable. One group found here is isopods of the family Cirolanidae. Also coming into this zone when it gets washed by waves are the various sand crabs and other animals that feed in a similar way. (See previous section.) The lowest zone—the infralittoral fringe or surf zone—is inhabited by the greatest number of species, including the large surf clams (*Tivela, Spisula*), sand dollars (*Dendraster*), various polychaete worms, a host of crustaceans, and the larger carnivorous snails (*Natica, Polinices, Bullia*). In the tropics, there is some tendency for this

lowest zone to be occupied by hippid crabs, while in temperate climes, the zone is inhabited by haustorid and other amphipods (Figures 6.46, 6.47, and 6.48).

In contrast to rocky shores where the fauna is sedentary or fixed, open sandy beaches are characterized by tidal migrations of many of the macrofauna. Any zonation pattern established by a researcher, therefore, reflects the faunal distribution at the time the samples were taken. Most macrofaunal migrations are simple movements of the animals that maintain them in the swash zone, where feeding conditions are good and predation pressure is low. These tidal migrations compress the zonation pattern at high tide, when macrofaunal organisms are concentrated, and spread them at low tide.

In addition to tidal changes in faunal composition on open sand beaches, there are seasonal changes. The most dramatic of these changes are seen on the beaches of India where Ansell et al. (1972) and McLusky et al. (1975) have shown that most of the fauna disappears from the intertidal zone during the monsoon. Another seasonal change, occurring mainly on temperate-zone beaches, is onshore and offshore movements in winter, which may be an adaptation to avoid winter storms. Long-term changes may also occur and may be dramatic—for example, periodic mass mortalities.

In direct contrast to experimental work on rocky shores, which has demonstrated the importance of physical factors, competition, and predation in determining zonation patterns, the explanations for the establishment of the pattern on open sandy beaches have not been subjected to rigorous experimental analysis. As a result, we know little of the factors or the interactions among them that establish the sandy beach distribution patterns.

FIGURE 6.46 Generalized pattern of distribution of organisms on sandy beaches.

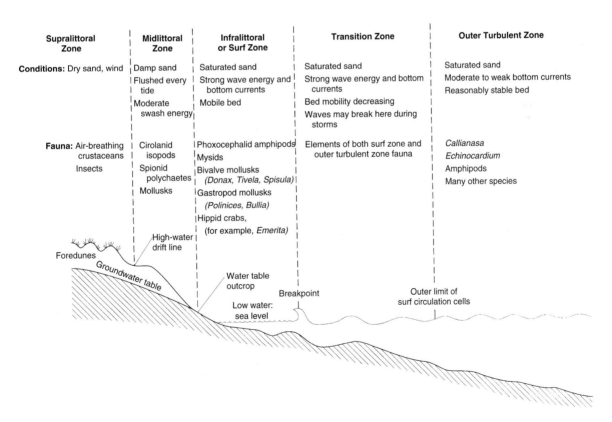

Temperate | Tropical

Supralittoral fringe

Midlittoral zone

Infralittoral fringe

Moon snail

Large clam

Sand dollar

Cirolanid isopods

Talitrid amphipods

Ghost crab

Supralittoral Zone	Midlittoral Zone	Infralittoral or Surf Zone	Transition Zone	Outer Turbulent Zone
Conditions: Dry sand, wind	Damp sand Flushed every tide Moderate swash energy	Saturated sand Strong wave energy and bottom currents Mobile bed	Saturated sand Strong wave energy and bottom currents Bed mobility decreasing Waves may break here during storms	Saturated sand Moderate to weak bottom currents Reasonably stable bed
Fauna: Air-breathing crustaceans Insects	Cirolanid isopods Spionid polychaetes Mollusks	Phoxocephalid amphipods Mysids Bivalve mollusks (*Donax, Tivela, Spisula*) Gastropod mollusks (*Polinices, Bullia*) Hippid crabs, (for example, *Emerita*)	Elements of both surf zone and outer turbulent zone fauna	*Callianasa* *Echinocardium* Amphipods Many other species

Foredunes

Groundwater table

High-water drift line

Water table outcrop

Low water: sea level

Breakpoint

Outer limit of surf circulation cells

FIGURE 6.47 Generalized scheme of zonation on sandy shores. (After "Sandy Beach Ecology," A. McLachlan, in *Sandy Beaches as Ecosystems*, edited by A. McLachlan & T. Erasmus, pp. 321–380. Copyright © 1983 Dr. W. Junk, with kind permission from Kluwer Academic Publishers.)

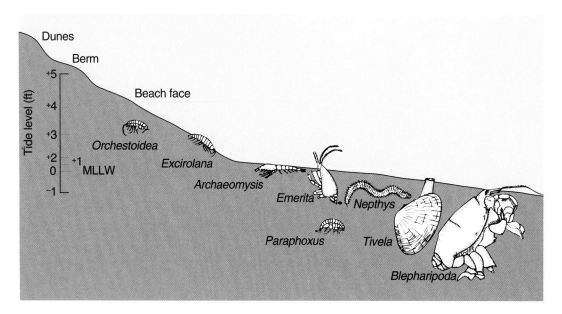

FIGURE 6.48 Zonation of the numerically dominant invertebrate genera on a central California intertidal sandy beach: *Orchestoidea*, *Paraphoxus* (Crustacea, Amphipoda); *Excirolana* (Crustacea, Isopoda); *Archaeomysis* (Crustacea, Mysidacea); *Emerita*, *Blepharipoda* (Crustacea, Decapoda); *Nepthys* (Annelida, Polychaeta); *Tivela* (Mollusca, Bivalvia). (Not to scale.)

The extremely crowded conditions for organisms observed on rocky shores are not found on sandy beaches. Indeed, the fauna is relatively sparse and does not occupy all the available space. Also, the three-dimensional nature of the sedimentary substrate, coupled with the fact that many macrofaunal species burrow into it, minimizes interference competition. Moreover, it is difficult to conceive of organisms overgrowing others in this environment, where animals move freely vertically in the substrate, as well as horizontally on the surface. This concatenation of conditions immediately suggests that competition for space is not a major contributor to the observed distribution pattern. Competition for food also appears negligible, since the sparse populations and abundant plankton make competition unlikely. The fact that most sandy beach animals are opportunistic feeders also suggests that competition is not important. Similarly, there are relatively few indigenous invertebrate predators, and it seems unlikely that they are responsible for any major distributional pattern. This observation is supported by the fact that, on the few occasions that predators have been experimentally excluded from sediments, diversity has increased, not decreased, as on rocky shores. The effect, however, of large vertebrate predators, such as birds and fishes, remains to be investigated.

Little is known of the infralittoral, or surf-zone, fish fauna and its potential effects on open-beach community structure. Nor are things likely to change in this regard, due to the danger and difficulty of sampling in the surf zone. What we do know is that the fish assemblages are dynamic, have few resident species, and are variable in time and space. Most species are opportunistic feeders. These generalities were confirmed in a four-year study by Gibson et al. (1993) on the fish fauna of a Scottish sandy beach. The researchers found a clear seasonal cycle in both abundance and species composition. Very few species were year-round residents, and most were present only from spring to fall. There was also a marked annual cycle of species numbers and abundance. Recruitment of young produced during the year appeared to be the main reason for the increase in both number of species and numbers of individuals in the spring. Seasonal and annual changes in these parameters resulted from subsequent mortality of these recruits rather than either immigration or emigration by adults.

A similar situation prevails for birds. Here, again, our lack of knowledge stems primarily from a lack of direct studies and experimental manipulations. Indirect work, particularly in South Africa, suggests, however, that birds may be significant predators, removing 10–49% of the macrobenthic invertebrate production.

In the absence of indigenous large plants, open sand beaches have food sources derived mainly from the surrounding sea. The major determinants of the amount of food on the beach are the proximity and size of the food source and the characteristics of the surf zone delivering the food to the beach. The primary types of food available on open sand beaches include particulate organic matter, carrion, large dead algae, dissolved organic material, plankton, and resident benthic microflora. Open-sand macrofaunal food chains generally include filter feeders, detritus feeders, scavengers, and carnivores. Filter feeders are more abundant on open sand beaches, detritus feeders on the more protected beaches. Scavengers and carnivores are rarely abundant. Fishes, birds, and certain large crustaceans dominate the top of open-sand food chains (Figure 6.49). In general, open sand beaches have a gradient in which terrestrial predators (mainly birds) dominate the upper shore zones and marine predators (fish) the lower shores. These distributions are summarized in Table 6.1.

■ COMMUNITY ORGANIZATION ON PROTECTED SAND FLATS

In contrast to the situation on open sand beaches, protected sand flats are accessible for ecological investigations and experimental manipulations. As a result, we know more about the structuring of the communities of protected sand flats. Both physical and biological factors are important.

Physical factors play a significant role in the organization of sand-flat communities in a number of ways. First, the grain size of the sand may set limits on the organisms that can inhabit the area.

Communities dominated by suspension feeders and those dominated by deposit feeders tend to be separated spatially due to trophic group amensalism (see p. 204). Also, deposit feeders tend to be more abundant in fine sediments, where the particles have more organic material and more nutrients for them, whereas suspension feeders are most abundant in clean coarse sand. These patterns seem to hold true for the intertidal sand as well as for the subtidal sand, as described in Chapter 5, but it is well to remember that many exceptions to this hypothesis exist.

Other direct effects on sand flats are the result of seasonal weather changes. For example, in the New England intertidal sand flats, a number of dominant epifaunal invertebrates are affected by seasonal changes in the water temperature. Whitlatch (1982) reports that, as the water temperature declines in the fall, many crustacean species migrate to deeper waters, where they become inactive over winter. The same is true for the dominant gastropod, the mud snail *Ilyanassa obsoleta*. Since *Ilyanassa* has an impact on algal mats, the stability of the sediment, and, subsequently, the abundance of infauna, this seasonal migration can have significant effects on community structure. (See Chapter 5, pp. 202–204.) In addition, particularly in high latitudes, the formation of sea ice and its scouring effect on the intertidal flats create changes in the community. (See Chapter 5, pp. 243–248, on polar communities.)

The prime biological factors that structure sand-flat communities are predation, competition, and disturbance. Current information suggests that, of the three, either predation or predation combined with disturbance is especially significant. A number of sand-flat epifaunal predators may have significant effects on the community. For example, on the U.S. east coast, the moon snail *Polinices duplicatus* is an active predator of bivalves, particularly the soft-shell clam *Mya arenaria*. Wiltse (1980) has demonstrated that when *Polinices* is excluded from areas, not only does the clam population rise, but so do the populations of other infauna. Apparently, this snail affects community structure not only by feeding on clams, but also by disturbing the rest of the infauna as it plows through the substrate. Similar findings have been reported by Peterson and Peterson (1979), who caged out predators in North Carolina tide flats. Another group of predators on the North American east coast consists of various large crabs, such as the green crab, *Carcinus maenas*, in New England and the blue crab, *Callinectes sapidus*, in the Middle Atlantic states. Virnstein (1977) has demonstrated that *Callinectes* is a voracious predator that digs in the sediment to locate its prey; exclusion of the crab results in a great increase in infaunal density. In Chesapeake Bay, clams recruit to all shallow-water soft-substrate habitats every spring in extremely high densities of thousands of recruits per square meter. By the end of each summer, however, blue crab predation eradicates all clams from all areas except those with coarse sand or sea grass, which, according to Lipcius and Hines (1986),

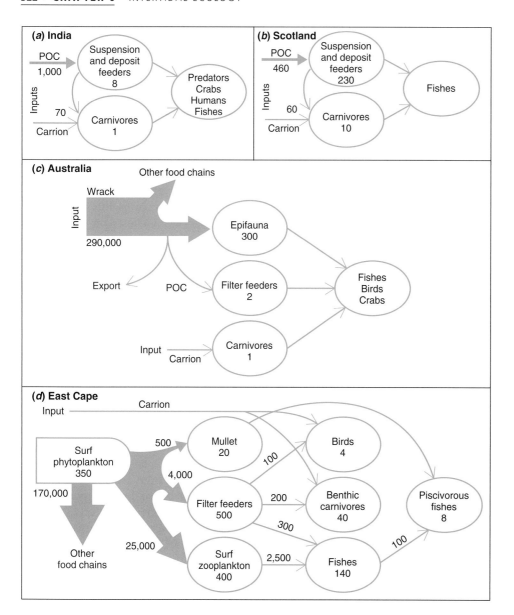

FIGURE 6.49 Macroscopic food chains and carbon flow in several different open sandy-beach systems. (a) India. This is an example of a low-energy, mostly reflective beach with negligible surf and no drift line. Macrofauna are dominated by particulate feeders. Crabs, fish, and humans are the main predators. (b) Scotland. An example of a low-energy dissipative beach. The macrofauna are dominated by particulate feeders, but the dominant predators are fishes, particularly juvenile flatfishes. (c) Western Australia. Example of a low-energy reflective beach without a surf zone, but with large inputs of dead algae (wrack). The fauna here consists of large numbers of surface dwellers that consume wrack, small populations of hippid crabs and donacid bivalves in the midlittoral zone, and oxypodid crabs at the drift line. Predators are few. (d) South Africa, East Cape. Example of a high-energy beach tending toward the dissipative type and characterized by large numbers of surf diatoms supporting a rich fauna and more complex food chains. (All values are given in grams of carbon per meter squared per year.) (Modified from *Ecology of the Sandy Shores*, A. C. Brown and A. McLachlan. Copyright © 1990 with permission from Elsevier Science.)

TABLE 6.1

Major Beach Types by Climate Zone, Conditions, and Macrofaunal Composition

TYPE	LATITUDE	PREDOMINANT SOURCES	MACROFAUNAL BIOMASS	DOMINANT TROPIC GROUP	PREDATORS
High turnover, particulate-based food chain	Tropical	Particulates	Low to moderate	Filter feeders	Invertebrates, fishes
Low turnover, particulate-based food chain	Temperate	Particulates	Low to high	Filter feeders	Fishes, birds invertebrates
Macrodebris-based food chain	Temperate	Stranded macrophytes	Moderate	Scavengers, grazers	Birds
Sediment-based food chains in sheltered beaches	Temperate or tropical	Deposited detritus microorganisms and benthic microflora	Low to high	Deposit feeders	Birds, fishes, invertebrates
Carrion-based food chains in extremely exposed steep beaches	Temperate or tropical	Stranded carrion	Low	Scavengers, predators	Birds, crabs

Source: From "Sandy Beach Ecology" A. McLachlan in *Sandy Beaches as Ecosystems,* pp. 321–360, 1983. Copyright © 1983 Dr. W. Junk. Reprinted by permission.

limit the foraging ability of the crabs. A seasonal predator on eastern seaboard tidal flats is the horseshoe crab *Limulus polyphemus,* which digs distinctive pits when searching for infaunal prey. Woodin (1978) has demonstrated that *Limulus* also significantly reduces other infaunal organisms.

Other large east-coast invertebrate predators include several species of whelks (*Busycon* spp.) and tulip shells (*Fasciolaria*). The whelks, in particular, are devastating to the sand-flat clam populations because they can consume all sizes. Unlike the barnacles and some mussels of the rocky intertidal zone, the clams have no refuge in size, but do find refuge living in the dense root mats of sea grasses.

Despite the massive effects of the larger predators, there are areas that are refuges from predation and disturbance by these animals. Woodin (1978), for example, has demonstrated that, by forming upright tubes in the sand, the polychaete worm *Diopatra cuprea* of the eastern United States intertidal sand flats effectively deters both *Limulus* and *Callinectes* to the extent that the infaunal abundances are greater in and around *Diopatra* tubes. All of the above strongly suggest that predation, together with its accompanying disturbance, is the main biological structuring force in intertidal sand flats.

Another significant biological process determining community structure is, surprisingly, deposit feeding.

In North Carolina, one of the more abundant deposit feeders is the hemichordate worm *Balanoglossus aurantiacus.* This animal, which Peterson and Peterson (1979) have called a "funnel feeder," digs a U-shaped burrow and then collapses the head end. Thus, sediment continues to fall down to form a funnel-shaped depression directed at the head of the worm. The worm then ingests the sediment, consuming the smaller infaunal organisms and causing the death of those it does not digest. The densities of these large worms and their continual reworking of the sand mean that they keep the numbers of infaunal organisms low, as well as exclude potential competitors. In North Carolina, the effect of *Balanoglossus* is heightened by the presence of yet another sediment processor—the sea cucumber *Leptosynapta tenuis,* which operates at the sediment surface.

Adult–larval interactions probably also have an effect on the composition of sand-flat communities. These interactions should probably be categorized as predation because the interaction consists of deposit or suspension feeders ingesting the settling or newly settled larvae. Woodin (1976) suggests that when adults are in dense assemblages, such interactions could be important, but we currently lack any experimental studies to document this hypothesis.

The role of competition in structuring sand-flat communities is different from that observed on rocky shores. The three-dimensional nature of the protected sand shores, the dynamic relationship between the substrate and the animals living therein, and the interactions between juveniles and adults all tend to reduce the severity of direct competition. The best examples of such competition in sand flats involve mobile animals that live on the sediment surface. Levinton et al. (1985) have demonstrated that competition between two mud snails on Long Island Sound determines their intertidal distribution there. *Hydrobia totteni*, which occupies the whole intertidal zone in Europe, is confined to the high intertidal zone on Long Island Sound because it is an inferior competitor to the native American mud snail *Ilyanassa obsoleta*. *Hydrobia* avoids *Ilyanassa* by living in the high intertidal zone, which *Ilyanassa* does not inhabit. A similar competitive interaction involves *Ilyanassa* and the introduced European snail *Littorina littorea*. According to Brenchley and Carlton (1983), *Ilyanassa* avoids contact with *Littorina* because *Littorina* destroys its egg capsules; this leads to spatial separation of the two.

Although less common, competitive interactions leading to different depth distributions do occur in protected sand beaches. One good example comes from South Carolina, where two amphipod species are found at different depths. *Acanthohaustorius* lives in the surface oxidized layer of sediment, while *Pseudohaustorius* lives in deeper, anoxic sediments. Through a series of experimental manipulations, Grant (1981) demonstrated that both species prefer to occupy the oxidized upper layer, but that the presence of the superior competitor *Acanthohaustorius* displaces *Pseudohaustorius* to the deeper sediments (Figure 6.50). Although this example of competition leads to physical depth displacement of one of the species resulting from competition for space, competition among the more sedentary infaunal organisms, such as bivalve mollusks, is usually reflected in reduced growth rates and diminished fecundity rather than displacement or death, suggesting that competition for space is not a significant factor in the distribution of these animals.

In direct contrast to the results of removing predators in the rocky intertidal zone, removing predators in soft substrates does not lead to increased competition and competitive exclusion. This is not to say that

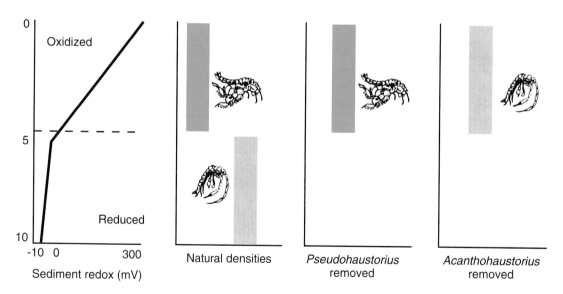

FIGURE 6.50 Competitive depth displacement of two species of amphipods under natural conditions and with each species separately removed. (Modified from *The Ecology of Atlantic Shorelines*, M. D. Bertness, p. 281. Copyright © 1999 Sinauer Associates, Inc. Reprinted by permission.)

competition does not occur, for competition for space in sediments has been found by Levinton (1977) and Peterson (1977). The competition, however, generally results in lower growth rates, not mortality, and so does not seem to be a structuring force that determines the makeup of a community.

Another interesting contrast with rocky intertidal areas is provided by Peterson (1991), who notes that the major predators on rocky shores are the sedentary, slow-moving starfish and mollusks that are limited to the lower parts of the intertidal zone by their lack of mobility and need for coverage by water. In contrast, protected sand flats and mudflats are dominated by highly mobile predators, such as crabs, fishes, and birds, that have feeding times less affected by the tides and that are capable of ranging throughout the intertidal zone.

We can summarize this section by stating that the macrofauna of intertidal sand flats displays high temporal and spatial variability. Indeed, the total numbers of species and individuals may vary by several orders of magnitude within and between years. This high variability is due to the high predation pressure and biological disturbance that control the infaunal communities composed largely of weedy, opportunistic species.

Muddy Shores

Sharp boundaries exist between rocky shores and sandy shores, making them easily defined and recognizable. Boundaries between sand flats and muddy shores are not as easily defined. Indeed, it is not even possible to draw sharp boundaries between the two, because, as shores become more protected from wave action, they become finer grained and accumulate more organic matter; thus, they become more "muddy." Sand and mud shores are, therefore, the opposite ends of a continuum, with the sand beaches having larger grain sizes and the muddy areas having the finest grain sizes.

Since sharp boundaries do not exist between sandy and muddy shores, and since one blends into the other along a gradient of increasing protection from wave action, the fauna and flora of muddy shores also show a change from organisms typical of open sand beaches to those typical of muddy shores along the same gradient.

This section will discuss the features characteristic of typical muddy shores—that is, those at the opposite end of the continuum from the open sand beaches. It should be borne in mind, however, that various transitional communities exist that will show mixtures of organisms from both extremes. Since muddy shores are also composed of sediments, similar factors apply as they did with protected sandy shores and will not be repeated.

Muddy substrates are also characteristic of estuaries and salt marshes (Chapter 8). Most of the muddy shores of the world are associated with estuaries and similar embayments; hence, many of the types of organisms and adaptations are similar between estuaries and salt marshes. To avoid the problem of overlap in discussing these two areas, this section will treat the infaunal organisms of mudflats, as well as their adaptations, feeding biology, and trophic structure. The same considerations also will apply to the estuarine infauna of Chapter 8. The role of other organisms, productivity, and ecological relationships are discussed in that chapter as well.

■ PHYSICAL FACTORS

The major contrast with open sand beaches is that muddy shores cannot develop in the presence of significant wave action. Therefore, muddy shores are restricted to intertidal areas that are completely protected from open-ocean wave activity. Muddy shores are best developed where there is a source of fine-grained sediment particles and are located in various partially enclosed bays, lagoons, harbors, and especially estuaries. (See Chapter 8.)

Since these areas develop where water movement is minimal, the slopes of muddy shores are much flatter than those of sandy beaches. This is why muddy shores are often referred to as mudflats. Muddy shores are more stable than sand substrates and more conducive to the establishment of permanent burrows.

The very fine particle size, coupled with the very flat angle of repose of these sediments, means that water in the sediments does not drain away and is held within the substrate. The long retention time for water, combined with a poor interchange of the interstitial water with the seawater above, and a high internal bacterial population usually result in complete depletion of the oxygen in the sediments below the first few centimeters of the surface. Thus, **anaerobic** conditions prevail within the sediment, and this is one of the most important characteristics of the muddy shore that separate it from sand flats.

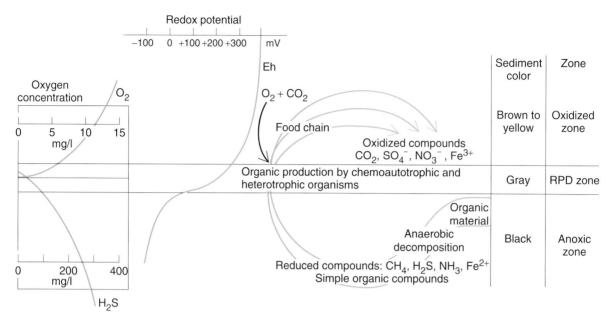

FIGURE 6.51 Diagrammatic representation of the physical and chemical characteristics of sediments across the redox discontinuity layer and the biological processes occurring in each.

Between the upper aerobic layer and the lower anaerobic layer is a transition zone called the **redox potential discontinuity (RPD) layer**. This zone is characterized by a rapid change from a positive redox potential (E_h), as measured by an electrode, to a negative redox potential. Below the RPD zone, sediments are anaerobic, and organic compounds are decomposed by anaerobic bacteria. Above the RPD zone, decomposition is by aerobic means. The RPD layer is usually characterized by the gray color of the sediments, the oxidized layer above is usually brown or yellow, and the anoxic layer is black (Figure 6.51). The RPD zone is significant biologically for several reasons. First, reduced compounds diffuse upward, and as soon as oxygen is available, bacteria oxidize these compounds. The oxidized end products—including CO_2, NO_3, and SO_4—in turn are incorporated into bacterial biomass and form the basis of new food chains. Some compounds, however, diffuse downward below the RPD zone and are used by the anaerobic bacteria. These bacteria, in turn, produce more reduced compounds, which, in their turn, complete the cycle and release phosphate, important for new plant growth. Also present in the RPD zone are the chemoautotrophic bacteria. By oxidizing the reduced compounds and fixing CO_2, these bacteria produce more organic material. It is little wonder that high numbers of organisms inhabit the RPD zone.

Muddy shores tend to accumulate organic material. There is thus an abundant potential food supply for the resident organisms, but the innumerable small organic particles "raining" down on the mud flat also have the potential to clog respiratory surfaces.

■ **ADAPTATIONS OF ORGANISMS**

As we noted regarding sandy shores, the surface of muddy shores is often barren, since few animals inhabit the surface of the mud flat. Most organisms inhabiting these areas are adapted to burrowing into and through soft substrate, or else they inhabit permanent tubes in the substrate, which they are able to form because of the increased stability of the fine sediments. In contrast to the open sand beaches, muddy shores advertise the presence of organisms on the surface by the presence of various holes of differing sizes and shapes.

A second major adaptation concerns the anaerobic conditions in the substrate. If organisms are to survive in the substrate, either they must be adapted to live under anaerobic conditions or they must have some way of bringing the overlying surface water with its oxygen supply down to them. Since most

FIGURE 6.52 Oxygen dissociation curves for the blood of *Arenicola* (A) and *Nephthys* (NV). (Modified from "Observations on the Respiratory Physiology and on the Hemoglobin of the Polychaete Genus *Nephthys* with special reference to *N. hombergi*," J. D. Jones, *Journal of Experimental Biology*, Vol. 32, pp. 110–125, 1954. Reprinted by permission of the Company of Biologists, Ltd.)

multicellular organisms cannot survive without oxygen, the latter is the more common adaptation. It is to obtain this oxygen-rich surface water and food that the various burrows, holes, and tubes appear on the surface of the mudflat.

Although most mudflat organisms are intolerant of completely anaerobic conditions, many have adaptations that permit them to exist at lower oxygen tensions than similar forms that live on sand beaches. The most common adaptation to a low oxygen supply is the development of a carrier (for example, hemoglobin) that will pick up oxygen at concentrations well below that of pigments in other organisms (Figure 6.52). Another adaptation is the use of glycogen stores for anaerobic metabolism during periods when there is little or no oxygen. Since wave action is essentially absent on mudflats, there is no need to develop either rapid burrowing or a heavy body to maintain the organism's position, as is the case on open sandy shores.

■ TYPES OF ORGANISMS

In contrast to sand beaches, muddy shores often develop a substantial growth of various plants. On mudflats, the most abundant plants are diatoms, which live on the muddy surface and often give a brownish color to the surface at low tide. Other plants include large macroalgae, such as species of *Gracilaria* (red algae), *Ulva*, and *Enteromorpha* (green algae). These algae often undergo seasonal cycles of abundance, becoming common in the warmer months and virtually disappearing in colder months.

Other areas, particularly the lowest tidal levels, may be covered with seagrasses, such as the genus *Zostera*. (See Chapter 5.) As a result of the occurrence of these primary producers, there is substantial primary productivity in mud flats.

Mudflats also contain large numbers of bacteria, which feed on the copious amount of organic matter that is present. Bacteria are the only abundant organisms found in the anaerobic layers of the mud shore and constitute a significant biomass. Among the dominant bacteria inhabiting this area are a number capable of using the potential energy of the various reduced chemical compounds that are abundant here. These **chemoautotrophic**, or **sulfur, bacteria** obtain energy through the oxidation of a number of reduced sulfur compounds, such as various sulfides (for example, H_2S). The bacteria are thus primary producers of organic matter, as green plants are in other areas. The bacteria produce organic matter by means of energy obtained from the oxidation of reduced sulfur compounds, whereas plants produce organic matter via energy obtained from sunlight. (See also discussion of these bacteria at hydrothermal vents in Chapter 4.)

Since these autotrophic bacteria are located in the RPD layer of the mud, mudflats are unique among marine environments in that they have two separate layers in which primary productivity occurs: the surface, where diatoms, algae, and marine grasses carry on photosynthesis; and a deeper layer, where bacteria conduct chemosynthesis.

The dominant macrofaunal groups on muddy shores are the same as those encountered on sand beaches and sand flats, namely, polychaete worms, bivalve mollusks, and small and large crustaceans, but of different sorts.

■ FEEDING BIOLOGY AND TROPHIC STRUCTURE

Because of the greater amount of organic matter present in and on muddy shores, and because of increased productivity due to both bacteria and plants, there is vastly more food available on muddy shores than on sand beaches. This bounteousness permits more large organisms to live on mud shores, and, indeed, mudflats may be very densely populated.

The dominant feeding types on mudflats are deposit feeders and suspension feeders, a situation somewhat different from that on open sand beaches. Deposit feeders are particularly abundant due to the large amount of organic material and the large populations of bacteria in the sediments. Deposit-feeding polychaetes are represented by such genera as *Arenicola* and *Capitella*. Both feed by burrowing through the substrate, ingesting it, and digesting the organic matter contained within, including bacteria. The undigested material is passed out of the animal through the anus. *Arenicola* spp., however, form a U-shaped burrow in which one arm of the U is a permanently open shaft to the surface, while the other arm is filled with sediment that the worm ingests (Figure 6.53). The worm consumes sediment from the filled shaft, passes the sediment through its gut, and then moves up in the open burrow to defecate at the surface, leaving characteristic mounds. Capitellids, by contrast, form no permanent tubes, but move like aquatic earthworms through the surface layers of the substrate, ingesting it as they proceed. Other deposit feeders include the surface-feeding terebellid polychaete worms and the burrowing hemichordates.

Deposit-feeding bivalves are also common in these areas. Temperate-zone mudflats often have large numbers of small tellinid clams of the genus *Macoma* or *Scrobicularia*. These bivalves have siphons that are separated from each other. Each clam lies buried in the substrate with its siphons extended to the surface. There, the intake siphon is moved over the surface like a vacuum cleaner, ingesting organic particles and bringing them down to the clam for digestion (Figure 6.54).

(a)

(b)

FIGURE 6.54 (a) *Macoma nasuta* in the substrate. (b) *Macoma nasuta* feeding with its in-current siphon. (Modified from "Life in Sandy Shores," A. E. Brafield, *Studies in Biology*, No. 89, 1978. Reprinted with the permission of Cambridge University Press.)

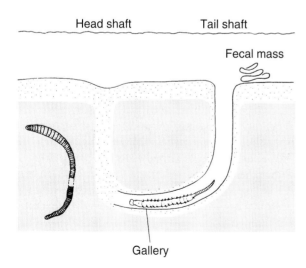

FIGURE 6.53 Two common polychaete worms of mudflats: *Capitella* (left), burrowing through the substrate, and *Arenicola* (right), in its U-shaped burrow.

Suspension feeders include various other species of clams, some crustaceans, and numbers of polychaetes. Suspension-feeding mechanisms are no different from those observed in other areas. There is, however, some question as to how exclusively these forms are feeding on suspended plankton. The fine-sediment particles are easily stirred up on mudflats and remain resuspended in the water. Suspension feeders may take in various amounts of these particles in addition to plankton organisms. Most of the suspension feeders on mudflats also partially take in resuspended sediments; hence, they are actually feeding on both deposited and suspended material. In general, deposit feeders are more common in fine-grained sediments, and suspension feeders become more abundant in coarser sediments, where there is little organic matter (Figure 6.55).

The major carnivores on mud shores are often the fishes, which feed when the tide is in, and the birds,

which feed when the tide is out (Figure 6.56). Indigenous mudflat predators include a few polychaete worms (*Glycera* spp.), moon snails (*Polinices, Natica*), nemertean worms, and crabs.

Despite the relative abundance of plant material on muddy shores, there are very few herbivores. Most plant material finds its way into the food chain only after it has been broken down into small pieces to enter the deposit-feeding food web as detritus. Thus, the trophic structure of a mudflat is often built up from two bases: a detritus–bacteria base and an autotrophic base. The detritus-base web is derived from plants and other organic sources and includes the bacteria that live on the detrital particles. The particles can either be taken in directly by the large macrofaunal invertebrates or be consumed by bacteria. An additional bacterial component at the base of this trophic pyramid is the sulfur bacteria.

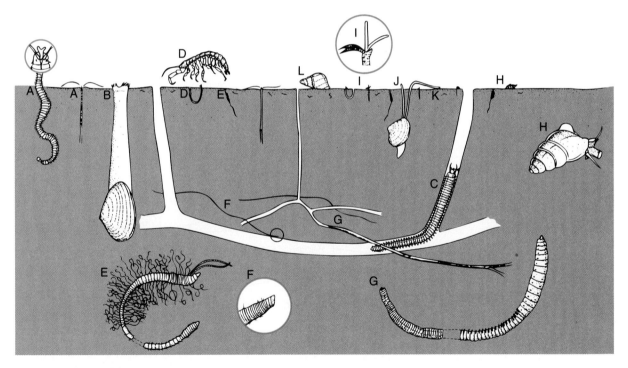

FIGURE 6.55 Some representative New England mudflat benthic invertebrates. Suspension feeder: (B) *Mya arenaria* (soft-shelled clam). Surface deposit feeders: (A) *Polydora ligni* (spionid polychaete); (D) *Corophium* spp. (gammaridean amphipod); (H) *Hydrobia totteni* (hydrobid gastropod); (I) *Streblospio benedicti* (spionid polychaete); (J) *Macoma balthica* (bivalve); (L) *Ilyanassa obsoleta* (mud snail). Burrowing omnivore: (C) *Nereis virens* (nereid polychaete). Burrowing deposit feeders: (E) *Tharyx* sp. (cirratulid polychaete); (F) *Lumbrineris tenuis* (lumbrinerid polychaete); (G) *Heteromastus filiformis* (capitellid polychaete); (K) oligochaete. (From R. Whitlatch, 1982, *The Ecology of New England Tidal Flats: A Community Profile*, U.S. Fish and Wildlife Service, Biological Services Program.)

FIGURE 6.56 Birds feeding on a mudflat. (Photo courtesy of David Weintraub.)

Bacteria are consumed commonly by various nematode worms, which occupy a position similar to that of herbivores in other areas. Bacteria are also consumed by deposit feeders as they ingest the organic particles on which the bacteria are found. Nematodes and deposit feeders are, in turn, consumed by various carnivores, including predatory invertebrates (moon snails and polychaetes such as *Glycera*), as well as birds and fishes (Figure 6.57).

The second food web is based on the microscopic diatoms as autotrophs. The diatoms are consumed by several different polychaetes, mollusks, and crustaceans (Figure 6.58). These, in turn, are consumed by the large predatory birds and fishes.

■ ZONATION AND COMMUNITY STRUCTURE

Little information exists regarding the zonation of mudflats. The very gentle slope of these areas means that the intertidal zone is often extensive, more so than rocky or sandy shores. The upper area, the supralittoral fringe, is frequently inhabited by various species of crabs, many of which burrow into the substrate. The extensive midlittoral area is the home of most of the common species of clams and polychaetes. There is no sharp boundary with the infralittoral fringe, and similar organisms are encountered there.

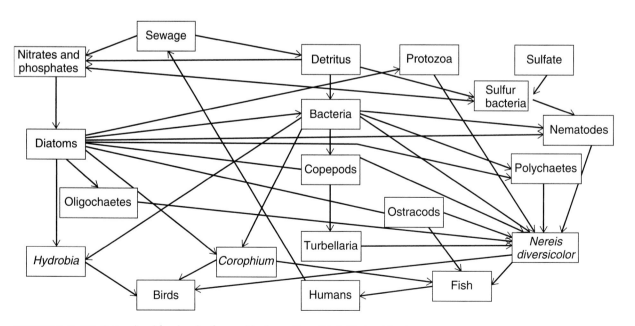

FIGURE 6.57 Generalized food web of a muddy shore. (From *Life in Mud and Sand*, S. K. Eltringham, 1971, Crane Russak.)

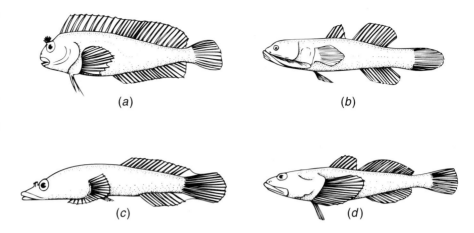

FIGURE 6.58 Examples of rocky intertidal fishes from the four dominant families: (a) Blenniidae (blennies). (b) Gobiidae (gobies). (c) Gobiesocidae (clingfishes). (d) Cottidae (cottids). (Modified from "The Biology of Behavior of Littoral Fish," R. N. Gibson, *Oceanography and Marine Biology: An Annual Review*, Vol. 7, pp. 367–410, 1969.)

The role of physical factors and biological factors in structuring these communities has been studied in a number of places. Structuring forces are similar to those acting in sand flats. Various biological factors seem significant, with predation being perhaps the most important. Many of the predators are similar to those described for sand flats, and sand flats and mudflats seem to be similarly organized.

Intertidal Fishes

Whereas a great deal of research has been concentrated on the ecology of invertebrates and plants of the intertidal zone on all three types of shores, there are relatively few accounts of the fishes of these areas or of the role they may play in community organization as grazers or predators. Since fishes are often present in considerable numbers in the intertidal zone, and since it is known that they have a significant effect on other communities, such as coral reefs (see Chapter 9), and in various subtidal communities (Chapter 5), it is reasonable to assume that future research may clarify their role here as well.

Because of the turbulent environment, most rocky intertidal fishes are small. Their body shape is usually compressed and elongate (Blenniidae, Pholidae) or depressed (Cottidae, Gobiesocidae)—shapes that allow them to inhabit holes, tubes, crevices, or depressions for protection against both desiccation and wave action (Figure 6.58). Most also lack swim bladders and are closely associated with the substrate. Many of these fishes are adapted to withstand greater ranges of salinity and temperature than their subtidal relatives. A few are even adapted to spend some time out of the water. (See Chapter 9 on *Periophthalmus*).

Most rocky intertidal fishes in the temperate zone are visual carnivores, again suggesting a potentially significant role in intertidal community organization.

The life-history patterns of the few species that have been investigated are similar. Eggs are demersal and laid on stones, rocks, or submerged vegetation. Often, the eggs are guarded by the male. The eggs hatch into planktonic larvae after a few weeks. Depending on the species, the planktonic period varies and may be as long as two months. During this period, the larvae gradually acquire adult features and finally become demersal. Life spans of the adults are generally short, from two to ten years, and sexual maturity occurs in the first or second year.

On mud and sand flats, skates and rays, as well as various species of flatfishes, are significant predators of infaunal organisms. Rays are particularly common on intertidal flats. Their method of foraging, by excavating depressions or pits, causes widespread mortality among the infaunal organisms. Their effect extends beyond the species they consume: Digging has uncovered many small infaunal species that are either killed by the disturbance or consumed by the smaller fishes that follow the rays. Flatfishes also may prey on various infaunal invertebrates, but their role in community structure seems more species specific. Some have a minimal effect because they feed on few important infaunal species, whereas others reduce both the abundance and the diversity of species in a community. Virnstein (1977) demonstrated that the hogchoker (*Trinectes maculatus*) had a minimal effect on the infauna of Chesapeake Bay, whereas the spot (*Leiostomus xanthurus*) significantly reduced infaunal numbers. Some intertidal fishes are migratory, moving either tidally, diurnally, or seasonally.

Birds

At low tide, a considerable variety of birds are often associated with the intertidal zone. Birds are particularly conspicuous on intertidal sand flats and mudflats, much less so in rocky intertidal areas. Sandy and muddy intertidal flats are exploited by many different shorebird species (Figure 6.59), mainly predators of infaunal invertebrates. Given their abundance and high food requirements, one might wonder how so many predatory species can coexist in a single flat without depleting the infauna or undergoing severe competition. There are several explanations. In the temperate zone, most of these birds are migratory and are present only during certain portions of the year. While they are

present, they may locally deplete the infauna, but it has time to recover during the periods when they are absent. Another explanation is that tidal flats are not uniform in community structure, as we have seen. Different infaunal associations are found in different areas, due to differences in grain size and various succession patterns that follow disturbances. The infauna is also segregated by depth in the sediment, as well as by tide level. A variety of different microhabitats exists. If the bird species differ in preference for food or for microhabitat, then they may be separated in space or time, thereby reducing competition and the chance that resources will become depleted. Indeed, this seems to be the case. For example, Harrington and Schneider (1978) have shown that sanderlings prefer sandy

FIGURE 6.59 New England tidal-flat birds, showing general feeding positions of the various groups. (From R. Whitlatch, 1982, *The Ecology of New England Tidal Flats: A Community Profile*. U.S. Fish and Wildlife Service, Biological Services Program.)

substrates, while dowitchers are more abundant on mud. Similarly, temporal segregation has been demonstrated among sanderlings and semipalmate plovers. As Harrington et al. (1974) found, sanderlings follow the water's edge as the tide recedes, whereas semipalmate plovers forage in the middle intertidal zone. Evans et al. (1979) have shown that competition between knots and dunlins is reduced because knots eat mollusks, whereas dunlins eat polychaetes. Morphological features, including different leg and bill lengths, also influence where the birds can feed. Finally, behavior patterns dictate what food resources may be exploited.

Given that birds are present and abundant and that they do consume intertidal invertebrates, how important are they in structuring communities? It seems, on the basis of most studies, that they can be quite important. Evans et al. (1979), for example, reported that wintering shorebirds were responsible for removing 90% of the *Hydrobia* snail population and 80% of the nereid polychaetes from an English mudflat. In the same vein, Harrington and Schneider (1978) reported that the whole group of foraging shorebirds eliminated 50–70% of the invertebrate population on the east coast of the United States. Finally, Quammen (1984) demonstrated that shorebirds had a significant effect on the densities of infaunal invertebrates in mudflat areas, but little or no effect on sand-flat invertebrate densities, on the Pacific coast of North America.

A special case may exist on "bird islands"—isolated small islands on which birds congregate in great numbers for breeding. In such situations, as Bosman and Hockey (1986) have noted, the greater nutrient input to the intertidal zone via bird guano leads to enhanced algal production and, if there are predators of the herbivorous molluscan grazers, such as oyster catchers, may lead as well to an intertidal zone dominated by foliose algae. We can summarize the situation by saying that bird predation may be intense and significant, particularly at certain times, and most certainly plays an important role in structuring tidal-flat communities (Figure 6.60).

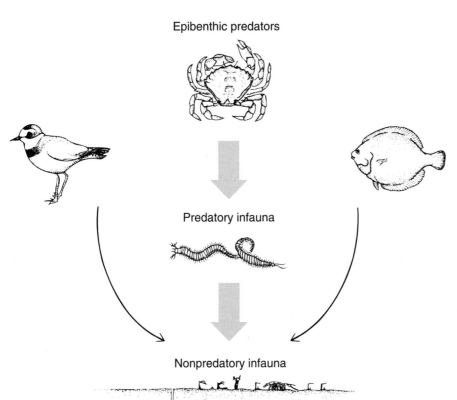

Epibenthic predators

Predatory infauna

Nonpredatory infauna

FIGURE 6.60
Diagrammatic representation of the interactions between various epibenthic predators and the infauna of sedimentary shores. Broad arrows indicate relations that may be particularly important. (From "Role of Predatory Infauna in Structuring Marine Soft-bottom Communities," G. Ambrose, *Marine Ecology Program Services,* Vol. 17, pp. 102–115, 1984. Reprinted by permission of Inter-Research.)

Summary of Key Concepts

- The intertidal zone accounts for the smallest area of the world's oceans, but is the best known and most accessible to humans.

- The intertidal zone has the greatest variation in environmental factors of any marine area.

- The intertidal zone is inhabited primarily by marine organisms.

- The tide is the daily predictable rise and fall of sea level and is the most important factor influencing life in the intertidal zone.

- Tides are formed due to the interaction of the gravitational attraction of the sun and the moon on the Earth and the centrifugal force generated by the rotating Earth–moon system.

- Extreme spring tides occur when the sun and moon are directly aligned, whereas minimal neap tides occur when the sun and the moon are at right angles to each other.

- Differences in the number of high and low tides per day and in their heights are due to the peculiarities of the basins in which the tides occur.

- Tides with a single high and low per day are called diurnal tides; tides with two highs and lows per day are called semidiurnal tides.

- Tides can have two direct effects on organisms. The first involves the duration of exposure to the air. Since the organisms are primarily marine, the longer they are exposed to the air, the greater is the chance that they will become desiccated beyond their limits of tolerance or that they will encounter a lethal temperature. Intertidal organisms differ in their tolerance to exposure, and this difference contributes to their patterns of distribution.

- The second effect is the result of the time of day that exposure to air occurs. Exposure to air during midday in the tropics could lead to lethal temperatures; exposure at night in the cold temperate zones in winter could lead to the organism's freezing to death.

- The great predictability of tides induces certain rhythms, such as feeding and reproduction, in intertidal organisms.

- Air temperatures always have a greater range than water temperatures, and extremes may either kill organisms or weaken them, making them susceptible to death by other factors, such as desiccation.

- Wave action directly affects organisms in two ways. First, on sedimentary shores it moves the entire substrate around, in the process smashing organisms and tearing them apart. Second, waves throw water higher on the shore than the water level due to tides; therefore, wave action allows marine organisms to live higher than the tides would permit them to.

- Salinity changes in the intertidal zone that can affect organisms are due primarily to the flooding of the zone with fresh water from runoff or heavy rains.

- In the cold temperate zones in winter, catastrophic mortality in the intertidal zone can be caused by ice scour.

- Adaptations of intertidal organisms are primarily for avoiding stresses of daily exposure to air.

- Organisms exhibit various methods of avoiding water loss, including moving into moist habitats at low tide, shielding soft tissues from the environment by closing their shells, clamping down into home scars, and secreting a mucous covering. Others simply tolerate a great water loss during low tide and rehydrate during high tide.

- Intertidal organisms reduce their heat gain from the environment in several ways, including possessing a large body size and reducing the surface area of their body that is in contact with the substrate.

- Intertidal organisms lose heat from their bodies through radiation from ridges, sculpturing on their shells, having light colors, and the evaporation of water from their bodies.

- Adaptations to reduce mechanical wave stress include streamlined shapes that minimize exposure to the lift and drag forces of waves, flexibility and bending in algae, and strong attachments to the substrate.

- Motile animals avoid wave stress by moving into sheltered areas.

- Most intertidal animals have respiratory surfaces that are adapted to extract oxygen from seawater; thus, either the animals protect those surfaces during exposure to the air or the animals' organs are modified to extract oxygen from air. Intertidal fishes may employ cutaneous respiration.

- Feeding in intertidal animals usually occurs only when they are submerged, thereby protecting the fleshy parts of the animal from desiccation.

- Most intertidal animals have planktonic larvae, and breeding cycles are synchronized with the tides.

- There are significant differences between the rocky shores of the Atlantic and Pacific coasts of North America. In the Atlantic, rocky shores are generally absent south of New England and are replaced by sedimentary environments, whereas the Pacific coast is dominated by rocky shores, and sedimentary environments are uncommon. Air temperature ranges are greater on the Atlantic shores than the Pacific, and Atlantic intertidal organisms in New England and Canada are regularly killed by freezing or ice scour, both of which rarely occur in the Pacific intertidal zone. Also, the Pacific intertidal zone was spared the widespread extinction that occurred in the Atlantic intertidal zone during the last ice age.

- The number of species and species diversity are greater in the Pacific intertidal zone than in the Atlantic.

- A characteristic feature of the rocky intertidal zone throughout the world is that there is a distinct horizontal banding, or zonation, of organisms, with the bands succeeding each other vertically on the shore.

- The zonation pattern observed led to a "universal" classification scheme that established a standard format for describing shore zonation. This scheme has three main divisions in the intertidal zone: the uppermost is the supralittoral fringe; the middle is the midlittoral zone; and the lowest is the infralittoral fringe. The boundaries of these divisions are defined by the limits of distribution of the dominant organisms.

- The most obvious explanation for the occurrence of zonal patterns on rocky shores is that the patterns are the result of tidal action and reflect the different tolerances of the organisms to increasing exposure to the air and the resulting desiccation and temperature extremes. Many studies in different areas, carried out during different seasons, and under different conditions of tides, wave action, and temperature have produced evidence which suggests that physical factors are generally the determinants of the upper limits of intertidal organisms.

- Whereas early work on the causes of zonation focused on physical factors, more recent work has established the importance of biological factors, including competition, predation, grazing, and larval recruitment in the establishment of zonal patterns.

- In the rocky intertidal zone, there is intense competition among organisms for space, and several studies of competition among animals and plants have clearly demonstrated that the zonation of intertidal organisms is at least partially a function of biological competition.

- Competition and the outcome of competitive interactions may be modified by the biological factor of predation.

- On both the Atlantic and Pacific rocky coasts of North America, certain competitive dominant animals— *Mytilus californianus* on the Pacific coast and *Mytilus edulis* on the Atlantic coast—have the ability to outcompete all other animals and take over the substrate. They are prevented from doing this by predation by starfishes: *Pisaster ochraceus* in the Pacific and *Asterias forbesi* and *A. vulgaris* in the Atlantic. Thus, predatory activity by top predators permits other, less competitive animals to continue to exist on primary substrate in the intertidal zone. Because of their ability to influence the structure of the entire intertidal community, such predators have been called keystone species.

- Some prey animals escape predation in various ways that contribute to zonal patterns. On the Pacific coast, the barnacle *Balanus cariosus* escapes predation by the snail *Nucella* by growing too large to be attacked, while *Balanus glandula* escapes by settling so high in the intertidal zone that *Nucella* can reach the barnacle only at the near-certain risk of excessive desiccation.

- Some anomalies in the keystone species concept have caused a number of researchers to question its usefulness. Their doubts have led to a redefinition of a keystone species as a single predator that controls the abundance of a prey species that in turn can exclude all other species and dominate the community. If more than a single predator is involved in preventing competitive domination, then the predation is diffuse, and if predation alone cannot control the abundance of the competitive dominant, it is called weak predation.

- The keystone species concept remains valid for some temperate shores, but is not a universal concept, and its occurrence is variable within the range of the species. It may be replaced under certain environmental conditions by other community regulatory mechanisms, such as diffuse or weak predation, larval settlement, and physical factors.

- A central theme of rocky intertidal ecology is that, whenever predation is reduced, competition will be increased, but a reduction in compatition is not the inevitable result of the removal of predators.

- The relative importance of biological and physical factors on rocky shores varies over relatively short distances.

- Grazers play an important role in regulating the upper and lower limits of algal species.

- Grazing affects a number of parameters, including algal zonation, species diversity, patchiness, and succession.

- Algae have defenses against grazers, including laying down calcium carbonate in the plants' tissues, clustering reproductive sites underneath and away from grazers, producing toxic or noxious chemicals, and having evolved growth forms that are resistant to grazers.

- Some algal species have two growth forms that vary in their susceptibility to grazing.

- The importance of larval settlement to the structure of intertidal communities has become clear over the past decade.

- Larval recruitment is variable in time and space and can significantly affect intertidal community structure.

- Spatial and temporal fluctuations in larval recruitment lead to predictable variations in the importance of competition, predation, and grazing in communities.

- Studies of community structure on tropical intertidal shores are relatively uncommon. On the Pacific coast of Panama, intense consumer pressure by a variety of fish, crab, and snail consumers, as well as harsh physical conditions, lead to an intertidal zone dominated by bare space and heavily defended algal crusts. This general pattern may be common on tropical rocky shores.

- Small-scale spatial patchiness is a basic feature of rocky intertidal habitats and is caused by physical or biological factors that create an open space for recolonization, produce a different substrate, or alter physical conditions in an area.

- Successional sequences in the rocky intertidal zone are governed by both the inhibition and facilitation models and depend on the time of initial disturbance that frees the space.
- Rocky intertidal communities are best thought of as mosaics of patches in different stages of recovery.
- Community structure changes horizontally on wave-exposed gradients.
- In general, the high intertidal zone tends to have the distribution pattern of its characteristic species set by physical factors, whereas biological factors become more important in setting limits in the low intertidal zone. However, the two kinds of factors do not act in isolation, and the final distribution patterns observed on either a large or small scale are often the result of the interaction of two or more of either or both kinds of factors.
- Certain conditions affecting life in tide pools differ markedly from conditions in the surrounding intertidal zone.
- Tide pools are refuges from desiccation for intertidal organisms, but can have rapid changes in temperature, salinity, and oxygen concentrations that may strongly affect the organisms.
- Zonation patterns have been described for tide pools both within single pools and among pools at different levels on the shore. In general, numbers of species or organisms decrease with increasing height of the tide pool.
- The role of biological interactions in determining the community structure of tide pools is poorly documented.
- Perhaps the most important physical factor governing life on exposed sand beaches is wave action.
- Beaches are defined by three factors—particle size, wave action, and slope—interrelated such that defining two fixes the third.
- Particle size influences water retention on the beach.
- The slope of a beach is the result of the interaction among particle size, wave action, and swash and backwash water.
- A second important physical factor on sand beaches and flats is wave-induced motion of the substrate.
- On the basis of a continuum of grain size, slope, and wave action, a number of different types of beach have been described. Representing the extremes of the continuum are the dissipative and reflective beaches.
- The abundance and diversity of macrofaunal organisms on open beaches are more strongly correlated with particle size and slope than with wave action.
- Because of the ceaseless movement of the surface layers of sediment on sand beaches, few large organisms occupy those layers.
- Sand beaches and flats have a smooth, uniform profile and lack the topographic diversity of rocky shores.
- Whereas wave action on open sand beaches precludes most organisms from inhabiting surface layers, the

much-reduced wave action on protected sand flats makes them suitable for habitation by epifaunal organisms.
- Because of its insulating qualities and high water content, sand is an excellent buffer against changes in temperature and salinity.
- Oxygen may become depleted in soft-sediment substrates, especially in fine-grained sediments, where water exchange is reduced.
- Two main adaptations to continued substrate movement by animals inhabiting open sand beaches are deep burrowing and rapid burrowing.
- Other adaptations include smooth shells and reduced spines to aid in burrowing, tidal migrations to keep the animals in the swash zone, the avoidance of diurnal predators by becoming nocturnal, and the development of meshes that filter out sand to prevent the clogging of respiratory surfaces.
- On open sand beaches, the types of organisms that are conspicuous by their absence are large plants and sessile animals.
- The dominant animals on sand beaches and sand flats are polychaete worms, bivalve mollusks, and crustaceans.
- There is little primary productivity on open sand beaches: Animals found there are mainly filter feeders, detritus feeders, and scavengers.
- On protected sand flats, there is substantial primary productivity and all types of feeding animals are present.
- The zonation pattern of organisms on open sand beaches is neither as clearly defined nor as well understood as that of organisms on rocky shores, largely because of the researcher's inability to set up field experiments on such beaches.
- Open sand beaches generally have three zones: an upper supralittoral fringe inhabited by amphipods or ghost crabs, a broad midlittoral with sand crabs and isopods, and an infralittoral fringe with large number of species.
- On open sand beaches, physical factors such as wave action, particle size, and beach slope probably generate most of the patterns that are observed.
- On protected sand flats, due to trophic group amensalism, there is a tendency to have a complementary distribution of communities dominated by suspension feeders and deposit feeders, with the deposit feeders more abundant in fine sediments and filter feeders in coarser sediments.
- The prime biological factors that structure sand-flat communities are predation, competition, and disturbance. Of these, predation or predation combined with disturbance is especially significant.
- The role of competition in structuring sand-flat communities is different from that observed on rocky shores. The three-dimensionality of the substrate and the interactions between juveniles and adults reduce the severity of direct competition.

- Although not common, competitive interactions leading to different depth distributions do occur in protected sand beaches.

- In contrast to removing predators from rocky intertidal areas, removing them from soft substrates does not lead to increased competition or competitive exclusion.

- The macrofauna of intertidal sand flats displays high temporal and spatial variability, due to the high predation pressure and biological disturbance that control the infaunal community structure and the opportunistic nature of many infaunal organisms.

- There are no sharp boundaries between sand and mud shores, and they are at opposite ends of a continuum.

- Muddy shores are restricted to intertidal areas protected from ocean waves.

- Muddy shores have very fine particles that greatly restrict the interchange of water and, with the high numbers of bacteria, create anaerobic conditions in the substrate below a shallow surface aerobic layer.

- Between the aerobic and anaerobic layers is a transition area called the redox potential discontinuity (RPD) layer, characterized by a rapid change from positive to negative redox potential.

- One of the primary adaptations of mudflat organisms is the ability to burrow into the substrate or to form permanent tubes that reach the surface.

- Another adaptation of mudflat organisms is being able to live under conditions of low oxygen or having some way of bringing oxygenated water down to themselves.

- Mudflats often develop a substantial growth of various plants and contain large numbers of bacteria, especially in the RPD layer. Among the bacteria in this layer are chemolithoautotrophic, or sulfur, bacteria, that are primary which use the energy of reduced sulfur compounds to produce organic material.

- The dominant large animals on mudflats are polychaete worms, bivalve mollusks, and crustaceans.

- The dominant feeding types on mudflats are deposit feeders and suspension feeders. The major predators are fishes and birds.

- Community structuring factors in mudflats are similar to those in sand flats.

- There are relatively few accounts of the role of fishes in the rocky intertidal communities, but most appear to be visual carnivores, suggesting a potentially significant role in intertidal community organization.

- On sand flats and mudflats, skates and rays, as well as flatfishes, are significant predators of infaunal animals. They also have wider community effects by excavating pits, thereby causing widespread mortality among the other infaunal organisms.

- Bird predation on the intertidal may be intense and significant, particularly at certain times of the year, and it plays an important role in structuring tidal-flat communities.

Review Questions

ESSAY Develop complete answers to these questions:

1. An important theme of intertidal ecology is that release from predation pressure permits an increase in competition. Discuss this concept, presenting evidence to support your answer.

2. Discuss differences in the forces that structure the rocky intertidal communities of temperate and tropical waters.

3. Why must a researcher who is studying zonation on a sandy beach take repeated samples at many different times over the course of the study? Discuss.

4. How does predation modify the structure of zonation in the rocky intertidal? Discuss.

5. Discuss how variable recruitment can influence the organization of rocky-shore communities.

6. How does small-scale patchiness develop? What forces come into play in developing patches in the rocky-shore community?

7. What is meant by the term *succession*? Does succession occur in the rocky intertidal zone? Why or why not? Defend your answer.

8. What adaptations allow organisms to survive in the anaerobic mudflat regions? Discuss and give examples.

Bibliography

Ambrose, W. G. 1984. Role of predatory infauna in structuring marine soft bottom communities. *Mar. Ecol. Progr. Ser.* 17:102–115.

Ansell, A. D., P. Sivades, B. Narayanan, and A. Trevallion. 1972. The ecology of two sand beaches in southwest India. II. Observations on the populations of *Donax incarnatus* and *Donax spiculurn. Mar. Biol.* 17:318–322.

Bertness, M. D. 1984. Habitat and community modification by an introduced herbivorous snail. *Ecology* 65(2):370–381.

Bertness, M. D. 1989. Intraspecific competition and facilitation in northern acorn barnacles. *Ecology* 70(1): 257–268.

Bertness, M. D. 1999. *The ecology of Atlantic shorelines.* Sunderland, MA: Sinauer Associates, Inc.

Bertness, M. D., S. D. Garrity, and S. C. Levings. 1981. Predation pressure and gastropod foraging patterns: A latitudinal pattern. *Evolution* 35(5):995–1007.

Bertness M. D., and G. Leonard. 1997. The role of positive interactions in communities: Lessons from the intertidal. *Ecology* 78: 1976–1989.

Bertness, M. D., G. H. Leonard, J. Levine, P. Schmidt, and A. O. Ingraham. 1999. Testing the relative contribution of positive and negative interactions in rocky intertidal communities. *Ecology* 80: 2711–2726.

Bosman, A. L., and P. A. R. Hockey. 1986. Seabird guano as a determinant of rocky intertidal community structure. *Mar. Ecol. Progr. Ser.* 32 (2–3):247–257.

Boyle, P. R. 1969. The survival of osmotic stress by *Sypharochiton pelliserpentis. Biol. Bull.* 136:154–165.

Brafield, A. E. 1978. *Life in sandy shores.* Studies in Biology no. 89. London: Edward Arnold.

Brenchley, G. A., and J. T. Carlton. 1983. Competitive displacement of native mud snails by introduced periwinkles in the New England intertidal zone. *Biol. Bull.* 165:543–558.

Bridges, C. A. 1988. Respiratory adaptations in intertidal fishes. *Amer. Zoo.* 28:79–96.

Brown, A. C., and A. McLachlan. 1990. *Ecology of sandy shores.* New York: Elsevier.

Bruno, J. F. 2000. Facilitation of cobble beach plant communities through habitat modification by *Spartina alterniflora. Ecology* 81(5):1179–1192.

Butman, C. A. 1992. Larval settlement of soft-sediment invertebrates: The spatial scales of pattern explained by active habitat selection and the emerging role of hydrodynamical processes. *Oceanography and Marine Biology Annual Review* 25: 113–165.

Carefoot, T. 1977. *Pacific seashores, a guide to intertidal ecology.* Seattle: University of Washington Press.

Castenholz, R. W. 1961. The effect of grazing on marine littoral diatom populations. *Ecology* 42:783–794.

Chapman, A. R. O. 1990. Effects of grazing, canopy cover and substratum type on the abundances of common species of seaweeds inhabiting littoral fringe tide pools. *Bot. Mar.* 33:319–326.

Chapman, A. R. O., and C. R. Johnson. 1990. Disturbance and organization of macroalgal assemblages in the Northwest Atlantic. *Hydrobiologia* 192:77–121.

Chia, F. S. 1973. Sand dollar: A weight belt for the juvenile. *Science* 181:73–74.

Connell, J. H. 1961. The influence of interspecific competition and other factors on the distribution of the barnacle *Chthamalus stellatus. Ecology* 42:710–723.

Connell, J. H. 1970. A predator–prey system in the marine intertidal region. I, *Balanus glandula* and several predatory species of *Thais. Ecol. Monogr.* 40:49–78.

Connolly, S. R., and J. Roughgarden. 1998. A latitudinal gradient in Northeast Pacific intertidal community structure: Evidence for an oceanographically based synthesis of marine community theory. *Amer. Nat.* 151:311–326.

Davis, P. S. 1969. Physiological ecology of *Patella.* III. Desiccation effects. *Mar. Biol. Assoc. U.K.* 49:291–304.

Dayton, P. K. 1971. Competition, disturbance and community organization: The provision of and subsequent utilization of space in a rocky intertidal community. *Ecol. Monogr.* 41:351–389.

Dayton, P. K. 1973. Dispersion, dispersal and persistence of the annual intertidal alga *Postelsia palmaeformis* Ruprecht. *Ecology* 54(2):433–438.

Dayton, P. K. 1975. Experimental evaluation of ecological dominance in a rocky intertidal algal community. *Ecol. Monogr.* 45:137–159.

Denny, M. W., T. L. Daniels, and M. A. R. Koehl. 1985. Mechanical limits to size in wave-swept organisms. *Ecol. Monog.* 55: 69–102.

Dethier, M. N., and D. O. Duggins. 1988. Variation in strong interactions in the intertidal zone along a geographical gradient: A Washington–Alaska comparison. *Mar. Ecol. Progr. Ser.* 50:97–105.

Doty, M. S. 1946. Critical tide factors that are correlated with the vertical distribution of marine algae and other organisms along the Pacific coast. *Ecology* 27:315–328.

Eltringham, S. K. 1971. *Life in mud and sand.* New York: Crane Russak.

Evans, P. R., D. M. Henderson, T. J. Knights, and M. W. Pienkowski. 1979. Short-term effects of reclamation of part of Seal Sands, Teesmouth, on wintering waders and shelduck. I, Shorebird diets, invertebrate densities and the impact of predation on the invertebrates. *Oecologia* 41:183–206.

Foster, B. A. 1971. On the determinants of the upper unit of intertidal distribution of barnacles (Crustacea, Cirripedia). *J. Anim. Ecol.* 40:33–48.

Francis, L. 1973. Intraspecific aggression and its effect on the distribution of *Anthopleura elegantissima* and some related anemones. *Biol. Bull.* 144:73–92.

Frank, P. W. 1965. The biodemography of an intertidal snail population. *Ecology* 46:831–844.

Gaines, S. D., and M. D. Bertness. 1992. Dispersal of juveniles and variable recruitment in sessile marine species. *Nature* 360: 579–580.

Gaines, S. D., and M. D. Bertness. 1993. Measuring the connection between dispersal and population dynamics: Why ecologists must learn to integrate. *Ecology* 74: 2430–2435.

Gaines, S. D., and J. Lubchenco. 1982. A unified approach to marine plant–herbivore interactions. II. Biogeography. *Ann. Rev. Ecol. Syst.* 13:111–138.

Gaines, S. D., and J. Roughgarden. 1987. Fish in offshore kelp forests affect recruitment to intertidal barnacle populations. *Science* 235:479–481.

Garrity, S. D. 1984. Some adaptations of gastropods to physical stress on a tropical rocky shore. *Ecology* 65: 559–574.

Gibson, R. N. 1969. The biology and behavior of littoral fish. *Oceanog. Mar. Biol. Ann. Rev.* 7:367–410.

Gibson, R. N., A. D. Ansell, and L. Robb. 1993. Seasonal and annual variations in abundance and species composition of fish and macrocrustacean communities on a Scottish sandy beach. *Mar. Ecol. Progr. Ser.* 98:89–105.

Grant, J. 1981. Dynamics of competition among estuarine sand burrowing amphipods. *J. Exp. Mar. Biol. Ecol.* 49:255–265.

Harrington, B. A., and D. C. Schneider. 1978. *Studies of shorebirds at an autumn migration stopover area.* Final Report for U.S. Fish and Wildlife Service, Migratory Bird and Habitat Research Laboratory, Laurel, MD.

Harrington, B. A., S. K. Groves, and N. T. Houghton. 1974. *Season progress report. Massachusetts shorebird studies.* Contract 14-16-008-687. U.S. Fish and Wildlife Service, Manomet, MA.

Haven, S. B. 1971. Niche differences in the intertidal limpets *Acmaea scabra* and *Acmaea digitalis* (Gastropoda) in central California. *Veliger* 13:231–248.

Hawkins, S. J. 1981. The influence of season and barnacles on the algal colonization of *Patella vulgata* exclusion areas. *J. Mar. Biol. Assoc. U.K.* 61:1–15.

Hawkins, S. J., and R. G. Hartnoll. 1983. Grazing of intertidal algae by marine invertebrates. *Oceanog. Mar. Biol. Ann. Rev.* 21:195–282.

Hay, C. 1979. Some factors affecting the upper limit of the southern bell kelp *Durvillaea antarctica* (Chamisso) Hariot on the New Zealand shores. *J. Roy. Soc. N.Z.* 9:279–289.

Hay, M. E. 1981. The functional morphology of turf forming seaweeds: Persistence in stressful marine habitats. *Ecology* 62:739–750.

Hay, M. E. 1986. Associational plant defenses and the maintenance of species diversity. *Amer. Nat.* 128:617–641.

Hedgpeth, J., ed. 1957. *Treatise on marine ecology and paleoecology. Vol. 1, Ecology.* Washington, DC: Geol. Soc. Amer. Memoir 67.

Himmelman, J. H., and D. H. Steele. 1971. Food and predators of the green sea urchin *Strongylocentrotus droebachiensis* in Newfoundland waters. *Mar. Biol.* 9:315–322.

Jones, N. S. 1948. Observations and experiments on the biology of *Patella vulgata* at Port St. Mary, Isle of Man. *Liverpool Biol. Soc. Proc. Trans.* 56:60–77.

Jones, N. S., and J. M. Kain. 1967. Subtidal algal colonization following the removal of *Echinus*. *Helg. Wiss. Meeresunt.* 15:460–466.

Kanwisher, J. 1957. Freezing and drying in intertidal algae. *Biol. Bull.* 113:275–285.

Kanwisher, J. W. 1960. Freezing in intertidal animals. *Biological Bulletin* 116: 258–264.

Kennish, R., G. A. Williams, and S. Y. Lee. 1996. Algal seasonality on an exposed rocky shore in Hong Kong and the dietary implications for the herbivorous crab *Grapsus albolineatus*. *Mar. Biol.* 125:55–64.

Kooistra, W. H. C. F., A. M. T. Joosten, and C. van den Hoek. 1989. Zonation patterns in intertidal pools and their possible causes: A multivariate approach. *Bot. Mar.* 32:9–26.

Kozloff, E. N. 1983. *Seashore life of the northern Pacific Coast.* Seattle: University of Washington Press.

Levinton, J. S. 1977. Ecology of shallow water deposit-feeding communities. In *Ecology of the marine benthos*, edited by B. C. Coull. Columbia: University of South Carolina Press, 191–227.

Levinton, J. S., and G. R. Lopez. 1977. A model of renewable resources and limitation of deposit feeding benthic populations. *Oecologia* 31:177–190.

Levinton, J. S., S. Stewart, and T. H. DeWitt. 1985. Field and laboratory experiments on interference between *Hydrobia totteni* and *Ilyanassa obsoleta* (Gastropoda) and its possible relation to seasonal shifts in vertical mudflat zonation. *Mar. Ecol. Progr. Ser.* 22:53–58.

Lewis, J. R. 1964. *The ecology of rocky shores.* London: The English Universities Press, Ltd.

Lipcius, R. N., and T. Hines 1986. Variable functional responses of a marine predator in dissimilar homogeneous microhabitats. *Ecology* 67(5): 1361–1371.

Littler, M. M., and D. S. Littler. 1980. The evolution of thallus form and survival strategies in benthic marine macroalgae: Field and laboratory tests of a functional form model. *Amer. Nat.* 116:25–44.

Lowe, G. C., Jr., and E. R. Cox. 1978. Species composition and seasonal periodicity of the marine benthic algae of Galveston Island, Texas. *Contrib. Mar. Sci. Univ. Texas* 21:9–24.

Lubchenco, J. 1978. Plant species diversity in a marine intertidal community: Importance of herbivore food preference and algal competitive abilities. *Amer. Nat.* 112(983):23–39.

Lubchenco, J. 1980. Algal zonation in the New England rocky intertidal community: An experimental analysis. *Ecology* 61:333–344.

Lubchenco, J. 1982. Effects of grazers and algal competitors on fucoid colonization in tide pools. *J. Phycol.* 18:544–550.

Lubchenco, J., and S. D. Gaines. 1981. A unified approach to marine plant–herbivore interactions. I. Populations and communities. *Ann. Rev. Ecol. Syst.* 12:405–437.

Lubchenco, J., and B. Menge. 1978. Community development and persistence in a low rocky intertidal zone. *Ecol. Monogr.* 59:67–94.

Lubchenco, J., B. A. Menge, S. D. Garrity, P. J. Lubchenco, L. R. Ashkenas, S. D. Gaines, R. Eralet, J. Lucas, and S. Strauss. 1984. Structure, persistence, and role of consumers in a tropical intertidal community (Taboguilla Island, Bay of Panama). *J. Exp. Mar. Biol. Ecol.* 78:23–73.

MacGinitie, G. E., and N. MacGinitie. 1949. *Natural history of marine animals.* New York: McGraw-Hill.

Mann, K. H. 1972. Macrophyte production and detritus food chains in coastal waters. *Mem. Inst. Ital. Idrobiol.* 29(suppl.):353–383.

McLachlan, A. 1983. Sandy beach ecology. In *Sandy beaches as ecosystems*, edited by A. McLachlan and T. Erasmus. The Hague: Dr. W. Junk, 321–380.

McLusky, D. S., S. A. Nair, A. Stirling, and R. Bharoava. 1975. The ecology of a central west Indian beach with particular reference to *Donax incarnatus. Mar. Biol.* 30:267–270.

Menge, B. A. 1976. Organization of the New England rocky intertidal community: Role of predation, competition and environmental heterogeneity. *Ecol. Monogr.* 46(4):355–393.

Menge, B. A. 1991. Relative importance of recruitment and other causes of variation in rocky intertidal community structure. *J. Exp. Mar. Biol. Ecol.* 146:69–100.

Menge, B. A. 1995. Indirect effects in marine rocky intertidal interaction webs: Patterns and importance. *Ecol. Monogr.* 65:21–74.

Menge, B. A., and J. Lubchenco. 1981. Community organization in temperate and tropical rocky intertidal habitats: Prey refuges in relation to consumer pressure gradients. *Ecol. Monogr.* 51:429–450.

Menge, B. A., E. L. Berlow, C. A. Blanchette, S. A. Navarrete, and S. B. Yamada. 1994. The keystone species concept: Variation in interaction strength in a rocky intertidal habitat. *Ecol. Monogr.* 64:249–286.

Menge, B. A., J. Lubchenco, L. R. Ashkenas, and F. Ramsey. 1986. Experimental separation of the effects of consumers on sessile prey in the low zone of a rocky shore in the Bay of Panama: Direct and indirect consequences of food web complexity. *J. Exp. Mar. Biol. Ecol.* 106:225–269.

Metaxas, A., and R. E. Scheibling. 1993. Community structure and organization of tide pools. *Mar. Ecol. Progr. Ser.* 98:187–198.

Morgan, S. D. 2001. The larval ecology of marine communities. In *The ecology of marine benthic communities*, edited by Bertness, M. D., S. D. Gaines, and M. Hay. Sinauer Associates, Sunderland, MA.

Moran, M. J. 1980. The ecology and effects on prey of the predatory gastropod, *Morula marginalba.* Ph.D. dissertation, University of Sydney.

Newell, R. C. 1970. *Biology of intertidal animals.* New York: Elsevier.

Newell, R. C. 1979. *Biology of intertidal animals.* Faversham, Kent, U.K.: Marine Ecological Surveys, Ltd.

Paine, R. T. 1966. Food web complexity and species diversity. *Amer. Nat.* 100:65–75.

Paine, R. T. 1969. A note on trophic complexity and community stability. *Amer. Nat.* 103:91–93.

Paine, R. T. 1974. Intertidal community structure: Experimental studies on the relationship between a dominant competitor and its principal predator. *Oecologia* 15:93–120.

Paine, R. T. 1979. Disaster, catastrophe and local persistence of the sea palm *Postelsia palmaeformis. Science* 205:685–687.

Paine, R. T. 1988. Habitat suitability and local population persistence of the sea palm *Postelsia palmaeformis. Ecology* 69(6):1784–1794.

Paine, R. T., and R. L. Vadas. 1969. The effects of grazing by sea urchins, *Strongylocentrotus* spp., on benthic algal populations. *Limn. Oceanog.* 14:710–719.

Paine, R. T., J. C. Castillo, and J. Cancimo. 1985. Perturbation and recovery patterns of starfish dominated intertidal assemblages in Chile, New Zealand and Washington State. *Amer. Nat.* 125(6):679–691.

Palumbi, S.R. 1984. Tactics of acclimation: Morphological changes of sponges in an unpredictable environment. *Science* 225:1478–1480.

Peterson, C. H. 1977. Competitive organization of the soft bottom macrobenthic communities of southern California lagoons. *Mar. Biol.* 43:343–359.

Peterson, C. H. 1991. Intertidal zonation of marine invertebrates in sand and mud. *Amer. Sci.* 79(3):236–249.

Peterson, C. H., and N. Peterson. 1979. *The ecology of intertidal flats of North Carolina: A community profile.* Slidell, LA: U.S. Fish and Wildlife Service, Office of Biological Services FWS/OBS-79/39.

Pinada, J. 1994. Spatial and temporal patterns in barnacle settlement along a southern California rocky shore. *Marine Ecology Progress Series* 107: 125–138.

Quammen, M. 1984. Predation by shorebirds, fish and crabs on invertebrates in intertidal mudflats: An experimental test. *Ecology* 65:529–537.

Ricketts, E., and J. Calvin. 1985. *Between Pacific tides.* 5th ed. Rev. by D. W. Phillips. Stanford, CA: Stanford University Press.

Sanford, E. 1999. Regulation of keystone predation by small changes in ocean temperature. *Science* 283:2095–2097.

Sanford, E., M. S. Roth, G. C. Johns, J. P. Wares, and G. N. Somero. 2003. Local selection and latitudinal variation in a marine predator prey interaction. *Science* 300: 1135–1137.

Segal, E., and P. Dehnel. 1962. Osmotic behavior in an intertidal limpet, *Acmaea limatula. Biol. Bull.* 122:417–430.

Sousa, W. 1980. Experimental investigations of disturbance and ecological succession in a rocky intertidal algal community. *Ecol. Monogr.* 49:227–254.

Southward, A. J. 1965. *Life on the seashore.* Cambridge, MA: Harvard University Press.

Southward, A. J., and C. E. Southward. 1978. Recolonization of rocky shores in Cornwall after use of toxic dispersants to clean up the Torrey Canyon spill. *J. Fish. Res. Bd. Canada* 35:682–706.

Stanley, S. 1969. Bivalve mollusk burrowing aided by discordant shell ornamentation. *Science* 166:634–635.

Steneck, R. S. 1982. A limpet–coralline alga association: Adaptations and defenses between a selective herbivore and its prey. *Ecology* 63:507–522.

Stephenson, T. A., and A. Stephenson. 1949. The universal features of zonation between tidemarks on rocky coasts. *J. Ecol.* 37:289–305.

Stimson, J. 1970. Territorial behavior in the owl limpet, *Lottia gigantea. Ecology* 51:113–118.

Sutherland, J. P. 1990. Recruitment regulates demographic variation in a tropical intertidal barnacle. *Ecology* 71:955–972.

Timko, P. 1976. Sand dollars as suspension feeders: A new description of feeding in *Dendraster excentricus. Biol. Bull.* 151(1):247–259.

Trussell, G. C., A. S. Johnson, S. G. Rudolph, and E. S. Gilfillan. 1993. Resistance to dislodgement: Habitat and size-specific differences in morphology and tenacity in an intertidal snail. *Marine Ecology Progress Series* 100:135–144.

Trussell, G. C., P. Ewanchuk, B. R. Silliman, and M. D. Bertness. 2004. Trophic cascades in rocky shore tide pools: distinguishing lethal and nonlethal effects. *Oecologia.* 139: 427–432.

Ulanowicz, R. E., and J. H. Tuttle. 1992. The trophic consequences of oyster stock rehabilitation in Chesapeake Bay. *Estuaries* 15:298–306.

Underwood, A. J. 1980. The effects of grazing by gastropods and physical factors on the upper limits of distribution of intertidal macroalgae. *Oecologia* 46:201–213.

Underwood, A. J., and E. J. Denley. 1984. Paradigms, explanations and generalizations in models for the structure of intertidal communities on rocky shores. In *Ecological communities: Conceptual issues and the evidence,* edited by D. R. Strong, D. Simberloff, G. Abele, and A. B. Thistle. Princeton, NJ: Princeton University Press, 151–180.

Underwood, A. J., and P. Jernakoff. 1981. Effects of interactions between algae and grazing gastropods on the structure of a low shore intertidal algal community. *Oecologia* 48:221–233.

Vermeij, G. J. 1971. Temperature relations of some tropical Pacific intertidal gastropods. *Mar Biol.* 10: 308–314.

Virnstein, R. W. 1977. The importance of predation by crabs and fishes on benthic infauna in Chesapeake Bay. *Ecology* 58:1199–1217.

Voss, G. L. 1976. *Seashore life of Florida and the Caribbean.* Miami: E. A. Seeman.

Wethey, D. S. 1984. Thermal effects on the distribution of barnacle populations in New England. *Biol. Bull.* 59:160–169.

Wethey, D. S. 1985. Catastrophe, extinction and species diversity: A rocky intertidal example. *Ecology* 66:445–456.

Whitlatch, R. 1982. *The ecology of New England tidal flats: A community profile.* Washington, DC: U.S. Fish and Wildlife Service, Biological Services Program FWS/OBS-81/01.

Wiltse, W. I. 1980. Effects of *Polinices duplicatus* (Gastropoda: Naticidae) on infaunal community structure at Barnstable Harbor, Massachusetts. *Mar. Biol.* 56:301–310.

Woodin, S. A. 1976. Adult–larval interactions in dense infaunal assemblages: Patterns of abundance. *J. Mar. Res.* 34:25–41.

Woodin, S. A. 1978. Refuges, disturbance and community structure: A marine soft bottom example. *Ecology* 59:274–284.

CHAPTER 7

MEIOFAUNA

In addition to the large macrofaunal organisms that inhabit sedimentary shores, there is another world there, made up of organisms that occupy the microspaces between particles or that live on the individual particles. These are the **meiofaunal** or **interstitial** organisms. The existence of this unique association of organisms was not realized by biologists until the twentieth century, when European scientists began to investigate these areas. Since then, many scientists throughout the world have made studies of this unique habitat. The first studies were primarily taxonomic, to determine what kinds of organisms were present and to name them. More recent ecological studies have given us a better understanding of the peculiar conditions under which these animals live.

An interstitial faunal and floral assemblage has been found in intertidal and subtidal sedimentary substrates throughout the world, in both fresh water and salt water. Major interstitial groups have been discovered by Coull et al. (1977) at water depths as great as 5,000 m. Studies of these associations are, however, unequally distributed. The marine interstitial fauna is better known than the freshwater interstitial fauna, and the organisms and associations of the European seas are the best known of the various geographical areas. The Pacific coast of North America is practically unknown and unstudied with respect to interstitial fauna, whereas the Atlantic coast of North America has been the site of numerous studies.

In this chapter, we will investigate the principles that govern meiofaunal life in marine sediments.

■

■ DEFINITIONS

As with any discipline or subdiscipline, there are a few technical terms to be learned. *Interstitial* is a general term that refers to the spaces between the sedimentary particles and is also used to refer to the organisms living there. *Meiofauna*, however, is the preferred term for the latter usage. This term refers to all animals that pass through a 0.5-mm screen, but are retained on a 62-μm screen. A synonym is **meiobenthos**. Since the term *meiofauna* defines a size class, but does not specify a spatial location or how the organisms move within the habitat, and since the interstitial space is a function of the size of the particles, a few other terms have been used to refine the classification and indicate location. **Endobenthic** refers to meiofaunal-sized organisms that move within the sediment by displacing particles (i.e., the organisms are bigger than the interstitial space). **Mesobenthic** refers to meiofaunal organisms living and moving within the interstitial space, and **epibenthic** refers to those meiofauna living at the sediment–water interface. On sedimentary shores, the mesobenthic organisms are virtually identical to the meiofauna. In this chapter, we deal primarily with meiobenthos operating as mesobenthic animals. In very fine sediments, however, meiobenthic organisms, such as nematodes, kinorhynchs, copepods, and others, displace sediment grains as they move; thus, they operate as endobenthic animals.

Environmental Characteristics

The conditions that affect the meiofauna are somewhat different from those which affect the macrofauna in the same area. Perhaps the most important factor influencing the presence, absence, and types of interstitial organisms is **grain size**, which is of paramount importance in determining the amount of interstitial space

available for habitation. The coarser the grain size, the greater is the volume of the interstitial space and the larger are the interstitial organisms that can inhabit the area. Conversely, the finer the grain size, the lesser is the space available and the smaller the organisms must be to inhabit the area. Grain size acts as a barrier to the movement of meiofaunal organisms. Certain organisms will be confined to sediments of a certain grain size or to areas of a given environment simply because they are too large to fit between the grains of adjacent areas and, therefore, cannot disperse through the whole area. Meiofaunal organisms show a definite zonation based on grain size. For example, sediments in which the median particle size is less than 125 μm are dominated by burrowing organisms, such as kinorhynchs, whereas sediments with particle sizes above 125 μm are dominated by forms that move in the interstices.

Grain size is also of great importance because it controls the ability of the sediment to retain and circulate water. Meiofaunal organisms are aquatic and require the presence of water in the interstices of the substrate in order to survive (Figure 7.1). If the

FIGURE 7.1 A transect of a Danish sand beach showing gradients of a) water, b) salinity, and c) oxygen at low tide. (From "Diurnal and Annual Variations of Temperature & Salinity of Interstitial Water in Sandy Beaches," B. O. Jansson, *Ophelia*, Vol. 4, pp. 173–201, 1967. Copyright © 1967 Apollo Books. Reprinted by permission.)

(a)

(b)

(c)

grain size becomes too coarse, the water will not be held in the sand by capillary action when the tide ebbs and will drain away, leaving only a very thin layer coating the grains. By contrast, fine-grained sediments can hold considerable water in the interstices through capillary action.

The presence of water circulating through the pore spaces in the sediment is important because this water movement is responsible for renewing the oxygen supply. Circulation is best in coarse-grained sediments and is reduced in fine-grained ones. If the grain size becomes too fine, as in mud shores, circulation practically ceases. The result is the creation of anaerobic layers in the sediment. (See the discussion of the RPD layer in Chapter 6, pp. 325–326.) Maximal numbers of meiofauna have been found where sand grain diameters were between 0.175 and 0.275 mm. Fenchel (1978) reports that the interstitial fauna tends to disappear when the median grain size diameter drops below 0.1 mm.

In addition to grain size, the mineral nature of the grains is of importance in determining the composition of the meiobenthos. Different associations of organisms may be found in siliceous sand as opposed to carbonate sand.

As the preceding paragraphs suggest, oxygen is an important factor in this environment. Practically all marine sediments have an oxidized layer at the surface, beneath which lies a completely anoxic layer of different chemical composition. Thus, meiofaunal organisms living below a certain depth will encounter oxygen-free conditions. The thickness of the oxygenated layer depends on a number of factors, such as grain size, the amount of organic material, water turbulence, and bacterial metabolism.

Temperature is another environmental factor of importance in determining the presence or absence, and the distribution, of meiofaunal organisms. Temperature ranges are most extreme in intertidal beaches and in the upper layers of the beach. Temperature changes are minimal in subtidal sediments and in intertidal sediments below 10–15 cm in depth (Figure 7.2). The surface layer of sand on a beach insulates the lower layers, dampening any significant temperature changes. Within the surface layers, however, the temperature may change markedly, depending on the air temperature, the effect of wind and rain, the amount of sunlight hitting the surface, and the temperature of the seawater. In cold temperate and polar

FIGURE 7.2 Graph showing the change in temperature at different depths in a Danish sand beach during a spring day. (From "Diurnal and Annual Variations of Temperature & Salinity of Interstitial Water in Sandy Beaches" B. O. Jansson, *Ophelia*, Vol. 4, pp. 173–201, 1967. Copyright © 1967 Apollo Books. Reprinted by permission.)

regions, the temperature may be low enough to freeze the upper layers. Some meiofaunal organisms (e.g., turbellarian *Coronhelmis lutheri*, gastrotrich *Turbanella hyalina*, oligochaete *Marionina southerni*, and harpacticoid copepod *Parastenocaris phyllura*) have adapted to survive despite being frozen. Because of the range of temperatures that may occur in the upper layers, the fauna inhabiting these areas may be different from those of lower layers, including those with a considerable tolerance for temperature change (eurythermal organisms).

Another factor acting on the meiobenthos, particularly of intertidal beaches, is **salinity**. Reduced salinities may occur in intertidal beaches as a result of freshwater runoff over the beach during low tide or heavy rainfall. As noted in Chapter 6, such salinity changes are usually confined to the upper layers of the beach, because the lower layers retain elevated levels of salt water through capillary action. Since fresh water is less dense than salt water, fresh water cannot penetrate below the point where marine water is held by the capillary action. This means that only the uppermost layers are subject to changes in salinity. Organisms inhabiting these layers are usually adapted to tolerate such changes (e.g., down to 15 psu).

A final environmental factor is **wave action**, which acts both intertidally and subtidally, but is most dramatic intertidally. Breaking and moving waves resuspend the upper layers of sand and completely remove or deposit large amounts of sand seasonally. Whenever the beach is churned up, all of its internal space is rearranged, and the organisms themselves are moved about. This process has profound effects on organisms that depend on staying in the space between grains. For those organisms, the interstitial space is constantly being rearranged, and they are in constant danger of being thrown out into the open water, where the risk of predation is much greater. Special adaptations for coping with this risk are known. (See pp. 351–352.)

Before leaving the realm of environmental factors, some mention should be made of light. Because light rarely penetrates more than 5–15 mm into sediments, microscopic plants should generally be restricted to the very surface layers. Diatoms, however, have been found at depths considerably below that level, raising speculation about how they manage to survive. Most of the meioflora and microflora are attached to the surface of the sand

grains and form an important food source for a variety of meiofaunal grazers.

Composition of the Interstitial Assemblages

The types of organisms constituting the meiofauna are drawn from a broad range of invertebrate phyla. Some are represented by one or a few species, whereas others are abundant both in individuals and in species. Their small size invites comparison to plankton organisms. In general, meiofaunal organisms have a size range similar to that of some of the

FIGURE 7.3 Types of interstitial ciliate protozoans of the phylum Ciliophora. (a) *Loxophyllum vermiforme*, with a flattened body. (b) *Tracheloraphis remanei*, with a cylindrical body. (c) *Remanella faurei*, with a flattened, elongated body. (d) *Remanella caudata*, with an elongated body and tail. (e) *Geleia gigas*, with a threadlike body. (From "The Interstitial Fauna of Marine Sand," B. Swedmark, *Biological Review*, vol. 32, pp. 1–42, 1964. Copyright © 1964. Reprinted by permission of Cambridge University Press.)

smaller mesoplankton and the microplankton. (See Chapter 2.)

Those invertebrate phyla which normally have small bodies are, of course, preadapted to live in the small spaces between sand grains and are represented by many species and individuals. The protistan phylum Ciliophora is represented abundantly by a large number of species. These are organisms diverse in form and, interestingly enough, often larger than the metazoan animals found with them (Figure 7.3). The class Turbellaria, of the phylum Platyhelminthes, is also abundantly represented by numerous small, flat, and elongate worms (Figure 7.4). Most abundant of all, perhaps, are the ubiquitous roundworms, the nematodes (Figure 7.5).

Several inconspicuous invertebrate phyla are well represented in the meiofauna, probably because they are normally small. These include the phyla Gastrotricha, Kinorhyncha, Tardigrada, and Rotifera (Figure 7.6). The Gastrotricha are particularly numerous in marine sediments (Figure 7.7) and Rotifera in fresh water. Certain phyla with very few living representatives, such as Priapulida, Loricifera, and Entoprocta, are also found in the meiofauna (Figure 7.8).

The worms of the phylum Annelida, with their elongated shape, are well suited to the interstitial environment and are abundant. Representatives in the meiofauna are from the classes Oligochaeta and Polychaeta (Figure 7.9).

FIGURE 7.4 Examples of some turbellian flatworms from an interstitial marine sand beach in Florida. (a) A member of the family Macrostomidae. (b) A member of the family Kalyptorhynchidae. (c) *Polystylophora* sp. (d) *Proschizorhyncus* sp. (e) *Cicerina* sp. (f) *Nematoplana* sp. (From Louise F. Bush, "Characteristics of Interstitial Sand Turbellaria," *Trans. Amer. Microscopical Society*, Vol. 87, No. 2, pp. 244–251, 1968. Reprinted by permission.)

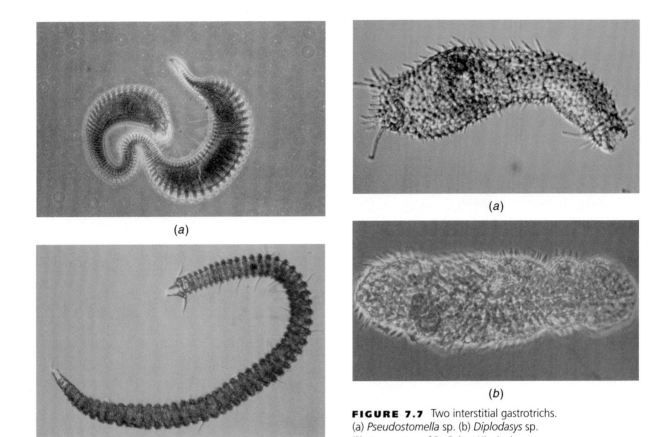

(a)

(b)

FIGURE 7.5 Two interstitial nematodes. (a) *Epsilonema* sp. (b) *Prototrichoma* sp. (Photos courtesy of Dr. Robert Higgins.)

(a)

(b)

FIGURE 7.7 Two interstitial gastrotrichs. (a) *Pseudostomella* sp. (b) *Diplodasys* sp. (Photos courtesy of Dr. Robert Higgins.)

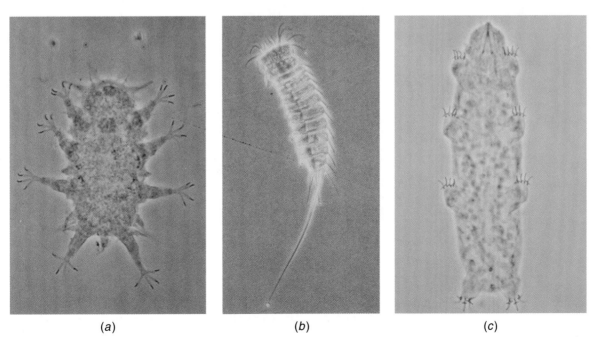

(a)   (b)   (c)

FIGURE 7.6 Representative interstitial Kinorhyncha and Tardigrada. (a) *Haleshiniscus* sp. (Tardigrada). (b) *Centroderes spinosus* (Kinorhyncha). (c) *Anisonyches diakidius* (Tardigrada). (Photos courtesy of Dr. Robert Higgins.)

(a)

(c)

(b)

FIGURE 7.8 Meiofaunal representatives of three small phyla. (a) *Macabbaeus tentataculus* (Priapulida). (b) *Nanoloricus* sp. (Loricifera). (c) *Loxosoma* sp. (Entoprocta). (Photos courtesy of Dr. Robert Higgins.)

FIGURE 7.9 Some representative interstitial annelids of the class Polychaeta. (a) *Mesonerilla intermedia*. (b) *Paranerilla limnicola*. (c) *Nerillidium simplex*. (d) *Diurodrilus* sp. (e) *Trilobodrilus* sp. (From N. C. Hulings, ed., 1971, *Proceedings of the First International Conference on Meiofauna*, Smithsonian Contributions to Zoology no. 76.)

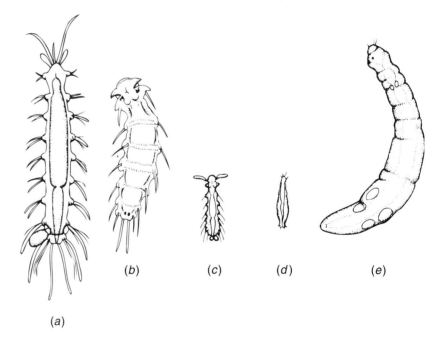

(b)   (c)   (d)   (e)

(a)

Crustaceans are represented primarily by an abundance of harpacticoid copepods and ostracods and a few other small groups, such as Mystacocarida (Figure 7.10).

Phyla that normally consist of large organisms or of sedentary or sessile forms are poorly represented in the meiobenthos. In the case of large-bodied phyla, this is due to the body size, which does not allow the organisms to fit into the small spaces. These groups are, therefore, represented by only a few rather aberrant forms. Examples of such phyla are Echinodermata and Cnidaria

(a) (b)

FIGURE 7.10 Some crustacea of the interstitial fauna. (a) *Derocheilocaris typica* (Mystacocarida). (b) Harpacticoid copepod. (Photos courtesy of Dr. Robert Higgins.)

(Figure 7.11). Phyla of sessile or sedentary habit have few representatives, because the dynamic nature of the substrate makes it difficult to remain permanently fixed in one place. Moreover, most of these phyla are suspension feeders that depend on filtering organisms out of the water they live in. In the interstitial environment, there is relatively little water to filter and there are very few organisms in the water. Most organisms are attached to, or adhere to, the particles. The sessile groups represented by a very few species include the bryozoans and ascidians (Figure 7.12).

Absent from the interstitial fauna are the phyla Phoronida, Echiura, Porifera, Ctenophora, Hemichordata, and Chaetognatha.

Sampling and Extracting Meiofauna

Because interstitial organisms are very small and occupy a unique habitat, special techniques are needed to obtain samples of them. A great variety of sampling and extracting techniques have been employed by meiobenthologists over the years, and none is standard. Most techniques are, however, simple and may be grouped into general classes.

The small size of the organisms means that only a small sample of the substrate is required. This is most conveniently obtained by taking a core. Coring devices are usually hollow plastic or metal cylinders that may be pushed into the sand by hand, closed off at one or both ends, and removed from the beach (Figure 7.13). The enclosed sedimentary column may

then be extruded and the organisms extracted. It is also possible to take a small volume or core from a larger grab or dredge sample taken for macrofauna. Since the meiofauna exists throughout a number of vertical layers of the sand, it is best to take a core that extends some distance vertically into the substrate. The majority of the meiofaunal individuals will, however, be concentrated in the top 10–15 cm, so a depth of 25–30 cm is adequate to sample all but a few forms.

The diameter of the coring device is not critical, but those in common use range in size from 2–4 cm. Larger sizes produce such large samples that the extraction process is prolonged and difficult.

Once the sample is in hand, it is necessary to extract the organisms. Meiobenthic animals are reluctant to leave the sediment and must be removed by special methods in order to be studied. A number of different extraction techniques have been used over the years, and each gives somewhat different results. Note that the fauna may be extracted alive or may be preserved. In most cases, it is preferable to extract the animals alive, because live animals are much more easily seen and because groups such as flatworms and gastrotrichs are virtually unidentifiable when preserved.

The simplest method of extraction is to scoop up a bucket of sand and let it stagnate until organisms are driven to the surface by the declining oxygen concentration. They may then be collected by pouring off the surface water through a 62-μm screen. Another simple technique is to treat a sample of sediment with fresh water, to osmotically shock the animals to release from the particles, and then to pour off the water through a 62-μm screen. Perhaps the most used technique is to treat the sample with

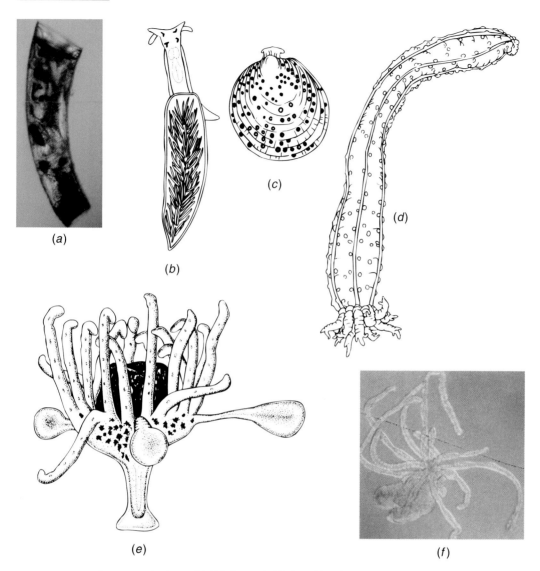

FIGURE 7.11 Representative interstitial Mollusca, Cnidaria, Echinodermata, and Brachiopoda. (a) *Caecum* sp. (Mollusca, Gastropoda). (b) *Hedylopsis brambeli* (Mollusca, Gastropoda). (c) *Gwynia capsula* (Brachiopoda). (d) *Labidoplax buskii* (Echinodermata). (e) *Stylocoronella riedli* (Cnidaria, Scyphozoa). (f) *Halammohydra* sp. (Cnidaria, Hydrozoa). (a, f, photos courtesy of Dr. Robert Higgins; b–e From *Proceedings of the First International Conference on Meiofauna*, Smithsonian Contributions to Zoology, No. 76, 1971.)

an isotonic solution of MgCl$_2$, which will anesthetize the animals. Then they can be collected by pouring the MgCl$_2$ solution through a fine net. Even more efficient is the method of **elutriation**. In this technique, a sample of sediment is introduced into a container with openings at the top and bottom. The sediment is constantly stirred up by water entering from below. As the sediment moves about, the organisms are dislodged and carried over and out with the water through the upper opening. This water is filtered on a very fine mesh, which then catches the organisms (Figure 7.14). With this technique, one must remember to use only filtered water at the input. Otherwise, the sample will be contaminated by plankton organisms. The technique can be employed with either living or preserved material.

FIGURE 7.12 Sample interstitial Bryozoa. *Monobryozoon ambulans* (Bryozoa). (From *Proceedings of the First International Conference on Meiofauna*, Smithsonian Contributions to Zoology, No. 76, 1971.)

Adaptations

Most of the morphological adaptations observed in the interstitial fauna correlate closely with a few aspects of the environment. In the dynamic environment, the individual sediment grains of the upper layers are constantly being resuspended and deposited; thus, the location and amount of interstitial space are constantly changing. To survive here, meiofaunal organisms must remain in these spaces in the sediment and must avoid both being crushed by moving sand particles and being thrown out into the plankton, where their chances of survival would be reduced.

Perhaps the most obvious adaptation to such an environment is that of *size*: All meiofaunal organisms are very small. Furthermore, in this fauna, one finds the smallest representatives of most large phyla. For example, mollusks, echinoderms, and annelids are usually fairly large animals. All have representatives in the meiobenthos, but they are remarkable for their small size. The holothurian *Leptosynapta minuta* is 2 mm, the gastropod *Caecum glabrum* also 2 mm (see Figure 7.11), the polychaete *Diurodrilus minimus* but 350 μm (see Figure 7.9), and the hydroid *Halamohydra* 1 mm (see Figure 7.11). In the case of the ciliate protozoans, however, the size is generally larger than that of ciliates living in other environments. At the same time, these larger ciliates are considerably flattened and elongated, which allows them to fit more readily into the interstices. (See Figure 7.3.) Thus, in this case, larger really means longer.

Another adaptation concerns the *shape* of the body. Most of the meiofaunal organisms have elongated or vermiform bodies—even those organisms which are normally not vermiform. (See Figure 7.3.) Another type of body shape observed is represented by organisms that are very flat. Flat organisms fit into the narrow spaces well, and they also present a greater surface

FIGURE 7.13 A simple coring device for obtaining samples of interstitial organisms. (Photo by James Nybakken.)

FIGURE 7.14 An elutriation apparatus for separating interstitial fauna from the sand. (Photo by James Nybakken.)

area for hanging onto the grains. This latter adaptation has survival value, in that it permits the organism to stay on the sediment grain during resuspension. Along with small body size, there is a reduction in the complexity and number of the body organ systems in metazoan animals. For example, in the polychaetes, the complex pharyngeal apparatus found in some macrofaunal polychaetes often disappears, as do the parapodial gills and even the kidneys (nephridia). In the meiofaunal hydroids, the number of tentacles is reduced from that observed in forms not living interstitially.

One might expect that organisms living in an environment in which the substrate was often rearranged would be at risk of being crushed as sediment grains banged into each other. In response to that possibility, many meiofaunal animals have developed various types of reinforcement to their body walls. Skeletons of spicules are found in such

ciliates as *Remanella*, such turbellarian flatworms as *Acanthomacrostomum spiculiferum*, and the acochlidiacean mollusks *Rhodope* and *Hedylopsis*. Gastrotrichs, by contrast, have an armor of scales, while the nematodes depend on a thick, heavy cuticle. Those delicate animals which lack a protective armor have developed great abilities to extend and contract the body quickly, a mechanism that also permits escape from crushing.

Perhaps because the environment is so dynamic, we find that most of the inhabitants are free moving. This is the case even for groups that are normally sessile, such as the Bryozoa and Ascidiacea. It is possible, however, that since all the common ways of extracting the meiofauna depend on the fauna's leaving the substrate, there may be other sessile forms that we have not yet observed.

Two final types of adaptation may be noted. Both enable the meiofaunal organisms to remain in the sediment during the resuspension process. Many species of various phyla, such as flatworms and gastrotrichs, have adhesive organs that fix them to the sediment grains. Crustaceans and tardigrades use hooks or claws on their limbs. **Statocysts**—organs that detect gravity and help organisms differentiate up from down—are also common. Such sense organs enable any organism to determine quickly which way to move during resuspension; thus, they can return to the sediment rather than be thrust up into the open water.

■ LIFE HISTORY

As yet, we know relatively little of the life histories of many meiofaunal organisms. There are two groups of interstitial organisms. **Temporary meiofauna** consists of the newly settled juveniles of the macrofaunal organisms, and **permanent meiofauna** consists of organisms that spend their entire lives in the interstices of the sediment. The temporary meiofaunal organisms are not considered in this text.

Since the permanent meiofaunal organisms are so small, the number of gametes that they can produce at any one time is limited. In contrast to the macrofauna, which often produce thousands or hundreds of thousands of eggs, egg production by meiobenthic organisms is almost always below 100 per individual and usually ranges between 1 and 10!

When the number of eggs is that small, a species cannot afford to lose many, or there will be no next generation. As a result, these animals have evolved a number of adaptations to ensure the survival of the few propagules. First, there are adaptations which ensure that fertilization will occur. One way is to have copulation in which the sperm are directly transferred to the female. Such is the case with harpacticoid copepods. Another way is to package all the sperm in a single unit called a **spermatophore** and attach it to the female where it will be available to fertilize the eggs as they issue. Such a mechanism occurs in certain polychaete worms and in the acochlidiacean mollusks. Finally, some species are hermaphroditic and have both male and female systems in the same individual, perhaps ensuring self-fertilization. Among these organisms are gastrotrichs, some hydroids (*Otohydra*), and some polychaetes (*Protodrilus*).

Once fertilization has occurred, it is important that the embryos be protected from loss. Usually, macrobenthic organisms produce larvae that enter the plankton in order to obtain maximum dispersal. However, the plankton community is notoriously dangerous, and most of the embryos are destroyed via predation. To compensate, the benthic animals using this route must produce large numbers of larvae. Since the meiofauna cannot produce large numbers, it would not be a good survival strategy for them to put larvae into the plankton—and in fact, that is what is observed: Very few meiobenthic animals have planktonic larvae, and those which do are the ones that produce the largest numbers of eggs. Instead, most meiofaunal organisms have brood protection. The eggs are kept with the female until the young have reached a sufficient size to be self-sustaining. That selection pressure for brood protection is high in this environment can be deduced by observing that brood protection occurs here in groups of invertebrates in which it is otherwise virtually unknown, such as cnidarians (*Otohydra*) and gastrotrichs (*Urodasys viviparus*).

For those few that do not have brood protection, additional safeguards are found. For example, eggs that are spawned either are sticky, so as to attach quickly to sand, or are confined in cocoons that are sticky. In any case, the development is usually direct; that is, the egg hatches as a juvenile and does not have a free-swimming larva, or a larva is produced that is nonpelagic and that remains in the interstices of the sediment.

Seasonal reproductive periods have been observed in many species. Studies by Gerlach (1971) suggest that generation times of interstitial organisms range from a few days to over a year. The average number of generations per year is three.

Ecology

Coull and Bell (1979) report that, on a worldwide basis, the average number of meiofaunal organisms per square meter is 10^6 and the biomass between 1 and 2 g/m^2. Both numbers of organisms and biomass are highest in intertidal areas and decrease with increasing depth in the oceans. The most numerically abundant taxa in all environments are nematodes and harpacticoid copepods.

The composition of the meiofauna varies both vertically and horizontally in the substrate. The factors creating this zonation are grain-size differences and physiochemical factors, particularly temperature, oxygen, and salinity. Characteristic patterns of assemblages of species are correlated with these factors. Most meiofaunal organisms are concentrated in the upper sediment layers, particularly in muddy sediments. Coull and Bell (1979) found that as much as 94% of all meiofauna were in the uppermost 1 cm of such sediments. In sand beaches, meiofauna have been found as deep as 90 cm. Subtidally, perhaps 95% of the meiobenthos occur in the upper 7 cm. The variability in vertical extent seems related to the oxygen concentration; most taxa are present only in aerobic sediments. There is often a definite vertical zonation (Figure 7.15), which may, however, also be due to predation. Coull et al. (1989) reported that vertical cores taken where the fish *Leiostomus xanthurus* was feeding showed that the meiofauna was significantly reduced in the upper 2 cm.

There is a specialized faunal assemblage that is confined to the anoxic layers, the **thiobios**. Characteristic of this assemblage are ciliates of the orders Trichostomatida, Heterotrichida, and Odontostomatida, of which the latter contains only anaerobic species and is confined to this area. Zooflagellates, certain nematodes, turbellarians, gnathostomulids, rotifers, and gastrotrichs are also present, but it is not known whether these forms are permanent members of the community or transients from the upper oxidized zones.

Vertical migration of the organisms also occurs, usually triggered by temperature or salinity changes. Vertical migration is especially prominent in beaches in the temperate zone. The same organisms are often found deeper during the winter months and shallower during the summer. Strong tidal or wave action following storms may also trigger a migration, which may occur on a much shorter time scale—on the order of a few days or a tidal cycle.

FIGURE 7.15 Vertical zonation of interstitial animals in a Scandinavian beach. (1) Microdasyoid, *Urodasys*; (2) gastrotrich, *Chaetonotus*; (3) nematode; (4) turbellarian, *Coelogynopora*; (5) tardigrades; (6) harpacticoid, *Cylindropsyllis*; (7) archiannelid, *Nemillidion*; (8) hydroid, *Halammohydra*. Ciliates: (9) *Pseudoprorodon*; (10) *Helicoprorodon*; (11) *Tracheloraphis*; (12) *Loxophyllum*; (13) *Litonotus*; (14) *Dilepthus*; (15) *Lynchella*; (16) *Chlamydodon*; (17) *Prontonia*; (18) *Frontonia*; (19) *Blepharisma*; (20) *Condylostoma*; (21) *Diophrys*; (22) *Strombidium*; (23) *Aspidisca*; (24) *Paraspathidum*; (25) *Mesodinium*; (26) *Remanella*; (27) *Kentrophorus*; (28) *Cardiostomum*; (29) *Homalozoon*; (30) *Lacrymaria*; (31) *Geleia*; (32) *Sonderia*; (33) *Metopus*; (34) *Caenomorpha*; (35) *Saprodinium*; (36) *Myelostoma*; (37) *Parablepharism*.

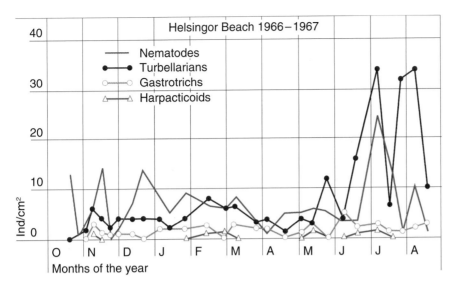

FIGURE 7.16 Changes in abundance of the major interstitial organisms with season in a Danish beach. (From "The Ecology of Marine Microbenthos" Fenchel et al., *Ophelia*, Vol. 6, pp. 1–182, 1969. Copyright © 1969 Apollo Books. Reprinted by permission.)

There are also seasonal changes in the abundance of meiofaunal organisms. Certain species are more abundant in one season than another (Figure 7.16). For example, on Danish beaches, Muus (1967) found that copepods were at a minimum in winter and peaked in numbers in the summer. Fenchel (1978) has noted that there is an annual change in the densities of the dominant groups in temperate shallow-water areas. Copepods are succeeded by oligochaetes, nematodes, ostracods, and turbellarians. This temporal succession appears to be regulated by primary productivity and predation. Not only do abundances vary seasonally, but so does the species composition, with different species becoming abundant at different times of the year. This variation is probably correlated with changes in the physiochemical factors mentioned previously, but it may also be due to migration.

Meiofauna are sensitive to both human-induced and natural perturbations. Studies by Coull and Palmer (1984) recorded an immediate decrease in the abundance and diversity of constituent meiofauna following a disturbance. In such perturbations, the crustaceans are the most affected, the nematodes the least affected. Recovery rates from perturbations are variable, taking from only hours to months. Because of their large numbers, stationary life habits, short generation time, intimate association with sediments, and rapid response to a disturbance, meiofaunal organisms would seem to be ideal organisms to use as indicators of pollution. However, the great difficulty in identifying these tiny animals, the lack of experts in the field, and the high expense of processing meiofaunal samples has precluded their widespread use for that purpose. In a review of the effects of pollution on meiofauna, Coull and Chandler (1992) noted that species diversity consistently decreased under polluted conditions, but the abundances of individuals of the major taxa increased in half the studies and decreased in the others.

Meiofaunal organisms are patchily distributed horizontally, even when the grain distribution of the sediment is homogeneous. We are ignorant of the causes of such patchiness, but various researchers have suggested several reasons. Prominent among these suggestions is the idea that food sources are patchy, which leads to concentration of the meiofauna. Others have suggested that the occurrence of macrofaunal organisms or their constructions, such as burrows and tubes, increases the heterogeneity of the environment and the patchiness of the animals' distribution. Finally, since most meiofaunal organisms lack a larval dispersal mechanism, reproduction may result in aggregations of juveniles.

There is little information on the long-term variability of meiofauna. The few studies have produced conflicting results. The earliest, a nine-year study by McIntyre and Murison (1973), suggested that annual variations in meiofauna abundance were small, whereas a five-year investigation led Coull and Bell (1979) to the opposite conclusion. More recently, in an 11-year study of two sites, one sand and one mud, Coull (1985a) demonstrated that the variability in the mud site was twice that of the sand, and the year-to-year variability was greater than the seasonality, but no long-term cycles were found.

Trophic relationships among the components of the interstitial fauna are not well known. It has been determined that certain flatfishes, such as *Pleuronectes platessa, Platichthys tiesus, Microchirus boscanion*, and *Limanda limanda* in European waters, actively prey upon meiofauna when they are young and recently metamorphosed. McIntyre (1969) reports that the stomachs of these fishes contained numbers of harpacticoid copepods, ostracods, oligochaetes, and nematodes. More recently, Nelson and Coull (1989) demonstrated that the juveniles of the fish *Leiostomus xanthurus* preyed preferentially on the meiofaunal copepods in contrast to the more abundant nematodes. Other macrofaunal predators on meiofauna include shrimps, larger polychaetes, and hydroids. Macrofaunal deposit feeders, such as polychaetes and holothurians, must also destroy numbers of meiofauna existing on the particles they consume.

In addition to macrofaunal predators, there are interstitial predators, including certain nematodes, turbellarians, some tardigrades, and hydroids, such as *Protohydra*. The latter is a particularly voracious feeder on copepods and nematodes, according to Muus (1966). Predation by both meiofaunal and macrofaunal predators reduces the populations of other interstitial organisms, but it is not known how extensive this reduction is.

In addition to predators, the interstitial fauna contains three other feeding types: herbivores, detritus feeders, and suspension feeders. While some turbellarians are predators, others are scavengers, feeding on dead meiofaunal and macrofaunal organisms. Harpacticoid copepods are primarily herbivores, feeding on the few attached diatoms or on detritus. Ostracods also ingest diatoms. Detritus feeders include gastrotrichs, some nematodes, and archiannelid polychaetes. Suspension feeders are quite rare, due to the absence of an interstitial plankton.

A considerable degree of food specialization exists among the meiofauna. Fenchel (1968), studying the interstitial ciliates, found a surprising degree of food specificity for various bacterial and algal taxa. In addition, he showed that, in certain habitats, whole groups of species subdivided their food resource according to size. A similar food niche separation has been reported by Tietjen and Lee (1977) in nematodes.

What controls meiofaunal communities? Again, we lack sufficient data to come up with an answer. In a study of two sites in South Carolina over an 11-year period, Coull (1985b) suggested that organisms in fine-grained sediment (mud) were controlled by predators, while in the nearby sand habitat the abundances of organisms were controlled by physical factors.

■ MEIOFAUNA–MACROFAUNA INTERACTIONS

What is the significance of the meiofauna to the general ecology of benthic sedimentary systems, and what, if any, is the relationship of the meiofauna to the macrofauna? These questions have generated both considerable interest and conflicting opinions over the years. The average meiofaunal biomass is from one-tenth to one-fifth that of the resident macrofauna. In certain areas, such as the intertidal sand and mudflats and in the deep sea, the ratio is closer to 1 to 1 (Platt, 1981). This suggests that, in these latter areas, meiofauna plays an important role in energetics. Because of their small size, higher metabolic activity, and high turnover rate, productivity of meiofaunal organisms, per unit of biomass, is greater than that of the macrofauna. Therefore, even in systems dominated by macrofauna, total organic meiofaunal production approaches that of the macrofauna. Clearly, the meiofauna represents an important reservoir of organic material. Furthermore, through their metabolic secretions, meiofaunal animals stimulate bacterial production in the sediments.

What is the ultimate fate of this production? Is it merely recycled within the meiofaunal system, or is there a linkage with the macrofauna and higher trophic levels? Earlier workers, such as McIntyre and Murison (1973), thought that the meiobenthic system was separate, its production unavailable to higher trophic levels of the macrofauna. It now appears likely that these two systems are linked to some degree. For example, nonselective deposit feeders of the macrofauna cannot avoid consuming meiofauna, and if the larger animals' digestion is equally nonselective, then meiofauna must contribute to their nutrition. Also, the resuspension of bottom particles, which are captured by large suspension feeders, would lead to a similar macrofaunal ingestion. Macrofaunal predators that selectively feed on meiofaunal organisms seem rare, but Feller and Kaczynski (1975) have demonstrated that juvenile salmon are predators of meiobenthic copepods, and Aarnio et al. (1998) have shown that the macrofaunal priapulid *Halicryptis spinulosa* significantly reduces meiofauna organisms in the Baltic Sea. Bell and Coull (1978) have demonstrated

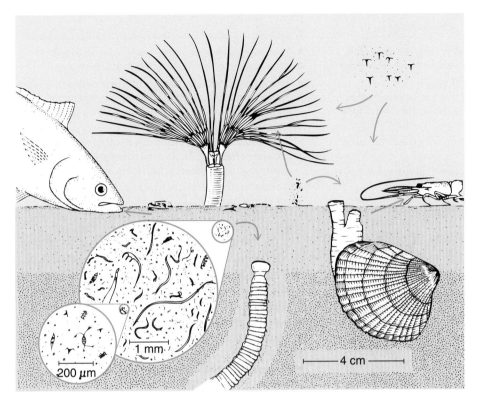

FIGURE 7.17
Hypothetical benthic food web, illustrating possible connections between the meiofauna and macrofauna. (From "Meiofaunal Dynamics and the Origin of the Metazoa," H. M. Platt, 1981, pp. 207–216, Natural History Museum.)

that the grass shrimp *Palaemonetes pugio* not only feeds on meiofaunal organisms, but also can control the meiofaunal population and community structure. Other fishes, particularly young stages of flatfishes, gobies, and mullet, have been found to contain large quantities of meiofaunal organisms in their guts. Thus, the meiofaunal organisms probably fit into the macrofaunal food web, as illustrated in Figure 7.17.

The extent to which meiofauna enters the macrofaunal food web varies with the type of sediment. It is likely that meiofauna is more significant as food to macrofauna in muddier sediments than in sands, partly because in muds the meiofaunal organisms are concentrated in the top layers, where they are more accessible, whereas in sand they may be much deeper and less accessible to many macrofaunal organisms. An important question is "Since the macrofauna interact with the meiofauna, what effects does the macrofauna have on the meiofaunal community structure? Little is known at present, but Bell and Coull (1978) have demonstrated that, in at least one case, meiofaunal numbers, particularly of the slower-moving wormlike meiobenthos, could be significantly reduced, suggesting a predator-controlled system.

■ SPECIES RICHNESS, DISTRIBUTION, AND DISPERSAL

One of the more remarkable features of the meiofauna is its species richness. Fenchel et al. (1967) found as many as 70 species of meiofaunal organisms in 50 cm^2 of sand, and they regularly obtained 30 to 50 species of ciliate protistans alone. Even more striking is Gerlach's (1965) discovery that the Arctic interstitial fauna is rich in contrast to the impoverished macrofauna at that latitude. On the basis of the limited studies available, it appears that there is no gradient in species richness in this fauna as one moves from polar regions to the tropics. This observation contrasts with the common distribution pattern of most macrofaunal organisms, where lower latitudes tend to have greater richness than higher latitudes.

Zoogeographically, the meiofauna have a pattern in which the common genera are often cosmopolitan, but the species are not. As a result, several different zoogeographical regions or provinces can be recognized.

Because most meiofaunal organisms lack a motile larval phase, they disperse mainly by the transport of adults by waterfowl, rafting on floating substrates (including icebergs), and turbulent water (Gerlach, 1977).

Summary of Key Concepts

- The meiofauna, or interstitial organisms, are those tiny animals which occupy the microspaces between particles or which live on individual particles.
- Environmental conditions affecting the meiofauna are somewhat different from those affecting the macrofauna in the same area.
- Perhaps the most important factor influencing meiofauna is grain size.
- Coarse grain size means more interstitial space; fine grain size means less interstitial space.
- Coarse-grained sediments hold less water by capillary action than do fine-grained sediments.
- The circulation of water and oxygen is greater in coarse sediments and reduced in fine-grained sediments.
- In very fine grained sediments, such as mud, water circulation practically ceases, the interstitial habitat becomes anoxic, and the meiofauna disappears.
- Temperature is another factor of importance in determining the presence, absence, and distribution of meiofaunal organisms. Temperatures are most variable in the surface layer of sediment, and the variability decreases with depth.
- Reductions in salinity due to flooding with fresh water usually affect only the uppermost layers of sediment, because the lower layers retain seawater by capillary action.
- Wave action churns up the sediment, constantly rearranging it and increasing the chances of throwing the meiofauna into the open water.
- The types of organisms constituting the meiofauna include a broad range of invertebrate phyla dominated by those phyla which normally have small-enough sizes to fit between the sediment grains.
- Phyla that consist primarily of large, sedentary, or sessile organisms are poorly represented in the meiofauna.
- Special techniques, including osmotic shock, anesthetization with magnesium chloride, seawater ice, and elutriation, are required to extract the meiofauna from the sediment.
- The most obvious adaptation of meiofaunal organisms is that they are very small in size.
- Another adaptation is that these animals tend to be flattened and elongated or vermiform in shape.
- An adaptation to the dynamics of disturbance in the sediment is the presence of skeletons that offer some protection against crushing.

- Most interstitial animals are free moving and have various adaptations to cling to the sediment grains.
- Temporary meiofauna consists of newly settled juveniles of macrofaunal organisms, whereas permanent meiofauna consists of organisms that spend their entire lives in the interstices of the sediment.
- Due to their small size, meiofauna produce very few eggs and ensure their fertilization and survival by various mechanisms that retain the embryos in the interstitial space.
- Worldwide, the average number of meiofaunal organisms per square meter is 10^6.
- The composition of the meiofauna varies both horizontally and vertically in the substrate, with most concentrated in the upper layers.
- Vertical zonation of meiofauna in the sediment appears to be related to oxygen concentration or predation.
- Vertical migration triggered by temperature or salinity changes is especially prominent in temperate-zone beaches.
- Seasonal changes in the abundances of organisms and species composition may be regulated by primary productivity and predation.
- Meiofauna are sensitive to both natural and human-induced perturbations.
- Meiofauna are distributed patchily, even where the sediment grain size is homogeneous; we are ignorant of the causes of this patchy distribution.
- There is little information on the long-term variability of meiofaunal communities, and the information that is available is conflicting.
- Trophic relations among the components of the interstitial fauna are not well known.
- Predators on the meiofauna include certain fishes and macrofaunal deposit feeders that kill meiofaunal organisms when they ingest the sediment. There are also meiofaunal predators.
- The meiofauna also contains herbivores, detritus feeders, and suspension feeders.
- The significance of meiofauna to macrofaunal communities and food webs appears to depend on the type of sediment. Meiofauna is more significant to macrofauna in finer sediments, probably partly due to the concentration of the meiofauna in the top layers, where they are more accessible.
- The meiofauna is species rich, and there is no gradient in richness with change in latitude.

Review Questions

ESSAY Develop complete answers to these questions:

1. How do temperature and salinity influence the distribution and types of interstitial organisms? Present evidence to support your answer.

2. What evidence supports the hypothesis that the selection pressure on meiofaunal females to brood their offspring is high?

3. Why might one expect a diel rhythm of vertical migration of the meiofauna in the sand? How could you test to see whether this rhythm is an endogenous behavior?

4. Discuss the evidence in support of the hypothesis that meiofauna and macrofauna interact.

5. What are the most commonly found adaptations to the interstitial environment?

6. Why don't meiofaunal organisms generally produce planktonic larvae?

Bibliography

Aarnio, K., E. Bonsdorff, and A. Norkko. 1998. Role of *Halicryptus spinulosus* (Priapulida) in structuring meiofauna and settling macrofauna. *Mar. Ecol. Progr. Ser.* 163:145–153.

Bell, S. S., and B. C. Coull. 1978. Field evidence that shrimp predation regulates meiofauna. *Oecologia* 35:141–148.

Bush, L. 1968. Characteristics of interstitial sand turbellaria: The significance of body elongation, muscular development and adhesive organs. *Trans. Amer. Microsc. Soc.* 87:244–251.

Coull, B. C. 1973. Estuarine meiofauna: A review: Trophic relationships and microbial interactions. In *Estuarine microbial ecology*, edited by L. H. Stevenson and R. R. Colwell. Columbia: University of South Carolina Press, 499–512.

Coull, B. C. 1985a. Long-term variability of estuarine meiobenthos: An 11-year study. *Mar. Ecol. Progr. Ser.* 24:205–218.

Coull, B. C. 1985b. The use of long-term biological data to generate testable hypotheses. *Estuaries* 8:84–92.

Coull, B. C., and S. S. Bell. 1979. Perspectives of marine meiofaunal ecology. In *Ecological processes in coastal and marine systems*, edited by R. J. Livingston. New York: Plenum Press, 189–216.

Coull, B. C., and G. T. Chandler. 1992. Pollution and meiofauna: Field, laboratory and mesocosm studies. *Oceanog. Mar. Biol. Ann. Rev.* 30:191–271.

Coull, B. C., and M. A. Palmer. 1984. Field experimentation in meiofaunal ecology. *Hydrobiologia* 19:1–19.

Coull, B. C., R. L. Ellison, J. W. Fleeger, R. P. Higgins, W. D. Hope, W. D. Hummon, R. M. Rieger, W. E. Sterrer, H. Thiel, and J. H. Tietjen. 1977. Quantitative estimates of the meiofauna from the deep sea off North Carolina, U.S.A. *Mar. Biol.* 39:233–240.

Coull, B. C., M. A. Palmer, and P. E. Meyers. 1989. Controls on vertical distribution of meiobenthos in the mud: Field and flume studies with juvenile fish. *Mar. Ecol. Progr. Ser.* 55:133–139.

Feller, R. J., and V. W. Kaczynski. 1975. Size selective predation by juvenile chum salmon (*Oncorhynchus keta*) on epibenthic prey in Puget Sound. *J. Fish. Res. Bd. Canada* 32:1419–1429.

Fenchel, T. 1968. The ecology of marine microbenthos. II. The food of marine benthic ciliates. *Ophelia* 5:73–121.

Fenchel, T. 1969. The ecology of marine meiobenthos. IV. Structure and function of the benthic ecosystem: Its chemical and physical factors and the microfaunal communities with special reference to the ciliated protozoa. *Ophelia* 6:1–182.

Fenchel, T. 1978. Ecology of micro- and meiobenthos. *Ann. Rev. Ecol. Syst.* 9:99–121.

Fenchel, T., B. O. Jansson, and W. von Thun. 1967. Vertical and horizontal distribution of the metazoan microfauna and of some physical factors on a sandy beach in the northern part of the Oresund. *Ophelia* 4:227–243.

Gerlach, S. A. 1965. Uber die Fauna in der Gezeitenzone von Spitzbergen. *Bot. Bothab. Acta Univ. Gothab.* 3:3–23.

Gerlach, S. A. 1971. On the importance of marine meiofauna for benthos communities. *Oecologia* 6:176–190.

Gerlach, S. A. 1977. Means of meiofauna dispersal. *Mikrofauna Meeresboden* 61:89–103.

Higgins, R. P., and H. Thiel, eds. 1988. *Introduction to the study of meiofauna*. Washington, D.C.: Smithsonian Institution Press.

Hulings, N. C., ed. 1971. *Proceedings of the first international conference on meiofauna*. Washington, D.C.: Smithsonian Contributions to Zoology no. 76.

Hulings, N. C., and J. S. Gray, eds. 1971. *A manual for the study of meiofauna*. Washington, D.C.: Smithsonian Contributions to Zoology no. 78.

McIntyre, A. D. 1969. Ecology of marine meiobenthos. *Biol. Rev.* 44:245–290.

McIntyre, A. D., and D. J. Murison. 1973. The meiofauna of a flatfish nursery ground. *J. Mar. Biol. Assoc. U.K.* 53:93–118.

Muus, B. J. 1967. The fauna of Danish estuaries and lagoons. *Meddr. Danm. Fisk.-og Havunders* 5:1–316.

Muus, K. 1966. Notes on the biology of *Protohydra leuckarti* Greef (Hydroidea, Protohydridae). *Ophelia* 3:141–150.

Nelson, A. L., and B. C. Coull. 1989. Selection of meiobenthic prey by juvenile spot (Pisces): An experimental study. *Mar. Ecol. Progr. Ser.* 53:51–57.

Platt, H. M. 1981. Meiofaunal dynamics and the origin of the metazoa. In *The evolving biosphere*, edited by P. L. Foray. British Museum (Natural History) and Cambridge University Press, U.K., 207–216.

Pollock, L. W. 1970. Distribution and dynamics of interstitial Tardigrada at Woods Hole, Massachusetts, U.S.A. *Ophelia* 7(2):145–165.

Sterrer, W., and P. Ax, eds. 1977. *The meiofauna species in time and space*. Mainz, West Germany: Akademie der Wissenschaften und der Literatur.

Swedmark, B. 1964. The interstitial fauna of marine sand. *Biol. Rev.* 39:1–42.

Tietjen, J. H., and J. J. Lee. 1977. Feeding behavior of marine nematodes. In *Ecology of marine benthos*, edited by B. C. Coull. Columbia: University of South Carolina Press, 21–35.

Vernberg, W., and B. C. Coull. 1981. Meiofauna. In *Functional adaptations of marine organisms*, edited by F. J. Vernberg and W. B. Vernberg. New York: Academic Press, 147–177.

CHAPTER 8

ESTUARIES AND SALT MARSHES

Whereas the intertidal zone is the place where terrestrial and the marine habitats meet, an estuary is a place where fresh water and salt water meet. It is a transition zone between the two aquatic ecosystems. Estuaries have been, and remain, in close association with humans, because many of the major cities of the world sit on estuaries. Consequently, estuaries are one of the marine systems most heavily affected by humans. We will discuss the various special physical and chemical conditions in an estuary and how these create a rigorous environment in which relatively few species can persist.

◾

Types of Estuaries

There are many definitions of an estuary, because several geomorphological features of coastlines, such as lagoons, sloughs, fjords, and other shallow embayments, are often considered estuaries. A simple definition is that an **estuary** is a partially enclosed coastal embayment where fresh water and seawater meet and mix. This definition implies the free connection of the sea with the freshwater source, at least during a part of the year. It excludes permanently isolated coastal water impoundments, as well as such isolated brackish or saline bodies of water as the Caspian Sea, Aral Sea, and Great Salt Lake.

As a result of an area's geomorphology and geological history, and the prevailing climatic conditions, there may be different types of estuaries (Figure 8.1), each displaying somewhat different physical and chemical conditions. Perhaps the most common type is the **coastal-plain estuary**. These estuaries were

formed at the end of the last ice age when the rising sea level invaded low-lying coastal river valleys. Estuaries such as the Chesapeake Bay and the mouths of the Delaware and Hudson rivers in the United States and the Cornwall and Devon estuaries in Great Britain are examples of this type. Similar to the coastal-plain estuary is the **tectonic estuary**, which forms when the sea reinvades the land due to subsidence of the land, not as a result of a rising sea level. A good example is San Francisco Bay. A third type of estuary is the **semienclosed bay**, or **lagoon**. Here, sandbars build up parallel to the coastline and partially cut off the waters behind them from the sea. A shallow lagoon forms behind the sandbars and collects the freshwater discharge from the land. The water in such lagoons varies in salinity, depending on climatic conditions, whether any major river flows into the lagoon, and the extent to which the bars restrict access to seawater. Such estuaries are common in North Carolina, along the Texas and Florida Gulf coasts, in northwestern

Europe (the Netherlands), and parts of Australia. A final category of estuary is the **fjord**—a valley that has been deepened by glacial action and is then invaded by the sea. Fjords are characterized by a shallow sill at their mouths that restricts the interchange of water between the deeper waters of the fjord itself and the sea. Often, these deeper waters are stagnant. Fjords are abundant on the coasts of Norway, Chile, Scotland, Alaska, New Zealand, and British Columbia.

Estuaries may be classified in yet another way, depending on the manner in which the salinity gradients are formed. In most estuaries, there is a gradient in salinity from full seawater (33–37 psu) at the mouth to fresh water at the upper reaches. Fresh water is less dense than seawater, and where the two meet, the fresh water will "float" on the seawater. Mixing occurs where the two come in contact, but the extent of the mixing varies with many other environmental factors, including the shape of the basin, the nature of the tide in the area, river flow, and rainfall.

In estuaries where there is substantial freshwater outflow and reduced evaporation (typical temperate-zone estuaries), the fresh water moves out over the top of the salt water, mixing with it near the surface, reducing the salinity there, and leaving the deeper waters more saline. In such a situation, a cross section of the estuary shows isohalines (lines of equal salinity), which extend upstream at the bottom (Figure 8.2). At any given point on the estuary, a vertical column of water has its highest salinity at or near the bottom and its lowest at or near the surface. This is a **positive estuary**, or **salt-wedge estuary**, also called a **river-dominated** or **stratified estuary**. Such estuaries form a continuum—from those with little mixing and prominent salt wedges, through those with partial mixing and lesser salt wedges, to **homogeneous**, **marine-dominated**, and **neutral estuaries**, in which either complete mixing or an evaporation rate equal to the freshwater inflow gives similar salinities from surface to bottom at any point. Where an estuary fits

FIGURE 8.1 Types of estuaries. (a) Bay or lagoon lying behind a barrier beach: the major lagoons of the North Carolina coast. (b) Tectonic estuary: San Francisco Bay. (c) Boat at Milford Sound in Fiordland, New Zealand. (a, after C. Peterson and N. M. Peterson, 1979, *The Ecology of Intertidal Flats of North Carolina: A Community Profile*, U.S. Fish and Wildlife Service, Biological Services Program FWS/OBS-79/39. b, after T. J. Conomos, ed., 1979, *San Francisco Bay, the Urbanized Estuary*, courtesy of Pacific Division—American Association for the Advancement of Science. c, Copyright Peter Hendrie/Getty Images/The Image Bank.)

(a)

(b)

(c)

FIGURE 8.1 continued

FIGURE 8.2 Positive and negative estuaries. (a) Positive (salt-wedge) estuary. Freshwater runoff is significant throughout the year and is greater than evaporation. Seawater enters along the bottom and gradually mixes with outward-flowing fresh water. (b) Negative estuary. Freshwater flow is diminished or absent during part of the year. Seawater enters along the surface. Evaporation is greater than runoff, so salinity increases as one moves up the estuary. The hypersaline water sinks and flows out below the incoming seawater.

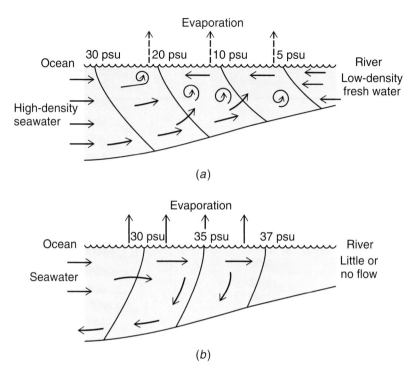

in this continuum depends not only on the amount of mixing of the water masses, but also on the tidal regime, the geometry of the estuarine basin, and the river flow. Moreover, the tidal regime and river flow may be further altered seasonally. In general, positive estuaries are common and neutral estuaries are rare. Galveston Bay, Texas, and Alligator Harbor, Florida, are examples of neutral estuaries.

In desert climates, where freshwater input to the estuary is small and the rate of evaporation is high, a **negative**, or **evaporite**, **estuary** results. In a negative estuary, the incoming salt water enters at the surface and is somewhat diluted by mixing with the small amount of fresh water. The high evaporation rate, however, causes this surface water to become hypersaline. More dense than seawater, hypersaline water sinks to the bottom and moves out of the estuary as a bottom current. A salinity profile of such an estuary is the reverse of that of the positive estuary; that is, a negative estuary has its highest salinity at the top and its lowest at the bottom (Figure 8.2). A final category is the **seasonal**, or **intermittent**, **estuary**, formed in areas where there is a marked wet and dry season (a Mediterranean climate). In the rainy season, these estuaries have freshwater input and are open to the ocean. In the

dry season, they have no freshwater input, may become dry or stagnant, and are often cut off from the ocean by seasonal sandbars. Salinity in seasonal estuaries thus varies not spatially but temporally.

Physical Characteristics of Estuaries

The physical–chemical regime of estuaries is highly variable in many parameters, creating a stressful environment for many organisms. These stresses limit the numbers and kinds of species living in estuaries.

■ SALINITY

The dominant feature of the estuarine environment is fluctuation in salinity. By definition, a salinity gradient exists at all times in an estuary, but the pattern of that gradient varies with the seasons, the topography of the estuary, the tides, and the amount of fresh water in the estuary. We have already discussed the patterns of salinity distribution in salt-wedge, partially mixed, and negative estuaries. There are, however, other factors that alter salinity patterns. The tide is one such factor. Where the tidal range is significant, high tide drives the salt water further up the estuary, displacing

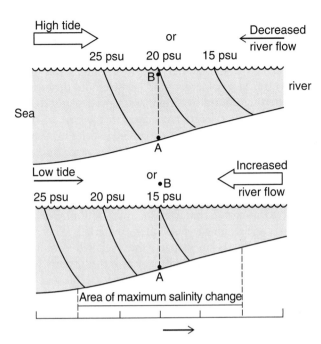

FIGURE 8.3 Change in salinity in an estuary with change in tide level or change in river discharge. Point A is on the bottom and B is intertidal. Because point B is covered only at high tide, it is inundated only with high-salinity water and is unaffected by low-salinity water at low tide. Point A, in contrast, is covered by water of different salinities at different tidal levels.

the isohalines upstream. Low tides, by contrast, displace the isohalines downstream. As a result, a certain area of the estuary is subject to a salinity regime that changes with each tide (Figure 8.3). This area has the maximum salinity fluctuation. The time scale of the salinity fluctuations is such that the salinity range over a 6- to 12-hour period equals or exceeds the entire annual range of salinities for some areas, even within an estuary.

A second force is the Coriolis effect. The rotation of the Earth deflects flowing water. In the Northern Hemisphere, this effect deflects outflowing fresh water in north–south-oriented estuaries to the right as one looks down the estuary toward the sea. Salt water flowing into the estuary from the ocean is also displaced to the right as one looks from the sea toward the estuary. The opposite is true in the Southern Hemisphere. As a result, two points, each on opposite sides of the estuary equidistant from the mouth, may have predictably different salinities (Figure 8.4).

Seasonal changes in salinity in the estuary are usually the result of seasonal changes in evaporation, freshwater flow, or both. In areas where freshwater discharge is reduced or absent for part of the year, higher salinities (salt wedges) may be found further upstream. With the onset of increased freshwater flow, the salinity gradients are moved downstream toward the mouth of the estuary. At different seasons

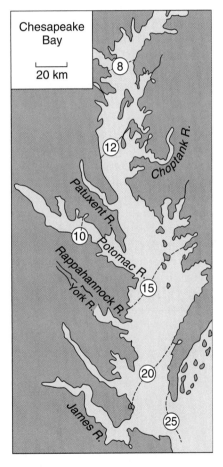

FIGURE 8.4 Result of Coriolis effect on salinity in the north–south-oriented Chesapeake Bay estuary. Note the deflection of the isohalines to the right in this Northern Hemisphere estuary. Circled numbers refer to salinity values in parts per thousand. (After *Ecology of Estuaries*, D. S. McLuskey. Copyright © 1971 Heinemann Educational Publishers. Reprinted by permission of Heinemann Educational Publishers, a division of Reed Educational & Professional Publishing, Ltd.)

of the year, therefore, a given point in the estuary may exhibit different salinities.

Thus far, we have considered only salinity changes in the water column. In the substrate, a different situation may prevail. Since estuarine substrates are often sand and mud, water is held in the interstices between the particles. This interstitial water, which originates from the overlying water, undergoes changes in salinity much more slowly than the overlying water does, because of the slow interchange between the two. The interstitial water and the surrounding mud and sand are, therefore, buffered with respect to the overlying water. Organisms dwelling in the substrate are subject to less drastic salinity changes than are organisms in the water column. This buffering effect is probably most effective in the lower intertidal zone. Indeed, variability in interstitial salinity increases with tidal height, because interstitial water at higher elevations has more opportunity to become diluted during rains and is subject to more evaporation in dry weather (Figure 8.5).

■ SUBSTRATE

The substrate in estuaries is highly variable and dependent on both the geological history of a habitat and contemporaneous sediment transport processes. Fjord estuaries that have been carved out of rock by receding glaciers have a bedrock substrate with a sediment cover that is dependent on the sediment supplied by the adjacent watershed. Estuaries in which receding glaciers left cobble deposits, such as Narragansett Bay, Long Island Sound, and Puget Sound, typically have a cobble substrate covered to varying degrees by sediments. To at least some degree, most estuaries have soft, muddy substrates derived from sediments carried into the estuary by both seawater and fresh water. Wind (aeolian) transport of larger sand particles into the estuary is often significant in certain areas, particularly coastal lagoons behind barrier beaches. In the case of fresh water, rivers and streams carry silt particles in suspension. When these suspended particles reach and mix with seawater in the estuary, the presence of the various ions in the seawater causes the particles to flocculate, creating larger, heavier particles that then settle out, forming the characteristic mud bottom. Seawater also carries much suspended material. When the seawater enters an estuary, the sheltered conditions reduce the water motion that has kept the particles in suspension. As a result, the particles settle out and contribute to the forming of the mud or sand substrate. The relative importance of freshwater-borne or marine-borne particles to the development of the muddy substrate varies both from estuary to estuary and geographically in general.

Particle deposition is further controlled by currents and particle size. Large particles settle out faster than small particles, and strong currents keep particles in suspension longer than weak currents. Where strong currents prevail, the substrate will be coarse (sand or gravel), as only large particles settle out; where waters are calm and currents weak, fine silt will settle out. Thus, both seawater and fresh water drop their coarse sediments first, the former at the mouth of the estuary and the latter in the upper reaches of the river itself.

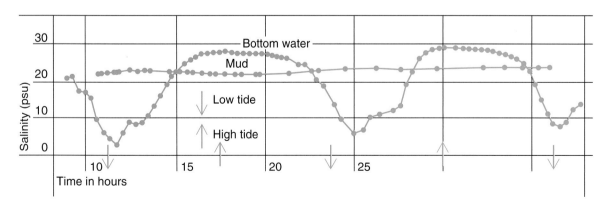

FIGURE 8.5 Comparison of salinity fluctuations in the water column with that interstitially in the bottom mud. Data from the Pocasset River estuary in Massachusetts. (Reprinted with permission from "Salinity Measurements in Estuaries," by P. C. Mangelsdorf, in *Estuaries*, Pub. No. 83, edited by G. H. Lauff, 1967, pp. 71–79. Copyright © 1967 American Association for the Advancement of Science.)

The area where the two waters mix is dominated by fine silt (mud), resulting from decreased water movement and the intermixing of the two water masses. Among the particles that settle out in estuaries is much terrestrial and marine organic matter. As a result, the substrate that accumulates is rich in organic material that may serve as a large food reservoir for estuarine organisms. The large surface area relative to volume of the very small particles means that there is a very large area for bacteria to inhabit.

Catastrophic events may also play a significant role. Massive storms and their accompanying floods may deposit or remove vast amounts of sediment in estuaries, causing severe mortality of organisms. For example, Zedler (1982) reported 4 cm of silt deposited during a 1980 flood on the Tijuana estuary in southern California, and 40% of the low-tide volume of Magu Lagoon was lost during the same flood. Similarly, Andrews (1973) reported that the massive flooding of fresh water into Chesapeake Bay on the U.S. east coast following Hurricane Agnes in 1972 caused a severe reduction in salinity and catastrophic losses of organisms.

■ TEMPERATURE

Water temperatures in estuaries are more variable than in nearby coastal waters. In part, this is because there is usually a smaller volume of water in an estuary and a larger surface area; therefore, the water heats up and cools down more rapidly under prevailing atmospheric conditions. (Fjords, being deep and with a large volume, do not show this variation.) Another reason for the variation is freshwater input. Fresh water in rivers and streams is more subject to seasonal changes in temperature than seawater is. Rivers in the temperate zones are colder in winter and warmer in summer than the adjacent seawater. When these fresh waters enter the estuary and mix with the seawater, they alter the temperature of the seawater. As a result, estuarine waters are colder in winter and warmer in summer than surrounding coastal waters. The time scale is of interest, in that, with the changing tides, a given point in an estuary will show a large temperature variation as a function of the difference between the temperature of the seawater and that of the river water.

Because fresh waters exhibit the greatest temperature ranges, as one progresses up the estuary, the annual temperature range becomes greater. Similarly, the temperature range is least at the entrance to the estuary, where mixing with fresh water is minimal. On a short-term basis, the central area of an estuary may show the greatest change with changing tide.

Temperature also varies vertically. Surface waters have the largest temperature range, deeper waters the smallest. In salt-wedge estuaries, this vertical temperature difference also reflects the fact that surface waters are dominated by fresh water, whereas the deeper waters may be completely or predominantly marine.

■ WAVE ACTION AND CURRENTS

Estuaries are surrounded by land on three sides. This means that the distance over which a wind can blow over water to create waves is minimal, at least in comparison with oceans. Since the height of waves depends on the *fetch*, or the open water distance over which the wind can blow, a small area of water can generate only small waves. The shallow depth of the water in most estuaries also precludes development of large waves. The narrowness of the mouth of the estuary, coupled with the shallow bottom, rapidly dissipates the effects of any waves entering the estuary from the sea. As a result of these processes, estuaries are generally places of relatively calm water. Reduced wave action also promotes the deposition of fine sediments, allowing the development of rooted plants, which further stabilizes these sediments.

Currents in estuaries are caused primarily by tidal action and river flow. Currents are generally confined to channels, and within these channels, velocities up to several knots can occur. Any object that occludes tidal or river flows in estuaries, such as an island or the mouth of a bay, will lead to predictable areas of accelerated flow. The highest water velocities occur in the middle of channels, where the frictional resistance from the bottom and the banks is lowest. The flow regime in estuaries strongly dictates the area's sediment and larval delivery patterns. Although the estuary is an overall place of sediment deposition, channels and other habitats in which currents are accelerated are areas of erosion. Leonard et al. (1998) found that areas of accelerated flows in estuaries are also areas with high larval recruitment of organisms (including planktonic larva) as well as areas of high productivity. This is because high fluid flows lead to an increase in the flux of planktonic larvae, which are particulate food of filter feeders, and to increased gas exchange in benthic plants and algae. Whenever currents change position, new channels are quickly eroded and old channels filled, particularly in inter-tidal areas, where a natural cycle of erosion and deposition occurs. However, in most estuaries, deposition exceeds erosion and there is a net accumulation of silt. In intermittent estuaries, where the mouth is often closed off from the sea during the dry part of the year, water movement is severely reduced at that time, leading to stagnation, a reduced oxygen content, algal blooms, and fish kills.

Most estuaries have a continual input of fresh water at the head. A given amount of this fresh water moves down the estuary, mixing with seawater to a greater or lesser degree. A volume of this water is eventually discharged from the estuary at its mouth or evaporated to compensate for the next volume introduced at the head or origin. The time required for a given mass of fresh water to be discharged from the estuary is called the **flushing time** and can be a measure of the stability of the estuarine system. Long flushing times are important to the maintenance of estuarine plankton communities and flushing characteristics can regulate the retention and export of planktonic larvae of an estuary.

■ TURBIDITY

Because of the large number of particles in suspension in estuarine waters, at least at certain times of the year, the turbidity of the water is high. The highest turbidities occur during times of maximum river flow. Turbidity is generally at a minimum near the mouth, with its full seawater, and increases with distance inland. The more an estuary approaches a lagoon morphology, the more turbidity is a function of plankton concentration or wind speed (producing resuspension).

The major ecological effect of turbidity is a marked decrease in light penetration. This, in turn, decreases photosynthesis by phytoplankton and benthic autotrophs, thereby reducing productivity. Under conditions of severe turbidity, phytoplankton production may be negligible, and the major production of organic matter is by emergent marsh plants. (See pp. 393–395.)

■ OXYGEN

The regular influx of fresh water and salt water into estuaries, coupled with the shallowness, turbulence, and mixing caused by wind, usually means that there is an ample supply of oxygen in the water column. Since the solubility of oxygen in water decreases with increased temperature and salinity, the precise amount of oxygen in the water will vary with those parameters. In salt-wedge estuaries—or, indeed, any deep estuary during the summer, when a thermocline can develop and where there is vertical salinity stratification—there is often little interchange between the oxygen-rich surface waters and the deeper layers. This isolation of the deep waters from oxygen interchange, coupled with high biological activity and a slow renewal or flushing rate, may deplete the oxygen in the bottom waters.

Oxygen can be severely depleted in the substrate as well. The high organic content and high bacterial populations of the sediments exert a large oxygen demand on interstitial water. Since a fine particle size restricts the interchange of interstitial water with the water column above, oxygen is quickly depleted in muddy sediments. Consequently, muddy estuarine sediments are often anoxic below the first few centimeters—unless they have large particles or large numbers of burrowing animals, such as the ghost shrimp *Callianassa* and the hemichordate worm *Balanoglossus*, which by their activities oxygenate lower sediment layers.

Human-caused events that can have devastating anoxic effects on estuarine bottom communities have become common throughout the world over the past few decades and are a serious threat to North American estuaries. These events are caused by the increased use of artificial nitrogen fertilizers that find their way into estuaries as runoff. In combination with high summer water temperatures, increased levels of nitrogen lead to large volumes of anoxic water at the bottoms of estuaries. Significant events with anoxic effects that kill important components of estuarine ecosystems are now common during the warm summer months in Narragansett Bay and Chesapeake Bay and in the Gulf of Mexico in the plume of the Mississippi River. Fish, oyster reefs, and clams have all been shown to suffer mass mortality from events causing these severe anoxic effects. The continued use of artificial fertilizers, together with global warming, will make these problems worse in the next century and beyond.

The Biota of Estuaries

■ FAUNAL COMPOSITION

There are three types of fauna in estuaries: marine, freshwater, and brackish-water, or estuarine, fauna.

The marine fauna makes up the largest group in terms of numbers of species and includes two sub-groups. **Stenohaline** marine animals are marine forms that are either unable or barely able to tolerate changes in salinity. These organisms are usually restricted to the mouths of estuaries, where the salinity is generally 25 psu or above. They are often the same species found in the open sea. The second subgroup, called **euryhaline** marine animals, consists of animals that are capable of tolerating varying amounts of reductions in salinity below 30 psu. Such species can penetrate varying distances up the estuary. Most tolerate salinities down to 15–18 psu, with a few hardy species tolerating levels down to 5 psu (Figure 8.6).

The brackish-water, or true estuarine, species are found in the middle reaches of the estuary in salinities between 5 and 18 psu and are not found in fresh water or in full seawater. Examples of these animals are the polychaete *Nereis diversicolor*, oysters (*Crassostrea, Ostrea*), clams (*Scrobicularia plana, Macoma balthica, Rangia flexuosa*), small gastropods (*Hydrobia*), crabs (*Callinectes*), and shrimps (*Palaemonetes*) (Figure 8.7). Some of these estuarine genera may be limited in a seaward direction, not by physiological tolerances, but by biological interactions, such as competition and predation; hence, in any given estuary, this component may not be easily defined.

Another component has its origin in fresh water. These animals usually cannot tolerate salinities much above 5 psu and are restricted to the upper reaches of the estuary. (See Figure 8.6.)

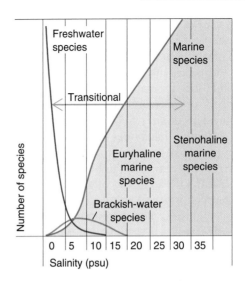

FIGURE 8.6 Numbers of species in each of the three major components—marine, freshwater, and brackish-water species—of estuarine faunal and their distribution with salinity. (After *Ecology of Estuaries*, D. S. McLuskey. Copyright © 1971 Heinemann Educational Publishers. Reprinted by permission of Heinemann Educational Publishers, a division of Reed Educational & Professional Publishing, Ltd.)

Finally, a transitional component includes those organisms, such as migratory fishes, which pass through the estuary on their way to breeding grounds, either in fresh water or salt water. Common examples are salmon (*Salmo, Oncorhynchus*) and eels (*Anguilla*). Also included are forms that spend only part of their lives in the estuary—usually the juvenile stage—and the adults are found at sea. Good examples of this latter group are the various shrimps of the family Penaeidae

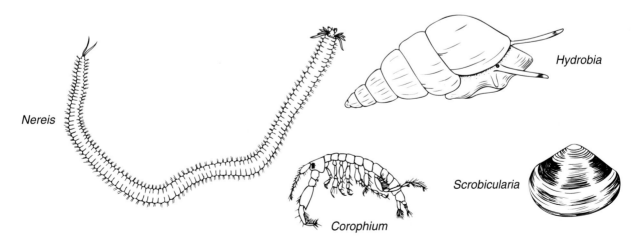

FIGURE 8.7 Some typical estuarine animals. (Not to scale.)

(*Penaeus setiferus, P. aztecus, P. duorarum*), which form the basis of the Gulf of Mexico shrimp fishery and whose young inhabit estuaries. The transitional fauna also contains forms that enter the estuary only to feed (many birds and fishes, for example).

The number of species inhabiting estuarine systems, as noted by Barnes (1974), is significantly lower than the number inhabiting nearby marine or freshwater habitats. Most true estuarine organisms are derived from marine, not freshwater stocks; the situation is similar to that in the other transitional zone—the intertidal, which is also populated mainly with marine organisms, not terrestrial forms. In contrast to marine species in the intertidal zone, however, the number of true estuarine species is small. (See Figure 8.6.) Why are there so few estuarine species? The most common explanation is that fluctuating environmental conditions—mainly salinity—are so extreme that few species have evolved the necessary physiological specializations to exist there. An alternative explanation is that estuaries have not had a long enough geological history to permit a complete estuarine fauna to develop. These two hypotheses are not mutually exclusive, so both could be operating. In the final analysis, which hypothesis is responsible for the low diversity of estuaries is not currently known.

■ ESTUARINE VEGETATION

Like the macrofauna, the macroflora of estuaries is limited. Most of the permanently submerged portions of estuaries consist of mud substrates that are unsuitable for macroalgal attachment. In addition, the highly turbid water restricts light penetration to a narrow upper layer. As a result, the deeper layers of the estuary are often barren of macrofloral life. The upper layers of water and the intertidal zone have a limited number of plants. In the lower reaches of the estuary at and below mean low water, there may be beds of seagrasses (*Zostera, Thalassia, Cymodocea*). (Considered subtidal communities, these beds are examined in Chapter 5.)

The intertidal mudflats are inhabited by a limited number of green algal species, common among which are the genera *Ulva, Enteromorpha, Chaetomorpha,* and *Cladophora*. These algae are often seasonally abundant, disappearing during certain times of the year (Figure 8.8).

Estuarine mudflats often have an abundant diatom flora. According to Lackey (1967), benthic diatoms are typically more abundant in estuaries than are their planktonic relatives. Many are motile

Cladophora sericea

Chaetomorpha linum

Enteromorpha intestinalis

FIGURE 8.8 Typical estuarine algae. (Not to scale.)

and undergo a rhythmic migratory pattern, moving up to the surface or down into the mud, depending on the illumination. Estuarine muds also provide a suitable habitat for filamentous cyanobacteria of several types. These autotrophs may form mats up to 1 cm thick on the mud surface and on the bases of various emergent vascular plants.

In terms of biomass, the dominant vegetation in estuaries is emergent plants—generally, long-lived flowering plants that root in the upper intertidal zone and form the characteristic salt-marsh communities that fringe estuaries throughout the temperate zones of the world. The dominant genera include *Spartina* and *Salicornia*. (See later in this chapter.) In the tropics, another community, the mangrove forest, is found. (See Chapter 9.)

A final estuarine component is bacteria. Both the water and the mud of estuaries are rich in bacteria because of the abundance of organic matter that is subject to decomposition. Wood (1965) has shown that estuarine waters contain hundreds of times more bacteria than seawater, and the upper layers of mud contain more than a thousand times more bacteria than the overlying water. Densities of bacteria of 100–400 million per gram have been reported in estuarine muds.

■ ESTUARINE PLANKTON

The estuarine plankton is also species poor and follows the same trend as the macrofauna and macrovegetation. Diatoms frequently dominate the phytoplankton, but dinoflagellates may achieve dominance during the warmer months and may remain dominant in some estuaries at all times. Dominant diatom genera include *Skeletonema*, *Asterionella*, *Chaetoceros*, *Nitzchia*, *Thalassionema*, and *Melosira*. Abundant dinoflagellate genera include *Gymnodinium*, *Gonyaulax*, *Peridinium*, and *Ceratium*. The phytoplankton may also be temporarily enriched by the resuspension of characteristic bottom-dwelling diatoms. High turbidity and rapid flushing may restrict phytoplankton numbers and productivity in some estuaries. Where turbidity is low and flushing time long, however, diverse populations and relatively high productivity may result. Consequently, depending on prevailing conditions, estuaries may differ considerably in both phytoplankton numbers and phytoplankton productivity.

Zooplankton in estuaries mirrors the phytoplankton in being limited in species composition, which also varies seasonally and with salinity gradients up the estuary. The few true estuarine zooplankters occur in larger, more stable estuaries, where salinity gradients are less variable. Shallow, rapidly flushed estuaries are inhabited mainly by a typical marine zooplankton assemblage carried in and out with the tide. Characteristic estuarine zooplankters include species of the copepod genera *Eurytemora*, *Acartia*, *Pseudodiaptomus*, and *Centropages*; certain mysids, such as species of the genera *Neomysis*, *Praunus*, and *Mesopodopsis*; and certain amphipods, such as species of *Gammarus*. The estuarine zooplankton average about 1 ml/m^3 displacement volume, somewhat greater than the concentration in adjacent coastal waters.

Adaptations of Estuarine Organisms

The variable physical nature of the estuarine habitat—especially its fluctuating salinities, temperatures and oxygen levels—makes it a particularly stressful and rigorous habitat. For organisms to survive and successfully colonize this area, they must possess certain adaptations.

■ MORPHOLOGICAL ADAPTATIONS

Few morphological adaptations among estuarine organisms can be attributed simply to living under conditions of fluctuating temperature and salinity. Most are adaptations to a given habitat, such as burrowing into mud. (See Chapter 6 for a discussion of adaptations of organisms living in mud.) Mud-dwelling organisms, whether estuarine or not, often have fine fringes of hair, or setae, that guard the entrances to their respiratory chambers to prevent clogging by silt particles. Such a situation prevails for estuarine crabs and many bivalve mollusks.

Other morphological changes in estuarine organisms, according to Remane and Schieper (1971), include a smaller body size than relatives living in full seawater and a reduced number of vertebrae among fishes. There are also certain morphological adaptations among vascular plants living in estuaries and salt marshes. All such plants are relatively small, perhaps the result of long-term selection stemming from water stress. All have a special tissue, the **aerenchyma**, that supplies oxygen to the roots, which are embedded in anoxic mud. This tissue may also bring oxygen indirectly to animals in the mud. Many plants have mechanisms for dealing with salinity as well. Some, such as *Spartina alterniflora*, have **salt glands** for eliminating excess salt; thus, they maintain a stable water balance. Others have extensive stores of carbohydrates in their roots, to provide a source of sugar to cope with stressful conditions. Abundant stomata and thin cuticles are suggestive of high transpiration rates. Another adaptation of vascular plants living in high-saline estuarine soils is the **succulence strategy**—a mechanism for maintaining high water concentrations in their tissues to buffer against water loss by osmosis. Finally, a reduced leaf area, few stomata, and photosynthetic stems further reduce water loss.

■ PHYSIOLOGICAL ADAPTATIONS

Adaptations required for estuarine life include those which maintain the ionic balance of body fluids in fluctuating external salinities. **Osmosis** is the physical process in which water passes through a semipermeable membrane that separates two fluids of different salt concentration, to move from areas of lower to higher concentration. The ability to control the concentration of salts or water in internal fluids is called **osmoregulation**. Most marine organisms cannot control their internal salt content and are **osmoconformers**. Their ability to penetrate estuaries is limited by their tolerance for changes in their internal fluids. The inability to cope with variation in salinity is the reason that starfish, sea urchins, and sea anemones are uncommon in estuaries. **Osmoregulators** are organisms that have physiological mechanisms to control the salt content of their internal fluids. Most estuarine animals either are osmoregulators or are osmoconformers able to function with fluctuating internal salt concentrations.

Penetrating into an estuary means encountering water with lowered salinity. Since the internal salt concentration of marine species is higher than that of the estuarine water, water tends to move across their membranes into their bodies to equalize the concentrations. Regulation means excreting the excess water without losing salts or excreting water and salts and replacing lost salts via an active uptake of ions from the environment. For freshwater animals, which move from a more dilute to a less dilute medium when entering estuaries, the reverse is true.

Osmoregulatory ability is found in several taxa of estuarine animals, primarily polychaete worms, mollusks, and crustaceans. The field of osmoregulation has been the subject of intensive research for many years, resulting in a large literature. Detailed coverage is beyond the scope of this introductory book, but students interested in more in-depth studies may refer to the literature cited at the end of the chapter. Suffice it to say that osmoregulation can operate in three ways: (1) Animals may move water, (2) they may move ions, or (3) they may adjust their internal water–ion balance. In advanced invertebrates and vertebrates, the osmoregulatory organ is generally the kidney, where excess water is excreted and needed ions resorbed. Among many invertebrates, gills are the most common osmoregulatory structures. However, special cells may also exist in other parts of the body, particularly for taking up or removing specific ions.

In certain soft-bodied forms, such as polychaete worms, an osmoregulatory mechanism is developed, but is relatively slow to respond. These animals can tolerate wide ranges of internal ion concentrations, at least for short periods. For example, placed in dilute seawater, *Nereis diversicolor* takes up water due to osmosis. After some time, however, it begins to osmoregulate, and according to McLusky (1971), the internal water content falls to near that found when the worm was in full seawater (Figure 8.9). Not only is this osmoregulation delayed; it does not come into effect until the salinity of the external environment falls below a certain level. That is, *Nereis* remains an osmoconformer as salinity decreases, and the animal regulates only when the external salinity falls below a certain level (Figure 8.9).

Bivalve mollusks and gastropods are generally poor osmoregulators and respond to drastic decreases in salinity by closing up in their shells to prevent excessive dilution of internal fluids.

Among higher crustaceans, such as crabs, osmoregulation is well developed in some species. Greatly restricted body permeability due to the exoskeleton and the ability to regulate ion concentrations of internal fluids are probable reasons for these animals' success in estuaries. Among vertebrates, osmoregulation is primarily through water removal by the excretory organs, coupled with ion uptake from the environment to replace lost salts. In invertebrates, other mechanisms may be involved. For example, the cells of crustaceans engage in osmoregulation through changes in the concentration of amino acids. This "intracellular osmotic regulation" results from movements of free amino acids in and out of individual cells to equalize the osmotic concentration in the cells with the extracellular bathing fluid, preventing excess water from building up and possibly rupturing the cells (Figure 8.10).

Vascular plants growing under estuarine conditions also display a set of physiological adaptations to osmotic stress. These rooted plants must maintain high concentrations of osmotically active substances to counter the water-retaining capacity of the surrounding medium. Generally, when the osmotic potential (basically, an increase in ion concentration) of the water in the soil around the roots increases, the plants respond by increasing the ion concentration of their tissues. Such an osmotic adjustment is the plant equivalent of osmoregulation in animals. In saline environments, salts are continuously transported into the plant. To prevent the salts from reaching a lethal level, most plants have mechanisms to get rid of the excess salt. Such mechanisms may include salt-secreting glands, leaching by rains of salts deposited

FIGURE 8.9 Changes in the body fluids of *Nereis diversicolor* with changes in salinity. (a) Weight changes occurring after the animals are transferred to 20% seawater. The initial weight increase is due to the intake of water by osmosis. The weight decreases when the animal osmoregulates. (b) Osmotic concentration of the body fluids in relation to salinity change. The straight line on the graph indicates osmoconformity. (From *Ecology of Estuaries*, D. S. McLuskey. Copyright © 1971 Heinemann Educational Publishers. Reprinted by permission of Heinemann Educational Publishers, a division of Reed Educational & Professional Publishing, Ltd.)

FIGURE 8.10 Graph showing the change in salinity of the blood of the crab *Australoplax tridentata* with a change in the salinity of the external medium. (Modified from *Ecology of Estuaries*, D. S. McLuskey. Copyright © 1971 Heinemann Educational Publishers. Reprinted by permission of Heinemann Educational Publishers, a division of Reed Educational & Professional Publishing, Ltd.)

on external surfaces by transpiration, and the shedding of leaves loaded with excess salts. Nonsucculent halophytes lose salts by some sort of secretion, whereas succulents release salts by shedding leaves.

■ BEHAVIORAL ADAPTATIONS

Behavioral adaptations are widespread in estuarine organisms. Among invertebrates, burrowing into the mud is a common adaptation to living in soft sediments. Although this adaptation is certainly not exclusive to estuarine animals, it has two beneficial effects for such organisms. First, for species with limited osmoregulatory abilities, living in the mud means exposure to interstitial water, which has less variation in both salinity and temperature than open water has. Living in the mud thus affords a means of reducing salinity and temperature changes. Second, burying oneself in the substrate means that the animal is less likely to be consumed by surface- or water-dwelling predators, such as birds, fishes, or crabs.

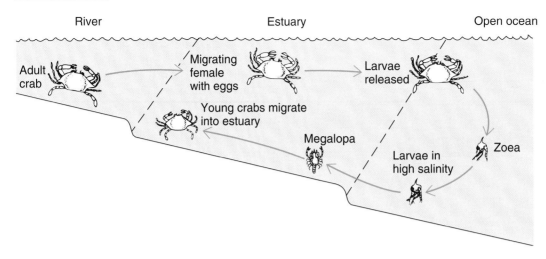

FIGURE 8.11 The life cycle of the blue crab *Callinectes sapidus* in estuaries of the Atlantic coast of the United States.

Many adult estuarine crabs are able to live successfully in low salinities due to their developed osmoregulatory abilities, but their eggs and young often lack such abilities. Therefore, many such crabs exhibit a specific migratory pattern, moving from the estuary to the adjacent sea for the breeding season. One example of this type of behavior is that of the crab *Eriocheir sinensis*, which has been introduced into Europe from its native China. Although adults live hundreds of miles up rivers, *E. sinensis* migrates back to the sea to breed and to spend the early part of its larval life. Another example is the blue crab *Callinectes sapidus*, which lives as an adult in estuaries of the American east and Gulf coasts, including Chesapeake Bay. The females migrate to waters of higher salinity to hatch the young. Once the young crabs pass through their larval stages, they migrate back up the estuary (Figure 8.11).

In contrast, certain other species take advantage of the abundant food and the relative scarcity of predators and use estuaries as a nursery for their young. Most of these species are fishes. Examples include mullet (*Mugil* sp.), striped bass (*Roccus saxatilis*), and flounder (*Platichthys flesus*), which enter the estuary as juveniles and migrate back to the sea after they mature. Various shrimp (*Penaeus*) also use the estuary as a nursery.

Ecology of Estuaries

Having considered the major physical–chemical features of estuaries, the common organisms, and the major adaptations of those organisms, it is now possible to consider certain aspects of the ecology of estuaries.

■ PRODUCTIVITY, ORGANIC MATTER, AND FOOD SOURCES

If we ignore salt marshes for the moment, the primary productivity of estuaries resides in their phytoplankton, benthic diatoms, and sea grasses, as well as in various algal mats. When all four sources are considered together, however, primary productivity by algae has traditionally been considered very low. Estuaries, however, have large amounts of organic material, large numbers of organisms, and high secondary productivity. If algal primary productivity is low, where does this organic material originate? Classically, it was believed that the major source of primary productivity in estuaries resided not in the estuaries themselves, but in the emergent plants of the salt marshes surrounding the estuaries. Teal's (1962) oft-cited study of a Georgia salt marsh found a net productivity of 6,850 kcal/m^2/year, but the various algae contributed only 1,600 kcal/m^2/year. Algae may, however, be more important in some areas, such as the Pacific coast of North America, of which Zedler et al. (1978) have shown that the algae contribute significantly to primary productivity. In addition, studies by Haines (1979) in the Georgia marshes suggested that the detritus from benthic and planktonic algae forms the bulk of the organic detritus. Haines found that most salt-marsh plant detritus was accumulated and consumed in the marsh itself, leaving algal detritus as the major material exported. In this scenario, the marsh is primarily a nursery for fishes, crabs, and shrimps, which are later exported to coastal waters.

Primary productivity in and around the estuary, however, is not the only source of organic material: As we noted previously, estuaries act as sinks for organic material brought down by rivers and in from the sea. It is difficult to assess the role of primary productivity within the estuarine system in contributing to the total organic production, for several reasons. First, estuaries have few herbivores that feed directly on plants. Most of the plant material, therefore, must be broken down to detritus before entering the various food webs. This breakdown process involves bacterial action, which is an additional complicating factor. Also, once the plant material has become detritus, it is not easily distinguished from other organic detritus brought into the system by the river and the sea. Some fraction of the great amounts of organic material that move through the estuary is produced in the estuary (autochthonous), but some is transported in and is the result of primary production elsewhere (allochthonous). The latter includes organic matter transported from the surrounding salt marsh. Currently, the relative amount contributed by each source is under debate.

McLusky (1989) has considered the various sources of carbon "food" in the temperate estuaries of Europe and North America. From his studies, he has separated estuaries into two major types that are at opposite ends of a spectrum and are named for their primary geographical locations. The **European** (for example, in Britain and the Netherlands) **estuary** is dominated by large mudflats and is relatively barren of large plants, but has large populations of surface benthic diatoms that produce most of the primary productivity along with phytoplankton that inhabit the water column. These estuaries derive most of their energy from outside, in the form of organic matter imported from the sea or river. They are net recipients of energy. Their ability to support large populations of consumers stems not from their productivity, but from the fact that they are sinks for imported energy. At the opposite extreme is the **American estuary** (for example, in Louisiana, Florida, and Georgia), dominated by extensive stands of *Spartina* grass and other salt-marsh plants surrounding less extensive mudflats. Phytoplankton and benthic diatoms are still significant primary producers here, but are overshadowed by the huge productivity of *Spartina* and the salt-marsh plants. This type of estuary produces an excess of carbon, which is then exported to the surrounding sea, usually as detritus.

The detritus forms a substrate for a rich bacterial and algal growth, which is an important food source for various suspension- and detritus-feeding animals. When one refers to the detritus food source of estuaries, however, the term *detritus* is used in the widest sense, to refer to organic particles, bacteria, algae, and even associated protozoans. This detritus forms an accumulation of food for estuarine organisms. As an example of the amount of food, Odum and de la Cruz (1967) noted that open seawater contains 1–3 mg of dry organic matter per liter, whereas drainage waters in estuaries may contain up to 110 mg of dry-weight organic matter per liter.

Recent studies have begun to call into question the long-standing assumption that consumers do not play a large role in the control of estuarine productivity. As mentioned earlier, estuarine ecologists have long thought that the production in estuaries is derived from seagrass and salt marsh plants and is controlled largely by physical factors such as soil salinity, soil oxygen levels, and the availability of nutrients—a type of control called *bottom-up control*. Recent studies, however, have suggested that *top-down* consumer control has been overlooked in these habitats. For example, Jefferies (1997) found that goose populations are expanding because of agricultural food subsidies stimulated by the use of artificial fertilizers and are destroying extensive areas of arctic salt marshes, marsh snail grazing in southeastern U.S. marshes from Virginia to Georgia has been found by Silliman and Bertness (2002) to exert strong top-down control on production, the introduced nutria, a rodent, has been shown by Gough and Grace (1998) to limit the production of salt marshes on the U.S. Gulf Coast, and a superabundant burrowing crab has been found by Bortolus and Iribarne (1999) to limit the production of marsh plants on the Atlantic coast of South America. These recent results, to be discussed in more detail latter in the chapter, suggest that estuarine productivity patterns and food webs are more complex than is currently understood. Moreover, Jackson (1997) has suggested that the seagrass meadows thought to historically be at the base of many estuarine food webs may be a human artifact stemming from the removal of large vertebrate consumers such as turtles, seals, and dugongs that either controlled populations of grazers or grazed themselves on seagrasses.

Current productivity patterns in estuaries have also been strongly influenced by eutrophication, which in estuaries is driven both by the increased use of artificial nitrogen fertilizers and by shoreline development, which increases the flow of human waste products into the estuaries. In New England, shoreline development—particularly the removal of natural vegetation buffers on the shoreline that intercept and absorb runoff—has been shown to increase the abundance of fast-growing weedy algae. These ephemeral algae blooms form massive rafts 1–2 feet thick, covering the bottoms of large eutrophic basins. According to Valiela et al. (1992), the algal rafts suffocate the plants and animals on the underlying sediment, leaving almost entirely bare anoxic substrate (Figure 8.12).

In Chesapeake Bay a different scenario, but with a similar outcome, is occurring. There, increased eutrophication is leading to increased production of planktonic diatoms, dramatically decreasing light transmission through the water column. As a consequence, seagrass beds have died off in all but very shallow waters, because the insufficient light fails to match their photosynthetic output to their metabolic demands (Orth and Moore, 1983). Clearly, estuarine productivity patterns and food webs are being heavily affected by humans.

■ COMMUNITY STRUCTURE OF ESTUARIES

The factors that contribute to the spatial structuring or patterning of estuarine communities have not

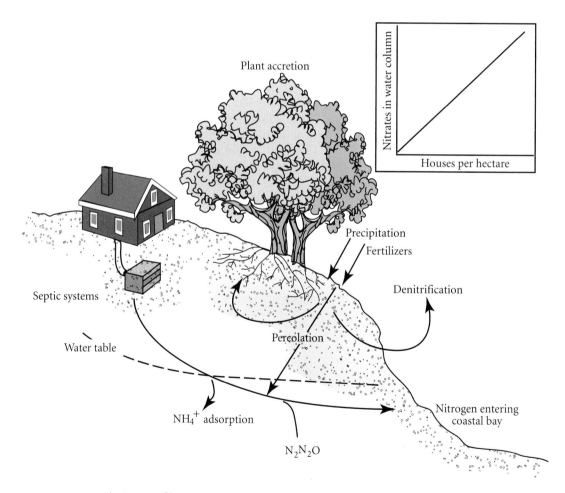

FIGURE 8.12 The impact of local eutrophication on shoreline communities. Increased nitrogen supplies lead to weedy green seaweeds dominating shallow-water habitats. Lost are seagrass beds and most of the organisms, such as clams and snails, that typically live buried in the substrate. (From *The Ecology of Atlantic Shorelines*, M. D. Bertness, Figure 2.1, p. 54. Copyright © 1999. Used with permission of Sinauer Associates, Inc.)

received much attention. Moving along the shores of estuaries, one finds conspicuous horizontal patterns in the distribution of plants and animals, but surprisingly few studies have examined the causes of these patterns. In most cases, the distributions have been assumed to be a simple reflection of the physical tolerances of different organisms expressed across a strong physical gradient, but this assumption has rarely been tested. One of the patterns that is conspicuous in estuaries worldwide is the shift one sees in the species composition of marsh plants that are important to the economy of the estuary as one moves from full-strength seawater to fresh water along estuarine salinity gradients. Salt marshes are dominated by a characteristic suite of species, freshwater marshes are characterized by an entirely different set of species, and, along estuarine salinity gradients, the plant community gradually shifts from one assemblage to another. While estuarine ecologists had long held that salt marsh and freshwater marsh plants were each adapted to the unique conditions in their respective habitats, recent evidence has suggested otherwise. When the known variables are quantified, it turns out that virtually all marsh plants grow best in fresh water. Moreover, when salt marsh plants are transplanted to freshwater marshes, they do better there in the absence of the native freshwater plants, but succumb to competition from freshwater natives when transplanted to freshwater habitats with those plants present. In contrast, freshwater plants died when transplanted to salt marshes with or without neighbors present. These findings (Crain et al. 2004), suggest that freshwater marsh plants are competitively dominant to salt-marsh plants and that salt marsh plants are competitively excluded from freshwater marshes. Thus, salt marshes may be viewed as competitive refuges for salt marsh plants. How important competitive interactions are in influencing the segregation of estuarine animals across estuarine salinity gradients has not been examined.

A second factor that appears to be generally important in generating sharp spatial patterns along estuarine salinity gradients in estuaries is current patterns. Sanford et al. (1994) and Leonard et al. (1998) found that any obstruction to flow in tidal estuaries—such as the mouths of embayments, or narrows caused by islands or other geological features that occlude flow—result in accelerated water movement. Accelerated flow increases the delivery of larvae to a site, increases the flux of particulate food

to filter feeders, and increases the efficiency of gas exchange on leaf surfaces. The last of these in turn increases the photosynthetic and metabolic rates of vascular plants and algae and typically decreases the importance of consumers, because of both a decreased locomotor ability of predators in rapidly moving water and a decreased ability of predators to find prey because olfactory signals are lost in the turbulent flow of fast-moving water. As a result, estuarine habitats with high flow not only are habitats with less deposition of sediment and a coarser substrate, but are generally communities with high densities of organisms with fast growth rates and low predation. Seagrass beds, oyster beds, and mussel beds are typically characteristic features of high-flow habitats in estuaries. In contrast, areas with slower water movement generally have the opposite characteristics: low larval delivery, low particulate food delivery, low gas exchange rates, and high consumer pressure because predators can easily move through the habitat and locate prey. On rocky shores in tidal estuaries on the coast of Maine, high-flow habitats are dominated by mussel beds, while low-flow habitats are dominated by a canopy of impalatable long-lived algae with a heavily grazed understory dominated by bare substrate (Leonard et al., 1998).

■ FOOD WEBS

Estuarine food webs have long been thought to be primarily detritus based. The relatively low water-column primary productivity, the lack of abundant herbivores, and the presence of large amounts of detritus have led to this conclusion. While the claim is currently being debated, it is a convenient starting point. Later in the chapter, we will discuss recent evidence which suggests that consumers play a much stronger role in estuarine food webs than is currently recognized.

To consider detritus the base of estuarine food webs is actually misleading, because it implies that detritivores actually digest organic detritus. Instead, most detritus is digested by bacteria and other microorganisms, and the macroorganisms that consume detritus actually feed, and are dependent on, those microorganisms to break down the detritus into a usable food resource. This chain of events should be kept in mind in the discussion that follows.

Detritus may be consumed directly, either while it is in suspension in the water column by various suspension-feeding benthic invertebrates or, more commonly, by direct consumption of the substrate by deposit feeders. Suspension or filter feeders, such as the clams *Cardium, Mya*, and *Mercenaria*, may obtain a portion of their food from suspended detritus along with the phytoplankton and zooplankton they ingest. Direct detritus consumers or deposit feeders include certain clams, such as *Scrobicularia* and *Macoma*, that have separated siphons and feed by "vacuuming up" the detrital particles from the substrate via their inhalant siphons. (See Figure 6.54.) Some clams and worms use both feeding modes—suspension feeding in high flows when most small food particles are in suspension, and deposit feeding in low flows when most small food particles settle out on the bottom.

Polychaete worms are abundant in estuaries, as are many deposit feeders. Members of the families Spionidae, Terebellidae, and Ampharetidae spread long tentacles over the mud surface and collect and transport food particles via ciliary mucous tracts. Others, such as the families Capitellidae and Arenicolidae, directly ingest the substrate, digesting organic material and bacteria during the passage of the substrate matter through their guts. Bacteria may also be consumed directly by small protozoans and nematodes, both of which occur in large numbers in the upper layers of the mud. Other animals, such as amphipods of the genus *Corophium*, sort particles with their appendages, while small gastropods, such as *Hydrobia* and *Cerithidea*, rake the surface of the mud for food particles (Figure 8.13). The feeding and burrowing of all these organisms continually rework the substrate.

The large number of deposit feeders and suspension feeders are, in turn, consumed by both vertebrate and invertebrate predators. The main invertebrate predators are crabs and shrimps; certain polychaetes, such as the genera *Glycera, Nephtys*, and *Nereis*; and various predatory gastropod mollusks, such as *Polinices, Aglaja, Chelidonura*, and *Busycon*. Of these predators, the blue crab *Callenectes* and the whelk *Busycon* are particularly dominant predators in the extensive estuaries of the southeastern coast of North America. Lipcius and Hines (1986) and Peterson (1982) have suggested that both play a keystone role in controlling community structure of the benthic communities in those estuaries.

Fishes and birds are also dominant predators in estuaries. Hartley (1940) found that in European estuaries the plaice (*Pleuronectes platessa*) feeds on polychaetes, while the flounder (*Platichthys flesus*) feeds on mysids, shrimps, and amphipods. Darnell (1961) found that in Louisiana estuaries the three most important larger invertebrate species that were heavily grazed by fishes were the clam *Rangia cuneata* and two crabs: *Rithropanopeus harrisii* and *Callinectes sapidus*. Different fish species concentrate on different types of animals, and feeding habits change with age. A common developmental pattern

FIGURE 8.13
Cerithidea snails lay on a mud flat in Australia. (Copyright Brandon D. Cole/Corbis.)

in estuarine fishes is to progress from eating zoo-plankton, to detritus, to macroinvertebrates, to even other fishes. Much of the productivity of estuaries that ends up in fishes is ultimately lost to the estuarine system when the fishes move offshore as adults. Kneib (1997) calls this movement of resources from shallow water to offshore habitats a *trophic relay*.

Estuarine birds include ducks, geese, and a host of shorebirds, wading birds, gulls, and terns. The majority of the shorebirds feed on the infaunal invertebrates of mudflats by probing at low tide. (See also Chapter 6, pp. 332–333.) The variety of bill lengths among these birds enables different species to feed at different levels in the mud (Figure 8.14). Birds follow the rise and fall of the tide during daylight hours to obtain their food. Observations and analyses of stomachs of these birds suggest that they are capable of consuming enormous numbers of invertebrates. Studies by Drinnan (1957), Goss-Custard (1969), and Prater (1972) have respectively reported that populations of oyster catchers (*Haematopus ostralegus*) consume 315 *Cardium*, redshanks (*Tringa totanus*) consume 40,000 *Corophium*, and knots (*Calidris canutus*) consume 730 *Macoma* per day. The effect of this predation on the populations of infaunal invertebrates is

not known, but various scientists have suggested that birds take 4–20%.

There has been considerable interest over the past three decades in understanding the role played by large predators in the soft-bottom communities of estuaries. Peterson (1979) reviewed early studies and found that, when large predators are removed from soft-bottom communities, the density of benthic organisms increases, but competitive exclusion leading to dominance by a competitive dominant does not occur. This situation is in contrast to that on rocky shores, where the removal of a predator commonly leads to a competitive dominant emerging, occupying all space, and forcing other species out. Why should such a different result occur here? Peterson (1979) has suggested that interference competition, which operates in the rocky intertidal zone, cannot operate in soft muds because the organisms living there cannot obtain enough purchase to push away or crush another organism and because the organisms can exploit the third dimension (depth) of the soft substrate. Competitive exclusion by overgrowth, common on rocky shores, is also precluded because of high mobility of the sediment. Adult–larval interactions, such as the consumption of settling larvae by

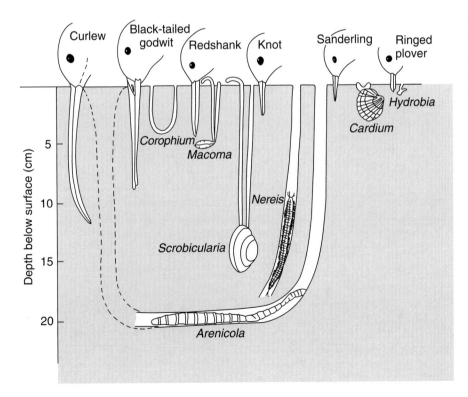

FIGURE 8.14 Diagram showing the bill length of some common European shorebirds in relation to the depth of some common estuarine invertebrates. (After *The Biology of Estuarine Animals*, J. Green, 1969.)

adult filter and deposit feeders, also may be of suffi-cient intensity that they keep densities below a popu-lation's carrying capacity, preventing competitive exclusion even in the absence of predation. (See also the discussion of intertidal mudflats in Chapter 6.)

It is important to note that modern estuarine food webs have long been dramatically affected by humans and that current American estuarine food webs are probably very different than they were four centuries ago, before Europeans began colonizing the Western Hemisphere. Valiela et al. (1992) and Orth and Moore (1983) found that eutrophication can lead to algal blooms and virtually preclude seagrasses from living in many estuaries. Overexploitation of oysters and other filter feeders has probably also had a mas-sive, little-understood impact on our estuaries, say Ulanowicz and Tuttle (1992). Before Europeans colo-nized North America, oyster reefs were a dominant feature of the estuaries on the east coast of North America south of Cape Cod. Narragansett Bay, Long Island Sound, and Chesapeake Bay, for example, were dominated by oyster reefs that likely filtered the entire water column in those systems every day. By filtering the water column, oysters probably left the water clear and with good light transmission. The result was

likely seagrass-dominated systems with vastly differ-ent food webs and dynamics than those we see in contemporary estuaries that have been massively impinged upon by humans.

In sum, estuarine food webs have traditionally been thought to have a number of energy pathways generated primarily from a detrital base and to ex-port production to other marine habitats. The pres-ence of abundant food and the fact that estuaries afford a refuge from marine predators have led to estuaries being used as nurseries for the juveniles of many animals that spend their adult lives elsewhere. Those same reasons are also why migratory adults of many species of birds and fishes use estuaries as feed-ing grounds. Estuaries are one of the marine habitats that have been most heavily affected by humans. The trophic interactions of estuaries are summarized in simplified form in Figure 8.15.

■ PLANKTON CYCLES

Although nutrient levels in estuaries are high, they are not well balanced; nitrogen concentration is often low and may even occasionally be limiting to estuar-ine phytoplankton. As we observed about the open sea (Chapter 2), there is also an annual cycle of

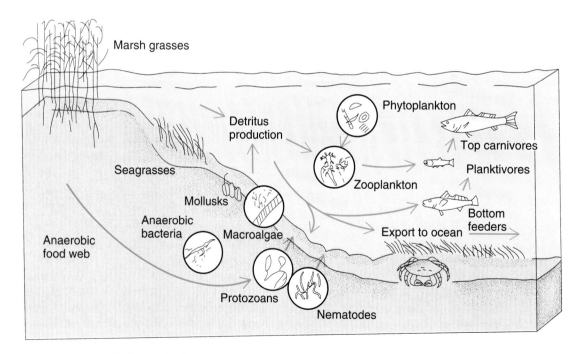

FIGURE 8.15 The food web of a typical estuary, showing some of the major aquatic trophic groupings. (After *Estuarine Ecology*, Day et al., Copyright © 1989 John Wiley & Sons, Inc. Reprinted by permission.)

phytoplankton in estuaries. In the temperate zones, low phytoplankton populations usually occur in late fall and winter and are due to reduced light and high turbidity caused by high river runoff and turbulence. The reduced populations of phytoplankton are followed by a later winter diatom bloom, which ends in late spring, often not by zooplankton grazing, but probably through the depletion of nitrogen sources, leading to the accumulation of diatoms on the mud surface. Phytoplanktonic populations remain low in the summer due to low nutrient levels and grazing, but there may be occasional blooms of dinoflagellates, leading to red tide situations. A large fall bloom occurs in some estuaries that rivals the spring bloom, whereas in others it is absent or reduced.

In comparison to the relative abundance of nutrients, the phytoplankton in estuaries has a relatively low annual rate of production. McLusky (1989) suggests that the low rate stems from three factors: (1) the shallow water, which is much less than the optimal depth for maximum photosynthesis; (2) the turbidity of the water, which limits light penetration; and (3) the continual removal of the phytoplankton caused by the rapid flushing rate of the estuary.

The zooplankton of estuaries may also be limited or reduced, is either by turbidity, which, by limiting the

phytoplankton, limits the food available to the zooplankton, or by currents, which carry the zooplankton out to sea. Elimination by currents can be reduced by the zooplankton remaining close to the bottom.

The zooplankton cycle in estuaries is variable and, as Deevey (1960) found, may change drastically from year to year. The zooplankton cycle does not closely follow the phytoplankton cycle. Populations of zooplankton peak later in the spring than the phytoplankton, and their numbers remain high throughout the summer, declining through the fall to winter lows. Zooplankton in estuaries consumes only 50–60% of the net phytoplankton production, leaving a significant portion available to the bottom suspension feeders. In shallow estuaries, the suspension-feeding benthos competes directly with the zooplankton for food, a situation that is not present in offshore waters.

Salt Marshes

Bordering temperate and subpolar estuaries and protected marine shores and embayments throughout the world are special plant associations known as salt marshes (Figure 8.16). Salt marshes develop wherever sediment accumulates and is bound together and stabilized by vascular plants forming a transition area

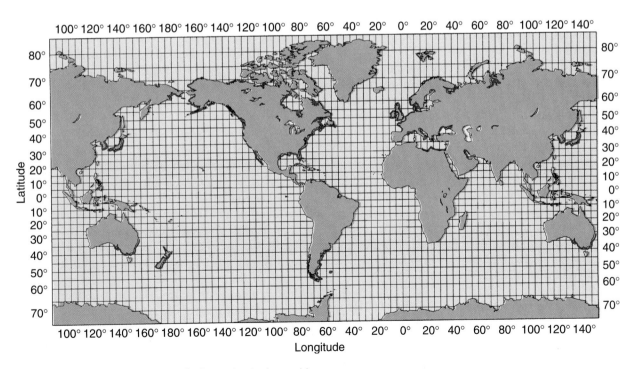

FIGURE 8.16 The distribution of salt marshes in the world.

FIGURE 8.17 Areal extent of salt marshes on the Atlantic coast of the United States. (After "Mangals and Salt Marshes of the Eastern United States," R. J. Reimold, in *Wet Coastal Ecosystems*, ed., V. J. Chapman, 1977, pp. 157–166. Copyright © 1977 with permission from Elsevier Science.)

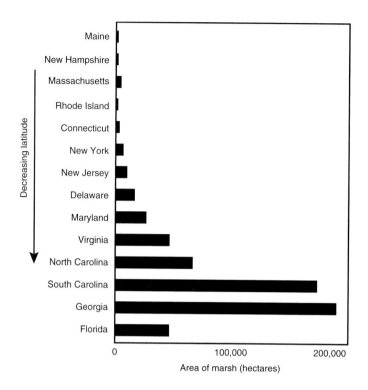

between aquatic and terrestrial ecosystems. Such marshes dominate the intertidal Atlantic coast of North America south of New England (Figure 8.17), as well as the northern coast of the Gulf of Mexico and much of the Atlantic coast of South America. They are a less common component of intertidal habitats on the Pacific coasts of North and South America.

Salt marshes are particularly important systems ecologically. They buffer shorelines from storm damage and erosion and serve as biochemical filters on runoff water entering estuaries. They also serve as important nursery grounds for the juveniles of many marine fish and crustaceans (Figure 8.18). In spite of their ecological value, salt marshes have long been abused by humans. In New England, early colonists ditched and drained salt marshes for livestock grazing and haying; since then, the marshes have commonly been filled over for development. Most European marshes and marshes on the Pacific coast of North America have also been heavily influenced by human activities.

■ DEFINITION AND CHARACTERIZATION

Salt marshes are communities of emergent herbs, grasses, or low shrubs rooted in soils alternately inundated and drained by tidal action. Salt marshes exist mainly at the higher tidal levels in areas of protected water and most often in association with estuaries. Salt marsh plants are **halophytes**, meaning that they can grow in soils with a high salt content. The dominant plants are herbaceous angiosperms. Because the upper portions of the plants are above water even during periods of high tide, the association between salt marshes and the plants that grow there has both terrestrial and aquatic components.

■ ENVIRONMENTAL CHARACTERISTICS

Salt marshes are rigorous physical environments that show wide variation in environmental factors, particularly salinity and the availability of oxygen in the soil. The salt marsh shares with the estuary a fluctuating salinity that results from the interaction between river flow and marine water. Because the salt marsh is in the intertidal zone, however, variations in salinity may be more sudden and extreme than those acting in the waters of the estuary. For example, a sudden rainstorm at low tide may reduce surface salinities to near zero, whereas the return of the high tide may inundate the marsh with nearly full strength seawater. It is not uncommon for marshes to experience salinities varying from 20 to 40 psu in a single tidal cycle. Moreover, during extended periods of

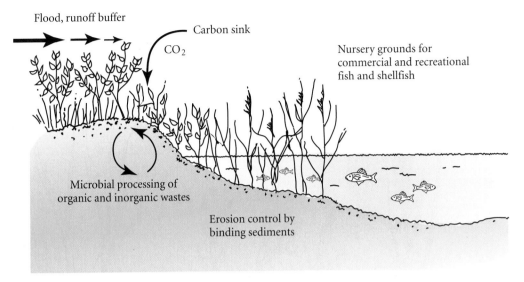

Flood, runoff buffer

Carbon sink

CO_2

Nursery grounds for commercial and recreational fish and shellfish

Microbial processing of organic and inorganic wastes

Erosion control by binding sediments

FIGURE 8.18 Environmental services provided by salt marshes. (From *The Estuarine Ecosystem*, 2/e, D. S. McLuskey, copyright © 1989 Chapman and Hall, with kind permission from Kluwer Academic Publishers.)

slack tides in the warm summer months, much of the water in high marsh soils can evaporate, exposing marsh plants to salinities well over 100 psu.

Marsh sediments can also be highly anoxic. The high productivity of marshes leads to sediments with high organic content and high microbial activity that depletes the sediments of oxygen. As a consequence, in addition to needing to cope with high soil salinities, the plants that live in marshes must have morphological or physiological mechanisms for dealing with anoxic soils.

■ COMPOSITION AND DISTRIBUTION

One of the most characteristic features of salt marshes is that they are species poor. Because of high soil salinities and anoxic soils, relatively few plants and animals live in salt marshes. Furthermore, these tolerant organisms show a high degree of taxonomic similarity worldwide. The dominant plants of salt marshes are grasses of the genus *Spartina* and species of the genera *Juncus* and *Salicornia* (Figure 8.19). Characteristic animals of marine origin are various species of crabs (*Uca, Hemigrapsus, Sesarma*), mussels (*Geukensia*), certain snails (*Littorina, Cerithidea, Melampus*), and smaller crustaceans such as amphipods and shrimp.

Insects represent the major terrestrial animal component of salt marshes and may live there permanently. There are also terrestrial animals,

including raccoons, that forage in marshes. At high tide, marine and estuarine animals enter the salt marsh, and at low tide, terrestrial species forage in the marsh (Figure 8.20). Despite the taxonomic similarity of salt marsh species over broad geographic areas, certain combinations of topographic features and species distributions have produced several different types of salt marsh. In the United States, there are two major groups: those of the Atlantic and Gulf coasts, on the one hand, and those of the Pacific coast, on the other. On the sediment-rich east and Gulf coasts, salt marshes are far more extensive than west coast marshes and occupy large areas of gently sloping coastline that surrounds the numerous broad estuaries and shallow bays associated with the offshore continental shelf. The geologically young Pacific coast, by contrast, has fewer rivers and bays, and steep coastal mountains lead directly to a narrow continental shelf. Consequently, river mouths are narrow, and the extent of both estuaries and salt marshes is restricted.

■ MARSH DEVELOPMENT AND ZONATION

Like coral reefs, salt marshes are biogenic structures. Built and maintained by salt marsh plants, the salt marsh has a general appearance that is similar throughout the world. Salt marshes are broad, flat expanses of low herbaceous, grassy, or shrubby plants of similar form, cut by a dendritic arrangement of

FIGURE 8.19 Some dominant salt marsh emergent plants.

Spartina alterniflora

Salicornia

Juncus gerardi

channels leading to larger tidal creeks (Figure 8.21). These creeks in turn lead to mudflats that are devoid of macrophytes or to the open waters of the estuary. Within the marsh, there may be scattered open areas (pans) and shallow pools. On the upland border of the marsh, terrestrial shrubs and trees replace the halophytes. In contrast to the rocky intertidal zone, where the majority of organisms are marine in origin, salt marsh plants are mainly terrestrial.

On the Atlantic coast of North America, marsh development begins with the colonization of intertidal sediment by smooth cordgrass, *Spartina alterniflora*. Cordgrass spreads vegetatively by rhizomes, dampens waves, slows water movement, enhances sedimentation, and binds sediment particles into a dense mat with its roots. Young marshes therefore usually consist of a monoculture of *S. alterniflora* (Figure 8.22). Over time, with the accumulation of belowground root material and sedimentation, marshes accrete (grow) vertically and, when mature,

consist of a high marsh and a low marsh cut by tidal creeks or drainage channels. The high marsh habitat is thus built by successful low marsh development. This evolution is readily seen in a cross section of a New England marsh where low-marsh cordgrass root debris, or **peat**, underlies the plants in the high marsh (Figure 8.23). The high marsh is flooded only by the highest tides, whereas the low marsh and drainage channels are flooded daily by the tides. As the high marsh develops, other salt marsh plants become established, and vertical zonation patterns emerge due to the changing physical conditions across the intertidal zone. In the northern marshes of the Atlantic coast, maturation leads to the accumulation of belowground plant debris or peat, which, along with sedimentation, adds to the vertical accretion of the marsh. Peat in northern marshes is the consequence of high-latitude marsh plants seasonally storing their biomass below ground during the winter and cooler temperatures limiting belowground decomposition.

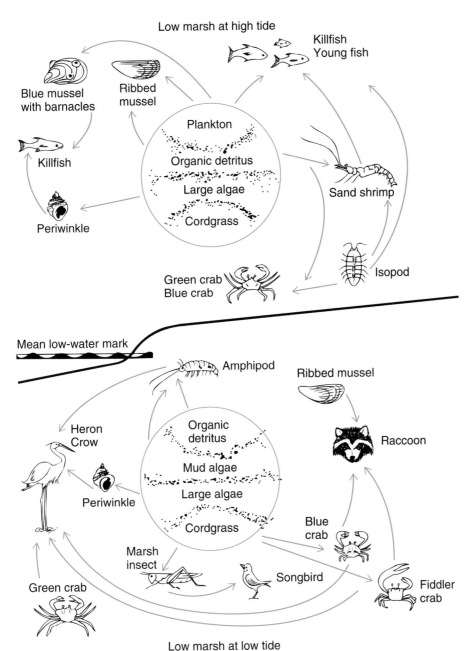

Low marsh at high tide

Low marsh at low tide

FIGURE 8.20 Characteristic animals present in a salt marsh at low and high tides on the Atlantic coast of North America.

Peat does not occur in lower latitude marshes where marsh plants do not seasonally shift their biomass below ground to overwinter; hence, peat accretes only by sedimentation. Peat was used extensively for fuel in Europe in the 17th and 18th centuries.

A cross section of a typical mature marsh from the tidal creek to the true terrestrial vegetation reveals a zonation pattern. This pattern varies in detail and species composition from one geographical area to another because of the tidal regime, drainage, slope, climate, and other factors, but a basic common zonation pattern for the Atlantic coast of North America exists and is described next.

The lowest zone is the creek bank and bottom, usually barren of macrophytic vegetation and containing mainly infaunal estuarine and marine animals. In New England, the low marsh—the area flooded daily by the tides—is dominated by the

FIGURE 8.21 Tidal creeks penetrating into a coastal salt marsh on the lower Coastal Plain of Georgia, U.S.A. (Copyright James Randklev/Getty Images/The Image Bank.)

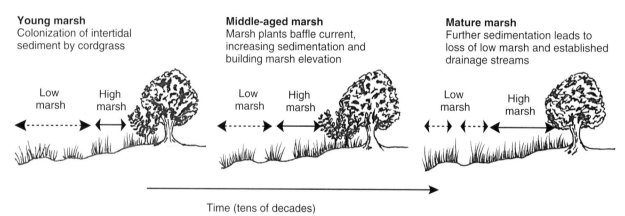

Young marsh
Colonization of intertidal sediment by cordgrass

Low marsh   High marsh

Middle-aged marsh
Marsh plants baffle current, increasing sedimentation and building marsh elevation

Low marsh   High marsh

Mature marsh
Further sedimentation leads to loss of low marsh and established drainage streams

Low marsh   High marsh

Time (tens of decades)

FIGURE 8.22 Marsh development. (After "Coastal Salt Marshes," R. W. Frey & P. B. Basan, in *Coastal Sedimentary Environments*, ed., R. A. Davis, pp. 225–301, 1985. Copyright © 1985 Springer-Verlag. Reprinted by permission.)

cordgrass *Spartina alterniflora* (Figure 8.24a). This zone is usually characterized by lush, tall cordgrass on the seaward margin and a shorter growth form of the same species immediately landward, where the soil is anoxic and accumulates peat. The next zone, the seaward border of the high marsh, is dominated by salt marsh hay, *Spartina patens*. Progressing landward, the next high marsh zone is dominated by the black rush *Juncus gerardi*. Black rush, a perennial with a dense root mass, is more intolerant of salty soils and flooding than *S. patens* is. Bordering the terrestrial vegetation at the upper limit of the marsh in New England is the marsh elder *Iva frutescens*, a woody shrub. This species is highly intolerant of tidal flooding and grows only where the soil is well drained. In addition to these dominant species, there

are some less abundant stress-tolerant, fugitive species that are found in small patches, usually inhabiting disturbed areas within the aforementioned zones. These species include the spike-grass *Distichlis spicata* and the glassworts *Salicornia europaea* and *Salicornia biglovii*.

Southern Atlantic marshes have a somewhat different set of plants and zonation pattern (Figure 8.24b). These marshes are subject to higher rates of evaporation, and hence higher salinities, as a result of exposure to more intense solar radiation. Hence, as Pennings and Bertness (1998) report, they are characterized by a plant assemblage that is more salt tolerant. The high salinities also foster the development of vegetation-free areas in the marsh, called **salt pans**. In New England marshes, vegetation growth is highly

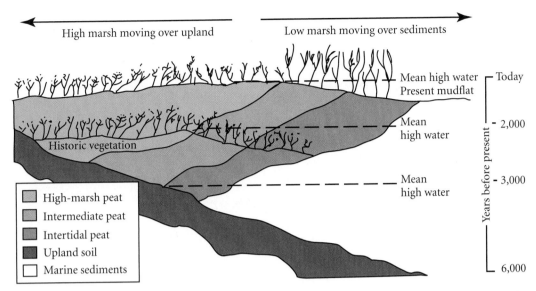

High marsh moving over upland   **Low marsh moving over sediments**

Mean high water
Present mudflat

Historic vegetation

Mean high water

Mean high water

Years before present

Today

- 2,000

- 3,000

- 6,000

High-marsh peat
Intermediate peat
Intertidal peat
Upland soil
Marine sediments

FIGURE 8.23 Redfield model of the development of New England salt marshes, illustrating how high-marsh vegetation overgrows low-marsh vegetation during the development of the marsh. (From *The Ecology of Atlantic Shorelines*, M.D. Bertness, Figure 7.4, p. 317. Copyright © 1999. Used with permission by Sinauer Associates, Inc.)

seasonal, from May to September, and during the fall and winter the emergent portions of the vegetation are composed of the dead remains from the growing season. This seasonally produced debris can form into mats over the winter and either be rafted by the tides onto the upper marsh, where it can smother other marsh plants, or be exported from the marsh in a seasonal pulse. In southern marshes, vegetation growth occurs throughout the year and the plant debris has no seasonal pulse.

The seaward border of the southern Atlantic low marsh that is flooded daily is dominated by tall cordgrass, *Spartina alterniflora*, a pattern similar to that in New England marshes (Figure 8.24b). Landward from this border is an extensive stand of the short form of the cordgrass, which displaces the marsh hay *Spartina patens*. Although here marsh hay is not the dominant plant that it was in New England, it is still found in patches. The highest parts of southern salt marshes are dominated by the black needle rush *Juncus romerianus*, which displaces both *Spartina patens* and *Juncus gerardi* from these high marsh habitats, which they occupied in New England. Between the short cordgrass zone and the *J. romerianus* zone is a distinct intermediate zone, which typically contains permanent salt pans that have extremely high soil salinities (often greater

than 100 psu), precluding plant colonization except on the fringes. Whereas tidal flushing at low elevations and rain and runoff at high elevations prevent accumulations of soil salts in those areas, the absence of such flushing, combined with high evaporation, leads to the high soil accumulations and salt pans. Around the borders of the salt pans is a characteristic suite of usually salt tolerant plants forming the salt meadow zone (Figure 8.24b). Among these plants are stunted forms of the black needle rush, spike grass (*Distichlis spicata*), and the glassworts *Salicornia europaea*, *S. virginica*, and *S. biglovii*, the latter of which seems to be the most salt tolerant and can be found growing in the middle of salt pans.

On the West Coast of the United States, native marshes tend to have a narrow fringe of *Spartina* in the low marsh, followed by a broad expanse of *Salicornia* occupying most of the midintertidal zone. The highest intertidal area is occupied by a more diverse assemblage, of which *Jaumea*, *Distichlis*, and *Limonium* are common (Figure 8.25). Adam (1990) notes that, in contrast to intertidal rocky shores, where the reverse is true, competition among the various plant species is a significant factor governing the uppermost zonal boundaries of plants in salt marshes, while the lower limits are set by physiological tolerances for submersion in salt water.

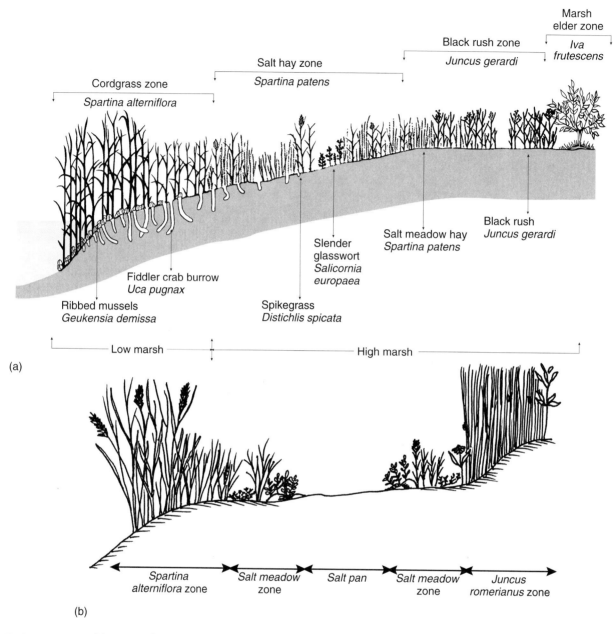

FIGURE 8.24 Diagrammatic representation of the zonation patterns of salt marshes of the Atlantic coast of the United States. (a) Zonation in a New England marsh. (b) Zonation in a southern marsh. (a, Copyright © Vergi Kask. Reprinted by permission. b from *The Ecology of Atlantic Shorelines*, M. D. Bertness, Figure 7.12, p. 326. Copyright © 1999. Used with permission by Sinauer Associates, Inc.)

■ CAUSES OF ZONATION

The physical and biological interactions that generate the striking zonation patterns observed in salt marshes have not been investigated as systematically throughout the world as they have been in the rocky intertidal zone. The most research has been done on the extensive marshes of the Atlantic and Gulf coasts of North America. They are discussed here as an example of how biological and physical factors interact to cause zonal patterns. The similarity of salt

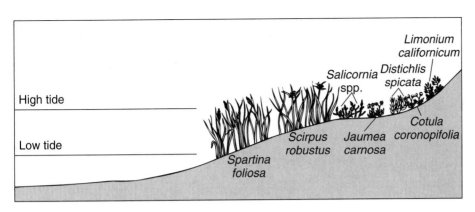

FIGURE 8.25 Generalized zonation of a Pacific coast salt marsh based on the marshes of San Francisco Bay.

marshes throughout the world suggests that it is reasonable to assume that these examples are generalizable to the organization and community structure of salt marshes in other parts of the world.

As noted earlier in New England the tidal salt marshes are characterized by a distinct zonation. (See Figure 8.24a.) The highest zone is dominated by the marsh elder *Iva frutescens*. The next highest zone is occupied by the black rush *Juncus gerardi* and is flooded only at the highest tides. Beginning at the mean highest tide line, the black rush gives way to a broad zone of *Spartina patens*, salt meadow hay, that extends down to the mean high water line. Also found in this zone are patches of the spike grass *Distichlis spicata* and the glasswort *Salicornia europaea*. Below the mean high water line *Spartina patens* is replaced by the cordgrass, *Spartina alterniflora*.. Surrounding the roots of the cordgrass are large numbers of burrows of the fiddler crab *Uca pugnax*. The cordgrass finally gives way to open mudflats, marking the end of the marsh. At this edge, large numbers of ribbed mussels, *Geukensia demissa*, live at the bases of the cordgrass.

In Chapters 5 and 6, we saw that much of the zonation of intertidal and subtidal organisms on rocky substrates could be explained by biological interactions—mainly competition, grazing, and predation. Salt marshes differ from the rocky subtidal and intertidal habitats in that grazers often have only a limited effect on the dominant plants. While recent research by Silliman and Bertness (2002) and by Jefferies (1997) has shown that grazers are much more important in marshes than was generally believed, grazers do not frequently appear to be important in generating the striking vertical zonation of salt-marsh plants. Instead, plant competition and

physical stresses seem typically to be responsible for generating the strong zonation patterns in salt marsh plant communities. We discuss this finding next, before examining the role that consumers play in these communities.

In New England marshes, at the uppermost part of the high marsh, the marsh elder *Iva frutescens* outcompetes *Juncus gerardi*, displacing it to a lower position in the high marsh. However, *J. gerardi* in turn outcompetes all other marsh plants. *J. gerardi* does this in two ways, as Bertness (1991a) found from removal and transplant experiments. First, *J. gerardi* has a very dense root mat, whereas *Distichlis* and *Spartina alterniflora* both have belowground runners. Various studies have demonstrated that dense root mats win out in competition with runners. Second, *J. gerardi* emerges early in the spring and begins growing before any other marsh plant does, thereby defeating the others, including *S. patens*, which also possesses a dense root mat. *Spartina patens* is restricted to the lower parts of the high marsh by competition with *J. gerardi*, but *S. patens*, in turn, is competitively dominant over *S. alterniflora*.

Since *S. alterniflora* is competitively inferior to both *J. gerardi* and *S. patens*, how does it manage to survive? The answer has to do with the sediment conditions and the morphology of *S. alterniflora*. As Bertness (1991b) has shown, the low marsh where *S. alterniflora* is dominant has waterlogged, anoxic sediment that *S. patens* cannot tolerate. *Spartina alterniflora*, however, is able to thrive in the anoxic sediment because it has well-developed aerenchyma tissue that provides oxygen to its roots an as well as efficient anaerobic metabolism. Cordgrass is usually taller in a band nearest the water and the open mudflat, whereas, in the upper reaches where it meets the *S. patens* band, it is stunted—the result of poor drainage and the accumulation of peat in the soil. Near the water, there is good drainage and little peat, and cordgrass grows well. Further inshore, peat has accumulated in the soil, poor soil drainage leads to increasingly anoxic soils, and cordgrass growth is stunted. *Spartina alterniflora*, therefore, degrades its habitat as it moves outward to colonize the open mudflat.

Our discussion so far has focused on the plant zonation only of the major dominants; we have not discussed the patchy occurrence of two other plants, *Distichlis spicata* and *Salicornia europaea*, that are competitively inferior to the dominants. How can they exist when they are outcompeted by the two species of *Spartina* and the *Juncus*? The answer lies with disturbance of the dominants and the salinity tolerances of the two fugitive marsh plant species.

In New England salt marshes, plants deal with harsh winter physical conditions by storing above-ground resources in overwintering belowground roots and rhizomes. Thus, in the fall, the cordgrass, *Spartina alterniflora*, reallocates aboveground resources to belowground storage structures, leaving standing dead plant skeletons composed primarily of cellulose. Over the winter months, these standing dead plants begin to decompose, fall over, and form large free-floating mats of plant debris, commonly called *wrack*, that can be over 100 m^2 in surface area. During spring tides and storms, this wrack is often transported to the high marsh, where it is stranded. In high New England marshes, stranded wrack can cover small areas, but it is also not unusual to find areas of many hundreds of square meters covered with a thick (up to 30 cm) blanket of wet, decomposing wrack. Typically, wrack transported to high elevations by extreme spring tides is stapled in place by emerging high marsh plants and remains in place for over a year until it decomposes or is washed away by spring tides and storms, revealing bare substrate where all the underlying plants have been killed. With the increasing temperatures of late spring and summer, the sediment in these bare patches heats up, causing the water in the surface layers to evaporate and the salinity to increase as much as 30 times that of the surrounding sediment, which is densely covered with marsh plants. This increase in salinity occurs at middle elevations in the high marsh that is normally dominated by either *S. patens* or black rush. At higher elevations, rainwater dilutes the salinity, while lower in the *S. alterniflora* zone frequent inundation by seawater prevents the salt from accumulating.

The hypersaline conditions of the bare patches prevent the competitive dominants, *S. patens* and *Juncus*, from invading. *Salicornia europaea*, however, thrives in hypersaline conditions, germinating and taking over patches by its seeds. The plant's dominance of patches, however, is short lived because *Distichlis* invades patches vegetatively by sending out runners from plants in nearby less saline environments. *Distichlis* can invade not only because it is salt tolerant, but because it receives its water through the runners from plants outside the hypersaline patch. As the spike grass takes over patches, it shades the sediment, reducing evaporation and decreasing salinity. After a few years, when the salinity has returned to normal, the competitive dominants displace *Salicornia* or *Distichlis* (Figure 8.26).

The salt marshes of the southern Atlantic and Gulf coasts are not subject to destruction by winter ice, nor do they typically produce massive rafts of wrack that may settle and destroy patches of high-marsh plants. However, they are subject to higher solar radiation, leading to the presence of more or less permanent salt pans. Thus, the differences in zonation patterns observed between southern marshes and northern marshes is tightly linked to their differing climatic conditions.

POSITIVE PLANT–ANIMAL INTERACTIONS   Because salt-marsh habitats are physically stressful on the plants that live in them, and because some of the invertebrates that live in the marshes can alleviate some of the stresses on marsh plants, positive plant–animal interactions are common in salt-marsh communities.

In southern New England, the ribbed mussel *Geukensia demissa* lives in dense aggregations on the seaward border of salt marshes attached to the stems and rhizomes of cordgrass. On the seaward border of these marshes, cordgrass growth is limited by disturbance and by nitrogen, and mussels help alleviate both of these problems. The mussels filter food from the water column and deposit high-nitrogen fecal matter on the roots and rhizomes of cordgrass and in the sediment. In essence, the mussels operate as nutrient transformers, filtering otherwise unavailable nutrients out of the water column and depositing them on the roots of the marsh plants (Figure 8.27). Bertness (1984) suggests that the resulting production of the cordgrass can be increased by 50% in a single season. In addition, the mussels buffer the

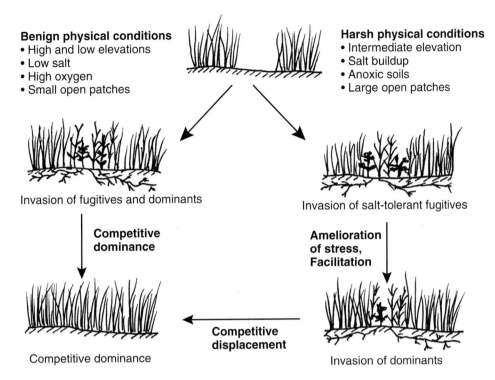

Benign physical conditions
• High and low elevations
• Low salt
• High oxygen
• Small open patches

Harsh physical conditions
• Intermediate elevation
• Salt buildup
• Anoxic soils
• Large open patches

Invasion of fugitives and dominants

Invasion of salt-tolerant fugitives

Competitive dominance

Amelioration of stress, Facilitation

Competitive dominance

Competitive displacement

Invasion of dominants

FIGURE 8.26 Alternative succession patterns in bare patches of the high marsh as a result of differing physical conditions. (From *The Ecology of Atlantic Shorelines*, M. D. Bertness, Figure 7.30, p. 348. Copyright © 1999 Sinauer Associates, Inc. Reprinted by permission.)

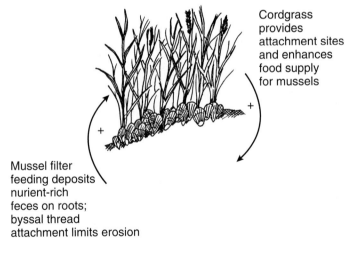

Cordgrass provides attachment sites and enhances food supply for mussels

Mussel filter feeding deposits nurient-rich feces on roots; byssal thread attachment limits erosion

FIGURE 8.27 Facultative mutualism between the filter-feeding marsh mussel *Geukensia demissa* and cordgrass on the Atlantic coast of North America. ("Ribbed Mussels and the Productivity of *Spartina alterniflora* in a New England Salt Marsh," M. D. Bertness, *Ecology*, Vol. 65, pp. 1794–180, 1984. Copyright © 1984 Ecological Society of America. Reprinted by permission.)

seaward edge of the cordgrass against physical disturbance by binding sediment and cordgrass roots with their byssal threads; thus, the seaward edge of the marsh is protected from wave- and current-induced erosion. Indeed, the mussels could not exist there if the cordgrass did not provide the anchoring material of roots for the byssal threads of the mussels. Since mussels depend on marsh cordgrass for attachment, the relationship between ribbed mussels and cordgrass in New England is best described as a facultative mutualism wherein both species benefit from, but are not dependent on, the association. Dame

(1993) has shown that other filter-feeding bivalves, such as oysters and other species of mussels, play a similar role in other marsh systems.

Another facultative mutualistic interaction is found between the cordgrass, *Spartina alterniflora*, and the fiddler crab, *Uca pugnax*, in southern New England marshes (Bertness, 1985). As mentioned earlier, cordgrass production in these marshes can be limited by poorly drained, waterlogged, and anoxic soils. The fiddler crab is a deposit feeder that feeds on the surface of the sediment and can be found at densities of over 150 burrows/m^2. The crab burrows in the mud around the roots of the cordgrass. Both the burrowing and the animal's deposit feeding aerate the soil, increase the decomposition of peat, and slow its accumulation (Figure 8.28), thereby enhancing the productivity of the cordgrass. At the same time, the presence of the roots and rhizomes of the cordgrass in soft mud stabilizes the mud enough to facilitate fiddler crab burrowing.

POSITIVE PLANT FEEDBACK The physical stresses in salt marshes can often be alleviated by the plants themselves, leading to positive feedback among plants playing a leading role in the organization and structure of salt marshes. Whereas we have highlighted the roles that competition and physical stress play in generating zonation patterns in the marshes, positive interactions that ameliorate the physical

FIGURE 8.28 As a by-product of its burrowing and deposit-feeding behavior, the mud fiddler crab *Uca pugnax* positively influences cordgrass growth on marshes on the Atlantic coast of North America by oxygenating the soil and increasing the rate of belowground decomposition. (Copyright Mark Bertness.)

environment are extremely important in all salt marshes. As already described, salt marshes are biogenic systems built and maintained by marsh plants. Providing habitats for other organisms, foundation marsh plants such as *Spartina alterniflora* on Atlantic coast marshes are responsible for initially stabilizing habitats with mobile substrate. This initial stabilization stimulates sedimentation and stabilizes the accumulated sediment, initiating the marsh accretion process and buffering the habitat for further colonization. Bruno (2000) has experimentally shown that, once cordgrass has colonized, the establishment of other halophytic marsh plants is dependent on the cordgrass buffering propagules and emerging seedlings from wave stress. Cordgrass arenchyma also oxygenate waterlogged low marsh soil and lead to positive feedback between cordgrass density and primary production.

At intermediate elevations of high salt marshes, where tidal flushing and rainwater runoff are reduced, high soil salinities can occur in the absence of plant cover. Because plant cover reduces the evaporation of water from the soil surface, there is positive feedback among plant neighbors that buffers them from salinity stress. In southern New England marshes, a host of positive interactions occurs at these elevations. Bertness and Hacker (1994) and Hacker and Bertness (1999) found that the woody shrub *Iva frutescens* forms a zone of short (25–50 cm tall) plants at this elevation in *Juncus* vegetation, and it and most of the other forbs that live within the *Juncus* zone at these elevations are entirely dependent on *Juncus* alleviating salt stress for them to live in this habitat. In addition, the insects that live in this zone are also all dependent on *Juncus* facilitation. When *Juncus* was experimentally removed from the system, all these plants and insects died, demonstrating that the most species-rich areas of these marshes are entirely dependent on positive interactions. Also, positive interactions between *Iva frutescens* and its neighbors at the seaward edge of the zone ensure *Iva*'s survival. In general, salt marsh communities appear to be hierarchically organized, with positive feedback by foundation species building and maintaining a habitat that is physically suitable for the rest of the community, while stress gradients, competition, and other biological forces generate the community structure on the biogenic surface.

In sum, plant competition and the steep environmental stress gradient in salt marshes leads to the striking spatial segregation, or zonation, of plants across salt marshes, with natural disturbances providing competitive refuge for weedy ephemeral plants to persist. Since the harsh physical stresses of salt marshes can be ameliorated by both marsh plants and animals, positive interactions are also important in the structure and organization of marshes. Without the amelioration of habitat afforded by the plants that build salt marshes, complex marsh communities would not exist.

■ PRODUCTIVITY

The primary producers of the marsh are the emergent marsh plants and the various microalgae that live on the surface of the plants and the mud. In contrast to productivity in the water and mudflats of estuaries, productivity is very high in marshes, especially for the rooted emergent plants.

Although salt marshes have been extolled as one of the most productive of all plant communities, it is now clear that the methods and assumptions used to obtain the primary production figures were often flawed or had various limitations, making them suspect or causing widely varying estimates. The figures presented here may be revised in the future and are presented only to approximate the relative production among different areas.

In considering primary productivity, it is most convenient to divide producers into two groups: vascular emergent plants and various algae. The most complete data on productivity exist for the vascular plants, especially for the east coast marshes. The major primary producers in these marshes are the various species of *Spartina*. Teal (1962) has shown that Georgia marshes have an average net productivity of 1,600 g C/m^2/year. This figure decreases as one proceeds northward: In New Jersey, Good (1965) reported only 325 g C/m^2/year. On the Gulf coast marshes, the primary productivity of vascular plants ranges from 300–400 to more than 3,000 g C/m^2/year, according to Turner (1976). On the Pacific coast, Atwater et al. (1979) reported San Francisco marshes at 50–1,500 g C/m^2/year of net production (compare with Table 2.2, p. 00), and Seliskar and Gallagher (1983) estimated 100–1,000 g C/m^2/year for the marshes of the Pacific Northwest. Silliman and Bortolus (2003) have convincingly argued that marsh productivity figures have been systematically underestimated wherever grazers are important, but unrecognized, regulators of marsh plant productivity.

Algal productivity estimates are fewer, more variable, and unavailable for many areas. Studies of the Atlantic coast by Teal (1962), Gallagher and Daiber (1974), and Van Raalte et al. (1976) have provided productivity estimates from 20 to 25% of that of the emergent vegetation. Zedler (1982), however, has shown that algal productivity in southern California marshes is 40–60% of that of the emergent vegetation. Seliskar and Gallagher (1983) estimate algal production in the Pacific Northwest to be considerably less: 100 g C/m^2/year.

Studies by Valiela and Teal (1974) and Mendelssohn (1979) indicate that salt marshes, in general, appear to be nitrogen limited, a property that can have dramatic consequences. Levine and colleagues (1998) demonstrated that, in New England marshes, nitrogen fertilization can completely change the competitive interactions described earlier that lead to sharp zonation in these communities. Under experimental conditions of high nitrogen, *Spartina alterniflora* and *Distichlis spicata* outcompete *Juncus gerardi* and *Spartina patens* and invade and dominate the high marsh. These findings suggest that the historical pattern of zonation actually seen is due to persistent low nitrogen levels in the marshes (Figure 8.29) and that these marshes would undergo predictable shifts in species composition under conditions of human-induced eutrophication.

Zedler (1982) has suggested that the reduced productivity of vascular plants and the enhanced productivity of algae observed in southern California marshes, but not in the east coast marshes, may be because the California marshes surround intermittent estuaries in which the soils are hypersaline during the summer growth period. The hypersalinity reduces the growth of the vascular plants, leading to a relatively open canopy; hence, more light penetrates to the substrate, enhancing algal production. Zedler also points out that this system may be more energy efficient, since algae can be consumed directly, whereas the vascular plants must be attacked by decomposers first, thereby adding another step to the food chain and reducing the total energy available to higher trophic levels.

Over the years, a prevailing view has developed among biologists that, because of the high productivity of marshes and because the productivity is little used or even underused by resident animals, salt marshes are an outwelling source of nutrition that sustains the productivity of estuarine and nearshore waters, particularly fisheries. The view that marsh

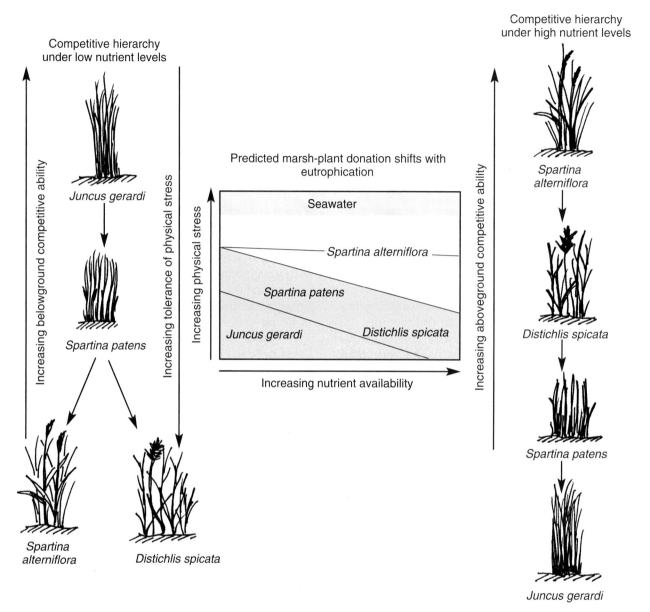

FIGURE 8.29 Change in competitive relations among marsh plants under high and low nitrogen levels. Competitive dominance at low nitrogen levels falls to plants that exploit belowground resources, whereas competitive dominance at high nitrogen levels falls to plants that are the better aboveground competitors. (After "Nutrient Availability and the Zonation of Marsh Plant Communities," J. S. Brewer & M. D. Bertness, *Journal of Ecology*, Vol. 86, pp. 285–292, 1998. Copyright © 1998 Blackwell Science, Ltd. Reprinted by permission.)

plants are not heavily utilized by grazers was based on early estimates by Smalley (1960) and Teal (1962) of herbivore damage and on arguments that marsh plants were heavily defended, of poor nutrient quality, and impalatable. (For a review, see Valiela, 1984.)

While little real direct data backed up this claim, the concept was widely supported by Teal (1962) in his suggestion that 45% of marsh production was removed, permitting that fraction to be used to support organisms outside the marsh. Since that time, however, additional studies and the use of new methodologies have led to a modification of this view. Indeed, there now seems to be agreement that salt marshes do export dissolved and particulate organic matter, but that the magnitude of the exported material and its importance to productivity in adjacent waters are variable and depend on a number of biological and geographical features. For example, Nixon (1980) showed that, where marsh area is large relative to the area of nearshore water, the organic supplement from the marsh may amount to about 50% of the open-water phytoplankton production. Such areas would be like those found off Georgia. In other cases, such as Chesapeake Bay and most New England marshes, the organic supplement is 10% or less. Furthermore, there is no evidence that the exported carbon provides significantly greater production of fish or shellfish than is found in coastal areas without marshes. We do not know why the export of organic matter does not lead to increased production at the higher trophic levels, but Nixon suggests that it may be due to such factors as sedimentation, the low nutrition of outwelled particles, or the rapid respiration of carbon by bacteria.

Marshes also export dissolved nitrogen and phosphorus. According to Nixon (1980), however, these nutrients do not have significant effects on primary productivity in nearshore waters, since primary productivity is no higher in areas where marshes occur than in areas where they do not. On the Pacific coast, marshes are small and are adjacent to nutrient-rich, upwelling ocean waters. Any outwelling of carbon or nutrients, therefore, would be insignificant to ocean productivity in these areas. Wherever consumers are important regulators of marsh production, they of course also influence the importance of outwelling to adjacent systems.

We can summarize this section by saying that our understanding of the productivity of salt marshes and the fate of production there, including export to estuaries and nearshore coastal waters, has been vigorously discussed for over 40 years, but that there are still many unknowns. There is agreement that marshes can export carbon, but the fraction exported varies with geographical, biological, and physical features.

■ CONSUMER CONTROL IN SALT-MARSH SYSTEMS

Salt-marsh plant communities have long been thought of as detritus-based communities in which the majority of the primary production of marsh plants is not consumed by grazers, but dies, is attacked by microbes, and enters the food chain as detritus. The paradigm that salt-marsh systems are detritus-based food webs is grounded in work performed nearly 50 years ago on Sapelo Island, Georgia. Among early salt-marsh ecologists, Teal (1962) noted that marsh plants were not heavily grazed, that they appeared to be relatively impalatable, and thus that most plants entered salt-marsh food chains as detritus. This paradigm was widely and uncritically accepted for decades. Over the past decade, however, it has become clear that salt-marsh systems can be under strong consumer control. Initial evidence for this hypothesis came from systems in which human intervention could be seen as leading to consumer impacts, but most recently, evidence has accumulated suggesting that natural marsh systems may, in general, be strongly influenced by consumers. In a study by Furbish and Albano (1984), introduced feral horses on the barrier islands of Maryland and Virginia were shown to have decreased marsh-plant primary production and increased plant species diversity. Another study by Gough and Grace (1998) showed that the introduced rodent nutria did the same on Gulf coast marshes of Louisiana. Both of these examples, however, involves a human introduction. Probably the most stunning example of natural consumer control of marsh production is the impact of snow geese on the arctic marshes of Hudson Bay. Robert Jefferies of the University of Toronto began working with snow geese herbivory in the Canadian arctic in the early 1970s. At that time, he found that snow geese had a positive effect on arctic marsh plants. Geese would arrive in their arctic summer breeding ground and feed on marsh plants by grubbing roots and rhizomes. In the process of foraging they would both thin the marsh plants and defecate on them, leading to an increase in the productivity of the plants the following year. Removing geese with cages reduced the plant biomass. Over the next 30 years, the snow geese population of North America began to grow dramatically due to agricultural nitrogen subsides at temperate latitudes. Whereas snow geese populations in the 1970s had reached low numbers because the North American wetlands that were the geese feeding grounds along their flyway were disappearing, over the next few decades snow geese shifted their temperate zone food source to golf courses and agricultural fields, which were rich in nitrogen because of the heavy use of artificial nitrogen fertilizers. This shift dramatically increased the size of the snow geese population, which then had a devastating impact on arctic marshes of Hudson Bay (Figure 8.30). Geese have destroyed literally hundreds of thousands of acres of the Hudson Bay wetland, the largest wetland in the world. The damage is severe, and recovery will be virtually impossible because the geese grub plant roots, removing the positive feedbacks between the arctic marsh plants and the physical conditions of arctic marsh soil. The system is relatively fragile to begin with: The soil is poorly developed, and the dominant marsh plants are restricted to only the top 2–5 cm of the soil surface. Once the geese removed these marsh plants by grubbing their roots, the positive effects that the plants have on growing conditions in the soil are removed. Because of the ensuing increased evaporation, the soils become highly saline, and without plants increasing the oxygen content of the soil with their parenchyma, the soil becomes anoxic. As a result, none of the plants that make up this community can sexually recruit by seed into the community. Reinvasion can occur only very slowly at the edges of still-vegetated habitats, aided by the amelioration of the remaining vegetation. The area destroyed by geese covers thousands of acres and, due to the slow pace of recovery, these areas of runaway consumption are essentially permanent features of the landscape. This example of the role that grazers can play in marshes, however, still involves human intervention. Jefferies (1997) discusses the ecological damage to the area wrought by the geese.

More recently, it has become clear that grazing by the common marsh snail, *Littoraria irrorata*, on the Atlantic shores of North America is an example of a natural grazer that can have dramatic runaway consumption effects on native salt marsh vegetation. This snail had long been considered a detritivore that was attracted to areas of dead and dying cordgrass. Recent work by Silliman (1998), however, found that the snail was capable of strongly suppressing cordgrass production, and work by Silliman and Bertness (2002) showed that, in habitats where snail recruitment was high, the snail could turn tall, productive stands of cordgrass into mudflats in less than a growing season. *L. irrorata* has this large top-down impact on cordgrass both by ingesting cordgrass when the snails are at high density and by rendering the plant more susceptible to

(a)

(b)

(c)

FIGURE 8.30 Over the past two decades, increasing populations of snow geese have destroyed large areas of the marshes of the Hudson Bay, one of the worlds largest wetlands. (a) Population increases in snow goose population of North America over the past 50 years. (b) Geese exclusion cages, showing the results of runaway herbivory by geese in Hudson Bay. (c) Map showing the massive spatial scale of the destruction (in red) of Hudson Bay marshes over the past 20 years. (Photograph and image courtesy of R. Jefferies.)

fungal infection through grazing scars. Ultimately, heavily grazed cordgrass succumbs to such infections. *Littoraria* populations are kept in check by their predators, blue crabs, turtles, and fishes that prevent the snails from living on the seaward border of most marshes. On snailless borders, lush cordgrass growths appear, restricting high snail densities to higher elevations, where, as expected, the snails strongly suppress cordgrass growth. This situation is a classic example of a trophic cascade where consumers—blue crabs, turtles etc.—control grazer densities and, as a by-product, enhance marsh grass growth (Figure 8.31). These findings also tie the harvesting of blue crabs, one of the principle predators of the snails, to the health of southern marshes. Bortolus and Iribarne (1999) found similar strong consumer effects on natural marsh systems driven by the superabundant grapsid crab in the salt marshes of Argentina and Brazil on the Atlantic coast of South America.

■ TROPHIC INTERACTIONS AND FOOD WEBS

From the preceding discussion of the factors influencing estuarine and salt marsh production, it should be clear that estuarine trophic interactions and food webs are far more complex than once thought. The notion that these systems are simple ones in which consumers are unimportant, and the idea that such systems can be easily characterized as detritus-based food webs is naïve and overly simplified. Estuarine and marsh systems have also been so intensively modified by human activities that characterizing their natural condition is difficult. Ulanowicz and Tuttle (1992) have shown that overharvesting filter feeders such as oysters that once kept the water column clean enough for sea grass to flourish in deeper water has obviously had an enormous effect on estuarine and salt marsh trophic interactions and food webs, and Jackson (1997) and Jackson et al. (2001) have found that depleting the large consumer populations of seals, fishes, crabs, turtles, and lobsters that once may have exerted strong control over primary production or grazer populations has had the same effect. Moreover, as Orth and Moore (1983) and Valiela et al. (1992) have shown, the heavy impacts of eutrophication in most estuaries and salt marshes has further simplified these systems and limited our insight into their natural dynamics. With these constraints and limitations in mind, we can now say something about the interactions and food webs of salt marshes.

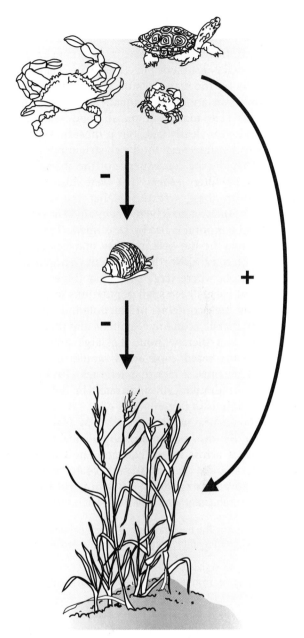

FIGURE 8.31 Cartoon illustrating consumer control of vegetation in the extensive salt marshes of Georgia and the Carolinas. If not kept in check by its predators, the snail, *Littoraria*, is capable of destroying these marshes due to runaway consumption. The fishes, crabs, and turtles that consume these snails are therefore responsible for the existence of many of the marshes. (From Silliman, B. R. and M. D. Bertness, "A Trophic Cascade Regulates Salt Marsh Primary Production," *Proceedings of the National Academy of Sciences (USA)*, vol. 99, pp. 10500–10505, 2002. Reprinted by permission of Brian Silliman.)

Traditionally, the food webs of salt marshes were thought to be based primarily on the consumption of detritus (and associated bacteria) and perhaps various algae. Since most of the detrital consumers distinguish their food only by particle size, a certain amount of live benthic algae is also consumed. These detrital consumers, plus the few insects feeding directly on the marsh plants, are, in turn, consumed by a variety of both aquatic and terrestrial carnivores. That many salt marsh food webs are detritus based is clearly true, but it must be kept in mind that the superficial food web structure of salt marshes may be the consequence of the absence of predators or filter feeders that were historically important in these systems or that the current structure of the system is being maintained at larger spatial and temporal scales by consumers. This, for example, may be the case in many marshes along the Atlantic coast of North America in Georgia and the Carolinas, where large predators (blue crabs, turtles, and fishes) keep snail populations in check, resulting in high cordgrass production and a detritus-based marsh economy. Silliman and Bertness (2002) suggest that, without the large predators, these detritus-based food webs would have no source of detritus, a fact that Jefferies (1997) has shown to be dramatically illustrated in the arctic marshes of Hudson Bay.

Predators appear also to generally play a large role in shaping the distribution and abundance patterns of organisms across marsh habitats. The two best-studied examples are on Atlantic coast marshes, where, as Stiven and Gardner (1992) found, predation by the blue crab *Callinectes sapidus* restricts the ribbed mussel *Geukensia* to high marsh habitats in southern marshes and, as Hamilton (1976) observed, the same predation excludes marsh periwinkles from places where they cannot migrate out of the water at high tide (Figure 8.32). Similarly, Van Dolah (1978) and Vince et al. (1976) showed that killifish (*Fundulus heteroclitis*) predation limits the snail *Melampus*, juvenile fishes, and various amphipods to the high marsh in association with dense vegetation (Figure 8.33).

Competitive interactions among animals in salt marshes have received relatively little attention and remain a fertile area for future work. The few studies that exist suggest that competition may be important in setting distribution patterns in marshes. For example, Race (1982) has demonstrated that the introduced snail *Ilyanassa obsoletus* has eliminated the native snail *Cerithidea californica* from mudflats in San Francisco Bay. *C. Californica* is now found only in the upper portions of the marsh, because it is more tolerant of desiccation than *Ilyanassa* is. Similarly, Sliger (1982) found that the two common marsh crabs *Pachygrapsus crassipes* and *Hemigrapsus oregonensis* were segregated vertically. *Pachygrapsus* appeared to be the dominant crab and had eliminated *Hemigrapsus* from higher tidal areas. *Hemigrapsus*, however, had a refuge in the lower intertidal areas, because it could tolerate silty water. Its gills were not so easily clogged by silt as those of *Pachygrapsus* were. Lee (2002) has shown that, on Atlantic coast marshes, marsh periwinkes (*Littoraria*) competitively dis-

FIGURE 8.32 The migration of the marsh periwinkle *Littoraria irrorata* up the cordgrass blades to avoid predation by blue crabs (*Callinectes sapidus*). (From *The Ecology of Atlantic Shorelines*, M. D. Bertness, Figure 35, p. 355. Copyright © 1999 Sinauer Associates, Inc. Reprinted by permission.)

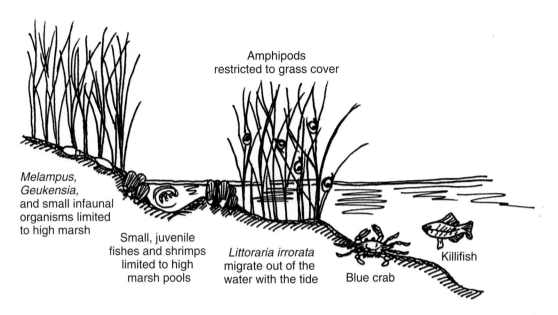

FIGURE 8.33 Representation of the effects of predation on some of the salt marsh inhabitants. (From *The Ecology of Atlantic Shorelines*, M. D. Bertness, Figure 48, p. 370. Copyright © 1999 Sinauer Associates, Inc. Reprinted by permission.)

place the pulmonate snail *Melampus* to live exclusively at extremely high marsh elevations.

As already mentioned, interactions in salt marshes and salt marsh food webs are also strongly influenced by non-consumer plant–animal interactions because salt marshes are physically and biologically stressful habitats for marsh organisms. Still, the organisms themselves can alleviate many of the stresses. Marsh plants are beneficial to animals in a number of ways other than providing the primary production that supports higher trophic levels. For example, marsh plants provide cover that reduces predation. Small invertebrates, such as crabs, snails, and insects, are protected from predation from birds and fish if they are beneath the canopy of marsh plants. Willason (1980) found that the shelter of the marsh was important to the survival of the crab *Pachygrapsus crassipes*, and Boland (1981) showed that most predatory birds used open mudflats rather than marsh areas. In southern California, there is evidence that the native hornsnail (*Cerithidea californica*) is more heavily preyed on in open mudflats than in areas within the marsh. Plants also stabilize the substrate and provide attachment surfaces for many animals. Without *Spartina*, for example, the mussel *Geukensia* would have nothing to attach its byssal

threads to, barnacles would not have a surface to cement to, and terrestrial insects and pulmonate snails would have no vertical substrate to climb to escape inundation at high tide. Marsh plants also reduce wave stress and solar radiation and thus ameliorate underlying disturbances, temperatures, and salinities, decreasing these stresses on understory organisms.

In turn, animals have many positive nonconsumer effects on the plants. For instance, the burrowing of crabs aerates otherwise stagnant mud and benefit the roots of the marsh plants. Some types of burrowing, however, may damage plants. For example, Josselyn (1983) suggests that burrowing by the isopod *Sphaeroma* may lead to erosion of the bank and the loss of parts of the marsh as they collapse into tidal creeks. In contrast, filter-feeding bivalve mollusks, such as the oyster *Crassostrea virginica* and the ribbed mussel *Geukensia demissa*, transfer nutrients from the water column to rooted vascular plants, thus coupling the water-column and marsh food webs. All of these positive indirect effects help to increase marsh plant production.

Salt marsh communities are relatively simple communities. In spite of their simplicity, however, the dynamics and structure of salt marshes are generated and maintained by a balance of negative consumer and competitive processes and positive habitat-ameliorating interactions.

■ HUMAN IMPACTS ON SALT MARSHES

At the beginning of this section, we mentioned that, historically, humans have had a heavy impact on salt marshes. Human impacts on marshes have been, and continue to be, severe. Early colonists in New England diked, drained, and ditched marshes to improve the yield of salt hay and to convert the marshes to dry land for development. Ditching of marshes continued almost to the end of the 20th century, to control insects. A worse problem during the 19th and 20th centuries, however, was the filling of salt marshes for shoreline development. Railroads, highways, housing developments, and parking lots have displaced

marshes throughout North America, and, as Nichols et al. (1986) observes, San Francisco Bay the largest estuarine salt marsh system on the Pacific coast, has been reduced from an areal extent of 2,200 km^2 in 1850 to less than 125 km^2 today, due to human development. Marsh losses in New England have been equally dramatic. Less obvious human impacts on salt marshes are those mediated by human modification of consumers that play a large role in marshes. We have already outlined two striking examples: (1) the heavy use of artificial nitrogen fertilizers in North America, triggering a snow goose population explosion that is rapidly destroying the arctic marshes of Hudson Bay by runaway consumption, and (2) the overharvesting of blue crabs and other predators of marsh snails on the Atlantic coast of

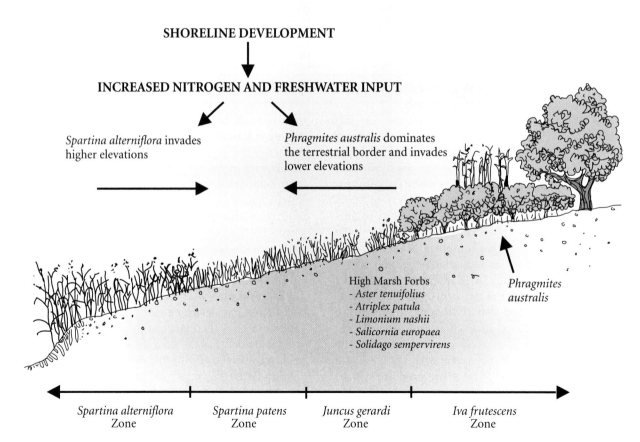

FIGURE 8.34 Cartoon illustrating the consequences of shoreline development on the vegetation in New England salt marshes. (From M. D. Bertness et al., "Anthropogenic Modification of New England Salt Marsh Landscapes," *Proceedings of the National Academy of Science* 99 (3):1395–1398, 2002. Copyright © 2002 National Academy of Sciences, U.S.A. Used with permission.)

North America, which may threaten marshes in this region with runaway top-down control by snails.

North American marshes are also currently being threatened by an invasion of exotic plants that is dramatically and perhaps irreversibly changing the marshes. On the east coast of North America, the reed *Phragmites australis* is invading salt marshes and displacing native marsh communities from Maine to Chesapeake Bay. *Phragmites* has been a member of the east coast plant assemblage for thousands of years, but, as Saltonstall (2001) explains, a European genotype was introduced to North America in the 19th century. This exotic genotype has invaded freshwater marshes throughout the continent. Over the past two decades, shoreline development in New England salt marshes has led to the invasion of *Phragmites*. By removing the woody vegetation that normally buffers marshes from runoff water, shoreline development increases the nitrogen supply to marshes and decreases soil salinities. Under such circumstances, *Phragmites* becomes the competitive dominant plant in these marshes, where it moves from being a minor component at the terrestrial border of the marsh to taking over the high marsh completely and displacing the entire high marsh

plant assemblage. According to Bertness et al. (2002), the addition of nitrogen also stimulates cordgrass to expand its range shoreward. Together, the two effects result in *Phragmites* and cordgrass driving the entire native community to extinction (Figure 8.34). This scenario does not appear to be happening in southern marshes, because of the presence of large competitively dominant native plants in these marshes. A similarly disturbing plant invasion, however, is occurring on the Pacific Coast of North America, where the Atlantic cordgrass, *Spartina alterniflora*, has been introduced intentionally. In California, Ayres et al. (2003, 2004) have shown that Atlantic cordgrass is hybridizing with the native cordgrass, creating a superhybrid cordgrass that is displacing the native species, aggressively invading lower elevations, and displacing mudflats that are crucial foraging habitats for migratory shorebirds. In the more Northern marshes of Washington State, there is no native cordgrass species, but Atlantic cordgrass is invading and dominating low intertidal mudflats and displacing the feeding grounds of migratory shorebirds. Clearly, human-mediated impacts on North American salt marshes are seriously threatening the very existence of these ecologically valuable habitats.

Summary of Key Concepts

- An estuary is defined as a partially enclosed coastal embayment where fresh water and seawater meet and mix.

- Based on geology and geomorphology, there are four basic types of estuary: coastal plain, tectonic, semienclosed bay or lagoon, and fjord.

- Based on salinity gradients, there are three groups of estuaries: positive or salt wedge, negative or evaporite, and seasonal or intermittent.

- The dominant feature of the estuarine environment is its fluctuation in salinity.

- The pattern of water-column salinity distribution in an estuary may be altered by the tidal regime, the Coriolis effect, and seasonal changes in evaporation, freshwater flow, or both.

- Salinity changes in the substrate are less than in the water column, so organisms in the substrate are partially buffered.

- Most estuaries have soft, muddy substrates derived from sediments carried into the estuary by both seawater and fresh water.

- Reduced water motion and the presence of various ions in the water causes silt particles to flocculate and settle out.

- Catastrophic events such as massive storms and accompanying floods may produce large depositions or removals of sediments in an estuary.

- Large amounts of particulate organic matter settle out in an estuary, which can then serve as a food reservoir for estuarine organisms.

- Water temperatures in estuaries are more variable than in coastal waters, due to a smaller volume and larger surface area for evaporation.

- The temperature range increases as one progresses up the estuary.

- The temperature also varies vertically, with the temperature at the surface showing a greater range than that in deeper waters.

- Wave action is minimal in estuaries, due to their shallow depth and short fetch.

- Currents in estuaries are caused primarily by tidal action and river flow, with the highest velocities occurring in the middle of channels, where the frictional resistance from bottom and side banks is lowest.

- The time required for a given mass of fresh water to be discharged from the estuary is called the flushing time.

- Because of particles in suspension, turbidity is high in estuaries, decreasing light penetration and reducing primary productivity.

- Oxygen is usually in ample supply in the water column, but is severely depleted in the substrate.

- There are four types of fauna in estuaries: marine, freshwater, estuarine, and transitional.

- The largest group of species consists of marine organisms and includes both stenohaline and euryhaline species.

- True estuarine organisms are found in the middle reaches of the estuary, where the salinity is between 5 and 18 psu.

- Freshwater species are intolerant of salinities above 5 psu and are confined to the upper reaches of the estuary.

- The transitional fauna includes those animals which pass through the estuary to breeding grounds either in fresh or salt water and those which spend only part of their life cycle in the estuary.

- The number of species inhabiting estuaries is significantly lower than the number inhabiting nearby freshwater or marine habitats.

- Deeper substrates of estuaries are often barren of plant life, but seagrass beds, macroalgae, and diatoms may dominate shallow and intertidal habitats.

- The dominant vegetation in the intertidal habitats of estuaries are emergent flowering plants that build salt marshes.

- Bacteria are extremely abundant in both the water column and the substrate of estuaries.

- Estuarine plankton is species poor, with diatoms dominating the phytoplankton and copepods the zooplankton.

- There are few morphological adaptations among estuarine animals that can be attributed simply to living in estuaries.

- Among plants, morphological adaptations include aerenchyma tissue, salt glands, a high lignin content, abundant stomata, thin cuticles, and a reduced leaf area.

- Osmosis is the physical process in which water passes through a semipermeable membrane that separates two fluids with different salt concentrations. The water moves from the fluid with the lower concentration to that with the higher concentration, thereby tending to equalize the salinities.

- Organisms that can control their internal salt concentrations are called osmoregulators; those which cannot are called osmoconformers.

- Osmoregulation can operate in three ways: Animals may move water, they may move ions, or they may adjust their internal water–ion balance.

- Most salt marsh plants have mechanisms to get rid of excess salts, such as salt-secreting glands, the deposition of salt on external surfaces, and the shedding of leaves loaded with excess salt.

- A common behavioral adaptation in estuarine organisms is burrowing into the substrate, exposing them to much less variation in salinity and temperature and decreasing their exposure to surface predators.

- Many adult estuarine crabs can live successfully in estuaries, but the eggs and young cannot, necessitating a migration back to the sea for breeding purposes.

- Many species take advantage of the estuary's food supply and scarcity of predators to use the estuary as a nursery for their young.

- The primary productivity of estuaries resides in the phytoplankton, benthic diatoms, seagrasses, and various algal mats; of these, the diatoms and algal mats are the most significant.

- Classically, it was believed that the major source of primary productivity in estuaries resided not in the estuaries themselves, but in the emergent plants of the surrounding salt marsh.

- Algae may be more important as primary producers in certain areas, particularly in southern California.

- Estuaries serve as a sink for organic material brought within them by the rivers and the sea. Thus, some of the great amount of organic material that moves through the estuary is produced there, and some is transported in from elsewhere.

- A European estuary is dominated by relatively barren mudflats, with primary productivity centered in the benthic diatoms and phytoplankton. An American estuary is dominated by extensive stands of emergent salt marsh plants around the mudflats. These plants are the major primary producers in the estuary.

- The estuarine food web has a number of energy pathways, generated primarily from a detrital base. These pathways have a somewhat incomplete use of energy that allows export to other areas, primarily by fishes and birds.

- The incidence of omnivory is generally higher in estuaries than in other systems.

- There is an annual cycle of phytoplankton in estuaries, with a late-winter diatom bloom ending in late spring through the depletion of nitrogen. Populations remain low through the summer, due to low nutrient levels and grazing, but the fall may bring another bloom.

- The zooplankton cycle in estuaries is variable and does not closely follow the phytoplankton cycle, as the zooplankton population peaks later than the phytoplankton population and remains high during the summer.

- Salt marshes are communities of herbs, grasses, and shrubs rooted in soils alternately inundated and drained by tidal action. All these plants are halophytes growing on soils with a high salt content.

- The salt marsh is a rigorous environment with wide variations in salinity and soil oxygen.

- Salt marshes are species poor, and the dominant plants show a high degree of similarity over a wide geographic area.

- The major terrestrial animals of marshes are insects, and marshes are invaded at high tide by marine and estuarine animals.

- Two major groups of marshes are found in North America. The Atlantic and Gulf coast marshes are extensive, occupying gently sloping substrates, and are dominated by various species of *Spartina*; the Pacific coast marshes occupy much smaller areas, characterized by broad bands of *Salicornia*, with *Spartina* of lesser importance.

- The general appearance of a salt marsh is similar throughout the world. It is a broad, flat expanse of low herbaceous, grassy, or shrubby plants of similar form, cut by dendritic channels leading to tidal creeks that in turn lead to bare mudflats or the open waters of an estuary.

- On the Atlantic coast of North America, marsh development begins with the colonization of the intertidal sediment by the smooth cordgrass *Spartina alterniflora*; in time, other plants become established, and a vertical zonation pattern emerges due to the changing physical conditions across the intertidal zone.

- The zonation pattern in New England marshes has the lowest zone dominated by *Spartina alterniflora*, followed by a zone of *Spartina patens*, and next a zone of *Juncus gerardi*, followed at the uppermost edge by a band of *Iva frutescens*, bordering the terrestrial vegetation.

- Southern Atlantic marshes have the seaward edge of the marsh dominated by *Spartina alterniflora*, extending landward and displacing most of the *Spartina patens*, which is still found in patches. The highest parts of the southern marshes are dominated by *Juncus romerianus*, and between this zone and the *Spartina* zone is a distinct intermediate zone with permanent bare salt pans of exceedingly high salt content bordered by a suite of very salt tolerant plants.

- Pacific coast marshes tend to have a narrow seaward fringe of *Spartina*, followed by a broad expanse of *Salicornia* occupying most of the midintertidal zone. The highest zone is occupied by a more diverse assemblage.

- New England salt marshes have their primary space dominated by dense stands of perennial, turf-forming plant species which, through competitive interactions, form monospecific bands that succeed each other across the intertidal zone.

- Competitively inferior fugitive plant species are rare in New England salt marshes, occurring only in patches created in the turf species when mats of rafted vegetation have been deposited on the dominants, killing them and creating hypersaline conditions through increased exposure of sediment to the evaporation of water. The very harsh physical environment of the hypersaline bare patches cannot be tolerated by the competitively superior plants until that environment has been altered by the fugitive species.

- *Spartina alterniflora* forms a mutualistic association with both mussels and fiddler crabs.

- The primary producers of marshes are emergent plants and various microalgae that live on the surfaces of the plants and substrate.

- Vascular plant productivity in marshes is high, but varies with geographic locality, while algal productivity is less in all regions.

- In most marsh systems, productivity is nitrogen limited.

- The reduced productivity of vascular plants and the enhanced algal productivity seen in southern California marshes may be due to the open canopy, which permits more light to penetrate to the algae on the substrate.

- Herbivory on marsh plants is much higher than has been appreciated.

- Salt marshes do export dissolved and particulate organic matter, but the magnitude of this exported material and its importance to the production in adjacent waters are variable and depend on a number of geographical, biological, and physical features.

- Marsh plants are beneficial to the animals in the marsh in a number of ways, including providing cover that reduces predation, stabilizing the substrate and providing a surface for attachment, and reducing wave and current action.

- Animals, in turn, have effects on the plants, including aerating the substrate by burrowing, grazing off of epiphytic algae, providing nutrients through fecal deposition, and, in the case of mussels and oysters, stabilizing the substrate.

- Competitive interactions among animals in the salt marsh have been little studied and remain a fertile area for future work.

- Although marshes experience a number of natural and human disturbances, those caused by humans often have the most profound effects and can threaten the existence of the communities inhabiting the marsh.

Review Questions

ESSAY Develop complete answers to these questions:

1. Why are there relatively few truly estuarine species? Discuss.

2. Compare and contrast the European estuary with the American estuary in terms of the plants that dominate the mudflats and the relative inputs and outputs of carbon.

3. What generally occurs when predators are excluded from soft-bottom invertebrate communities? Discuss.

4. In what ways can eutrophication affect estuarine communites?

5. How do physical and biological factors interact to create zonal patterns in the New England salt marshes?

6. Why are positive plant–animal and plant–plant interactions so common in salt marshes?

7. Why is it that stenohaline animals are located deeper in the water column the farther up an estuary they are found?

8. How can the disruption of habitat in coastal wetlands and estuaries affect organisms that live in the subtidal or intertidal regions? In coastal waters? In continental shelf waters? In the open ocean?

Bibliography

Adam, P. 1990. *Saltmarsh ecology.* New York: Cambridge University Press.

Andrews, J. D. 1973. Effects of tropical storm Agnes on epifaunal invertebrates in Virginia estuaries. *Chesapeake Sci.* 14:223–234.

Atwater, B. F., S. G. Conard, J. N. Dowden, C. W. Hedel, R. L. McDonald, and W. Savage. 1979. History, land forms and vegetation of the estuary's tidal marshes. In *San Francisco Bay, the urbanized estuary,* edited by J. Conomos. San Francisco: Pacific Division AAAS, 347–386.

Ayres, D. A., P. Baye, and D. R. Strong. 2003. *Spartina foliosa:* A common species on the road to rarity? *Madrono* 50:209–213.

Ayres, D. A., D. L., Smith, K. Zaremba, S. Klohr, and D. R. Strong. 2004. Spread of exotic cordgrasses and hybrids (*Spartina sp.*) in the tidal marshes of San Francisco Bay. *Biological Invasions,* in press.

Barnes, R. S. K. 1967. The osmotic behaviour of a number of grapsoid crabs with respect to their differential penetration of an estuarine system. *J. Exp. Biol.* 47:535–551.

Barnes, R. S. K. 1974. *Estuarine biology.* Studies in Biology, no. 49. London: Edward Arnold. (Distributed in U.S. by Baltimore: University Park Press.)

Bertness, M. D. 1984. Ribbed mussels and the productivity of *Spartina alterniflora* in a New England salt marsh. *Ecology* 65:1794–1807.

Bertness, M. D. 1985. Fiddler crab regulation of *Spartina alterniflora* production on a New England salt marsh. *Ecology* 66: 1042–1055.

Bertness, M. D. 1991a. Interspecific interactions among high marsh perennials in a New England salt marsh. *Ecology* 72(1):125–137.

Bertness, M. D. 1991b. Zonation of *Spartina patens* and *Spartina alterniflora* in a New England salt marsh. *Ecology* 72(1):138–148.

Bertness, M. D. 1992. The ecology of a New England salt marsh. *Amer. Sci.* 80:260–268.

Bertness, M. D. 1999. *The Ecology of Atlantic shorelines.* Sunderland, MA: Sinauer Associates, Inc.

Bertness, M. D., and S. D. Hacker. 1994. Physical stress and positive associations among plants. *American Naturalist* 144:363–372.

Bertness, M. D., and S. W. Shumway. 1993. Competition and facilitation in marsh plants. *Amer. Nat.* 142(4):718–724.

Bertness, M. D., P. Ewanchuk, and B. R. Silliman. 2002. Anthropogenic modification of New England salt marsh landscapes. *Proceedings of the National Academy of Science* 99(3):1395–1398.

Boland, J. M. 1981. Seasonal abundance, habitat utilization, feeding strategies and interspecific competition within a wintering shorebird community and their possible relationships with the latitudinal distribution of shorebird species. Master's thesis, South Dakota State University.

Bortolus, A., and O. Iribarne. 1999. Effects of the SW Atlantic burrowing crab *Chasmagnathus granulata* on a Spartina salt marsh. *Marine Ecology Progress Series* 178: 78–88.

Bruno, J. F. 2000. Whole-community facilitation through substrate stabilization by the intertidal grass *Spartina alterniflora. Ecology* 81: 1179–1192.

Callaway, J. C., and M. N. Josselyn. 1992.The introduction and spread of smooth cordgrass (*Spartina alterniflora*) in South San Francisco Bay. *Estuaries* 15: 218–226.

Chapman, V. J., ed. 1977. *Ecosystems of the world.* 1. *Wet coastal ecosystems.* New York: Elsevier.

Conomos, J., ed. 1979. *San Francisco Bay, the urbanized estuary.* San Francisco: Pacific Division AAAS.

Crain, C. M., B. R. Silliman, S. L. Bertness, and M.D. Bertness. 2004. Mechanisms of the spatial segregation of plants across estuarine salinity gradients. *Ecology.* (In press.)

Dame, R. F. 1993. The role of bivalve filter feeding material fluxes in estuarine ecosystems. In *Bivalve filter feeders in estuarine and coastal ecosystem processes,* edited by R. F. Dame, pp. 245–269. Springer-Verlag, New York.

Darnell, R. 1961. Trophic spectrum of an estuarine community based on studies of Lake Ponchartrain, Louisiana. *Ecology* 45(3):553–568.

Day, J. W., C. A. S. Hall, W. M. Kemp, and A. Yañez-Arancibia. 1989. *Estuarine ecology.* New York: Wiley.

Deevey, G. B. 1960. The zooplankton of the surface waters of the Delaware Bay region. *Bull. Bing. Ocean. Coll.* 17(2):54–86.

Drinnan, R. E. 1957. The winter feeding of the oyster catcher (*Haematopus ostralegus*) on the edible cockle (*Cardium edule*). *J. Anim. Ecol.* 26:441–469.

Frey, R. W., and P. B. Basan. 1985. Coastal salt marshes. In *Coastal sedimentary environments,* edited by R. A. Davis. New York: Springer-Verlag, 225–301.

Furbish, C. E., and M. Albano. 1994. Selective herbivory and plant community structure in a mid-Atlantic salt marsh. *Ecology* 75:1015–1022.

Gallagher, J. L., and F. C. Daiber. 1974. Primary production of edaphic algal communities in a Delaware salt marsh. *Limn. Oceanog.* 19:390–395.

Good, R. E. 1965. Salt marsh vegetation, Cape May, New Jersey. *N.J. Acad. Sci. Bull.* 10(1):1–11.

Goss-Custard, J. D. 1969. The winter feeding ecology of the redshank *Tringa totanus. Ibis* 111:338–356.

Gough, L., and J. B. Grace. 1998. Herbivore effects on plant species density at varying productivity levels. *Ecology* 79: 1586–1594.

Green, J. 1968. *The biology of estuarine animals.* Seattle: University of Washington Press.

Hacker, S. D., and M. D. Bertness. 1999. The role of positive interactions in the plant species diversity of salt marsh plant communities. *Ecology* 80:2064–2073.

Haines, E. B. 1979. Interactions between Georgia salt marshes and coastal waters: A changing paradigm. In *Ecological processes in coastal and marine systems*, edited by R. J. Livingston. New York: Plenum Press, 35–46.

Hamilton, P. V. 1976. Predation on *Littorina irrorata* by *Callinectes sapidus*. *Bull. Mar. Sci.* 26:401–409.

Hartley, P. H. T. 1940. The saltash tuck-net fishery and the ecology of some estuarine fishes. *J. Mar. Biol. Assoc.* 24:1–68.

Jackson, J. B. C.1997. Reefs since Columbus. *Coral Reefs* 16 (Supplement) 23–32.

Jackson, J. B. C., and 17 others. 2001. Historical overfishing and the recent collapse of coastal ecosystems. *Science* 293: 629–638.

Jefferies, R. L. 1997. Long-term damage to sub-arctic coastal ecosystems by geese: Ecological indicators and measures of ecosystem dysfunction. *Disturbance and recovery in Arctic lands: An ecological perspective*, edited by R. M. M. Crawford, pp.151–166. Kluwer Academic, Boston.

Josselyn, M. 1983. *The ecology of San Francisco Bay tidal marshes: A community profile*. U.S. Fish and Wildlife Service Division of Biological Services FWS/OBS-83/23.

Kinne, O. 1963. The effects of temperature and salinity on marine and brackish water animals. *Oceanog. Mar. Biol. Ann. Rev.* 1:301–340.

Kneib, R. T. 1997. Tidal marshes offer a different perspective on estuarine nekton. *Annual Review of Oceanography and Marine Biology* 35: 1–120.

Lackey, J. 1967. The microbiota of estuaries and their roles. In *Estuaries*, edited by G. H. Lauff. Publ. no. 83. Washington, DC: AAAS, 291–302.

Lauff, G. H., ed. 1967. *Estuaries*. Publ. no. 83. Washington, DC: AAAS.

Lee, S. 2002. Competitive interactions among snails on a Georgia salt marsh. Undergraduate thesis, Brown University.

Leonard, G., J. M. Levine, P. Schmidt and M. D. Bertness. 1998. Flow-generated bottom-up forcing of intertidal community structure in a Maine estuary. *Ecology* 79: 1395–1411.

Levine, J., S. Brewer, and M. D. Bertness. 1998. Nutrient availability and the zonation of marsh plant communities. *J. Ecol.* 86:285–292.

Lipcius, R. N., and T. Hines. 1986. Variable functional responses of a marine predator in dissimilar homogeneous microhabitats. *Ecology* 67 (5): 1361–1371.

Long, S. P., and C. F. Mason. 1983. *Saltmarsh ecology*. London: Blackie and Son, Ltd.

McLusky, D. S. 1971. *Ecology of estuaries*. London: Heinemann.

McLusky, D. S. 1981. *The estuarine ecosystem*. New York: Wiley.

McLusky, D. S. 1989. *The estuarine ecosystem*, 2d ed. New York: Chapman and Hall.

Mendelssohn, I. A. 1979. Nitrogen metabolism in the height forms of *Spartina alterniflora* in North Carolina. *Ecology* 60: 106–112.

Nichols, F. H., J. E. Cloern, S. N. Luoma, and D. H. Peterson. 1986. The modification of an estuary. *Science* 231:567–573.

Nixon, S. W. 1980. Between coastal marshes and coastal waters—a review of twenty years of speculation and research on the role of salt marshes in estuarine productivity and water chemistry. In *Estuarine and wetland processes*, edited by P. Hamilton and K. B. MacDonald. New York: Plenum Press, 437–525.

Nixon, S. W. 1982. *The ecology of New England high salt marshes: A community profile*. Biological Report 81(55). U.S. Fish and Wildlife Service, Washington, DC.

Odum, E. P., and A. A. de la Cruz. 1967. Particulate detritus in a Georgia salt marsh–estuarine ecosystem. In *Estuaries*, edited by G. H. Lauff. Publ. no. 83. Washington, DC: AAAS, 383–388.

Odum, H. T., B. J. Copeland, and E. A. McMahan. 1974. *Coastal ecosystems of the United States*. 4 vols. Washington, DC: The Conservation Foundation.

Orth, R. J., and K. A. Moore. 1983. Chesapeake Bay: An unprecedented decline in submerged aquatic vegetation. *Science* 222:51–53.

Pennings, S. C., and M. D. Bertness. 1999. Using latitudinal variation to examine effects of climate on coastal salt marsh pattern and process. *Proceedings of the OECD Workshop on Global Change and Wetlands. Journal of Biogeochemistry* 3:100–111.

Peterson, C. H. 1979. Predation, competitive exclusion and diversity in the soft bottom benthic communities of estuaries and lagoons. In *Ecological processes in coastal and marine systems*, edited by R. J. Livingston. New York: Plenum Press, 233–264.

Peterson, C. H. 1982. Clam predation by whelks (*Busycon* spp.): experimental tests of the importance of prey size, prey density, and seagrass cover. *Marine Biology* 66:159–170.

Peterson, C., and N. M. Peterson, 1979. *The ecology of intertidal flats of North Carolina: A community profile*. U.S. Fish and Wildlife Service, Biological Services Program. FWS/OBS-79/39.

Pomeroy, L. R., and R. G. Wiegert, eds. 1981. *The ecology of a salt marsh*. New York: Springer-Verlag.

Prater, A. J. 1972. Ecology of Morecambe Bay. III. The food and feeding habits of knot (*Calidris canutus*) in Morecambe Bay. *J. Appl. Ecol.* 9:179–194.

Race, M. 1982. Competitive displacement and predation between introduced and native mud snails. *Oecologia* 54:337–347.

Redfield, A. C. 1965. Ontogeny of a salt marsh. *Science* 147: 50–55.

Reid, G. K., and R. D. Wood. 1976. *Ecology of inland waters and estuaries*. 2d ed. New York: Van Nostrand.

Reimold, R. J. 1977. Mangals and salt marshes of the eastern United States. In *Wet coastal ecosystems*, edited by V. J. Chapman. Amsterdam: Elsevier, 157–166.

Reimold, R. J., and W. H. Queen, eds. 1974. *Ecology of halophytes*. New York: Academic Press.

Remane, A., and C. Schlieper. 1971. *Biology of brackish water*, 2d ed. Stuttgart: Schweizerbartsche.

Saltonstall, K. 2002. Cryptic invasion by a non-native geno-type of the common reed, *Phragmites australis*, into North America. *Proceedings of the National Academy of Science* 99: 2445–2449.

Sanford, E., D. Bermudez, M. D. Bertness, and S. D. Gaines. 1994. Flow, food supply and acorn barnacle population dynamics. *Marine Ecology Progress Series* 104: 49–62.

Seliskar, D. M., and J. L. Gallagher. 1983. *The ecology of tidal marshes of the Pacific Northwest Coast: A community profile.* U.S. Fish and Wildlife Service Division of Biological Services, Washington, DC FWS/085-82/32.65.

Silliman, B. R. 1998. Top down control of *Spartina alterniflora* growth: Salt marsh periwinkles eat more than just detritus. Master's thesis, University of Virginia, Charlottesville.

Silliman, B. R., and M. D. Bertness. 2002. A trophic cascade regulates salt marsh primary production. *Proceedings of the National Academy of Science* 99 (16):10500–10505.

Silliman, B. R., and A. Bortolus. 2003. Underestimation of *Spartina alterniflora* production in Western Atlantic salt marshes. *Oikos.* 143:549–555.

Sliger, M. 1982. Distribution and microhabitat selection of *Hemigrapsus oregonensis* (Dana) and *Pachygrapsus crassipes* (Randall) in Elkhorn Slough, Monterey County, California. Master's thesis, California State University, Hayward.

Smalley, A. E. 1960. Energy flow of a salt marsh grasshopper population. *Ecology* 41:672–677.

Stickney, R. R. 1984. Estuarine ecology of the southeastern United States and the Gulf of Mexico. College Station: Texas A & M University Press.

Stiven, A. E., and S. A. Gardner. 1992. Population processes in the ribbed mussel *Geukensia demissa* (Dillwyn) in a North Carolina salt marsh tidal gradient: Spatial pattern, predation, growth and mortality. *J. of Experimental Marine Biology and Ecology* 160: 81–102.

Stout, J. P. 1984. *The ecology of irregularly flooded salt marshes of the northeastern Gulf of Mexico: A community profile.* U.S. Fish and Wildlife Service Biological Report 85 (7.1).

Teal, J. 1962. Energy flow in the salt marsh ecosystem of Georgia. *Ecology* 43:614–624.

Teal, J. 1985. *The ecology of New England low marsh habitats: A community profile.* Washington, DC: U.S. Fish and Wildlife Service.

Turner, R. E. 1976. Geographic variations in salt marsh macrophyte production: A review. *Contrib. Mar. Sci.* 20:47–68.

Ulanowicz, R. E., and J. H. Tuttle. 1992. The trophic consequences of oyster stock rehabilitation in Chesapeake Bay. *Estuaries* 15: 298–306.

Valiela, I. 1984. *Marine ecological processes.* New York: Springer-Verlag.

Valiela, I., and J. M. Teal. 1974. Nutrient limitation of salt marsh vegetation. In *Ecology of halophytes*, edited by R. J. Reimold and W. H. Queen, pp 547–563. New York: Academic Press.

Valiela, I., P. Peckol, C. D'Avanzo, C. H. Sham, and K. Lajtha. 1992. Couplings of watersheds and coastal waters: Sources and consequences of nutrient enrichment in Waquoit Bay, Massachusetts. *Estuaries* 15:443–457.

Valiela, I., J. M. Teal, and W. Sass. 1975. Production and dynamics of salt marsh vegetation and the effects of experimental treatment with sewage sludge: Biomass, production and species composition. *J. Appl. Ecol.* 12:973–982.

Van Dolah, R. F. 1978. Factors regulating the distribution and population dynamics of the amphipod *Gammarus palustris* in an intertidal salt marsh community. *Ecological Monographs* 48: 191–217.

Van Engel, W. A. 1958. The blue crab and its fishery in Chesapeake Bay. *Commercial Fisheries Rev.* 20(16):6–17.

Van Raalte, C. D., I. Valiela, and J. M. Teal. 1976. Production of epibenthic salt marsh algae: Light and nutrient limitation. *Limn. Oceanog.* 21:862–872.

Vince, S., I. Valiela, N. Backus, and J. M. Teal. 1976. Predation by the salt marsh killifish *Fundulus heteroclitus* (L.) in relation to prey size and habitat structure: Consequences for prey distribution and abundance. *Journal of Experimental Marine Biology and Ecology* 23:255–266.

Waisel, Y. 1972. *Biology of halophytes.* New York: Academic Press.

Wiegert, R. G., and B. J. Freeman. 1990. *Tidal salt marshes of the southeast Atlantic coast: A community profile.* Biological Report 85(7.29). Washington, DC: U.S. Fish and Wildlife Service.

Willason, S. W. 1980. Factors influencing the biology and coexistence of grapsoid crabs, *Pachygrapsus crassipes* (Randall) and *Hemigrapsus oregonensis* (Dana) in a California salt marsh. Master's thesis, University of California, Santa Barbara.

Wood, E. J. F. 1965. *Marine microbial ecology.* London: Chapman and Hall.

Zedler, J. B. 1982. *The ecology of southern California coastal salt marshes: A community profile.* U.S. Fish and Wildlife Service Biological Services Program FWS/OBS-81/54.

Zedler, J., T. Winfield, and D. Mauriello. 1978. Primary productivity in a southern California estuary. *Coastal Zone* 3:649–662.

Zobell, C. E., and C. B. Feltham. 1942. The bacterial flora of a marine mud flat as an ecological factor. *Ecology* 23:69–78.

CHAPTER 9

TROPICAL COMMUNITIES

For color, beauty of form and design, and variety of life, perhaps no natural areas in the world can equal coral reefs. Their beauty has long fascinated generations of scientific and lay people alike. By contrast, few marine environments are more ecologically valuable, but have been less appreciated than mangrove forests. These two very different communities are characteristic of vast tropical regions of the world and must be considered in order to understand shallow-water tropical ecosystems. In this chapter, we discuss the distribution and structure of these assemblages and then examine the organisms that inhabit them.

Coral Reefs

Over a vast region (millions of square miles) of the tropics, the shallow inshore waters are dominated by coral reefs, and such reefs are often used to define the limits of the tropical marine environment (Figure 9.1). Coral reefs have been estimated to occupy about 600,000 square miles of the surface of the Earth— about 0.17% of the total area of our planet.

Karlson (1999) notes that coral reefs have the greatest diversity per unit area of any marine ecosystem with respect to higher taxa, and it has been estimated that perhaps 4–5% of all species, or about 91,000, are found on coral reefs. Coral reefs are globally significant in that about half the calcium that enters the world's oceans each year is taken up and bound into the reefs as calcium carbonate ($CaCO_3$). Since each bound calcium atom requires the incorporation of a molecule of carbon dioxide, reefs also remove about 700 billion kilograms of carbon per year. All these facts mean that coral reefs play a significant role in the continuing health of the marine environment and, indeed, our entire planet.

Coral reefs are also among the most productive systems in the marine environment and have existed on the planet for hundreds of millions of years. Despite their long history and status as the most massive structures built by living organisms, the thin layer of living coral tissue seems particularly sensitive to natural and human disturbances that today appear to be converging and challenging the continued survival of these beautiful ecosystems.

Like salt marshes and mangroves, coral reefs are biogenic structures built entirely by biological activity. Coral reefs are essentially massive deposits of $CaCO_3$ produced primarily by corals (phylum Cnidaria, class Anthozoa, order Scleractinia), with lesser additions from calcareous algae and other $CaCO_3$-secreting

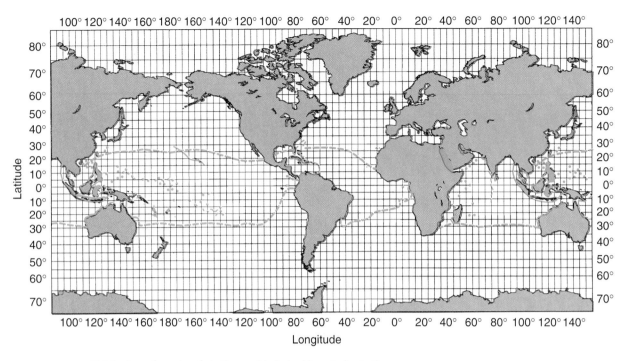

FIGURE 9.1 Distribution of coral reefs in the world; dashed lines indicate the 20°C isotherm.

FIGURE 9.2
Underwater view of a Pacific Ocean coral reef. (Photo courtesy of Dr. Gary Williams.)

organisms, such as mollusks, sponges, and segmented worms (Figures 9.2 and 9.3). Although corals are found throughout the world's oceans in polar and temperate waters, as well as in the tropics, it is only in the tropics that coral reefs develop. This is because there are two different types of corals. **Hermatypic** corals are corals that produce reefs; **ahermatypic** corals do not form reefs. Ahermatypic corals are distributed worldwide; hermatypic corals are found only in the tropical regions. Besides these two distinguishing features,

FIGURE 9.3 Photo showing the crowding of corals on a coral reef in Palau. (Photo courtesy of Dr. Gary Williams.)

hermatypic corals have small symbiotic algal cells called **zooxanthellae** living in their tissue, whereas most ahermatypic corals do not have these cells. Zooxanthellae enhance the growth of hermatypic corals, as well the corals' ability to secrete the calcium carbonate skeletons of the organisms, and thus are critical for reef growth and development.

■ REEF DISTRIBUTION AND LIMITING FACTORS

Six major physical factors limit coral reef development: temperature, depth, light, salinity, sedimentation, and emergence into air. A glance at a map of the distribution of coral reefs in the world immediately shows that nearly all are found within the 20°C surface isotherm. (See Figure 9.1.) Hermatypic corals can maintain themselves for varying periods at temperatures somewhat below this, but, according to Wells (1957), no reefs develop where the annual mean minimum temperature is below 18°C. Optimal reef development occurs where the mean annual water temperatures are about 23–25°C, and some coral reefs can tolerate temperatures up to about 36–40°C. Of note is the fact that reefs are reduced or absent from large areas on the west coast of South and Central America and also from the west coast of Africa, both of which are areas well within the tropics. The reason for the great reduction in, or absence of, reefs in these areas is that the west coasts of both of these continents are areas of strong upwellings of cold water, which reduces the temperature of the shallow inshore waters below that required for reef development in many places. Furthermore, both coasts have strong north-flowing cold currents—the Peru Current on

the South American coast and the Benguela Current off West Africa—that keep the temperature down. The situation is also complicated by the influx of large amounts of fresh water from the Niger and Congo rivers in Africa and the Amazon in South America.

Coral reefs are limited by depth. Coral reefs do not develop in water that is deeper than about 50–70 m, and most reefs grow in less than 25 m of water. This explains why coral reefs are restricted to the margins of continents or islands. The peculiar feature of the development of atolls, which rise out of water many kilometers deep, will be discussed later. (See pp. 412–414.) The depth restriction is due to the light requirements of hermatypic corals. Light is one of the most important factors limiting coral reefs: Sufficient light must be available to allow photosynthesis by the symbiotic zooxanthellae in coral tissue. Without sufficient light, photosynthesis is reduced and, with it, the ability of the corals to secrete calcium carbonate and produce the reefs. Yentsch (1966) asserts that the compensation point for most reef corals (the point at which photosynthesis is reduced too much for coral growth) seems to be the depth at which light intensity is reduced to 1–2% of surface intensity.

Another factor that restricts coral reef development is salinity. Hermatypic corals are true marine organisms and, like most cnidarians, are intolerant of salinities deviating significantly from that of normal seawater (32–35 psu). Wherever inshore waters are subject to continuing influxes of fresh water from river discharge so that the salinity is lowered, reefs will be absent. Such is the case along large portions of the Atlantic coast of South America, where the Amazon and Orinoco rivers discharge a huge volume of fresh

FIGURE 9.4 Summary of physical factors acting on coral polyps and coral reefs that may limit their distribution.

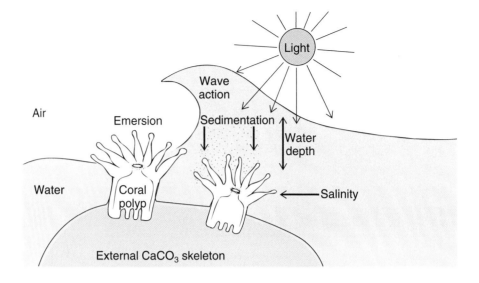

water. (See Figure 9.1.) On a smaller scale, this phenomenon occurs in many areas of the tropics, where rivers and streams running out to sea cause breaks in reef development. At the other extreme, coral reefs occur even in regions of elevated salinity, such as the Persian Gulf, where reefs flourish at 42 psu.

Sedimentation is often correlated with freshwater runoff carried by rivers and streams. Sediment, both in the water and settling out on the coral reefs, has an adverse effect on corals. Many corals can remove limited amounts of sediment by trapping it in mucus and carrying it off by ciliary action. Most hermatypic corals, however, cannot withstand heavy sedimentation, which overpowers their ciliary-mucus cleansing mechanism, clogs their feeding structures, and smothers them. Sediment in the water (turbidity) also reduces the light necessary for photosynthesis by the zooxanthellae in coral tissue. As a result, coral reef development is reduced or eliminated in areas of high turbidity. While hermatypic coral species may vary in their tolerance for sedimentation, few can tolerate high sedimentation rates.

In general, coral reef development is greatest in areas subject to moderate wave action. With their dense skeletons of calcium carbonate, coral colonies can be quite resistant to damage by wave action, depending on their colony morphology. At the same time, wave action provides a constant source of fresh oxygenated seawater and prevents sediment from settling on the colony. Wave action is also responsible for renewing the plankton, which is food for the coral colony.

Finally, coral reefs are limited in an upward direction by exposure to air during low tide. Loya (1976) found that, whereas mucus secretion prevents the reefs from dehydrating for short periods of an hour or two out of water, most corals are killed by longer exposure to air, limiting their upward growth to the level of the lowest tides. The factors that limit coral growth are summarized diagrammatically in Figure 9.4.

■ STRUCTURE OF CORALS

Since the prominent members of coral reefs are corals, it is necessary to understand their anatomy. Corals are members of the phylum Cnidaria, which includes such diverse forms as jellyfishes, hydroids, and sea anemones. Corals and sea anemones are members of the same taxonomic class, Anthozoa; they differ primarily in that corals secrete an external calcium carbonate skeleton and anemones do not. Corals may be colonial or solitary, but almost all hermatypic corals are colonial, with individual coral animals, or **polyps**, occupying little cups, or **corallites**, in the massive skeleton (Figures 9.5, 9.6, and 9.7). The calcium carbonate skeletons of reef-building corals are external. Coral polyps divide asexually on the surface of the skeleton, creating a sheet of genetically identical polyps, which secrete the colony's skeleton at their base. Each cup or corallite of the skeleton contains and is secreted by one polyp and has a series of sharp bladelike **sclerosepta** rising from the base. The pattern of these sclerosepta differs from species to species and is often used as a taxonomic

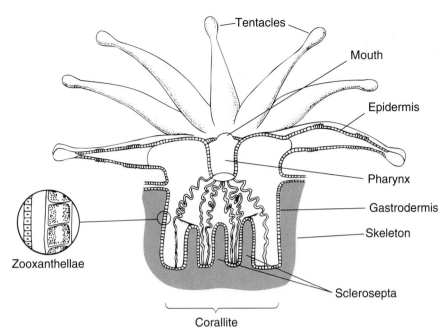

FIGURE 9.5 Anatomy of a coral polyp.

Tentacles

Mouth

Epidermis

Pharynx

Gastrodermis

Skeleton

Zooxanthellae

Sclerosepta

Corallite

FIGURE 9.6 A single coral colony, showing the characteristic appearance when the polyps are expanded. (Photo courtesy of Drs. Lovell and Libby Langstroth.)

FIGURE 9.7 Expanded coral polyps. (Photo courtesy of Drs. Lovell and Libby Langstroth.)

tool to distinguish among coral species. The sclerosepta of the skeletal cup alternate with the internal septa of the gastrovascular cavity of the coral polyp. Differences in the morphology of coral colonies are determined by patterns of polyp growth and division. Each polyp is a three-layered animal with an outer epidermis separated by a fibrous mesoglea from an internal gastrodermis. Around the mouth is a series of tentacles that have batteries of stinging capsules, or nematocysts, that the animals use to capture their zooplankton food (Figure 9.8). Symbiotic zooxanthellae, which are actually dinoflagellates (see Chapter 10), are found intracellularly in the gastrodermal layer. Coral colonies grow by having the polyps bud off new polyps asexually. New colonies are established either sexually by the fragmentation of skeletal pieces or sexually through the production and settlement of a planktonic **planula** larva.

■ TYPES OF REEFS

Coral reefs occur in different sizes and shapes, resulting from particular hydrological and geological conditions in tropical seas. This variety has led scientists

FIGURE 9.8 Orange tube coral, *Tubastrea coccinea*, showing the clusters of nematocysts in the form of dots on the tentacles. (Photo courtesy of Dr. Diane Nelson.)

to classify reefs into various schemes, often generating a bewildering array of names and terms. In an attempt to keep things simple and prevent excessive terminology, we choose to limit the number of types to those which are most common. In general, coral reefs are grouped into one of three categories: **atolls**, **barrier reefs**, and **fringing reefs** (Figure 9.9). Atolls are usually easily distinguished because they are modified horseshoe-shaped reefs that rise out of very deep water far from land and enclose a lagoon, which may contain **lagoon reefs** or **patch reefs**. With few exceptions, atolls are found in the Indo-Pacific area and develop on the top of seamounts. Barrier reefs and fringing reefs, by contrast, tend to grade into each other, are not readily separable, and develop on the edge of continental landmasses. Some scientists would prefer to group the two forms into a single category. Both occur adjacent to a landmass, with a barrier reef being separated from the landmass by a greater distance and deeper water channel than a fringing reef (Figure 9.9b). Fringing reefs and barrier reefs are common throughout the coral reef zones in

all oceans. The largest barrier reef is the Great Barrier Reef of Australia, which stretches for almost 2,000 km along the eastern coast of Australia, from near New Guinea to just north of Brisbane. The second-largest barrier reef is that off the Yucatan Peninsula in Belize.

■ ORIGIN OF REEFS

Different types of reefs and reefs in different oceans may have diverse origins and histories. The greatest interest in the origin of reefs has centered on atolls. For many years, humans speculated as to how such reefs could develop out of deep water, miles from the nearest emergent land. Interest was heightened when it was discovered that reef corals could not live deeper than 50–70 m. This state of affairs led to the development of several theories concerning the origin of atolls. Only one need be discussed here: the **subsidence**, or **compensation, theory**. This explanation for the origin of atolls was first proposed by Charles Darwin following his five-year voyage on the *Beagle*, during which time he had the opportunity to study reefs in several areas. According to Darwin's subsidence theory, atolls

(a)

FIGURE 9.9 Aerial photo of an atoll (a) and a barrier reef (b) (a, Paul Chesley/Getty Images Inc./Stone Allstock. b, Art Wolfe/Getty Images Inc./Stone Allstock.)

(b)

are created when fringing reefs begin to grow on the shores of newly formed volcanic islands that have pushed to the surface from deep water. These islands often begin to subside or sink under their own weight, and if the subsidence is not too fast, reef growth will keep up with it, forming a barrier reef next and, finally, an atoll as the island disappears beneath the sea (Figure 9.10). When the island has disappeared, corals continue to grow on the outside and keep the reef at the surface. On the inside, where the island used to be, quiet water conditions and high sedimentation prevail. These conditions prevent continued vigorous

FIGURE 9.10 Geological evolution of a coral atoll according to the subsidence hypothesis of Darwin. (After *General Zoology*, 6th ed., Stoyer et al., 1979.)

coral growth; hence, a lagoon develops. This theory links all three types of reef in a geological sequence, but is not an explanation for the development of all fringing and barrier reefs.

Since the current surface features of atolls give no evidence of a volcanic base, in the years after the development of Darwin's theory other explanations were offered, and the whole concept of the origin of atolls became embroiled in the "coral reef problem." If Darwin's theory was correct, then drilling down through the current atoll reefs would yield layer after layer of limestone until, finally, volcanic rock would be encountered. Because the technology of the time was incapable of drilling to the base of atoll reefs, the problem had to wait until the middle of the twentieth century for its resolution. In 1953, borings at Eniwetok Atoll in the Marshall Islands penetrated 1,283 m of reef limestone and then hit volcanic rock. This was the evidence that Darwin's theory was substantially correct. The theory has been strengthened by the discovery of flat-topped submerged mountains, or **guyots**, that, at present, have their tops many hundreds or thousands of meters below the ocean surface, but on whose surfaces reside the remains of shallow-water corals. Evidently, these mountains sank too fast for reef growth to keep within the lighted zone.

Although the subsidence theory links all three types of reef in a successional sequence, not all barrier reefs and fringing reefs can be explained by such a mechanism. Indeed, the reasons barrier and fringing reefs occur around continental margins and high nonvolcanic islands are simply that these areas offer suitable environmental conditions for the growth of reefs and a suitable substrate on which to begin that growth. The extensive reefs around the Indonesian Islands, the Philippines, New Guinea, Fiji, and most of the Caribbean islands and Florida are there because a suitable substrate in shallow water existed on which their growth could be initiated. In none of

these areas are large land areas subsiding, nor will these reefs ultimately become atolls.

Whereas atolls are often very old structures (for example, the 1,283 m of reef limestone at Eniwetok represents a reef growth of about 60 million years), other reefs—particularly those in the Atlantic Ocean—may be much younger. These reefs are all geologically very recent, most dating only from the last glacial age, or 10,000–15,000 years before the present. Perhaps even more remarkable is the finding cited by Sale (1991) and Anderson (1990) that the world's largest reef system, the Great Barrier Reef of Australia, is at least 40 times younger than previously thought, with new estimates of its age ranging from 500 thousand years to as young as 9 thousand years.

Reef growth is not simply a product of deposition by calcium carbonate–secreting organisms; it also involves bioeroding organisms and geochemical processes. Scoffin (1972) and Adey (1977) have shown that calcium carbonate–secreting organisms provide the calcium carbonate building blocks required for reef formation, while bioeroding organisms deposit fine calcium carbonate sediments necessary for the lithification of reef sediments or the precipitation of magnesium calcite to form fossilized reef sandstone.

■ COMPOSITION OF REEFS

Although corals are the major organisms that form the basic reef structure, there is a bewildering array of other organisms associated with reefs, such that these areas are perhaps the most diverse and species-rich areas that exist in the marine environment today. Members of practically all phyla and classes may be found on coral reefs. It is not practical to mention all of these forms in this text, nor do we even know the roles played by many of them. It is worth pointing out some of the more important and abundant groups, however.

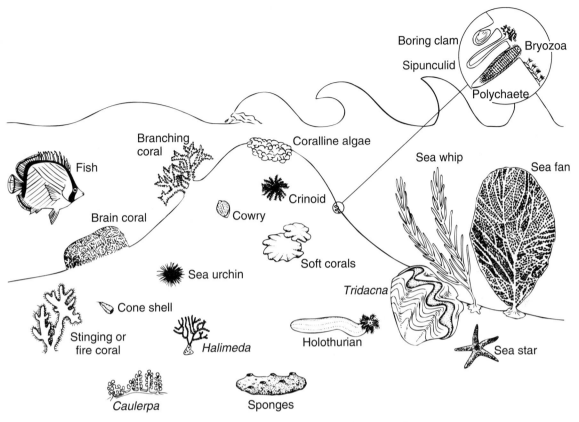

FIGURE 9.11 Some of the dominant and conspicuous components of a coral reef.

In addition to the stony corals (order Scleractinia), certain other cnidarians contribute strongly to reefs. Among these organisms are those anthozoan relatives of the corals called **gorgonians** (order Gorgonacea), commonly called sea fans and sea whips (Figure 9.11; see also Figure 9.3). Gorgonians have internal skeletons consisting of a flexible axial protein rod and spicules. They are particularly conspicuous on Atlantic reefs. The soft corals (order Alcyonacea) are common on Indo-Pacific reefs and may be more abundant than stony corals in some areas of the reef. They are rare, however, on Atlantic reefs. The hydrocoral *Millepora* (order Hydrocorallina), called "fire coral" for its powerful nematocysts, is a conspicuous member of Atlantic reefs.

Coralline algae are extremely important in constructing and maintaining reefs. These red algae precipitate $CaCO_3$, as do the corals, but are encrusting, spreading out in thin layers over the reef and cementing the various pieces together. By this action, the united pieces form a strong bulwark that resists destruction from the ocean waves. These plants form the reef's **algal ridge**, which is the most rapidly calcifying zone on the reef. Some green algae grow erect and secrete $CaCO_3$. Together with the corals, the coralline red and calcareous green algae make up the three dominant $CaCO_3$-secreting organisms of the reef. Much of the sand in coral reef systems is derived from the breakup of these erect calcified algae, such as the green alga genus *Halimeda*. Other free-living, noncoralline algae are inconspicuous on reefs, but a significant algal flora exists just below the surface layers of $CaCO_3$ in the coral colonies themselves.

Sponges are common on reefs and also play a major role in modern reef construction and ecology. Bakus (1964) and Wulff (1997) found that large barrel-shaped and encrusting sponges, often brightly colored to advertise their chemical defenses to potential predators, are common on open surfaces on reefs and drab-colored encrusting and cryptic sponges typically fill the cracks and crevices of reefs. The soft-tissue biomass of sponges in coral reefs can be even greater than the soft-tissue biomass of corals, and since sponges are extremely effective filter feeders, sponge feeding may be important in filtering the water mass above reefs and in keeping light transmission high, ensuring that reef-building corals receive enough light to grow well. As Wulff and Buss (1979) observed, sponges are also important in the role they play binding substrate together, in terms of both preventing reef structures from collapsing under their own weight or from storm and consumer damage and holding reef structures in place long enough for lithification—the chemical process that converts the $CaCO_3$ secreted by reef organisms into the limestone that underlies reefs—to take place. Finally, boring sponges are one of the most important bioeroders of reefs. Bioerosion is a critically important process in coral reefs, both because it can destroy $CaCO_3$ skeletons that contribute to reef growth and because it creates the fine $CaCO_3$ sediments necessary for reef lithification and limestone deposition.

Significant contributions to the $CaCO_3$ deposits on reefs are also made by mollusks of various types. The most conspicuous and important are the giant clams (*Tridacna, Hippopus*), which, as Salvat and Richard (1985) found, can occur at densities as great as 60 per square meter in the South Pacific. Typically, however, as a study by McMichael (1974) indicated, their densi-

FIGURE 9.12 A giant clam, *Tridacna maxima*, on a Palau coral reef. The patterned mantle tissue, which contains the symbiotic zooxanthellae that furnish the clam with a portion of its nutrition, is expanded. (Photo courtesy of Drs. Lovell and Libby Langstroth.)

ties rarely exceed 1 per square meter (Figure 9.12). Giant clams are absent from Atlantic reefs.

Echinoderms—primarily sea urchins, sea cucumbers, starfishes, and feather stars (crinoids)—are another abundant and conspicuous $CaCO_3$-secreting group on reefs. Whereas their role in contemporary reef ecosystems (except for the sea star *Acanthaster* and certain urchins) is incompletely understood (Figures 9.13 and 9.14), they have been important

FIGURE 9.13 The starfish *Fromia monilis* on a soft coral on the Great Barrier Reef. (Photo courtesy of Dr. Diane Nelson.)

FIGURE 9.14 A crinoid of the genus *Comanthus* on a Great Barrier coral reef. (Photo courtesy of Dr. Diane Nelson.)

FIGURE 9.16 A hawkfish swimming among the branches of a coral reef in Palau. (Photo courtesy of Drs. Lovell and Libby Langstroth.)

FIGURE 9.15 Moorish idols, *Zanclus canescens*, on a coral reef in Palau. (Photo courtesy of Drs. Lovell and Libby Langstroth.)

contributors to $CaCO_3$ deposition on reefs. Although various crustaceans and polychaete worms are abundant on reefs, they are often inconspicuous, and little is known about their functions in the reef ecosystem.

The final large conspicuous group on reefs is the fishes (Figures 9.15–9.18). The number of fish species present on coral reefs can be in the hundreds or, occasionally, thousands and makes the coral reef one of the richest environments for fishes on Earth. The fishes contribute considerably to the biological structure and integrity of the reefs.

Bacteria are also very abundant on reefs and are responsible for part of the high productivity of the reef, as well as the decomposition and quick cycling of organic matter.

■ CORAL DISTRIBUTION AND REEF ZONATION

The largest number of reef coral species and genera occur in an area of the Indo-Pacific that includes the Philippine Islands, the Indonesian Archipelago, New Guinea, and northern Australia (Figure 9.19). Veron (1986) believes that, within the Indo-Pacific, 88 genera and 500 species of corals occur. The number of genera and species declines in all directions away from this area of central diversity, but most central Indo-Pacific reefs have 20–40 genera of corals. In contrast to the Indo-Pacific, Goreau and Wells (1967) list only 36 genera and 62 coral species in the Atlantic Ocean. A dominant reef-building genus in both oceans, *Acropora*, has as many as 200 species in the Indo-Pacific, but only three in the Atlantic. Some important reef-building genera of the Pacific, such as *Pocillopora*, *Pavona*, and *Goniopora*, are absent from the Atlantic. Thus, the species richness of Atlantic reefs is much less than that of Pacific reefs. In either ocean, however, as one moves away from the centers of the distribution, the number of coral species declines and is dramatically reduced at the edges of the coral reef zone. For example, the reefs at Midway Island in the Hawaiian chain in the northern Pacific have only 7 coral genera, according to Grigg (1983), while in the Atlantic, the most northern reefs in Bermuda have only 13 coral genera and those in Jamaica have 24.

Coral reefs are large and complex associations of organisms living in a number of different types of habitats, all present in the same system. For example,

FIGURE 9.17 Queen angelfish, *Holocanthus ciliaris*, on a Caribbean coral reef. (Photo courtesy of Dr. Diane Nelson.)

FIGURE 9.18 A lionfish, *Pterios volitans*, on a Pacific coral reef. (Photo courtesy of Drs. Lovell and Libby Langstroth.)

there are areas of hard substrate on which many sessile organisms attach and that are analogous to rocky shores. At the same time, there are areas of sand that require a different set of adaptations, much as we saw for soft bottoms. (See Chapter 5.) Similarly, there are areas of heavy wave action and strong currents and areas of virtual calm, where water movement is minimal. Still other areas may have a lush growth of calcareous green algae (*Halimeda*) mimicking seagrass beds. In other areas, there are mangroves or actual seagrass beds. All of these types of habitat are found in a relatively small area. Therefore, one of the reasons for the great diversity of life in coral reefs is the great diversity of habitats. As might be expected, these different habitats, as well as others not mentioned here,

FIGURE 9.19 The number of genera of reef corals occurring in various tropical regions. (After "Distribution of Reef-Building Corals," J. E. N. Veron, *Oceanus,* Vol. 29, No. 2, p. 27, 1986. Copyright © 1986 Woods Hole Oceanographic Institution.)

vary in their extent or presence on reefs in different geographic areas; hence, it is difficult to describe a typical zonation pattern for reefs that will be generally applicable. On all reefs, gradients in water motion (waves, currents) and light penetration (depth, turbidity) are the two most significant physical factors determining zonation. When these factors are coupled with the various biological interactions (competition, predation, grazing), conspicuous zonal patterns emerge. Broad, large-scale zonation patterns are often the result of physical factors, whereas small-scale "patchiness" is frequently the result of biological interactions.

On atolls, reef development and reef zonation are strongly influenced by wave action, a product of the prevailing wind. Therefore, even though these reefs are essentially horseshoe shaped, a cross section shows different zones, depending on whether the reefs face the prevailing wind and waves or lie in their lee. Figure 9.20 is a diagrammatic cross section of a typical atoll and indicates the major features of zonation.

Approaching an atoll from the sea on the windward side, we see the first zone of the reef, the **outer seaward slope**, where living coral becomes abundant at a depth of about 50 m. Corals in this area are few and often delicate. At about 15 m, there is frequently a terrace or step in the otherwise steep outer slope;

from here to the surface, coral growth is lush. This is also the area where wave action is the most severe. At the surface of the water is the **windward reef margin**, which is characterized by the development of coral–algal spurs or buttresses extending into the waves and down to the terrace. These spurs or buttresses alternate irregularly with deep channels or grooves into which water surges onto the reef and through which water and associated debris leave the reef (Figure 9.21). This is the so-called **spur-and-groove**, or **spur-and-buttress, zone**. The sides of the grooves or channels and the spurs themselves support a lush growth of the dominant reef-forming corals, such as *Acropora*. Because environmental conditions are optimal, it is here that coral reef accretion occurs most rapidly. Immediately behind the buttress zone is a smooth, coral-free ridge of encrusting coralline algae called the **algal ridge**. This ridge receives the full impact of the wave energy and, therefore, is virtually free of any organisms except the encrusting coralline algae, which are usually of several genera, including *Porolithon*, *Hydrolithon*, and *Lithothamnion* or related genera—hence the origin of the alternative name for this zone: the **lithothamnion ridge**. Algal ridges are responsible for the asymmetry that is characteristic of many atolls. Typically, the windward side exhibits the widest islands and has the algal ridge. There is a correlation between the development of an algal ridge and

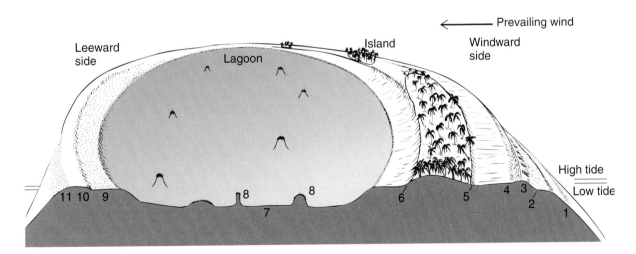

FIGURE 9.20 Diagrammatic cross section of a typical atoll: (1) outer seaward slope; (2) windward reef margin with spur-and-buttress zone; (3) algal ridge; (4) the reef flat; (5) seaward beach of the island; (6) lagoon beach of the island; (7) lagoon floor; (8) lagoon reefs; (9) leeward reef flat; (10) leeward reef margin; (11) leeward reef slope. (Modified from "The Biology of Coral Reefs," C. M. Yonge, 1963, *Adv. Mar. Biol,* Vol. 1, pp. 209–260. Copyright © 1963 Academic Press, Ltd. Reprinted by permission.)

FIGURE 9.21 Generalized sketch of the spur-and-buttress zone of a windward reef. (AR) Algal ridge; (B) buttresses or spurs; (G) grooves; (SC) surge channels; (T) 18-m terrace. (After "The Biology of Coral Reefs," C. M. Yonge, 1963, *Adv. Mar. Biol*, Vol. 1, pp. 209–260. Copyright © 1963 Academic Press, Ltd. Reprinted by permission.)

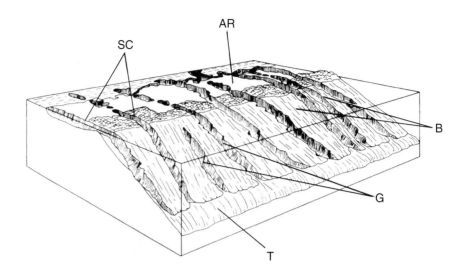

the degree of fetch and wind duration, so that many Pacific reefs have such ridges while many Caribbean reefs do not. The grooves of the buttress zone, often called **surge channels**, penetrate behind the algal ridge as deep channels or as blowholes where the channel has been roofed over by algal or coral growth, leaving only one or a few openings on the inside of the ridge through which water may enter or leave. The blowholes may offer a spectacular display as the water is forced into the tubes at the outer slope of the reef and explodes as a geyser on the inside. This whole area—algal ridge and buttress zone—is extremely dangerous. Wave action is violent, and a step into a blowhole or surge channel could mean disaster.

Behind the algal ridge is a very shallow **reef flat** zone—a complex area with strong gradients in such environmental factors as temperature, turbidity, and exposure, moving from the algal ridge to the shore of the atoll island. These gradients, coupled with differences in depth and type of substrate (coral rock, sand), provide many habitats that subdivide this zone. Near the algal ridge, conditions for coral growth are good, and a variety of corals is found. This is another area with active accumulation of reef material. Conditions for coral growth depart from optimal as one approaches the island, and corals become less and less abundant. Microatolls often develop in deeper areas of the middle reef flat. They are larger colonies of a massive coral, such as some species of the coral *Porites*, in which the top of the colony has been killed by exposure to air or excess sedimentation. The death of the top of the colony is followed by erosion of the animals' skeletons, so the

microatoll becomes a small basin filled with water in which other organisms may grow (Figure 9.22). Microatolls support a variety of organisms, including giant clams (*Tridacna, Hippopus*).

The reef flat zone ends at the **seaward beach** of the atoll island. On the lagoon side of the island is a narrow beach, followed by the quiet waters of the lagoon, which has two major zones: the **lagoon reefs** and the **lagoon floor**. Lagoon reefs are found around the perimeter of the lagoon and also exist as patch reefs or pinnacles rising from the lagoon floor to the surface. The depth of the lagoon is usually less than 50 m and, therefore, within the depth range of coral growth. Conditions within the lagoon are not as good for coral growth as on the windward side, because wave action and circulation are not as great and sedimentation is greater. As a result, coral growth is often restricted to areas where conditions are better, leading to the existence of patch reefs and pinnacles and leaving large areas of the lagoon floor free of coral growth. Sedimentation leads to large areas of sand, and the lagoon floor may have extensive beds of seagrasses (*Thalassia, Cymodocea*) or of green algae, such as *Caulerpa* and *Halimeda*.

The leeward side of the lagoon is bordered by the **leeward reef flat**, which is generally narrower than the seaward reef flat and is separated from the lagoon reefs by a barren boulder zone or by an area of poor coral growth dominated by isolated colonies of *Porites*. The seaward margin of this flat is similar to the windward flat. At the leeward reef margin, the algal ridge is usually poorly developed or absent, and there are no surge channels or buttresses. At the

FIGURE 9.22 A microatoll on the Great Barrier Reef. (Photo by James Nybakken.)

seaward edge, extending down to 15–20 m, coral growth is lush and diverse, dominated by the branching arms of *Acropora*. In addition to the massive corals seen on the seaward reef margin, numerous branching species are found here, due to the absence of violent wave action. Below 15–20 m, the leeward reef slope is similar to the outer seaward slope.

Fringing reefs and barrier reefs can be considered truncated portions of the atoll zonation scheme. Fringing reefs and barrier reefs on the windward sides of islands have a zonation like that of the windward side of atolls, and those on the leeward side have a zonation similar to that of the leeward reef of atolls. The major difference between the two is that barrier reefs—especially large ones such as the Great Barrier Reef—may have an extensive lagoon between the reef flat and the shore of the continent or island, and that lagoon may develop additional coral reefs.

There is also conspicuous depth zonation of corals on most reefs (Figure 9.23). In shallow water, corals

SEA   REEF CREST   REEF FLAT

Beach

Sea grass bed

Small corals in pools and channels

Heavy branching corals

Massive corals

14
16

Zonation of a coral reef

FIGURE 9.23 Depth zonation of corals on a typical reef. (From *Invertebrate Zoology*, 6/e, HSIE 6th Edition by Ruppert. © 1994. Reprinted with permission of Brooks/Cole, a division of Thomson Learning: www.thomsonrights.com. Fax 800-730-2215.)

FIGURE 9.24 The elkhorn branching coral, *Acropora palmata* (Photo courtesy of Mark Bertness).

with a high colony surface area and small polyps dominate. Among these corals are the highly branching ones, such as the elkhorn (Figure 9.24) and staghorn corals that dominated Caribbean reefs for centuries until about 20 years ago. Porter (1976) has shown that these corals rely heavily on their zooxanthellae and have plantlike (arborescent) growth morphologies to maximize light interception. In deeper water, massive mound-shaped and platelike morphologies, often with large polyps, dominate (Figure 9.25). These corals rely more on capturing prey than on zooxanthellae for their nutrition. There is even morphological variation within coral species that reflects this general pattern: Lasker (1981) found that the Caribbean coral *Montastrea cavernosa* has smaller polyps and more zooxanthellae when found at shallow depths and larger polyps and a more predaceous lifestyle in deeper waters (Figure 9.26).

■ ATLANTIC AND INDO-PACIFIC REEFS

As we have already indicated, there are many differences between Atlantic and Indo-Pacific reefs. Among these are differences not only in structure and zonation features, but also as regards the presence, absence, and abundance of various fauna and flora.

The physical conditions under which reefs grow in both the Atlantic and the Pacific are similar with respect to temperature, salinity, and turbidity, but the number of coral genera and species is very different between the two areas and is not limited to corals; it also occurs among most of the reef components, including mollusks, fishes, and crustaceans. For example, there are perhaps 5,000 species of mollusks on Indo-Pacific reefs, but only 1,200 on Atlantic reefs. Similarly, there are 2,000 species of Indo-Pacific reef fishes compared with 600 Atlantic species. On a more detailed note, on the one hand, Atlantic reefs commonly have large numbers of sea fans and whips (gorgonians), whereas these are much less common in Indo-Pacific reefs. On the other hand, Indo-Pacific reefs have large numbers of soft, or alcyonacean, corals, such as *Sarcophyton* and *Lobophyton*, whereas these are inconspicuous in the Atlantic. The soft corals *look* like corals with calcium carbonate skeletons, but lack a hard skeleton. Atlantic reefs also lack the giant clams (*Tridacna*, *Hippopus*) and the octocorals *Heliopora* and *Tubipora*. There are no anemone fishes (clown fishes, *Amphiprion*) in the Atlantic, although the Atlantic does have fishes that associate with anemones. Similarly, crabs and shrimps that guard certain corals against predators are absent from the Atlantic, as is the coral-eating starfish *Acanthaster planci*.

Dominance is also different in the two regions. Atlantic reefs were historically dominated by the elkhorn coral, *Acropora palmata*, and the fire coral *Millepora*

FIGURE 9.25 A typical massive coral with large polyps (Photo courtesy of Mark Bertness).

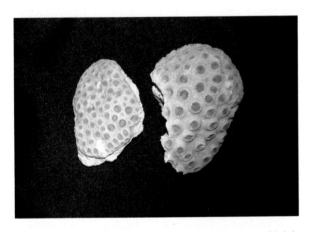

FIGURE 9.26 Shallow-water (left) and deepwater (right) ecomorphs of the common massive Caribbean coral *Montastrea cavernosa* (Photo courtesy of Mark Bertness).

complanata, a hydrocoral. *Millepora* does not dominate any Indo-Pacific reef. Rhythms differ in that most Atlantic corals are nocturnal, but in the Pacific, many, such as *Porites, Goniopora,* and *Favia,* are diurnal.

There are also differences in relative abundances of some of the noncoral species. Sponge biomass, for example, as Wilkinson (1987b) notes, is 2–10 times

as great on Caribbean reefs as on the Great Barrier Reef, and few, if any, Caribbean sponges have the phototrophic symbionts present in many Pacific sponges. Similarly, Sammarco (1987) found that, whereas echinoids are the major grazers in the Atlantic, fishes fill that role in the Pacific.

Not only are there differences in species composition and relative abundances between the two areas, but also there are differences in age and reef morphology (construction). Atlantic reefs usually rest on shallow banks or platforms, probably the result of erosion by wave action during the Pleistocene period, when sea level was much lower. Borings through Atlantic reefs, therefore, do not show the vast depth of reef limestone that indicates the great age of the Pacific atoll reefs. Why this great discrepancy in age? No one knows for certain, but two explanations may be offered: First, the Atlantic is a geologically more recent ocean; hence, there has been less time for reefs to develop. Second, in the ice age, the Atlantic may have become too cold for reefs to survive; thus, they died and were replaced only when the seas warmed following the retreat of the glaciers.

Zonation is also somewhat different. Atlantic reefs have their buttress zones deeper than the Pacific reefs, and active coral growth extends to 100 m in the Atlantic, but only to 60 m in the Pacific. Also, atoll islands are often absent in the Atlantic, but reef flats and shallow lagoons are common. Finally, the percentage of cover by corals differs: In the Atlantic, corals cover less than 60% of the substrate, as opposed to 80–90% in the Pacific.

■ PRODUCTIVITY

Coral reefs are truly oases in a watery desert. As we noted in Chapter 2, except in upwelling areas, tropical marine waters are extremely poor in nutrients; hence, they have very low productivity. Coral reefs, however, abound with life, and all studies done on them have indicated high productivity. Most studies, including that of Atkinson (1992), have estimated gross primary productivity to be 1,500–5,000 g C/m^2/year. This figure is in contrast to a gross primary productivity in the open tropical oceans of 18–50 g C/m^2/year. How can coral reefs maintain that level of production in such nutrient-poor areas? This is less hard to explain for reefs bordering high islands or continents, where much nutrient material might become available through runoff from the land, than it is for atolls isolated far from land. Indeed, in the latter case, we do not yet know all the reasons for the high productivity. While there is little or no disagreement among scientists that coral reefs *are* extremely productive, there is still no consensus about the underlying factors that sustain their high productivity. The problem revolves around the necessary source or sources of the limiting nutrients nitrogen and phosphorus. Some researchers have suggested that reefs need a sustained external source of nutrients and have invoked such processes as upwelling in the water column and the leaching of groundwater. Other researchers, including Sorokin (1990), dispute the need for external sources of nutrients and suggest that tight recycling by biological processes such as filter feeding can maintain high enough fluxes of the nutrients to sustain the reef's productivity. Still others, such as Atkinson (1992), have suggested that a high flow rate of the tropical oligotrophic water across the reef is sufficient to account for the high productivity without invoking outside sources or even recycling.

In the absence of any consensus about nutrient fluxes, it is still possible to suggest other reasons that reefs are so productive. First, the amount of photosynthetic tissue present on a reef and capable of photosynthesis is very large, much larger than in a similar area of the open ocean. Photosynthetic organisms present on a reef include zooxanthellae, prokaryotic symbionts in sponges, crustose coralline algae, turf algae, large macroalgae, seagrasses, and phytoplankton. Each of these organisms varies in productivity and in the amount of cover they provide on reefs (Table 9.1), but together they form a significant biomass. Indeed, the biomass of these photosynthestic organisms is greater than the animal biomass, even if less conspicuous. Of these autotrophic components, it appears that the ubiquitous small turf algae are the most significant primary producers (Table 9.1). Reefs also act as sinks for any nutrients brought in from the outside. This feature ensures that the nutrients remain within the reef system and are not lost to deeper, offshore waters. It also means that any plankton from the open sea that impinges on the reef remains there, as do the nutrients brought with them.

Another factor that aids in the productivity of coral reefs is the close coupling of photosynthetic organisms with either the animals or the matrix of the reef. The best example is the location of the zooxanthellae in the coral polyps. Residing in coral tissue means that the algae will not be washed from the reef. It also means that any nutrients produced by the coral (NO_3, PO_4) as a result of its metabolism will be available directly to the algae without cycling it outside into the water, where it might be lost. The same is true of the cyanobacteria in sponges, and similarly, most of the free-living algae on the reef contain calcium carbonate. They are too heavy to be moved easily off the reef, or they are fused into the reef structure by the coralline algae; hence, their nutrients remain on the reef to nourish additional generations. Rapid cycling is also aided by the large bacterial populations on reefs. These bacteria quickly break down dead material, making the nutrients formerly enclosed within them available. For all these reasons, and probably for others yet to be discovered, coral reefs have very high productivity.

■ BIOLOGY OF HERMATYPIC CORALS

Hermatypic corals are the dominant group involved in reef formation and maintenance. Understanding their biology is key to understanding of the ecology of reefs.

NUTRITION   Reef corals have long been known to be carnivorous animals, similar in that respect to most other cnidarians. They have tentacles studded with stinging capsules, or nematocysts, used to capture small planktonic organisms. Since reef corals live in colonies of hundreds or thousands of individuals, and since the number of colonies is also huge, the total area for feeding is enormous. In addition to their tentacles and nematocysts, the outer epidermis of corals is ciliated and produces mucus. This ciliary mucous

TABLE 9.1

Primary Productivities of Component Autotroph Types on Coral Reefs and the Distribution of These Communities on Reefs

AUTOTROPH TYPE	RANGE OF PRODUCTIVITY (g C/m²/day)	APPROXIMATE AREA/ COVER ON REEFS %
Benthic algae	0.1–4	0.1–5
Turf algae	1–6	10–50
Zooxanthellae	0.6	10–50
Sand algae	0.1–0.5	10–50
Phytoplankton	0.1–0.5	10–50
Seagrasses	1–7	0–40

Source: From Larkham, 1983.

mechanism is generally used by corals to rid themselves of sediment settling on the surface. In some corals, it has been modified for feeding as well. Those species using mucous mechanisms to capture plankton tend to have smaller, shorter tentacles than those which feed directly on plankton by snaring them with their tentacles. Food organisms are detected by chemoreception.

Since corals exist in such numbers on reefs, and since the surrounding ocean is notoriously poor in plankton, how is it possible for the corals to obtain enough food? This question has puzzled generations of coral reef biologists, but the answer is now becoming clear. Rigorous plankton sampling carried out in studies by Johannes et al. (1970) and Porter (1974) just offshore on reefs both during the day and at night has revealed an abundance of plankton. A large fraction of this plankton, however, is not from the open ocean, but is indigenous to the reef itself. This indigenous plankton is composed primarily of demersal meroplankton, which spends the daylight hours on the sea bottom and rises into the waters over the reef at dusk. This may also explain why many corals are closed during the daylight hours, expanding to feed only during the hours of darkness. Johannes et al. (1970) calculated that demersal plankton is sufficient to satisfy only 5–10% of the corals' total food requirements. Where does the remainder of their food come from? Only one source seems possible: the zooxanthellae found in the coral tissues.

The role of zooxanthellae in coral biology has long fascinated biologists. Early on, it was learned that corals were carnivores and could not digest plant tissue. Furthermore, experimental studies showed that, if large-tentacled corals were starved or kept in the dark, they would expel their zooxanthellae, but could continue to live for a few weeks without them. This was possible because these corals could capture zooplankton and use them. The experimental observations suggested that zooxanthellae were really unnecessary to the coral. Work by Johannes (1974), however, found that if corals were not fed, but were kept in light, they gained weight, and radioactive tracer studies performed by Muscatine and Cernichiari (1969) showed that organic compounds fixed by the zooxanthellae in photosynthesis were transferred to the coral, to be subsequently eaten. Thus, whereas the role of zooxanthellae in nutrition is now accepted, we still do not know how important zooxanthellae are for all coral species. As Porter

(1976) and Wellington (1982b) have noted, in general, although corals are primarily **phototrophic** (dependent on zooxanthellae), corals with large polyps and tentacles are less dependent on zooxanthellae for nutrition than species with smaller polyps.

Zooxanthellae are acquired by corals either directly from the parent or indirectly from the environment. In the case of asexually reproducing corals, zooxanthellae are directly transmitted during the division or fragmentation process. Sexually reproducing corals, however, may transmit the parent zooxanthellae directly to the eggs or brooded larvae; alternatively, the larvae or young corals may have to acquire zooxanthellae from the environment after being released from the parent. Which mechanism is employed is generally species specific.

Originally, it was thought that there was only a single species of zooxanthellae, but recent molecular taxonomic work has revealed at least 16 species. Zooxanthellae in corals grow rapidly and fix carbon at a rapid rate. Rapid growth continues until the animals reach a certain density, at which point they stop reproducing and their growth slows as they release carbon compounds to the coral. If, at this point, higher concentrations of N and P are added, the zooxanthellae once again begin to reproduce and reach even higher densities. This finding suggests that corals are nitrogen limited. Corals seem able to limit the growth and densities of zooxanthellae through some unknown "host factor," causing the zooxanthellae to release the organic molecules they use for food. Recent research by Rowan (1998) has also revealed that different zooxanthellae may also be characteristic of different reef microclimates. Clearly, much more remains to be learned about coral zooxanthellae.

GROWTH AND CALCIFICATION Although there has been much interest in the growth rate of both corals and coral reefs for many years, it is only relatively recently that we have come to understand the growth of individual corals and reefs. The first requirement for active coral growth is light. If corals are shaded or denied access to light, they stop growing. If light is withheld long enough, they die. This requirement for light is based on the needs of the zooxanthellae, and, as Goreau (1961) has noted, it suggests the other important role of zooxanthellae in the life of corals: They significantly increase the calcification rate of corals and the growth rate of coral colonies. Exactly how zooxanthellae increase the rate of skeletal accretion is not fully understood, but it seems that light enhances oxygen production, which, in turn, stimulates coral metabolism and leads to increased $CaCO_3$ deposition. The latter again indicates the importance of zooxanthellae in coral reefs, where rapid calcification is needed to maintain the reefs against the various destructive forces acting against them.

The most direct way to measure coral growth—a method employed by Buddemeir and Kinzie (1976)—is to record a particular dimension of the coral and then repeat the measurement after an appropriate interval. A rapid and simple way to measure coral growth, used by Connell (1973), is to take a time series of photographs of permanent locations on a reef and then measure changes in coral size or cover over time (Figure 9.27).

Coral growth rates vary with species, age of the colony, and location on the reef. Young, small colonies tend to grow more rapidly than older, larger colonies. Corals with different morphologies also differ in growth rates. In general, branched or foliaceous plant-shaped corals tend to grow more rapidly than massive or mound-shaped corals. For example, Davis (1983) has shown that *Acropora*, a branched genus, grows 2.5–26.6 cm/year in the Caribbean. In contrast, *Montastrea annularis*, a massive, moundlike coral, had a linear growth of 0.81–2.5 cm/year. Despite much recent work that has given us good estimates of growth rates in some species, the fact that there is growth throughout the life of the colony and that growth rates in the larger, older colonies are variable and difficult to estimate means that we still do not understand the age structure of the colonies seen on a reef, nor do we know how long corals can live. What little information there is, primarily that gleaned by Connell (1973), suggests that most coral colonies on reefs are 10 years of age or younger. However, some of the large, massive corals, such as those forming microatolls, may be 100 years old or even older, suggesting that corals can live much longer than the ages generally attained by most colonies on the reef.

The growth form of a given coral species may also vary, depending on the location of the species on a reef. In contrast to the same species in shallow water, corals at greater depth are often thinner and more fragile, probably as a result of less calcification. Wave action forces branched species to have shorter and stubbier branches, and currents cause branching

forms to have a definite alignment of their branches. This ability of coral to have different skeletal shapes has resulted in much confusion among coral reef taxonomists, such that different growth forms of the same species have often been erroneously deemed different species.

SEXUAL MATURITY, REPRODUCTION, AND RECRUITMENT Corals reproduce both sexually and asexually. Asexual reproduction is accomplished when new individuals bud off the parent. Continued budding is the mechanism for increasing the size of the colony, but generally not for producing new colonies. Highsmith (1982) has observed that fragmentation is an important means of starting new colonies in some species. Jackson and Hughes (1985) found that, in a three-year period on a Jamaican reef, 38 new colonies of the coral *Leptoseris cucullata* were started by fragmentation, while only 16 were established by larval recruitment. Sexual reproduction in corals results in free-swimming planula larvae, which initiate new colonies when they settle (Figure 9.24). Most corals reach sexual maturity at 7 to 10 years of age. Corals are primarily hermaphroditic; a few are dioecious (gonochoric).

The time at which corals form gametes depends on the size and age of the parent colony. Fragmentation or partial destruction of a colony can affect its fecundity. Most corals are broadcast spawners, releasing eggs and sperm into seawater, but a smaller number are brooders, retaining fertilized eggs in the gastrovascular cavity until development has proceeded to the planular stage (Figure 9.28). Spawning is usually seasonal, with a single brief annual event. Broadcast spawners are regulated by the lunar cycle, and there are some interesting regional differences. For example, Szmant-Frolich (1985) found that many corals on the Great Barrier Reef spawn in late spring to early summer, after the October and November lunar cycles and often synchronously. In the Caribbean, the major spawning period is longer, lasting from July through September, and is asynchronous. Among the brooders, the release of planulae is also coordinated by the lunar cycle.

Planulae are released and swim in open water for an undetermined time, generally a few days to a week or more, before settling down and starting a new colony. There is still disagreement over the amount of time planula can remain in the plankton and, therefore, over the dispersal range of coral

(a)

(b)

(c)

(d)

FIGURE 9.27 Photographs of the same areas of a 1-m quadrat showing the changes in growth and occurrence of coral colonies over an eight-year period at Heron Island on the Great Barrier Reef. For each photograph, the square-meter frame, divided into 20 × 20 cm squares, was positioned on permanent stakes. (Only a portion of the 20 × 20 squares is shown.) Note the great changes that occurred in the growth and presence of corals over short intervals. (a) 1963. (b) 1965. (c) 1969. (d) 1971. (Photos courtesy of Dr. Joseph Connell, Dept. of Biological Sciences, University of California, Santa Barbara.)

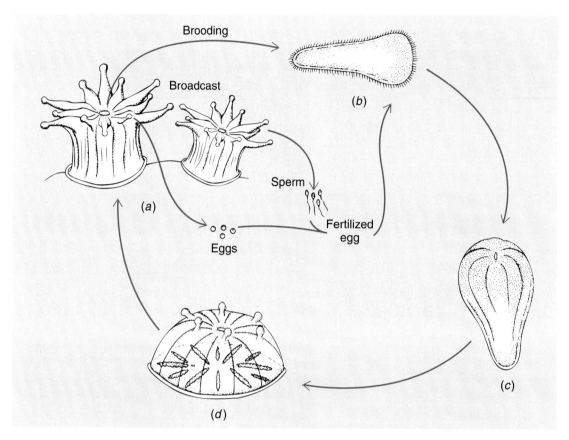

FIGURE 9.28 Reproduction of corals. (a) Adult polyp. (b) Planula (larva). (c) Later planula with developing septa. (d) Young polyp after attachment. (Not to scale; planula and young polyp are much smaller than the adult.) (Modified from *The Invertebrates*, Vol. 1, L. Hyman 1940.)

species. Since adult corals are fixed in place, the planula larvae are the only means of dispersal of the various coral species.

The rate of recruitment of new coral colonies to a reef system is variable in time and in space. In both the Great Barrier Reef and the Caribbean, *Acropora* is a dominant coral genus, but in 1985, Sammarco discovered that *Acropora* recruits are rare in the Caribbean, but abundant in the Great Barrier Reef. This difference could have important implications for the recovery of reefs from catastrophic events. In the Great Barrier Reef system, the higher number of recruits and higher survival rates may lead to faster recovery, whereas in the Caribbean, recovery may be much slower. Connell (1973) found that, on the Great Barrier Reef, coral recruitment is 0–13 new colonies/m²/year. The mortality of new coral colonies is thought to be very high.

■ SPECIES INTERACTIONS AND THE ECOLOGY OF REEFS

At this point, we have covered the basic biology of the signature organisms of reefs—the scleractinian corals—and have outlined general zonation patterns on reefs. Reefs, however, are much more than simple static systems: They are living systems that increase or decrease in size as a result of complex interactions among biological and physical forces. We next consider how these interacting factors produce the reef and influence its persistence. Since coral reefs are among the most complex systems in the marine environment, and since they have not received as much detailed ecological study as rocky shores, our knowledge of this aspect of the coral reef system is not as complete as we would like it to be. However, it is possible to suggest how factors *appear* to interact to maintain the system. The general picture now emerging is that the interactions responsible for generating patterns on reefs are similar to those responsible for generating the patterns we observed on rocky shores.

COMPETITION   One of the features of many coral reefs is that there is virtually no open space on them. The primary space is almost completely covered by the dominant corals or an algal turf (all the small, short macroalgae growing a few centimeters from the substrate). Because all the space is occupied, and because corals must have access to light to survive, we might expect competition among corals for space and light.

Competition does indeed occur among corals. Upright, branching corals grow more rapidly than encrusting or massive corals, and they often grow over encrusting corals shading them. In this kind of competition, called **exploitative competition**, the shaded part of the encrusting colony often dies. Since branching corals can grow faster, and since they shade slower growing massive and encrusting corals, how do the slow-growing corals manage to persist? There are a number of reasons, but the most intriguing is that uncovered by Lang (1973): Slow-growing massive corals can extend digestive filaments from their gastrovascular cavities and kill adjacent competing coral species (Figure 9.29). This

FIGURE 9.30 Competitive interaction between two massive brain coral species. (Photo courtesy of Dr. Ron Shimek.)

kind of interaction, called **interference competition**, allows slower growing, but more aggressive, corals to maintain space and outcompete more rapidly growing neighbors (Figure 9.30). As Lang (1973) has shown, corals capable of utilizing their digestive filaments in this way can be arranged in a competitive hierarchy in which the winner of any given encounter be predicted. This aggressive pecking order may contribute to maintaining local diversity on coral reefs, but the situation is actually much more complicated. For example, in the eastern Pacific, the slow-growing massive coral *Pavona* dominated the faster growing foliaceous coral *Pocillopora* in short-term experiments performed by Glynn (1974). Over time, however, *Pocillopora* develops tentacles 30 times their normal length and increases the potency of their nematocyst batteries. These "sweeper tentacles" contact *Pavona* colonies and inflict severe tissue damage. Wellington (1980) showed that this retaliation allows *Pocillopora* to repair the original damage, continue growing toward the *Pavona*, and eventually displace *Pavona*. Richardson et al. (1979) reported a similar system of sweeper tentacles in two other Caribbean corals, suggesting that the phenomenon is widespread. The competitive superiority of *Pocillopora* helps explain why it is the dominant coral on shallow reefs of the eastern Pacific.

Huston (1985) found that, at shallow depths, cover by coral or other organisms may approach 100% on many coral reefs. Under such crowded conditions, the rapidly growing branching corals reduce or eliminate the slower growing massive corals by

FIGURE 9.29 Photo showing the killing of portions of a tabular *Acropora* colony (left) by the brain coral *Ctenella chagius* (right), thereby preventing overgrowth by the faster growing *Acropora*. (Photo from C. R. C. Sheppard, 1979, Interspecific aggression between reef corals with reference to their distribution. *Mar. Ecol. Progr. Ser.* 1:237–247.)

overtopping them, thereby reducing their access to light and to water movement. Such exploitative competition may explain the dominance of *Pocillopora* at the shallowest depths in the eastern Pacific and *Acropora* in the Caribbean. If the slow-growing corals cannot gain access to full sunlight, how do they survive? One way, discovered by Huston (1985), is that they are more shade tolerant and grow at greater depths, where reduced competition under the low-light conditions permits their survival. A greater survival rate, in turn, leads to increased species diversity with depth, a relationship we see not only in corals, but in other members of the reef fauna as well.

As more studies have been performed on competitive interactions among coral species, many mechanisms of coral competitive interactions have been identified, including mesenterial filaments, sweeper tentacles, mucous secretions, overgrowth, overtopping, sweeper polyps, and waterborne chemicals. Interspecific competition is made even more complex because it can also vary geographically and temporally and can be influenced by other factors, such as predation, disturbance, and changing environmental conditions. For example, a hierarchical ranking of coral species based on the aggressive use of mesenterial filaments is different in Jamaica and Bermuda, and in the aforementioned *Pavona–Pocillopora* interaction, the hierarchical ranking by mesenterial filaments puts *Pavona* on the top, but if the ranking is by sweeper tentacles, then *Pocillopora* is ranked first. These categories may also describe competition between corals and other space-occupying invertebrates on the reef. For example, the inhibition of growth and tissue necrosis among scleractinian corals as a result of competitive encounters with soft corals seem to be mediated by waterborne chemicals, according to Sammarco and Coll (1992). Similarly, sponges have been known to overgrow corals, and chemicals produced by the sponges have been implicated as the competitive mechanism permitting the sponges to displace the corals.

Competition among corals and other reef taxa, particularly fast-growing algal species, is reduced by grazing fishes and invertebrates. Also, as Birkeland (1977) has shown, whereas algae can be competitively superior and overgrow juvenile coral colonies in shallow water, corals can maintain their space in deeper waters. The importance of algal grazers to corals has been shown by Brock (1979), Hatcher (1983), and Sammarco (1980) with experiments

demonstrating that algae can dominate corals in the absence of grazers.

Competitive interactions on reefs are not limited to those between scleractinian corals and algae. For example, Wahle (1980) reported that hydrocorals of the genus *Millepora* directed their growth toward adjacent gorgonians (*Plexaura homomalia* and *Briareum asbestinum*) and, over a period of months, completely overgrew them. Similarly, Karlson (1980) showed that the zoanthid *Palythoa caribaeorum* and the gorgonian *Erythropodium caribaeorum* overgrew the zoanthid *Zoanthus solanderi* in the Caribbean.

It is clear, reports Karlson (1999), that competitive interactions among corals and other clonal sessile invertebrate taxa are complex and involve multiple mechanisms. Moreover, these interactions can be variable in time and space and may be modified by disturbances, geography, and predation such that the resultant community structure is difficult to predict. Finally, many competitive interactions can result in standoffs, reversals, and even competitive coexistence, as well as competitive exclusion. Thus, the degree to which competition among corals, or between corals and other space-occupying invertebrates, controls community structure on reefs is presently unknown. Furthermore, competition is only one mechanism contributing to coral reef diversity, structure, and zonation patterns.

PREDATION Although many species of mobile invertebrates exist on coral reefs, one *sees* virtually no mobile invertebrates on the reef, except for a few large echinoderms (sea cucumbers, urchins, feather stars) and large mollusks (*Tridacna* and others). The reef seems dominated by the corals and fishes. This gap between appearance and reality arises because the other invertebrates are hidden. Bakus (1964) has suggested that the reason for this cryptic pattern is the intense predation pressure on the reef: Any soft-bodied invertebrate in the open would be quickly consumed. This hypothesis can be verified by turning over coral heads and exposing the fauna, which consists primarily of small mobile and sessile invertebrates and which immediately attracts fishes to the area to consume the exposed organisms. Larger echinoderms and mollusks are less vulnerable to predation.

The role of predation in determining the structure and composition of coral reefs is not as well studied or understood as it is in rocky intertidal habitats, but

FIGURE 9.31 Photo showing the large numbers of *Acanthaster planci* on a heavily infested reef in Palau in 1979. (Photo courtesy of Dr. Diane Nelson.)

we do have an increasing body of evidence of its importance. A surprising number of animals feed on live coral and can be classified as coral predators. Most of these predators, however, are small in comparison to coral colonies. Predation by these organisms resembles grazing by herbivores, wherein patches of coral polyps are removed, but the entire colony is not destroyed. If not too much tissue is removed, the coral may regrow polyps to cover the area, very much as grass regrows after grazing. The small predators in question include gastropod mollusks, at least one nudibranch mollusk (*Phestilla*), polychaete worms (*Hermodice*), and several crabs (for example, *Mithraculus*). Where these predators have been studied, it appears that most are dietary specialists with respect to the types of corals they consume. In general, the mollusk and echinoderm corallivores seem to prey on the fast-growing and dominant corals. Taken together, these predators usually do not appear to have a significant effect on coral colonies, nor do they seem able to affect a colony's community structure. However, as Glynn (1990) has noted, if the living corals on a reef are greatly reduced as a result of massive destruction by a physical event, such as a hurricane or an El Niño episode, the remaining living colonies may concentrate the now-starving corallivores, resulting in secondary mortality, which could have an even greater

effect than the initial destruction. Indeed, Knowlton et al. (1981) reported such a delayed-mortality event in the dominant Caribbean coral *Acropora cervicornis* following the destructive passage of a hurricane on the island of Jamaica.

Two other taxa of predators are, however, capable of destroying coral colonies and modifying the structure of reefs: the crown-of-thorns starfish, *Acanthaster planci*, and a variety of fishes. The crown-of-thorns starfish is very large and feeds on living coral tissue (Figure 9.31). Because of its great size, it is capable of destroying an entire colony during feeding. Normally, *Acanthaster* is rare on coral reefs in the Indo-Pacific (the animal is not found in the Atlantic), but when it is present in moderate to large numbers, it can have a heavy impact on reefs. For example, Glynn (1976) observed that, on eastern Pacific reefs, *Acanthaster* preferred to feed on the abundant, fast-growing *Pocillopora* colonies, but in the field, it fed most of the time on the slow-growing, nonbranching colonies in the deeper parts of the reef. This selective predation on nonpocilloporid corals reduced coral cover and diversity. The reason the preferred *Pocillopora* was not eaten is that the colonies harbored symbiotic shrimp (*Trapezia*, *Alpheus*) that repulsed the starfish and protected the colonies. *Acanthaster* may also cause severe changes in reef structure when a preceding event gives it access to otherwise inaccessible corals. On coral reefs of the eastern Pacific, for instance, the presence of the pocilloporid corals prevented *Acanthaster*'s access to the other corals because shrimp symbionts living in *Pocillopora*, plus the *Pocillopora*'s nematocysts, prevented the star from crossing the coral barrier.

On Indo-Pacific coral reefs, outbreaks of *Acanthaster* can cause large-scale coral mortality. For example, according to Endean (1973), starting in the 1960s *Acanthaster* underwent a population explosion on certain reefs in the western Pacific, with the population density of starfishes increasing tenfold to tens of thousands of starfish on each reef (Figure 9.31). In densities such as these, the starfishes can destroy the entire reef over a period of months. Even when *Acanthaster* populations are normal (20–40 individuals per hectare), they affect the relative abundance of corals, since *Acanthaster* feeds preferentially on fast-growing specimens.

Of course, *Acanthaster* is not without its own predators. In the Red Sea, for instance, puffer fish and triggerfish kill as many as 4 thousand of the sea stars per hectare per year and may be responsible for the low densities of *Acanthaster* there. Other small predators include the shrimp *Hymenocera* and the polychaete worm *Pherecardia*. Such predators may be able to prevent the population outbreaks seen in the western Pacific, but there is currently no evidence that they have done so. Birkeland reports that *Acanthaster* outbreaks appear to have long been part of the ecology of Indo-Pacific reefs.

Two groups of fishes actively graze on coral colonies. Some (corallivores), such as puffers (Tetraodontidae), filefishes (Monacanthidae), triggerfishes (Balistidae), and butterfly fishes (Chaetodontidae)(Figure 9.32a), consume entire coral polyps. In addition, a group of omnivores, such as surgeonfishes (Acanthuridae) and parrot fishes (Scaridae), remove coral polyps to obtain either the algae in the coral skeleton or various invertebrates that have bored into the skeleton (Figure 9.32b).

Corallivores that feed by biting off and ingesting pieces of coral skeleton tend to prey on fast-growing branching species, such as *Porites* and *Acropora* in the Atlantic and *Pocillopora, Acropora*, and *Montipora* in the Pacific. There is evidence that these corallivorous fish can influence community structure. For example, in experiments carried out by Neudecker (1977), *Pocillopora damicornis* colonies were severely damaged when transplanted to depths of 15–30 m, but

were unaffected by corallivorous fish at depths of 2–3 m. Similarly, Wellington (1982a) showed that *Pocillopora* grew better in damselfish territories, where it was protected from corallivorous fish by the aggressive territorial defense of the damselfish. These two cases—plus the observations that *Pocillopora* is rarely found deep on reefs and that, when caged away from predators, their colonies can grow faster at 15 m than at 2 m—are strong evidence that corallivorous fish determine community structure. The extent to which predation restricts corals to certain regions or is responsible for coral distribution patterns is, however, unknown at present. What is certain is that the two families of fishes, Acanthuridae (surgeonfishes) and Scaridae (parrot fishes), that do consume living and dead coral, are among the more abundant fishes on reefs. By passing coral skeletons through their guts, they are also significant producers of $CaCO_3$ sand and sediment on reefs.

Other carnivores may affect reef structure by preying on groups of organisms that sometimes compete with the corals. If carnivores preyed upon sponges, alcyonarians, gorgonians, and tunicates more than they preyed upon reef corals and coralline algae, it would give the latter an advantage in sequestering space. This hypothesis, however, currently remains speculative because of a lack of evidence.

GRAZING Aside from the abundant and conspicuous coralline algae on reefs, there is little evidence of other large algae or other plants on pristine reefs that have not been heavily affected by human activities. If you look closely, however, you will find a short algal turf or small clumps of algae in many parts of the reef. Algae can be major space competitors with corals on reefs, invading and establishing themselves more quickly than the corals. Since algae are not dominant on pristine reefs and obviously do not outcompete corals, what restrains them? The answer appears to be intense grazing pressure by fishes and invertebrates, primarily sea urchins. Many families of fishes, including Siganidae and Pomacentridae, as well as Acanthuridae and Scaridae, contain herbivorous grazers. Continued grazing by these fishes, plus grazing by abundant sea urchins, such as the black long-spined urchin *Diadema*, keep the algae reduced to a minimum and enhance the survival of coral recruits (Figure 9.32b). If the herbivores are either removed from reefs or prevented from grazing by caging, coral is rapidly

(a)

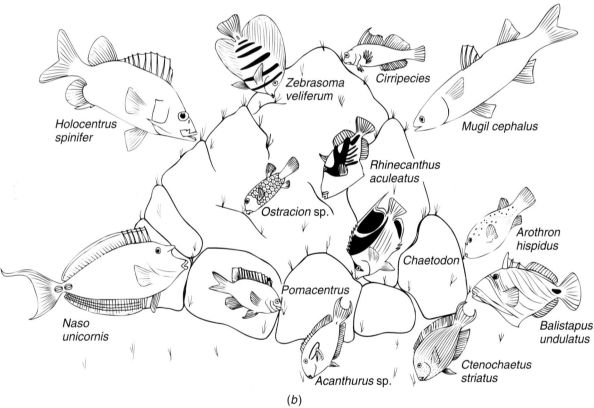

(b)

FIGURE 9.32 (a) Fishes that graze coral colonies: butterfly fish (Chaetodontidae), *Chaetodon, Colunula*; parrot fish (Scaridae), *Scarus*; triggerfish (Balistidae), *Balistapus*; puffer (Tetraodontidae), *Arothron*. (b) Herbivorous fishes of the reef: triggerfish (Balistidae), *Balistapus, Rhinecanthus*; squirrelfish (Holocentridae), *Holocentrus*; surgeonfish (Acanthuridae), *Acanthurus, Ctenochaetus, Zenbrasoma, Naso*; damselfish (Pomacentridae), *Pomacentrus*; butterfly fish (Chaetodontidae), *Chaetodon*; puffer (Tetraodontidae), *Arothron*; mullet (Mugilidae), *Mugil*; boxfish (Ostraciontidae), *Ostracion*; blenny (Blenniidae), *Cirripecies*. (Not to scale.) (Modified from "Ecological Relationships of the Fish Fauna on Coral Reefs of the Marshall Islands," R. W. Hiatt & D. W. Strasburg, *Ecological Monographs*, Vol. 30, pp. 65–127, 1960. Copyright © 1960 Ecological Society of America. Reprinted by permission.)

overgrown and killed by algae. Indeed, Mark Hay (1991) has noted that, not only are there more herbivorous fishes on coral reefs than in any other marine habitat, but their grazing rates often exceed the rates of grazing for any other habitat, terrestrial or marine. Herbivorous fish appear to have the ability to remove anywhere from 50 to 100% of total algal production. It has been suggested that the reason for the removal of such vast quantities of algae is that the algae are actually low in nutritive value for the fishes' needs; hence, the fishes must consume large amounts.

Grazing by herbivorous fishes and urchins directly affects algae, but affects corals indirectly. The effect of the grazers on the algae is to reduce the algal biomass, in some cases considerably. In an experiment performed by Sammarco et al. (1974) in the Virgin Islands, when *Diadema antillarum* were removed from an isolated reef, there was a tenfold increase in algal biomass within four months. There may also be more complex and subtle effects: As Sammarco et al. (1974) showed, grazing may prevent a single competitively dominant alga from taking over all the space and reducing the diversity.

Grazing can also result in shifts in abundance of groups of algal species the grazers prefer. This phenomenon is best exemplified by coralline and fleshy filamentous algae. The latter grow rapidly and may eliminate the slower-growing corallines if ungrazed. However, the fleshy filamentous algae are much more palatable and so are preferred by the grazers, so, in areas of heavy grazing, coralline algae increase at the expense of fleshy algae.

Given the intensity of grazing pressure, how do so many algae survive in these reef systems? There are a number of ways that algae deter various herbivores. One is to become less palatable. This can be done, as in the case of the coralline algae, by laying down $CaCO_3$ in the tissues. A second way is to produce chemicals that are unpalatable or noxious to the grazers. In another strategy, some algae reduce or avoid grazing by living in reef habitats that provide spatial or temporal refuges from grazers. In this respect, it is known that herbivory is highest in the shallow forereef and decreases markedly in the deeper forereef area, as well as on the sandy lagoon and deep sand plains. Residing in deeper water means that water temperatures are lower as well, and temperature affects the rate of grazing; thus, the algae that live here survive, and the reef becomes a mosaic of such refuges interspersed with heavily grazed areas, giving rise to different microhabitats that support different algal communities and that lead to a greater diversity of organisms.

The indirect effects of grazing on corals and coral growth produce a more complex situation. Although it is undoubtedly true that herbivores, in general, permit the survival of coral that would otherwise be eliminated by competition with the more rapidly growing fleshy algae, it is possible that they may have certain deleterious effects. For example, in contrast to fishes, sea urchins seem to be rather unselective grazers and thus may graze young coral colonies to destruction. At very high densities, urchins will graze on all organisms, not just algae; hence, as Sammarco et al. (1974) report, coral growth is prevented as well. At lower densities, urchins will remove algae and permit coral colonies to become established. Grazing by moderate densities of urchins clears areas for coral planulae to settle and contributes to the maintenance of coral reefs.

Within the tropics, the relative importance of grazing fishes and urchins to the maintenance and structure of reefs varies with human fishing pressure. Hay (1984) has demonstrated that fishing removes both herbivorous fishes and those fishes which prey on urchins. Thus, urchins become relatively more important on reefs that are subject to heavy fishing pressure. In general, moderate grazing prevents the dominant macroalgae from monopolizing the substrates and increases diversity. By contrast, heavy grazing leads to the destruction of not only the algae, but the corals as well, and *decreases* the diversity of the reef system.

A special case of the complex effect of grazers on coral is that of the damselfish (Pomacentridae). Damselfishes are territorial fishes that graze either selectively or nonselectively on the algae within their territory (Figure 9.33). They defend their "algal gardens" against other grazing fishes, and, as Ogden and Lobel (1978) report, they may even control the algal species composition in their territory by selective "weeding." According to Potts (1977), since the algae in damselfish territories are protected from heavy grazing by other grazers, they grow rapidly enough to kill the small settling corals by overgrowth and shading. Damselfish "gardens" may occupy significant portions of reef surfaces. Indeed, as Wellington (1982a) reports, up to 60% of shallow reef habitats may be occupied by damselfish territories, so these areas are important contributors to community

FIGURE 9.33 Three-spot dam-selfish in its territory. (Photo courtesy of Charles V. Angelo/ Photo Researchers, Inc.)

structure, as well as refuges for juvenile invertebrates and demersal plankton. All is not, however, that simple: Since damselfishes protect their territory from all grazing fishes, including grazing corallivorous fishes, it follows that, because massive corals recover from fish grazing and grow more slowly than branching foliose corals, damselfish contribute to the selective dominance of the reef areas by branching corals in the areas the damselfishes occupy. Kaufman (1977) has shown that, in the Caribbean, damselfishes eliminate the massive coral *Montastrea annularis* from the upper parts of the reef, where the damselfishes are most common. *Montastrea*'s demise then leads to dominance by the fast-growing coral *Acropora cervicornis*. In the eastern Pacific, there is an even more dramatic example of the effects of damselfishes on reef structures. According to Glynn and Wellington (1983), in this system the damselfishes also remove the dominant urchin, *Eucidaris thouarsii*, from their territories; thus, the fishes contribute to the survival of corals that might otherwise be destroyed by the grazing urchin.

The relative importance of fishes and urchins on reefs has been poignantly pointed up by the catastrophic die-off of virtually all of the *Diadema antillarum* in the Caribbean. Beginning at the Atlantic entrance to the Panama Canal and at Barbados, black-spined urchin populations began to disappear mysteriously starting in 1983. Over the next 13 months, the populations died off throughout the Caribbean. Lessios (1988) reported that no tropical western Atlantic area was spared and mortality was 93–99%. Mortality was followed by the nearly complete failure of urchin larval recruitment in 1984.

The immediate response to this loss of a major grazer was an increase in the algal biomass, most dramatically in areas where herbivorous fishes had been reduced by fishing pressure. In areas that experienced the largest algal blooms, the fleshy algae increased at the expense of the encrusting coralline forms. In the decade since the die-off, most Caribbean reefs have become dominated by fleshy seaweed. Coral cover has declined dramatically. To date, there has been little recovery of the urchin populations or signs of a return to coral dominance—see, however, Carpenter and Edmunds (2001). Most reef ecologists, including Jackson et al. (2001), believe that this massive shift in the Caribbean from a coral-dominated system to a seaweed-dominated one has been generated by the urchin die-off and the overfishing of reef fishes, which, together, have largely removed grazing pressure from these systems.

POSITIVE INTERACTIONS While it is easy to focus on the conspicuous roles played by competition for space and consumer pressure in shaping coral reef communities, one should not overlook the fundamental role played by positive interactions. Best described as facultative mutualisms or simply facilitations, positive interactions are also responsible for the presence and persistence of coral reefs. Like salt marshes on wave-protected temperate-zone shores, coral reefs are biogenic structures built and maintained by the reef organisms. Without the mutualism between hermatypic coral and zooxanthellae, most modern coral reefs would not exist, because coral growth and $CaCO_3$ deposition would not be sufficient to keep pace with rise in sea level. In addition, as we have already discussed, and as Castro (2000) observes, most mobile invertebrates and smaller organisms on coral reefs are entirely dependent for their survival on refuges provided by corals and other large sessile organisms. Finally, whereas coral reefs are found only in oligotrophic tropical waters of low productivity, they are responsible for generating remarkably high levels of primary production due to the success of zooxanthellae and also to the animal and plant assemblages that depend on the structure and resources provided by reefs. It is most accurate to think of coral reef communities as complex communities molded by competitive and consumer processes, but built on a foundation of positive interactions and feedbacks.

REGIONAL DIFFERENCES There are a number of interesting regional differences in species interactions. It is not known yet, however, how they relate to community structure and coral reef development. First, it is likely that herbivory is more intense on Pacific reefs than on those of the Atlantic. Fishes remove more algae from settling plates in the Pacific, and there are more bite marks per unit area of substrate in the Pacific. Also, at least before the massive die-off of *Diadema*, herbivorous sea urchins were more common in the Atlantic than in the Pacific. Vermeij (1978) has investigated the frequency of shell injuries to gastropod mollusks in the Pacific versus those in the Atlantic. He reports that these injuries are more common in the Pacific and suggests that the difference is the result of a higher rate of predation in Pacific reefs.

■ BACTERIA IN REEF SYSTEMS

The role of bacteria in coral reef systems is poorly understood. Only a few studies have been undertaken, primarily in the Great Barrier Reef system.

On the Great Barrier Reef, bacteria have been investigated in both the water column and the sediment. Bacterial production is dependent on the dissolved organic carbon (DOC) from the water

flowing over the reef and the particulate organic carbon (POC) from the breakdown of reef organisms. The fluxes of DOC and POC are great enough to permit large populations of bacteria to exist in both the water columns of the lagoons and the sediments. As Ducklow (1990) notes, bacterial biomass and production are orders of magnitude higher in the sediments than in the water column, and the ratio of total reef bacterial production to total reef primary production ranges from 0.6 to 1.0. Bacteria represent a huge energy resource for the reef community, and Sorokin (1990) suggests that a microbial loop system could be present, similar to that existing in the plankton (see Chapter 2), which could conserve energy in the reef system. A loop of this kind would explain the maintenance of high productivity on the coral reefs, but such a loop system has not yet been discovered, according to Ducklow (1990).

■ ROLE OF ALGAE IN REEF SYSTEMS

As we have remarked, algae tend to be small and inconspicuous on coral reefs, in marked contrast to the great kelp forests of temperate shores. However, through various biological interactions, algae are ecologically important to the community structure of the reef system, for a number of reasons.

First, encrusting red coralline algae, such as *Lithothamnion*, maintain the integrity of the reef by constantly cementing together various pieces of calcium carbonate; thus, they reinforce the reef against destruction by wave action and prevent individual pieces from being carried off the reef. The algal ridge is further responsible for breaking the velocity of waves and producing calmer conditions that allow other organisms to grow in the reef flat behind the ridge. A second group of calcareous algae comprises the greens, dominated by the genus *Halimeda*. These algae are significant contributors to the sand found in reefs, particularly in the lagoon area. They, therefore, create a special habitat.

Certain algae also bore into coral skeletons, but they do not appear to be significant in respect of bioerosion. Algae also create habitats in and around themselves and may furnish necessary shade from the hot tropical sun for certain organisms.

Finally, algae are important as primary producers in the reef system and as a food for various herbivores. These roles are summarized in Figures 9.34 and 9.35.

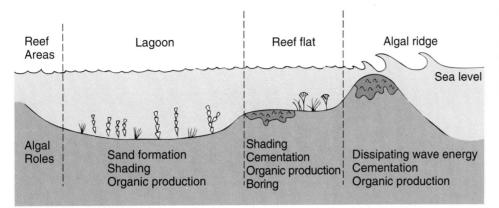

FIGURE 9.34 The various roles of free-living algae in different zones of a reef. ("Proceedings of the Second International Symposium on Coral Reefs," Vol. 1, M. Doty, 1974.)

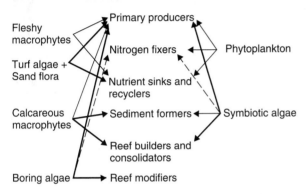

FIGURE 9.35 The main algal groups and their roles in the coral reef ecosystem. The thickness of the line indicates the relative importance of that algal group to the process listed. A dashed lines indicates a possible role for the group. (From "Reef algae," M. A. Borowitzka and A. W. D. Larkum, *Oceanus*, Vol. 29, No. 2, p. 52, 1986. Reprinted by permission of Woods Hole Oceanographic Institution.)

TABLE 9.2

Number of Fish Species in Several Coral Reef Areas

GEOGRAPHICAL AREA	NUMBER OF FISH SPECIES
Philippine Islands	2,177
New Guinea	1,700
Great Barrier Reef	1,500
Seychelles Islands	880
Marshall and Mariana Islands	669
Bahama Islands	507
Hawaiian Islands	448

Source: After Goldman and Talbot, 1976.

(Table 9.2). Atlantic reefs are also relatively species poor in comparison. The number of species found on a single reef in the Pacific is remarkable: 500 for one reef in the Great Barrier Reef system. How does this diversity of species come about, and how is it maintained? These are questions that have intrigued coral reef biologists for years, and they are still generating debate. We can only suggest some of the reasons here; more remain to be discovered.

One of the reasons for the high diversity of species on reefs is the great variety of habitats that exist there. Coral reefs encompass not only coral, but also areas of sand, various caves and crevices, areas of algae, shallow and deep water, and different zones progressing across the reef. This diversity of habitats alone can go far to explain the increased number of fishes. These habitats and their associated fish inhabitants are summarized in Figures 9.36 and 9.37.

The numerous habitats, however, are not enough to explain the high diversity of coral reef fishes, particularly in local areas. A number of experimental and theoretical studies have attempted to explain how so many species of reef fish can exist within local

■ ECOLOGY OF REEF FISHES

Coral reefs harbor more species and more diverse fish communities than any environment on Earth. It is impossible to conceive of a coral reef without the myriad brilliantly colored fishes gracefully moving among corals and associated organisms. Indeed, fishes are the most abundant and conspicuous large mobile organisms on a reef. Because of their abundance and pervasive presence, it seems obvious that they must contribute in significant ways to the reef ecosystem. We have noted some of these important structuring relationships in previous sections; in this section, we will review other aspects of their ecology.

The species richness of coral reef fishes is similar to that of corals. The central Indo-Pacific area of the Philippine and Indonesian archipelagos has the greatest number of species, and the number decreases in all directions away from this center

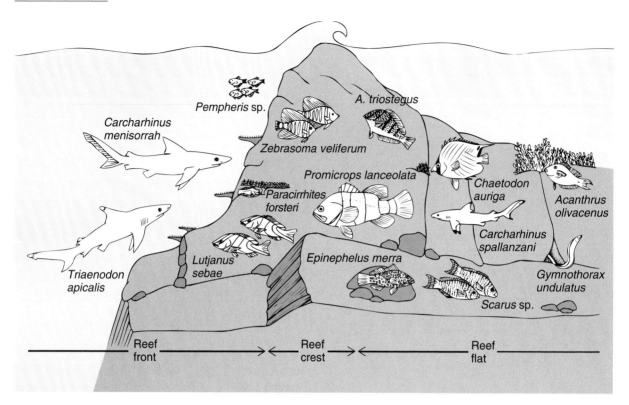

FIGURE 9.36 Generalized picture of the characteristic fish species and their habitats on a seaward coral reef in the Marshall Islands: Requiem sharks (Carcharinidae), *Triaenodon, Carcharhinus*; moray eels (Muraenidae), *Gymnothorax*; hawkfishes (Cirrhitidae), *Paracirrhites*; groupers (Serranidae), *Epinephelus, Promicrops*; snappers (Lutjanidae), *Lutjanus*; surgeonfishes (Acanthuridae), *Acanthurus, Zebrasoma*; parrot fishes (Scaridae), *Scarus*; butterfly fishes (Chaetodontidae), *Chaetodon*; sweepers (Pempheridae), *Pempheris*. (Not to scale.) (Modified from "Ecological Relationships of the Fish Fauna on Coral Reefs of the Marshall Islands," R. W. Hiatt & D. W. Strasburg, *Ecological Monographs*, Vol. 30, pp. 65–127, 1960. Copyright © 1960 Ecological Society of America. Reprinted by permission.)

communities. Four opposing theories, based on different aspects of recruitment or competition, have arisen over the last 25 years (Figure 9.38). The more classical view is that high diversity is the result of strong competitive interactions following recruitment that lead to a high degree of specialization. Each species has a specific set of adaptations that give it the competitive edge in at least one situation on a reef. That is to say, these fishes have narrower ecological niches, so more species can be accommodated in a given area. This model, devised by Smith and Tyler (1972), is called the **competition model**. It is partially supported by correlations between fish species diversity and habitat complexity, or "rugosity," uncovered by Risk (1972) and Luckhurst and Luckhurst (1978). In this model, recruitment is not significant in structuring the adult population, because it is assumed that there is always an excess of potential recruits. A second model is Sale's (1977, 1980) **lottery hypothesis**, which rests on the fact that most coral reef fishes produce large numbers of larvae that are dispersed into the plankton. It states further that fishes are not that specialized (that is, many similar species have the same requirements) and that there is active competition among the species. Local success and persistence result from chance as to which species of the planktonic larval pool occupies the vacant space. According to this hypothesis, competition is unimportant and recruitment dominates. The third view is that fish populations are not in equilibrium: Predation, catastrophe, and unpredictable recruitment ensure that populations never become large enough to undergo

FIGURE 9.37 Fishes associated with individual (a) branching and (b) plate-type coral colonies: Butterfly fishes (Chaetodontidae), *Chaetodon, Forcipiger*; damselfishes (Pomacentridae), *Chromis, Dascyllus*; wrasses (Labridae), *Epibulus, Thalassoma, Stethojulus, Gomphusus*; velvet fishes (Caracanthidae), *Caracanthus*; gobies (Gobiidae), *Paragobiodon*; puffers (Tetraodontidae), *Arothron*; filefishes (Monocanthidae), *Oxymonocanthus*; hawkfishes (Cirrhitidae), *Paracirrhites*; triggerfishes (Balistidae), *Balistapus, Rhinecanthus, Abalistes*; squirrelfishes (Holocentridae), *Holocentrus*. (Not to scale.) (Modified from "Ecological Relationships of the Fish Fauna on Coral Reefs of the Marshall Islands," R. W. Hiatt & D. W. Strasburg, *Ecological Monographs*, Vol. 30, pp. 65–127, 1960. Copyright © 1960 Ecological Society of America. Reprinted by permission.)

FIGURE 9.38 Classification of the four models offered to explain the diversity of species among coral reef fishes based on the importance of competition and recruitment processes. (From *The Ecology of Fishes on Coral Reefs*, edited by Peter F. Sale. Copyright © 1991 by Academic Press, reproduced by permission of the publisher. All rights of reproduction in any form reserved.)

	Postrecruitment (PR) competition	
	Intense	Weak
Recruitment modified by PR processes	1. Competition model (Smith and Tyler)	3. Predation disturbance models (for example, Talbot et al.)
Recruitment not modified by PR processes	2. Lottery model (Sale)	4. Recruitment limitation model (for example, Victor)

competitive exclusion, because they are kept below the numbers at which resources or food become limiting. This is the **predation-disturbance model**, formulated by Talbot, Russell, and Anderson (1978). Finally, the most recent explanation is the **recruitment limitation model**, discussed by Victor (1983) and Doherty (1982). This model argues that the larval supply is never sufficient for the adult population to reach its carrying capacity. The adult population reflects variation in larval recruitment and not postrecruitment events. In an empirical test of the model carried out by Doherty and Fowler (1994), 90% of the variation in abundance of the damselfish *Pomacentrus moluccensis* was explained by recruitment. However, further data are needed regarding other fishes. Evidence is currently inconclusive as to which of the four views is correct. We can say, however, that one consistent fact that has arisen in the 25 years of controversy is that fish recruitment is variable.

Perhaps as a result of the large number of species and the partitioning of the habitat, we find that, despite their obvious mobility, most reef fishes are restricted to certain areas of the reef and are very localized. Most do not migrate, and many of the smaller species, such as gobies, blennies, and damselfishes, defend their territories. However, certain of the larger fish do make diel migrations to feeding areas. The distances covered may range from only a few meters to several kilometers and are often predictable (Figure 9.39).

There has been considerable debate over whether populations of reef fishes are open or closed. In closed populations sexual recuits come from the local population, whereas in open populations recruits come from other adult populations. The lottery and recruitment limitation models of regulation of the reef fish population imply that larval fish populations are open. In open populations, recruitment is a stochastic or random process wherein larvae passively disperse in the plankton. Thus, these models decouple the production of larvae from larval recruitment. Recent evidence, however, has suggested that some reef fish have closed populations, whereby larvae have mechanisms of staying in or returning to the populations where they were produced. This self-recruitment hypothesis has been promoted by Jones et al. (1999) and Swearer et al. (1999). How common self-recruitment and closed populations are in reef fishes in not currently known, but is clearly important for the understanding, management, and conservation of coral reef fish.

One of the most interesting discoveries concerning the fishes of coral reefs is the differences seen in fish species between day and night. Most people see reefs during the day, when the majority of the fish species are visible. At night, these diurnal fishes seek shelter in the reef and are replaced by a smaller number of nocturnal species not seen during the day (Figures 9.40 and 9.41). Since some of the nocturnal species are ecologically similar to certain diurnal

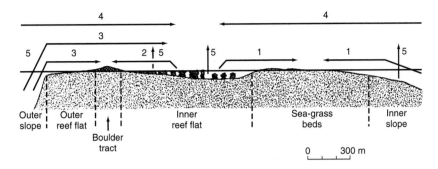

FIGURE 9.39 Patterns of movement of different groups of fishes within a fringing reef during an incoming tide. 1 and 2: from the inner reef flat to sea-grass beds and the rubble bank, respectively, to feed; 3: from the outer slopes to feed in shallower waters; 4: movement of pelagic fishes from deeper waters to feed on the inner reef flat; 5: general movement up into the water column. (From *The Ecology of Fishes on Coral Reefs*, edited by Peter F. Sale. Copyright © 1991 by Academic Press, reproduced by permission of the publisher. All rights of reproduction in any form reserved.)

species (for example, Apogonidae replace Pomacentridae), this is another way of permitting a greater number of species to exist on the reef without direct competition. Hobson (1968) notes that all nocturnal species of reef fishes are predaceous, but diurnal fishes span nearly all trophic modes (carnivores, planktivores, omnivores, etc.).

The feeding relationships of reef fishes are of some interest (Figure 9.42 and Table 9.3). The most abundant feeding type on reefs is the carnivores, which may constitute as much as 50–70% of the fish species. According to Goldman and Talbot (1976), many of these carnivores are not specialized to a given food source, but are opportunistic, taking whatever is available to them. They also feed on different prey at different stages in their life cycles. The fact that the majority of the fishes on reefs are generalized carnivores argues against the competition model of the species richness of reefs. Indeed, the lack of specialization makes it difficult to explain the species richness at all. This is not to say that no specialized carnivores exist; there are some. However, correlated with the generalized feeding habits of most carnivores, the numbers of true scavenger fishes are few because the carnivores also act as scavengers and pick up any recently dead organisms.

Herbivores and coral grazers make up the next largest group of fishes (about 15% of the species), and the most important of these are the families Scaridae and Acanthuridae (Figure 9.43). The remaining fishes are generally classed as omnivores or multivores and include representatives from virtually all families of fishes on the reef. Only a few fishes are zooplankton feeders, and these are mainly small schooling fishes of the families Pomacentridae, Clupeidae, and Atherinidae.

A side effect of the feeding of some fishes, such as grunts (haemulids), is to enhance the nutrition and growth of corals. Meyer et al. (1983) note that schools of certain fishes, which rest by day in coral heads, feed at night on seagrass beds and then return to defecate material rich in nitrogen and phosphorus into the coral heads.

When one considers the fact that most reef fish are unspecialized carnivores, it is easy to understand why most invertebrate organisms are hidden from view, since exposing a soft body on a reef would certainly invite instant attack by some carnivore. How, then, do a few invertebrates, notably the large sea cucumbers, remain abundant lying unprotected on reefs? The answer apparently is that those few which are visible have evolved defense mechanisms to deter potential predators. Tropical sea cucumbers, for example, asserts Bakus (1973), have two such mechanisms: They produce toxic substances that can kill fishes, and they have viscous, sticky strands (cuvierian organs) that can be extruded to literally tie up any potential predator.

The development of distasteful or toxic substances is not limited to invertebrates: Many of the fishes also produce toxic substances, in the form of (1) venom associated with various spines or poisonous material extruded onto the body surface (**crinotoxin**) or (2) flesh

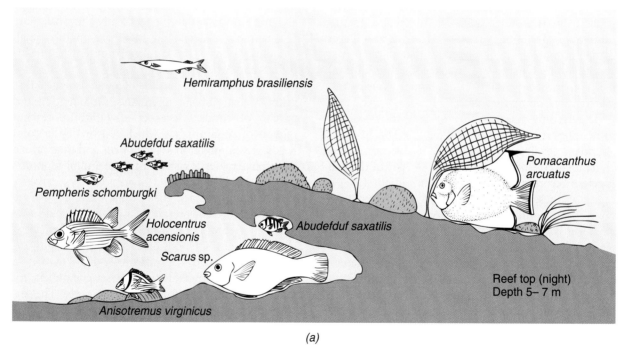

(a)

FIGURE 9.40 Day and night fish distribution on a Florida coral reef at two different depths. (a) Reef top at 5–7 m. (b) Deep reef at 25–30 m. Damselfishes (Pomacentridae), *Abudefduf, Chromis*; wrasses (Labridae), *Clepticus, Thalassoma, Lachnolaimus*; parrot fishes (Scaridae), *Scarus*; angelfishes (Chaetodontidae), *Pomacanthus, Holacanthus*; grunts (Pomadasyidae), *Haemulon, Anisotremus*; cardinal fishes (Apogonidae), *Apogon*; squirrelfishes (Holocentridae), *Holocentrus*; sweepers (Pempheridae), *Pempheris*; snappers (Lutjanidae), *Lutjanus*; halfbeaks (Hemiramphidae), *Hemiramphus*. (Data for the drawing based on W. A. Starck and W. P. Davis, 1966, "Night Habits of Fishes of Alligator Reef, Florida." *Ichthyolog. Aquar. J.* (38) 41:313–356.)

(b)

FIGURE 9.40 continued

FIGURE 9.41 A nocturnal reef fish, *Myripristis murdjan*. Note the large eyes. (Photo courtesy of Drs. Lovell and Libby Langstroth.)

FIGURE 9.42 The trophic relationships of coral reef fisheries. (After "Ecology of Fishes in Tropical Waters," R. H. Lowe-McConnell, 1977, *Studies in Biology*, No. 76, 1977. Reprinted with the permission of Cambridge University Press.)

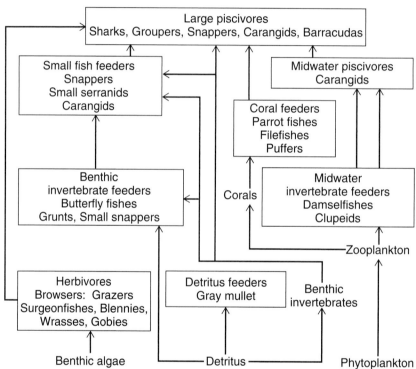

or internal organs that are toxic. Truly venomous fishes are rare on reefs and are confined mainly to the stonefishes (Synanceiidae) and scorpion fishes (Scorpaenidae), but a large number of fishes, including the abundant parrot fish (Scaridae), the wrasse (Labridae), and the surgeonfish (Acanthuridae), have toxic secretions on their outer surfaces. One explanation for the prevalence of these substances is that they deter predation by the abundant carnivores.

The final category, namely, toxic flesh or internal organs, has caused the most interest among humans, because eating tropical fishes with toxins in the flesh

or organs produces a serious disease called **ciguatera**, one of the most mysterious and puzzling diseases found in the tropics. Banner (1976) notes that the symptoms in humans are primarily neurological, including exhaustion, visual disturbance, inversion of senses (hot feels cold and vice versa), paralysis, loss of reflexes, and, finally, death by respiratory failure. The mystery came originally from several sources:

1. The symptoms are not consistent.
2. The fishes that have the toxin vary in toxicity and can include virtually any or all of the large reef

TABLE 9.3

Proportion of Fish Species in Different Trophic Categories from Seven Tropical Studies[a]

TROPHIC CATEGORY	H & S[b]	RAND[c]	G & T[d]	HOBSON	W & H	SANO	T & C
Herbivores	26	13	22	7	15	18	20
Planktivores	4	12	15	18	20	15	38
Benthic invertebrate feeders	49	44	27	56	53	41	33
Coral feeders	6	1		9	5	9	
Sessile animal feeders	8	6		13	3		
Mobile invertebrate feeders	35	37		34	45		
Omnivores	13	7		10	4	19	
Piscivores	10	25	38	7	8	4	8
Others (e.g., cleaners)				2		2	1

[a]Figures given are percentages of the number of species surveyed, unless otherwise noted. The category of benthic animal feeders is the sum of the three subcategories following and indented from it. The seven studies were H & S, Hiatt and Strasburg (1960); Rand, Randall (1967); G & T, Goldman and Talbot (1976); Hobson, Hobson (1975); W & H, Williams and Hatcher (1983); Sano, Sano et al. (1946b); and T & C, Thresher and Colin (1986).
[b]Hiatt and Strasburg (1960) classified some species into more than one category.
[c]Randall (1967) indicated that the survey was biased toward important sport and commercial species.
[d]Goldman and Talbot (1976) used only the four categories shown. Herbivores were placed into a category of grazers that included coral feeders. Figures given are percent weight of all groups, averaged over three types of habitat: fore reef, lagoon, and back reef.
Source: The Ecology of Fishes on Coral Reefs, edited by Peter F. Sale. Copyright © 1991 by Academic Press, reproduced by permission of the publisher. All rights of reproduction in any form reserved.

fishes, from carnivores to herbivores. A given species may be toxic on one reef, but not on another, or the same species may be toxic at one time of the year on a given reef and not at another.

3. There seems to be no repeatable pattern to the occurrence of the toxin.

We now know somewhat more about the disease. Perhaps most significant was the discovery in 1977 of the causative organism: the dinoflagellate *Gambierdiscus toxicus*. This organism is scarce on reefs with large amounts of living coral, but it proliferates after disturbances have created dead coral surfaces covered by filamentous or calcareous algae. The toxin, which is a polyether, has also been isolated and partially characterized. We are, however, still ignorant of most of the ecological aspects of *G. toxicus* and of the factors that trigger a bloom, as well as the mechanisms by which the toxin comes to reside in the flesh of certain carnivorous fish. Ciguatera, therefore, remains of considerable importance to the people of the tropics, since the fisheries of many areas cannot be developed until this toxicity problem is solved.

Apparently, coral reef fishes are also subject to occasional epidemics that cause mass mortality. Williams and Williams (1988) cite a Caribbean-wide mass die-off of fishes that was recorded in August and September of 1980 following the passage of a hurricane. However, this mass mortality event was not studied, so the nature of its cause(s) is unknown.

Another of the outstanding characteristics of reef fishes is their color. Why—especially given the great predation pressure—should they be so brightly colored? Biologists cannot agree as yet, and it is likely that color and pattern serve several different functions. One explanation is that the bright colors advertise the fact that the species is toxic or otherwise distasteful so that predators will ignore it (warning coloration). Such colors may also camouflage the species, either by breaking up the shape of the fish or making it appear as something else. The latter explanation especially has been used to explain the dramatic banding patterns seen in some reef fishes (Figure 9.44; see also Figures 9.15 and 9.18). In addition to their role in various predator–prey interactions, color and pattern may be important in visual communication, since these brightly colored fishes are highly dependent on vision. Color and pattern may serve in species and gender recognition as well and may be used in courtship and mating behaviors.

Despite the ubiquitous presence of fishes on coral reefs in seemingly large numbers, little is known about the population dynamics of reef fishes. We do know that the populations of individual species fluctuate over time and in different areas of the reef, but

CORALLIVORES

Chaetodontidae
(butterfly fishes)

Balistidae
(Triggerfishes)

Tetraodontidae
(puffers)

HERBIVORES

Pomacentridae
(damselfishes)

Scaridae
(parrot fishes)

Acanthuridae
(surgeonfishes)

Siganidae
(rabbitfishes)

FIGURE 9.43 Families of reef fishes that include the most corallivorous and herbivorous species. (From *The Life and Death of Coral Reefs*, C. Birkeland, editor, © 1997 Chapman & Hall, p. 232, Figure 10.1. Reprinted by permission.)

Lo vulpinus

Heniochus acuminatus

FIGURE 9.44 Reef fishes with dramatic banding patterns.

we are ignorant of the exact causes. Much of our lack of knowledge comes from the presently almost impossible task of following the dynamics of the larval populations in the plankton and their settlement patterns on the reef.

One of the most conspicuous aspects of the biology of many reef fishes is their schooling behavior. Tropical reef fishes are often found both in single-species schools and in mixed-species schools, ranging in size from dozens to thousands of individuals. The two most widely accepted functions of schooling behavior in reef fishes are as a defense against predators and as an advantage in foraging. Fishes in schools are thought to be less vulnerable to consumers, because predators have difficulty focusing on individual prey in large schools. Many fish respond to the presence of predators by joining schools. Schooling behavior can also increase the

foraging success of fishes. In shallow-water schools of herbivorous fishes, for example, members of schools feed freely in the territories of damselfishes, whereas solitary foragers are generally prevented from foraging in damselfish territories by their owners. Thus, notes Robertson (1976), schooling gives these fishes access to the rich algal gardens maintained by the territorial damselfishes.

A final aspect of reef fish ecology concerns the phenomenon of **cleaning behavior**, a specialized form of predation in which certain small fishes (*Labroides* spp.) or shrimps remove various ectoparasites from other, usually larger, fish species. Cleaning behavior is widespread and apparently occurs on all reefs. In exhibiting this behavior, the cleaner fishes (or shrimps) often set up "cleaning stations" where they appear to advertise their presence through bright, contrasting colors. The fish to be cleaned comes to the cleaning station area (often a prominent coral head or

FIGURE 9.45 Humpback cleaner shrimp on a tomato cod (*Lysmata ambionesis*). (Photo courtesy of Steve Wulper/ DRK Photo).

FIGURE 9.46 A cleaning shrimp, *Periclimenes yucatanicus*, on the tentacles of the anemone *Condylactis gigantea*. (Photo courtesy of Dr. Diane Nelson.)

boulder) and remains motionless as the cleaner moves over its body, removing parasites (Figures 9.45 and 9.46). The cleaners may even enter the mouth or gill chambers of the fish. A sort of truce among predators prevails at the cleaning stations, as the fishes that are being cleaned are vulnerable to predation. Fishes even "line up" at the stations, awaiting their turn for cleaning!

The importance of cleaning behavior to the fish population and the economy of the reef has not been established. Some biologists believe that if cleaners are removed from a reef, the fish fauna decreases, moving away or otherwise showing signs of distress. Other biologists have removed cleaners with no noticeable effect on the reef or reef fishes.

■ CORAL REEF CRYPTOFAUNA AND BIOEROSION

The term **cryptofauna**, applied to coral reefs, refers to all invertebrates that inhabit the coral substrates, boring into the $CaCO_3$, living in crevices and cracks, and occupying space under coral heads and boulders. According to Hutchings (1983), two groups exist: those which actually bore into the $CaCO_3$—the true borers—and those which are unable to bore into the coral rock, but live in preexisting holes, cracks, and crevices—the "opportunists" (Figure 9.47). True borers

of coral rock include sponges, bivalve mollusks, sipunculans, and some polychaete worms. The nestlers or opportunists represent most marine invertebrate phyla, but crustaceans and polychaete worms seem to be particularly common.

Very little is known about the structure, recruitment, or succession of these cryptofaunal communities. Hutchings (1983) believes the community to be dominated by filter and deposit feeders, and the animals living in such communities feed mainly on particulate food in the water column and organic detritus that has been collected in mucus secreted by the coral and moved off the colony. The filter and deposit feeders may play an important part in the coral reef system by using the large amounts of coral-produced mucus and recycling the organic material and nutrients trapped therein.

Colonial cryptofaunal organisms, including sponges, bryozoans, and ascidians, tend to dominate the surface fauna of the undersides of corals, caves, and overhangs. Their ability to reproduce asexually and to overgrow other potential space occupiers gives them a competitive edge in such areas. According to Jackson (1977, 1979), these fast-growing invertebrates

FIGURE 9.47
Diagrammatic representation of a variety of external and internal bioeroders of coral skeletons. (From *The Life and Death of Coral Reefs*, C. Birkeland, editor, © 1997 Chapman and Hall, p. 69, Figure 4.1, with kind permission from Kluwer Academic Publishers.)

Internal Bioeroders (Borers)
Microborers
A. Algae, fungi, bacteria

Macroborers

  B. sponges *(Clionidae)*
  C. Bivalves *(Lithophaga)*
  D. Barnacles *(Lithotrya)*
  E. Sipunculans *(Aspidosiphon)*
  F. Polychaetes *(Eunicidea)*

External Bioeroders *(Grazers)*

  G. Parrot fish *(Scaridae)*
  H. Puffer fish *(Arothron)*
  I. Hermit crab *(Aniculus)*
  J. Limpet *(Acmaea)*
  K. Urchin *(Diadema)*
  L. Chiton *(Acanthopleura)*

often undergo intense competition. Buss and Jackson (1979) cite overgrowth relationships among these organisms as forming competitive networks rather than hierarchical sequences. Owing to their indeterminate and rapid asexual growth, many of the cryptofauna have an advantage over the slower-growing ahermatypic corals and solitary organisms. Jackson (1977), however, observes that their competitive abilities are offset by an increased susceptibility to predation and disturbance.

Most certainly, the cryptofauna is significant in bioerosion of the reef (Hutchings 1986). The rates of bioerosion due to the activities of sponges, echinoids, and grazing mollusks have been assessed and, when added to that of various grazing fishes, suggest that bioerosion may be the most important of the processes that break down coral reefs. Stearn and Scoffin (1977) found that, in Barbados, corals and algae laid down about 160 metric tons of $CaCO_3$ annually over an area of about 1 hectare. In the same area, boring sponges, barnacles, and bivalves annually removed 1.5 metric tons from live corals and 23.5 metric tons from dead coral. Parrot fishes removed an additional 1 metric ton, and the urchin *Diadema* removed 163 metric tons. These figures suggest that the reef is eroding faster than it is built. A second study, however, carried out by Scoffin et al. (1980) suggested a higher calcification rate of 206 tons per year and a total bioerosion rate of 163 tons per year.

■ SUCCESSION AND STABILITY IN CORAL REEFS

Ecological succession refers to the change in community structure and composition of an area or habitat over time. Whereas early evidence suggested that orderly succession did not occur in coral reefs, current indications, primarily from two long-term studies in the Pacific, are that predictable species replacements do occur in some areas. In Guam, the coral reefs were devastated by *Acanthaster* in 1967–69, losing up to 90% of the coral cover. Colgan (1982) has shown that the recovery of the reefs over the next 11 years followed a definite successional sequence. At Heron Island on the Great Barrier Reef, Joseph Connell has studied the dynamics of the coral community since 1962, giving us the longest continuous time record for any coral reef. These data, too, report a predictable succession of species through time. However, other long-term studies of reefs in Hawaii by Grigg (1983) and Dollar and Tribble (1993) did not reveal any predictable replacements of species. These studies have suggested that local disturbance events can strongly influence successional events and that biological mechanisms such as competition, life-history attributes, inhibition or facilitation of colonization, and predation on eggs and larvae all influence the direction and rate of successional changes such that the entire process is highly variable geographically. Much more work is needed before we can make any definitive statements about succession in coral reefs.

Related to the question of succession is the stability of coral reefs. On short time scales, reefs appear to be quite stable but they are not necessarily so over the longer term. The structure, and even the existence, of reefs can be radically changed by natural catastrophic events such as hurricanes, El Niño–induced weather and oceanographic changes, *Acanthaster* outbreaks, and diseases affecting corals and other reef organisms. As with succession, the stability of reefs appears to be quite variable in time and space.

■ CATASTROPHIC MORTALITY AND RECOVERY OF REEFS

Although reefs appear to be large and very stable systems, existing, in the case of atolls, for millions of years, they do suffer large-scale destruction from various forces. Perhaps the major physical source of massive reef mortality is mechanical destruction by severe tropical storms. Hurricanes or typhoons are intense enough to destroy large areas of reef. Since most coral reefs lie within the zone most frequently traversed by typhoons or hurricanes, the chance that a reef or a section of a reef will be destroyed or severely damaged is great. Damage in such storms usually is due to coral colonies being toppled or uprooted and carried off the reef, thus freeing space for new occupants.

The second major source of catastrophic mortality of Pacific reefs, at least in recent times, is the population explosion of the sea star *Acanthaster planci*. Since 1957, when the animal's population explosions were first observed, *A. planci* has caused catastrophic mortality of reefs in a number of places in the western Pacific (Figure 9.48). This starfish's ability to destroy large areas of reef is remarkable. Chesher (1969) reports that, in Guam, 90% of the coral reef

FIGURE 9.48 Reported coral bleaching events, *Acanthaster* outbreaks, and mass coral mortalities from 1870 to 1990. (From *The Life and Death of Coral Reefs*, C. Birkeland, editor, © 1997 Chapman and Hall, p. 355, Figure 15.1, with kind permission from Kluwer Academic Publishers.)

along 38 km of shoreline was wiped out in just 2.5 years. Endean (1973) notes that, on the Great Barrier Reef, the bulk of the corals on an 8-km^2 reef were destroyed in 12 months. More recent surveys cited by Walbran et al. (1989) and Endean and Cameron (1990) have shown that about one-third of the reefs of the Great Barrier Reef have been ravaged. When these rates are multiplied by the number of reefs in the Pacific that were infested by *A. planci* in the 1960s (including 60% of the reefs of the Great Barrier Reef between 16 and 19°S), it becomes clear why there is concern about the survival of reefs.

What causes these population explosions, and are they natural phenomena? These are the two major questions that biologists have asked regarding the "*Acanthaster* problem," and their answers show how little we really know about how coral reefs function. Although there is no consensus concerning the answer to either question, there is evidence that the population explosions are natural phenomena. That similar outbreaks have occurred in the past is suggested by Walbran et al.'s (1989) findings of aggregations of *Acanthaster* skeletal remains in sediment cores. Also, Birkeland (1982) reports that *Acanthaster* is mentioned in folklore and in the memories of certain island peoples. Whether such explosions are natural to other reef systems is not known. Even if these earlier outbreaks were natural, the frequency and magnitude of recent ones could still have been caused by human factors. In any event, natural or human induced, what is the trigger? Examinations of infested reefs have produced no consistent pattern of human-induced or natural activity that indicates a single cause. Many causes have been suggested, including various dredging activities, which purportedly opened new space for the *A. planci* juveniles; some chemicals (e.g., pesticides), which supposedly freed *A. planci* from its normal population controls; the removal of important predators, which increased the survival rates of juveniles; and severe tropical storms, which presumably opened space for juvenile settlement or caused adults to aggregate.

Although no one knows for certain what the cause is, the two suggestions that have received most attention are the ones involving predators and the effects of storms. The predator-removal theory necessitates the presence of predators on adult *Acanthaster* that normally keep the population in check. Are there such predators? In the Indo-Pacific, there are at least three: a fish, the hump-headed wrasse, *Cheilinus undulatus*; the puffer fish, *Arothron hispidus*; and the large gastropod *Charonia tritonis*. In the eastern Pacific, the polychaete worm *Pherecardia striata* and the shrimp *Hymenocera picta* appear to be important predators. Of course, there are undoubtedly predators on the planktonic larvae and newly settled juveniles of *Acanthaster* as well, but they are little known or studied. Of the predators on the adults, attention has been focused mainly on *Charonia*, because it is a known specialist predator on starfishes. It is also in demand by shell collectors. Endean and Cameron (1990) have suggested that the removal of this relatively uncommon gastropod from the reefs by collectors could have allowed the population explosion to begin, but currently there is little evidence to support this hypothesis. If that explanation is correct, than *C. tritonis* is a keystone species in the same sense that *Pisaster ochraceus* is in the rocky intertidal zone. However, we know so little about the interactions on reefs that the *Acanthaster* population explosion certainly could not have been predicted ahead of time.

Another idea is Birkeland's (1982) runoff hypothesis, which suggests that juvenile recruitment of *A. planci* is enhanced by a combination of unusually low salinities, high nutrients, and high temperatures. In this scenario, the occurrence of a year or several years of abnormally heavy rainfall in an area, especially if it is coupled with human destruction of the native vegetation on adjacent land areas, causes increased runoff that provides enough nutrients to stimulate phytoplankton blooms and thereby furnishes food for large numbers of *Acanthaster* larvae. The increased survival of the larvae and their subsequent settlement on the reefs leads to outbreaks three years later. This model uses a natural explanation that fits the pattern of recent outbreaks, but currently there is little evidence to support it. Offered as an alternative to the runoff model is the adult-aggregation theory proffered by Dana et al. (1972) and Potts (1981), which says that the destruction of reef areas by typhoons causes starfishes to condense into aggregations that attack the remaining living coral en masse.

A third cause of catastrophic mortality is El Niño, a naturally recurring periodic phenomenon manifested as a massive influx of nutrient-poor warm water into the usually cold, nutrient-rich surface waters off the coasts of Ecuador and Peru. In fact, El Niño is but part of an interacting set of atmospheric and oceanographic conditions called the El Niño–Southern Oscillation (ENSO), which periodically alters the atmospheric pressure, winds, rainfall patterns, ocean currents, and sea level over large areas of the tropical and subtropical Pacific Ocean. These natural, unpredictable ENSO events vary greatly in intensity, areal extent, and ecological effects. The greater the intensity, the greater is the change in oceanographic and meteorological conditions, the larger is the area affected, and the greater is the effect on living organisms. Glynn (1988) reports that the 1982–83 El Niño was perhaps the strongest El Niño event of the twentieth century up to that date and caused unprecedented disturbances in marine populations and communities. It may have had a periodicity of 1 in 100 to 1 in 250 years and resulted in a rise of ocean surface temperatures to 30–32°C in the eastern Pacific, a temperature that was 2–4°C above normal and remained for several months. According to Lessios et al. (1983), this prolonged elevation of temperature, coupled, on the one hand, with changes in sea level up to 44 cm below normal in the eastern Pacific and, on the other, with unusual hurricane activity and hurricane paths, killed large tracts of reef corals in the Gulf of Panama, the Galápagos Islands, Tahiti, and the Society Islands, as well as in other places throughout much of the tropical Pacific. Mortality rates for coral ranged from 50 to 98%. Particularly devastated were the reefs in the Galápagos, where Glynn (1985) estimated that 95% of corals died, and those in the Gulf of Panama, where, again, Glynn (1985) found that 70–80% of the corals were destroyed. The 1997–98 El Niño had an intensity equal to the 1982–83 event.

The loss of the corals and the coral reef framework in areas such as the Society Islands and the Galápagos Islands had profound secondary effects. On the one hand, reported Glynn (1988), it eliminated habitats for numerous reef-associated species. Whole fish communities disappeared from areas of dead coral. On the other hand, in the eastern Pacific, the recruitment of sea urchins was enhanced and densities of *Diadema* in Panama and *Eucidaris* in the Galápagos rose. The activities of the urchins, in turn, caused increased bioerosion, which destroyed the dead reef framework. Even corals that survived the initial mortality by the change in physical factors were subjected to increased predation by surviving corallivorous sea stars and sea urchins when the protective shrimp and crab symbionts died or migrated. Hence, the eastern Pacific reefs have not shown much recovery in the more than 15 years since the 1982–83 El Niño.

A puzzling event that can cause catastrophic coral mortality is **coral bleaching** (Figure 9.49). Corals bleach when they expel the zooxanthellae that normally inhabit their tissues. Following the expulsion, corals become white—hence the name of the phenomenon. If the zooxanthellae are absent

FIGURE 9.49 Coral bleaching. (Copyright Constantinos Petrinos/Nature Picture Library)

for a significant time, the coral will die. Coral bleaching events have been reported on many reefs in both the Caribbean and the Indo-Pacific in the 1980s and 1990s (Figure 9.48). Usually, the bleaching affects most, but not all, coral species and is most severe in shallow-water areas. Widespread and severe coral bleaching occurred in the Caribbean in 1987–88 and in the eastern Pacific following the previously described 1982–83 El Niño. More recent bleaching events have been reported in Jamaica by Goreau (1990) and in Thailand, French Polynesia, and the Andaman Sea by Brown (1997).

The cause of bleaching is unclear, but most scientists attribute it to stress, particularly stress induced by water temperatures of 30°C or above. Bleaching is growing more common, and there is concern that the pattern of bleaching events may be an early warning of the onset of global warming. (See Chapter 11.) If bleachings continue to increase, the integrity of many reefs may be compromised because corals may not have time to recover and large reef areas may die.

The activities of humans can directly cause catastrophic mortality on reefs through dredging, pollution, and overfishing. The reefs in Kaneohe Bay, Hawaii, for example, were destroyed through sewage pollution. During World War II, the military closed off the lagoon at Palmyra Atoll, which resulted in the death of all lagoon reefs. In Jamaica, the spectacular die-off of the sea urchin *Diadema antillarum*, coupled with the chronic overfishing of most fishes, has caused the reefs there to become dominated by algae rather than coral. In certain areas of India, coral is actively mined for use in building, and this mining is also causing catastrophic mortality. In the Philippines, dynamiting for fishes is a major cause of destruction. Oil pollution has also been shown to affect corals adversely, with the branching species, such as *Acropora*, being more heavily affected than the massive species.

Two diseases that affect corals, **black-band disease** and **white-band disease**, named after the diagnostic symptoms they produce on infected coral colonies, may cause local die-offs, but have not as yet caused catastrophic mortality. Black-band disease has also been found in milleporinids and gorgonians. The cyanobacterium *Phormidium corallyticum* has been suggested as the cause of the disease, but it also provides a microhabitat for other bacteria that may be important. The cause of coral tissue lysis as the band

moves has yet to be determined. Not all corals are equally susceptible to black-band disease. Large brain corals are the most susceptible, whereas the branching corals are resistant. In white-band disease, the cause is unknown. White-band disease seems more prevalent in branching corals.

When catastrophic mortality occurs, how long does it take a reef to recover? Again, we are ignorant, but the little information we have on the recolonization of reefs destroyed by hurricanes suggests that it may take as long as 25 to 30 years before the reef has completely recovered. Estimates of recovery time from *Acanthaster* devastation range from 7 to 40 years, and recovery may take even more time, depending on whether the devastated reefs are reinfested with *Acanthaster* before they can recover. Endean and Cameron (1990) report that, in many parts of the Great Barrier Reef, reefs attacked and devastated by *Acanthaster* in the 1960s were reinfested in the 1980s, suggesting that complete recovery may take considerably more than 40 years. Glynn (1985) believes that, in the case of the 1982–83 El Niño, some areas may take 100 or more years to recover. The reasons for the widely varying estimates of recovery time are because recovery depends on several factors, including the extent of the initial destruction, the nearness of a source of larvae for recolonization, favorable water currents and conditions for establishment of the larvae, the time between major disturbances, and biological interactions. Some reefs may be hindered in their recovery by persistent grazing by large numbers of sea urchins or by the mats of algae and soft corals, which quickly take over the dead coral skeletons, rendering them unsuitable substrates for recolonization.

■ LONG-TERM DYNAMICS

Coral reefs throughout the world are suffering from an onslaught of both natural and human disturbances that, in concert over the last few decades, have degraded or destroyed many reefs. Extensive assessments carried out by 80 countries and the Global Coral Reef Monitoring Network has raised the estimate of the percentage of reefs destroyed throughout the world by late 2000 to 27%. Moreover, as Wilkinson (2002) notes, it has been predicted that, by 2030, close to 60% of the worlds coral reefs may be lost if current trends in climate change, habitat degradation, and overharvesting continue. Over the past decade, note Jackson et al. (2001) and Pandolfi et al.

(2003), it has been realized that all reefs have already been significantly degraded by human activities and that there are no pristine reefs. Caribbean reefs seem to have suffered disproportionately and are also the best documented. It seems appropriate to end this section on coral reefs with a brief discussion of the current concern for their long-term survival.

The island of Jamaica lies at the center of the coral reef diversity in the Atlantic, with more than 60 species of reef-building corals occurring along its reef-lined coast. The Jamaican reefs have also been the subject of intensive study going back to the 1950s, making them among the best understood reef systems in the world. Terry Hughes (1994) has documented what 40 years of human and natural disturbances have done to these reefs. His findings can serve here as an example of the kinds of changes that we may see in other parts of the coral reef areas of the world if we are not careful.

As Hughes points out, the amount of damage to the Jamaican coral reefs is enormous. Censuses taken from 1977 to 1993 along 300 kilometers of coastline reveal a decline in coral cover from an average of 52% to 3% and a concomitant increase in fleshy algae from a mean of 4% to 92%! In other words, there has been a phase shift from coral domination to algae domination in shallow waters. The classic zonation pattern of coral reefs first described by Goreau in the 1950s no longer exists.

The phase shift probably began in the 1960s, when the exponentially rising Jamaican human population greatly increased fishing pressure on the reefs. By the end of the decade, fishing had reduced the fish biomass by as much as 80% along the north coast. By 1973, the fishing pressure was perhaps two to three times the sustainable yield, and large predatory fish species were rare. Species composition changed, indicating severe overfishing. This paucity of fishes—particularly herbivorous fishes that compete with sea urchins, such as scarids and acanthurids, and predatory fishes that eat sea urchins, such as balistids—permitted the increase in the sea urchin *Diadema antillarum* such that populations on overfished reefs were much higher than on pristine reefs. During this time, the reefs appeared normal and no large algal blooms occurred as a result of the demise of the herbivorous fish because the abundant *Diadema* replaced the fish.

In 1980, Hurricane Allen, a class 5 storm, hit Jamaica after almost 40 years during which the island had been spared a major storm. Allen smashed the shallow-water reef corals, especially the dominant branching *Acropora cervicornis* and *A. palmata.* Immediately after the storm, there was a short-lived algal bloom, but it was not sustained, probably due to urchin grazing, and corals began to recolonize the reefs.

The recolonization was cut short by another natural disaster: Starting in 1982, a waterborne pathogen began to attack the *Diadema* populations, eventually causing mass mortality of the urchins throughout the entire Caribbean. *Diadema* populations were reduced by 99%. In the decade since the die-off, there has been no significant recovery.

Without either the herbivorous fishes or *Diadema* to keep the algae in check, all reefs have undergone an explosive algal bloom that has preempted virtually all space on them. The algae have overgrown and killed the coral colonies that survived the hurricane. The takeover by the algae was aided by a series of bleaching events in 1987, 1989, and 1990 that further reduced the remaining corals.

It now appears that this new pattern will persist even in the face of future hurricanes, because, while such storms will destroy the algae, they will also destroy the remaining corals, and the algae regenerate or recruit much more rapidly than the corals. The ultimate recovery of the corals depends on the return of significant numbers of herbivores to the reefs, probably both the fishes and the urchins. As Hughes concludes, to save the reefs for the future, it will be necessary to control overfishing and build the populations of herbivores. If this is not done, the prime coral reef area of the Atlantic faces a gloomy future.

The most recent assessments of the global status of coral reefs have warned that, without a serious international effort to limit human damage to coral reefs systems, global warming, overharvesting of reef organisms, shoreline development, and the continued use of artificial nitrogen fertilizers will together continue to rapidly degrade coral reef systems. Indeed, according to Jackson et al. (2001), Pandolfi et al. (2003), and Hughes et al. (2003), we could easily see the collapse of existing reef systems during our lifetime.

Mangrove Forests

Mangrove forest, or **mangal**, is a general term used to describe a variety of tropical inshore communities dominated by several species of trees or shrubs that grow in salt water. Mangroves are terrestrial flowering plants that have reinvaded the fringes of the sea. See Ellison et al. (1999) for a good evolutionary history of mangroves. The term **mangrove** refers to the individual plants, whereas *mangrove forest, mangrove swamp, tidal forest,* and *mangal* refer to the whole community or association dominated by these plants. Walsh (1974) maintains that 60 to 75% of the coastline of the Earth's tropical regions is lined with mangroves. Like salt marshes, their temperate-zone counterparts, mangrove forests are ecologically important communities in nearshore tropical and semitropical systems. Mangrove forests buffer tropical shorelines from erosion, serve as biochemical filters on terrestrial runoff, and are important nursery grounds for a wide variety of mobile tropical marine organisms ranging from crabs and shrimp to fish. Unfortunately, also like temperate-zone salt marshes, mangrove forests throughout the world are an extremely threatened ecosystem because of their vulnerability to a variety of human activities. Recognizable descriptions of mangal associations were made by the Greeks as early as 325 B.C. Mangrove communities dominate the same habitats in tropical and semitropical latitudes that are dominated by salt marshes at higher, temperate latitudes. Occurring in low-wave-energy, high-sedimentation shoreline habitat dominated by salty anoxic soils at low latitudes, mangroves replace, and appear to competitively displace, salt marshes at these lower latitudes. On the coast of Florida, mangroves often recruit to more northerly habitats and begin to displace salt-marsh vegetation, but die back during harsh winters. On the east coast of Australia, Santilan and Williams (1999) have found that global warming over the past decade has triggered an expansion of mangroves into temperate habitats, where they are displacing salt-marsh vegetation. Mangroves appear to be quite sensitive to freezing, which limits them from extending their range into temperate habitats. In this section, we examine the basic ecological factors governing mangrove communities.

■ STRUCTURE AND ADAPTATIONS

Mangrove forests are composed of trees and shrubs belonging to some 12 genera of flowering plants in eight different families (Table 9.4). The most important or dominant genera are *Rhizophora, Avicennia, Bruguiera,* and *Sonneratia.* Mangroves share a number of characteristics that allow them to live in shallow marine waters. They are, first of all, shallow rooted, with roots that spread widely or with peculiar prop roots that stem

TABLE 9.4

The Genera of Mangroves

Avicennia	Aegiceras
Suaeda	Aegialitis
Laguncularia	Rhizophora
Lumnitzera	Bruguiera
Conocarpus	Ceriops
Xylocarpus	Sonneratia

Source: After Lugo and Snedaker, 1974.

from the trunk or branches of the plan (Figure 9.50). The shallow roots of some mangrove species send up extensions, called **pneumatophores**, above the surface of the substrate that allow the roots to receive oxygen in the otherwise anoxic mud in which these trees grow (Figure 9.51). The aerial prop roots of other mangrove species are also adaptations for oxygenating roots in anoxic soils. Mangroves have numerous mechanisms to cope with the osmotic problem of living in saline soils, namely, that water is constantly diffusing out of the plants. To deal with this problem, some mangroves have salt glands, which help maintain their osmotic balance by excreting salt. (Recall the salt glands of seabirds in Chapter 3, pp. 131–132.) Other mangroves elevate their osmotic concentrations by storing amino acids in their internal fluids. Still others have a succulent strategy of possessing internal water storage tissue, and yet others exclude salt and separate fresh water from salt water at the roots by a reverse-osmosis process. Mangrove leaves are also conspicuously tough and have high concentrations of defensive compounds to limit herbivory.

According to Rabinowitz (1978), nearly all mangroves share two reproductive strategies: (1) vivipary, or the uninterrupted development of the embryo following fertilization, such that there is no "resting" or true "seed" stage; and (2) dispersal by water. Vivipary is particularly well developed in the red mangrove genus *Rhizophora* and can serve as a model. In *Rhizophora,* the seed, while still on the parent plant, germinates and begins to grow into a seedling without a resting stage. During this time, the seedling root elongates and the weight distribution changes, so that the root becomes heavier at the outer, free end (Figure 9.52). Eventually, seedlings drop from the parent plant and, because of their weight distribution, float upright in the water. Seedlings are then carried by water currents until they encounter shallow water. When they contact

FIGURE 9.50 Mangrove forest, showing typical root system and roots descending from the branches. (Photo by Dr. Roger Seapy.)

FIGURE 9.51 Pneumatophores on the roots of a mangrove tree. (Copyright Dr. Aaron Ellison.)

the bottom, they produce anchoring roots and develop into trees. The advantages of this system of reproduction are obvious for a plant living on the fringe of the sea. Having seedlings that are able to float means that they can be dispersed by water currents, while the fact that the seedling floats upright with its root below water means that, when water shallow enough for the mangrove to grow in is reached, the seed will ground itself (Figure 9.53). Viviparous mangrove propagules are also all relatively

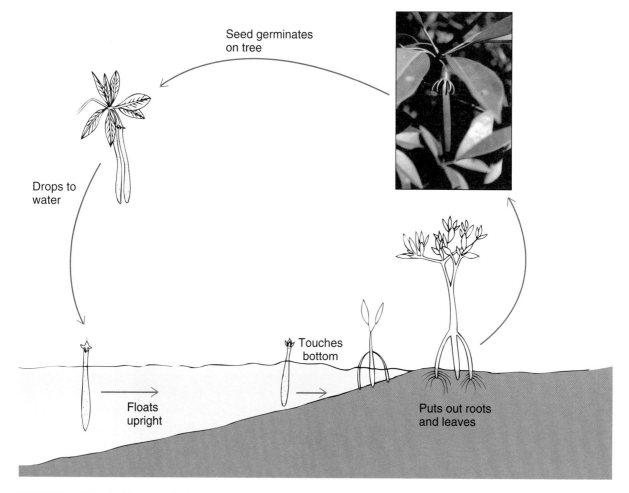

Seed germinates
on tree

Drops to
water

Touches
bottom

Floats
upright

Puts out roots
and leaves

FIGURE 9.52 The life cycle of a typical mangrove tree.

FIGURE 9.53 Mangrove seedlings in Belize (Photo courtesy of Mark Bertness).

large, and, as Farnsworth and Ellison (1997a) noted, a large parental investment is probably needed for them to cope with both the harsh physical conditions of anoxic saline soils found in mangrove habitats and the high herbivore pressure on seedlings.

■ **DISTRIBUTION**

Mangrove forests are distributed throughout the tropical and subtropical oceans of the world. They can grow only on shores that are sheltered from wave action; otherwise, the seeds would never be able to ground properly and put down roots. These shores may be directly along the lee sides of islands or island chains or on islands or landmasses behind protecting offshore coral reefs. Such shores are particularly well developed in estuarine areas of the tropics, where they reach their greatest aerial extent.

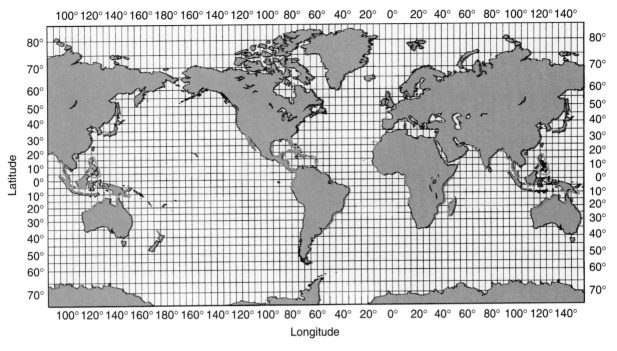

FIGURE 9.54 Distribution of mangrove forests in the world.

Mangroves occur over a larger geographical area than coral reefs and may be found in areas well outside the tropics, such as along the west coast of central and northern South America and Africa where reefs are rare or absent, and as far south as the north island of New Zealand and nearly to the mouth of the Rio de la Plata in Argentina (Figure 9.54). Mangrove forests may also penetrate some distance upstream along the banks of rivers; for example, McNae (1968) reports that they extend as far as 300 km along the Fly River in New Guinea. Mangroves are usually absent from atolls and isolated high islands such as Hawaii. (Mangroves were introduced into Hawaii in 1902.)

As with corals, the diversity of mangroves varies geographically. The largest number of species is in the Indonesia–Australia–New Guinea area; the eastern Pacific and the tropical Atlantic are represented by only a few species (Table 9.5).

■ PHYSICAL CONDITIONS OF MANGROVE FORESTS

Since mangroves can become established only where there is no significant wave action, the first physical condition noted in mangrove habitats is that water

TABLE 9.5

Species Diversity of Mangroves in Different Geographical Regions

REGION	NUMBER OF GENERA	NUMBER OF SPECIES
Australia and New Guinea	16	35
Asia, including Indonesia	17	39
East Africa and Madagascar	8	9
West Africa	3	5
West Atlantic, including the Caribbean	3	6
Eastern Pacific	4	7

Source: Species Diversity, Vol. 1: Ecological Communities, by Ricklefs and Schluter, p. 218. Copyright © 1993 University of Chicago Press. Reprinted by permission.

motion is minimal. This absence of vigorous water motion has significant effects. Slow water movement means that fine-sediment particles settle out and accumulate on the bottom. The result is an accumulation of mud; hence, the substrate in mangrove swamps is usually mud. These areas are, then, much like the previously discussed muddy shores and salt marshes of temperate shores, where poor interstitial circulation and high bacterial numbers lead to anoxic conditions. Such conditions are why mangroves have such shallow roots and pneumatophores. The soils of mangrove forests are low in oxygen, high in salt content, and fine grained, with a high organic content. Note that mangroves can grow on many types of substrates, so it is not unusual to see small islands of mangroves even on coral rock or sand. As Odum et al. (1982) report, the extensive mangrove forests, however, flourish only on mud or fine-grained sediment.

The low water movement in mangrove forests is further enhanced by the mangroves themselves. As noted earlier, many mangroves have peculiar prop (stilt) roots, which extend downward from the trunk and branches. These roots are often so numerous that they form an impenetrable tangle between the surface of the mud and the surface of the water. The dense root systems decrease water movement even more, and the finest particles settle out around the roots of the mangroves, creating accumulating layers of sediment. Once settled, sediment is usually not resuspended and transported out. Sediment thus accumulates rapidly in mangroves. Mangrove forests are significant producers of new coastal land at the expense of the sea. The rate at which new land forms can be remarkable. For example, the great Venetian traveler Marco Polo visited a port city called Palembang in Sumatra in 1292. At that time, the city was directly on the sea. At present, it lies 50 km inland! This movement, observes McNae (1968), translates into an accumulation of land at the rate of 73 m/year over the last 675 years. In other areas, the rate may be even faster (200 m/year in the Bodri Delta of east Java). This creation of land means that, as one progresses through a mangrove from the seaward side to land, there is a gradual change from marine to terrestrial conditions. The change is the main reason for the zonation observed in many mangrove forests.

Although mangroves are thought to be facultative halophytes that are capable of growing in freshwater habitats, mangrove forests do not develop in strictly freshwater environments, probably because, as Kuenzler (1974) speculates, they are not good competitors and high salinity is important in reducing competition from freshwater and terrestrial vascular plants. Crain et al. (in press) have recently demonstrated this relationship for salt marsh plants. These findings mean that both salt marsh and mangrove plants live in physically stressful anoxic and salty habitats not because they are habitats where the plants grow best, but as a refuge from competitively superior plants that dominate freshwater habitats in both temperate and tropical latitudes.

The final physical factor to be considered is the tide. Tidal ranges and types vary across the geographical range of the mangroves. Mangrove forests develop only in shallow water and intertidal areas and are thus strongly influenced by the tides. Whereas, according to Odum et al. (1982), tidal action is not a direct requirement for mangroves, it is indirectly important in a number of ways. First, tidal action helps exclude other, competitively superior vascular plants, permitting mangroves to exist. Second, tides bring salt water up the estuaries, permitting mangroves to penetrate inland. Tides also transport nutrients into, and export material out of, mangrove forests. Finally, the action of the tides prevents soil salinities from reaching lethal levels in areas of high evaporation, and the tides aid in dispersing propagules. Perhaps these are the reasons mangrove communities reach their greatest development where there are large tidal fluctuations. It is the tide and its vertical range that determine the periodicity of the inundation of the forests, which is of considerable importance in determining what kinds of mangrove associations develop in an area and which also may be responsible for the different types of zonation patterns observed.

■ ZONATION

Until recently, most studies of mangrove forests have been descriptive. As might be expected, these studies have produced different schemes that reflect the varying underlying conditions, such as tidal factors. No one has suggested a truly universal scheme such as that proposed for rocky shores (see Chapter 6), but a general scheme of wide applicability for mangrove forests of the Indo-Pacific region may be applicable.

The seaward area of most Pacific (Old World) mangrove forests is dominated by one or more species of *Avicennia*. This *Avicennia* fringe is usually narrow, because *Avicennia* seedlings do not grow well under conditions of shade or heavy siltation, which prevail in the interior of the forest. Associated with this zone and growing seaward of it are trees of the genus *Sonneratia*, which grow where they experience daily wetting (Figure 9.55).

Behind the *Avicennia* fringe is a zone dominated by one or more species of *Rhizophora*. These are the trees that most characterize mangrove communities; they have the arching stilt roots that make areas they inhabit such an impenetrable maze for humans.

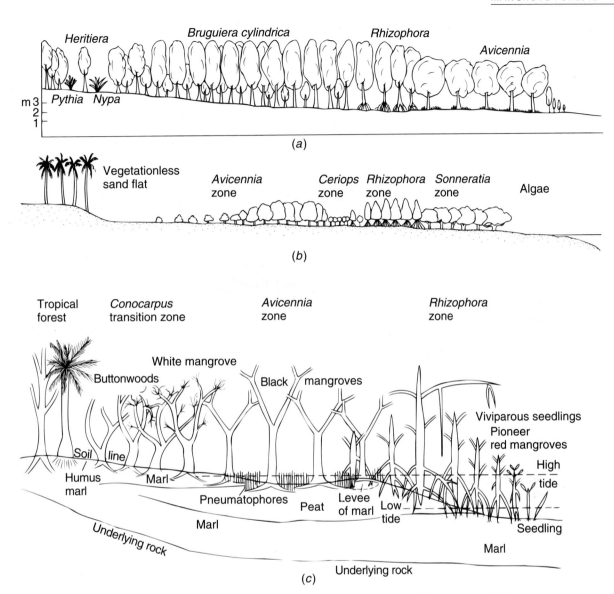

FIGURE 9.55 Diagrams of the zonation of mangroves in several areas. (a) Malaysia. (b) East Africa. (c) Florida. (a, b from *Ecosystems of The World*, Vol. 1, ed., V. J. Chapman, 1977. Copyright © 1977 V. J. Chapman. c, from "The Ecology and Geologic Role of Mangroves in Florida," J. H. Davis, Jr., 1940, Carnegie Institute of Washington, Publication 517, Tortugas Lab, Paper 32. Reprinted by permission.)

Species of *Rhizophora* are often quite tall and develop across a broad area of the intertidal zone, from levels flooded at all high tides to areas flooded only at the highest spring tides.

As one progresses landward, the next zone is the *Bruguiera* zone. Trees of the genus *Bruguiera* develop on more heavy sediment (clays) deposited at the high-water level. The final zone of the mangrove forest, which is not always present, is the *Ceriops* zone, an as-

sociation of small shrubs. This is a variable zone and may be merged with the trees of the *Bruguiera* zone.

In the Americas (New World), mangrove zonation is less pronounced, due in part to fewer mangrove species. For example, many Indo-Pacific regions have 30–39 mangrove species present, but Florida has just 3: red mangrove (*Rhizophora mangle*) (Figure 9.56), black mangrove (*Avicennia germinans*), and white mangrove (*Laguncularia racemosa*). Another species,

FIGURE 9.56 The red mangrove *Rhizophora* common in Florida and the Caribbean (Photo courtesy of Mark Bertness).

buttonwood (*Conocarpus erectus*), though not a mangrove, often occurs between mangroves and terrestrial vegetation. Zonation in the Americas places the *Rhizophora* mangal zone at the seaward edge, according to West (1977). This zone is followed inland by a broad band of *Avicennia* and then a belt of *Laguncularia* or *Conocarpus* (Figure 9.55c).

Zonation may also be limited by tidal action. Wherever the tidal range is small, the intertidal zone is also restricted, as are mangrove forests. The most extensive forests develop on shores that have a substantial vertical tidal range. This is also true of salt marshes, and the enhancing effect of a large tide on salt marshes has been called a *tidal subsidy*.

It is important to note that the preceding discussion gives only general plant zonation schemes and that not all mangrove forests will exhibit them. In fact, in some areas, zonation may be considerably abbreviated or represented by only a few individuals. This is particularly true at the limits of the distribution of the mangroves. Zonation may also be ambiguous where local conditions are such that evaporation of water from the soils makes them hypersaline. Hypersalinity can kill mangroves, creating bare areas. Zonation can also be difficult to detect or may not be present in mangroves where the gradient in physical conditions is not sharp or is heterogeneous and interrupted by local conditions.

In addition to the plant zonation often seen in mangrove forests, there is conspicuous zonation of invertebrate epiphytes on the roots of mangroves that live with their roots submerged in seawater. Vertical zonation on mangrove roots is similar to that found on rocky shores. Typically, barnacles, oysters, and periwinkle snails are found at high intertidal heights, while clonal organisms such as sponges and sea squirts dominate low intertidal and subtidal heights on mangrove roots. Interestingly, there is often striking horizontal zonation of epiphytes on mangrove roots. On the seaward edge of mangrove forests, the cover and density of invertebrates on submerged mangrove roots is typically high, whereas at the same tidal height in the interior of mangrove stands, submerged roots can be almost free of invertebrates. Bingham (1992) has shown this pattern to be due largely to larval settlement patterns. Once marine larvae are ready to settle and undertake an adult lifestyle, the larvae of most marine invertebrates will settle on the first suitable surface they encounter. In mangrove forests, this typically means that roots on the seaward edge are exposed to more larvae and have the highest densities of invertebrates. Because most larvae settle on the seaward edge of mangrove roots, few larvae get to roots on the interior of mangrove forests, so those roots remain relatively free of encrusting invertebrates.

■ ASSOCIATED ORGANISMS

Mangrove communities are unusual. Due to the height of the trees, true terrestrial organisms can occupy the portions that are never in water, while true marine animals occupy the bases that are in the water. Like salt marshes, mangrove forests, then, form a strange mixture of marine and terrestrial organisms and have been suggested to be an evolutionary pathway from land to sea and vice versa.

Most terrestrial organisms that live in mangroves show no special adaptations for life in mangrove forests, since they spend their lives out of marine waters in the upper reaches of the trees, although they may forage on the marine animals at low tide. Among the large variety of terrestrial arthropods in mangrove forests are herbivorous and carnivorous insects, but their significance to mangroves is little known. Farnsworth and Ellison (1997a) have shown that, before they are released by the parent tree, mangrove flowers and seeds are heavily damaged by consumers in mangroves throughout the world. These authors also found that 50–90% of mangrove seeds are destroyed by consumers (mostly insects and crabs) while the seeds are still attached to adult trees.

Marine organisms associated with mangroves are of two types: those which inhabit the hard substrate of the roots of mangroves and those which occupy the mud. Mangrove associations are different from those of muddy shores primarily because of the roots' vast, hard surface area, which is absent from typical mud shores. (See pp. 325–326.)

The dominant marine animals in mangrove forests are mollusks, crustaceans, sponges, sea squirts, and some peculiar fishes. Mollusks are represented by a number of snails, one group of which (Littorinidae) generally lives on the roots and trunks of the mangrove trees and the other, comprising mainly detritus feeders (Ellobiidae and Potamididae), on the mud at the base of the roots. Little is known of the contribution of these snails to the mangrove forest. A second group of mollusks includes the bivalves—especially oysters—which attach to the mangrove roots, can be common, and can have a significant biomass (Figure 9.57).

Mangrove forests are also inhabited by numerous large-sized crabs and shrimps. These animals, which include such common genera as *Uca*, the fiddler crabs; *Cardisoma*, the tropical land crabs; and various ghost crabs (*Dotilia, Cleistostoma*), excavate burrows in the soft substrate. The mangrove crab *Aratus pisonii* is abundant in Caribbean forests, where it lives on the roots just above the waterline. Also abundant in Caribbean mangroves is *Goniopsis cruentata*, which lives among the roots. This crab does not construct burrows, but may take refuge in burrows dug by other crabs. All of these crabs—*Uca*, for example—can be specialized to feed on the detrital particles found in the mud. Generally, they separate the organic detrital particles from the nonorganic matter by filtering the substrate through a set of fine hairs around the mouth. This type of feeding method is called *deposit feeding*. Other crabs are general omnivores and, as Smith (1992) has noted, have been shown to play an important role in mangrove forests as mangrove seed predators. Among these crabs are *Sesarma, Goniopsis*, and *Aratus*. All show varying degrees of adaptation toward a more terrestrial mode of life; the usual modification is a vascularization of the walls of the gill chambers so that they become more "lunglike." As we noted previously, mangroves are transitional areas between land and sea, and the crabs living in them certainly reflect this dualism in their partial adaptation to air breathing.

The burrows of many of these crabs, such as *Uca* and *Sesarma*, as well as those of mud shrimps, such as *Upogebia* and *Thalassina*, serve several functions, perhaps the foremost of which is that they provide refuges from predation and physical stress, as well as a place to breed and to store eggs safely. Mangrove crabs will typically retreat to their burrows both during physically harsh low tide exposures to terrestrial conditions and at any time during the tidal cycle when large predators are present. These behaviors suggest that burrows are important refuges from both physical stress and predators. Just as they do in salt-marsh ecosystems (Bertness, 1984a), burrowing crabs enhance the growth of mangroves by aerating mangrove sediments and allowing oxygen to penetrate more deeply into the substrate. In this manner, says Smith (1987), the crabs ameliorate anoxic conditions that can limit mangrove growth. The deposit feeding of many mangrove crustaceans, such as *Uca*, *Upogebia*, and *Thalassina*, is also likely responsible for aerating mangrove soil and thus benefiting the plants. The common Caribbean mangrove crabs *Aratus pisonii* and *Goniopsis cruentata* are not deposit feeders, but instead are omnivores with a diet that includes mangrove leaves and seeds. Thus, these crabs may be significant herbivores. Sousa and Mitchell (1999) and Smith (1992) report that some mangrove crabs are consumers of mangrove seedlings and play

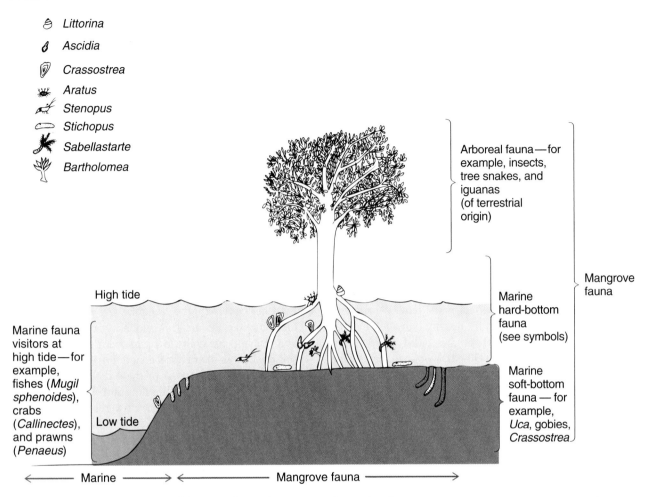

♮ *Littorina*

𝕝 *Ascidia*

Crassostrea

Aratus

Stenopus

Stichopus

Sabellastarte

Bartholomea

Arboreal fauna—for example, insects, tree snakes, and iguanas (of terrestrial origin)

Mangrove fauna

High tide

Marine hard-bottom fauna (see symbols)

Marine fauna visitors at high tide—for example, fishes (*Mugil sphenoides*), crabs (*Callinectes*), and prawns (*Penaeus*)

Low tide

Marine soft-bottom fauna — for example, *Uca*, gobies, *Crassostrea*

←——— Marine ———→ ←——————— Mangrove fauna ———————→

FIGURE 9.57 Representation of the macrofauna of a mangrove forest, showing the vertical distribution and ecological relationships. (Modified from *Marine Biology*, H. Friedrich, Sedwick & Jackson.)

a significant role in maintaining mangrove zonation by differential consumption of different species' seedlings. Crabs, however, are not the only organisms that feed on mangroves: In an experiment in Florida, Devlin (1999) found that the most significant consumer of mangrove propagules and seedlings in Florida is the beetle *Cocotrypes rhizophorae*, and there is even a report of mangrove foliage being consumed by green sea turtles!

A large numbers of marine invertebrates inhabit the prop roots of mangroves and form a colorful association of sponges, sea squirts, and oysters that appear to play a significance role in the ecology of mangroves. For example, Perry (1988), working on the red mangrove *Rhizophora mangle* in Costa Rica, experi-

mentally demonstrated some interesting ecological interactions among the mangrove, epifauna on mangrove roots, and various benthic and water-column predators. In this system, the isopod *Sphaeroma peruvianum* bores into mangrove roots, while barnacles (*Balanus* spp.) settle on the surface of the roots. Both cause a decrease in root growth rate, of 50% and 30%, respectively, as well as a decrease in root production of over 50% in both cases. However, not all roots are equally affected. Roots that are not in contact with the substratum are most affected, whereas those in contact suffer considerably less. The reason for this discrepancy is that those roots which are in contact with the substrate are accessible to a benthic predator, the hermit crab *Clibarnarius panamaensis*, which climbs the roots at high tide to feed on barnacles. The absence of isopods from certain roots can also be attributed to predation, as Perry (1988) demonstrated in studies of predator exclusion. A similar situation may occur with respect to algae. Taylor et al. (1986) found that, in mangroves in Belize, fleshy algae was common only on roots that were not in contact with the bottom. Roots that were in contact with the bottom were heavily grazed by benthic herbivores, such as urchins, keeping them clear of algae.

Sutherland (1980) studied the same red mangrove species studied by Perry in Venezuela, with quite different results. In Venezuela, the epifaunal organisms that occurred on mangrove roots were long lived and resistant to displacement by other organisms; predation had little effect on community development. Clearly, more work needs to be done to understand these assemblages.

Another important invertebrate on the roots of Caribbean red mangroves is the small isopod *Sphaeroma terebrans*. This isopod burrows into the prop roots of red mangroves, and the boring, together with secondary decomposition by bacteria and fungi, usually severs the prop root at the level of mean high tide. Rehm and Humm (1973) were the first to describe the phenomenon and suggested that the destruction of prop roots left the underlying sediment free to erode, allowing the shallow-rooted trees to fall into the water. In this way, the whole forest would slowly be cut down and destroyed. Since that time, however, other workers have taken issue with this ecocatastrophe view. Estevez and Simon (1975) have shown that the infestations of the isopod are patchy, limited in extent, and related to changes in salinity. Simberloff et al. (1978) suggested that, since the isopod boring stimulated the mangrove to produce more root growth and branching, which are potentially beneficial to mangroves, isopods may have a positive effect on the trees. More recently, Ellison and Farnsworth (1990) have demonstrated that sponges and perhaps other root epiphytes protect mangrove roots from isopod boring, so the dynamics of the mangrove root-fouling community may mediate the importance of isopod boring effects on mangroves.

Ellison et al. (1997) have shown experimentally that the epiphytic community on mangrove roots (Figure 9.58) plays an important role in the performance of red mangroves in the Caribbean. Most of the sessile and slow-moving invertebrates that live on the submerged portion of red mangrove roots are filter feeders making their living filtering food particles out of the water col-

FIGURE 9.58 Epiphytes on Caribbean red mangrove roots in Belize. (Copyright Dr. Aaron Ellison.)

umn. Among these creatures are sponges, sea squirts, and oysters. Onuf et al. (1977) have demonstrated that mangroves are nitrogen limited, so these filter feeders may serve as a transformer of nutrients from the water column, where they are thought to be unavailable to mangroves, to the sediment or root surface. Bertness (1984b) found the same relationship between marsh cordgrass and marsh mussels in salt marshes. Cordgrass is nitrogen limited, and mussels filter feed and deposit their nitrogen-rich feces directly on cordgrass roots, alleviating the nitrogen limitation and increasing cordgrass growth. Mangrove oysters that live directly attached to mangrove roots could be playing a similar role, but in most mangroves encrusting clonal filter feeders such as sponges and sea squirts are more abundant. Ellison and colleagues (1997) have demonstrated that, when sponges cover red mangrove roots, the roots respond by sending adventitious rootlets into sponge tissue, where they acquire nitrogen. Sponges, then, increase the growth of red mangroves. How common mangrove root epiphytes benefit growth in other mangrove assemblages is unknown.

Mangrove roots are not the only parts of the plant that are subject to consumption. Wada and Wower (1989) demonstrated that the ocypodid crab *Macrophthalmus quadratus* grazes on the pneumatophores of

FIGURE 9.59 An Indonesian mudskipper. (Photo courtesy of Dr. Greg Erickson.)

the mangrove *Sonneratia alba* and removes epiphytic algae and bark. Onuf et al. (1977) found that mangroves in Florida responded to nutrient enrichment derived from the guano of a bird-breeding colony by exhibiting higher growth rates compared with mangroves in nonenriched areas, but that the nutrient-enriched mangroves also had increased herbivory on the leaves by terrestrial herbivores. Underwood and Barrett (1990) have demonstrated an experimental relationship between the root-dwelling oyster *Crassostrea commercialis*, which resides on the roots of the mangrove *Avicennia marina*, and the presence of the littorinid snail *Bembicium auratum*; the oysters increase the distribution, abundance, and size of *Bembicium*. Underwood and Barrett ascribe this positive relationship to the ability of the oysters to provide a refuge from predation for the littorinid, thus protecting it from crustaceans and fish.

Mangroves also are nursery grounds for a variety of fishes and crustaceans—for example, penaeid shrimps and such fishes as mullet, which may spend the early part of their life cycles in mangroves before moving offshore. In addition, the prop roots of mangroves are known to be a nursery habitat for juvenile spiny lobsters (*Panulirus argus*) in the Caribbean, which may spend up to two years in that area.

Conspicuous by their size and abundance in the water and on the mud in Indo-Pacific mangrove forests are the small, big-eyed fishes of the genus *Periophthalmus* and their relatives. These fishes, collectively called mudskippers, are remarkable because they spend most of their time out of water, crawling around on the mudflats or even climbing mangrove roots (Figure 9.59)! In fact, they act much like frogs or toads. Mudskippers dig burrows in the mud for refuge and breeding. They are quite fast, moving over the exposed mudflats by "walking" on their strong pectoral fins or by using a series of skipping motions, or "bounds," in which the tail and caudal fins provide the thrust.

Aside from their peculiar unfishlike habit of crawling around on the mudflats, the other characteristic feature of these fishes is their modified eyes, which are set high on their heads and are arranged so that they focus best in air, not in the water. When mudskippers swim, their eyes protrude above the surface; hence, the fishes look like frogs. A final noteworthy adaptation is in the respiratory system: The gills are often reduced, and aerial respiration is accomplished by vascularized sacs in the mouth cavity and gill chambers.

■ **SUCCESSION AND MORTALITY**

Mangroves are subject to mortality from a number of different natural and human-induced causes. Since mangroves occur in physically stressful habitats for vascular plants, and since they ameliorate many of the stresses of their habitat to make it more habitable, recovery from disturbance can be a slow process. The prop roots and pneumatophores of mangroves increase the supply of oxygen to the soil, and shading by mangroves limits evaporation and salinity. Disturbances in mangrove forests can upset these positive feedbacks that are important for community structure and productivity. Disturbances will typically trigger recolonization by mangroves, but the process may be slow and may result in gradual change or succession.

Davis (1940), working in Florida, was the first to suggest a connection between zonation, succession, and changes in physical factors. In this now-classical view, Davis argued that the mangrove zonal patterns were equivalent to seral stages in succession. In his scheme, the seaward zone of red mangroves is the "pioneer," or initial, seral state. Farther landward, the black and white mangrove zones were regarded as progressively later stages of the successional sequence, leading eventually to terrestrial communities. This view, however, is not universally accepted. Other workers, such as Thom (1975), have demonstrated that mangrove zonation is mostly a response to external physical factors, rather than a successional sequence induced by the plants themselves. This view is supported by Rabinowitz (1975), who demonstrated in transplant experiments that various species of mangroves could grow well within any zone and that physical factors alone were not responsible for their absence. It now appears that the zonation of mangroves is controlled, not directly by physical factors, but by an interplay of these factors with interspecific competition and propagule dispersal and sorting. Unless there is heavy sedimentation, once a mangrove area reaches an equilibrium state, change occurs only if there is some sort of disturbance. That is, no further "succession" is induced simply by the plants themselves. Disturbances may be small (lightning strikes or droughts) or large (changes in sea level or hurricanes). Such perturbations set zonation back to an earlier state. Evidence from Florida, according to Odum et al. (1982), suggests that hurricane perturbations are frequent and regular enough to keep mangrove systems in a cyclical pattern of succession rather than in the classic facilitation sequence that would continually produce additional terrestrial forests as the climax stage. In this newer view, the role of mangroves in building land at the expense of the sea is seen as more passive than active, and various physical processes are the dominant forces determining the rate of growth or recession of mangrove shorelines. Mangroves have the

role of stabilizing sediments, the origin and major deposition of which are due to physical processes.

Perhaps the greatest causes of large-scale mortality are typhoons and hurricanes. These violent storms destroy large areas of mangrove forests by uprooting trees, by laying down massive sedimentation, or by altering the salinity of the water and soil. Hurricane Donna, which struck Everglades National Park in 1960, caused a mortality ranging from 25 to 75% of 100,000 acres of mangroves in the park. As with coral reefs, estimates of the mangroves' recovery time from such a massive mortality event are about 20–25 years (Figure 9.60).

Most other massive mortality of mangroves is due to human activity. Perhaps the greatest single destruction event occurred as a result of herbicide spraying on the mangrove forests of southeast Asia during the Vietnam war. According to Tschirley (1969), perhaps 100,000 hectares of mangrove forests were defoliated and destroyed by herbicides during the war. It is not known how long these forests will take to recover or whether they will regenerate at all in view of the extreme sensitivity of mangroves to residual herbicides.

Mangrove forests also have been destroyed by filling, dredging, and channelizing waters. Historically, mangrove forests probably occurred along the Persian Gulf, from which they are now absent due to cutting by early humans for firewood and for building boats. Ellison and Farnsworth (1996) report that, in the tropics today, mangroves around the world continue to be heavily exploited by humans. Human threats to mangroves, note Farnsworth and Ellison (1997b), vary in different parts of the tropics, but logging for firewood and charcoal, draining for land development (particularly for tourism) and partially destroying mangroves to build ponds for fish and shrimp culture or for salt production are the most pervasive threats (Figure 9.61).

As mentioned at the outset of this section, mangrove forests have demonstrably important value to humans. In addition to buffering tropical shorelines from erosion, like their temperate zone cousins, the salt marshes, mangrove forests serve as biochemical filters in processing runoff water entering nearshore tropical marine habitats and are nursery habitats for a wide variety of organisms, including penaeid shrimp, spiny lobsters, blue crabs, and various fishes, such as the mullet. The whole Tortugas pink shrimp industry of Florida, an 11-million-pound-per-year harvest, is dependent

FIGURE 9.60 Photo of dead mangroves destroyed in Everglades National Park by the hurricane of 1960. (Photo by James Nybakken.)

FIGURE 9.61 Harvesting a mangrove forest in Malaysia. (Copyright Dr. Aaron Ellison.)

on the mangrove system. Mangroves also shelter a number of endangered species, such as the American crocodile, the brown pelican, and the Atlantic Ridley sea turtle. Finally, the ability of mangroves to stabilize land in the face of tropical storms may be important in protecting inland habitats. All of these services are jeopardized if the global destruction of mangroves continues at its current pace.

Summary of Key Concepts

- Coral reefs are estimated to occupy 0.17% of the area of our planet.

- Coral reefs have the greatest higher taxonomic diversity of any marine ecosystem, comprising perhaps 4–5% of all species.

- Coral reefs are among the most productive systems in the marine environment and have existed for hundreds of millions of years, yet coral reefs are sensitive to a number of natural and human-made disturbances.

- Coral reefs, salt marshes, and mangrove forests are biogenic structures built by biological activity.

- Hermatypic corals have symbiotic zooxanthellae algae in their tissues and are the corals that produce reefs; ahermatypic corals lack zooxanthellae and do not produce reefs.

- Six major physical factors limit coral reef development: temperature, depth, light, salinity, sedimentation, and emergence in air.

- Coral reefs are limited to waters of full salinity that are not deeper than 50–70 m, do not fall below 18°C, and do not have excessive sediment.

- Corals are members of the phylum Cnidaria, and most hermatypic corals are colonial, with the various individual polyps occupying cups or corallites in the prodigious skeletal mass.

- Coral reefs may be grouped into three categories: fringing, barrier, and atoll.

- The subsidence, or compensation, theory of Darwin explains the existence of atolls and links fringing, barrier, and atoll reefs together in an evolutionary sequence, but the theory does not explain the origin of all fringing and barrier reefs.

- Although corals are the major organisms that form reef structures, members of practically all phyla and classes may be found on coral reefs.

- Among other organisms contributing strongly to the $CaCo_3$ budget of reefs are the coralline algae, the green algal genus *Halimeda*, the hydrocoral *Millepora*, and various mollusks, including the giant clams.

- Other prominent groups on coral reefs include the soft corals, gorgonians, echinoderms, sponges, crustaceans, polychaete worms, and fishes.

- The largest number of reef coral species and genera occurs in the Indo-Pacific in an area that includes the Philippines, Indonesia, New Guinea, and northern Australia; the number of species declines in all directions away from that area.

- Tropical Atlantic reefs have many fewer coral genera and species than the Indo-Pacific has.

- Coral reefs are complex associations of organisms with a high diversity of habitats, one reason for the high species diversity.

- The zonation across an atoll, beginning at the windward side and progressing across the atoll to the leeward side, includes an outer seaward slope with a windward reef margin; a spur-and-groove, or buttress, zone; an algal ridge; a reef flat zone; the seaward beach of the atoll island; the island; a lagoon with lagoon reefs and sea-grass beds; a leeward reef flat; and a leeward reef margin.

- Among the many differences between Atlantic and Pacific coral reefs are structure and zonation features, as well as differences in the presence, absence, and abundances of various faunal and floral components.

- Coral reefs maintain a high primary productivity in nutrient-poor tropical ocean waters, but there is little agreement among scientists as to the underlying sources of the nutrients that sustain the high productivity.

- Other factors that help sustain the high productivity of coral reefs include the large amount of plant tissue present on the reefs, close coupling of the photosynthetic organisms with the animals or matrix of the reef, and rapid cycling of nutrients by the large bacterial populations.

- Although reef corals are known to be carnivores similar to other cnidarians, they are also phototropic, depending upon the symbiotic zooxanthellae for much of their nutrition.

- Coral growth and calcification depend upon sufficient light for the zooxanthellae, which increase the rate of calcification and the growth of the coral colonies.

- The rate at which coral colonies grow differs with different species, with the age of the colonies, and in different areas of the reef.

- We are ill informed about the ages of corals, and what little information we have suggests that most coral colonies are 10 years or younger, but that some colonies can reach ages of 100 years or more.

- Corals reproduce both sexually and asexually. Asexual reproduction is usually by the budding off of new polyps from the parent and increases the size of the colony, but fragmentation of the parent colony can also lead to new colonies. Sexual reproduction results in a planula larva, which disperses from the parent to initiate a new colony.

- Most corals are hermaphroditic and may either brood the fertilized eggs to the planula stage or release the gametes to be fertilized externally in seawater.

- The rate of recruitment of new coral colonies to the reef system is variable in time and in space.

- Corals compete with each other for light mainly by overtopping and shading other colonies, a process that is a type of exploitative competition. Still other species maintain space on the reef by killing adjacent competing species with their nematocyst-studded mesenterial filaments, a process called interference competition.

- Corals using mesenterial filaments in interference competition can be arranged in a competitive hierarchy. Originally, it was suggested that this aggressive pecking order was a means of maintaining coral diversity, but coral competitive interactions are now known to be much more complicated.

- Another group of organisms for which competition for space is common is the cryptofauna.

- The role of predation in determining the structure and composition of coral reefs is not very well understood. Most predators of corals do not appear to have a significant effect on coral colonies or coral reef community structure. Most of these predators are small relative to the coral colony and do not remove enough coral tissue to kill the colony.

- Two taxa of predators are capable of destroying whole coral colonies and modifying the structure of reefs. The first is the starfish *Acanthaster planci*, which, when present in large numbers, can consume whole reefs. The second is the various corallivorous fishes.

- Algae are the major space competitors with corals on coral reefs and can outcompete them, but are generally kept from doing so by the intense grazing pressure of herbivorous fishes and urchins.

- Grazing can also produce shifts in the abundance of different groups of algae. Fleshy, filamentous algae grow faster and may eliminate the slow-growing coralline algae, but the fleshy algae are preferred by the grazers; thus, under heavy grazing pressure, the coralline algae survive.

- Algae deter herbivores by being less palatable through the deposition of $CaCo_3$, by producing noxious chemicals, and by settling into habitats that provide spatial or temporal refuges from grazers.

- Moderate grazing prevents monopolization of the substrate by algae and increases diversity on reefs, whereas heavy grazing destroys not only the algae, but corals as well, and decreases diversity.

- Damselfishes defend their territories against other grazing and corallivorous fishes and, in so doing, permit algae to overgrow, shade, and kill young, slow-growing, massive coral colonies while enhancing the growth of fast-growing branching corals, so that the latter dominate such territories.

- Herbivory and predation are more intense on Pacific reefs than on Atlantic reefs.

- Large numbers of bacteria exist on reefs, representing a huge energy source and suggesting the presence of a microbial loop system similar to that found in the plankton, but the existence of such a loop is still being debated.

- Algae are important to the reef in that they cement pieces of $CaCo_3$ together, produce sand, erode the reefs, are primary producers, and furnish habitats.

- Coral reefs harbor more species and more diverse fish communities than any environment on Earth, but only a fraction of this diversity can be explained by the large variety of habitats on the reef.

- Four competing hypotheses exist to explain the high local fish diversity of reefs: the competition model, the lottery hypothesis, the predator disturbance model, and the recruitment limitation model.

- There are marked differences between the fish species that are active on the reef during the day and those which are active at night.

- The largest trophic group of fishes on reefs consists of predators, constituting 50–70% of the species; most are unspecialized opportunists.

- The second-largest trophic group of fishes is the herbivores, while the remainder of the fishes are generally classed as omnivores with a few planktivores.

- Venomous fishes are rare on reefs, but many fishes have toxic secretions on their body surfaces.

- Ciguatera is a disease of humans caused by eating certain fishes whose flesh contains toxins produced by the dinoflagellate *Gambierdiscus toxicus*.

- Little is known of the population dynamics of reef fishes.

- Cleaning behavior is a specialized form of predation where small fishes or shrimps remove ectoparasites from other, usually larger, fishes.

- The cryptofauna are those invertebrates which inhabit the coral substrates, living in cracks, crevices, and caves and under and in coral heads and rock. These animals are important bioerosive forces on reefs.

- On short time scales, reefs appear to be stable, but not on longer time scales.

- There are three known natural causes of mass mortality on reefs: hurricanes or typhoons, population explosions of the corallivorous sea star *Acanthaster planci*, and the meteorological and oceanographic phenomenon known as El Niño. Another puzzling event, called coral bleaching, also causes significant coral mortality.

- Human activity may cause mass mortality of reefs by dredging, mining, overfishing, and pollution.

- White-band and black-band disease may cause local die-offs, but as yet they have not caused mass mortality.

- Recovery times for reefs that have undergone mass mortality range from 7 to 100 years.

- Coral reefs throughout the world are suffering from an onslaught of both natural and human disturbances that have degraded or destroyed the reefs. Many have not recovered in the decade or more since devastation, leading to considerable concern among humans about their survival.

- *Mangrove forest*, or *mangal*, is the term used to describe a variety of tropical inshore communities dominated by species of trees or shrubs that grow in salt water. These ecosystems are found on 60–75% of the tropical coastlines of the planet.

- Mangroves are shallow-rooted plants with widely spread roots and often with peculiar prop roots that stem from the trunk or branches, commonly with pneumatophores and with tough leaves and internal water storage tissue.

- Nearly all mangroves have vivipary and dispersal by water.
- Mangroves occur over a larger geographical area than coral reefs do, but are restricted to shores protected from wave action.
- Mangrove forests are areas of slow water movement, further slowed by the presence of the numerous roots of the mangroves themselves, thus accumulating sediment.
- The soils in mangrove forests are fine grained, low in oxygen, and high in organic and salt content.
- Mangrove forests are strongly influenced by the tides, which exclude other potential plant competitors; bring salt water further up estuaries, permitting mangroves to penetrate further inland; prevent soil salinities from reaching lethal levels; and disperse the mangrove propagules.
- The general zonation pattern for Pacific mangrove forests is to have a seaward area dominated by *Avicennia*, behind which is a zone of *Rhizophora*, then a *Bruguiera* zone, and, most landward, a *Ceriops* zone.
- Zonation is less pronounced in the Americas, where the *Rhizophora* zone is the most seaward, followed by *Avicennia* and then *Laguncularia* or *Conocarpus*.

- Zonation patterns may be interrupted by local conditions, mainly hypersalinity, which kills mangroves.
- Mangrove forests consist of organisms derived both from the marine environment and the terrestrial environment. Marine organisms include those which attach to the hard substrate of the roots and those which occupy the mud.
- Invertebrates inhabiting the mud often dig burrows that allow oxygen to enter more deeply into the substrate, ameliorating the anoxic conditions.
- Invertebrates inhabiting mangrove prop roots can both benefit and harm mangroves.
- Mangroves are nursery grounds for shrimps, for lobsters, and for fishes that live offshore as adults.
- Mangroves are subject to massive natural mortality, primarily by hurricanes and typhoons, and recovery times are in the same time range as for reef recovery.
- Most other mangrove mass mortality is due to human activity. Historically, the mangroves have been destroyed by filling, dredging, cutting for firewood, and building ponds for aquaculture.

Review Questions

ESSAY Develop complete answers to these questions:

1. Other than in the composition of coral species, how do coral reefs of the Indo-Pacific and Atlantic oceans differ? Discuss.

2. Examine the impact that coral predators have on coral reefs in terms of both destruction and stabilization.

3. Compare and contrast exploitative and interference competition in corals, providing examples of each to support your essay.

4. Explore the arguments that attempt to explain the high diversity of coral reef fish. Present evidence that supports each hypothesis.

5. Explore how mangals have both horizontal and vertical zonation patterns.

6. What role does the crown-of-thorns starfish (*Acanthaster*) play in the structural and biological stability of the coral reef?

7. Compare the primary productivity of coral reefs with that of open tropical oceans at the same latitude. Why do the open tropical oceans have such a low productivity? In contrast, why are the coral reefs so productive, especially in light of the fact that the waters of the coral reef are nutrient poor?

8. What environmental factors cause the destruction of mangals? What other forces are at work that destroy mangals?

Bibliography

Adey, W. H. 1977. Coral reef morphogenesis: A multidimensional model. *Science* 202:831–837.

Adey, W. H., I. G. MacIntyre, and R. Stuckenrath. 1977. Relict barrier system off St. Croix: Its implications with respect to late Cenozoic coral reef development in the Western Atlantic. *Proc. 3rd Int. Coral Reef Symp.* 2:15–22.

Anderson, I. 1990. Darwin may founder on the Great Barrier Reef. *New Scien.* (October):15.

Atkinson, M. J. 1992. Productivity of Enewetak Atoll reef flats predicted from mass transfer relationships. *Cont. Shelf Res.* 12(7/8):799–807.

Babcock, R. C. 1985. Growth and mortality in juvenile corals (*Goniastrea, Platygyra* and *Acropora*): The first year. *Proc. 5th Int. Coral Reef Symp.* 4:355–360.

Bakus, G. 1964. The effects of fish-grazing on invertebrate evolution in shallow tropical waters. Occ. paper no. 27, Allan Hancock Foundation.

Bakus, G. 1973. The biology and ecology of tropical holothurians. In *Biology and geology of coral reefs*, edited by O. A. Jones and R. Endean. Vol. II, *Biology 1.* New York: Academic Press, 325–367.

Banner, A. H. 1976. Ciguatera: A disease from coral reef fish. In *Biology and geology of coral reefs*, edited by

O. A. Jones and R. Endean. Vol. 3, *Biology 2*. New York: Academic Press, 177–213.

Bertness, M. D. 1984a. Fiddler crab regulation *Spartina alterniflora* production on a New England saltmarsh. *Ecology* 66: 1042–1055.

Bertness, M. D. 1984b. Ribbed mussels and the productivity of *Spartina alterniflora* in a New England saltmarsh. *Ecology* 65: 1794–1807.

Bingham, B. L. 1992. Life histories in an epifaunal community: Coupling of adult and larval processes. *Ecology* 73(6): 2244–2259.

Birkeland, C. 1977. The importance of rate of biomass accumulation in early successional stages of benthic communities to the survival of coral recruits. Vol. 1, *Biology. Proc. 3rd Int. Coral Reef Symp.* 15–21.

Birkeland, C. 1982. Terrestrial runoff as a cause of outbreaks of *Acanthaster planci* (Echinodermata: Asteroidea). *Mar. Biol.* 69:175–185.

Birkeland, C. 1989. The influence of echinoderms on coral reef communities. In *Echinoderm studies*, Vol. 3, edited by M. Jangoux and J. M. Lawrence. Rotterdam: A. A. Balkema, 1–80.

Birkeland, C., ed. 1997. *Life and death of coral reefs*. New York: Chapman and Hall.

Borowitzka, M. A., and A. W. D. Larkum. 1986 Reef algae. *Oceanus* 29(2)j 52.

Brock, R. E. 1979. An experimental study on the effects of grazing by parrot fishes and the role of refuges in benthic community structure. *Mar. Biol.* 51:381–388.

Brown, B. E. 1997. Disturbances to reefs in recent times. In *Life and death of coral reefs*, edited by C. Birkeland. New York: Chapman and Hall.

Brown, B. E., and J. C. Ogden. 1993. Coral bleaching. *Sci. Amer.* 268(1):64–70.

Buddemeir, R. W., and R. A. Kinzie, III. 1976. Coral growth. *Oceanog. Mar. Biol. Ann. Rev.* 14:183–225.

Buss, L.W., and J. B. C. Jackson. 1979. Competitive networks: Non-transitive competitive relationships in cryptic coral reef environments. *American Naturalist* 113: 223–234.

Castro, P., 2000. Biogeography of trapeziid crabs (Brachyura, Trapeziidae) symbiotic with reef corals and other cnidarians. In *The Biodiversity Crisis and Crustacea*, edited by J.C. von Vaupel Klein and F.R. Schram, *Proceedings of the Fourth International Crustacean Congress*, Amsterdam, vol. 2, Crustacean Issues, 12: 65–75.

Chapman, V. J., ed. 1977. *Ecostystems of the world*. Vol. 1, *Wet coastal ecosystems*. New York: Elsevier.

Chesher, R. H. 1969. Destruction of Pacific corals by the sea star *Acanthaster planci*. *Science* 165:280–283.

Colgan, M. W. 1982. Succession and recovery of a coral reef after predation by *Acanthaster planci*. *Proc. 4th Int. Coral Reef Symp.* 2:333–338.

Connell, J. 1973. Population ecology of reef building coral. In *Biology and geology of coral reefs*, edited by O. A. Jones and R. Endean. Vol. II. *Biology 1*. New York: Academic Press, 205–245.

Connell, J. H., T. P. Hughes, and C. C. Wallace. 1997. A 30-year study of coral abundance, recruitment and disturbance in space and time. *Ecol. Monogr.* 67:461–488.

Crain, C, B. R. Silliman, S. L. Bertness and M. D. Bertness in press. Mechanisms causing the spatial segregation of plants along estuarine salinity gradients. *Ecology*.

Crossland, C. 1952. Madreporaria, Hydrocorallinae, *Heliopora*, and *Tubipora*. *Sci. Repts. Great Barrier Reef Exped. 1928–29* 6:85–257.

Dana, T. F., W. A. Newman, and E. W. Fager. 1972. *Acanthaster* aggregations: Interpreted as primarily responses to natural phenomena. *Pac. Sci.* 26:355–372.

Davies, P. J. 1983. Reef growth. In *Perspectives on coral reefs*, edited by D. J. Barnes. *AIMS Contribution*, no. 200:69–106.

Davis, J. H., Jr. 1940. The ecology and geologic role of mangroves in Florida. Carnegie Inst. Wash. Publ. 517, Tortugas Lab. Pap. 32:303–412.

Devlin, D. J. 1999. A field experiment of predator strategies and mangrove resistance: The relationship between the red mangrove (*Rhizophora mangle*) and the scolytid beetle *Cocotrypes rhizophorae*. *Gulf Res. Rep.* 10:73.

Doherty, P. J. 1982. Some effects of density on the juveniles of two species of tropical, territorial damselfishes. *J. Exp. Mar. Biol. Ecol.* 65:249–261.

Doherty, P., and T. Fowler. 1994. An empirical test of recruitment limitation in a coral reef fish. *Science* 263:935–939.

Dollar, S. J., and G. W. Tribble. 1993. Recurrent storm disturbance and recovery: A long-term study of coral communities in Hawaii. *Coral Reefs* 12:223–233.

Ducklow, H. W. 1990. The biomass, production and fate of bacteria in coral reefs. In *Ecosystems of the world* 25, *Coral reefs*, edited by Z. Dubinsky. New York: Elsevier, 265–289.

Edmunds, P. J., and R. C. Carpenter. 2001. Recovery of *Diadema* leads to reduced macroalgal cover and increased abundance of juvenile coral on a Caribbean reef. *Proc. Natl. Acad. Sci. USA* 98: 5067–5071.

Ellison, A. M., and B. J. Farnsworth. 1990. The ecology of Belizean mangrove root-fouling communities: Epibenthic fauna are barriers to isopod attack. *J. Experimental Marine Biology and Ecology* 142:91–104

Ellison, A. M., and B. J. Farnsworth. 1996. Anthropogenic disturbance of Caribbean mangrove ecosystems: Past impacts, present trends and future predictions. *Biotropica* 28: 549–565.

Ellison, A. M., B. J. Farnsworth, and R. E. Merkt. 1999. Origins of mangrove ecosystems and mangrove biodiversity anomaly. *Global Ecology and Biogeography Letters* 8: 95–115.

Ellison, A. M., B. J. Farnsworth, and R. R. Twilley. 1997. Facultative mutualism between red mangroves and root-fouling sponges in Belizean mangal. *Ecology* 77: 2431–2444.

Endean, R. 1973. Population explosions of *Acanthaster planci* and associated destruction of hermatypic corals in the Indo-West Pacific region. In *Biology and geology of coral reefs*, edited by O. A. Jones and R. Endean. Vol. II, *Biology 1*. New York: Academic Press, 390–438.

Endean, R., and A. M. Cameron. 1990. *Acanthaster planci* population outbreaks. In *Ecosystems of the world* 25, *Coral reefs*, edited by Z. Dubinsky. New York: Elsevier, 419–437.

Estevez, E. D., and J. L. Simon. 1975. Systematics and ecology of *Sphaeroma* (Crustacea, Isopoda) in the mangrove habitat of Florida. In *Proceedings of the international symposium on the biology and management of mangroves*, edited by G. Walsh, S. Snedaker, and H. Teas. Gainesville: University of Florida, 286–304.

Fadlallah, Y. H. 1983. Sexual reproduction, development and larval biology in scleractinian corals. A review. *Coral Reefs* 2:129–150.

Farnsworth, E. J., and A. M. Ellison. 1997a. Global patterns of pre-dispersal propagule predation in mangrove forests. *Biotropica* 29: 316–330.

Farnsworth, E. J., and A. M. Ellison. 1997b. The global conservation status of mangroves. *Ambio* 26:328–334.

Glynn, P. W. 1972. Observations on the ecology of the Caribbean and Pacific coasts of Panama. In *The Panamic biota: Some observations prior to a sea level canal*, edited by M. L. Jones. *Bull. Biol. Soc. Wash.* 2:13–30.

Glynn, P. W. 1974. Rolling stones among the Scleractinia: Mobile coraliths in the Gulf of Panama. *Proc. 2nd Int. Coral Reef Symp.* 2:183–198.

Glynn, P. W. 1976. Some physical and biological determinants of coral community structure in the eastern Pacific. *Ecol. Monogr.* 46:431–456.

Glynn, P. W. 1984. Widespread coral mortality and the 1982–83 El Niño warming event. *Envir. Conserv.* 11:133–146.

Glynn, P. W. 1985. Corallivore population sizes and feeding effects following El Niño (1982–1983) associated coral mortality in Panama. *Proc. 5th Int. Coral Reef Symp.* 2:149.

Glynn, P. W. 1988. El Niño–Southern Oscillation 1982–1983: Nearshore population, community and ecosystem responses. *Ann. Rev. Ecol. Syst.* 19:309–345.

Glynn, P. W. 1990. Feeding ecology of selected coral reef macroconsumers: Patterns and effects on coral reef community structure. In *Ecosystems of the world* 25, *Coral reefs*, edited by Z. Dubinsky. New York: Elsevier, 365–400.

Glynn, P. W. 1997. Bioerosion and coral reef growth: A dynamic balance. In *The life and death of coral reefs*, edited by C. Birkeland. New York: Chapman and Hall.

Glynn, P. W., and G. M. Wellington. 1983. *Corals and coral reefs of the Galápagos Islands*. Berkeley: University of California Press.

Glynn, P. W., G. M. Wellington, and C. Birkeland. 1979. Coral reef growth in the Galápagos: Limitations by sea urchins. *Science* 203:47–48.

Goldman, B., and F. Talbot. 1976. Aspects of the ecology of coral reef fishes. In *Biology and geology of coral reefs*, edited by O. A. Jones and R. Endean. Vol. III, *Biology 2*. New York: Academic Press, 125–154.

Goreau, T. F. 1961. On the relation of calcification to primary production in reef building organisms. In *The biology of Hydra, etc.*, edited by H. M. Lenhoff and W. F. Loomis. Coral Gables, FL: University of Miami Press, 269–285.

Goreau, T. F., and J. W. Wells. 1967. The shallow water Scleractinia of Jamaica: Revised list of species and their vertical distribution range. *Bull. Mar. Sci.* 17(2):442–453.

Goreau, T. F., N. I. Goreau, and T. J. Goreau. 1979. Corals and coral reefs. *Sci. Amer.* 241(2):124–136.

Goreau, T. J. 1990. Coral bleaching in Jamaica. *Nature* 343:417.

Grigg, R. W. 1983. Community structure, succession and development of coral reefs in Hawaii. *Mar. Ecol. Progr. Ser.* 11:1–14.

Hatcher, B. G. 1983. Grazing in coral reef ecosystems. In *Perspectives on coral reefs*, edited by D. J. Barnes. Townsville, Australia: AIMS, 169–179.

Hay, M. E. 1984. Patterns of fish and urchin grazing on Caribbean coral reefs: Are previous results typical? *Ecology* 65:446–454.

Hay, M. E. 1991. Fish–seaweed interactions on coral reefs: Effects of herbivorous fishes and adaptations of their prey. In *The ecology of fishes on coral reefs*, edited by P. Sale. New York: Academic Press, 96–119.

Hiatt, R. W., and D. W. Strasburg. 1960. Ecological relationship of the fish fauna on coral reefs of the Marshall Islands. *Ecol. Monogr.* 30:65–127.

Highsmith, R. C. 1982. Reproduction by fragmentation in corals. *Mar. Ecol. Progr. Ser.* 7:207–226.

Hobson, E. S. 1968. *Predatory behavior of some shore fishes in the Gulf of California*. Bur. Sports Fish. and Wildlife (U.S.), Res. Rept. 73.

Hogarth, P. J. 1999. *The biology of mangroves*. Oxford: Oxford University Press.

Hughes, T. P. 1994. Catastrophic phase shifts and large scale degradation of a Caribbean coral reef. *Science* 265:1547–155.

Hughes, T. P., and 16 others. 2003. Climate change, human impacts, and the resilience of coral reefs. Science 301:929–933.

Huston, M. A. 1985. Patterns of species diversity on coral reefs. *Ann. Rev. Ecol. Syst.* 16:149–177.

Hutchings, P. A. 1983. Cryptofaunal communities of coral reefs. In *Perspectives on coral reefs*, edited by D. J. Barnes. Townsville, Australia: AIMS, 200–208.

Hutchings, P. A. 1986. Biological destruction of coral reefs: A review. *Coral Reefs* 4:239–252.

Jackson, J. B. C. 1977. Competition on marine hard substrata: The adaptive significance of solitary and colonial strategies. *Amer. Nat.* 111:743–767.

Jackson, J. B. C. 1979. Overgrowth competition between encrusting cheilostome ectoprocts in a Jamaican cryptic reef environment. *J. Anim. Ecol.* 48:805–823.

Jackson, J. B. C., and T. P. Hughes. 1985. Adaptive strategies of coral-reef invertebrates. *Amer. Sci.* 73:265–274.

Jackson, J. B. C. and 12 others. 2001. The future of coral reefs. *Science* 293: 629–634.

Johannes, R. E. 1974. Sources of nutritional energy for reef corals. *Proc. 2nd Int. Coral Reef Symp.* 1:133–137.

Johannes, R. E., N. T. Kuenzel, and S. L. Coles. 1970. The role of zooplankton in the nutrition of some scleractian corals. *Limn. Oceanog.* 15:579–586.

Jones, G. P., M. J. Milicich,, M. J. Emslie, and C. Lunow. 1999. Self-recruitment in a coral reef fish populations. *Nature* 401: 802–804.

Jones, O. A., and R. Endean, eds. 1973. *Biology and geology of coral reefs*. Vol. II, *Biology 1*. New York: Academic Press.

Jones, O. A., and R. Endean, eds. 1976. *Biology and geology of coral reefs*. Vol. III, *Biology 2*. New York: Academic Press.

Karlson, R. H. 1980. Alternative competitive strategies in a periodically disturbed habitat. *Bull. Mar. Sci.* 30:118–131.

Karlson, R. H. 1999. *Dynamics of coral communities*. Boston: Kluwer Academic Publishers.

Kaufman, L. 1977. The three spot damselfish: Effects on benthic biota of Caribbean coral reefs. *Proc. 3rd Int. Coral Reef Symp.* 1:559–564.

Knowlton, N., J. C. Lang, M. C. Rooney, and P. Clifford. 1981. Evidence for delayed mortality in hurricane damaged Jamaican staghorn corals. *Nature* 294:251–252.

Kuenzler, E. J. 1974. Mangrove swamp systems. In *Coastal ecological systems*, edited by H. T. Odum, B. J. Copeland, and E. A. McMahon. Vol. 1. Washington, DC: Conservation Foundation, 346–371.

Ladd, H. S., E. Ingerson, R. C. Townshend, M. Russell, and H. K. Stephenson. 1953. Drilling on Eniwetok Atoll, Marshall Islands. *Bull. Amer. Assoc. Petrol. Geol.* 37:2257–2280.

Lang, J. C. 1973. Interspecific aggression by scleractinian corals, 2. Why the race is not only to the swift. *Bull. Mar. Sci.* 23(2):260–279.

Lang, J. C., and E. A. Chornesky. 1990. Competition between scleractinian reef corals—a review of mechanisms and effects. In *Ecosystems of the world* 25, *Coral reefs*, edited by Z. Dubinsky. New York: Elsevier, 209–252.

Larkham, A. W. D. 1983. The primary productivity of plant communities on coral reefs. In *Perspectives on coral reefs*, edited by D. J. Batches. Townsville, Australia: AIMS, 221–230.

Lasker, H.R. 1981. Phenotypic variation in the coral *Montastrea cavernosa* and its effects on colony energetics. *Biol. Bull.* 160:292–302.

Lessios, H. A. 1988. Mass mortality of *Diadema antillarum* in the Caribbean: What have we learned? *Ann. Rev. Ecol. Syst.* 19:371–394.

Lessios, H. A., P. W. Glynn, and D. R. Robertson. 1983. Mass mortalities of coral reef organisms. *Science* 222:715.

Lobel, P. S. 1980. Herbivory by damselfishes and their role in coral reef community ecology. *Bull. Mar. Sci.* 30:273–289.

Lowe-McConnell, R. H. 1977. *Ecology of fishes in tropical waters*. Studies in biology no. 76. London: Edward Arnold.

Loya, Y. 1976. Recolonization of Red Sea corals affected by natural catastrophes and man-made perturbations. *Ecology* 57(2):278–289.

Luckhurst, B. E., and K. Luckhurst. 1978. Analysis of the influence of substrate variables on coral reef fish communities. *Mar. Biol.* 49:317–323.

Lugo, A. E., and S. C. Snedaker. 1974. The ecology of mangroves. *Ann. Rev. Ecol. Syst.* 5:39–64.

McMichael, D. F. 1974. Growth rate, population size and mantle coloration in the small giant clam *Tridacna maxima* (Roding) at One Tree Island, Capricorn Group, Queensland. *Proc. 2nd Int. Coral Reef Symp.* 1:241–254.

McNae, W. 1968. A general account of the fauna and flora of mangrove swamps and forests in the IndoWest Pacific region. *Adv. Mar. Biol.* 6:73–270.

Meyer, J. L., E. T. Schultz, and G. S. Helfman. 1983. Schools: An asset to corals. *Science* 220:1047–1048.

Motoda, S. 1940. The environment and life of massive reef coral, *Goniastrea aspera* Verrill, inhabiting the reef flat in Palao. *Palao Tropical Biol. Sta. Studies* 2:61–104.

Muscatine, L., and E. Cernichiari. 1969. Assimilation of photosynthetic products of zooxanthellae by a reef coral. *Biol. Bull.* 137:506–523.

Neudecker, S. 1977. Transplant experiments to test the effect of fish grazing on coral distribution. *Proc. 3rd. Int. Coral Reef Symp.* 1:317–323.

Newel, N. D. 1972. The evolution of reefs. *Sci. Amer.* 226(6):54–65.

Odum, H. T., B. J. Copeland, and E. A. McMahan, eds. 1974. *Coastal ecological systems of the United States*. Vol. I: B, *Natural tropical ecosystems of high diversity*. Washington, DC: Conservation Foundation, 346–514.

Odum, W. E., C. C. McIvor, and T. J. Smith, III. 1982. *The ecology of the mangroves of South Florida: A community profile*. U.S. Fish and Wildlife Service, Office of Biological Services FWS/OBS-81/24.

Ogden, J. C., and P. S. Lobel. 1978. The role of herbivorous fishes and urchins in coral reef communities. *Envir. Biol. Fishes* 3:49–63.

Onuf, C. P., J. M. Teal, and I. Valiela. 1977. Interactions of nutrients, plant growth and herbivory in a mangrove ecosystem. *Ecology* 58:514–526.

Pearson, R. G. 1975. Coral reefs, unpredictable climatic factors and *Acanthaster*. Crown-of-thorns starfish seminar proceedings. *Aust. Govt. Pub. Ser. Canberra*, 131–134.

Pandolfi, J. M., et al. 2003. Global trajectories of the long-term decline of coral reef ecosystems. *Science* 301:955–956.

Perry, D. M. 1984. Direct and indirect regulatory effects by root fauna on the growth rate of prop roots in the red mangrove. *Amer. Zool.* 24(3):126.

Perry, D. M. 1988. Effects of associated fauna on growth and productivity in the red mangrove. *Ecology* 69(4):1064–1075.

Porter, J. 1972. Predation by *Acanthaster* and its effect on coral species diversity. *Amer. Nat.* 106(950):487–492.

Porter, J. 1974. Zooplankton feeding by the Caribbean reef-building coral *Montastrea cavernosa*. *Proc. 2nd Int. Coral Reef Symp.* 1:111–125.

Porter, J. 1976. Autotrophy, heterotrophy and resource partitioning in Caribbean reef building corals. *Amer. Nat.* 110:731–742.

Potts, D. C. 1977. Suppression of coral populations by filamentous algae within damselfish territories. *J. Exp. Mar. Biol. Ecol.* 28:207–216.

Potts, D. C. 1981. Crown-of-thorns starfish—man induced pest or natural phenomenon? In *The ecology of pests*, edited by R. L. Kitching and R. E. Jones. Melbourne: CSIRO Publications Service, 55–86.

Rabinowitz, D. 1975. Planting experiments in mangrove swamps of Panama. In *Proceedings of the international symposium on the biology and management of mangroves*, edited by G. Walsh, S. Snedaker, and H. Teas. Gainesville: University of Florida, 385–393.

Rabinowitz, D. 1978. Dispersal properties of mangrove propagules. *Biotropica* 10:47–57.

Rehm, A. E., and H. J. Humm. 1973. *Sphaeroma terebrans*: A threat to the mangroves of southeastern Florida. *Science* 182:173–174.

Richardson, C. A., P. Dustan, and J. C. Lang. 1979. Maintenance of living space by sweeper tentacles of *Montastrea cavernosa*, a Caribbean reef coral. *Mar. Biol.* 55:181–186.

Risk, M. J. 1972. Fish diversity on a coral reef in the Virgin Islands. *Atoll Res. Bull.* 153:1–6.

Robertson, D. R., H. P. Sweatman, E. Fletcher, and M. Cleland. 1976. Schooling as a mechanism for circumventing the territoriality of competitors. *Ecology* 57: 1208–1220.

Rowan, R. 1998. Diversity and Ecology of zooxanthellae on coral reefs. *J. Phycology* 34: 407–417.

Rowan, R., N. Knowlton, A. Baker, and J. Jara. 1997. Landscape ecology of algal symbionts creates variation in episodes of coral bleaching. *Nature* 388:265–269.

Sale, P. F. 1977. Maintenance of high diversity in coral reef fish communities. *Amer. Nat.* 111:337–359.

Sale, P. F. 1980. The ecology of fishes on coral reefs. *Oceanog. Mar. Biol. Ann. Rev.* 18:367–421.

Sale, P. F., ed. 1991. *The ecology of fishes on coral reefs.* New York: Academic Press.

Salvat, B., and G. Richard. 1985. Takapoto atoll, Tuamotu archipelago. In *Fifth international coral reef congress, Tahiti 27 May–1 June, 1985.* Vol 1: *French Polynesian coral reef*, edited by B. Delesalle, R. Galzin, and B. Salvat. Antenne Museum Ephe: Moorea, French Polynesia, 323–378.

Sammarco, P. W. 1980. *Diadema* and its relationship to coral spat mortality: Grazing, competition and biological disturbance. *J. Exp. Mar. Biol. Ecol.* 45:245–272.

Sammarco, P. W. 1985. The Great Barrier Reef versus the Caribbean: Comparisons of grazers, coral recruitment patterns, and reef recovery. *Proc. 5th Int. Coral Reef Symp.* 4:391–397.

Sammarco, P. W. 1987. A comparison of some ecological processes on coral reefs of the Caribbean and the Great Barrier Reef. *Unesco Rept. Mar. Sci.*, no. 46: 127–166.

Sammarco, P. W., and J. C. Coll. 1992. Chemical adaptations in the Octocorallia: Evolutionary considerations. *Mar. Ecol. Prog. Ser.* 88:93–104.

Sammarco, P. W., J. S. Levinton, and J. C. Ogden. 1974. Grazing and control of coral reef community structure by *Diadema antillarum* Philippi (Echinodermata: Echinoidea): A preliminary study. *J. Mar. Res.* 32(1):47–53.

Santilan, N., and R. J. Williams. (1999). Mangrove transgression into salt marsh environments in south-east Australia. *Global Ecology and Biogeography*, 8:117–124.

Scoffin, T.P. 1972.The fossilization of Bermuda patch reefs. *Science* 178:1280–1282.

Scoffin, T. P., C. W. Stearn, D. Boucher, P. Frydl, C. M. Hawkins, I. G. Hunter, and J. K. MacGeachy. 1980. Calcium carbonate budget of a fringing reef on the west coast of Barbados. II, Erosion sediments and internal structure. *Bull. Mar. Sci.* 30:475–528.

Simberloff, D., B. J. Brown, and S. Lowrie. 1978. Isopod and insect root borers may benefit Florida mangroves. *Science* 201:630–632.

Smith, C. L., and J. C. Tyler. 1972. Space resource sharing in a coral reef fish community. *Bull. Nat. Hist. Mus. Los Angeles County* 14:125–170.

Smith, T. J. 1992. Forest structure. In *Tropical Mangrove Ecosystems*, edited by A. I. Robertson and D. M. Alongi, pp 101–136. American Geophysical Union, Washington, DC.

Sorokin, Yu. I. 1990. Aspects of trophic relations, productivity and energy balance in coral reef ecosystems.

In *Ecosystems of the world* 25, *Coral reefs*, edited by Z. Dubinsky. New York: Elsevier, 401–410.

Sorokin, Yu. I. 1993. *Coral reef ecology*. Ecological studies, vol. 102. New York: Springer.

Sousa, W. P., and B. J. Mitchell. 1999. The effect of seed predators on plant distributions: Is there a general pattern in mangroves? *Oikos* 86:55–66.

Stearn, C. W., and T. P. Scoffin. 1977. Carbonate budget of a fringing reef, Barbados. *Proc. 3rd Int. Coral Reef Symp.* 2:471–476.

Stehli, F. G., and J. W. Wells. 1971. Diversity and age patterns in hermatypic corals. *Sys. Zool.* 20:115–126.

Stoddard, D. R. 1969. Ecology and morphology of recent coral reefs. *Biol. Rev.* 44(4):433–498.

Sutherland, J. P. 1980. Dynamics of the epibenthic community of roots of the mangle, *Rhizophora mangle*, at Bahia de Buche, *Venezuela. Mar. Biol.* 58:75–84.

Swearer, S. E., J. E. Caselle, D. W. Lea, and R. R. Warner. 1999. Lartval retention and recruitment in an island population of a coral reef fish. *Nature* 402: 799–802.

Szmant-Froelich, A. 1985. Reproduction and recruitment of corals: Conclusion. *Proc. 5th Int. Coral Reef Symp.* 4:399–400.

Talbot, F. H., B. C. Russell, and G. R. V. Anderson. 1978. Coral fish communities: Unstable, high diversity systems? *Ecol. Monogr.* 48:425–440.

Taylor, P., M. Litter, and D. Litter. 1986. Escapes from herbivory in relation to the structure of mangrove island macroalgae communities. *Oceologia* 69:481–490.

Thom, B. G. 1975. Mangrove ecology from a geomorphic viewpoint. In *Proceedings of the international symposium on the biology and management of mangroves*, edited by G. Walsh, S. Snedaker, and H. Teas. Gainesville: University of Florida, 469–481.

Tschirley, F. H. 1969. Defoliation in Vietnam. *Science* 163:779–786.

Underwood, A. J., and G. Barrett. 1990. Experiments on the influence of oysters on the distribution, abundance and sizes of the gastropod *Bembicum auraturn* in a mangrove swamp in New South Wales, Australia. *J. Exp. Mar. Biol. Ecol.* 137(1):25–45.

Vaughn, T. W. 1915. The geologic significance of the growth rate of the Floridian and Bahamian shoalwater corals. *J. Wash. Acad. Sci.* 5:591–600.

Vermeij, G. J. 1978. *Biogeography and adaptation*. Harvard University Press, Cambridge, MA.

Vermeij, G. J. 1987. Interocean differences in architecture and ecology: The effects of history and productivity. *Unesco Rept. Mar. Sci.*, no. 46:105–126.

Veron, J. E. N. 1986. Distribution of reef-building corals. *Oceanus* 29:27–31.

Victor, B. C. 1983. Recruitment and population dynamics of coral reef fish. *Science* 219:419–420.

Wada, K., and D. Wower. 1989. Foraging on mangrove pneumatophores by ocypodid crabs. *J. Exp. Mar. Biol. Ecol.* 134(2):89–100.

Wahle, C. M. 1980. Detection, pursuit and overgrowth of tropical gorgonians by milleporid hydrocorals: Perseus and Medusa revisited. *Science* 209:689–691.

Walbran, P. D., R. A. Henderson, A. J. Timothy Jull, and M. J. Head. 1989. Evidence from sediments of long-term *Acanthaster planci* predation on corals of the Great Barrier Reef. *Science* 245:847–850.

Walsh, G. E. 1974. Mangroves, a review. In *Ecology of halophytes*, edited by R. J. Reimold and W. J. Queen. New York: Academic Press, 51–174.

Wellington, G. W. 1980. Reversal of digestive interactions between Pacific reef corals: Mediation by sweeper tentacles. *Oecologia* 47:340–343.

Wellington, G. W. 1982a. Depth zonation of corals in the Gulf of Panama: Control and facilitation by resident reef fishes. *Ecol. Monogr.* 52:223–241.

Wellington, G. W. 1982b. An experimental analysis of the effects of light and zooplankton on coral zonation. *Oecologia* 52:311–320.

Wells, J. W. 1954. *Recent corals of the Marshall Islands, Bikini and nearby atolls*. U.S. Geol. Survey Paper 260-I:385–486.

Wells, J. W. 1957. Coral reefs. In *The treatise on marine ecology and paleoecology*. Vol. I, *Ecology*. Geol. Soc. of Amer. Memoir 67, 609–631.

West, R. C. 1977. Tidal salt marsh and mangal formations of middle and south America. In *Wet coastal ecosystems*, edited by V. J. Chapman. New York: Elsevier, 193–213.

Wilkinson, C. 1987a. Interocean differences in size and nutrition of coral reef sponges. *Science* 236:1654–1657.

Wilkinson, C. 1987b. Sponge biomass as an indication of reef productivity in two oceans. *Unesco Rept. Mar. Sci.*, no. 46:99–104.

Wilkinson, C., ed. 2002. *Status of coral reefs of the world*. Australian Institute of Marine Science, Townsville, Australia.

Williams, E. H., and L. B. Williams. 1988. Caribbean marine mass mortality: A problem with a solution. *Oceanus* 30(4): 69–75.

Wulff, J. L. 1997. Mutualisms among species of coral reef sponges. *Ecology* 78: 146–159.

Wulff, J. L., and L. W. Buss. 1979. Do sponges help hold coral reefs together? *Nature* 281: 474–475.

Yentsch, C. S. 1966. Primary production. In *The encyclopedia of oceanography*, edited by R. W. Fairbridge. New York: Van Nostrand, 722–725.

Yonge, C. M. 1963. The biology of coral reefs. *Adv. Mar. Biol.* 1:209–260.

CHAPTER 10

SYMBIOTIC RELATIONSHIPS

One of the most striking features of marine organisms is the large number of close associations among many different species. These are a series of associations that are neither predator–prey nor herbivore–plant relationships in which one member consumes the other. These close relationships between unlike species generally are either harmless to either member or, more likely, beneficial to one or both. **Symbiosis**, the name given to such associations, means an interrelationship between two different species. Symbiotic relationships are found in the terrestrial environment as well as in the aquatic, but they appear disproportionately common, extensive, and well developed in the marine environment, such that they merit separate coverage in this volume.

■

■ COVERAGE AND DEFINITIONS

Symbiotic relationships cover a broad spectrum of associations, from random, casual, or facultative associations, through ever more obligatory groupings that benefit one or both members, and finally to those relationships which are parasitic. Such differences in the degree of association have led to the subdivision of symbiosis into more narrowly defined groupings. The term **commensalism** is often used to refer to an association that is clearly to the advantage of one member while not harming the other. **Inquilinism** is a special subdivision of commensalism in which one animal lives in the home of another, or in its digestive tract, without being parasitic. **Mutualism** is that form of symbiosis in which two species associate for their mutual benefit. In a mutualistic relationship, the partners are often called **symbionts.** In a commensal or inquiline relationship, however, the partner gaining advantage is called the **commensal** and the other the host.

The term *parasitism* generally refers to an association in which one species lives in or on another and draws nourishment from that species at its expense—that is, to its detriment. In other words, it is an association in which the advantage is solely to one member at the expense of the other. Parasitism is an extremely common association in both marine and terrestrial environments. It will not be discussed in this chapter, since it contains few features unique to the marine environment.

There are two broad groupings of nonparasitic symbiotic associations in the sea: those between algal cells and various invertebrate animals and those between various animals, both vertebrate and invertebrate. We shall consider these groupings in that order. A final category is that between various bacteria and animals, which has become particularly well publicized through the discoveries of organisms living in deep-sea thermal-vent systems. These symbioses were covered in Chapter 4.

Symbioses Among Algae and Animals

Most known symbiotic relationships in the sea between autotrophic organisms and animals are between unicellular algae or their chloroplasts and a wide variety of marine invertebrate animals. These symbiotic relationships are most common in tropical waters, but they also are prevalent in temperate oceans. They are, however, virtually absent from polar waters. Furthermore, the associations are, for obvious reasons, restricted to the shallow subtidal or intertidal areas, or to the uppermost layers of the pelagic realm where sufficient light for photosynthesis is present.

■ TYPES AND COMPOSITION OF THE ASSOCIATIONS

There are basically two types of symbiotic associations between autotrophs and invertebrates. The more common is to have the entire functioning autotrophic cell associated with the invertebrate animal. The second is to have only the functioning chloroplasts from the algal cells incorporated into the tissues of the invertebrate body.

The algal-cell symbionts have typically been classified into groups on the basis of their color. **Zooxanthellae** is the name given to brown, golden, or brownish-yellow cells, **zoochlorellae** to those which are green. A third, smaller group is blue or bluish green and has been called **cyanellae** (Cyanobacteria) (Table 10.1). These color groups, however, do not distinguish among the actual autotrophic species involved. Zooxanthellae, the most common in the seas of the world, are primarily species of dinoflagellates (Figure 10.1), but include a few diatoms and cryptophyceans, haptophyceans, and chrysophyceans. Schoenberg and Trench (1976) reported the most common zooxanthellae species to be the dinoflagellate (Dinophyceae) *Symbiodinium (Gymnodinium) microadriaticum*. In the past, the difficulty in identifying these symbiotic algae with traditional methods led researchers to embrace concepts as disparate as that of a single zooxanthellae species in all hosts to one according to which there is a different species in each host. With the use of modern molecular techniques, however—primarily RNA sequencing—Rowan and Powers (1991) were able to recognize 10 algal taxa among 22 hosts sampled. The zoochlorellae are much less common in the sea, but dominate symbiotic associations in fresh water. Among the marine associations, zoochlorellae include the algae *Platymonas (Tetraselmis) convolutae*, *Pedinomonas* of the Prasinophyceae, and *Chlorella* and *Chlamydomonas* of the Chlorophyceae. Cyanellae are all cyanobacteria (formerly called blue-green algae) and include the genera *Aphanocapsa* and *Phormidium*, and probably others as yet unidentified (Table 10.1). Cyanellae are most common as symbionts in sponges and planktonic diatoms.

The symbionts all occur inside the bodies of the animals, and most are found either within vacuoles

TABLE 10.1

Summary of the Types of Associations between Algae and Marine Invertebrates

ALGAL SYMBIONT GROUP	ALGAL TAXON	INVERTEBRATE ANIMAL HOST TAXA
Zooxanthellae	Dinophyceae	Prostista, Porifera, Cnidaria, Platyhelminthes, Mollusca
	Haptophyceae	Protista
	Chrysophyceae	Protista
	Bacillariophyceae	Platyhelminthes (*Convoluta convoluta*)
	Cryptophyceae	Protozoa
Zoochlorellae	Prasinophyceae	Platyhelminthes (*Convoluta roscoffensis*), Protista
	(?) Chlorophyceae	Cnidaria, Protista
Cyanellae	Cyanophyceae	Porfera, Protista, Echiura
Prochlorophyta	*Prochloron*	Urochordata
Chloroplasts	Chlorophyceae	Mollusca (Sacoglossa)

Source: Modified from "Symbiosis of Algae with Invertebrates" by D. C. Smith in *Oxford Biology Reader*, No. 43, p. 16.

FIGURE 10.1 Symbiotic zooxanthellae (dinoflagellates) from the anemone *Anthopleura xanthogrammica*. (Copyright Scott R. Santos.)

inside individual tissue cells or in various body spaces between or within tissue layers. Generally, the autotroph cells are restricted to certain tissues or areas of the host, and they grow and reproduce within the invertebrate without being digested.

Among those associations which involve only the chloroplasts and the invertebrate body, the chloroplasts are usually derived from the cells of larger green algae (*Codium, Caulerpa, Cladophora, Bryopsis*) first ingested by the animal in feeding and subsequently transferred from the digestive tract of the animal to other tissues.

Autotrophic symbiosis, either cellular or via chloroplasts, is widespread among various invertebrate groups. Symbiotic associations occur in Protista, Porifera, Cnidaria, Platyhelminthes, Mollusca, and Echiura. Within these phyla, report McLaughlin and Zahl (1966), about 130 genera have algal symbionts. Autotrophic symbiosis is most common among Protista and Cnidaria.

Although most algal symbionts can be readily identified in their hosts, it has been recently discovered that a number of protistans previously considered photosynthetic, such as the ciliate *Mesodinium rubrum* and the dinoflagellates *Peridinium balticum* and *Kryptoperidinium foliaceum*, are, in fact, in symbiotic associations in which the algal cells have lost their cellular integrity and do not appear as separate from the host body. These associations are examples of **cryptic symbiosis**.

■ ORIGIN OF THE ASSOCIATION

Although we will probably never know the exact origin of algal–invertebrate symbiotic relationships, it is possible to outline some suggestions. In all the phyla mentioned that contain symbiotic algae or chloroplasts, the final phase of digestion is intracellular. Individual cells take up particles from the stomach. It is thus possible to conceive of this association as arising through the ingestion by cells of the digestive tract of either intact algal cells or chloroplasts, taken in by the animal during feeding. If, then, the algal cells were resistant to digestive action or the animal lacked enzymes to digest plant cellulose, the basis for a new association would be laid. This association could then evolve into a more obligatory relationship, provided that the new association conveyed an enhanced survival value to both symbionts. Since these associations are common, they must have some selective value. (See pp. 482–483.)

■ DISTRIBUTION OF AUTOTROPH–INVERTEBRATE ASSOCIATIONS

Although there is neither space nor time to list and discuss all the species in which autotrophic symbionts occur, it is of use to discuss briefly the various groups of organisms in which this association is prevalent.

Beginning with protistans, symbiotic relationships are found in all the epipelagic planktonic Radiolaria, where the zooxanthellae occur in the outer frothy layer

FIGURE 10.2
Zooxanthellae and
zoochlorellae in various invertebrate taxa. (a) Zooxanthellae in the radiolarian
Acanthometra pellucida (left)
and *Sphaerozoum acuferum*
(right). (b) Zoochlorellae in
the acoel flatworm *Convoluta
roscoffensis*. (After *The Biology
of Marine Animals*, J. A. Colin
Nicol, 1960, Pittman Books.)

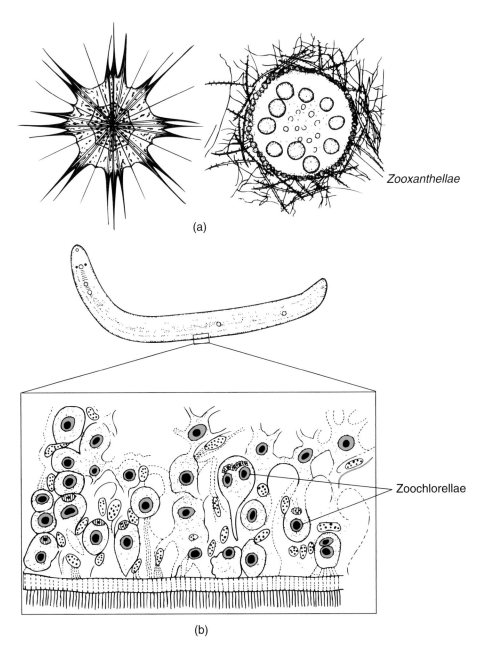

Zooxanthellae

(a)

Zoochlorellae

(b)

(Figure 10.2). Symbiotic relationships also are found in a number of planktonic Foraminifera (*Globigeri-noides*), in all the larger benthic species (*Elphidium, Nonion, Heterostregina*), and even in marine ciliates (*Paraeuplotes, Trichodina*).

Symbiosis in sponges is particularly important in species that inhabit coral reefs. The association is usually with cyanobacteria. According to Wilkinson (1983), sponges with such symbionts constitute 80% of the sponge species and biomass on certain reefs of the Great Barrier Reef. Zooxanthellae occur in *Cliona* and cyanellae in Demospongia.

Among the Cnidaria, symbiosis is extremely common, especially among tropical species. Virtually all tropical, shallow-water anemones, soft corals, sea fans, whips, and stony corals have symbiotic zooxanthellae in their tissues. Even certain tropical jellyfish (*Cassiopeia, Mastigius*) contain zooxanthellae.

Granule

Chloroplast

Nucleus

(a)

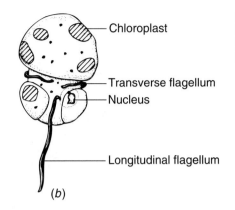

Chloroplast

Transverse flagellum
Nucleus

Longitudinal flagellum

(b)

FIGURE 10.3
Symbiotic (a) and free-living (b) zooxanthellae cells.

In temperate seas, the incidence is not as great, although many anemones, such as *Anthopleura*, retain the symbionts.

A peculiar symbiosis occurs in some compound ascidians (*Didemnum, Trididemnum*), wherein an unusual prokaryotic unicellular cyanobacterium called *Prochloron*, which combines features of both cyanobacteria and green algae, is embedded in the test or in the cloacal system.

Whereas a few symbiotic relationships occur in Ctenophora (*Beröe*), Annelida (*Eunice*), and Echinodermata (*Ophioglypha*), the remaining major reservoirs of symbiotic relationships are found in the Platyhelminthes and Mollusca. Among marine flatworms, the classic case is that of the genus *Convoluta*, which has been extensively investigated. Symbiotic relationships in the Mollusca are generally restricted to certain marine gastropods of the order Sacoglossa, in which the animals retain only the chloroplasts, and to the bivalve family Tridacnidae and the genera *Corculum* and *Fragum*.

■ MODIFICATIONS RESULTING FROM THE ASSOCIATION

The symbiotic association between the algae and the invertebrates is generally a very close one (mutualistic) and has resulted in significant anatomical and physiological changes in both the algal cells and the various invertebrate hosts.

Perhaps the most profound changes that occur as a result of the association are found in the algal cells. Most marine symbiotic algae are dinoflagellates. The symbionts, however, have lost their locomotory flagellae and the characteristic grooves around their body. Furthermore, their cell walls are much reduced in thickness. The zoochlorellae in the flatworm *Convoluta*

have lost even more, in that the cell wall disappears, as do the light-sensitive stigmata. These cells become little more than bags containing chloroplasts.

If the algal cells are removed from the host and are cultured outside the animal, they will develop the characteristic flagellae, cell walls, and other organs of a typical free-living form. It is apparent that the changes observed are a direct result of the symbiotic association (Figure 10.3).

In the case of the animal hosts, the changes vary with the type of organism and with the degree of interdependence established between the symbionts. The single universal modification is that all of these invertebrates live in shallow water, where they can obtain adequate light, so that the algae can carry on photosynthesis. Perhaps the least amount of modification occurs in the lower invertebrates, such as protistans and Porifera, in which the algae occur as simple inclusions in the cytoplasm of the animal. According to Wilkinson (1983), among sponges, those with symbiotic cyanobacteria enhance their light-gathering ability by morphological flattening, thereby increasing the area available to intercept light.

Among the Cnidaria, definite modifications begin to become apparent. In marine Cnidaria, the algae occur in the innermost of the cell layers, the gastrodermis (endoderm). The numbers of zooxanthellae in different corals vary, and those corals with the most seem to have a reduced tentacle size, indicating that they are less able to capture zooplankton food. Among certain soft corals of the family Xeniidae, the digestive regions of the animal are reduced and the animals are not responsive to animal food. In the jellyfish *Cassiopeia*, a behavioral modification is striking. These animals, rather than swimming in open waters as do most jellyfishes, lie upside down on the bottom in

FIGURE 10.4 *Cassiopeia* jellyfish in normal resting position upside down on the sea bottom. The brownish color is due to the symbiotic zooxanthellae. (Photo courtesy of Diane Nelson.)

shallow tropical waters, exposing their oral arms to the light to illuminate their algae (Figure 10.4). The oral arms are also much enlarged and expanded, to provide more area for habitation by the algal cells—truly a strange way of life for a jellyfish!

Among flatworms, the most studied case of symbiosis is that of *Convoluta roscoffensis*, a small animal inhabiting sand beaches on the Brittany coast of France. The alga inhabiting *Convoluta* is *Platymonas* (*Tetraselmis*) *convolutae*, which also occurs free living in the same area. *Convoluta* exhibit perhaps the most profound behavioral and life-history modifications in response to the alga, but little in the way of anatomical changes (Figure 10.5).

Convoluta roscoffensis live burrowed in the sand of beaches near the upper reaches of the tidal zone. Whenever the tide is in, the animals are found buried. When the tide recedes, the animals move up onto the surface of the sand, where they spread out to expose their symbionts to the light. When the tide begins to return, the vibrations trigger a burrowing response, so the animals are safely under the sand again before the rising tidal waters can sweep them out to sea. Even more significantly, as Smith (1973) notes, these animals apparently do not feed as adults. Furthermore, if the young worms do not ingest the algal symbionts upon hatching, they will not complete their development and will die even if they feed.

It is among the mollusks, however, that we find the most dramatic changes of all resulting from the association. Although there are thousands of species of

FIGURE 10.5 *Convoluta roscoffensis*, showing the green color of the symbiotic zoochlorellae. (Photo courtesy of D. P. Wilson/Science Source/Photo Researchers, Inc.)

mollusks in the seas of the world (mollusks are the second-largest phylum after the arthropods), symbiotic associations with algal cells have developed in only nine species. All are bivalve mollusks, seven in a single family, the Tridacnidae, which includes the giant clams. Six of the species are in the genus *Tridacna* itself, and the remaining one is in the genus *Hippopus*. The remaining two species, *Corculum cardissa* and *Fragum fragum*, are in the family Cardiidae.

All tridacnid clams are distributed only in the Old World tropics of the Indo-Pacific. They also include the largest bivalve mollusk in existence: *Tridacna gigas*, which has been recorded by Rosewater (1965) to reach 1.2 m in length and 263 kg in weight. The other species are more modest in size, but all are still large in comparison with most other bivalve species.

These tridacnid clams are inhabitants of coral reef areas, where they are found abundantly in the shallow, sunlit water (Figure 10.6). They usually have an uncharacteristic position for a bivalve mollusk, in that they either lie on the surface of the bottom or bore into coral or coral rock with the opening between the valves facing up toward the surface of the water. The valves usually gape widely, and within this opening is an extensive, brightly colored layer of tissue in which the symbiotic zooxanthellae are found. The brightly colored tissue exposed within the gape is siphonal tissue; the color results from the interaction of various pigments deposited there. The bright colors protect the tissues of the clam from the dam-

aging effects of sunlight, while they pass enough light to allow the zooxanthellae to photosynthesize.

What is truly remarkable about tridacnids, however, is the tremendous change their bodies have undergone from that of a typical clam to accommodate this symbiotic association. Bivalves generally rest with their foot either embedded in the substrate or held against it. In this position, a normal clam has its hinge uppermost. Obviously, this strategy would not do for a tridacnid, since such a position would not permit the tissue and zooxanthellae to be illuminated. The result, as Yonge (1975) has described, is that the entire tridacnid has undergone a tremendous rotation with respect to the foot, so the hinge comes to lie on the underside next to the foot and the opening of the shell faces upward (Figure 10.7). At the same time, the siphons and siphonal tissue underwent an expansion. The siphons grew and extended themselves, covering the length of the upward-facing opening and providing the expanded area for occupation by the zooxanthellae. As a result of this rotation and expansion, one of the tridacnid's shell-closing muscles was lost. Because of these profound anatomical changes, the tridacnids are markedly different in body orientation from any other bivalves. The changes can be attributed only to the association with the symbiotic algae.

Symbiotic relationships with chloroplasts have been reported by Greene (1970) in the marine gastropods of the order Sacoglossa (*Elysia, Placida, Placobranchus*; Figure 10.8). Animals of this order regularly

FIGURE 10.6 A giant clam, *Tridacna gigas*, with siphonal tissue expanded. (Photo courtesy of R. Shimek.)

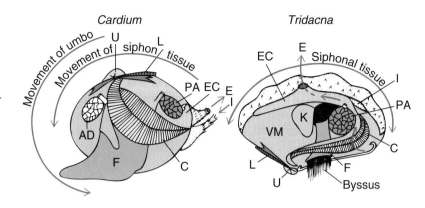

FIGURE 10.7 Comparison of the body orientation and internal anatomy of a typical clam, *Cardium*, with that of *Tridacna*. (AD) Anterior adductor; (C) ctenidium; (E) exhalent current; (EC) exhalent chamber; (F) foot; (I) inhalent current entering inhalent aperture; (K) kidney; (L) ligament; (PA) posterior adductor; (U) umbo; (VM) visceral mass. (Modified from "Symbiosis," *The Treatise on Marine Ecology and Paleontology*, Vol. 1, G. Thorson, 1957, ed. J. Hedgpeth, © 1957 Geological Society of America. Used with permission.)

consume the contents of algal cells, usually of the order Caulerpales. Algae of this order are particularly suitable for these suctorian molluscan feeders, because they have partially syncytial construction (that is, they lack cell walls). Some species of sacoglossans have evolved the ability to retain the chloroplasts in a functioning position on the dorsal surface. The only modifications so far observed are that they do not feed very often, and the tropical forms provide a screen of light-absorbing material above the chloroplasts to cut down the light intensity.

■ **VALUE OF THE ASSOCIATION**

Since symbiotic associations with algae are so common among invertebrates, and since they often result in pro-

found anatomical or behavioral changes in the partners, it only seems natural to assume that they must have some positive value to each of the two organisms. What might be the value of such associations?

We have already partially answered this question with respect to corals in Chapter 9, where we saw that the corals obtain food materials (see p. 425) from the zooxanthellae and also that the zooxanthellae enhance the ability of corals to lay down calcium carbonate (see p. 426). In turn, the zooxanthellae receive nutrients in the form of nitrates and phosphates produced in the metabolic processes of the coral—nutrients that are rare in external waters.

Among most invertebrates that have a symbiotic association with algal cells, the cells retain their

FIGURE 10.8 *Elysia hedgpethi*, a sacoglossan mollusk that contains chloroplasts obtained from the green algae *Bryopsis corticulans* or *Codium fragile*. (Photo by James Nybakken.)

integrity and are not digested by the animal for a source of nutrients. Any nutrients that do pass are in the form of chemical compounds. Energy-containing molecules, such as glycerol, produced by the zooxanthellae in photosynthesis pass to the animal. Nitrates and phosphates, nutrients the algae require, pass from the invertebrate to the algal cell.

A similar situation occurs between the sacoglossan mollusks and the symbiotic chloroplasts. Greene (1970) has demonstrated the translocation of organic material from chloroplast to animal. In the sacoglossans so far investigated, however, the life-time of the chloroplasts is much shorter, and the animals must periodically replenish their supply by ingesting the contents of algal cells.

With respect to the giant clams, however, Fankboner (1971) has suggested that, in addition to the reciprocal transference of nutrients, the animals may digest the zooxanthellae cells. Doing so would seem counterproductive, however, especially considering the great anatomical changes these animals have undergone to provide properly for their algal guests. Apparently, the clams can discern between healthy zooxanthellae cells and those which are degenerating or are senile. As a result, the clam transports only the senile or degenerating cells from the blood spaces of the outer mantle, where they photosynthesize, into the deeper tissues, where they are consumed by phagocytic blood cells. In this manner, the clams retain the symbiotic relationship. At the same time, they cull the unfit algal cells to obtain additional nutrients that would otherwise be lost when such cells die.

Other values may result from the association. As we have noted, most of these associations are tropical, and tropical waters are typically low in plant nutrients. It may be that the symbionts benefit from the association through access to a larger and more reliable source of nutrients in the metabolic products of the animal (NO_2^{2-}, PO_4^{3-}, CO_2) than they would obtain from the open water. Among the benthic Foraminifera, the partnership probably accounts for the twenty- to one-hundred fold increase in the calcification rate observed between those with symbionts and those without.

Tropical waters also are lower in oxygen than temperate waters are, because water at higher temperatures holds less oxygen. Since the photosynthetic process produces oxygen, it may be that the animals get additional amounts of oxygen produced by the symbiotic algae—a benefit that is especially valuable in the crowded conditions of the shallow waters of the tropics. Zooxanthellae also are protected from the actinic damage of the intense solar radiation of the tropics by living in the tissues of various marine animals.

Wilkinson (1983) has demonstrated that, in the symbiotic relationship between cyanobacteria and sponges, the cyanobacteria are photosynthetically active. Wilkinson (1983) also has estimated that 5–12% of the photosynthetically produced carbon is translocated to the host sponge. That this additional energy source is significant has been demonstrated by Wilkinson and Vacelet (1979), who showed that sponges grew faster in light than in the dark. It has also been suggested that the cyanobacteria may protect host-sponge cells from high light intensity.

■ ESTABLISHMENT AND TRANSMISSION OF ALGAL SYMBIONTS

Whereas each generation of sacoglossans must feed on algae to obtain its supply of symbiotic chloroplasts, the symbiotic association with zooxanthellae hosts is perpetuated from generation to generation in one of two ways: Either the zooxanthellae cells must be passed on directly from the parent to the eggs or larvae, or each new generation must reinfect itself with algal cells found in the surrounding environment.

The passage of algal symbionts directly to the next generation via the egg or larvae is the method used by most Cnidaria. In brooding corals, for example, the zooxanthellae enter the eggs or larvae at some time prior to their release from the parent. Among broadcast spawners, however, the zooxanthellae do not make their appearance, according to Babcock (1985), until 5 to 10 days after the larvae have settled. It is assumed, though without much evidence, that the giant clams transmit the zooxanthellae through the eggs, but it has also been suggested that the clams receive the dinoflagellates from the surrounding corals as well.

With other hosts, the perpetuation of a symbiotic relationship depends on the reinfection of each generation with algal symbionts obtained from the environment. Among protistans and Porifera, such reinfection is the common means of transmission. Each new generation obtains the algal cells by ingesting the free-living form of the symbiont. Surprisingly, this is also the case in *Convoluta roscoffensis*, where the association is a highly dependent one. Here, the algae-free larvae ingest the free-living

Platymonas (Tetraselmis) convolutae. To ensure that the larvae will be infected, however, a chemical has evolved which attracts the alga to the egg cases of *Convoluta,* so when the young hatch and begin to feed, the algae are present. Under such conditions, the new generation is virtually assured of reinfection.

A final possibility for transmission is the ingestion by the potential host of food organisms themselves containing symbiotic algae. Although this route has not yet been proven important, it remains an area requiring study.

Little is known about the transmission of cyanobacteria to succeeding generations of sponges, but cyanobacteria have been found in the eggs of some species, suggesting that these algae may be transmitted directly.

■ SYMBIOSIS IN THE DEEP SEA

Aside from the aforementioned symbiosis between various invertebrates and the chemosynthetic bacteria discussed in Chapter 4, various symbioses have been discovered between certain protistan eukaryotes

(a)

(b)

(c)

(d)

FIGURE 10.9 Scanning electron micrographs of protistan-prokaryote epibiotic symbioses from two deep-sea sulfidic sites in California. (a) Euglenoid from Monterey Bay cold seep with rod-shaped prokaryotes completely covering the cell. (b) Euglenoid from Santa Barbara Basin with rod-shaped prokaryotes lying in pellicular grooves at the posterior end of the cell. (c) Ciliate from Monterey Bay cold seep with rod-shaped prokaryotes completely covering the cell. (d) High magnification of specimen shown in (c) showing the prokaryote morphology and the pairing of cilia. F, flagella: P, prokaryote: C, cilia. Scale bars = 5 μm. (Photos courtesy of Kurt R. Buck, previously published in "Protistan-Prokaryotic Symbioses in Deep-Sea Sulfidic Sediments.")

and a number of prokaryotic microorganisms in the deep sea. These relationships have come to light only recently and remain relatively unexplored due to the difficulty of sampling deep-sea environments and of separating such organisms from the fine-grained sediments. Buck and Bernhard (2002) report that the great majority of protistans, mainly euglenoids and cilitates, taken from several sites in the Santa Barbara basin and from Monterey Bay in California had either epibiotic or endobiotic prokaryotic symbionts (Figure 10.9). The extent to which such symbiosis exists elsewhere in the deep sea is unknown, as is the nature of the relationship.

Symbioses Among Animals

We are concerned here with those special interactions in which members of different species are regularly associated with each other in nonparasitic relationships. Such symbiotic associations are widespread in the sea, primarily in the crowded reaches of the epipelagic, intertidal, and shallow subtidal zones. As with algal symbiosis, these relationships are somewhat more common or spectacular in the tropics, but they are also common in temperate seas.

■ TYPES OF ASSOCIATIONS

Symbiotic relationships among marine animals cover a broader spectrum than the strictly mutualistic associations we have seen among algae or chloroplasts and marine invertebrates. The simplest type of associations are commensal, wherein the "guest" lives on another "host" organism or in or on some construction of that organism, such as a tube or burrow. These associations are similar to epiphytic relationships among terrestrial plants, wherein the epiphyte uses the other plant as a substrate without actually taking sustenance from it. In such cases, the commensal usually gains in some measurable way from the association, and the host is not seriously inconvenienced.

Marine commensals that live on other invertebrates are called **epizoites**. Those which live inside other animals, but are not parasites, are called **endozoites**. Epizoites are abundant in marine waters, and many are probably not true commensals; the relationship is the result of organisms that normally settle on the substrate, but instead settle at random on the outside of a slow-moving or sessile invertebrate. These will not be considered further here. Other epizoites are highly specific, and a symbiotic relationship is certainly the case, with the "guest" somehow seeking out the correct

"host." These symbiotic epizoites are the most abundant group of commensals and are spread among the protistans and the phyla Cnidaria, Entoprocta, Annelida, Arthropoda, and Mollusca.

Many invertebrates harbor specialized ciliate protistans, either on their external surfaces, internally in the digestive tract, or among the gills. For example, the vorticellid ciliate *Ellobiophyra donacis* is restricted to the gills of the clam *Donax vittatus* (Figure 10.10), and the collared ciliate *Lobochona prorates* is found on the telson of the isopod crustacean *Limnoria tripunctata*. Many other examples are also known. Endozoic commensal ciliates are common as well, especially in the digestive tracts of larger animals. For example, Beers (1961) found that *Strongylocentrotus droebachiensis*, a large green sea urchin of the cold temperate waters of North America, has as many as seven species of ciliate protozoans in its gut. These symbionts are found nowhere else and die if removed.

Among the Cnidaria, the class Hydrozoa furnishes most of the cases of epizoites. The best studied is the case of two species of *Proboscidactyla*, two-tentacled hydroids, which always occur on the rim of the tubes of polychaete worms of the genera *Potamilla* and *Schizobranchia* in the North Pacific Ocean. Similarly, along the Pacific, Atlantic, and Gulf coasts, the hydroid genus *Clytia* occurs on clams of the genus *Donax*. Hydroids of the genera *Hydractinia* and *Podocoryne* are well-known ectocommensals on shells occupied by hermit crabs and on the snail *Cantharus*, respectively. Other hydroids are epizoic on gorgonians, pennatulids, and ascidians, but the closeness of the association is not always known.

Turbellarian flatworms are often endozoic in the digestive tracts of larger marine invertebrates and in the mantle cavities of various mollusks. The polyclad *Notoplana ovalis* occurs in the mantle cavity of the limpet *Patella oculis*, while the triclad *Nexilis epichitonius* occurs in the mantle cavity of the common Pacific coast chiton *Mopalia hindsii*. The rhabdocoels *Syndesmis dendrastorum* and *S. franciscanus* are endozoites in the guts of the sand dollar, *Dendraster excentricus*, and the sea urchins *Strongylocentrotus franciscanus*, *S. purpuratus*, and *S. droebachiensis*, respectively. Bryozoans offer two remarkable examples of symbiosis. Larvae of *Hippoporida* seek out a tiny shell occupied by a young hermit crab. The bryozoan colony grows in such a way that it forms a chamber that enlarges continuously with the aperture of the snail shell and keeps pace with the increase in size of the hermit crab, so the crab never needs to change

FIGURE 10.10 Some epizoites. (a)
Ellobiophyra donacis on the gills of the bivalve
mollusk *Donax*. (b) *Proboscidactyla stellata* on
the tube of a polychaete worm. ((a) from
E. Chatton and A. Lwoff. (1929), Contribution à
l'étude l'adaptation *Ellobiophrya donacis* Ch. Et Lw.,
péritriche vivant sur les de l'aephale *Donax viinius*
da Costa, *Bull. Biol.* 63, pp. 321–349. (b) after *The
Biology of Marine Animals*, J. A. Colin Nicol, 1960,
Pittman Books.)

(a)

(b)

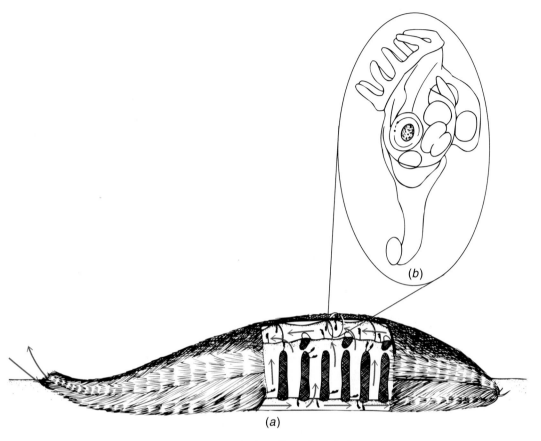

FIGURE 10.11 Epizoic entoprocta, *Loxosomella fauveli*, on the polychaete *Aphrodite aculeata*. (a) Position of the entoprocts in the respiratory channels of the worm, between the body wall and the thick feltlike setae covering the body surface. (b) Enlarged view of a single *Loxosomella fauveli* individual. (Modified from "Studies on Danish Entoprocta," C. Nielsen, *Ophelia*, Vol. 1, No. 1, pp. 1–76, 1964. Copyright © 1964 Apollo Books. Reprinted by permission.)

shells. The growth forms of the bryozoan take on peculiar spindle shapes that are known to Gulf coast beachcombers as "Texas longhorns"). A second bryozoan, *Triticella elongata*, is commonly found on the crab *Pinnixa*, which is commensal in the tubes of the polychaete *Chaetopterus*. The phylum Entoprocta is a small, little-known group, most of which, as Nielsen (1964) has documented, are epizoic commensals on other marine invertebrates. For example, they inhabit respiratory current areas of various polychaete worms (Figure 10.11).

Among the Annelida, a number of polychaete worms are epizoites with definite host specificity. The best studied is the hesionid *Ophiodromus pugettensis*, which occurs on the ambulacral grooves of the bat star, *Asterina (Patiria) miniata*, of Pacific coast waters (Figure 10.12). However, the whole group of so-called scale worms (Polynoidae) are all common epizoites on various other echinoderms and mollusks.

In the class Crustacea, the epizoic forms occur mainly within three groups: copepods, amphipods, and decapod crabs. Copepods associated with other marine animals are usually highly evolved and modified parasites, but a few are simply commensal, scampering over the outer surfaces of various invertebrates. Examples include several species of the genus *Hemicyclops*, which move on the surface of other, larger crustaceans. Others, such as *Paranthessius*, are found in the mantle cavity of various bivalves (*Tresus, Protothaca, Saxidomus*). One of the most specialized epizoic associations is that among several species of barnacles and various large whales. Two families of barnacles are found only on

FIGURE 10.12 The scale worm *Ophiodromus pugettensis* on the bat star *Asterina miniata*. Arrow points to worm. (Photo by James Nybakken.)

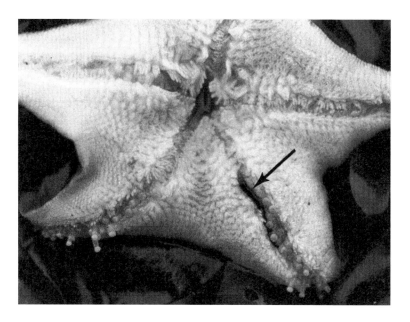

whales, and several are species specific. *Cryptolepas rachianectes* occurs only on the California gray whale, *Eschrichtius robustus*, and *Conchoderma auritum* attaches to *Coronula diadema*, which, in turn, is attached to the skin of whales (Figure 10.13). A curious form of epizoic relationship occurs between the amphipods of the suborder Hyperiidea and the large jellyfishes, where they apparently feed on the jellyfish. Hyperiids are often epizoic or endozoic on jellyfish. Also epizoic on jellyfishes are certain juveniles of various benthic crab species (for example, *Cancer jordani*).

Most mollusk epizoites are found among certain groups of clams. Generally, these are small species that attach via a foot or byssus to the outside of other invertebrates. Thus, *Montacuta ferruginosa* is found on the heart urchin, *Echinocardium cordatum; Orobitella rugifera*, attached via a byssus, is found on the underside of the burrowing shrimp *Upogebia stellata;* and *Mysella pedroana* is seen on the legs of the sand crab *Blepharipoda occidentalis*. Among the gastropods found on other invertebrates, the genus *Crepidula* is one of the most common (Figure 10.14).

The above is certainly not an exhaustive list of the various epizoic and endozoic symbiotic associations,

FIGURE 10.13 The whale barnacle *Cryptolepas rachianectes* and the cyamid amphipod *Cyamus scammoni* on the skin of the California gray whale, *Eschrichtius robustus*. (Photo by James Nybakken.)

FIGURE 10.14 Commensal slipper shells, *Crepidula adunca*, adhering to the black shell of the predatory gastropod *Mitra idae*. (Photo courtesy of Drs. Lovell and Libby Langstroth.)

FIGURE 10.15
The echiuran worm
Urechis caupo, to the
left, and some of the
symbionts that inhabit
its burrow. Shown are
the pea crab
Scleroplax granulata,
the scale worm
Hesperonoe adventor,
and the long-fingered
shrimp *Betaeus longi-
dactylus*. (Photo by
Dr. Richard Mariscal.)

but it gives an idea of the extent and diversity of such associations. Many, many more could be mentioned, but would not serve the purpose of this text.

A second type of symbiotic association is that of organisms associated with animals forming a tube or burrow. In this case, the commensal or guest occupies the tube or burrow constructed by the host, but is not necessarily in close association with the body of the host. These commensals use the tube or burrow as a refuge from predation. They may or may not also tap the food source of the host. This type of symbiotic relationship ranges from one that is simply fortuitous, as in the case of an animal diving into any convenient burrow to avoid a predator, to examples of obligatory associations, in which the commensals are not found other than in such tubes or burrows and are dependent on them for their existence. In the former category are certain small gobiid fishes, such as *Clevelandia ios*, which take refuge in any available tube or burrow, but emerge to feed. Obligate tube dwellers include many small crabs of the family Pinnotheridae, known as pea crabs. Many species of this family are known throughout the world, where they inhabit the tubes and burrows of polychaete annelids and other invertebrates, sometimes in species-specific associations.

One of the best-known examples of symbiosis with tube dwellers is that of the echiurid worm *Urechis caupo* on the Pacific coast of North America.

This animal creates a permanent U-shaped burrow in which four or more commensals can be found (Figure 10.15). At the upper end is the fish *Clevelandia ios*, which merely uses the tube as a refuge. Further in the tube and closely associated with the worm are the scale worm *Hesperonoe adventor* and the pinnotherid crab *Scleroplax granulata*. *Hesperonoe adventor* seems restricted to *Urechis* burrows and lives against the worm, snatching food from it. Although *Scleroplax* is restricted to tubes, it may also be found in the burrows of other animals. Another commensal is the small clam *Cryptomya californica*, which has very short siphons and inserts them into the burrow of *Urechis*, enabling the clam to live lower in the substrate than the length of its siphons would normally permit. Occasionally, the shrimp *Betaeus longidactylus* may inhabit the burrow as well.

A similar situation prevails with the tube-building polychaete *Chaetopterus*. This genus is virtually cosmopolitan, building tough parchment tubes in the muddy intertidal zone. The tubes are also inhabited by various species of pinnotherid crabs and scale worms, with the species varying with geographical location.

Another type of association concerns those invertebrates which live in the mantle cavities of various mollusks, a common site for commensals to occupy. For example, the nemertine *Malacobdella grossa* lives

FIGURE 10.16 Symbiotic relationships among crabs and anemones. (a) The anemone *Dardanus gemmatus* on a hermit crab. (b) The crab *Lybia tessellata* with anemones of the genus *Triactis* in its chelae. (Photos courtesy of (a) Mike Severns/Tom Stuck & Associates, (b) Andrew J. Martinez/Photo Researchers, Inc.)

(a)

(b)

in the mantle cavity of various clams on the Pacific coast. Pinnotherid crabs occupy the mantle cavity of numerous bivalves, including oysters, where they are considered a pest by oyster fishers. Polychaete worms inhabit mantle cavities of chitons and limpets. In the tropics, shrimps of the family Palaemonidae are often found in mantle cavities of large bivalves such as *Pinna*, *Atrina*, and *Tridacna*.

The remaining types of associations are mainly mutualistic rather than strictly commensal. The organisms involved display varying levels of behavioral and physiological modifications for the association.

One of the most obvious, but not well understood, associations occurs between various crabs and sea anemones. Many hermit crabs all over the world have anemones attached to their shells, and in some genera, such as *Dardanus*, the presence of anemones seems to be universal. Still other crabs, such as *Hepatus, Munidopagurus macrocheles*, and *Stenocionops furcata*, are found with anemones attached directly to their backs. The ultimate symbiotic relationship appears to be that evolved among a few crabs (*Lybia, Polydactylus*) which

carry anemones on their chelae and use them for defense or to capture food (Figure 10.16). Interestingly, most of the anemones involved in these associations are of three genera—*Calliactis, Paracalliactis*, and *Adamsia*—species of which are rarely found elsewhere than with crabs. This type of association must be mutualistic, because whenever the crab changes shells or molts its carapace, the anemone is usually transferred also. The transfer is effected by the manipulatory movements of the crab, but requires a degree of cooperation on the part of the anemone in loosening its grip on the shell or carapace and allowing itself to be physically moved.

Another type of association is that between other Cnidaria and fishes. Throughout the Indo-West Pacific in the coral reef areas, there is a striking association of small fishes of the genera *Amphiprion, Premnas*, and *Dascyllus* with various large anemones. These anemone fishes are able to live nestled among the tentacles of the anemones by preventing the discharge of the formidable nematocysts of the tentacles (Figure 10.17). Other fishes cannot prevent the nematocysts from

FIGURE 10.17 An anemone fish of the genus *Amphiprion* nestled in the tentacles of its anemone. (Photo courtesy of Drs. Lovell and Libby Langstroth.)

discharging, and the anemones can kill and eat fishes of a size similar to that of anemone fishes. A similar association seems to exist between the dangerous siphonophore called the Portuguese man-of-war (*Physalia*) and the little fish *Nomeus gronovii*, which swims among the tentacles bearing the powerful nematocysts. (See Figure 10.21). Still other juvenile fishes often congregate under the bells of the large scyphozoan jellyfishes (*Cyanea, Chrysaora*), where they presumably obtain protection from predation.

A similar situation prevails between fishes and the long-spined tropical sea urchin *Diadema*. As we noted in Chapter 9, these urchins are circumtropical in shallow water and have enormously long, thin spines that easily penetrate flesh. At least two tropical fishes—*Aeoliscus strigatus* and *Diademichthys deversor*—have adapted themselves to live among these spines, presumably for protection (Figure 10.18). The fishes also look like sea urchin spines and so are camouflaged.

Another type of association is that of fishes with other fishes, usually large, predaceous fishes. Pilot fishes (*Naucrates ductor*) and remoras (*Echeneis remora*) are always found with larger fishes, with marine vertebrates such as turtles, or even with moving, inanimate objects (Figure 10.19).

The final type of association is that known as cleaning behavior. As noted in Chapter 9, this is an association in which various species of fishes and shrimps actively attract large fishes so that the smaller ones can clean them of various ectoparasites. (See p. 446.)

■ **ORIGIN OF THE ASSOCIATION**

With such a diverse array of relationships, it seems unlikely that there is any single explanation for the origin of all the different types of associations. However, one factor common to all associations is that they have arisen in areas of the ocean that are the most crowded with life. It would, therefore, appear reasonable that epizoites in particular could have become established due to the great competition for space among the settling invertebrate larvae. As a result, many settled on other invertebrates, and the association evolved, in some cases, into more obligatory relationships in which commensal individuals gained some advantage over the free-living forms. This hypothesis is borne out by noting that many epizoites or burrow-inhabiting species are not strongly attracted to any given host species, while others are highly specific. Similarly, for many of the tube and burrow inhabitants, as well as those found in the mantle cavities of various mollusks, we can probably trace the origin of the relationship to the search for space. After all, tubes, burrows, and mantle cavities really represent a certain amount of potentially occupiable, semienclosed space.

The initial association of crabs with anemones may also have begun when the anemones began to reside on

FIGURE 10.18 *Diadema*, the long-spined sea urchin, with three shrimpfish, *Aeoliscus strigatus*, and two clingfish, *Diademichthys deversor*, sheltering among the spines. (After *Marine Animals, Partnerships and Other Associations*, R. V. Gotto, 1969. Copyright © 1969 R. V. Gotto.)

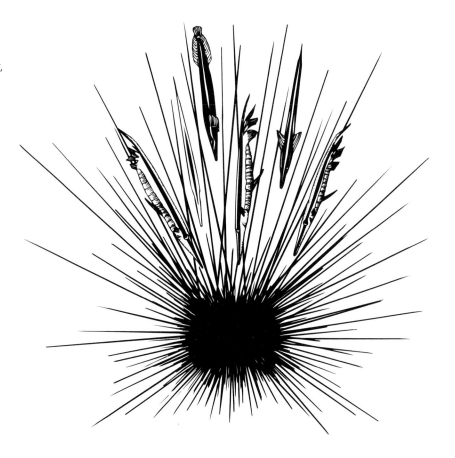

FIGURE 10.19 A remora, *Echeneis*, attached to a manta ray. (Photo courtesy of Jeff Rothmann Photography.)

shells already occupied by hermit crabs. Alternatively, the association could have arisen out of the habit of various crabs that pick up materials from the bottom and apply them to their carapaces as camouflage.

Many of the commensals that are epizoites or burrow or tube dwellers—especially the polychaete worms—tend to be from groups that show strong **thigmotaxis**; that is, they respond positively to stimulation by contact with other, solid bodies. Such a behavioral response would naturally preadapt them to move into tubes or burrows or into various cavities in the bodies of larger animals. Once the practice was established, selection would favor the development of chemosensory clues enabling these animals to find such burrows, tubes, or cavities, and more specific associations would follow.

Among the fishes and Cnidarians, the various types of symbiotic relationships seem to have their origin in the protection from larger predators that such associations confer. It does not appear that this relationship—especially between epipelagic fishes and Cnidarians—stems from crowding. Provided that the symbionts can prevent the nematocysts from discharging, the tentacles of anemones and jellyfishes represent a secure, formidable barrier to a large predator. However, in the case of large sharks, pilot fishes, and remoras, the association probably originated in the search for food in the epipelagic zone. Pilot fishes and remoras probably were originally attracted by the amounts of food dropped in feeding by the larger fishes, and they could have established themselves either by avoiding falling prey to the sharks or by being small enough not to be recognized as food by the larger fishes.

Cleaning behavior is a complex relationship involving many fish species, and its origin and ecological functions have been confounded by alternative hypotheses and factual inaccuracies, as Gorlick et al. (1978) have discussed. More recently, Grutter (1999) has demonstrated that, for at least one cleaner, *Labroides dimidiatus*, the association is a mutualistic one.

■ MODIFICATIONS DUE TO THE ASSOCIATION

As with algal–animal symbiosis, animal–animal relationships can result in anatomical, physiological, and behavioral modifications to one or both of the partners.

The fewest modifications occur among the epizoites—especially the facultative ones—and also are found in nonsymbiotic situations. Often, these animals have no anatomical modifications attributable to the symbiotic association, but those found with specific hosts usually have developed a means of recognizing the host from a distance. Such an ability to recognize hosts is even more highly developed among obligate symbionts of various types. But what is its basis? Investigators such as Davenport (1950) have shown that it is most likely a chemical, unique to the host, that is released into the water. The commensal has, in turn, developed chemosensory receptors that detect the chemical and enable the animal to home in on its host. If such symbionts are tested in the laboratory in an apparatus designed to give them

a "choice" of host, they show a remarkable ability to choose the host on which they are naturally found from among other closely related forms. Chemical attraction of this kind has been demonstrated in several scale worms (Polynoidae) commensal with various starfish species, in the peculiar fishes of the genus *Carapus* that live in sea cucumbers, and in the polynoid *Hesperonoe adventor* and its host, *Urechis caupo*. More recently, VandenSpiegel and colleagues (1998) have suggested that vision is the first sense that allows the shrimp *Synalpheus stimpsoni*, which lives on tropical commatulid crinoids, to find a commatulid, and then olfaction acts, permitting the shrimp to select the right commatulid species.

Another modification of commensals, and one that also aids in recognition of the host, is the ability of the larvae or young to select the appropriate substrate. In many epizoite species that settle onto their hosts from the plankton, metamorphosis and subsequent development do not occur unless the larva is in contact with the appropriate host. The substrate discrimination abilities of some of these symbionts are truly remarkable, with some even able to recognize substrate surface texture at the molecular level! It is not surprising, then, that whale barnacles can settle not just on whale skin, but only on certain species of whales. In this case, however, it is likely that a chemical unique to the host is also present. Similarly, the clam *Modiolaria* is able to recognize the structure of tunicin, the major chemical component of the external layer of the sea squirts with which it lives.

A basic modification of tube-dwelling symbionts is that they are often smaller or thinner and flatter than their free-living relatives. Thus, the pinnotherids, or pea crabs, are among the smallest of the marine crabs, and the shrimps that inhabit the mantle cavities of bivalve mollusks are also among the smallest of all shrimps. Similarly, the scale worms are often highly flattened in comparison with other polychaetes. In the same vein, fishes such as *Clevelandia ios*, which are commensal in tubes, and *Apogonichthys* and *Carapus*, in the internal spaces of clams and sea cucumbers, respectively, are very thin so that they can fit into those restricted areas. Finally, the pearlfishes of the genus *Carapus* show the greatest modifications in having lost their scales and pelvic fins and having shifted the anal opening far forward under the head. The latter change apparently is to ensure that defecation will occur outside the body of the sea cucumber. Furthermore, as

FIGURE 10.20 A pearlfish, *Carapus*, entering tail first into the cloacal opening of its holothurian host. (After *Marine Animals, Partnerships and Other Associations*, R. V. Gotto, 1969. Copyright © 1969 R. V. Gotto.)

Arnold (1957) has shown, these fishes exhibit behavioral modification in that they enter the cucumber tail first (Figure 10.20)! Such extensive modifications suggest a long history of adaptation.

Among anemones associated with crabs, a number of behavioral modifications occur. First, the anemones must respond positively to manipulations by the crab when it rushes to transfer the anemone to another shell or from its shed carapace to a new one. The anemone does this by releasing its grip on the old shell, so the crab can remove it with its claws. This level of "cooperation" requires a higher degree of nervous integration than we have believed could be accomplished by the extremely primitive, brainless nervous system of the anemone. Nor do we understand the process among the crabs that carry the anemones in their chelae. Here again, the anemones show a behavioral adaptation in that they allow themselves to be carried in the chelae and do not close up. In these cases, however, the crab also shows changes in that the chelae are modified to carry the anemones rather than to gather food. The crab extends the anemones when threatened or when gathering food. It uses its second pair of legs to transport food to the mouth, rather than the chelae, since the latter carry the anemones. Perhaps the most drastic changes undergone by the

anemone member of these crab–anemone systems is that exhibited by *Adamsia palliata*, which lives with the hermit crab, *Eupagurus prideauxi*. Here, the anemone steadily expands its basal disk to completely encircle the vulnerable abdomen of the hermit crab; thus, it eliminates the need for the crab to change its shell as it grows. The anemone then expands to allow for the growth of the crab and orients itself so that the oral disk and mouth are ventral to the crab. In that manner, it can obtain food from the crab.

The greatest morphological changes among fishes associated with other fishes occur in the remoras, which have their dorsal fin modified to form a large sucker, enabling them to remain attached to their hosts. (See Figure 10.19.) Pilot fishes (*Naucrates*) and the anemone fishes (*Amphiprion*) have no special morphological modifications. Among anemone fishes, however, there are definite behavioral and biochemical modifications. Studies have shown that anemone fishes of the genus *Amphiprion* must undergo a period of acclimatization to an anemone before they can dive into its tentacles with impunity. During this period, the fish is not capable of preventing the anemone from discharging its nematocysts. Schlichter (1976) demonstrated that the fish goes through a series of first brief and then longer encounters with the anemone. During this time, the fish coats itself with anemone mucus; thus, it tricks the anemone into discerning the fish as itself, and the anemone does not discharge its nematocysts. When the coating is complete, the acclimatization period is over and the fish can dive in among the tentacles without causing the nematocysts to discharge.

In addition to this fascinating coating process, anemone fishes have additional behavioral modifications. In contrast to most fishes, which swim away when approached by humans or other large potential predators, anemone fishes swim out of their anemones *toward* the potential predator! This behavior attracts the predator to them, at which time they do an abrupt about-face and dive into the anemone. Presumably, an unwary potential predator will chase them and be caught and consumed by the anemone. To further attract attention to themselves, anemone fish also have vivid color patterns, contrasting markedly with their background. (See Figure 10.17.)

In contrast to the anemone fishes, which develop immunity to the nematocysts of their host, the fishes (*Nomeus gronovii*) associated with the Portuguese man-of-war do not acclimate, nor do they ever

FIGURE 10.21
Portuguese man-of-war, *Physalia*, with the fish *Nomeus gronovii* sheltering among the tentacles. (After *The Biology of Marine Animals*, J. A. Colin Nicol, 1960, Pittman Books.)

attain complete immunity from the nematocysts (Figure 10.21). Instead, they appear to spend their lives playing a perpetual game of Russian roulette with the tentacles of the siphonophore, constantly moving to avoid contact with the lethal array of nematocysts! The same is true for the juveniles of various fishes that shelter below the umbrellas of large jellyfishes.

Among cleaning fishes, certain morphological modifications can be noted, such as pointed, narrow snouts and forcepslike teeth, adaptations for picking up small ectoparasites. Both cleaning shrimps and cleaning fishes also tend to have very bright and contrasting colors, which make them stand out against their background and advertise their presence. Certain cleaning shrimps—most notably the Pederson shrimp, *Periclimenes pedersoni*—have extraordinarily long antennae, which they wave about to further advertise their presence to the fishes.

Cleaning relationships also impose behavioral modifications on the fishes that are being cleaned. These animals must seek out and present themselves at the "cleaning station" and then remain motionless while the cleaner moves over their bodies. They must also open their mouths and spread their gills and opercula to allow the cleaner to enter the mouth and gill area. Of course, at the same time, the larger fish being cleaned must refrain from sudden inhalations, which might accidentally ingest the small cleaner!

■ VALUE OF THE ASSOCIATIONS

The symbiotic associations described in this chapter have obvious value to one or both of the partners, and the relative value of the relationship to each varies with the association.

With regard to the epizoites, the major value of the symbiotic relationship would seem to be with the epizoite rather than the host organism. For some epizoites, the value of the association lies in the fact that the host represents the only available suitable substrate in an otherwise unsuitable area. For example, the hydroid *Clytia bakeri*, which is epizoic on the clam *Donax gouldi* in sand areas in southern California, could not live in the area unless it had the clam to settle on, because it cannot exist on sand. There is, however, no obvious value to the clam. Other epizoites fix themselves onto the host to take advantage of feeding or respiratory currents produced by the host. This kind of relationship seems particularly common. The two-tentacled hydroids of the genus *Proboscidactyla* are fixed on the rims of the tubes of their annelid hosts to take advantage of the feeding tentacles and to remove food particles from the host. Similarly, the various epizoic entoprocts are usually found where there are respiratory currents. In this case, they do not steal food directly from the host, but filter the food from the moving current. Perhaps most epizoites benefit from the protection afforded them by the larger host; thus, the various flatworms occurring in the guts and mantle cavities of their larger hosts are protected from predation and, in the case of intertidal forms, from possible desiccation. Surely, the hyperiid amphipods found on the large scyphozoan jellyfishes would be much more subject to predation if they were free living in the plankton. Finally, the epizoite may also gain the advantage of movement—an advantage that is particularly significant if the epizoite is sessile and the host freely moving.

Commensals that inhabit tubes primarily gain protection from predation. This, for example, is the main benefit to the goby *Clevelandia ios*, which uses the tubes of *Urechis caupo*. Other commensals, however, such as the scale worms (Polynoidae), gain not only protection but food as they snatch morsels from their tube-dwelling hosts. Commensals in the mantle cavities of bivalve mollusks benefit from the protection of the bivalve shells, but they can also use the feeding and respiratory currents of the clam for their own needs. In a few cases, these symbionts tend toward parasitism in that they may consume pieces of the gills or mantle of their bivalve host.

Among the fishes associated with large jellyfishes, the Portuguese man-of-war, and the large sharks, the whole value of the association is to the fishes, in the form of protection from predation in an area (the epipelagic zone) notorious for great predation pressure. Pilot fishes and remoras are also protected from predation by their association with the large predaceous sharks, but they may obtain food as well, when the shark is feeding. The fishes associated with sea urchins and living in the respiratory trees of sea cucumbers are protected by their association, but their presence is no advantage to their hosts.

Among the mutualistic associations, the values of the association are more pronounced and, as the term implies, extend to the host as well. In the various crab-and-anemone associations, the crabs, as hosts, gain the advantage of the nematocysts of the anemones, either as protection against predators or else as an offensive weapon used to capture food. The anemones, while still sessile, gain the advantage of movement; some, such as those on hermit crabs, may also be able to obtain food fragments lost while the crab feeds. A similar situation prevails between anemone fishes and their anemones. The fishes gain protection, but they benefit their partners by attracting prey within reach of the tentacles and also by picking up pieces of food outside the reach of the anemone and depositing them in the anemone's mouth. Fautin (1991) has demonstrated that the fishes protect the anemone from predatory fishes. The fishes may also benefit the anemone by producing nutrients in the form of wastes that the anemone takes up.

In the case of cleaning behavior, the cleaner organisms' gain is in the form of food, whereas the large fishes cleaned presumably enjoy an increased measure of health due to the removal of the various ectoparasites.

■ LUMINESCENT BACTERIA

A final category of symbiosis is the curious relationship that has developed between various marine animals and luminescent bacteria. In this relationship, the bacteria are usually confined to a cavity in the body of the larger animal near the outer surface. The cavity is usually connected to the outside through an opening of some sort.

Such symbiotic relationships are confined to three groups of marine animals: fishes, squids, and pelagic tunicates. The relationship is most common among fishes and squids inhabiting the mesopelagic zone (see pp. 163–165), but it is still less common among these groups than is bioluminescence produced intrinsically without bacteria. The symbiotic bacteria are species of the genera *Vibrio* or *Photobacterium*. The fishes or squid of each generation apparently acquire their luminescent bacteria from the environment, where the bacteria are widespread. The various suggested values for the production of light by mesopelagic animals have been discussed in Chapter 4.

The relationship between the bacteria and the fishes or squid is a mutualistic one in which the bacteria obtain food from the larger animal and, in turn, the fishes or squid use the light produced by the bacteria for various defensive or offensive purposes, as discussed in Chapter 4.

The light produced by bacteria is usually continuous. As a result, the fishes and squids often develop elaborate modifications to control the light. These anatomical developments include the production of a reflecting layer behind the cavity to reflect the light outward, lenses to focus and concentrate the light, and most often, a screen or shade on the outside that can be raised or lowered to either "turn on" or "turn off" the light. The bacteria show no corresponding changes in behavior or morphology as a result of the association.

The origin of these elaborate associations is not well understood, and the bacteria apparently are not passed from one generation of fishes or squid through the eggs. It would appear that each generation somehow must reinfect its light organs from the external environment.

Summary of Key Concepts

- Symbiosis is the name given to close relationships between two different species.
- Symbiotic relationships cover a broad spectrum of associations, including commensalism, inquilinism, and mutualism.
- Basically, there are two groupings of nonparasitic symbiotic associations in the oceans: between autotrophs and invertebrates and between various animals.
- Most symbiotic relationships between autotrophs and animals are between unicellular algae or their chloroplasts and a wide variety of invertebrates.
- Symbiotic relationships are most common in the tropics, less common in temperate waters, and absent from polar waters.
- There are three groups of autotrophic cell symbionts: brown zooxanthellae, green zoochlorellae, and blue-green cyanellae.
- Algal symbionts are generally restricted to certain tissues or areas of the host.
- In cryptic symbiosis, the algal symbionts have lost their cellular integrity and do not appear separate from the host body.
- The origin of the symbiotic association is likely through ingestion of the algal cells or chloroplasts by cells of the digestive tract of the host.
- Algal–invertebrate associations are found most commonly in protistans, sponges, cnidarians, ascidians, flatworms, and mollusks and rarely in a few other higher taxa.
- Modifications in the algal symbionts include the loss of locomotory organs and reduced or absent cell walls.
- Among the animal hosts, a wide variety of anatomical and behavioral modifications are found, of which the most dramatic occur in the jellyfish *Cassiopeia*, the flatworm *Convoluta*, and the bivalves of the genus *Tridacna*.
- The value of the symbiotic association to the algae seems to be that they obtain nutrients from the host, which itself benefits by obtaining food molecules produced by photosynthesis in the algal cell. Perhaps the host also benefits by receiving additional oxygen.
- There are also associations between prokaryotic cells and protistans in the deep sea.
- Each new generation of host either receives the algal symbiont directly from the parent by passage via egg or larvae or must be reinfected from the environment.

- Symbiotic relationships among animals cover a broader spectrum than the strictly mutualistic associations seen among plants and animals.
- The simplest type of animal symbiosis is commensalism, wherein the "guest" lives on the host organism or in some construction of the host (usually a tube or burrow), gains in some way, and does not seriously inconvenience the host.
- Commensals that live on the outsides of other animals are called epizoites; those which live inside other animals, but are not parasitic, are called endozoites.
- Epizoites are the most abundant group of commensals and are found among protozoans, Cnidaria, Entoprocta, Annelida, Arthropoda, and Mollusca.
- Associations between hosts and commensals in which the latter occupy tubes or burrows constructed by the host range from simply fortuitous associations to obligatory ones.
- Most other types of associations among animals are mutualistic and include associations between anemones and crabs and between cnidarians and fishes.
- Cleaning behavior is an association in which various species of shrimp or small fishes clean larger fishes of ectoparasites.
- Although there is no single explanation for the origin of all the different animal–animal associations, the one factor common to all is that they have arisen in areas of the ocean most crowded with life; hence, epizoites might well have become established due to competition for space.
- Associations between crabs and anemones may have begun when anemones began to reside on shells already occupied by hermit crabs.
- Many tube- or burrow-inhabiting commensals show strong thigmotaxis, which would naturally preadapt them to move into tubes or burrows.
- Symbiotic relationships among fishes and cnidarians seem to have their origin in the protection from larger predators that such relationships confer.
- Cleaning behavior is a complex relationship whose origin is still debatable.
- Anatomical and behavioral modifications due to the association of animals with other animals vary dramatically, with the fewest modifications—oftentimes none—occurring among the epizoites.
- A common modification of obligate commensals is their ability to recognize the host from a distance, usually through chemosensory means.

- Some commensals have great powers of substrate recognition.
- A basic modification of tube- or burrow-dwelling symbionts is that they are often smaller or thinner and flatter than their free-living relatives.
- Anemones associated with crabs show a number of primarily behavioral modifications that allow the crab host to manipulate and move the anemone.
- The greatest morphological changes among fishes associated with other fishes occurs in the remoras, whose dorsal fin is modified to a sucker.
- Anemone fishes lack morphological modifications, but exhibit a definite set of behavioral and biochemical modifications.
- Cleaning fishes have long, pointed snouts and forcepslike teeth, and both cleaning fishes and shrimp have contrasting colors that make them stand out against the background.
- The fishes being cleaned must remain motionless during cleaning, with their mouths and opercula open to permit the cleaner to enter the mouth and gill area.

- The major value of the symbiotic relationship between epizoite and host would seem to be with the epizoite.
- Commensals inhabiting tubes or burrows primarily gain protection from predation, as do fishes associated with jellyfishes, whereas pilot fishes and remoras also are protected from predators, but may obtain food as well, when the host is feeding.
- In mutualistic associations such as those between crabs and anemones and between anemone fish and anemones, both partners benefit.
- In cleaning behavior, the cleaners gain food while the cleaned fishes enjoy better health due to the removal of their parasites.
- Symbiotic relationships between luminescent bacteria and fishes, squids, and tunicates are mutualistic, with the bacteria obtaining food from the host and the host using the light for various defensive and offensive purposes.

Review Questions

ESSAY Develop complete answers to these questions.

1. Compare and contrast chloroplast symbiosis with the endosymbiotic theory of the evolution of eukaryotic organelles.
2. Develop a table that presents the benefits accrued by host and guest in algal–invertebrate symbiosis. Evaluate the evidence that supports one of these benefits.
3. Speculate on why symbiosis is such a recurring and dominant theme in marine systems.
4. There appear to be many more symbiotic associations (for example, that between corals and their algae and that between the *Tridacna* and its algae) in marine systems than in freshwater systems. Speculate on why this might be so.
5. Discuss more than one evolutionary scenario for the development of the cleaning symbiosis exhibited by cleaner shrimp and cleaner fishes.

Bibliography

Arnold, D. C. 1957. Further studies on the behavior of the fish *Carapus acus. Staz. Zool., Naples* 23:91.

Babcock, R. C. 1985. Growth and mortality in juvenile corals (*Goniastrea, Platygyra,* and *Acropora*): The first year. *Proc. 5th Int. Coral Reef Symp.* 4:355–360.

Beers, C. D. 1961. The obligate commensal ciliates of *Strongylocentrotus droebachiensis.* Occurrence and division in urchins of diverse ages; survival in sea water in relation to infectivity. *Biol. Bull.* 121:69–81.

Buck, K. R., and J. M. Bernhard. 2002. Protistan–Prokaryotic symbiosis in deep-sea sulfidic sediments. In *Symbiosis: Mechanisms and model systems,* edited by J. Sackbach. Dordrecht, Netherlands, and Boston: Kluwer Academic Publishers, 507–517.

Cheng, T., ed. 1971. *Aspects of the biology of symbiosis.* Baltimore: University Park Press.

Cote, I. M. 2000. Evolution and ecology of cleaning symbiosis in the sea. *Oceanography and Marine Biology: An Annual Review* 38: 311–355.

Dales, R. P. 1957. Commensalism. In *The treatise on marine ecology and paleoecology,* edited by J. E. Hedgpeth. Vol. I, *Ecology.* Geol. Soc. of Amer. Memoir 67, 391–412.

Davenport, D. 1950. Studies in the physiology of commensalism. I, The polynoid genus *Arctonoe. Biol. Bull.* 98(2):81–93.

Fankboner, P. V. 1971. Intracellular digestion of symbiotic zooxanthellae by host amoebocytes in giant clams (Bivalvia: Tridacnidae) with a note on the nutritional role of the hypertrophied siphonal epidermis. *Biol. Bull.* 141:222–234.

Fautin, D. G. 1991. The anemone fish symbiosis: What is known and what is not. *Symbiosis* 10:23–46.

Gorlick, D. L., P. D. Atkins, and G. S. Losey. 1978. Cleaning stations as water holes, garbage dumps and sites for evolution of reciprocal altruism. *Amer. Nat.* 112 (984):341–353.

Gotto, R. V. 1969. *Marine animals, partnerships and other associations.* New York: Elsevier.

Greene, R. W. 1970. Symbiosis in sacoglossan opistho-branchs: Symbiosis with algal chloroplasts; transloca-tion of photosynthetic products from chloroplast to host tissue. *Malacologia* 10(2):357–380.

Grutter, A. S. 1999. Cleaner fish really do clean. *Nature* 398:672–673.

Henry, S. M. 1966. *Symbiosis.* Vol. I, *Associations of microorganisms, plants and marine organisms.* New York: Academic Press.

McLaughlin, J. J. A., and P. Zahl. 1966. Endozoic algae. In *Symbiosis,* edited by S. M. Henry. Vol. I, *Associations of microorganisms, plants and marine organisms.* New York: Academic Press, 257–297.

Nicol, J. A. C. 1960. *The biology of marine animals.* New York: Wiley.

Nielsen, C. 1964. Studies on Danish Entoprocta. *Ophelia* 1(1):1–76.

Rosewater, J. 1965. The family Tridacnidae in the Indo-Pacific. *Indo-Pacific Mollusca I* 6:347–396.

Rowan, R., and D. Powers. 1991. A molecular genetic classi-fication of zooxanthellae and the evolution of ani-mal–algal symbiosis. *Science* 251:1348–1351.

Schlichter, D. 1976. Macromolecular mimicry: Substances released by sea anemones and their role in the protec-tion of anemone fish. In *Coelenterate ecology and behavior,* edited by G. O. Mackie. New York: Plenum Press, 438–441.

Schoenberg, D. A., and R. K. Trench. 1976. Specificity of symbioses between marine cnidarians and zooxan-thellae. In *Coelenterate ecology and behavior,* edited by G. O. Mackie. New York: Plenum Press, 423–432.

Smith, D. C. 1973. Symbiosis of algae with invertebrates. *Oxford Biology Reader,* no. 43. London: Oxford University Press.

Smith, D. C., and A. E. Douglas. 1987. *The biology of symbio-sis.* Baltimore: Edward Arnold.

VandenSpiegel, D., I. Eekhaut, and M. Jangoux. 1998. Host selection by *Synalpheus stimpsoni,* an ectosymbiotic shrimp of comatulid crinoids, inferred by a field sur-vey and laboratory experiments. *J. Exp. Mar. Bio. Ecol.* 225:185–196.

Vernberg, W. B. 1974. *Symbiosis in the sea.* The Belle W. Baruch Library in Marine Science, no. 2. Columbia: University of South Carolina Press.

Wilkinson, C. R. 1983. Phylogeny of bacterial and cyanobacterial symbionts. In *Endocytobiology, endosymbiosis and cell biology,* edited by W. Schwemmler and H. E. A. Schenk. Vol. II. Berlin: Walter de Gruyter, 993–1002.

Wilkinson, C. R., and J. Vacelet. 1979. Transplantation of marine sponges to different conditions of light and current. *J. Exp. Mar. Biol. Ecol.* 37:91–104.

Yonge, C. M. 1957. Symbiosis. In *The treatise on marine ecology and paleoecology,* edited by J. E. Hedgpeth. Vol. I, *Ecology.* Geol. Soc. of Amer. Memoir 67, 429–442.

Yonge, C. M. 1975. Giant clams. *Sci. Amer.* 232:96–105.

CHAPTER 11

HUMAN IMPACTS ON THE SEA

The oceans are truly the last frontier for human exploration and exploitation on this planet. For years, the vast volume of the oceans, coupled with the fact that humans are terrestrial organisms who are naturally ill equipped to enter into water any more than superficially, kept the oceans and their communities safe from significant human interference or effects. As the human population of the world increased over the centuries, large land areas were significantly altered, but the oceans remained relatively untouched. The major usage of the sea during these centuries was as a source of food, but the primitive nature of gear and the limited range of vessels combined to make the effects of fishing relatively insignificant.

As we enter the twenty-first century, all this has changed. The explosion of technology has enabled human beings to penetrate all parts of the oceans. The rapidly increasing human population, coupled with sophisticated technology, has had significant effects on the ecology of the world's oceans in only a few short decades. The purpose of this chapter is to discuss and evaluate some of these effects in light of our knowledge of the functioning of the oceans' ecosystems.

Fisheries

The earliest use of the oceans by humans was probably for food. Early human populations living along the ocean borders captured various shore fishes, marine mammals, and shellfishes for consumption. This is recorded in the various large "shell mounds" excavated by archaeologists in different parts of the world. With the advent of vessels to venture onto the sea and the development of more refined nets, fishing became an important source of food to people living near the oceans. However, the gear and the vessels were so inefficient through much of human history that, even as late as the 1880s, such scientists as T. H. Huxley believed that the major fisheries were inexhaustible. This optimism however, does not mean, that there were no human impacts on the ocean, since, by the 1300s, inshore fish populations in northwestern Europe had been reduced and, in the 1700s and 1800s, the vast turtle populations in the Caribbean had been virtually eliminated. In the last decades of the twentieth century, however, the old ships and gear were replaced by much larger and more powerful vessels, more effective nets and traps, and electronic devices for detecting schools of fishes.

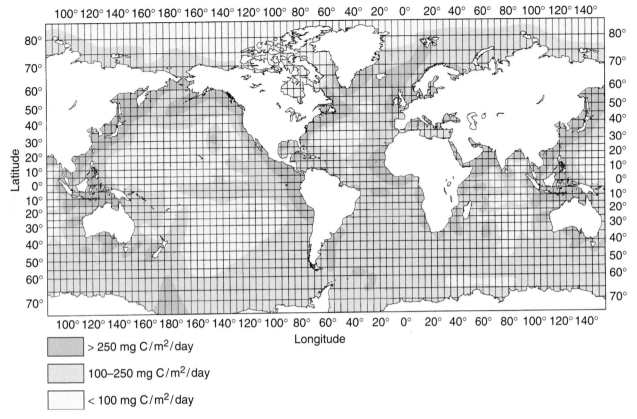

FIGURE 11.1 Geographical variations in the primary productivity of the world's oceans. Light blue areas are areas of lowest productivity (less than 100 mg C/m²/day), medium blue indicates medium productivity (100–250 mg C/m²/day), and dark blue signifies areas of highest productivity (more than 250 mg C/m²/day). (After *FAO atlas of the living resources of the sea*, 1972, Food and Agriculture Organization of the United Nations.)

The result has been a significant reduction in many fish populations and the disappearance or overexploitation of others at a time when increasing human populations are demanding more food. Since many believe that the oceans are the major food reservoir for upcoming generations of humans, the impact of fishing and its potential for sustaining future generations bear serious consideration.

■ **MAJOR FISHING AREAS**

The major fisheries are concentrated in the waters overlying the continental shelves around the world. The only major fisheries that operate in the open oceanic regions of the world are those for tuna and whales. There are several reasons for the concentration of fisheries in neritic waters. First, inshore waters have a much higher primary productivity than most open-ocean waters have (see pp. 72–75)

and, therefore, support larger populations of fishes at all trophic levels (Figure 11.1). Second, the bottom is fairly shallow on the shelf and is accessible to the various nets and traps used by humans to capture fishes. The deep ocean floor, by contrast, is so far from the surface that, even with our current sophisticated capture mechanisms, we cannot commercially fish there. Finally, the paucity of food in abyssal depths precludes any large fish population from existing there. Whatever fishes do live there cannot sustain a fishery for any length of time.

Although the fisheries are concentrated in the shelf regions, there is a very unequal distribution of the tonnage caught in various areas. The shelf areas of northwest Europe, those along the upwelling coast of western South America, and those off Japan produce the largest catches of fishes (Figure 11.2). The southern oceans and the tropics, with the exception

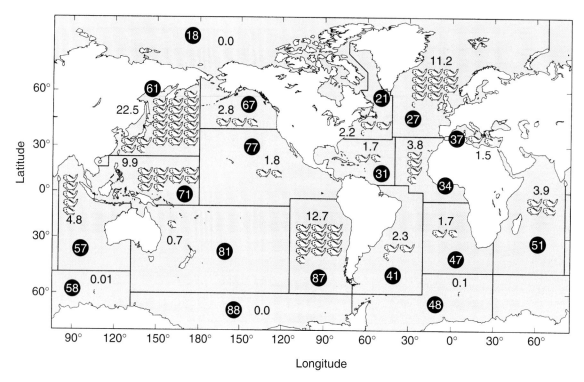

FIGURE 11.2 The major marine fishing regions of the world and their respective 1997 catches. Each symbol represents 1 million metric tons. The circled numbers refer to the region in the FAO division of the oceans. (Modified from the *FAO 2001 yearbook of fishery statistics*, Food and Agriculture Organization of the United Nations.)

of the west coast of South America, contribute much less to world fisheries.

■ MAJOR COMMERCIAL SPECIES

The commercially important marine animals are drawn from four groups: bony and cartilaginous fishes, marine mammals, mollusks, and crustaceans. Of these groups, the various fishes constitute by far the greatest tonnage. Among the thousands of species of marine fishes, only a very few make up the majority of catches in fisheries throughout the world. As Table 11.1 shows, aside from the "miscellaneous marine fishes" (see shortly), the herrings, sardines, and anchovies account for the largest tonnage of fishes. Of these, a single species, the Peruvian **anchoveta** (*Engraulis ringens*), has consistently accounted for almost half the catch and was the basis of the largest single fishery in the world, off Peru (Figure 11.3). Significant other fishes in this group are the menhaden, caught in fisheries off the Atlantic and Gulf coasts of the United States, and the herring, captured in

fisheries of northwestern Europe. Before its collapse in the 1940s and 1950s, the California sardine fishery was an important contributor to tonnage of fishes in the group. Herrings, anchovies, and anchovetas, all commonly called clupeoid fishes, are small, pelagic, and feed on a low trophic level, either directly on phytoplankton (*Engraulis*) or on herbivores, such as copepods. Most are not used as food for humans, but are reduced to protein meal for animal food.

The second-largest group of fishes landed consists of the gadoids, comprising various cod, haddock, pollock, and hake. These are bottom-dwelling (demersal) fishes, and major fisheries for them are on the shallow banks of the North Atlantic and North Pacific oceans (Figure 11.3).

The third-largest group comprises the scombroid fishes, generally known by the common name mackerel. Fishes of this group are common in both temperate and tropical waters, where they are fast-moving pelagic carnivores in shallow water. The newest editions of the FAO Fisheries Statistics now group these

TABLE 11.1

Worldwide Catches of Different Groups of Fishes and Shellfishes (in Millions of Metric Tons)

	1993	1994	1995	1996	1997	1998	1999	2000	2001
Herrings, sardines, anchovies	22.01	25.91	22.02	22.31	21.58	16.67	22.64	24.90	20.46
Cods, hake, haddock	9.93	9.73	10.72	10.74	10.23	10.33	9.34	8.65	9.22
Miscellaneous marine fishes	29.97	30.39	32.21	33.50	33.29	31.28	30.65	30.52	31.91
Tunas, bonitos, billfishes	4.57	4.65	4.73	4.63	4.85	5.81	5.97	5.87	5.82
Shrimps, prawns	2.08	2.25	2.30	2.45	2.54	2.75	3.02	3.07	2.95
Squid, cuttlefishes, octopus	2.71	2.77	2.88	3.06	3.32	2.86	3.60	3.66	3.35
Clams, cockles, ark shells	1.05	0.93	0.95	0.90	0.83	0.84	0.84	0.82	0.81
Flounder, halibut, soles	1.10	0.98	0.92	0.93	0.99	0.94	0.96	1.01	0.95
Oysters	0.16	0.17	0.19	0.19	0.19	0.16	0.16	0.28	0.20
Crabs	0.99	1.17	1.15	1.21	1.18	1.06	1.06	1.09	1.10
Mussels	0.26	0.28	0.24	0.20	0.22	0.24	0.22	0.24	0.26
Sharks, rays, chimeras	0.74	0.76	0.75	0.80	0.79	0.82	0.83	0.86	0.82
Scallops, pectens, etc.	0.49	0.61	0.50	0.48	0.48	0.55	0.61	0.66	0.70
Lobsters, squat lobsters	0.21	0.22	0.23	0.22	0.25	0.22	0.23	0.23	0.23
Krills, planktonic crustacea	0.09	0.08	0.12	0.10	0.08	0.08	0.10	0.10	0.10
Salmon, trout, smelt	0.98	0.99	1.15	1.03	0.92	0.89	0.91	0.81	0.89

Source: From *United Nations Food and Agriculture Organization Yearbook of Fishery Statistics*, 2001.

fishes with three other formerly separate groups (jacks, mullets, and sauries; redfishes, basses, and congers; and miscellaneous fishes) into a very large heterogeneous category called miscellaneous marine fishes. This classification, however, makes it impossible to sort out the the components, with the result that the category has now become the largest (Table 11.1). Closely related to mackerels are the various tunas, among the largest fishes in commercial fisheries. Widespread, fast-swimming carnivores of tropical and warm temperate seas (see Chapter 3 for a discussion of their biology), tunas are also the basis of the only major open-ocean fishery. Although their tonnage is not as great as that for other fishes, tunas are significant because they command a high price both as a canned fish product and as a fresh fish.

The flatfishes, such as halibut, sole, plaice, and flounder, are all well known and of considerable economic importance because they are premium food fishes. All are caught in fairly shallow waters, and many have been heavily exploited for years.

The sharks and other cartilaginous fishes have recently been more extensively exploited for food, and some have been demonstrated to have excellent flesh.

Salmon, trout, and smelt produce a respectable tonnage, but since trout are freshwater fishes, it is not possible to determine what fraction of the landings

consists of various salmons and what fraction comprises trouts. Also, salmon are currently being raised commercially in mariculture.

The major crustacean group in world fisheries is shrimp, with various species being caught in both warm and cool waters in all oceans. The other significant crustaceans are various species of crabs and lobsters. All are economically valuable and command high prices for direct human consumption.

The major mollusk group in terms of tonnage is the squid. Significant numbers of squids are caught in Japan, Europe, and California. The other major mollusk group commercially harvested includes the bivalves, such as oysters, mussels, and clams.

The major fishery for marine mammals has been that for various species of whales. Persistent overexploitation of the stocks in the twentieth century has, however, virtually doomed this fishery. Other harvestable mammals are mainly seals and sea lions.

Minor fisheries exist for various species of algae. Some are taken in Japan for human consumption, but most are collected to extract various products (algin derivatives) for use in industry.

■ SUSTAINABLE YIELD AND THE FUTURE

In 2001, the last year for which published FAO data are available, the total world fishery catch was

FIGURE 11.3 Some of the major commercial fishes of the world. (Not to scale.)

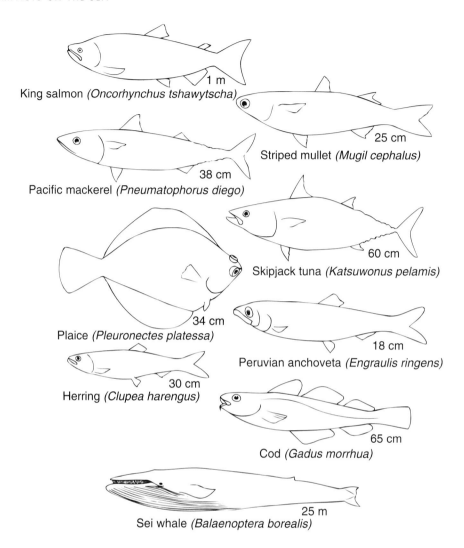

King salmon *(Oncorhynchus tshawytscha)* 1 m

Striped mullet *(Mugil cephalus)* 25 cm

Pacific mackerel *(Pneumatophorus diego)* 38 cm

Skipjack tuna *(Katsuwonus pelamis)* 60 cm

Plaice *(Pleuronectes platessa)* 34 cm

Peruvian anchoveta *(Engraulis ringens)* 18 cm

Herring *(Clupea harengus)* 30 cm

Cod *(Gadus morrhua)* 65 cm

Sei whale *(Balaenoptera borealis)* 25 m

92.4 million metric tons, of which 83.7 million metric tons were from marine fisheries. Weber (1994) cites 1989 as the peak year for marine fisheries landings. Since that time, the landings have actually dropped or, at best, remained the same (Figure 11.4). The fishes caught for human consumption represent perhaps 1% of all human food, but a significant 10% of the protein intake; hence, they are important.

Throughout the 1970s and 1980s, total marine landings increased steadily, finally peaking in 1989 (Figure 11.4). The increase was due primarily to heavy fishing for certain unexploited fishes, such as pollock and hake, that are used mainly for fish meal, and to the expansion of fisheries in Southern Hemisphere oceans. Since the human population is still increasing and is expected to reach nearly 7 billion early in the twenty-first century, and since many are anticipating that much of the food needed to feed these increased numbers will come from the sea, does the current decline or stagnation mean that we have reached the limit of the amount of fish protein we can harvest from the seas to meet the growing needs of a burgeoning human population? What are the facts with respect to current and potential amounts of food from the sea?

To answer these questions, it is necessary to look briefly at the situation of current fisheries. Of the present fishing effort, 90% is concentrated on the 7% of the world ocean area represented by the continental shelves. These shallow waters are the most productive ocean areas.

The great growth of the world fisheries in the 1950s and 1960s was due to two factors. The first was

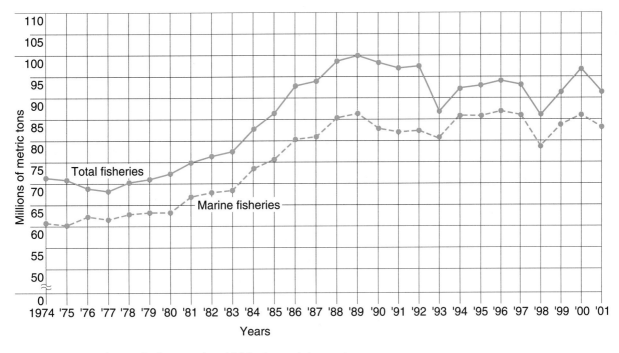

FIGURE 11.4 Changes in the annual world fisheries catch in recent years.

the discovery and exploitation of the anchoveta fishery of Peru, which sent that country from having less than 1% of world fishery landings in 1950 to 17% in 1967 and to a position as the top fishing nation in the world. The second factor was the introduction of modern fish-capturing and -handling gear, including factory ships, larger nets, and more accurate methods of locating fishes. Most of that effort was expended on the continental shelf regions around the world, with the result that, at present, this limited area has been fully exploited for its existing fish stocks and many, if not most, of these fisheries have been overfished. (See pp. 507–511.) It would appear, therefore, that little or no increase can be expected from continental shelf areas.

What about the remaining 93% of the world's oceans? Currently, the major oceanic fisheries of the world are for various tunas, swordfishes, billfishes, and so forth. In 2001, the landings of this group were 5.82 million metric tons, or only 6.7% of the total marine fishery landings that year. However, the 5.82 million figure is not an increase in landings over the previous years and suggests that we are approaching the limit in tunas that may be caught without depleting the stocks. Whereas 25 years ago the tuna stocks were considered nearly impossible to overfish, the

advent of purse seines dramatically changed that fact in a few years. Today, there is considerable concern about the continued ability of these fish stocks to sustain the current fishing pressure. Indeed, the bluefin tuna of the Atlantic, perhaps the world's most valuable fish, bringing $260 per kilogram in Japan, has had a population decline of 90% in the western Atlantic and has been suggested for listing as an endangered species.

Bottom fishing on the abyssal plains under the oceanic area is not feasible, since the diminished amount of energy reaching the abyssal floor does not allow the existence of substantial fish stocks. (See pp. 153–155.)

Furthermore, as we noted in Chapter 2, the surface waters of these open-ocean areas are impoverished with respect to nutrients and phytoplankton. They cannot sustain the populations of fishes, even in the surface waters, that the continental shelf waters are able to. Open oceans are "watery deserts" that, vast though they be, cannot match the production of the shallow coastal waters. In fact, we are approaching the limit of what we can obtain from the sea altogether.

What, then, is a reasonable estimate for the maximum sustainable yield of food that we can expect from the seas? By **maximum sustainable yield**, we mean the largest number of fishes that can be harvested year after year without diminishing the stocks. Perhaps the best estimate is based on total primary productivity of the oceans. As we have already said, most food chains and trophic levels depend on photosynthesis, the energy of which is fixed and finite in the world's oceans. By knowing the trophic level of the various fishes taken as food and the efficiency of energy transfer from level to level, it is possible to obtain a figure for maximum production. Such was the approach taken by Ryther (1969) to estimate potential yield.

To obtain his estimate, Ryther first divided the ocean into three areas, each representing waters with significantly different levels of primary productivity. The largest area (90%) was that of the open oceans of the world, but it had the lowest primary productivity (Table 11.2). The second-largest area (9.9%) was the neritic zone, and the final—highest in productivity but smallest in size (0.1%)—was the restricted area where upwelling occurs (Table 11.2). Next, Ryther estimated the efficiency of energy transfer among trophic levels in each division and, finally, the trophic level of the major fishes captured in each (Table 11.3).

Combining all the data, he arrived at a total world fish production of 24×10^7 tons ($= 240$ million tons). This figure, however, does not represent the total amount that may be taken: A significant fraction must be left as breeding stock to sustain future yields. Another large fraction is consumed by other carnivores in the sea. Ryther estimated these latter fractions to account for somewhat more than half of the total production, leaving about 100 million tons as the sustained yield for the world's oceans.

Other scientists, such as Cushing (1969) and Schaefer (1965), have also estimated the total harvestable amounts of fishes in the sea. Their estimates were made on the basis of primary productivity, extrapolating from current landings and taking into account trophic-level and ecological efficiencies. More recently, Pauly and Christensen (1995) estimated that, in the most intensively fished continental shelf regions, more than 20% of the primary productivity is needed to sustain the fisheries. Because scientists disagree on which figures to use for primary productivity, ecological efficiency, and average trophic level, their estimates have varied from a low of 60 million tons to a high of 290 million tons.

TABLE 11.2

Division of the Oceans into Provinces Based on Level of Primary Productivity

PROVINCE	PERCENTAGE OF THE OCEAN	AREA (km²)	MEAN PRODUCTIVITY (GRAMS OF CARBON PER m² PER YEAR)	TOTAL PRODUCTIVITY (10^9 TONS OF CARBON PER YEAR)
Open ocean	90.0	326.0×10^6	50	16.3
Coastal zone	9.9	36.0×10^6	100	3.6
Upwelling areas	0.1	3.6×10^5	300	0.1
Total				20.0

Source: After Ryther, 1969.

TABLE 11.3

Estimated Total Fish Production of the World's Oceans Based on the Three Provinces of Table 11.2

PROVINCE	PRIMARY PRODUCTION (TONS OF ORGANIC CARBON)	TROPHIC LEVELS	EFFICIENCY (PERCENT)	FISH PRODUCTION (TONS, FRESH WEIGHT)
Oceanic	16.3×10^9	5	10	16×10^5
Coastal	3.6×10^9	3	15	12×10^7
Upwelling areas	0.1×10^9	1.5	20	12×10^7
Total				24×10^7

Source: After Ryther, 1969.

Given the current situation of world marine fisheries in which we are seeing continued declines or stagnation in landings, and few, if any, unexploited fisheries, it appears at this time (2004) that we are either at or above the point of sustained yield for many of the world's fisheries. That we may well have overshot the limit on the number of fish we can take is manifested in the decline of various fisheries.

■ OVEREXPLOITATION

In recent years, there have been abundant examples of the decline of fish stocks of all types in all areas of the world. For example, Weber (1994) notes that productivity has declined in all but 2 of the 15 major marine fishing areas of the planet (Figure 11.5). In 4 of the most significant areas—the northwest, west central, and southeast Atlantic and the northeast Pacific—the catch has declined by 12% to 64% (Table 11.4). If one looks at individual species of fishes, one finds that 18 have dropped by more than 100,000 tons since their peak landings (Table 11.5). In the New England region alone, the total harvest of groundfish (cod, haddock, flounder, etc.) has declined from near 800,000 tons to about 100,000 tons. Indeed, what has been one of the world's greatest fishing grounds—the Grand Banks and Georges Bank, off northeastern North America—is, according to Safina (1995), essentially closed due to the collapse of the populations, making them essentially commercially extinct. A similar massive decline in the rockfish populations on the Pacific Coast of North America may also precipitate a closure. A major part of the problem has been the failure to realize that the world's fisheries yield the only wildlife still hunted as a major source of food for humans. Therefore, their populations are dependent upon the integrity of the ecosystem in which they occur and cannot be artificially increased beyond some natural level. Most, if not all, other human food comes from some sort of farming wherein populations of selected species are manipulated for maximum yield outside of any natural ecosystem.

Many of the declines in fisheries are clearly due to overexploitation, but others have more complex origins. All indicate that we are exceeding the maximum amount of fishes we can take from the oceans and that we have often mismanaged fisheries, to the point where we have not achieved a maximum sustainable yield. This recognition of the significance of overexploitation is manifest in the U.N. Food and Agriculture Organization's conclusion that the present way we operate the world's fisheries cannot be sustained. In the United States, the National Academy of Sciences reported in 1995 that human actions have caused such drastic effects in the marine fauna and flora as to endanger the functioning of whole ecosystems. Another disturbing factor is the article by Jackson et al. (2001) concluding that overfishing is not a new phenomenon. Using paleoecological, archaeological, and historical data, these researchers report that overfishing actually preceded

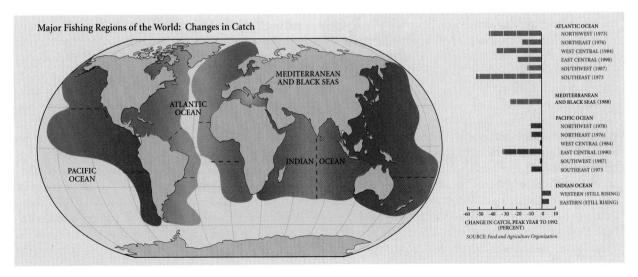

FIGURE 11.5 Changes in peak catch of fishes worldwide from peak year to 1992. (Copyright *Scientific American*. Reprinted by permission.)

TABLE 11.4

Change in Catch for Major Marine Fishing Regions, Peak Year to 1997

REGION	PEAK YEAR	PEAK CATCH (MILLION TONS)	1997 CATCH (MILLION TONS)	CHANGE (PERCENT)
Atlantic Ocean				
Northwest	1973	4.4	2.05	−53
Northeast[a]	1976	13.2	11.66	−12
West Central	1984	2.6	1.8	−31
East Central	1990	4.1	3.6	−12
Southwest	1997	2.6	2.6	0
Southeast[a]	1973	3.1	1.1	−64
Mediterranean and Black Seas[a]	1988	2.1	1.5	−28
Pacific Ocean				
Northwest	1988	26.4	24.6	−7
Northeast[a]	1987	3.4	2.8	−17
West Central	1997	8.9	8.9	0
East Central	1981	1.9	1.6	−16
Southwest	1991	0.9	0.8	−12
Southeast	1994	20.2	14.4	−29
Indian Ocean				
Western	Still rising		4.1	+6[b]
Eastern	Still rising		3.9	+5[b]

[a]Rebounding from a larger decline. [b]Average annual growth since 1988.

Note: Percentages calculated before rounding off catch figures. The catch in the Antarctic is at 356,000 tons, down from a peak of 653,000 tons in 1982, primarily because of reduced interest in krill. The catch in the Arctic is zero.

Source: From Weber, 1994 and *FAO Yearbook,* 1997.

all other human degradation of the shallow-water marine habitats by centuries. According to them, the magnitude of the removal of large animal biomass means that, throughout the world in shallow coastal waters, such species are at present effectively ecologically extinct, and since their absence predates scientific investigation, we do not know how the natural systems might have operated.

Another cause of the overexploitation can be found in economics. Hall (1999) points out that, whereas the value of the world's fishery catch is about $70 billion, the cost of catching the fishes is $124 billion. The deficit is met by various government subsidies of fuel costs, price controls, low-interest loans for gear, etc., thus encouraging more entries into an overcrowded field, a situation economists recognize as *overcapitalization*. As a result, between 1970 and 1990, the world's fishing fleet grew at twice the rate of the total global fisheries catches.

The signs of overfishing are normally a decline in the average size of fishes and an increase in the effort needed to land the same amount of fishes. The age structure of the fishes also changes. In some cases, overexploitation has been compounded by certain natural or human-induced changes in the environment that also reduced the stocks. In most situations, however, overexploitation is the direct result of human activity. We discuss three examples next: one with environmental consequences affecting the fishery, one in which the causes are uncertain, and one related directly to humans.

The small clupeid fish known as the Peruvian anchoveta (*Engraulis ringens*) is a filter feeder that occurs in large schools in the upwelling water mass of the Peru–Chile Current off the west coast of South America, where it feeds on the abundant plankton. This fish occurred in tremendous numbers in these cold waters before the advent of human fishing. The fish's major predators were various species of marine birds, collectively called guano birds, that lived by the millions along the continent's shores and produced huge quantities of

TABLE 11.5

Fishery Declines of More than 100,000 Tons, Peak Year to 1992

SPECIES	PEAK YEAR	PEAK CATCH (MILLION TONS)	1992 CATCH (MILLION TONS)	DECLINE	CHANGE (PERCENT)
Pacific herring	1964	0.7	0.2	0.5	−71
Atlantic herring	1966	4.1	1.4	2.6	−63
Atlantic cod	1968	3.9	1.2	2.7	−69
Southern African pilchard	1968	1.7	0.1	1.9	−94
Haddock	1969	1.0	0.2	0.8	−80
Peruvian anchovy (anchoveta)[a]	1970	13.1	5.5	7.6	−58
Polar cod	1971	0.35	0.02	0.33	−94
Cape hake	1972	1.1	0.2	0.9	−82
Silver hake	1973	0.43	0.05	0.38	−88
Greater yellow croaker	1974	0.20	0.04	0.16	−80
Atlantic redfish	1976	0.7	0.3	0.4	−57
Cape horse mackerel	1977	0.7s	0.4	0.3	−43
Chub mackerel	1978	3.4	0.9	2.5	−74
Blue whiting	1980	1.1	0.5	0.6	−55
South American pilchard	1985	6.5	3.1	3.4	−52
Asian pollock	1986	6.8	5.0	1.8	−26
North Pacific hake	1987	0.30	0.06	0.24	−80
Japanese pilchard	1988	5.4	2.5	2.9	−54
TOTALS		51.48	21.67	30.01	−58

[a]The catch of the Perusian anchovy hit a low of 94,000 tons in 1984, less than 1 percent of the 1970 level, before climbing up to the 1992 level.
Source: FAO. From Weber, 1994.

droppings or guano at their nesting areas. The guano was harvested for fertilizer.

The anchoveta population depends on abundant plankton, which, in turn, depends on the nutrient-rich upwelling water. In certain years, the upwelling in this area is blocked by an influx of warm surface water known as the El Niño Current. (See Chapter 9, pp. 450–451.) When this occurs, plankton disappear and the population of anchoveta drops, as does the number of birds; hence, there is a natural abundance cycle.

Direct human exploitation of the anchoveta began in the 1950s, and the fishery increased at a phenomenal rate until the late 1960s. The catch peaked in 1970 at 13 million metric tons, more than one-sixth of the entire world's landings!

This huge catch was accompanied by a decline in bird numbers, since the birds had fewer fishes to eat. The fishery experienced one drop in numbers in 1965 as a result of the occurrence of El Niño, but this temporary setback was made up the following year.

Fishery experts had estimated a maximum sustainable anchoveta yield of 9.5 million metric tons. This amount was exceeded in 1967, when over 10 million tons were landed. By 1970, more than 13.1 million metric tons were landed, but the size of the fishes was decreasing, and it was apparent to experts that the fishery was in trouble and that too few fishes were escaping to provide the breeding stock. As might easily have been forecast, the crash came in 1972, aided by another El Niño, and landings dropped to 4.7 million tons. The year 1973 was even worse, with only 1.7 million tons landed. The anchoveta fishery has not fully recovered. Even in 1997, only 7.7 million tons were landed (up from 5.5 million tons in 1982; see Table 11.5). Sadly, due to overfishing, a once huge resource has been reduced far below its sustainable yield, and the human population of the world has been denied a significant protein supply.

Much more mysterious is the demise of the California sardine fishery. Like the anchoveta, the California sardine, *Sardinops sagax*, is a small plankton-feeding clupeid fish. Its range is from Mexico to southern Alaska. In the 1930s, this species was the basis of the greatest fishery in North America. During the 1936–37 season, fishermen landed 650,000 tons in California alone, the major fishing region. The decline began in the 1940s, when fishermen saw the age structure of captured sardines begin to change to older age groups, reflecting the decimation of the younger fishes and their failure to recruit to the fishery. Despite the establishment of the California Cooperative Oceanic Fisheries Research Program in 1949, the fishery was commercially dead by the early 1950s. In Monterey, California, a center of the industry, the famed Cannery Row of author John Steinbeck closed up. The fishery has not recovered since that time, although 1981 saw the start of an encouraging increase in the numbers of fishes. The population now approaches commercial levels and some limited fishing has begun.

Although most people attributed the demise of the fishery over the 10-year period from 1940 to 1950 to overfishing, it seems apparent now that other factors were also at work. The decline of the fishery began with the onset of a 15-year period of unusually cold water in the sardines' habitat. Marr (1960) stated that this cold water affected the sardine population adversely, either directly through a reduction in breeding or indirectly through an effect on the food supply. In either case, the result was a catastrophic decline in numbers. If overfishing were the only reason for the demise, one would have expected that, in the more than 35 years during which there has been no fishing, the stocks would have recovered or at least returned to their known breeding grounds. Because this has occurred very slowly, it is strongly suspected that other factors were involved.

Finally, from an examination of the scales of sardines preserved in layers of sediment in certain anoxic basins, Soutar and Isaacs (1969) have presented evidence that the California sardine naturally undergoes huge cyclic changes in numbers, and the cycles range from 500 to 1,700 years. If this is correct, the California sardine would be present in copious numbers for a certain period of years and then vanish for a longer period before reappearing. During its absence, the niche is filled by another clupeid, the anchovy. Currently, the increase in anchovy numbers would seem to substantiate this theory. If it is true, Cannery Row might again bustle with the activities of fish rendering in only 500 years or so!

The final example of the demise of a fishery is much simpler: It is the direct result of human overfishing. Here, we refer to the whaling industry. The ignominious end of the world's whale fishery is now being observed.

In contrast to most animals taken in various fisheries, whales are warm-blooded mammals that have a long life span and a very low reproductive rate. This means that they are very slow to build up their numbers once reduced and are readily susceptible to extinction.

Large-scale whaling began in the North Atlantic in the sixteenth century. At first, methods were primitive and refuges from whalers existed for most species. However, the advent of long-range vessels and, especially, the harpoon gun with explosive charge spelled the end of whale stocks in the North Atlantic and North Pacific by 1900.

Attention then turned to the large stocks of baleen whales in the Antarctic region, where most whaling of the twentieth century has been concentrated. Coincident with the exploitation of these southern stocks was the use of huge factory ships, more speedy catcher boats, and airplanes and helicopters to find the whales. All these technological advancements led to a greater slaughter of whales in the first 60 years of the twentieth century than had been achieved by all whaling in all preceding centuries. Rounsefell (1975) notes that, in approximately 60 years of Antarctic whaling, perhaps 1.3 million whales were taken, and Hall (1999) gives a figure of 1.48 million.

In an attempt to regulate this fishery, the International Whaling Commission (IWC) was set up in 1948. The commission was to obtain data on whale stocks and to set catch quotas for the various species to conserve the stocks. Unfortunately, because the commission was given no enforcement powers, had no independent means of obtaining data on the status of the various whale stocks outside of what the whalers reported, and was subject to intense political pressure by the whaling nations that were members of it, its attempts to conserve stocks has been a farce. The commission has consistently set its quotas too high and refused to heed the advice of its own and outside experts. As a result, the stocks of whales have continued to decline. Today, some species are on the verge of extinction, and others are so reduced in numbers that they are commercially extinct. Along with the demise of the whales has come the demise of the whaling industry: When the IWC was set up in 1948, its 20 member nations were whalers; by 1979, only two—Russia and Japan—remained as international whalers.

The dramatic decline in this fishery can be appreciated from a few examples. In the 1930–31 season in the Antarctic, 25,000 blue whales (*Balaenoptera musculus*) were taken, but by the 1963–64 season, only

112 were taken. In 1968, the IWC reluctantly gave complete protection to the species. Declining blue-whale stocks forced whalers to change to fin whales (*Balaenoptera physalus*) after World War II, and from 1955 to 1961, more than 25,000 were taken each season. By 1965, the number had dropped to less than 2,500, and by the 1974–75 season, it had fallen to less than 1,000. Sei whales (*Balaenoptera borealis*) went from 25,453 in 1964–65 to 3,859 in 1974–75.

Clearly, the unrealistically high quotas set by the IWC and the greed of the whaling nations have reduced the whale stocks to the point where it is, or soon will be, uneconomical to continue whaling. Still, even if all whaling is stopped now, as many countries have urged, it will take many species years—perhaps centuries—to recover, because of their low fecundity and the long time necessary for them to reach reproductive size. That there is some hope for recovery of at least some of these leviathans lies in the recovery of the California gray whale (*Eschrichtius robustus*). This species was reduced to only a few individuals in the early decades of the twentieth century, but since complete protection in 1938, they have now increased in numbers to more than 8,000.

The collapse or decline of fisheries may be caused indirectly by other activities of humans. The collapse of the salmon fishery in Washington, Oregon, and British Columbia is one example. In this case, logging and dam building destroyed or silted up the spawning streams on which the stocks of salmon depended. The stocks may also have been overfished.

■ ECOSYSTEM CHANGES

Overfishing may also result in changes in the larger ecosystems or communities in which the target fish reside. For example, the demise of the North Atlantic cod and haddock fishery resulted in an increase in the number of dogfish sharks and skates that now fill that niche; the increase may contribute to additional declines in cod and haddock, as these sharks feed on the young of those fishes (Figure 11.6). In the northeast Pacific, heavy fishing of the Alaska pollock resulted in a decrease in the Steller's sea lion (*Eumetopias jubatus*) population from 300,000 individuals in 1960 to 66,000 in 1990, most likely because the Alaska pollock was the chief food of the sea lions.

Fishing has certain other effects on marine communities, which may be changed directly or indirectly through the removal of certain species or the alteration of their habitats by the fishing gear. Fishing affects the community directly by removal of species. One of the effects of this is the reduction in species diversity particularly on a local level. Decreases in species diversity may also come about indirectly, through alterations in predator–prey relationships and through the removal of individuals of fished species. This loss of species richness has been best documented in the case of coral reef fisheries. Although fishing reduces species diversity, it does not appear to cause species to become extinct. Fishing may also alter the size structure, reproduction, and life-history characteristics of the fished populations. Since most fishing selectively removes the larger specimens of a target species, the species' reproduction and subsequent recruitment may be affected if the major contributors to the reproductive pool are the larger individuals or if reproductive age is not reached until a large size is attained. Such size selection may also alter the sex ratios in the target species and curtail the reproductive life span, particularly in hermaphroditic species and gonochoristic species in which one sex is larger than the other. For example, Jennings and Kaiser (1998) note that about 50% of tropical reef fish are hermaphroditic, usually sequentially, meaning

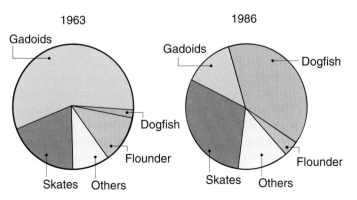

FIGURE 11.6 Percent composition of the demersal fishes of Georges Bank in 1963 and 1986. (From *Advancements in Marine Biology*, Vol. 34, p. 281, Fig. 26, © Academic Press, London.)

that they change from one sex to another as they grow. This means that if the larger individuals are either male or female and are systematically removed by fishing, the reproductive potential of the species is reduced, leaving a preponderance of just one sex in the fished population. Similarly, Smith and Jamieson (1991) report that, in a trap fishery for Dungeness crab in British Columbia, the size of females is smaller than that of males, so that the females are effectively barred from the fishery due to size limits on the traps. Thus, males are rare in the population, reducing mating and thus egg production.

Fishing can also alter the structure of the fished communities themselves. Perhaps the best and most recent example is the change in the composition of the fish communities of the Georges Bank fishing grounds in the North Atlantic. As Sherman (1991) reported, during the period from 1963 to 1986 the total biomass of the demersal fish declined precipitously and the composition of the community changed as well, as the desirable gadoids and flounders were fished out and replaced by spiny dogfish and skates (Figure 11.6).

When overfishing is coupled with natural changes in the environment, the effects may extend well beyond the immediate area. A good example of this effect is the relationship between changes in oceanographic conditions in the Barents and Norwegian seas and the fisheries for cod, herring, and capelin, as reported by Blindheim and Skjoldal (1993). In this system, intensive fishing of herring coupled with lower sea temperatures in the 1960s reduced herring predation on juvenile capelin and led to heightened capelin recruitment. However, in the 1980s the sea temperatures rose again, this time favoring cod recruitment, but not capelin. In 1984 and 1985, capelin recruitment failed, and this failure, coupled with intensive fishing, led to the collapse of the capelin stocks. Herring biomass was still low, and hence those other species which fed on herring and capelin had few alternative prey. Among the organisms affected were cod, seabirds, and seals, which began to starve and whose mortality rates increased (Figure 11.7).

Another source of destruction in fisheries is the **bycatch**—the nontarget organisms and undersized target organisms that are captured by the fishing gear. Most bycatch arrives on the fishing vessels dead or dying and is discarded back overboard, either because it is illegal to land the animals or because there is no market for them (Figure 11.8). Alverson et al. (1994) have estimated this bycatch at 27 million tons annually, or about 31% of the total world marine fisheries catch at that time. Bycatch thus represents a significant destructive impact on marine communities and is particularly devastating if it involves mammals, seabirds, turtles, and sharks, which have long life spans, experience long times before reaching reproductive age, and produce few young. The best-known example of mammal bycatch is the tuna purse seine fishery in the eastern tropical Pacific. The tuna schools are often associated with porpoises, and when the nets are set, porpoises are captured and drowned. Dayton et al. (1995) note that, up to 1987, perhaps 6 million porpoises were taken, substantially reducing populations. Since that time, various changes have been introduced into the fishery that have markedly reduced porpoise deaths.

Perhaps even more devastating than the tuna purse seine fishery was the bycatch associated with the high-seas drift-net fishery. At this fishery's height, in the 1960s, Japanese, Korean, and Tawainese boats deployed about 295,000 kilometers of nets in its waters. The nets tangled and killed thousands of seabirds and mammals. Brothers (1991) observes that, given the small population of albatrosses worldwide, the most devastating bycatch was the estimated 44,000 of them killed in the southern ocean by the Japanese longline fishery. Not all fisheries are equal in the amount of bycatch; the ratio of discard to landings is the highest for the shrimp and prawn trawl fisheries (Figure 11.9). The extensive bycatch of marine birds and mammals by the high-seas drift-net fisheries, particularly those of the North Pacific, finally caused such a stir that the nets were banned.

In connection with the preceding discussion, we must remember that the fishing gear itself has a deleterious effect on the habitat. Large trawling nets with heavy steel doors and chains that dig into the ocean bottom destroy large numbers of invertebrates that constitute the infaunal community structure. For example, Bergman and van Santbrink (2000) report that, in the North Sea, a single pass of a trawl caused direct mortality to 5 to 40% of the initial densities of gastropods, echinoderms, crustaceans, and annelids. As a result, the whole benthic community may change (Figure 11.10). The size of the area that can be so affected by the gear is enormous in some heavily fished areas. For instance, researchers have demonstrated that every square meter of the seabed in the North Sea is affected by fishing gear at least once a year. Dayton et al. (1995) report that some

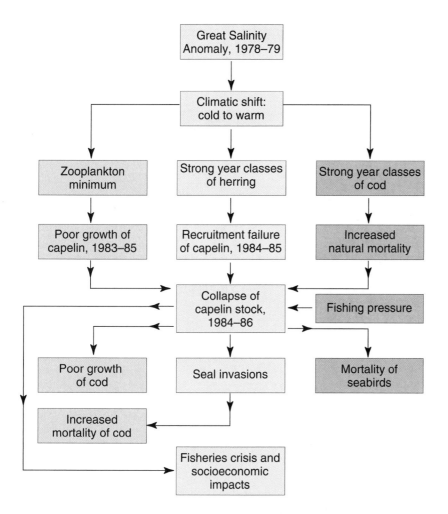

```
                    ┌──────────────────┐
                    │  Great Salinity  │
                    │  Anomaly, 1978–79│
                    └──────────────────┘
                             │
                             ▼
                    ┌──────────────────┐
                    │  Climatic shift: │
                    │   cold to warm   │
                    └──────────────────┘
```

FIGURE 11.7 Interactions among the organisms and the environment in the Barents Sea and Norwegian Sea ecosystems from the late 1970s to the mortality of birds and seals in the 1980s. (Modified from *Advancements in Marine Biology*, Vol. 34, p. 267, Fig. 19, © Academic Press, London.)

FIGURE 11.8 Fishing bycatch. (Copyright © Mark Conlin/ www.norbertwu.com.)

FIGURE 11.9 Ratio between bycatch and landed weight in 11 groups of fisheries. (Modified from *Advancements in Marine Biology*, Vol. 34, p. 268, Fig. 2 © Academic Press, London.)

FIGURE 11.10 Photos showing the difference in habitat (a) before and (b) after trawling gear has passed over an area. (From S. Simpson, 1998, Fishing trawlers scrape rock bottom. *Science News* 154:388. Photos courtesy of *Science News*.)

areas of the North Sea are trawled 0.5 to 7 times each year and areas of Long Island Sound and Narragansett Bay three times per year. Watling and Norse (1998) equate the effects of bottom trawling with clear-cutting of forests in the terrestrial environment and suggest that such trawling reduces the structural diversity of the bottom and therefore the biological diversity of the bottom communities. These same researchers also suggest that the area of the continental shelf of the world so affected is 150 times the forest area clear-cut yearly in the terrestrial environment. Recovery of the bottom communities destroyed or disturbed by dredging and trawling is affected by the life-history traits of the benthic organisms, coupled with the time between one dredging or trawling event and the next one, a period called the *return interval*. For some organisms, particularly the longer lived structural-forming ones that have a long recovery period, when the return interval is shorter than the time for recovery, the bottom community may be permanently changed.

Another type of human activity that is removing organisms from marine habitats and certainly contributing to community change is the collection of live marine organisms for the aquarium trade. This activity has the same effect as fishing, since it removes organisms permanently. The aquarium trade is centered primarily on coral reefs, because of their colorful inhabitants. Wheeler (1996) has indicated that the numbers of corals, fishes, and invertebrates collected for the aquarium trade is massive, and the United States is by far the largest consumer. According to Wheeler (1996), the worldwide trade in marine fishes alone is 18–60 million specimens. Although some of these fishes are aquarium bred, most—probably between 90 and 99%—are collected from the wild. Another unfortunate aspect of this trade is that as many as 90% of these animals die in transit or within a year or two in captivity. While several methods are employed to capture live fishes, one of the most common is to spray cyanide on the reef to stun the fishes. Unfortunately, cyanide also kills corals at concentrations hundreds of times lower than that used to collect the fishes, thus leading to high coral mortality. Worse, cyanide is poisonous to fishes, often causing delayed mortality only after the live fishes have been shipped. If delayed mortality is taken into account, only about 1 fish out of 10 collected survives. The effect of this collection on marine communities, especially coral reefs, is little known due to poor or nonexistent records, lack of enforcement of regulations (if any), and remoteness of collecting areas. The good news is that the aquarium trade is striving to improve survivability and working toward captive breeding and improved husbandry that will eventually reduce the numbers harvested from the wild.

Finally, there is the problem of lost gear. Every year, fishermen lose gear. This, however, does not mean that the gear stops fishing: Lost nets or traps may continue to capture marine organisms for varying periods (Figure 11.11). This phenomenon is known as **ghost fishing**. The extent of the problem can be appreciated from a few examples. Dayton et al. (1995) report that, on Georges Bank, 236 grapnel tows retrieved 341 actively fishing ghost nets, while an ROV survey off California found that 1% of the bottom was covered with lost gear, much of it still fishing. In Norway, nets that had been lost in 1983 were found to be still fishing in 1990. In the aforementioned high-seas drift-net fishery of the North Pacific, some 30,000–40,000 kilometers of nets were set per day, and the daily loss rate was estimated to be 20%.

SOLUTIONS

If we are serious about saving our fisheries and about protecting their marine habitats and their

FIGURE 11.11 Grey reef shark (*Carcharhinus amblyrhynchos*), dead after being entangled in an abandoned fishing net. (Photo courtesy of Norbert Wu/Peter Arnold, Inc.)

biodiversity, we must understand that the world's fisheries are dependent upon the integrity of their environment. This means that we must adopt a new approach to the management of the fisheries which employs ecological principles such as that called for by a National Marine Fisheries Advisory Panel report to Congress in 1999 and reiterated in the Pew Trust report of Dayton, Thrush, and Coleman in 2002. Basically, we must adopt an ecosystem approach that takes into consideration a more thorough understanding of the various biological and physical mechanisms that drive population dynamics, structure communities, and ensure biodiversity. Since time is also of importance, we must be able to make decisions without having the luxury of perfect knowledge. Also, it is likely that a part of this change in approach will be the implementation of marine protected areas and no-take zones to help ensure recruitment and to protect critical spawning areas.

TRAGEDY OF THE COMMONS

In 1968, Garrett Hardin published an article in *Science* magazine entitled "The Tragedy of the Commons," in which he pointed out that freedom of access to a common resource ultimately brings ruin to the resource. The core of Hardin's argument is that whenever there is a resource held in common by a large number of individuals or political entities, such as countries, each will try to maximize its own gain from that resource. Not to do so will allow another to do so, and the negative aspects of acting selfishly do not devolve on the one acting alone, but on all who share the resource.

Unfortunately, such an analysis is made by every entity sharing the resource. As a result, the resource is destroyed, and therein lies the tragedy.

The plight of the whale fishery can be likened to such a tragedy. What happened to it may also be applicable to other open-ocean fisheries, such as tuna, or potential fisheries, such as krill (see pp. 516–517).

Open-ocean fisheries lie in international waters beyond the control of any one country. Presumably, these fisheries are held in common for the good of the world's people. Since they are owned by no one, they and their resources are accessible by all. Countries with the means to exploit these areas do so because it serves their own interests. They do not have to answer to another nation or regulatory agency, and the tragedy is that, if they do not catch the fishes, another nation will. This short-term greed has destroyed the whale fishery and may do so next to tuna. The recent interest on the part of some countries in mining deep-ocean manganese nodules may also fall into this category.

■ **REGULATION**

The decline of many world fisheries due to common access and the pressures of an increased demand for food by an ever-increasing human population have led to friction among fishing nations and various attempts to regulate fishery resources. As long as the fishery exists on the continental shelf, where individual nations have jurisdiction by international convention, the fishery can be controlled by a single nation. In the past, the area of an individual nation's jurisdiction was 12 miles (19.3 km). Some, however, have acted unilaterally to extend these limits out to 50 or 200 miles or even beyond. These new limits have caused several confrontations, such as the "cod war" between Iceland and Britain, the seizure and fining of many U.S. tuna boats by Ecuador and Peru, and, more recently, the confrontation between Canada and Spain over fishing the depleted stocks on the Grand Banks. Following the third United Nations Conference on the Law of the Sea, a standard 200-mile-wide fishing and economic zone (called the "Exclusive Economic Zone") was established for the waters of each coastal nation. This means that individual nations now have full control of all fishery activity within 200 miles of their shores.

International agreements among nations for the control of fisheries have proved more difficult to bring about and implement effectively, as we witnessed in the case of the IWC. However, there are

some outstanding examples that have worked. One of the oldest is the International Fur Seal Treaty, established in 1911 among the United States, Russia, Japan, and Canada to regulate exploitation of the northern fur seal (*Callorhinus ursinus*). This treaty has been an outstanding success in regulating fur seal numbers to achieve a sustained yield. (The more recent declines in fur seal populations stem from large losses suffered by the adults in encountering the huge, miles-long gill nets set for salmon and tuna in the open Pacific.) Another success has been the International Halibut Commission, established in 1924 between Canada and the United States to regulate the then-declining Pacific halibut fishery. This was the first fish species to be conserved internationally. After establishment of the commission, the stocks of halibut recovered. Since these early attempts, other international regulatory agencies have been established, including the Atlantic and Baltic commissions, the International Tuna Commission, and the International North Pacific Fisheries Commission (INPFC). Many of these more recent regulatory agencies have had a beneficial effect on certain fisheries, but problems remain. The major one is that the various commissions have limited jurisdiction geographically or are limited to certain species. For example, the waters around Japan are not in the jurisdiction of the INPFC, and the Halibut Commission regulates only that species.

It is well to remember that not all commissions and regulatory agencies are successful. (Witness the IWC.) Perhaps the most dramatic recent failure of a regulatory agency is the International Commission for the Conservation of Atlantic Tuna (ICCAT). Despite the existence of this regulatory body, Atlantic bluefin tuna populations have continued to drop, 90% since 1975 for those which breed in the western Atlantic and 50% for those which breed in the Mediterranean.

To effect a truly worldwide regulation of fisheries, the current commissions or others will have to expand their jurisdiction to include all species in their areas. Those areas where no regulatory agencies exist, mainly in the Southern Hemisphere, will have to be brought under control. Also, regulatory agencies increasingly will have to take a whole-ecosystem approach.

■ **NEW FISHERIES**

Given that the current world fisheries catch is either declining or plateauing in many long-established fisheries, are there any prospects for finding substantial new fisheries? The answer may be yes, but whether these new fisheries will be able to contribute significantly to the human food supply is much less certain.

Now that the stocks of baleen whales have been reduced to the point of extinction in Antarctic waters, a relatively new fishery has developed for the **krill** (euphausiids) that formerly constituted the food of the whales. Krill exist in large numbers in the seas around Antarctica and are large enough (6 cm) to be captured without resorting to very small mesh plankton nets (Figure 11.12). The most conservative estimates of stocks of krill are 50 million tons, but other estimates range up to 100 million tons, according to Idyll (1970). In 1977, the harvest of a krill fishery was estimated at as much as 150 million tons. The disparity in these early estimates is because we really do not have reliable estimates of krill population sizes, where they live, or their local distribution. We also lack information on the age structure of the populations and where they spend the winters. (See

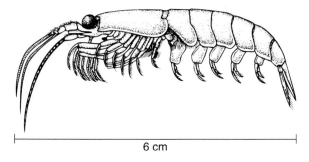

6 cm

FIGURE 11.12 Krill, or *Euphausia superba*, of the southern ocean.

pp. 254–256). Because of this absence of basic biological knowledge, the most recent estimates of the annual production of krill reported by Nicol and de la Mare (1993) range from 75 million tons to 1.35 billion tons, depending on what parameter is used to make the extrapolation (Figure 11.13). There is little doubt, however, that the potential for this fishery is enormous—certainly the largest of any unexploited fishery in the world today. Much remains to be learned about the krill fishery and what effect the taking of huge numbers of the animals may do to the remainder of the fragile Antarctic ecosystem (see also Chapter 5 pp. 254–256).

Another area of potential exploitation, until recently relatively untapped, is the area of the continental slopes at depths of 300–1,000 m. Of late, substantial stocks of blue whiting have been discovered in this zone off the British Isles. A similar area is being exploited off the Pacific coast of North America for grenadiers (Pacific roughy) and "rock cod" (*Sebastolobus*). The zone also contains substantial numbers of squid.

Rodhouse (1995) has suggested that there are significant underexploited squid resources in the southern oceans surrounding Antarctica. He estimates that the various higher vertebrate predators in these oceans take about 34 million tons of squid, some of which would be suitable for human food if they are palatable and have behavior patterns that would allow them to be harvested commercially.

Finally, the entire Indian Ocean has a low level of fisheries development by current standards, producing only about one-fifth the catch per unit area of the

From biomass estimates	135–1350
From estimates of primary production	100–500
From estimates of predator requirements	470
From the abundance of krill larvae	75

Estimates of annual production of Antarctic krill (millions of metric tons)

FIGURE 11.13 Various estimates of the annual production of krill, showing the widely varying numbers, depending on the source. (After "Ecosystem Management and Antarctic Krill," S. Nicol and W. de la Mare, *American Scientist*, Vol. 81, No. 5, pp. 36–47. Copyright © 1993 American Scientist/Elyse Carter. Reprinted by permission.)

Atlantic and Pacific. It would appear that further efforts in these waters would increase the catches without overexploiting the stocks. On the basis of primary production alone, Idyll (1970) estimated that the Indian Ocean can produce 10 million metric tons of fishes, as opposed to 7.96 million tons harvested in 1997.

■

WE CAN SUMMARIZE *this section by saying that the catches of marine fisheries have peaked and are now stagnating or declining, and many world fisheries are showing signs of overexploitation due to unregulated access and overfishing. Furthermore, only limited amounts of increases can be expected from fishing stocks and underexploited areas, with the possible exception of the Antarctic krill. An estimate of a total sustainable yield of 100 million metric tons or less for the world's oceans appears reasonable. With catches now at about 83 million tons, it is unlikely that a significant increase in food can come from harvesting the wild populations in the sea. Therefore, the seas cannot, and will not, be the source of substantial amounts of food for an increasing human population.*

■

Mariculture

As the limit has come into sight for the harvesting of wild populations of marine animals and plants in fisheries, humans have begun to culture various marine and estuarine species in situations that are analogous to terrestrial-based agriculture. Such a practice is termed **aquaculture** and involves humans rearing selected aquatic plants and animals under controlled conditions to increase the amount of food available to humans compared with that obtained through traditional fishing. In this section, we are concerned only with marine aquaculture, or **mariculture**.

■ HISTORY AND EXTENT

Despite evidence that marine organisms were cultured at least as far back as ancient Rome, aquaculture through the centuries has never become a significant contributor to human food. Food from the seas remained the province of the hunters and gatherers—the fishermen. Only in the twentieth century has aquaculture received wide recognition and undergone significant development. This development is partly in response to the pressure for more human food,

especially protein, and partly from the decline of wild stocks. Mariculture encompasses marine organisms raised not only for human food, but also for other products, such as pearls and terrestrial animal food.

Mariculture is practiced throughout the world in the tropical and temperate zones, but is more technologically advanced in the industrialized nations of the Northern Hemisphere. At present, it is confined to shallow coastal embayments or artificial ponds.

■ SPECIES CULTIVATED

Few marine organisms are currently cultivated, and most are either the type that fetches a premium price on the market, which economically justifies the high cost of their culture, or the type that produces large amounts of biomass under intensive culture. In the former category, we could include certain species of shrimps and abalone cultured in Japan. In the latter category are oysters and mussels. As might be anticipated, estuarine and bottom-dwelling species are more commonly cultivated than pelagic, open-ocean species.

Fishes that are successfully cultured commercially include several species of Pacific salmon (*Oncorhynchus*), Atlantic salmon (*Salmo*), mullets (*Mugil, Chanos*), yellowtail (*Seriola*), and flatfishes (*Pleuronectes, Solea*). Crustaceans cultured for market include various species of shrimps and prawns (*Penaeus, Metapenaeus, Leander*). Among the mollusks, oysters (*Crassostrea, Ostrea*) and mussels (*Mytilus*) are the subject of intensive commercial culturing (Figure 11.14). Other mollusks raised include several clams (*Anadara, Tapes, Mactra, Mercenaria, Meretrix*) and the abalone (*Haliotis*). Compared with the total list of fishes, crustaceans, and mollusks taken for food in various fisheries, this is a very short list. Other species, such as lobsters, scallops, and crabs, have been cultured experimentally, but at present, no commercial industry exists, suggesting that there is considerable room for future expansion.

Mariculture of certain marine algae is particularly significant in China and Japan. The red algae *Porphyra yesoensis* and various brown *Laminaria* species are commonly cultivated. In China, about 4.8 million tons of brown and red algae are produced annually.

■ YIELDS

Even though aquaculture has lagged far behind agriculture in development, its potential for the production of human food is considerable. For example, Bardach (1968) has noted that the unmanaged common oyster

FIGURE 11.14 Pearl oyster culture racks in Japan. (Photo courtesy of George Gerster/ Photo Researchers, Inc.)

grounds along the East Coast of the United States yield only about 10 kg of oysters per hectare, while the intensive culture of oysters in hanging racks in Japan yields 58,000 kg/hectare. Even larger yields are possible with mussels. In Spain, an intensive culture on hanging racks produces 300,000 kg/hectare!

A similar situation occurs with fishes. In Indonesia, the milkfish (*Chanos chanos*) is cultured in brackish ponds, but is not subjected to such intensive care as in Taiwan. Yield is 400 kg/hectare in Indonesia, but 2,000 kg/hectare in Taiwan.

Presently, the world yield from aquacultural practices is probably over 4 million tons, but most is from fresh water. Indeed, in 1992, two aquacultured freshwater fish—the silver carp and the grass carp—were included among the top ten fish, by weight, landed in the world. This was the first time that any cultured fish had made that listing. While freshwater aquaculture is becoming significant, mariculture currently contributes little human food yield. However, the science of mariculture is as yet at a low level of development. If the previous examples are any indication of what can be expected, the future yield could be substantial; however, it is not great for reasons discussed in the next section.

■ PROBLEMS WITH, AND RESTRICTIONS ON, MARICULTURE DEVELOPMENT

Mariculture is not a panacea for the problem of feeding the human race, nor is it likely to contribute significantly to food supplies in the immediate future. The reasons for this are numerous; space permits mentioning only the more significant.

First, marine animals require the water in which they live to have the proper physical and chemical characteristics. These are difficult to maintain in ponds or tanks and require complex filtering and water treatment to remove wastes and other toxic materials. Attempts to raise marine organisms on a large scale mean an often economically crippling investment in equipment to maintain the water quality.

Another drawback is that many marine species go through a series of larval stages, each requiring different conditions and food, before attaining adult size. To rear these forms successfully is often too costly or simply impossible in captivity. Still other species are pelagic in the open ocean and will not survive when confined to smaller spaces. Since we cannot control animals in the pelagic areas, we cannot culture there.

Then there is the problem of various diseases and parasites, which proliferate under captive or crowded conditions. Last, but certainly not least, in the United States there is the problem of government-imposed regulations and permits that are often extremely costly both in time and money. For example, in Maine and California, the permitting process may take up to six or seven years to complete and cost over $100,000, on average.

Present and future mariculture appears most practical for species that have fewer or no larval stages, that are nonpelagic, and that live and reproduce well in crowded conditions. Furthermore, our inability to control conditions or organisms in the open ocean means that future mariculture and increased yields must come from sheltered inshore bays, lagoons, and estuaries. Since the areas suitable for such culturing are a very minor part of the total world oceans, even at maximum culturing intensity, we cannot expect substantial quantities of food from mariculture in the near future. Recently, experiments aimed at growing out pelagic fish species in cages in the open ocean have had some success, so this approach may be viable in the future.

The foregoing notwithstanding, there are definitely some success stories in mariculture. Perhaps the one that is best known to people in the north temperate industrialized countries is salmon. Both Atlantic and Pacific salmon are currently cultured on a large scale in North America, Europe, and Chile such that the fish are now available year round and at low cost. However, this kind of success seems to come with a caveat: Hites et al. (2004) report that such farmed salmon have significantly higher concentrations of contaminants than do the fishes caught in the wild, probably due to the feed.

Pollution

The seas have been considered the ultimate dumping grounds for the wastes of human societies. In the same way that fishery stocks were once considered inexhaustible, so we have felt until recently that the immense volume of the world's oceans had an infinite capacity for absorbing all of our waste. In other words, dilution was the solution to pollution. Lately, we have realized that however large they are, the oceans are not infinite in their capacity to absorb wastes. We have also learned that, even in very small amounts, some of our wastes have significant effects on communities and species. This section discuss briefly a few of the more significant sources of marine pollution and their effects or potential effects on marine ecosystems. (For more extensive information on pollution, see the bibliography at the end of the chapter.)

■ OIL

Extensive media coverage over the last 25–35 years has made the oil pollution problem in the seas widely known to the public. Oil pollution in the seas results primarily from the spillage of crude oil from offshore drilling platforms or from accidents involving tankers. What is less well known is that a few marine environments are subject to natural oil contamination (Coal Oil Point in Santa Barbara, California, for example). One of the largest natural seeps was that reported by Harvey et al. (1979), who discovered a seep 1.5 km wide, extending for hundreds of kilometers off the island of Trinidad in the Caribbean Sea. This seep was found at a depth of 100 m, had a thickness of 100 m, and contained in excess of 1 megaton of oil.

Crude oil released into the sea usually floats, although some components may sink, and after a time, the evaporation of certain fractions may cause even more to sink. Away from land, the floating oil probably has little effect on the majority of planktonic and nektonic organisms. The lone exceptions are the marine birds and certain mammals, such as sea otters. When birds contact the oil, they become coated with it, and their feathers lose their insulating qualities. As a result, most die of exposure in the water or are unable to feed, which ends in the same way. Oil is, therefore, devastating to marine birds, causing significant losses.

Inshore, the oil may coat the shallow subtidal and intertidal zone and smother the communities that live there. The devastation is nearly complete initially, but recovery usually occurs with time. More serious than the oil itself have been the various chemicals, such as detergents, used to break up or disperse the oil in the water. As Nelson-Smith (1973) notes, in the Torrey Canyon disaster of 1967, these chemicals were shown to have caused more mortality of marine organisms than the oil itself.

Oil disasters have become more numerous with the increased oil demand of the industrialized world, the necessity of transporting oil from distant sources, and the increased numbers of offshore drilling rigs.

The immense damage caused by oil disasters involving the giant supertankers is revealed by the loss of the *Amoco Cadiz* off the Brittany coast of France in March 1978. In this calamity, 220,000 tons of oil spilled and one-third reached shore to contaminate 320 km (192 miles) of shoreline. The spill contaminated or destroyed major commercial marine products, including oyster beds, mussel beds, and lobster holding pens. By 1985, many of the areas affected by the spill appeared to the casual observer to be completely recovered. However, investigations such as that done by a joint United States–France commission consisting of the U.S. National Oceanic and Atmospheric Administration (NOAA) and the French Centre National pour l'Exploitation des Océans (CNEXO) have shown that some differences still exist between communities in the spill area and those present in the same area prior to the spill. This is particularly true for the bays and estuaries where studies, such as those by Riaux-Gobin (1985) and Boucher (1985), suggest that the hydrocarbons have affected the long-term trends of community structure, particularly with respect to microalgae and nematodes.

The great size of the area that can be affected by a single oil spill from a modern, large oil tanker was brought home to the world through the extensive media coverage of the disaster of the *Exxon Valdez*. Just past midnight on March 24, 1989, the tanker *Exxon Valdez* ran aground in Prince William Sound,

Alaska, spilling 260,000 barrels of Alaskan crude oil. The oil eventually was distributed by winds and currents over 30,000 km^2 of the coastal waters of Alaska (Figure 11.15). According to Platt et al. (1990), at the time of the spill this area was occupied by about 1 million seabirds, of which about 400,000 were killed. Of these, 70% were murres—certainly a significant loss. There was, however, no massive loss of fishes.

Undoubtedly the greatest losses from the spill were among the birds and marine mammals. It has been hard, however, to pinpoint exactly how large the losses were, because of the lack of recent census data on those animals immediately before the spill. In the case of killer whales (*Orcinus orca*), before the spill there were an estimated 297 individuals in Prince William Sound in 11 resident pods. Following the spill, some pods showed decreases in numbers, but

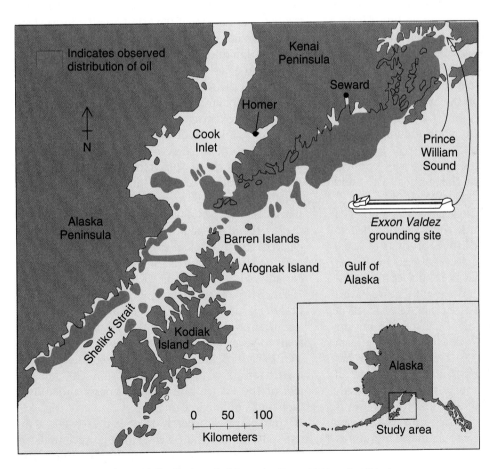

FIGURE 11.15 Observed distribution of oil from the *Exxon Valdez* oil spill between March 24 and May 20, 1989. (From "Immediate Impact of the Exxon Valdez Oil Spill on Marine Birds," J. F. Platt et al., *Auk*, Vol. 107, pp. 387–397, 1990 American Ornithologists' Union. Reprinted by permission.)

we are unsure of the source of the mortality, since no carcasses were ever recovered. For sea otters (*Enhydra lutris*), our data are somewhat better. Prior to the spill, the population size was estimated to be about 5,000 individuals in Prince William Sound. Following the spill, 878 carcasses were recovered; 123 others were brought in alive, but died in treatment; and 37 were treated and farmed out to aquaria. Therefore, about 1,000 animals were removed from the population as a result of the spill. What we do not know is how many additional animals whose carcasses were not recovered may have been lost. In the case of the harbor seals (*Phoca vitulina*), the prespill population was estimated at 2,500–3,000 individuals and was declining about 10% annually. Following the spill, the decline went to 35% in oil-affected sites, and about 200 individuals were killed.

Among the algae and invertebrates, the intertidal forms suffered most from the spill, but the results varied with the site, the tidal height, and the kind of organism. High intertidal algae, limpets, and mussels were most affected. The losses from the smothering effects of the oil were compounded by the hot water and steam used to clean the rocky intertidal zone. This use of hot water made the intertidal zone cosmetically more appealing, but the destruction it brought actually delayed the recolonization of those shores compared with those which were left to be cleaned by natural sources. Interestingly enough, there is little evidence that the subtidal areas and communities suffered much from the spill.

Before the *Exxon Valdez* oil spill, most studies of the ecological effects of such spills were limited to short-term monitoring and laboratory tests of acute toxicity to laboratory animals. However, the *Exxon Valdez* spill environment has been subject to continued study for 14 years, with the result that we have a much better understanding of the long-term effects of oil in the marine environment and on the marine ecosystem. In a summary review of these long-term studies, Peterson et al. (2003) report that there has been unexpected multiyear persistence of oil and oil residues in the ecosystem in surprising amounts, leading to chronic exposure to a number of organisms and resulting in long-term impacts on various algae, invertebrate, and vertebrate populations. The authors report that the long-term impacts fall into three categories. The first is the long-term exposure to persistent oil of species associated with shallow, relatively undisturbed sediments in which the oil was

sequestered from loss and weathering and has remained. For example, this type of exposure has affected the sea otter population in Prince William Sound that feeds on sediment-associated prey such as various clams. In those areas which became heavily oiled, sea otters had not yet recovered to their prespill numbers by the year 2000. The second category is sublethal exposure leading to compromised health, growth, and reproduction. For example, pink salmon embryos subjected to sublethal chronic oil exposure experienced stunted growth and a survival rate at sea of only half that of those not so exposed. Sublethal exposure also affected certain shorebirds, such as the black oyster catcher, which suffered reduced breeding rates and higher chick mortality where foraging territories were heavily oiled. The final category is cascading indirect effects, which are transmitted via various biological interactions that go beyond acute mortality. In the case of Prince William Sound, the oil spill led to the initial loss of the alga *Fucus gardneri* which was the most significant provider of cover for other organisms in the intertidal zone. The death of *F. gardneri* led to a cascade of events, including the loss of important gastropod grazers and predators and blooms of ephemeral green algae. Further cascading followed, and *Fucus* was inhibited from returning, since its recruits settle under the adult canopy to avoid desiccation. These indirect effects influenced succession on the rocky intertidal shores for over a decade. Peterson et al. (2003) have concluded that in the future it will be necessary to expand the time frame and scope of oil pollution studies beyond that of the initial acute toxicity stage, to encompass delayed effects on the various components of the ecosystem. Their comparison of the old and emerging paradigms is given in Table 11.6.

Since the *Exxon Valdez* spill, there have been relatively few massive oil spills from tankers. The most significant was that of the tanker *Prestige*, which split apart in a storm on November 19, 2002, off the Galician coast of Spain. This spill released about 79,000 metric tons of oil, which affected some 900 kilometers of the French and Spanish coasts and caused perhaps $1 billion in damage.

Until relatively recently, there had been no extensive spills from offshore wells, although such wells are potentially more devastating than spills from ships, since the amount of oil gushing out is not limited by the hold of the ship. Such a disaster happened in the Gulf of Mexico, where, in 1979, the Ixtoc I well blew

TABLE 11.6

Changing Paradigms in Oil Ecotoxicology, Moving from Acute Toxicity Based on Single Species Toward an Ecosystem-Based Synthesis of Short-Term Direct Plus Longer-Term Chronic, Delayed, and Indirect Impacts

OLD PARADIGM	EMERGING APPRECIATION
Physical shoreline habitat	
Oil that grounds on shorelines other than marshes dominated by fine sediments will be rapidly dispersed and degraded microbially and photolytically.	Oil degrades at varying rates, depending on the environment, with subsurface sediments that are physically protected from disturbance, oxygenation, and photolysis retaining contamination by only partially weathered oil for years.
Oil toxicity to fish	
Oil effects occur solely through short-term (~4-day) exposure to water-soluble fraction (1- to 2-ringed aromatics dominate) through acute narcosis mortality at parts-per-million concentrations.	Long-term exposure of fish embryos to weathered oil (3- to 5-ringed PAHs) at ppb concentrations has population consequences through indirect effects on growth, deformities, and behavior, with long-term consequences on mortality and reproduction.
Oil toxicity to seabirds and marine mammals	
Oil effects occur solely through short-term acute exposure of feathers or fur and resulting death from hypothermia, smothering, drowning, or the ingestion of toxics during preening.	Oil effects also are substantial (independent of means of insulation) over the long term through interactions between natural environmental stressors and the compromised health of exposed animals, through chronic toxic exposure from ingesting contaminated prey or during foraging around persistent sedimentary pools of oil, and through the disruption of vital social functions (caregiving or reproduction) in socially organized species.
Oil impacts on coastal communities	
Acute mortality through short-term toxic exposure to oil deposited on shore and on the shallow seafloor or through smothering accounts for the only important losses of shoreline plants and invertebrates.	Cleanup attempts can be more damaging than the oil itself, with impacts recurring as long as the cleanup (including both chemical and physical methods) continues. Because of the pervasiveness of strong biological interactions in rocky intertidal and kelp forest communities, cascades of delayed, indirect impacts (especially of trophic cascades and biogenic habitat loss) expand the scope of injury well beyond the initial direct losses and thereby also delay recoveries.

Source: From Peterson et al, *Science,* Vol. 302, 12/19/2003 American Association for the Advancement of Science. Used with permission.

and spewed out from 450,000 to 1.4 million tons of oil in 295 days before being capped. The large range in the amount of oil spilled is due to uncertainties involved in estimating the flow from the well. This was the world's greatest oil spill; it is also one of the least studied, so data on its effects are not available.

As if accidental oil spills were not enough of a problem, 1991 saw a massive deliberate oil spill in the northern Persian Gulf, the result of the Iraqi army releasing stored Kuwaiti crude oil from tanker loading piers. The extent and damage of this spill remains questionable in view of the small number of follow-up studies that were carried out, as well as a relative ignorance about the condition of the communities in the Persian Gulf prior to the spills. During the war, early reports suggested that perhaps 30,000 birds had

been killed, about 20% of the mangroves had been covered with oil, and up to 50% of the coral reefs had been affected. According to Gerges (1993), follow-up studies by NOAA in 1992, a year after the war, indicated that 6–8 million barrels of oil had probably been spilled into the sea, producing an oil slick that contaminated a large portion of Saudi Arabia's northern coastline, but with much less effect on the Iraqi, Kuwaiti, and Iranian coasts. The oil pollution was exacerbated by the destruction of Iraqi sewage treatment plants, which resulted in the release of perhaps 50,000 m^3 per day of raw sewage into the marine waters of Kuwait. Surprisingly, even given this volume of pollutants, the follow-up studies by NOAA at 150 bottom sites a year later found oil at only three and no demonstrable direct effects on most coral reefs, except for three in Kuwaiti waters, where numerous corals were bleached or dead. Significant mortalities among corals off Saudi Arabia in 1992 were attributed to natural causes. What effect these pollutants may have on commercial fisheries or other communities in the Persian Gulf remains to be assessed.

Only a few disasters per year on the scale of the Iraqi oil release could add up to significant destruction, particularly of shallow-water communities. With both oil shipping and offshore drilling increasing, we can expect more such disasters in the future. A more optimistic view of the fate of oil in the sea comes from a comprehensive review by the National Academy of Sciences (1985), which concluded that biological impacts vary considerably. Low-energy environments, such as estuaries, salt marshes, and intertidal and subtidal sedimentary areas, appear most vulnerable and also show the slowest rates of recovery. There appears to be little detectable long-term effect on pelagic communities, save for birds, and in all cases examined, mechanisms for oil degradation should allow eventual recovery. Given the new data on the chronic effects of oil over a decade in the case of the *Exxon Valdez* spill, it is obvious that we do not yet know what the time span of recovery really is.

■ SEWAGE AND GARBAGE

The discharge of human sewage and garbage into coastal waters is practiced throughout the world. The sewage may or may not have had some treatment before discharge. Sewage adds a large volume of small particles to the water and also large amounts of nutrients. When the sewage is discharged in small volumes and with adequate diffusing pipes, it is difficult to detect any long-term effect on the communities of the open coast. Discharged in large volumes and in semienclosed embayments, sewage can have a devastating effect. Two examples should suffice.

In southern California, the Los Angeles area discharges in excess of 330 million gallons of sewage per day at the White's Point outfall off the Palos Verdes Peninsula. Extensive studies in and around this outfall and others in the area over a period of years by personnel of the Southern California Coastal Water Research Project (Bascom, 1978) have revealed that sewage has caused significant degradation in benthic invertebrate communities in areas near the outfall, compared with similar areas some distance away. In addition, the kelp beds in the vicinity of the outfall have disappeared, sea urchins have increased

markedly in number, and diseased fishes are more prevalent. About 4.6%, or 168 km^2, of the 3,640-km^2 southern California mainland shelf has been changed or degraded as a result of sewage discharge from four major outfalls (Figure 11.16).

Once known for its flourishing coral reefs, Kaneohe Bay on the Hawaiian island of Oahu is a large, shallow embayment with restricted water circulation. Banner (1974) reports that, since World War II, the area around the bay was subjected to a tenfold increase in population. As a direct result of this urbanization, the bay was subjected to massive domestic sewage discharges, along with significant siltation from runoff during storms. The result was the total destruction of the coral reef communities in two-thirds of the bay and their replacement by a massive, smothering growth of the green alga *Dictyosphaeria cavernosa* (Figure 11.17). By 1978, the sewage discharges were eliminated from the bay, and in a resurvey of the area in 1983, Maragos et al. (1985) reported a remarkable recovery of corals and water clarity, suggesting an early recovery of the reefs in this bay. However, Smith et al. (2002) have noted that an alien Philippine red alga, *Kappaphycus* sp., seems now to be invading the reefs.

In addition to sewage, large amounts of garbage are dumped into the sea each year from shore and from ships. What effect this material may have on the marine communities is unknown. Evidence of the extent of the garbage pollution is found in the occurrence of small bits of plastic in plankton tows taken in many parts of the North Atlantic and Pacific Oceans and in the digestive tracts of fishes, diving seabirds, and marine mammals. As yet, there is little scientific evidence that these plastic bits cause mortality, although it has been suggested that they cause intestinal blockage. More ominous is that they do not readily decompose, and the increasing use of plastics by modern human societies probably means increasing amounts in the oceans.

The whole spectrum of the offal of civilization characterizes the material dumped into the waters off New York City. Included are dredge spoils, sewage, chemicals, garbage, and various construction materials, the amounts of which, poured into the open coastal waters, is of such volume that it is visible even from satellites, according to Gunnerson (1981). As regards sewage alone, 127 municipal discharges contribute 2.6×10^9 gallons per day, or more than 100 times that of the White's Point outfall in the

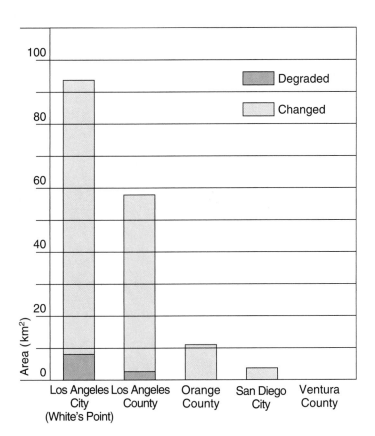

FIGURE 11.16 Changed or degraded areas near southern California's major sewer outfalls. (After W. Bascom, ed., 1978, Coastal water research project annual report, Los Angeles.)

Los Angeles area. The results of 35 years of dumping have been significant. Oxygen levels near zero have occurred over extensive bottom areas off New Jersey and have led, on numerous occasions, to mass fish and shellfish mortalities, as well as to chemical and bacteriological contamination, of fishes and shellfishes. Some fisheries have even had to be abandoned, according to Gunnerson (1981). Furthermore, even though most of the waste is dumped many miles offshore, certain materials, such as plastic, tar, grease balls, and even such medical wastes as hypodermic needles, have come ashore to contaminate beaches throughout the area. Clearly, human activities have had, and will continue to have, a significant effect on the marine ecosystem.

■ **CHEMICALS**

More insidious than oil or sewage, which are at least visible, are the various invisible toxic chemicals produced by the industrialized nations. Often transferred through the food chains in the sea, these chemicals find their way into the oceans' ecosystems and exert their effects in animals and places removed in time and space from the source. Because of this phenomenon, it is difficult to pin down the effects of a given chemical, especially if they turn up years later. Also, certain marine organisms enhance the toxic effects of many chemicals because they accumulate the substances in their bodies far above levels found in the surrounding water. Another factor that increases the effects of chemicals on living systems is **biological magnification**, in which the concentration of the chemical increases in the bodies of organisms with each succeeding trophic level. The chemicals are not metabolized in an organism; therefore, the amount accumulated in the tissues remains there. When several such individuals are consumed by a carnivore of the next trophic level, the carnivore gains the chemicals from all the individuals, increasing the concentration of the chemicals in the carnivore's body. If the food chains are long, significant levels of the chemical may accumulate in the top carnivore. As we saw in Chapter 1, marine food chains are long, so top carnivores in marine systems often have heavy loads of these chemicals. Humans also consume marine organisms, mainly from the higher

FIGURE 11.17 Kaneohe Bay, Hawaii, showing the extent of destruction of coral and overgrowth by the alga *Dictyosphaeria* at its maximum extent before cleanup was initiated. The darker blue portion indicates the area of destruction. (Modified from *Proceedings on the Second International Coral Reef Symposium*, Vol. 2, pp. 685–702.)

trophic levels. Chemicals have already had an effect on certain of our fisheries, such as those for swordfishes and tuna, in which levels of mercury have often been higher than the safe levels established by the FDA for humans, causing fishes to be rejected for consumption. Even more tragic is the case of Minamata disease, documented by Goldberg (1974).

In the late 1930s, the Chisso Corporation of Japan established a factory on the shores of Minamata Bay to produce vinyl chloride and formaldehyde. By-products of this plant contained mercury and were discharged into the bay. Through biomagnification, the marine fishes and shellfishes accumulated high concentrations of the toxic compound methylmercury chloride. The fishes and shellfishes were, in turn, consumed by the human inhabitants of the area. About 15 years after the dumping of mercury into the bay began, a strange, permanently disabling neurological disorder began to appear among the inhabitants, especially the children. This disorder was called **Minamata disease**. The cause was diagnosed as mercury poisoning in 1959, but it was not until the early 1960s that the active mercury compound was identified and the link to the factory discharge established. It took even longer—until the 1970s—for Japan to halt the discharge of mercury into the ocean and for other nations to blacklist mercury dumping and establish standards for acceptable levels in food.

A second, well-documented example is the case of the pesticide dichlorodiphenyltrichloroethane (DDT), the first of the synthetic pesticides of a class known as chlorinated hydrocarbons. DDT was put into worldwide use in 1945 and was hailed as a boon to humankind for its effectiveness in destroying a wide range of insect pests while remaining relatively nontoxic to humans. It has, however, a few attributes that were to bring about serious environmental consequences. First, it is a remarkably stable compound in natural systems. DDT and dichlorodiphenyl-trichloroethylene its first breakdown product, (DDE), persist for years; just how long is unknown. DDT is relatively insoluble in water, but soluble in fats or lipids, and it adheres strongly to particles.

Although DDT is not used in the marine environment, it has entered marine food webs through runoff from land, precipitation, and dumping. As a result, by the 1960s, DDT had been found in marine organisms as remote as Antarctica. The insidious effects of DDT in marine food webs became apparent in the 1960s, when certain top carnivores were found to have high levels in their tissues. DDT enters the marine food chains at the level of plankton, in which the chemical adsorbs on the surfaces of the planktonic organisms or is dissolved in their lipids. Since DDT is not metabolized, biomagnification concentrates it through succeeding trophic levels to the top carnivore. Concentrations of DDT in certain marine fishes, such as mackerel, were found to exceed the permissible 5 ppm established by the FDA as safe for human consumption, and many catches had to be destroyed.

The most serious effects, however, occurred among marine birds. Once DDT reached certain high levels in birds, it interfered with calcium deposition in their eggshells. As a result, thin eggs were laid, which broke open when the birds incubated them. The greatest amount of attention was focused on the brown pelican (*Pelicanus occidentalis*), in which successful breeding virtually disappeared in the main rookeries on the islands off southern California in the late 1960s and early 1970s. As Goldberg (1976) notes, the birds' reproductive failure was traced directly to the high levels of DDT in their tissues. What was curious here was that most of southern California is urbanized and had not used much DDT—certainly not as much as that used in agricultural areas. The waters around Los Angeles, however, had concentrations of DDT averaging 370 ppm, as opposed to 1 ppm off Baja California. Where did the chemical come from? The source was discovered to be the White's Point sewer outfall. It turned out that a single chemical plant that produced most of the world's DDT was located in Los Angeles and was dumping its

wastes into this outfall at the rate of 100 tons a year, a figure ten times as high as the total amount of DDT carried per year by the Mississippi River into the Gulf of Mexico. The decline of the brown pelican was thus linked through the marine food web to a toxic chemical discharged from a single source. Subsequent cessation of DDT dumping in Los Angeles has been followed by the slow recovery of the pelican.

These are only two examples of the problems associated with toxic chemicals. They emphasize the insidious nature of the often much-delayed effects and the difficulty of tracing the source of a problem through complex food webs about which we know relatively little. Since modern industrial states now produce thousands of actually or potentially toxic chemicals each year, of which many will enter the oceans, how many more of these disasters will we see in the future?

■ RADIOACTIVE WASTES

Since 1944, the world's oceans have been receiving an input of radioactive wastes generated from the production of nuclear weapons and electric power. Major sources of input have been fallout from nuclear weapons tests, releases or dumping of wastes from nuclear fuel cycle systems (mainly power or reprocessing plants), the sinking of nuclear powered submarines, and certain accidents. In the last category, one of the most significant contributors was a 1964 incident involving an aerospace nuclear generator that reentered the atmosphere when the launch malfunctioned and that deposited into the oceans a quantity of plutonium-238 equal to half the total oceanic deposits of the isotope that were already there. In the category of reprocessing plants, the Sellafield nuclear reprocessing plant in England is the world's largest and oldest operating nuclear power plant. Over the years, various accidents, chronic emissions, and overflows of nuclear wastes stored at the plant have caused the nearby Irish Sea to lay claim to be the most radioactive sea on the planet. In a similar vein, the purposeful discharge of radionuclides from the Hanford Reservation in Washington State to the Columbia River from the 1940s to 1971 created a large plume extending from the mouth of the river to 350 miles offshore. The radionuclides were transported via sinking particles to the sediment, where they were actually used by oceanographers to trace sediment movements on the continental shelf.

There are two main categories of radioactive wastes: **Low-level wastes** contain less than 1×10^{-8} curies of radioactivity per gram: **high-level wastes** contain more than 1×10^{-8} curies per gram.

At present, the dumping of high-level radioactive wastes in the oceans is prohibited by international agreement, but the dumping of low-level wastes is permitted. Low-level ocean dump sites have been designated and used in the northeast Atlantic (one site), off the coast of Europe (three sites), off the U.S. east coast (one site), and off the U.S. Pacific coast (one site). From 1946 to 1970, the United States dumped approximately 107,000 containers of low-level radioactive wastes at the two sites off our coasts, according to Park et al. (1983). Currently, the only active site is the one off the east coast. Whereas low-level radioactive waste seems manageable, there is considerable concern about high-level wastes, such as those generated from discarded fuel assemblies from nuclear generators or those which arise from various reprocessing operations. These wastes usually have long half-lives and powerful ionizing radiation. Such wastes must be safely isolated from the biosphere for periods that are geological in scale.

If a radioactive waste can be contained for ten times the half-life of the radionuclide, the radioactivity will be reduced by 1,024 times. Plutonium 239, a common radionuclide of spent nuclear fuel, has a half-life of 24,100 years. This means that it must be isolated from the biosphere for 241,000 years.

There is much debate over what to do about the disposal of these high-level nuclear wastes. This debate and the need for its resolution have been heightened by the proliferation of nuclear power plants and weapons, generating more wastes, and by the fact that the United States also needs to decommission and safely dispose of three or four nuclear submarines per year for the next 25 or more years. The problem has become more acute with the demise of the Cold War and the decommissioning of more weapon systems.

Since high-level radioactive wastes remain toxic to humans for many thousands of years, it is necessary to find a disposal site that will be stable for at least that length of time and that also will be beyond the reach of human tampering for the same length of time. Given the ingenuity of humans and our present technological prowess, there are few places on the planet that meet these requirements. One such place is the deep ocean, well away from the edges of the tectonic plates. The abyssal plains of the deep ocean lie under 4–5 km of water, beyond any regular current human activity in the ocean. Furthermore,

the bottom is covered with a thick layer of sediment that has accumulated over a long time and has remained stable. If the radioactive waste could be placed in this sediment, there is a good chance that it would remain safely away from the realm of human activities for the required time. Even if wastes should leak from the containers after some years, the nuclides would adhere to the sediment and probably remain in place (Figure 11.18).

Because of the foregoing advantages, the deep-sea bed has been suggested as the one area of the planet where we could safely isolate these toxic wastes for thousands of years. However, in evaluating the deep sea for feasibility of high-level waste disposal, we need to consider the consequences of leakage or failure of the containment vessels. It is, therefore, necessary to understand the fate of escaped radionuclides once they are out of the vessel. Such an understanding requires a considerable knowledge of both the physical and biological mechanisms that influence the movement of radionuclides. Since, as we saw in Chapter 4, the deep sea is the least understood of all the ocean areas of the planet, there is obviously a great deal of information that we need. As we grapple with the problem of the safe disposal of radioactive wastes, we should encourage the research necessary to answer the important questions, lest we again fail to make an intelligent choice and resort to expediency.

■ THERMAL POLLUTION

The siting of electrical generating plants along the seacoasts to take advantage of the marine waters for cooling purposes has led to the perturbation of marine communities by heated water discharge. This **thermal pollution** of the marine environment has been little investigated; thus far, the effects appear restricted to the communities immediately adjacent to the discharge. The damaging effects of thermal discharges are most pronounced in the tropics, where the organisms are living naturally at water temperatures close to their thermal maximum. It takes little increase in temperature to stress such communities. An example of the effects of a thermal discharge on tropical communities is described by Banus and Kolehmainen (1976). In Guayanilla, on the south coast of Puerto Rico, a power-generating station discharges heated water into a lagoon. The discharged water is 8–10°C higher than the ambient seawater and produces summer water temperatures of 39°C in the lagoon. This temperature is high enough to affect

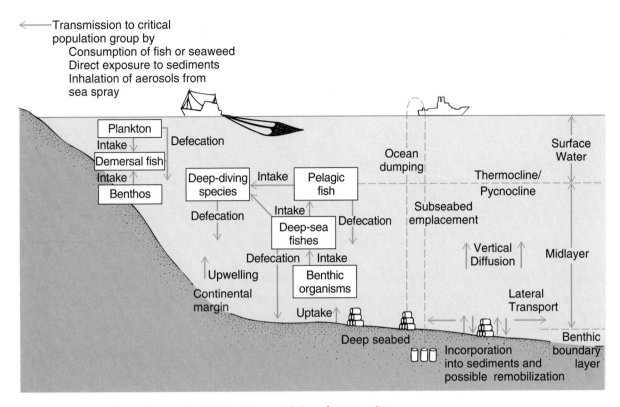

←—— Transmission to critical
population group by
    Consumption of fish or seaweed
    Direct exposure to sediments
    Inhalation of aerosols from
    sea spray

FIGURE 11.18 Physicochemical and biological transmission of waste radionuclides from deep-sea disposal sites. (Modified from *Wastes in the Ocean*, Vol. 3, P. K. Park et al. Copyright © 1983 John Wiley & Sons, Inc. Reprinted by permission.)

the reproduction of the dominant mangrove trees in the bay, and no young rooted and growing mangrove seedlings have been found in the lagoon.

Eutrophication

Eutrophication refers to the release of excess nutrients into coastal waters. These nutrients come from the fertilizers spread on agricultural land that make their way to the sea via rivers and streams. The results of such movement are red tides, various yellow and green foams, slimes, and slicks—all of which indicate excess phytoplanktonic growth. Although, on a small scale, certain of these blooms are natural, they have become more common and extensive in recent years. For example, in 1988 a yellow-brown slick off the Scandinavian countries threatened the entire Norwegian salmon mariculture industry and led to the most massive peacetime maritime rescue

operation in the history of Norway, according to Cherfas (1990). In another incident in 1978, the German North Sea coastal beaches were inundated by a meter-deep, foul-smelling foam created by surf beating on a massive outbreak of a gelatinous encapsulated alga of the genus *Phaeocystis*. Aside from presenting aesthetic considerations, many of these massive outbreaks are increasingly toxic, contributing further to the disaster.

The significance of eutrophication in the oceans is most dramatically demonstrated by the presence of some 50 oxygen-starved coastal dead zones in various parts of the oceans (Figure 11.19). **Dead zones** result when oxygen in ocean waters is reduced below the level necessary to sustain marine life. The largest such zone in the Western Hemisphere is located every summer off the Louisiana and Texas coasts and spreads with each advancing year. In 1989, according to Malakoff (1998), this zone covered 9,000 km^2, but by 1997 it had expanded to 16,000 km^2. In 2002, it

FIGURE 11.19 Worldwide distribution of dead zones created by human activity. (Based on Diaz and Rosenberg, 1995, Marine Benthic Hyporia: A Review of Its Ecological Effects and the Behavioral Responses of Macrofauna. *Oceanogr. Mar. Biol. Ann. Rev.* 33:245–303.)

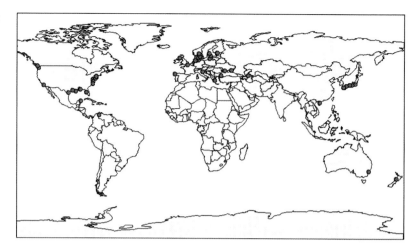

FIGURE 11.20 Extent of the dead zone off Louisiana and Texas. (Figure from "Dead Zone Grows" by Hold, *Science*, Vol. 297:1119, 8/16/2002. Copyright © 2002 American Association for the Advancement of Science. Used with permission.)

reached 22,000 km^2 (Figure 11.20). The Louisiana dead zone is the result of the massive amounts of nutrients carried off the agricultural lands of the Mississippi River drainage. During the spring and summer, the nutrient-laden warm fresh water of the Mississippi flows out over the more dense Gulf of Mexico seawater, forming a lidlike layer. Fueled by the nutrients, the phytoplankton bloom and attract zooplankton grazers. Large amounts of dead phytoplankton, zooplankton, and fecal pellets then sink to the bottom, where proliferating oxygen-consuming bacteria consume them and reduce the oxygen below the level necessary to sustain multicellular organisms. The dead zone is generally confined to shallow water up to 60 meters deep, but it is not confined to just the water immediately adjacent to the bottom and may affect as much as 80% of the water column, according to Malakoff (1998). The mobile organisms in the dead zone are driven away, but the sedentary

and sessile species, such as clams, starfishes, and worms, are killed. One of the largest concerns is that, as this dead zone spreads or becomes more persistent, it will destroy one of the richest fishing grounds in the Gulf of Mexico. To solve this problem, it will be necessary to somehow reduce the amount of fertilizers that leach from agricultural lands throughout some of the richest farmlands of the United States, a task that not only is daunting, but certainly cannot be achieved quickly.

How do we know that these blooms are not natural events and that their apparent increases are not just the result of better observation and quick communication common to modern times? Studies in Japan and Europe, areas with good historical records, show that such phenomena have increased markedly in the past 20 years. The good side to this is that the increased numbers of algal cells may absorb more carbon dioxide and retard the greenhouse effect.

(See the next section.) However, at present, no one knows whether this is true or not. Another unanswered problem concerns dimethyl sulfide (DMS), a natural compound produced by phytoplankton. When the phytoplankton die, DMS is released into the atmosphere, where it may be changed to sulfuric acid and contribute to acid rain.

Biological Invasions

Another human impact concerns the introduction of **alien animals and plants**. The devastating effects of alien animals and plants on native fauna and flora are well known and documented for the terrestrial environment. We know, for example, of the loss of the American chestnut through a disease imported from Europe and, more recently, the demise of our native elm trees through Dutch elm disease. Equally dramatic has been the devastation of European rabbits on the flora and fauna of Australia. What about introductions into marine waters?

Our knowledge of the effects of introducing an alien marine species is now well known. The major intentional introductions of marine species have been those of commercial or sport fishing value. Thus, the eastern (*Crassostrea virginica*) and Japanese (*Crassostrea gigas*) oysters were introduced into the bays and estuaries of the Pacific coast of North America to provide the basis of an oyster industry. Similarly, the striped bass (*Roccus saxatilis*) was introduced into the Pacific coast from the Atlantic and has become the basis of a valuable sports fishery.

Along with these purposeful introductions, however, have come other, unintentional species. For example, along with the introduced oysters came many invertebrate species attached to their shells. Around the world, other inadvertent introductions, such as fouling organisms on ships' hulls, have occurred as a result of transport. Increasingly through the twentieth century, the transport of organisms in ships' ballast water has led to many additional invasions of exotic species. Some estimates now suggest that as many as 3,000 species are transported daily in ballast water. Numerous additional global transport vectors, such as the commercial bait industry, the aquarium trade, and the global movement of live seafood, exacerbate the problem. The number of these species and their cumulative effects on indigenous communities are now beginning to be understood. That alien species may have profound effects is

witnessed by the Black Sea ecosystem crash discussed in the next paragraph. Another example is found in studies of introduced species on the central California coast. Carlton (1975, 1989) has shown that 150–200 marine invertebrates from various parts of the world have become established in the bays and estuaries of the Pacific coast. In the case of San Francisco Bay, the majority of the present shallow-water invertebrate fauna has been introduced, and it is now nearly impossible to know what the indigenous communities of this great bay were like. Another alien invasive species is the alga *Caulerpa taxifolia*. This alga is native to the tropics, but in 1984 it was observed in the Mediterranean Sea, where it has subsequently become a major problem, replacing native algae and invading and killing the native *Posidonia* seagrass beds. *Caulerpa* is also toxic or inedible to various herbivores, thus driving them away as well. It occupied only about an acre in the Mediterranean in 1984, but by 2000 had exploded to cover more than 30,000 hectares of shallow seafloor. In 2000, *Caulerpa* was also discovered in a bay in Southern California, where drastic efforts are now underway to destroy it.

How serious a threat these introductions are to marine communities around the world remains to be seen, but if the results we have observed in terrestrial communities are any guide, the threat could be considerable. As an example of the potential seriousness of the threat, we can mention the case of the ctenophore *Mnemiopsis leidyi*. In 1982, according to Travis (1993), this voracious planktonic carnivore was introduced into the Black Sea, probably from a ship that had taken on ballast water off the east coast of North America, where *Mnemiopsis* is native. *Mnemiopsis* underwent an almost unbelievable population explosion in the Black Sea. At times, it constituted as much as 95% of the sea's entire biomass! This gelatinous carnivore feeds on zooplankton, including the eggs and larvae of fishes, and is the prime suspect in the precipitous decline of the Black Sea fisheries. However, it is not the only reason for the decline of the Black Sea ecosystem, according to Kideys (2002): During the 1970s and 1980s there also was a great increase in the nutrient input to the Black Sea from the major rivers that led to eutrophication which spawned algal blooms, changed the dominant phytoplankton to favor flagellates, and resulted in hypoxia and mortality to benthic and fish populations. The appearance of *Mnemiopsis* completed the demise of the ecosystem, since the ctenophore fed on zooplankton as well as the eggs and larvae of one of the dominant target fishes of fishermen: the anchovy *Engraulis encrasicolus*. The sea's nutrient input was stabilized and even decreased in the late 1980s, leading to a decline in the phytoplankton population and the reemergence of diatoms as dominants, indicating a recovering ecosystem. Many species of zooplankton and fishes also began to reappear. All this happened despite the continued presence of *Mnemiopsis*. The question of how to combat this alien was still under debate at the time, but none of the options was without risk. Perhaps the best option was to introduce a predator or parasite of that animal. The problem was that we knew so little about the species and its predators and parasites that we could not be certain that they, too, might not become a problem if introduced. The *Mnemiopsis* problem was solved by the appearance in the Black Sea in 1997 of another ctenophore of the genus *Beroë*, which also probably arrived in ballast water taken in off the Atlantic coast of the United States. *Beroe* is a voracious predator of *Mnemiopsis* and caused a crash in its population, after which *Beroe* itself also crashed, indicating its dependence on *Mnemiopsis*. Since that time, the ecosystem of the Black Sea has continued to change. Is this a case of two wrongs making a right? Recently, *Mnemiopsis* emigrated into the Mediterranean Sea and may threaten its fisheries next. The potential for such introductions in the future is great. If, for example, a sea-level canal is constructed in Central America, allowing free interchange of the tropical Atlantic and Pacific faunas, massive changes may occur. The giant offshore oil-drilling rigs are another threat, accumulating a whole marine fauna in one area during drilling. When they are moved, often to an entirely different area, the fauna is brought along intact.

Miscellaneous Human Impacts

To provide docking facilities for commercial and pleasure vessels, humans have engaged in a great deal of dredging of estuaries and bays. Such activities destroy large areas of productive habitat for marine organisms and may have effects beyond the immediate area if nursery grounds for commercial species caught offshore are destroyed. Not only are communities wiped out by being physically removed, but the increased load of silt suspended in the water reduces light—and hence photosynthesis—and clogs the feeding and respiratory surfaces of many inverte-

brates, leading to their deaths. Such activity has destroyed many productive areas worldwide, and the expanding human population portends even more destruction in the future.

It may seem strange to suggest that dams built on freshwater rivers and streams can affect the oceans, but by preventing the flow of fresh water into the oceans, these structures have had effects in certain areas. Probably the best example is the Aswan Dam on the Nile River in Egypt. Prior to its construction in 1965, the Nile carried 43 billion cubic meters of water into the Mediterranean Sea. Now it carries only 3 billion. The dam has indirectly caused the erosion of the Nile delta and the loss of the fisheries of the eastern Mediterranean. For example, the catch in the sardine fishery in the eastern Mediterranean fell from an average of 18,000 tons per year in the 1960s before the dam was built to 600 tons in 1969 and has remained low in subsequent years, according to Weber (1994).

A final consideration is that of human trampling of intertidal shores. Whenever large numbers of humans visit a rocky intertidal shore, their peregrinations across the substrate trample various of the organisms. Some, such as small barnacles, may be well protected against damage, but other forms may be killed or severely damaged by being stepped on. This problem is especially severe on shores popular with the public and where there is no respite or season of relief from human visitors.

Marine Diseases

Within the last few decades, there has been an increase in mortality of many taxa of marine organisms, apparently due to disease. In addition, there seems to have been an increase not only in the frequency of epidemics, but in the occurrence of new diseases as well. Many of these outbreaks have occurred in shallow waters that are heavily affected by humans, suggesting human activity as a factor. Unfortunately, we know so little about diseases in the marine environment that we are, in many cases, unable to assess whether the epidemics are natural or whether they have resulted from some human intervention or activity in the marine environment. We have also made little progress in identifying the causes of many of the diseases. For example, the massive die-off of the dominant Caribbean sea urchin, *Diadema antillarum*, in the 1980s (see pp. 452–453) was one of the best-studied epidemics, and yet we still have not identified the pathogen. Similarly, the near elimination of the two dominant species of the coral *Acropora* at many locations in the Caribbean at the same time was by an agent as yet unidentified. Harvell et al. (1999) indicate that, currently, of the dozen or more coral diseases in the Caribbean, only three have a known causative agent. Nor are mass mortalities limited to the Western tropical Atlantic: Pathogen-caused epidemics have been reported in various parts of the temperate North Atlantic and the tropical Pacific and have affected a wide range of both animal and plant taxa. Concomitant with the apparent increase in disease has been an increase in toxic algal blooms that has affected both mammal and fish populations. For example, the toxic dinoflagellate *Pfisteria piscicida* was determined to be the cause of massive fish kills on the Atlantic coast of North America, according to Burkholder et al. (1992).

The question to be asked is whether this apparent increase in disease is real or whether it is an artifact of our increased scrutiny of the oceans. Several lines of evidence, primarily from coral reefs, suggest that the increase is real. First, our monitoring of coral reefs in Florida indicates that we are seeing diseases we have not seen before. Second, coral reefs are long-lived structures, and if these diseases had been in existence before, causing changes in community structure, we would see them in the fossil record. Finally, because of the massive mortality they cause, if they had occurred in previous decades, we certainly would have detected them.

If disease is increasing and some diseases are new, where did they come from? Current evidence indicates that most have come about not by new pathogenic organisms, but by existing organisms shifting hosts. Evidence for this **host shift** comes from several sources. In the Antarctic, the appearance of canine distemper virus in crabeater seals is thought to have come from contacts with sled dogs used in several Antarctic expeditions. In the Caribbean, a fungal pathogen that has attacked various species of sea fans has been identified as *Aspergillus sydownii*, which has somehow spilled over from its usual terrestrial reservoir. Still other diseases appear to be new. For example, phocine distemper virus, previously unrecognized, was determined to have caused mass mortality among harbor seals (*Phoca vitulina*) and gray seals

(*Halichorus gryphus*) in northwestern Europe in the 1980s, while two other new viruses—dolphin morbillivirus and porpoise morbillivirus—were found to be the cause of worldwide mortalities of dolphins, porpoises, and other cetaceans.

Why have we had these increasing outbreaks? Although we are uncertain as yet, correlative data suggest two conditions favoring outbreaks. The first is the climate-induced changes that have affected marine ecosystems. Although there is little or no direct proof, periodic climatic changes such as those produced by El Niño–Southern Oscillation events cause extreme stress on many marine communities, and some of the results thereof, such as coral bleaching, may be accelerated by pathogens that are themselves favored by such events. Correlative evidence of this relationship has been found in the variation in the occurrence of Dermo, a disease of the eastern oyster *Crassostrea virginica* caused by the protozoan *Perkinsus marinus*. Dermo outbreaks occur when La Niña conditions prevail and drop during El Niño events. The other condition favoring disease outbreaks is direct and indirect human activity. Humans have transported many marine species out of their native regions and introduced them into other areas. Doing so usually means transporting their pathogens as well. The best examples of the transfer of pathogens are the human-facilitated epidemics documented in the shrimp and bivalve mollusk culturing industries. Sometimes, the disease has even invaded wild populations. Another example of an outbreak brought about by human activities is that caused by human-produced pollutants which enter the oceans. These pollutants, from pesticides to sewage, may compromise the immune systems of marine animals or facilitate the movement of terrestrial pathogens to marine organisms.

In closing this section, we note that humans themselves are not immune to pathogens in seawater, since many terrestrial pathogens can persist for varying amounts of time in marine environments and can infect humans who are exposed to seawater. Perhaps the best example of this phenomenon is the human disease cholera. The organism responsible for cholera is *Vibrio cholerae*, which is associated with marine plankton. The incidence and severity of human epidemics of cholera have been strongly correlated with seawater temperature, turbidity, and plankton blooms.

Drugs from the Sea

Many marine organisms produce chemicals that are not found, or are rare, in terrestrial organisms. These chemicals are used by the marine organisms mainly to defend against predators. Certain of the compounds have biomedical properties that may be useful in combating human disease. Organisms especially productive of such biomedical compounds include sponges, cnidarians, tunicates, and opisthobranch mollusks. Among the useful chemicals are a shark-repelling saponin; ichthyotoxin, from the fish *Pardachirus marmoratus*; antitumor chemicals from the sponges *Haliclona* and *Halichondria*; and a potent nonprotein toxin, palytoxin, from the soft coral *Palythoa*. Perhaps the most interesting is a steroid called squalamine, found in the dogfish shark (*Squalus acanthias*). Stone (1993) reports that squalamine is effective in killing fungi that cause dangerous infections in AIDS and cancer patients. Over 1,700 compounds with some biomedical property have been reported in marine organisms in the last ten years, and more are certain to come with additional testing of organisms. The main stumbling block thus far to the commercial pharmaceutical development of these compounds has been the inability to raise most of the marine organisms in large numbers in captivity.

Global Warming and Changes in Sea Level

Certainly, one of the most talked-about subjects in science today, among not only scientists, but the public, politicians, and governments as well, is **global warming**, or the **greenhouse effect**, and how it will influence the climate of our planet. The term *global warming* refers to the increase in the temperature of the planet caused primarily by the increase in the atmosphere of certain gases, such as carbon dioxide, methane, nitrous oxide, and chlorofluorocarbons. These gases have the ability to trap the radiant energy from the sun, leading to increases in atmospheric temperatures, which, in turn, may cause changes in climate, alter rainfall and storm patterns, change ocean current patterns, and increase the rate at which the sea level rises.

There seems little doubt that the "greenhouse" gases have increased over the last 100 years, and much of the increase is due to humans burning fossil fuels and wood. Certainly, the increase in chlorofluorocarbons is directly attributable to humans, since this gas is a human invention that is used in refrigeration systems. The other gases are derived from natural sources. Exacerbating the problem has been the wholesale cutting of the world's forests, which would otherwise serve as a major sink and absorb excess carbon dioxide. The other major sink for carbon dioxide is the world's oceans. (Recall the carbon cycle, discussed in Chapter 1.) It is both the increase in the rate at which the sea level rises and the potential of the ocean to absorb greenhouse gases that directly concern us here. As global temperatures rise, the oceans of the Earth will expand because of the melting of glaciers and ice caps, the most significant of which are those of Greenland and Antarctica, and because of the thermal expansion of the ocean water. Depending on the amount of warming, sea levels could rise as much as 5–7 ft over the next century, an amount in stark contrast to the rise of 0.5 ft in the last 100 years. If the "worst-case" rise were to occur, there would be massive inundations of low-lying coastal areas, such as Louisiana, Cape Cod, Florida, Egypt, and Bangladesh, and many low-lying island archipelagos, such as the Maldives and Marshall Islands, might be completely obliterated.

If these unprecedented disasters could occur, why is there such controversy concerning global warming? The answer stems from our own lack of knowledge. Whereas we can chart with relative accuracy the trends in atmospheric gases, the rise (or fall) in the sea level, and the advance or retreat of glaciers over the last 100–200 years, that time frame is geologically insignificant and significantly variable to the extent that it cannot serve as a basis for predicting long-term changes. Besides, we know that the planet has undergone various sea-level and climatic changes in the past, but these changes have usually proceeded very slowly, over thousands or millions of years, not within a hundred or so years, as now seems to be the case. The real question, and the source of the controversy, is whether the changes we are now observing are simply natural variations in an otherwise stable climate since the end of the last ice age or whether they are the forewarnings of a long-term change. In spite of our current lack of the necessary long-term data, there now seems to be a growing consensus among scientists, according to Kerr (1990), that global warming is a reality.

There are a few startling examples from the marine environment which suggest that long-term changes are occurring and may be related to global warming. They also give an inkling as to what we might expect if trends continue. The first example concerns the depletion of the ozone layer over Antarctica precipitated by the chlorofluorocarbons. Smith et al. (1992) report that the thinning of the ozone layer over the Antarctic has led to an increase in ultraviolet light impinging on the Antarctic Ocean. This resulted in photoinhibition of the phytoplankton and a decrease in primary productivity of from 6 to 12%, which, in turn, decreased the phytoplankton's uptake of carbon dioxide. The results of this study suggest that, should the ozone layer be depleted over much larger areas of the planet, even less carbon dioxide would be absorbed, thus intensifying global warming.

In the second example, Roemmich and McGowan (1995) report on a 40-year study of the zooplankton in the waters off southern California. During this time, the temperature of the surface water warmed by about 1.5°C, resulting in increased stability of the thermocline such that wind-driven upwelling was displaced upward into shallower water and there were fewer nutrients (NO_3, PO_4) for the phytoplankton and, consequently, lower productivity. The latter, in turn, led to decreases in the zooplankton biomass by 70% between 1951 and 1993. The future consequences of such an observed trend depend on whether it is part of a natural cycle. If it is, the trend may reverse itself. However, if it is a manifestation of the longer human-induced global warming and there is a further 1–2°C temperature rise in ocean waters around the world in the next 40–50 years, causing increases in stratification, the effect on marine ecosystems could be devastating.

Given the fact that greenhouse gases are continuing to increase, and given the known fact that if this trend continues there will be a considerable rise in sea level, is there anything that humans can do to forestall such a calamity? The answer is yes, but it will require massive action on the part of all nations on the planet. Forests must be prevented from being cut (indeed, more need to be planted to absorb the extra carbon dioxide), chlorofluorocarbons must be taken out of use, and fossil-fuel usage must be reduced. Finally, there is the intriguing suggestion by Martin et al. (1990) that much of the excess carbon dioxide could be removed from the atmosphere to the ocean sink if a few hundred thousand tons of

iron could be made available to the phytoplankton of the equatorial Pacific Ocean, the southern Ocean, and the subarctic Pacific Ocean. All of these areas have been puzzling to oceanographers, because they have a high nutrient content, but low levels of chlorophyll. This enigma has led scientists to ask why there is such low biomass in areas with abundant nutrients and light. If Martin is correct, these phytoplankton are limited in growth and photosynthesis by the lack of trace amounts of iron in the water. If iron were added, maintains Martin, the increase in photosynthetic activity would be great enough to draw down the carbon dioxide from the atmosphere.

Initially, Martin's hypothesis was experimentally supported only by laboratory measurements with seawater samples that were uncontaminated with metals. Additional experiments performed by Coale (1991) and Johnson et al. (1994) with deck-board bottle enrichments of surface waters from the subarctic Pacific, the southern Ocean, and the equatorial Pacific also confirmed the hypothesis. However, all of these experiments were strongly criticized because of the effects of containment and the failure to take into account the complexities of the natural environment and the other members of the pelagic community.

Finally, in a unique series of trials known as the IronEx experiments, scientists from the Moss Landing Marine Laboratories led two international expeditions to the equatorial Pacific to test Martin's hypothesis in a series of open-ocean enrichment experiments. When approximately half a ton of iron was added to 100 km² of the nitrate-rich waters, phytoplankton biomass increased by a factor of 30, depleting carbon dioxide to 60% of its initial value, and the ocean turned green for miles (Coale et al., 1998). Whereas the confirmation of the iron hypothesis has suggested to some industrialists and politicians that we should use this technique to control global carbon dioxide, the scientific community is urging caution and asserting that much more needs to be understood about the oceans before large-scale fertilization is undertaken.

Concluding Remarks

For too long, the seas of the world have been considered an inexhaustible source of food, with an infinite capacity to absorb and purify our wastes and to supply all the raw materials we need to maintain an industrial society. It is now apparent that none of these assumptions is true and that, at the current level of technological development, the human population has the ability to inflict massive destruction on the seas, just as it has done on land. At present, the seas remain in good condition relative to the land, and we cannot afford to permit them to be degraded in the same way if we wish to continue our tenure as a species on this planet. We must employ ecological principles, such as those outlined in this text, to ensure that the potential of the oceans is realized without degradation, or else we shall relive yet another tragedy of the commons.

Summary of Key Concepts

- The explosion of technology and the rapidly increasing human population have had significant effects on the ecology of the world's oceans in recent years.

- Technological advances in fishing gear and ships have resulted in a significant reduction in, and overexploitation of, many fish stocks.

- The major fisheries are concentrated in the waters overlying the continental shelves, but the tonnages caught in different geographical areas are considerably unequal.

- The commercially important marine animals are drawn from four groups: bony and cartilaginous fishes, marine mammals, mollusks, and crustaceans; the greatest tonnage comes from fishes.

- In 2001, the total marine fisheries landings were 83.6million metric tons, but this figure is lower than the peak catch, which was reached in 1989.

- Currently, most of the fish stocks on the continental shelf areas of the world are fully exploited or over-exploited.

- Open-ocean waters are impoverished with respect to production and cannot sustain the population of fishes that the continental shelves do.

- The maximum sustainable yield is the largest number of fishes that can be harvested year after year without diminishing stocks.

- The estimated sustainable yield from the world's oceans varies between 60 million tons and 290 million tons because of disagreements among scientists on the proper figures to use for primary productivity, ecological efficiency, and the average trophic level.

- With few exceptions, productivity has declined in all of the major fishing areas of the planet due to overexploitation, often compounded by certain natural changes in the environment.

- The decline of the Peruvian anchoveta fishery was due to overfishing combined with environmental changes related to El Niño; the demise of the whale stocks was due entirely to overfishing, and the failure of the California sardine fishery remains a mystery.

- Overfishing may also result in changes in the larger ecosystems or communities in which the target species reside.

- Fishing affects the community directly by the removal of species, thus reducing species diversity; fishing may also alter the size structure, reproductive, and life-history characteristics of the populations targeted.

- Fishing may also alter the species structure of the affected communities.

- Overfishing coupled with environmental changes may have effects extending well beyond the immediate area.

- Bycatch is also a source of destruction to marine communities.

- Fishing gear itself has a deleterious effect on the habitat.

- Capturing fishes for the aquarium trade has resulted in significant losses, particularly to coral reefs.

- Ghost fishing by lost gear can continue to deplete organisms.

- To save our fisheries, we must adopt a new ecosystem approach.

- Open-ocean fisheries resources, presumably held in common for the good of the world's people, are at risk of succumbing to the tragedy of the commons.

- International agreements among nations for the control of fisheries have proved difficult to bring about and implement effectively, but they will be necessary to establish a truly worldwide regulation of fisheries.

- Potential new fisheries include the Southern Ocean (for krill and squid), the area of the continental slope at 300–1,000 meters, and many areas in the Indian Ocean.

- Mariculture is the rearing of marine plants and animals under controlled conditions to provide increased amounts of food for humans.

- Few marine organisms are cultivated, and most of these are the type that either fetch a premium price on the market or produce large amounts of biomass under culture.

- Although yields from mariculture can be very high, the science is currently at a low level of development; thus, mariculture presently contributes relatively little to human food.

- Mariculture development is restricted by several factors, including the difficulty of maintaining the proper physical and chemical characteristics of the water, the presence of many larval forms, diseases and parasites, and the cost in money and time to obtain permits.

- In recent years, we have come to realize that the oceans are not infinite in their capacity to absorb human wastes and wastes generated by human activity.

- Oil pollution usually results from large tanker accidents, leading to an oil spill, or from blowouts occurring in offshore drilling rigs.

- Oil is devastating to marine birds and mammals and can suffocate intertidal and shallow subtidal communities.

- The area affected by a spill from a large tanker can be extensive, as witnessed by the *Exxon Valdez* spill in Alaska, and can also chronically affect an area for years.

- Oil seems to have little effect on pelagic communities, save for birds.

- Discharge of human sewage to the marine environment in small amounts and with adequate diffusion seems to cause no effects on marine communities on the open coast, but in large volumes and in semienclosed embayments such discharge may be devastating.

- Large amounts of human garbage are dumped into the ocean each year, but there is little scientific data regarding their effects on marine communities.

- Various chemicals produced by the industrialized nations find their way into marine ecosystems; the chemicals are often transferred through marine food chains and exert effects in animals and places removed in time and space from the source.

- Biological magnification is the process whereby the concentration of chemicals in the bodies of organisms increases with each succeeding trophic level.

- Major sources of radioactive materials in the oceans have been fallout from nuclear testing, releases or dumping of wastes from nuclear fuel cycle systems, and certain accidents.

- Currently, dumping of high-level radioactive wastes into the oceans is prohibited, but dumping of low-level wastes is still permitted.

- High-level radioactive wastes must be isolated from the biosphere for geological time scales; the deep ocean floor has been suggested as the safest place to put such wastes.

- Thermal pollution by discharges of heated water is most pronounced in the tropics and is usually restricted to the communities immediately adjacent to the discharge.

- Eutrophication—the release of excess nutrients into coastal waters—leads to extravagant phytoplankton growth, resulting in red tides, various foams and slicks, and, most devastating, the depletion of oxygen, creating dead zones.

- Dredging of estuaries and bays for boating facilities destroys large areas of productive habitat for marine organisms.

- Dams built on freshwater rivers affect the oceans by reducing the sediment load, leading to the erosion of deltas and a reduction in the number of fisheries.

- Alien marine animals and plants have been introduced into various marine habitats around the world, both intentionally and unintentionally. The number of unintentional introductions and their effect on indigenous communities is little known, but potentially great.

- In recent years, there has been an increase in the mortality of many marine taxa from diseases. Both the frequency of epidemics and the occurrence of new diseases have risen.

- Some of the new diseases are a result of host shifts by the pathogens involved.

- Some of the recent epidemics have come about as a result of climate-induced changes in marine ecosystems, whereas others have been due to human activity.

- The incidence and severity of the human disease cholera have been strongly correlated with seawater temperature, turbidity, and plankton blooms.

- Certain marine organisms have been found to contain chemicals that have proved useful in combating human diseases.

- Global warming, or the greenhouse effect, has the potential to create a significant rise in sea level and also to change marine ecosystems.

Review Questions

ESSAY Develop complete answers to these questions:

1. What is meant by the term *maximum sustainable yield*? How does the notion relate to the biological potential of a fish? Elaborate.

2. Be it resolved that the solution to pollution is dilution. Present evidence to support your argument.

3. Describe the concept of the tragedy of the commons in terms of two different marine issues, two different terrestrial issues, two different atmospheric issues, and two different freshwater issues.

4. Why did the total catch in marine fisheries decline in the 1990s, even though new fisheries resources, such as krill, were (and still are) being exploited?

5. Discuss ways human efforts to increase food production on land have reduced the potential yields of food from the sea.

Bibliography

Alverson, D. L., M. H. Freeberg, J. G. Pope, and S. A. Murawski. 1994. A global assessment of fisheries by-catch and discards. FAO Fisheries Technical Paper 339.

Banner, A. H. 1974. Kaneohe Bay, Hawaii: Urban pollution and a coral reef ecosystem. In *Proceedings of the 2nd Int. Coral Reef Symp.* 2:685–702.

Banus, M. D., and S. E. Kolehmainen. 1976. Rooting and growth of red mangrove seedlings from thermally stressed trees. In *Thermal ecology*, Vol. 2, edited by G. W. Esch and R. W. McFarlen. Washington Technical Information Center of the Energy Research and Development Administration, 46–53.

Bardach, J. E. 1968. Aquaculture. *Science* 161:1098–1106.

Bardach, J. E., J. H. Ryther, and W. O. McLarney. 1972. *Aquaculture.* New York: Wiley.

Bascom, W., ed. 1978. Coastal water research project annual report. Los Angeles.

Bergmann, M. J. N., and J. W. van Santbrink. 2000. Fishing mortality of populations of megafauna in sandy sediments. In *Effects of fishing on non-target species and habitats*, edited by M. J. Kaiser and S. J. de Groot. Oxford, UK: Blackwell Scientific, 49–68.

Blindheim, J., and H. R. Skjoldal. 1993. Effects of climatic changes on the biomass yield of the Barent's Sea, Norwegian Sea and West Greenland large marine ecosystem. In *Large marine ecosytems: stress, mitigation and sustainability*, edited by L. M. Alexander and B. D. Gold. AAAS, 185–189.

Boucher, G. 1985. Long-term monitoring of meiofauna densities after the *Amoco Cadiz* oil spill. *Mar. Pollution Bull.* 16:328–333.

Brothers, N. 1991. Albatross mortality and associated bait loss in the Japanese longline fishery in the Southern Ocean. *Biological Conservation* 55:255–268.

Burkholder, J. M., E. J. Noga, C. H. Hobbs, H. B. Glasgow, and S. A. Smith. 1992. New "phantom" dinoflagellate is the causative agent of major estuarine fish kills. *Nature* 358:407–410.

Busch, L. 1991. Science under wraps in Prince William Sound. *Science* 252:772–773.

Carlton, J. 1975. Introduced intertidal invertebrates. In *Light's manual: Intertidal invertebrates of the central California coast*, 3d ed., edited by R. I. Smith and J. Carlton. Berkeley: University of California Press, 17–25.

Carlton, J. 1989. Man's role in changing the face of the ocean: Biological invasions and implications for conservation of near-shore environments. *Cons. Biol.* 3(3):265–273.

Carlton, J. 1996. Marine bioinvasions: The alteration of marine ecosystems by nonindigenous species. *Oceanography* 9: 36–43.

Carlton, J. 1996. Pattern, process and prediction in marine invasion ecology. *Biological Conservation* 78: 97–106.

Carlton, J. and J. Geller. 1993. Ecological roulette: The global transport of nonindigenous marine organisms. *Science* 261: 78–82.

Cherfas, J. 1990. The fringe of the ocean—under siege from the land. *Science* 248:163–165.

Coale, K. H. 1991. The effects of iron, manganese, copper and zinc on primary production and biomass in plankton of the subarctic Pacific. *Limn. Oceanog.* 36:1851–1864.

Coale, K. H., K. S. Johnson, S. E. Fitzwater, S. P. G. Blain, T. P. Stanton, and T. L. Coley. 1998. IronEx I, an in situ iron-enrichment experiment: Experimental design, implementation and results. *Deep-Sea Res. II* 45:919–945.

Culley, M. 1971. *The pilchard, biology and exploitation.* III: *The California sardine.* New York: Pergamon, 143–176.

Cushing, D. H. 1969. Upwelling and fish production. FAO Fisheries Technical Paper 84, 1–40.

Cushing, D. H. 1975. *Fisheries resources of the sea and their management.* Oxford, UK: Oxford University Press.

Cushing, D. H. 1977. *Science and the fisheries.* Studies in Biology, no. 85. Baltimore: Edward Arnold.

Dayton, P. K., S. F. Thrush, M. T. Agardy, and R. J. Hoffman. 1995. Environmental effects of marine fishing. *Aquat. Cons. Mar. Freshwat. Ecosyst.* 5:202–232.

Dayton, P. K., S. F. Thrush, and F. C. Coleman. 2002. Ecological effects of fishing in marine ecosystems of the United States. Washington, Pew Oceans Commission

FAO. 2002. *2001 Yearbook of fisheries statistics*, Vol.92. Rome: Food and Agriculture Organization of the United Nations.

Fautin, D., ed. 1987. *Biomedical importance of marine organisms.* Memoirs of the California Academy of Sciences, no. 13.

Gerges, M. 1993. Rehabilitating ROPME's ecosystems. *Siren* 48:1–4.

Goldberg, E. D. 1974. Marine pollution: Action and reaction times. *Oceanus* 18(1):6–18.

Goldberg, E. D. 1976. *The health of the oceans.* Paris: UNESCO Press.

Gunnerson, C. 1981. The New York Bight ecosystem. In *Marine environmental pollution*, edited by R. A. Geyer. Vol. 2, *Dumping and mining.* New York: Elsevier, 313–378.

Hall, S. J. 1999. *The Effects of Fishing and Marine Ecosystems and Communities.* Oxford, UK: Blackwell Scientific, Ltd.

Hardin, G. 1968. The tragedy of the commons. *Science* 162(3859):1243–1248.

Harvell, C. D., K. Kim, J. M. Burkholder, R. R. Colwell, P. R. Epstein, D. J. Grimes, E. E. Hofman, E. K. Lipp, A. D. M. E. Osterhaus, R. M. Overstreet, J. W. Porter,

G. W. Smith, and G. R. Vasta. 1999. Emerging marine diseases—climate links and anthropogenic factors. *Science* 285:1505–1510.

Harvey, G. R., A. G. Requejo, P. A. McGillivary, and J. M. Tokar. 1979. Observations of a subsurface oil-rich layer in the open ocean. *Science* 205:999–1001.

Hess, W. 1978. The *Amoco Cadiz* oil spill. A preliminary scientific report. NOAA/EPA.

Hites, R. A. et al. 2004. Global assessment of organic contaminants in farmed salmon. *Science* 303: 226–229.

Holmes, R. 1994. Biologists sort the lessons of fisheries collapse. *Science* 264:1252–1253.

Idyll, C. P. 1970. *The sea against hunger.* New York: Crowell.

Jackson, J. B. C. et al. 2001. Historical overfishing and the recent collapse of coastal ecosystems. *Science* 293: 629–638.

Jennings, S., and M. J. Kaiser. 1998. The effects of fishing on marine ecosystems. In *Advances in marine biology*, edited by J. H. S. Blaxter, A. J. Southward, and P. A. Tyler. 201–352.

Johnson, K. S., K. H. Coale, V. A. Elrod, and N. Tindale. 1994. Iron photochemistry and bioavailability in the equatorial Pacific. *Mar. Chem.* 46:319–334.

Kideys, A. E. 2002. Fall and rise of the Black Sea ecosystem. *Science* 297: 1482–1484.

Kerr, R. A. 1990. New greenhouse report puts down dissenters. *Science* 249:481–482.

Loughlin, T. R., ed. 1994. Marine mammals and the *Exxon Valdez*. San Diego: Academic Press.

Malakov, D. 1998. Death by suffocation in the Gulf of Mexico. *Science* 281:190–191.

Maragos, J., C. Evans, and P. Holthus. 1985. Reef corals in Kaneohe Bay six years before and six years after termination of sewage discharges. *Proc. 5th Int. Coral Reef Symp.* 2:236.

Marine pollution. 1974. A series of articles in *Oceanus* 18(1).

Marr, J. C. 1960. The causes of major variations in the catch of the Pacific sardine, *Sardinops caerulea*. Spec. Sci. Rept., U.S. Fish and Wildlife Ser. 208:108–125.

Martin, J. H., S. E. Fitzwater, and R. M. Gordon. 1990. Iron deficiency limits phytoplankton growth in Antarctic waters. *Global Biogeochem. Cycl.* 4:5–12.

McVay, S. 1966. The last of the great whales. *Sci. Amer.* 215:13–21.

National Academy of Sciences. 1985. *Oil in the sea, inputs, fates and effects.* Washington, DC: National Academy Press.

Nelson-Smith, A. 1973. *Oil pollution and marine ecology.* New York: Plenum Press.

Nicol, S., and W. de la Mare. 1993. Ecosystem management and the Antarctic krill. *Amer. Sci.* 81:36–47.

Park, P. K., D. R. Kester, I. W. Duedall, and B. Ketchum. 1983. *Wastes in the ocean.* Vol. 3, *Radioactive wastes and the ocean.* New York: Wiley.

Pauly, D., and V. Christensen. 1995. Primary production required to sustain global fisheries. *Nature* 374:255–257.

Peterson, C. H. 2001. The *"Exxon Valdez"* oil spill in Alaska: Acute, indirect and chronic effects on the ecosystem. *Advances in Marine Biology* 39:1–103.

Peterson, C. H. et al. 2003. Long term ecosystem response to the *Exxon Valdez* oil spill. *Science* 302:2082–2086.

Platt, J. F., C. J. Lensink, W. Butler, M. Kendziorek, and D. R. Nysewander. 1990. Immediate impact of the *Exxon Valdez* oil spill on marine birds. *Auk* 107:387–397.

Riaux-Gobin, C. 1985. Long-term change in microphytobenthos in a Brittany estuary after the *Amoco Cadiz* oil spill. *Mar. Ecol. Progr. Ser.* 24:51–56.

Rodhouse, P. G. 1995. Southern ocean cephalopod resources. *Bull. Malacological Soc. London* 25:1, 3.

Roemmich, D., and J. McGowan. 1995. Climatic warming and the decline of zooplankton in the California Current. *Science* 267:1324–1326.

Rounsefell, B. A. 1975. *Ecology, utilization and management of marine fisheries.* St. Louis: Mosby.

Russell-Hunter, A. D. 1970. *Aquatic productivity.* New York: Macmillan.

Ryther, J. H. 1969. Photosynthesis and fish production in the sea. *Science* 166:72–76.

Safina, C. 1995. The world's imperiled fish. *Scientific American* November 1995: 46–53.

Schaefer, M. B. 1965. The potential harvest of the sea. *Trans. Amer. Fish. Soc.* 94:123–128.

Scheuer, P. J. 1990. Some marine ecological phenomena: Chemical basis and biochemical potential. *Science* 248:173–177.

Sherman, K. 1991. The large marine ecosystem concept: Research and management strategy for living marine resources. *Ecol. App.* 1:349–360.

Smith, B. D., and G. S. Jamieson. 1991. Possible consequences of intensive fishing for males on the mating opportunities of Dungeness crabs. *Trans. Amer. Fish. Soc.* 120:650–653.

Smith, J. E., L. C. Hunter, and C. M. Smith. 2002. Distribution and reporductive characteristics of nonindigenous and invasive marine algae in the Hawaiian Islands. *Pacific Science* 56:299–315.

Smith, R. C., B. B. Prezelin, K. S. Baker, R. R. Bidigare, N. P. Boucher, T. Coley, D. Karentz, S. MacIntyre, H. A. Matlick, D. Menzies, M. Ondrusek, Z. Wan, and K. J. Waters. 1992. Ozone depletion: Ultraviolet radiation and phytoplankton biology in Antarctic waters. *Science* 255:952–959.

Soutar, A., and J. D. Isaacs. 1969. History of fish populations inferred from fish scales in anaerobic sediments off California. *Calif. Mar. Res. Comm., CalCOFI Rept.* 13:63–70.

Stone, R. 1993. *Deja vu* guides the way to new antimicrobial steroid. *Science* 259:1125.

Travis, J. 1993. Invader threatens Black, Azov Seas. *Science* 262:1366–1367.

Thrush, S. F., J. E. Hewitt, V. J. Cummings, P. K. Dayton, M. Cryer, S. J. Turner, G. A. Funnel, R. G. Budd, C. J. Milburn, and M. R. Wilkinson. 1998. Disturbance of the marine benthic habitat by commercial fishing: Impacts at the scale of the fishery. *Ecol. App.* 8:866–879.

Watling, L., and E. A. Norse. 1998. Disturbance of the seabed by mobile fishing gear: A comparison to forest clearcutting. *Conservation Biology* 12(6):1180–1197.

Weber, P. 1994. Net loss: Fish, jobs and the marine environment. Worldwatch paper 120.

Wheeler, J. A. 1996. The marine aquarium trade: A tool for coral reef conservation. A report for the Sustainable Development and Conservation Program. University of Maryland, College Park.

TAXONOMIC APPENDIX

WITH THE DISCOVERY of many new organisms in the later part of the twentieth century, along with knowledge of their chemical structure and metabolism, the task of organizing this increase in diversity into an acceptable heirarchy based on shared features and probable relationships became more complicated. Beginning with Linnaeus and into the early twentieth century, all life-forms were grouped into two kingdoms: plant and animal. Then, in the middle of the twentieth century, increasing knowledge of the diversity of microbial organisms, coupled with the understanding of their molecular structure and functioning, made it apparent that the two-kingdom system required changing. The first generally acceptable higher classification was formulated in 1969 by Robert Whittaker, who proposed a five-kingdom system that recognized two basically different types of

cell: prokaryotic and eukaryotic. All prokaryotic cells were bacteria and were set apart in the Kingdom Monera. Eukaryotic unicellular organisms that were either autotrophic or heterotrophic were placed in the Kingdom Protista. The multicellular eukaryotes were divided into three kingdoms: Plant (autotrophic), Fungi (saprotrophic), and Animal (heterotrophic). With the discovery and recognition of the structure and physiology of a group of organisms that were widely distributed in extreme environments, such as hot springs and ice, and that had cellular structures which did not fit into either a prokaryotic or eukaryotic type of cell, further changes were suggested. These organisms were first called archaeobacteria, but with the recognition of the magnitude of their differences, they were segregated out as a separate kingdom, Archaea. Further knowledge of the various protis-

TABLE 1

(a) A five-kingdom system

Monera	Protista	Plantae	Fungi	Animalia

(b) An eight-kingdom system

Bacteria	Archaea	Archaezoa	Protista	Chromista	Plantae	Fungi	Animalia

(c) A three-domain system

DOMAIN BACTERIA	DOMAIN ARCHAEA	DOMAIN EUKARYA (Eukaryotes)								
		Archaezoa	Euglenozoa	Alveolata	Stramenopila	Rhodophyta	Plantae	Fungi	Animalia

TABLE 2

A Comparison of the Three Domains of Life

	DOMAIN		
CHARACTERISTIC	BACTERIA	ARCHAEA	EUKARYA
Nuclear envelope	Absent	Absent	Present
Membrane-enclosed organelles	Absent	Absent	Present
Peptidoglycan in cell wall	Present	Absent	Absent
Membrane lipids	Unbranched hydrocarbons	Some branched hydrocarbons	Unbranched hydrocarbons
RNA polymerase	One kind	Several kinds	Several kinds
Initiator amino acid for start of protein synthesis	Formylmethionine	Methionine	Methionine
Introns (noncoding parts of genes)	Rare	Present in some genes	Present
Response to the antibiotics streptomycin and chloramphenicol	Growth inhibited	Growth not inhibited	Growth not inhibited
Histones associated with DNA	Absent	Present	Present
Circular chromosome	Present	Present	Absent
Ability to grow at temperatures >100°C	No	Some species	No

tan groups also suggested changes, which led to an eight-kingdom scheme. Finally, because of the fundamental differences in the three basic types of cell, there should be a "superkingdom" classification above that of the kingdom, termed Domain. In this system, there are three domains, as shown in Table 1.

In this taxonomic outline, we consider only taxa that are mentioned in the book. We omit taxa that are entirely parasitic and any others not present in marine systems. We also refrain from placing the viruses into the scheme, since marine viruses are so poorly known that any attempt to classify them is premature.

Since we do not distinguish or use the various subdivision of the taxonomic hierarchy in the Domains Bacteria (except for the Cyanobacteria) and Archaea in this book, we will not discuss them further here.

Domain Bacteria

Heterocysts

Kingdom Cyanobacteria

This kingdom includes bacteria that have chlorophyll a and that employ two photosystems to split water and produce oxygen. These bacteria may be single-celled or multicellular. *Trichodesmium* is an example.

The Domain Eukaryota or Eukarya

This domain includes organisms mentioned in this book that belong to the Kingdoms Plantae, Animalia, or Protista (or its subdivisions as kingdoms in the eight-kingdom and three-domain system of classification). Fungi certainly occur in the oceans of the world, but thus far they have been virtually ignored in marine ecology, and their taxonomic breakdown is not included here. The protistans include both autotrophic and heterotrophic organisms. Because in the past, the photosynthetic protistans have been claimed by the botanists and the heterotrophs by the zoologists, there have been considerable disagreements over their classification, and confusion still reigns. Here, we consider the autotrophs and heterotrophs together in a single classification system. Because a complete classification of the protistans would encompass 45 phyla, we have omitted those which have not been mentioned in the text.

The protistans can be subdivided into six kingdoms, of which three play major roles in the marine environment and can be briefly mentioned here.

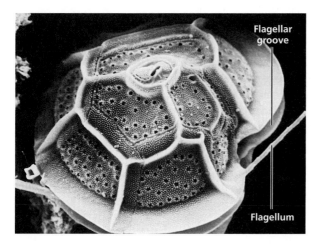

Flagellar groove

Flagellum

Kingdom Alveolata

This kingdom includes the dinoflagellates and the ciliates, both of which are significant in the marine environment. They are characterized by cells with subsurface cavities (alveoli).

■ PHYLUM SARCOMASTIGOPHORA

These are protistans with flagellae, pseudopodia, or both as locomotory organs and usually with a single type of nucleus. *Ceratium, Peridinium, Dinophysis, Gonyaulax, Globigerina, Acanthometron,* and *Acanthochiasma* are representative organisms.

■ PHYLUM CILIOPHORA

These protistans are characterized by having simple or compound cilia as locomotory organs and two types of nuclei. *Loxophyllum, Tracheoloraphis, Remanella,* and *Geleia* are members.

Kingdom Stramenopila

This kingdom includes the diatoms and the brown algae and is characterized by unicellular and multicellular cells, often with flagellae of unequal length and with rows of fine hairlike projections.

■ PHYLUM CHROMOPHYCOTA

This taxon includes the diatoms, which possess an external, two-part silicate skeleton. *Thalassiosira, Chaetoceros, Pseudoeunotia, Fragilaria,* and *Nitzchia* are examples.

■ PHYLUM PHAEOPHYCOTA

This is the phylum of the brown algae, which are multicellular photoautotrophs with cell walls of cellulose and other polysaccharides. *Laminaria, Nereocystis, Pterogophora, Macrocysti, Fucus,* and *Sargassum* are representative.

Kingdom Rhodophyta

This kingdom includes the red algae, with a very divergent morphology, chloroplasts containing various mixtures of chlorophylls, carotenoids, and pigments, and with various methods of spore and gamete formation. *Lithothamnion*, *Phymatolithon*, *Clathromorphum*, and *Corallina* are representatives.

Kingdom Chlorophyta

This kingdom includes the green algae, with chlorophyll a and b like those of higher plants, and cell walls of cellulose. *Ulva* and *Enteromorpha* are examples.

Kingdom Animalia

This kingdom currently comprises some 33 phyla, of which one is exclusively terrestrial and three are exclusively parasitic. That leaves 29 phyla to consider here. However, we discuss only those phyla which are ecologically significant. The kingdom includes all eukarya that are multicellular in structure and heterotrophic in nutrition. With the exception of certain taxa in the Phylum Chordata, all phyla are invertebrate organisms in that they lack an internal vertebral column as adults.

■ PHYLUM PORIFERA

The sponges are characterized by two loose cellular layers: an outer perforated layer and an internal layer of unique flagellated collar cells (choanocytes). Between the layers is a noncellular layer containing a skeleton of calcium carbonate or siliceous spicules and various spongin fibers. All sponges are sessile and lack true tissue layers and organ systems. There are four classes of sponges, but in marine waters the dominant ones belong to the Classes Demospongiae and Hexactinellidae.

In scyphozoans, the medusa stage is dominant and most are large pelagic animals in the open ocean. Anthozoans include the anemones and corals, and the polyp stage is the only one present.

CLASS SCYPHOZOA: *Pelagia*

CLASS ANTHOZOA: *Metridium, Anthopleura, Ptilosarcus.*

CLASS HYDROZOA: *Crassota, Muggiaea, Nanomia.*

■ PHYLUM CTENOPHORA

Ctenophores are similar to cnidarians in external appearance, but lack nematocysts and possess batteries of sticky cells called *colloblasts* for capturing prey. Most are planktonic and, unlike cnidarians, are never polymorphic.

CLASS TENTACULATA: *Peurobrachia*

CLASS NUDA: *Beroe*

■ PHYLUM CNIDARIA

Cnidarians are radially symmetrical animals possessing nematocysts, which are intracellular structures containing an inverted hollow tube that everts upon discharge to deter predators or to capture prey. As with the sponges, cnidarian bodies are in two layers separated by a thin or thick middle layer, here called the mesoglea. Many cnidarians exist in more than one body form. Two are recognized: the polyp, which is cyclindrical, is usually attached, and bears tentacles around the mouth at the free end; and the medusa, which is umbrella shaped or hemispherical, and has tentacles around the margin of the bell and a mouth that hangs down from the underside of the bell. The mouth leads into a single, blind gastrovascular cavity. In some cnidarians, the life cycle of the species alternates between the polyp and medusa form. There are three classes: Hydrozoa, Scyphozoa, and Anthozoa.

■ PHYLUM PLATYHELMINTHES

The flatworms have three germ layers and organ systems, but lack a true body cavity or coelom. As opposed to organisms in the preceding phyla, flatworms are bi-laterally symmetrical. Most species in the phylum are parasitic, but the Class Turbellaria is free living and well represented in the marine environment. Their significance in marine systems is unknown.

CLASS TURBELLARIA: *Convoluta*

▪ PHYLUM NEMERTEA

Nemertines are long thin worm like animals that have an eversible proboscis that lies in a cavity separate from the digestive tract. They are abundant in marine habitats but little known ecologically.

▪ PHYLUM NEMATODA

Nematodes are probably one of the most abundant phyla in the world in terms of numbers of individuals and are found everywhere in terrestrial, marine, and freshwater habitats. The total number of species is unknown, as is the animals' role in marine communities.

▪ PHYLUM ROTIFERA

Rotifers are pseudocoelomate animals of microscopic size usually recognized by the presence at the anterior end of a wheel like structure—the corona—and a distinctive muscular structure—the mastax—in the digestive tract.

▪ PHYLUM GASTROTRICHA

The gastrotrichs are perhaps the smallest of the metazoan animals and are characterized by a ciliated ventral surface, a cuticle of scales and prominent adhesive tubes. They are found interstitially. *Pseudostomella* and *Diplodasys* are examples.

▪ PHYLUM KINORHYNCHA

These pseudocoelomates have a triangular or oval body in cross section that is divided into 13 segments of which the first is an eversible spiny head. All are benthic animals inhabiting marine sediments. *Centorderes* is an example.

▪ PHYLUM LORICIFERA

This phylum was only discovered in 1983, is microscopic in size and inhabits coarse marine sand. The adults are sessile but the larvae move. They live interstitially. *Nanoloricus* is an example.

▪ PHYLUM TARDIGRADA

Tardigrades are characterized by a more or less oval body that has 4 pairs of short, retractable stubby limbs each terminated by either 4 claws or sticky pads. *Haleschiniscus* and *Anisonyches* are examples.

▪ PHYLUM ANNELIDA

The annelids are strongly metameric animals, in that the body is composed of a linear array of segments behind the head, and each segment contains a sheathing set of muscles that enclose a fluid-filled true body cavity, or coelom. A ventral nerve cord connects to a dorsal brain in the head. The annelids are characterized by a closed circulatory system and a set of paired excretory organs in each segment. A complete digestive tract runs from the anterior mouth to the posterior anus. Most annelids possess small chitinous structures called setae, which are employed in different ways in the different taxa. There are three classes of Annelids, but only one—the Class Polychaeta—is significant in marine ecosytem structure and function.

CLASS POLYCHAETA: *Nereis, Glycera, Echone, Clymenella, Owenia, Diopatra, Tomopteris, Nepthys.*

▪ PHYLUM ECHIURA

This is a small phylum related to annelids and characterized by having a body in two parts, a trunk and a muscular proboscis. They show no signs of metamerism but have a large trunk coelom. *Urechis* is an example.

▪ PHYLUM MOLLUSCA

With over 100,000 species, the molluscs are the second-largest phylum of the Animal Kingdom. Molluscs are characterized by a fleshy dorsal mantle that encloses most of the visceral organs and externally secretes a shell, if the animal produces one, and by a ventral muscular foot. The two structures unique to molluscs—the radula and the ctenidia—are not possessed by all species. With the exception of the Class Cephalopoda, molluscs are slow-moving animals functioning primarily with ciliary-mucous mechanisms. Seven classes are recognized, but only three are currently recognized as ecologically significant in structuring the various marine communities.

CLASS GASTROPODA: *Acmaea, Acanthina, Batillaria, Cerithidea, Collisella, Littorina, Lottia, Mitra, Nassarius, Nucella, Olivella, Polinices, Armina.*

CLASS BIVALVIA: *Calyptogena, Cryptomya, Donax, Macoma, Mya, Mytilus, Siliqua, Tridacna.*

CLASS CEPHALOPODA: *Abrailiopsis, Chiroteuthis, Architeuthis, Histioteuthis, Pterigioteuthis, Todarodes, Vampyroteuthis.*

▪ PHYLUM ARTHROPODA

This is the largest phylum of the Animal Kingdom. The total number of species in this taxon is really unknown, but has been estimated at anywhere from several million to more than 20 million. About a million species have been formally described. The phylum is found in all habitats on Earth, but the greatest number of species is found in the Class Insecta, which is almost exclusively terrestrial and fresh water in distribution. The marine waters are dominated by the Subphylum Crustacea, which is found in all marine habitats. The phylum is characterized by the possession of both an external jointed skeleton composed of chitin and protein and a segmented body consisting of a variable number of segments, or somites. Primitively, each segment had one pair of jointed appendages, but the loss or fusion of somites has reduced this condition in most living arthropods and has also led to a loss of appendages.

■ **SUBPHYLUM CRUSTACEA**

Arthropods with two pair of antennae.

CLASS PHYLLOPODA: *Evadne, Podon.*

CLASS MAXILLOPODA: *Acartia, Balanus, Calanus, Calocalanus, Chthamalus, Cryptolepas, Cypridina, Gausia, Gigantocypris, Neocalanus, Semibalanus.*

CLASS MALACOSTRACA: *Acanthohaustorius, Ampelisca, Alicella, Aratus, Blepharipoda, Cancer, Carcinus, Cyamus, Cystisoma, Emerita, Eriocheir, Euphausia, Gecarcinus, Gnathophausia, Jasus, Lybia, Meganyctiphanes, Notostomus, Pachygrapsus, Periclimenes, Pontoporeia, Portunus, Pseudohaustorius, Sergestes.*

■ **PHYLUM BRACHIOPODA**

The brachiopods are bivalved animals that are attached to the substrate by a stalk and possess a feeding structure called a lophophore. *Gwynia* is an example.

■ **PHYLUM BRYOZOA**

The bryozoans are another lophophorate phylum but the animals are colonial and sessile with the individual zooids usually less than 1 mm in size, but the colonies are macroscopic and many are polymorphic.

■ **PHYLUM ENTOPROCTA**

Entoprocts are solitary or colonial pseudocoelmate animals with a body situated on a muscular stalk and with a ring of tentacles surrounding the mouth and anus. *Loxosomella* is an example.

■ **PHYLUM ECHINODERMATA**

This phylum includes some of the most familiar of marine animals—sea stars, brittle stars, sea urchins, and sea cucumbers, to name a few. The phylum is defined by an internal skeleton, the elements of which are single crystals of calcium carbonate, and a complex compartmentalized coelom, including a fluid-filled water vascular system used for various functions, such as locomotion and the capture of prey. Echinoderms lack a defined head and centralized nervous system.

CLASS ASTEROIDEA: *Asterias, Asterina, Crossaster, Dermasterias, Heliaster, Hippasteria, Mediaster, Odontaster, Perknaster, Solaster, Stichaster.*

CLASS OPHIUROIDEA: *Amphiodia, Ophiomusium.*

CLASS ECHINOIDEA: *Dendraster, Diadema.*

CLASS HOLOTHUROIDEA: *Cucumaria, Molpadia, Peniagone.*

CLASS CRINOIDEA

■ **PHYLUM CHAETOGNATHA**

Chaetognaths are flattened arrow-shaped transparent predators abundant in the plankton of the oceans. The phylum is small, consisting of about 100 species, and is entirely marine. *Sagitta* is an example.

■ PHYLUM CHORDATA

All chordates are characterized by three features: pharyngeal gill slits, a dorsal hollow nerve cord, and a dorsal internal supporting rod called a *notochord*. These features are present at some time in the life cycle of the animals. By far the largest subphylum in terms of numbers of species is the Subphylum Vertebrata, which contains the familiar fishes, amphibians, reptiles, birds, and mammals. The taxa of this subphylum are well known to all and need not be elaborated further here. However, the subphylum Urochordata comprises the pelagic and benthic animals known as *tunicates*, which are prominent marine organisms.

■ SUBPHYLUM UROCHORDATA

Chordates without a dorsal nerve cord or notochord as adults.

CLASS ASCIDIACEA: *Aplidium*

CLASS LARVACEA: *Oikopleura*

CLASS THALIACEA: *Doliolium, Salpa.*

■ SUBPHYLUM VERTEBRATA

Chordates with a vertebral column.

CLASS AGNETHA

CLASS CHONDRICHTHYES

CLASS OSTEICHTHYES

CLASS AMPHIBIA

CLASS REPTILIA

CLASS AVES

CLASS MAMMALIA

The Kingdom Plantae

Plants are autotrophic organisms possessing advanced tissue differentiation and an alteration of generations between a haploid (gametophyte) and diploid (sporophyte) phase.

■ PHYLUM ANTHOPHYTA

These are seed plants with the seeds borne within fruits.

CLASS MONOCOTYLEDONES: *Zostera, Thalassia, Halodule, Syringodinium, Posidonia, Spartina, Juncus.*

CLASS EUDICOTYLEDONES: *Iva, Avicennia, Rhizophora, Ceriops, Conocarpus, Laguncularia, Bruguiera.*

GLOSSARY

abyssal zone The bottom zone of the oceans at depths between 4,000 and 6,000 m.

abyssal gigantism Phenomenon observed among several crustacean groups in which the animals' general size increases with increasing depth.

abyssal plain That area of the deep ocean floor lying between 4,000 and 6,000 m.

abyssalpelagic zone Pelagic aphotic zone lying between the bathypelagic and hadalpelagic zones.

acidity A measure of the concentration of hydrogen ions in a solution.

adenosine triphosphate The energy storage molecule of most living systems.

advection Movement of seawater horizontally or vertically, as in a current.

aerenchyma Tissue of thin-walled cells and spaces found in stems and roots of certain marsh plants; serve to transfer oxygen.

aerobic Condition in which oxygen is present.

ahermatypic coral A nonreef-producing coral without the symbiotic zooxanthellae in the tissues.

algal ridge Coral-free ridge of encrusting coralline algae lying immediately behind the buttress zone.

alien animal Animal that is not indigenous to an area.

alkalinity A measure of the concentration of hydroxyl ions in a solution.

allochthonous Of foreign origin; transported into an area from outside.

alternate stable states The occurrence in a single habitat of several different persistent communities in which the structure of each is the result of historical events.

altricial Pertaining to young birds that are hatched in an immature state and that require extended parental care.

American estuary An estuary dominated by extensive stands of emergent vegetation surrounding less extensive mud flats.

anaerobic Condition where oxygen is absent.

anchor ice Ice that forms around any convenient nucleus in the area below the permanent pack ice in Antarctica; it tends to carry organisms out of the area to be incorporated into the sea ice above.

anchoveta Common name for *Engraulis ringens*, the Peruvian anchovy.

anoxic Without oxygen.

anticyclonic Moving in a clockwise direction.

aphotic Without light; that area of the oceans without light.

apogee Farthest point of the moon from the Earth.

aquaculture The culture of aquatic organisms.

aquatic Living or existing within or on water.

Archaea One of the three Domains of life; prokaryotic.

area hypothesis The larger the area of a habitat the more species it will have.

asexual In reproduction, without involving sex.

atoll A modified ring-shaped coral reef arising from deep water far from continental landmasses and enclosing a shallow lagoon.

ATP *See* adenosine triphosphate.

autochthonous Formed or occurring in the place where it is found.

autotrophic Capable of producing one's own energy resources.

auxospore In diatoms, a reproductive cell that reestablishes the initial size of the species.

backwash Swash water that flows back down the beach.

bacterioplankton Bacteria that live in the plankton, usually of a size range from 0.2 to 2.0 μm (picoplankton).

baleen Horny material growing in comblike, fringed units from the upper jaws of whales of the order Mysticeti.

barrier reef Coral reef adjacent to landmasses and separated from them by a lagoon or channel of variable extent.

bathyal Bottom zone encompassing the continental shelf down to about 4,000 meters.

bathypelagic zone Pelagic aphotic zone lying between the mesopelagic and abyssalpelagic zones.

bathyscaphe A free-moving submersible designed to carry human observers into the deep sea.

bay A partially enclosed inlet of the ocean.

bends A disease occurring in humans who breathe compressed air in deep water. The bends result from a rapid decrease in pressure, causing nitrogen bubbles to form in the internal tissues and blood vessels.

benthic Pertaining to the area of the sea bottom or to organisms that occur on the sea bottom.

benthos Those organisms living on or in the sea bottom.

bet hedging A theory of life-history evolution in which the traits displayed by the organisms result from the fluctuations in juvenile survivorship probabilities relative to those of the adult; opposite of "r-K" selection theory.

binocular vision Type of vision providing depth-of-field focus due to overlap of the field of vision of two closely set eyes.

biochemistry A scientific discipline concerned with the study of the chemistry of living organisms and their products.

bioerosion The dissolution or breakdown of substrates, usually calcium carbonate, by a variety of living organisms.

biogeochemical cycle Cyclical movement of an element or a compound through living organisms and the nonliving environment.

biological conditioning Excretion by organisms of certain organic compounds into the water column that make the water either more or less suitable for themselves while enhancing or inhibiting other organisms.

biological interaction In ecology, the mutual or reciprocal action or influence between organisms. Predation, competition, and grazing are three kinds of biological interaction.

biological magnification Increase in concentration of chemicals in the bodies of animals with increasing trophic level.

bioluminescence The production of light by living organisms.

bioturbator Organism that is capable of modifying the local environment to favor or preclude other organisms from living there.

black band disease A coral disease characterized by a black band of dying tissue on infected corals.

blade In a kelp plant, the flattened part of the plant that terminates the stipe.

blubber The layer of lipid that serves as an insulating layer under the skin of whales and other marine mammals.

blue-green algae Group of prokaryotic photosynthetic organisms of the Domain Bacteria.

boiling point Temperature at which a liquid changes to a gas.

bottom-up process Control of community structure by environmental factors or prey.

bradycardia Slowing of the heart rate.

buffer A chemical solution that resists or dampens changes in pH with the addition of acids or bases.

bycatch Unintended capture of non-target organisms by fishing gear.

^{14}C method Method of measuring primary productivity by means of the radioactive isotope ^{14}C.

carbohydrate Biochemical compounds composed of carbon, hydrogen, and oxygen, often in the ratio of one carbon to two hydrogens and one oxygen.

Carnivora An order of mammals adapted to feed on other animals.

carnivore An animal that consumes other animals as food.

caudal Pertaining to the tail or posterior end of an organism.

caudal peduncle In fishes, that part of the body immediately in front of the tail.

cellulose A long-chain carbohydrate composed of repeated units of the sugar glucose.

centrifugal force Force that pulls objects out from a center of rotation.

Cetacea The order of mammals that contains the whales and porpoises.

chemoautotrophic bacteria Those bacteria able to obtain energy and therefore synthesize organic material through the oxidation of reduced sulfur compounds.

chitin Biochemically, a polymer of the carbohydrate glucosamine that forms the hard outer integument of crustaceans and other marine invertebrates.

chlorophyll maximum Depth in the ocean at which the concentration of chlorophyll reaches its highest level on a per volume basis.

ciguatera Disease of humans caused by consumption of tropical fishes with toxins in their flesh and organs.

cleaning behavior Special category of symbiosis in which large animals, usually fishes, permit themselves to be cleaned of various parasites by smaller fishes or invertebrates.

climax In ecology, the final stage of a successional sequence that is able to persist in the absence of environmental change.

clone A group of genetically identical individuals of a plant or animal species produced by asexual reproduction from a single individual.

coastal front Flow of water over the shallow continental shelf whereby friction with the bottom causes vertical turbulence that transfers nutrients and changes the chemistry of the water, leading to changes in the plankton.

coastal plain estuary Estuary formed by the flooding of a low-lying coastal valley due to a rising sea level.

coccolithophore Small, unicellular, flagellated alga, usually with an external covering of small pieces of calcium carbonate.

cohesion Mutual attraction of similar molecules, overcoming the external forces that would break them.

cold seeps Areas in the deep sea where reduced carbon or sulfur compounds exit from the rocks at ambient temperatures.

commensal In a symbiotic relationship, the partner that gains the advantage.

commensalism A symbiotic relationship between two different species in which one benefits and the other is neither benefited nor harmed.

compensation depth Depth at which the processes of photosynthesis and respiration are equal.

compensation intensity In the water column, that point at which light intensity is equal to 1% of the surface intensity.

compensation theory *See* subsidence theory.

competition The interaction among organisms for a necessary resource that is in short supply.

competition model Hypothesis that explains the high diversity of coral reef fishes in terms of a high diversity from intense competition leading to a high degree of niche specialization.

competitive exclusion The ecological principle that states that complete competitors cannot coexist.

competitive interference Exclusion of one species by another species through interruption of its normal activities.

consumer pressure Relative ability of a consumer to determine the structure of a community.

continental shelf The shallow underwater extension of a continent; usually limited in depth to 200 m.

continental slope The steeply descending bottom between the edge of the continental shelf and the continental rise; the ocean bottom between the depths of 200 and 3,000 m.

continental rise The gently sloping area between the continental slope and the abyssal plain.

copepod Small crustacean of the order Copepoda; the dominant macroscopic planktonic herbivore.

copepodid The larval stages of a copepod that follow the nauplius and metanauplius stages and that precede the adult.

coral bleaching Expelling of zooxanthellae from the tissues of reef-building corals.

coral reef Massive limestone structure built up through the cementing and depositional activities of anthozoans of the order Scleractinia and certain other invertebrate and algal species.

coralline algae A group of red algae common on coral reefs and that lay down calcium carbonate in their tissues.

corallite The cup into which a coral polyp fits.

corallivore An animal that consumes corals.

Coriolis effect The deflection imparted to moving water masses due to the spinning of the Earth on its axis; the deflection is to the right in the Northern Hemisphere and to the left in the Southern Hemisphere.

cosmopolitan In biogeography, an organism that is distributed throughout the world in suitable habitats.

crinotoxin A poisonous material secreted onto the surface of an organism.

critical depth The depth at which photosynthesis in the water column is equal to respiration in the water column; the depth to which phytoplankton cells may be mixed and still spend sufficient time above the compensation depth so that respiration and photosynthesis are equal.

critical tide level Points on the shore at which a small change in vertical distance results in a disproportionate change in exposure time to air.

cropper In the disturbance theory of Dayton and Hessler, a deep-sea animal that consumes living and dead animals smaller than itself.

cryptic coloration Color, hue, or pattern that mimics the background.

cryptic symbiosis Symbiotic association between an algal cell and a nonphotosynthetic organism in which the algal species' cellular integrity is so reduced that the alga appears to be an integral part of the host.

cryptofauna Fauna living on or in coral substrates.

current Water movements that result in the horizontal transport of water masses.

cyanellae Symbiotic blue-green algae.

cyanobacteria *See* blue-green algae.

Cyanophyceae A class of blue-green algae.

cyclonic Moving in a counterclockwise direction.

dead zones Areas of the ocean, usually in shallow water, where eutrophication has reduced the oxygen content below the level necessary to sustain most marine life.

decomposer An organism that breaks down dead protoplasm, freeing simple chemical substances for use by other organisms.

deep scattering layer (DSL) Concentration of midwater organisms that reflect sound waves produced by sonar devices.

deep sea That area of the ocean beyond the continental shelf and below the lighted zone.

deep water mass One of several water masses below the thermocline.

demersal Living close to the bottom of the sea.

density The mass per unit volume of a substance (physics); the number of individuals per unit area (biology).

deposit feeder An animal that feeds by consuming particles on or in the substrate.

desiccation The process of losing water.

diatom Microscopic autotrophic organism of the algal class Bacillariophyceae characterized by being enclosed in a two-part siliceous capsule.

diel Daily, or once every 24 hours.

diffuse predation Condition in which total predation is strong and capable of controlling the abundance of a competitively dominant species, but also in which the predation is spread over more than one predator.

dinoflagellate Microscopic organism of the class Dinophyceae possessing two locomotory flagellae.

disphotic zone The area of the ocean between the euphotic and aphotic zones; the disphotic zone has insufficient light for photosynthesis, but sufficient light for detection and response by animals.

dissipative beach A flat beach, maximally eroded, with fine sediments and with wave energy consumed in a surf zone away from the beach proper.

disturbance theory A hypothesis explaining the high diversity in the deep sea which suggests that extreme predation pressure prevents any species from becoming numerous enough to eliminate others by competition, therefore perpetuating a large number of species and high diversity.

diurnal Occurring daily or relating to daytime.

diurnal tide Tide with a single high and low each day.

divergence Surface horizontal flow of water away from a central area, permitting water to upwell.

diversity *See* species diversity.

DOM Abbreviation for dissolved organic matter.

dominant The numerically abundant species in a community.

drag Resistance to movement through a medium.

dysphotic *See* disphotic.

echolocation *See* sonar.

ecological succession An orderly process of community change controlled through modification of the physical environment.

ecology Discipline in biology that treats the spectrum of relationships existing between organisms and their environment and among organisms.

ecotone An area of intergradation between two biological communities or associations.

ectothermic Cold blooded; unable to regulate the body temperature.

eddy A circular movement of water.

Ekman spiral The change in direction and velocity of a current as one progresses downward in the water column.

electrical conductivity Relative ability of a material to allow the passage of electricity.

electron A negatively charged elementary particle in all matter.

element A substance composed of atoms having the same atomic number.

elutriation Separation of interstitial fauna from the sand through constant stirring with water and subsequent filtering on a fine mesh.

endobenthic Pertaining to meiofaunal-sized organisms that move within the sediments, displacing particles.

endothermic Warm blooded; able to regulate the internal body temperature.

endozoite An animal living symbiotically inside another animal.

energy The capacity to do work.

epibenthic Organisms living at the sediment–water interface.

epicercal tail A caudal fin of fishes with the dorsal lobe larger than the ventral lobe.

epifauna Benthic organisms that live on or move over the substrate surface.

epipelagic *See* photic.

epiphytic Pertaining to plants that are not parasitic, but that are attached to other plants.

epizoite An animal that is not parasitic, but that lives attached to another animal.

equilibrium species A species that has a life history characterized by long life, a long development time to reach maturity, a low death rate, and few reproductive periods per year.

estuary A partially enclosed coastal embayment where fresh water and seawater meet and mix.

euphotic *See* photic.

European estuary An estuary dominated by large mudflats and relatively barren of large plants, but with large populations of surface benthic diatoms.

euryhaline Able to tolerate wide fluctuations in salinity.

eutrophic Containing abundant nutrient material.

eutrophication The process of excessively increasing nutrient levels in the oceans through natural or artificial means.

evaporative estuary *See* negative estuary.

exploitative competition Competition among different species or members of the same species for a necessary resource that is in short supply.

extinction coefficient Ratio between the intensity of light at a given depth and the intensity at the surface.

femtoplankton A size class of plankton ranging from 0.02 to 0.2 μm.

fetch The distance over which the wind can blow.

filter feeder Animals that obtain their food by filtering particles out of the water column.

fin A flattened appendage of an aquatic animal used in locomotion or maneuvering in the water.

fishery The place or act of taking fishes or other sea organisms for human use.

fjord Estuary occurring in deep, drowned valleys, originally cut by glacial action; characterized by the presence of a shallow sill at the mouth, restricting interchange between surface water and deeper fjord waters.

flipper The pectoral or pelvic appendage of a marine mammal or reptile.

flushing time Time required for a given mass of fresh water to be discharged from an estuary.

food chain The pathway that transfers energy from a given autotrophic organism through a series of consumers.

food web The combination of all food chains in a given community or ecosystem.

form resistance Condition in which drag is proportional to the cross-sectional area of the object in contact with the water.

frictional resistance Condition in which drag is proportional to the amount of surface area in contact with the water.

fringing reef A coral reef that forms immediately adjacent to a landmass.

fugitive species *See* opportunistic species.

fundamental niche The total of all the ranges of all the biological and physical factors within which a species can exist; a multidimensional vector space.

gametophyte In algae, the haploid organism that produces gametes.

gas bladder Structure on the dorsal side of the body cavity of bony fishes that contains gas and is used by the fishes to regulate buoyancy.

geomorphology Discipline dealing with the form and configuration of the surface of the Earth.

ghost fishing Continued capture of marine organisms by fishing gear that has been lost in the ocean.

global warming Increase in average global temperature due to the buildup of gases such as CO_2, CH_4, and chlorofluorocarbons in the atmosphere.

gorgonians Anthozoans of the subclass Octocorallia that have spiculate internal skeletons; commonly called sea fans or sea whips.

grab A collecting device that picks up a volume of the bottom; usually deployed on a cable from a vessel.

gravity The attraction of terrestrial bodies toward the center of the Earth (the Earth's gravitational attraction).

grazer An animal that feeds on autotrophic organisms or other sessile animals.

grazing Feeding on vegetation or sessile colonial animals either by consuming the whole food organism or by cropping all or part of the surface growth.

greenhouse effect *See* global warming.

gross photosynthesis The total amount of photosynthesis before subtracting losses due to respiration.

gross primary production The total amount of organic material fixed in the primary production process.

guild Collection of species that make their living in the same way.

guyot *See* seamount.

gyre Circular motion of water in the major ocean basins.

habitat The place where an organism can be found.

hadal Pertaining to parts of the ocean bottom below 6,000 m.

hadalpelagic Deepest pelagic aphotic zone lying below the abyssalpelagic zone.

halophyte A plant adapted to grow in soils with high salt concentrations.

haptophytes Chlorophyll-bearing organisms of the class Haptophyceae.

harem In mammals, a group of females controlled by a single male during the breeding season.

heat Energy that moves from a higher to a lower temperature.

heat capacity Amount of energy required to raise the temperature of 1 g of material 1°C.

heat of vaporization Amount of energy required to evaporate a unit mass of a substance at constant temperature and pressure.

herbivore An animal that consumes autotrophic organisms.

hermatypic coral A reef-building coral with symbiotic zooxanthellae in its tissues.

heteropods Pelagic molluscs of the prosobranch order Heteropoda.

heterotrophic Obtaining nourishment from external sources of food.

high-level radioactive wastes Those wastes generating greater than 1×10^{-8} curies of ionizing radiation.

holdfast In kelp, that part of the plant which fixes it to the substrate.

holoepipelagic Pertaining to nektonic animals that spend their entire lives in the open ocean.

holoplankton Planktonic organisms that spend their entire lives in the plankton.

homeothermic *See* endothermic.

homeoviscous adaptation An adaptation to high pressure by incorporating more fluid lipids into biological membranes.

homogeneous estuary An estuary with complete mixing such that the salinity is similar from surface to bottom.

host The major provisioning partner of a symbiotic or commensal relationship.

host shift Movement of a pathogen from a natural host to another that was not previously a host.

hydrodynamic Pertaining to the physical features of water motion.

hydrogen bond Weak bond formed between a hydrogen atom in one molecule and an electronegative atom of the same or another molecule.

hydrographic Pertaining to the physical and chemical features of the oceans.

induced drag Disruption of laminar flow patterns around objects by throwing the water into vortices or eddies.

infauna Benthic organisms that live in or burrow through the bottom sediment.

infaunal predator A predator that lives within a sedimentary environment.

infralittoral fringe In the Stephenson universal zonation scheme, the lowest zone on the shore, bounded above by the upper limit of laminarians and below by the lowest tide level.

infralittoral zone *See* infralittoral fringe.

inquilinism Subcategory of commensalism in which one animal lives in the home or digestive tract of another without being parasitic.

insolation Direct exposure to the sun's rays.

interference competition Condition in which one species competes with another by directly interacting with it in some way.

intermittent estuary *See* seasonal estuary.

internal waves Waves occuring at the interface between water masses of different densities.

interstitial Pertaining to the space between adjacent particles in a sedimentary bottom.

intertidal zone Benthic area lying between the extremes of high and low tides.

interzonal fauna Mesopelagic animals that migrate vertically into other zones.

ion A positively or negatively charged atom or molecule produced through the loss or gain of one or more electrons.

isohaline A line joining points of equal salinity concentration.

isotherm A line joining points of equal temperature.

iteroparous Referring to organisms that reproduce more than once during their life spans.

keel A median longitudinal ventral ridge in certain nektonic fishes.

kelp Collective name for various large brown algae.

kelp bed A community dominated by large brown algae in which the plants do not form a surface canopy.

kelp forest A community dominated by large brown algae in which the plants form a surface canopy.

key industry species *See* keystone species.

keystone species A species that is disproportionately important in the maintenance of community integrity and without which drastic alterations of the community would occur.

krill Crustaceans of the Order Euphausiacea.

lagoon (a) Type of estuary formed through the cutting off of inshore waters by the buildup of sandbars parallel to the shore; (b) a shallow stretch of water separated from the open ocean by a coral reef or island.

lagoon reef *See* patch reef.

Langmuir convection cell A localized area of water a few meters wide and hundreds of meters long in which there is a local circulation pattern created by the wind, causing water to upwell and diverge from the center and converge and sink at the boundaries.

larva An independent morphological stage of an animal that develops from a fertilized egg and must undergo a greater or lesser amount of change before assuming adult features.

larval settlement When the larvae of an organism, living in the plankton, settles into its adult habitat.

latent heat of fusion Increase in heat content that accompanies the conversion of a unit mass of substance from a solid to a liquid state.

law of the minimum Law stating that, of those substances necessary to the survival of an organism, the one that is present in the smallest quantities or is reduced to a minimum first will limit the growth and survival of the organisms, even if all others are plentiful.

lecithotrophic larvae Planktonic larvae that do not feed in the plankton.

leeward The side of an island or reef protected from winds and waves.

Leibig's law of the minimum *See* law of the minimum.

light Electromagnetic radiation with wavelengths between 400 and 700 nanometers.

light/dark-bottle method Method of measuring primary productivity by employing two identical bottles, one transparent and the other opaque. Photosynthetic organisms in the transparent bottle undergo both respiration and photosynthesis while those in the dark bottle undergo only respiration. Subtracting the oxygen content of the two bottles gives a measure of photosynthesis.

lithothamnion ridge Synonym for algal ridge.

littoral zone *See* intertidal zone.

local distribution *See* patchiness.

lottery hypothesis Hypothesis explaining the high diversity of coral reef fishes as deriving from the high diversity due to chance as to which species from the larval pool occupies the vacant space.

low-level radioactive wastes Those wastes which generate less than 1×10^{-8} curies of ionizing radiation.

luciferase The enzyme that catalyzes the production of bioluminescence in the presence of luciferin and ATP.

luciferin The protein that produces bioluminescence in the presence of ATP and luciferase.

mackerel Common name for several species of economically important fishes of the family Scombridae.

macrofauna Benthic organisms more than 1 mm in size.

macroplankton Plankton with a size range from 2 to 20 cm.

Madreporaria The order of the class Anthozoa that contains the corals.

mangal *See* mangrove forest.

mangrove Common name for any of several species of inshore tropical trees or shrubs that dominate the mangal associations.

mangrove forest A variety of tropical inshore communities dominated by several species of shrubs or trees that grow in salt water.

mariculture The culture of marine organisms.

marine-dominated estuary *See* homogeneous estuary.

marine snow Name given to the various types of amorphous particulate matter found in the water column and derived from living organisms.

mass strandings When large numbers of whales beach themselves.

maximum sustainable yield The largest number of fishes that can be harvested year after year without diminishing the stock.

megaplankton Plankton with a size range above 20 cm.

meiobenthos A synonym for meiofauna.

meiofauna Benthic organisms between 0.5 mm and 62 μm in size, often deemed identical to interstitial fauna.

melon The large lipid-filled body in the head of a whale that serves as an acoustical lens.

meroepipelagic Pertaining to open-ocean nektonic organisms that spend part of their lives in the open ocean and part in other areas.

meroplankton Planktonic organisms that spend only part of their life cycles in the plankton.

mesobenthic Meiofaunal-sized organisms living and moving within the interstitial space.

mesopelagic zone Uppermost pelagic aphotic zone.

mesoplankton Plankton with a size range between 0.2 and 20 mm.

mesopsammon Organisms living between sand grains in fresh water.

metabolism The sum total of physical and chemical processes that occur in a living organism.

metabolite A biochemical produced by an organism and secreted into the surrounding medium.

metamorphosis The process of structural transformation in an animal whereby the animal changes from larva to adult.

microbial loop That part of the planktonic food web made up of bacteria and various nanoplanktonic organisms and that recovers and cycles dissolved and particulate carbon back into the larger planktonic food web.

microfauna Benthic organisms below the size of 0.1 mm.

micrograzer hypothesis *See* mixing and micrograzer hypothesis.

microplankton Plankton with a size range from 20 to 200 μm.

midlittoral zone In the Stephenson scheme of zonation, that area between the upper limits of laminarians and the upper limits of barnacles.

migration Periodic movement of animals from one place to another.

mimic An organism that assumes the shape, pattern, or behavior of a different species, usually for protection.

Minamata disease Crippling neurological disorder in humans caused by mercury poisoning.

mixed tide Tide consisting of a mixture of diurnal and semidiurnal tides.

mixing and micrograzer hypothesis Explanation for the lack of a spring phytoplankton bloom in the subarctic Pacific Ocean due to the continued presence of protistan micrograzers.

multivore *See* omnivore.

mutualism That form of symbiosis in which two species associate to their mutual benefit.

myoglobin An iron-containing protein compound in muscle that acts as an oxygen carrier and reserve oxygen supply.

nanoplankton Plankton with a size range from 2 to 20 μm.

nauplius First larval form of most crustaceans; characterized by possessing three pairs of appendages.

negative estuary Estuary in which evaporation is high and freshwater input low, so that salt water enters at the surface, evaporates, becomes hypersaline, and sinks, forming an outflowing bottom current.

nekton Pelagic animals that are powerful enough swimmers to move at will in the water column.

neritic Of, relating to, or constituting the water mass overlying the continental shelves.

net community photosynthesis The amount of photosynthesis in excess of respiration by both autotrophic and heterotrophic organisms.

net plankton Organisms caught in a standard zooplankton net (with a mesh above 0.2 mm).

net production Amount of total production left after losses from respiration (or that amount left to support other trophic levels).

neutral In chemistry, having equal numbers of OH^- and H^+ ions; a pH of 7.

neutral estuary *See* homogeneous estuary.

new production *See* net community photosynthesis.

niche The role of an organism in a community.

Niño, El An episodic increased warming of the equatorial Pacific Ocean due to atmospheric changes; El Niño can cause large-scale changes in oceanographic and atmospheric conditions that may lead to the catastrophic mortality of many marine organisms.

nonpelagic development Development without a planktonic period.

nucleus The central, dense, positively charged part of an atom.

oceanic Pertaining to the open-ocean water masses away from the continental shelves.

oligotrophic Containing little nutrient material.

omnivore An organism that consumes both plant and animal material. (*See* multivore.)

ooze A deep-sea sediment composed, at least partially, of the skeletal remains of certain pelagic organisms.

opportunistic species A species that has a life history characterized by a short life span, a short development time to reach maturity, a high death rate, and many reproductive periods per year.

osmoconformer An organism unable to regulate its internal fluid and salt balance and, therefore, one with a varying internal salt concentration.

osmoregulation The physiological activity of an organism that maintains its internal salt and fluid balance within a narrow, acceptable range.

osmoregulator An organism that has physiological mechanisms to control the salt content of its internal fluids.

osmosis The movement of water across a semipermeable membrane separating two solutions of differing solute concentrations. In osmosis, the water moves from the more dilute solution to the more concentrated one.

oval Special gas absorptive organ in physoclist fishes.

oxygen minimum zone An area in the ocean, usually between 500 and 1,000 m, in which oxygen values approach zero.

parallel bottom community The concept that similar sediment types at similar depths around the world contain similar communities of organisms in which the dominant animals are closely comparable taxonomically and ecologically.

parasitism Association in which one species lives on, or in, another, drawing nourishment from that species at its expense or to its detriment.

patch reef A small isolated reef structure existing in the lagoon behind the main coral reef.

patchiness An ecological condition in which organisms occur in isolated groups within a larger contiguous suitable habitat.

peat Organic matter in the soil of salt marshes accumulated from the root masses of marsh plants.

pectoral Pertaining to the chest region of the body.

pelagic Pertaining to the area of the open water of the world's oceans or to organisms that occur in the water column.

pelvic Pertaining to the hip region of the body.

perigee Nearest point of the moon from the Earth.

period Time required for two wave crests to pass a fixed point.

permanent meiofauna Organisms that spend their entire lives in the interstices of sediment.

Petersen grab A quantitative grab designed by C. G. Joh. Petersen to take a fixed area of soft bottom, usually a tenth or one-half of a square meter.

pheromone A chemical secreted by a species that influences the behavior of others of the same species.

photic Lighted; pertains to certain (epipelagic) areas of the ocean.

photophore A complex organ of certain deep-sea animals that produces light and regulates its emission and use.

photosynthesis Process occurring in autotrophic organisms whereby they use the energy of sunlight to synthesize energy-rich organic compounds from carbon dioxide and water.

phototaxis Movement of an organism in response to light.

phototrophic Name applied to organisms that use light as a source of energy or to an animal that is dependent upon photosynthetic zooxanthellae symbionts for energy.

physoclist In bony fishes, those fishes with no open duct between the gas bladder and the digestive tract.

physostome Fishes that have an open duct between the gas bladder and the esophagus.

phytoplankton Planktonic autotrophic organisms.

picoplankton Plankton in the size range from 0.2 to 2.0 μm.

Pinnipedia A suborder of mammals containing the seals, sea lions, and walruses.

plankton Those organisms floating free or drifting in the open water of the oceans and having their lateral and vertical movements determined by the motion of the water.

plankton net Fine-mesh conical net dragged in the water to collect plankton.

planktotrophic larvae Larvae that feed in the plankton.

planula The sexually produced larva of many cnidarians, including corals.

pneumatocyst The gas-filled floats on certain kelps.

pneumatophore Extension above the substrate of the roots of mangroves to permit aeration.

POC Abbreviation for particulate organic carbon.

poikilothermic *See* ectothermic.

polar The Arctic or Antarctic zone.

pollution Degradation of the natural environment.

polyp A sessile or sedentary cnidarian with a cylindrical body and, usually, tentacles surrounding a mouth at the free end.

positive estuary Estuary with substantial freshwater input and reduced evaporation to the extent that freshwater flows out at the surface and salt water flows in at the bottom; also known as a salt-wedge estuary.

positive interaction interaction between species that enhances community structure.

practical salinity unit (PSU) The total amount of dissolved material in seawater, expressed as parts per thousand.

preadaptation Possession by a species or group of species of features that favor the easy occupation of a new habitat.

predation Consumption of an individual of one species by an individual of another species.

predator An animal that feeds on another animal.

predator-disturbance model Hypothesis explaining the high diversity of coral reef fishes as deriving from the fact that the fish populations never reach equilibrium and therefore do not undergo competitive exclusion from predation, catastrophe, or unpredictable recruitment.

primary productivity The rate of formation of energy-rich organic compounds from inorganic materials.

prochlorophytes Tiny (0.6–0.8-μm) chlorophyll-bearing organisms that resemble cyanobacteria, but possess a different combination of pigments.

protein Complex biochemical compounds of large molecular weight composed of joined amino acids.

psammon An obsolete term referring to organisms living interstitially.

pteropods Pelagic molluscs of the opisthobranch orders Thecosomata and Gymnosomata.

pycnocline That area in the water column where the highest rate of change in density occurs for a given change in depth.

realized niche That portion of the fundamental niche which the species actually occupies or uses within its habitat.

recruitment limitation model Hypothesis explaining the high diversity of coral reef fishes as deriving from variation in larval recruitment.

red tide The name given to massive blooms of dinoflagellates in which the concentrations of the organisms are such that the water becomes red brown in color. Toxins are often secreted in sufficient quantities to affect other marine organisms.

redox potential discontinuity (RDP) layer Zone in muddy shores characterized by a rapid change from positive to negative redox potential; the boundary layer between aerobic and anaerobic sediments.

reflective beach A beach with no surf zone, a steep slope, and waves that break directly onto the sand and are then reflected back, producing large swashes.

refractive index For a medium, the ratio of the speed of light in a vacuum to that in the medium.

rete mirabile A large network of anastomosing small blood vessels serving a number of functions in marine vertebrates.

reverse migration Opposite of vertical migration; animals migrate downward at night and toward the surface during the day.

rhodopsin A light-sensitive pigment found in the rods of the eye.

rings *See* eddy.

river-dominated estuary *See* salt-wedge estuary.

rod The sensory body of the retina that is most responsive to low light levels.

salinity The total amount of dissolved material (salts) in seawater.

salt glands Special glands in plants and animals for eliminating excess salt.

salt marsh Community of emergent vegetation rooted in soils alternately inundated and drained by tidal action.

salt pans Vegetation-free areas in salt marshes caused by high soil salinity.

salt-wedge estuary Estuary with a pronounced difference in salinity between surface and deeper waters, with the latter being more saline. (*See also* positive estuary.)

Scleractinia Synonym for Madreporaria.

sclerosepta The calcium carbonate septa in the skeletons of corals.

seagrass Collective name for marine flowering plants of the families Potamogetonaceae and Hydrocharitaceae.

seamount Extinct volcanos that rise abruptly from the deep sea floor but do not reach the surface.

seasonal estuary Estuary having freshwater inflow and mixing with ocean water only for part of a year; the rest of the year it is a stagnant lagoon.

seasonal succession The change of dominant species in the phytoplankton with succeeding seasons of the year.

sedentary Characterized by remaining in one place for long periods or having limited movement.

sediment destabilizer Organisms whose activities cause the substrate to move, become resuspended, erode, or otherwise change.

sediment stabilizer Organisms whose activities bind the substrate particles or otherwise enhance substrate stability.

semelparous Breeding or reproducing only once during a lifetime.

semidiurnal tide A tide with two highs and two lows per lunar day.

semienclosed bay *See* lagoon (a).

sere One of the stages in a successional sequence leading to a climax community.

sessile Pertaining to an organism that is fixed in place.

shelf *See* sublittoral zone.

silicoflagellate Flagellated photosynthetic organism of the class Chrysophyta with an internal skeleton of silicon dioxide.

Sirenia The order of the class Mammalia that contains the sea cows (dugongs and manatees).

slough *See* lagoon (a).

sonar A method of using sound waves to locate objects in the water column.

species diversity In ecology, a numerical measure combining the number of species in an area with their relative abundance.

species richness The number of species in a given area.

spermaceti organ The melon of the sperm whale.

spermatophore A special packet that contains sperm.

sporophyte In algae, the diploid plant that produces spores.

spring bloom Dramatic increase in the population of photosynthetic organisms in the surface waters of the North Atlantic in the spring as a result of the abundant nutrients and absence of grazers.

stability–time hypothesis Theory seeking to explain the occurrence of high diversity among communities of the marine environment; the hypothesis states that high diversity occurs because highly stable environmental conditions have persisted over long periods, allowing numerous highly specialized species to evolve.

standing crop The biomass of an organism or a group of organisms present per unit volume or per unit area at a given point in time.

statocyst An organ that detects gravity.

stenohaline Able to tolerate only a narrow range of salinity changes.

stipe That part of a kelp plant between the blade and the holdfast.

stratified estuary *See* salt-wedge estuary.

sublittoral zone Bottom zone of the continental shelf.

submarine ridge Submerged ranges of mountains, usually marking the boundaries of crustal plates.

subsidence theory Darwin's theory as to the origin of coral atolls whereby atolls begin as fringing reefs around volcanic islands; subsequent subsidence of the island, coupled with upward reef growth, then produces first a barrier reef and, finally, with the disappearance of the volcano, an atoll.

succession The process of community change through time. The facilitation model of succession is the classical definition of the term, whereby there is an orderly sequence of seral stages, with the organisms of each modifying the environment and making it less suitable for them and more suitable for organisms of the next stage. The end point or climax perpetuates itself. The inhibition model of succession postulates no organism as competitively superior and asserts that those organisms which first occupy an area remain until a disturbance removes them. A third model, the tolerance model, is intermediate between the other two.

succulence strategy Adaptation of vascular plants living in highly saline soils whereby the plants maintain high tissue water concentrations to act as a buffer against water loss by osmosis.

sulfur bacteria *See* chemoautotrophic bacteria.

supralittoral fringe In the Stephenson universal zonation scheme, the highest zone on the shore, bounded below by the upper limit of barnacles and above by the upper limit of *Littorina*.

surface tension With respect to water, the attraction of the molecules of water at the surface (the air–water interface) so that the surface resists penetration by small objects.

surge channel Deep channels in the windward face of a coral reef through which water moves in and out of the reef.

suspension feeding Feeding by filtering particles out of the water column.

swash Water that runs up the face of a beach after a wave breaks.

swim bladder *See* gas bladder.

symbiont Each of the partners in a mutualistic relationship.

symbiosis The interrelationship between two different species.

tectonic estuary Estuary formed when the sea reinvades the land due to subsidence of the land as a result of earthquake activity.

temporary meiofauna Meiofaunal-sized organisms that are the juveniles of macrofaunal organisms.

terrigenous Ocean sediments derived from the land.

territoriality A pattern of behavior of an individual or a group that actively keeps other members of the same or different species out of an area.

territory An area occupied and protected by a species against others of its own species, usually for breeding or feeding purposes.

thalassopsammon The interstitial organisms in marine waters.

thermal pollution Degradation of the environment through increases in temperature.

thermal stratification Situation prevailing at all times in the tropics and during the summer in the temperate zones wherein the uppermost water mass decreases in density due to increased temperature and becomes isolated from the more dense, colder water below.

thermocline That portion of the water column wherein temperature changes most rapidly with each unit change in depth.

thermohaline circulation Circulation pattern in the oceans that produces both horizontal and vertical surface and subsurface currents circulating throughout the world's oceans

thigmotaxis Behavioral or motional response triggered by contact with a solid body.

thiobios Meiofauna that live in anoxic sediments.

tide The periodic rise and fall in the surface water of the oceans due to the gravitational attraction of the sun and moon and the rotation of the Earth.

tide pool Area in the rocky intertidal zone that retains some volume of water at low tide.

top-down processes Processes whereby predators control community structure.

toxin A chemical substance produced by an organisms and that is poisonous to other organisms.

trenches The narrow, steep-sided depressions in the ocean floor usually lying between 6,000 and 10,000 m in depth.

trophic cascade An indirect interaction chain in which an apex predator's impact penetrates through several or all trophic levels.

trophic group amensalism Exclusion of one group of organisms by modification of the environment by another.

trophic level Level in ecological system of classification of organisms according to their means of obtaining nutrition; the basic level is that of the autotrophs, the second is that of the herbivores, and the succeeding levels are those of carnivores.

trophic structure The total number of feeding levels through which energy passes in an ecosystem.

trophogenic zone The area in the ocean above the compensation point.

tropholytic zone The area in the ocean below the compensation point.

tsunami A very long period wave generated by an underwater earthquake, landslide, or volcanic eruption; also known as a tidal wave.

tubular eyes Eye in the form of a short cylinder and containing two retinas found in some deep sea fishes.

tuna The common name for any of several large pelagic fishes of the family Thunnidae.

turbidity Condition of reduced visibility in water due to the presence of suspended particles.

twilight zone *See* mesopelagic zone.

ultraabyssal *See* hadal.

ultraplankton Plankton less than 2 μm in size.

upper water mass The well-mixed water layer above the thermocline.

vernal Pertaining to the spring time.

vertical migration Diurnal vertical movement of pelagic organisms in the water column toward the surface at night and downward to the depths during the day.

viroplankton Planktonic organisms in the size range of 0.02–0.2 micrometer.

virus Infectious agent composed of a protein sheath around a nucleic acid core characterized by a lack of an independent metabolism. The absence of such a mechanism makes a virus totally dependent upon living organisms for reproduction.

viscosity That property of a liquid which resists movement through it.

vitamin An organic compound essential to certain metabolic processes in organisms.

wave With respect to the oceans, a disturbance that moves through the water but does not cause particles of the water to advance with it.

wavelength The horizontal distance between the tops of successive wave crests.

weak predation Condition wherein the total effect of predation on a competitive dominant prey is such that predation alone does not control the abundance of that prey.

whalebone *See* baleen.

white band disease A coral disease characterized by a conspicuous white band of dying tissue on infected corals.

windward The side of an island or reef that faces the prevailing wind.

zonation System of prominent horizontal bands of organisms that succeed each other vertically.

zoochlorellae Symbiotic green algae.

zooplankton Planktonic animals.

zooxanthellae Dinoflagellates symbiotic in the tissues of various marine invertebrates; symbionts that are brownish in color.

INDEX